TOPOGRAPHIC MAP SYMBOLS

VARIATIONS WILL BE FOUND ON OLDER MAPS

Hard surface, heavy duty road, four or more lanes ......

Hard surface, heavy duty road, two or three lanes ....

Hard surface, medium duty road, four or more lanes ....

Hard surface, medium duty road, two or three lanes .....

Improved light duty road..........................

Unimproved dirt road and trail......................

Dual highway, dividing strip 25 feet or less............

Dual highway, dividing strip exceeding 25 feet.........

Road under construction..........................

Railroad, single track and multiple track ....

Railroads in juxtaposition ....

Narrow gage, single track and multiple track...........

Railroad in street and carline......................

Bridge road and railroad ....

Drawbridge, road and railroad ....

Footbridge ....

Tunnel, road and railroad ....

Overpass and underpass ....

Important small masonry or earth dam ....

Dam with lock.................................

Dam with road.................................

Canal with lock................................

Buildings (dwelling, place of employment, etc.)..........

School, church, and cemetery......................

Buildings (barn, warehouse, etc.)....................

Power transmission line...........................

Telephone line, pipeline, etc. (labeled as to type).........

Wells other than water (labeled as to type). .... o Oil..... o Gas

Tanks; oil, water, etc. (labeled as to type)............. • ● ● ⊘ Water

Located or landmark object; windmill ....

Open pit, mine, or quarry; prospect .... ✕ ...... ✗

Shaft and tunnel entrance.........................

Horizontal and vertical control station:

    Tablet, spirit level elevation...................... BM △ 5653

    Other recoverable mark, spirit level elevation .... △ 5455

Horizontal control station: tablet, vertical angle elevation VABM △ 9519

    Any recoverable mark, vertical angle or checked elevation △ 3775

Vertical control station: tablet, spirit level elevation ..... BM ✕ 957

    Other recoverable mark, spirit level elevation .... ✕ 954

Checked spot elevation .... ✕ 4675

Unchecked spot elevation and water elevation .... ✕ 5657 870

Boundary, national ....

    State.....................................

    County, parish, municipio......................

    Civil township, precinct, town, barrio.............

    Incorporated city, village, town, hamlet...........

    Reservation, national or state...................

    Small park, cemetery, airport, etc...............

    Land grant...............................

Township or range line, United States land survey ......

Township or range line, approximate location ....

Section line, United States land survey ....

Section line, approximate location ....

Township line, not United States land survey ....

Section line, not United States land survey ....

Section corner, found and indicated .... + ... +

Boundary monument: land grant and other........... □ ...... □

United States mineral or location monument .... ▲

Index contour .... Intermediate contour..

Supplementary contour Depression contours ..

Fill................ Cut...............

Levee.............. Levee with road.....

Mine dump.... Wash.............

Tailings........... Tailings pond.......

Strip mine......... Distorted surface....

Sand area......... Gravel beach.......

Perennial streams .... Intermittent streams..

Elevated aqueduct.... Aqueduct tunnel.....

Water well and spring. o Disappearing stream..

Small rapids......... Small falls ....

Large rapids........ Large falls ....

Intermittent lake..... Dry lake..........

Foreshore flat...... Rock or coral reef...

Sounding, depth curve. 10 Piling or dolphin..... o

Exposed wreck....... Sunken wreck......

Rock, bare or awash; dangerous to navigation ....

Marsh (swamp)....... Submerged marsh

Wooded marsh...... Mangrove

Woods or brushwood. Orchard.........

Vineyard........... Scrub...........

Inundation area...... Urban area ....

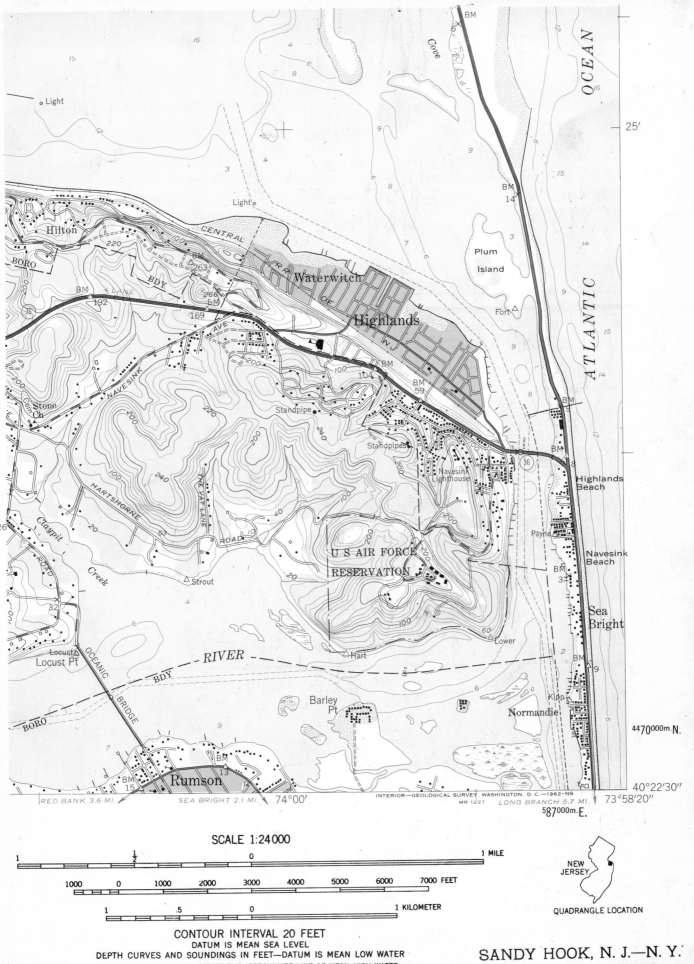

SCALE 1:24 000

CONTOUR INTERVAL 20 FEET
DATUM IS MEAN SEA LEVEL
DEPTH CURVES AND SOUNDINGS IN FEET—DATUM IS MEAN LOW WATER
SHORELINE SHOWN REPRESENTS THE APPROXIMATE LINE OF MEAN HIGH WATER
THE MEAN RANGE OF TIDE IS 3.8 FEET

NEW JERSEY

QUADRANGLE LOCATION

SANDY HOOK, N. J.—N. Y.
N4022.5—W7358.33/7.5x9.17

1954

Arthur N. Strahler
Alan H. Strahler

Modern Physical Geography

JOHN WILEY & SONS, NEW YORK
Santa Barbara / Chichester / Brisbane / Toronto

Cover photo A portion of the South Island of New Zealand, shown in Landsat false-color imagery. Snow-covered surfaces of the lofty Southern Alps (left) appear in pure white. Between the mountains and the sea is a lowland the Canterbury Plain, intensively used for sheep and cattle grazing, grain farming, and fruit orchards. Green vegetation appears as red in this image. Snow-fed rivers (blue) cross the plain to reach the Pacific Ocean. At the lower right, two deeply eroded extinct volcanoes form Banks Peninsula, lying just south of the coastal city of Christchurch. Width of the area shown is about 120 km. (NASA Image No. 2192-21265, courtesy of General Electric, Beltsville, Maryland.)

Throughout this book, illustrations with credit reading *Based on Goode Base Map* use as the base map Goode Map No. 201HC World Homolosine, copyrighted by the University of Chicago. Used by permission of the Geography Department, University of Chicago.

Library of Congress Cataloging in Publication Data:

Strahler, Arthur Newell, 1918-
Modern physical geography.

Bibliography: p.
Includes index.
1. Physical geography. I. Strahler, Alan H., joint author. II. Title.

GB54.5.S8   910'.02   77-20242
ISBN 0-471-01871-6

Printed in the United States of America

10 9 8 7 6 5 4 3 2 1

Biographical Note

ARTHUR N. STRAHLER (b. 1918) received his B.A. degree in 1938 from the College of Wooster, Ohio, and his Ph.D. degree in geology from Columbia University in 1944. He is a fellow of the Geological Society of America and the Association of American Geographers. He was appointed to the Columbia University faculty in 1941, serving as Professor of Geomorphology from 1958 to 1967 and as Chairman of the Department of Geology from 1959 to 1962. He is the author of several widely used textbooks of physical geography, environmental science, and the earth sciences.

ALAN H. STRAHLER (b. 1943) received his B.A. degree in 1964 and his Ph.D. degree in 1969 from The Johns Hopkins University, Department of Geography and Environmental Engineering. His published research lies in the fields of quantitative plant geography and forest ecology. He is Assistant Professor in the Department of Geography of the University of California at Santa Barbara. From 1973 to 1974 he held an appointment as Bullard Forest Research Fellow at Harvard University. He is the coauthor of textbooks on physical geography and environmental science.

Preface

In recent years, major changes have taken place in the content and curriculum of physical geography. New topics have been brought into the fold of physical geography, including cycling of energy and materials in ecosystems, plate tectonics, and remote sensing of the earth's varied landscapes. Important advances have also occurred as traditional topics have been approached in new ways with fresh ideas. For example, conventional classification systems of soils and climate used in past decades have been superseded by new systems based on a more thorough understanding of principles of soil science and the soil-water balance. In addition, the field of physical geography has acquired a new relevance through its multidisciplinary treatment of natural environmental processes and their relationship to Man's activity. The concept of the environmental region as a geographic area with its own unique problems and potentials for development has provided an essential focal point for students of physical geography.

In incorporating these advances into a physical geography text for the 1980s, our book emerged as a new work — not merely a revision of its predecessors, *Physical Geography*, Fourth Edition, and *Introduction to Physical Geography*, Third Edition. A new title was clearly in order, and we chose *Modern Physical Geography*.

The first eight chapters of *Modern Physical Geography* follow a well-established pattern of topics relating to the earth-sun relationships, cartography, and basic principles of atmospheric science. These topics have received careful updating of facts and improvements in presentation. Our treatment of the earth's radiation balance in Chapter 4 has been strengthened through new global maps and graphs showing albedo, insolation, longwave radiation, and net radiation. Basic principles of remote sensing are also presented in this chapter, where principles of electromagnetic radiation are a common theme.

Chapters 9 and 10 deal with climate and climate classification systems. The first of these chapters uses a well-established, descriptive approach to climate classification through recognition of distinctive temperature and precipitation regimes, illustrated by familiar climographs of representative stations. The Köppen system is included for optional use. Our second chapter places the descriptive

climate classification on a quantitative basis, using Thornthwaite's principles of the soil-water budget. This new approach adds greatly to the understanding of global patterns of soils and natural vegetation. Boundaries used for the climate classes are now based on station data of annual potential evapotranspiration, soil-water deficit and surplus, and mean monthly values of soil-water storage. The availability of both water and heat to plants is particularly well expressed through the structure of the new climate classification system. We thank Professor Donald R. Currey for reviewing the new text on climate classification.

Chapters 11 to 16 deal with two closely interwoven subject areas: pedology and biogeography. We have treated this section as a progression of topics, each one building logically on the foundations laid by those topics previously covered. Thus, we begin in Chapter 11 with the parent mineral matter of the soil. Silicate minerals and their alteration into clay minerals, stable oxides, regolith, and sediments are explained here because these materials provide inorganic substances that are cycled in the biosphere and involved in soil-forming processes. Chapter 12 traces the flow of energy and matter through the biosphere. These concepts of ecology are vital to an understanding of the global differentiation of both soils and natural vegetation; they also form the basis for evaluating agricultural ecosystems and their management. Chapter 13, devoted to basic soil science, uses modern concepts of the polypedon, horizon development, cation exchange, base status of soils, soil-temperature and soil-water regimes, and pedogenic processes. Thus the framework is established for Chapter 14, in which we present the U.S. Comprehensive Soil Classification System (CSCS) and Soil Taxonomy. Adoption of this modern system of soil classification is perhaps the most significant point of departure of *Modern Physical Geography* from its predecessors. We are grateful to Dr. Roy W. Simonson and Professor Henry D. Foth for their careful reviews of these chapters on soil science.

Chapters 15 and 16 present fundamentals of biogeography with particular emphasis on plant geography and the classification and global distribution of biomes and plant formation classes. A new world map of natural vegetation uses data compiled by Professor S. R. Eyre.

Chapters 17 to 23 provide a systematic treatment of geomorphic processes and the common landforms shaped by the agents of denudation. An important innovation is the modern treatment of fluvial denudation using a model based on the concept of exponential decay and the role of spasmodic isostatic uplift.

Chapters 24, 25, and 26 form a geologic group in which plate tectonics is used as the basis for explaining volcanic and tectonic landforms and the denudational landforms controlled by rock structure. Chapter 24 describes the lithosphere and explains plate tectonics, including such topics as plate subduction, continental rupture and collision, and suturing. Chapter 25 describes landforms produced by tectonic and volcanic activity and the natural hazards of volcanic eruptions and earthquakes. Some instructors may wish to cover Chapters 24 and 25 before beginning the study of geomorphology (Chapter 17). Chapter 26, devoted to structural control of landforms, makes use of plate tectonics to identify the various structural types in terms of their origin and position with respect to active plate boundaries, continental sutures, stable plate interiors, and mantle hot spots. The use of plate tectonics in classifying landforms controlled by structure is an important step forward in modern physical geography.

Our final pair of chapters (Chapters 27 and 28) represent a major innovation in physical geography. In establishing the concept of global environmental regions, we not only perform an integrative function that is the hallmark of geography, but we also provide physical geography with a useful purpose in the management of human affairs. An environmental region, based primarily on a single climatic regime, is endowed with a unique combination of soils, vegetation, hydrologic system, and geomorphic process. These factors are brought together in the description of a given environmental region, together with an evaluation of the opportunities and restraints which that particular physical environment imposes on human activity and agricultural productivity. Thus, the concept of global environmental regions can provide the scientific base for various branches of human geography and for studies of environmental management.

Adoption of the metric system has required extensive revision of many graphs and maps taken from the authors' published textbooks on physical geography and environmental science. Metric units are used throughout the text, with English units in parentheses where visualization of dimensions is helpful.

New guidelines issued by publishers call attention to the need to avoid sexist expressions in textbooks. Sexism in its broader sense means any arbitrary stereotyping of males and females by reason of their gender. As authors of a geography textbook, we are faced by the problem of selecting an acceptable word to denote collectively the individuals of the human race, including persons of both sexes. The word *man* is now regarded by many persons as closely associated with a male human being and is no longer broad enough in meaning to be applied to persons of either sex or to the human race. To avoid lengthy and often awkward alternatives that waste space, we use the proper name *Man,* always capitalized, to denote collectively all individuals or any group of individuals of the human race. *Man* is the English translation of the Latin name *Homo* for the genus to which all individuals of the human race belong. We feel that in this usage *Man* is no more a sexist word than, for example, *American, Canadian,* or *Navajo,* all of which are proper names for groups of persons including both sexes. We also capitalize two terms used frequently in environmental topics: *Man-made* and *Man-induced.* Except in those cases where the work of an individual geographer is cited, we have been careful to avoid using either feminine or masculine pronouns.

We wish to thank Paul A. Lee, geography editor at Wiley, for coordinating the activities of many individuals involved in production of a new textbook. His expertise and long experience in the fields of professional geography and publishing have proved invaluable in guiding our work through the long succession of complex stages from manuscript to bound copies.

Arthur N. Strahler
Alan H. Strahler

Contents

ix

APPENDIXES

COLOR PLATES

Introduction

Physical geography is an area of study that brings together and interrelates the important elements of Man's physical environment. Physical geography draws on several natural sciences for its subject matter, among them sciences of the atmosphere (meteorology, climatology), oceans (oceanography), solid earth (geology), soils (soil science), vegetation (plant ecology, biogeography), and landforms (geomorphology). But physical geography is much more than a collection of topics drawn from other sciences; it weaves that information into patterns of interaction with Man in a way not expressed within each of the contributing sciences. As a branch of geography, physical geography also emphasizes spatial relationships — the systematic arrangements of environmental elements into regions over the earth's surface and the causes for those patterns.

The focus of physical geography is on the **life layer**, a shallow zone of the lands and oceans containing most of the world of organic life, or biosphere. The quality of the life layer is a major concern of physical geography; "quality" means the sum of the physical factors that make the life layer habitable for all forms of plants and animals, but most particularly for Man. We use "Man," the first letter capitalized, as the English equivalent of genus *Homo,* thus referring collectively to all individuals of the human race. Man's full Latin name is *Homo sapiens,* the only surviving species of that genus. The surface of the lands is the home of Man because this animal requires a terrestrial rather than an aquatic environment.

The quality of the physical environment of the lands is established by factors, forces, and inputs coming from both the atmosphere above and the solid earth below. The atmosphere dictates climate, which governs the exchange of heat and water between atmosphere and ground. The atmosphere also supplies vital elements — carbon, hydrogen, oxygen, and nitrogen — needed to sustain all life of the lands. The solid earth forms the stable platform for the life layer and is also shaped into landforms. These landscape features — mountains, hills, and plains — bring another dimension to the physical environment and provide varied habitats for plants. The solid earth is also the basic source of many nutrient elements, without which plants and animals cannot live. These elements pass from rock into the shallow soil layer, where they are held in forms available to organisms.

Acting together, the inputs of energy and materials into the life layer from atmosphere and solid earth determine the quality of the environment and the richness or poverty of the organic life it can support. Thus an understanding of physical geography is vital to planning for survival of the earth's rapidly expanding human population. Survival will depend not only on how much fresh water and food is available; it will also depend on whether the environment is protected from forms of pollution and destruction that will reduce the capacity of the land to furnish those necessities. Here we touch on another of the important goals of physical geography: to evaluate the impact of Man on the natural environment. Environmental science, the study of interaction between Man and the environment, is gaining wide recognition today; many persons think of it as a new discipline. Actually, geographers have been investigating environmental science for many decades. Physical geography has always been at the heart of environmental studies because it is strongly oriented toward the interaction between Man and environment.

Our plan of study of physical geography is to begin with the atmosphere and the ways in which it furnishes light, heat, and water to the life layer. An evaluation of climate follows; here we develop background information for a system of global environmental regions. At this point, the concept of a global water balance is introduced.

Place-to-place variations in the availability of water to plants are studied through the soil-water balance, an accounting system on which a rigorous and useful climate classification system can be built.

Several chapters devoted to soils and natural vegetation follow. We start with a study of the parent mineral matter of the soil, which requires an overview of rocks and rock-forming minerals. Plants interact with the mineral soil, leading us to investigate the flow of energy and materials through the biosphere. From soil science we derive an understanding of the factors that underlie the global distribution of various types of soils. Biogeography, a branch of physical geography, adds an understanding of the facts that determine the global patterns of natural vegetation.

Vegetation and soils are also influenced by landforms — the relief features of the earth's land surfaces. Thus our next group of chapters investigates solar-powered processes that shape the land surface. These processes are weathering, mass wasting, and the activities of running water, waves and currents, glacial ice, and wind.

The systematic study of physical geography is completed by an investigation of deep-seated processes within and below the earth's crust by which entire continents are shaped. These processes produce major subdivisions of continents that show striking differences in relief and pattern. The recent development of a unifying geological theory, called "plate tectonics," provides geographers with a means of setting up a world system of structural types, each one strongly influencing the development of landforms.

Our final chapters bring the elements of physical geography into a synthesis through the recognition of a dozen global environmental regions. Each region has a unique combination of climate, vegetation, and soils, along with some distinctive kinds of landforms. We will find that Man has occupied and utilized certain environmental regions to great advantage, where the elements of physical geography have conspired to favor abundant food production through agriculture. Certain other environmental regions, as yet sparsely populated, are unfavorably endowed with opportunities for food production because of climates too dry, too cold, or too wet; because of soils too poor in plant nutrients; or because of terrain too rugged to farm. Through the study of environmental regions, you can gain a perspective of great value in assessing Man's continuing efforts to expand food production into these less promising environments.

The Geographic Grid and Its Projections

1

Our search for an understanding of Man's physical environment starts with an inquiry into things astronomical. The most basic environmental controls relate to two facts: First, the earth is approximately spherical in form; second, the spherical earth is in motion, spinning on an axis and, at the same time, traveling in a nearly circular path around the sun. Let us look into these facts to derive significant interpretations relating to the quality of the earth's surface environment.

A Spherical Earth

In this age of orbiting satellites, the spherical form of the earth is such an obvious fact that we may have difficulty imagining ourselves living in ancient times when the extent and configuration of the earth were entirely unknown. In that day, to sailors of the Mediterranean Sea, on their ships and out of sight of land, the sea surface looked perfectly flat and seemed to be terminated by a circular horizon. Based on this perception, the sailors might well have inferred that the earth has the form of a flat disk and that, if they traveled to its edge, their ships might fall off. Even so, these sailors could have sensed optical phenomena suggesting that the sea surface is not flat, but curved so as to be upwardly convex like a small part of the surface of a sphere. Even without optical instruments we can get some inkling that the earth's surface curves downward away from us through the observation that sunlight illuminates the tops of high clouds and high mountain summits after sunset and before sunrise.

Rigorous and convincing proof that the earth is spherical in form is not, however, as evident as it might seem if we limit ourselves to observations available to humans before the age of space vehicles. Consider the fact that a few survivors of Magellan's crew sailed completely around the globe and returned to their starting point in Spain. This discovery voyage, 1519–22, might seem at first thought to have proved the earth to be spherical; instead, it merely ascertained that the earth is a solid figure rather than a flat disk terminating in a sharp edge. Circumnavi-

gation could be performed on a cubical or cylindrical earth, or on an earth of any irregular solid form. So let us look for some possible proofs of earth sphericity available to earthbound humans. At least three basic experimental programs can achieve convincing results.

One proof of the earth's sphericity may be had from observations at sea. As a passing ship recedes farther and farther into the distance, it appears to sink slowly beneath the water level (Figure 1.1). Seen through binoculars or a telescope, the sea surface will appear to rise until the decks are awash, then gradually to submerge the funnel and the masts, leaving finally only smoke visible above the horizon. The explanation obviously is that the sea surface curves downward away from us. To prove that this curvature is spherical would require numerous observations in which measurements were made of the amount of apparent sinking of a vessel per unit of distance in many

Figure 1.1 **Because the sea surface is curved, a distant ship seen through a telescope appears to be partly submerged.**

3

different directions away from the observing point.

A second proof is found by observing many lunar eclipses, at which times the earth's shadow falls on the moon. The edge of the earth's shadow appears as an arc of a circle on the disk of the moon. It can be shown by geometrical proof that a sphere is the only solid body that will always cast a circular shadow. At the times of successive eclipses, the earth is rarely oriented in just the same position. No matter what earth profile is cast on the moon, the circular shadows are all alike; therefore the earth must be spherical.

A third proof uses a simple principle of astronomy, known in ancient times and used effectively by the Arabs as early as the ninth century A.D. An observer at the north pole always sees the North Star, Polaris, in the zenith point in the sky because that star lies in line with the earth's axis of rotation (Figure 1.2). As the observer travels southward, Polaris appears to shift its position toward the horizon, so that halfway between the north pole and the equator (at latitude 45°N), the star lies halfway between the zenith and the horizon. Upon approaching the equator, the observer would find Polaris located close to the horizon. A series of measurements of the angle between the horizon and Polaris would show that the angle always decreases by 1° for each 111 km (69 mi) of southward travel. This observation would prove that the travel path has followed the arc of a circle. Repetition of these observations along many different north-south lines (meridians) would establish that the northern hemisphere is one-half of a true sphere. A similar set of observations could be carried out

in the southern hemisphere, using a minor star in line with the south pole. In this way the earth could be proved to be spherical. This principle is actually used in navigation by the stars (celestial navigation). When we consider that, for more than two centuries, positions of vessels at sea have been accurately determined innumerable times by use of this basic method, the spherical form of the earth has obviously been demonstrated beyond question.

Measuring the Earth's Circumference

Scholars among the ancient Greeks believed the earth to be spherical. Pythagoras (540 B.C.) and associates of Aristotle (384–322 B.C.) held this view. They also speculated on the length of the earth's circumference, but with highly erroneous guesses. It remained until about 200 B.C. for Eratosthenes, head of the library in Alexandria, Egypt, to perform a direct measurement of the earth's circumference based on a sound principle of geometry. He observed that on a particular date of the year (close to summer solstice, June 21) at Syene, a city located on the upper Nile River far to the south, the sun's rays at noon shone directly on the floor of a deep vertical well. In other words, the sun at noon was in the zenith point in the sky, and its rays were perpendicular to the earth's surface at that point on the globe (Figure 1.3). At Alexandria, on the same date, the rays of the noon sun made an angle with respect to the vertical. The magnitude of this angle was one-fiftieth of a complete circle, or $7\frac{1}{5}°$.

Eratosthenes needed only to know the north-south distance between Syene and Alexandria to calculate the earth's circumference; he would simply multiply the ground distance by 50 to obtain the circumference. In those days distances between cities were only crude estimates, based on travelers' reports. Eratosthenes took the distance to be 5000 stadia. It is not easy to rate his results in terms of accuracy because we are not sure of the length equivalent of the distance unit (the stadium) that he used. If it were the Attic stadium, equivalent to 185 m (607 ft), the circumference in modern units would come to about 43,000 km (26,700 mi). Considering that the true circumference is close to 40,000 km (25,000 mi), Eratosthenes' results seem offhand to be remarkably good.

From Eratosthenes' classic experiment, it is an easy step to design an astronomical method of measuring the earth's figure using star positions instead of the sun. We need only to select a north-south line, whose length can be measured directly on level ground by surveying means. This line should be several tens of kilometers long. At the ends of the line, the angular position of any selected star can be measured at its highest point above the horizon or with respect to the vertical, using a level bubble or a plumb bob to establish a true horizontal or vertical reference. The difference in angular positions of the star will be equal to the arc of the earth's circumference lying between the ends of the measured line. This very procedure is believed to have been followed by ninth-century Arabs. Their measurements were probably much more accurate than those of Eratosthenes but, because the units of measure are not known in modern equivalents, their work cannot be checked.

Figure 1.2 **The height of the North Star above the horizon depends on one's position in the northern hemisphere. Because of the star's great distance from earth, its light rays are shown as parallel lines in the drawing.**

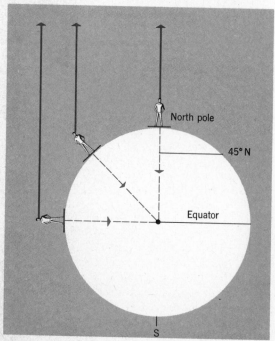

North pole

45° N

Equator

S

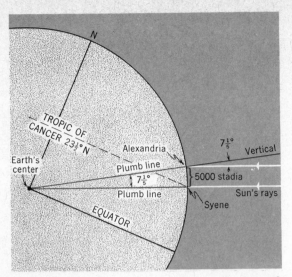

Figure 1.3 **Eratosthenes' method of measuring the earth's circumference.**

In the many centuries following Eratosthenes' work, Western science lay in stagnation. Then, about 1615, Willebrord Snell, a professor of mathematics at the University of Leiden, developed methods of precise distance and angular measurement, which he applied to the problem of the earth's circumference. His work presaged the era of scientific geodesy ("geodesy" comes from the Greek word meaning "to divide the earth") and led to remarkably accurate measurements of the earth's figure about a century later.

Gravity in the Environment

From the standpoint of life on earth, what is significant about the fact that the earth's form closely approximates that of a true sphere? One answer is, of course, "gravity." **Gravity** is the force acting on a small unit of mass at the earth's surface, tending to draw that mass toward the earth's center. Gravity is a special case of the phenomenon of **gravitation,** the mutual attraction between any two masses — special in the sense that gravity refers to a unit of mass so small in comparison with the earth's mass that the attraction of the unit mass for the earth can be disregarded. Gravitational attraction varies inversely as the square of the distance separating the centers of two masses. So gravity depends on the distance of any particle of matter from the earth's center of mass. That center lies close to the geometrical center of the sphere.

Perhaps you recall a principle of geometry to the effect that a sphere is defined as a solid on which all surface points lie equidistant from a common point: the center of the sphere. Because of this principle, gravity turns out to be an almost constant value at all points at sea level over the entire globe. This is a fact of fundamental importance to all life forms on earth. Life has evolved through geologic time under the influence of a value of gravity uniform over the earth and probably little changed during the major evolutionary period of a billion years or so.

Gravity thus represents the lowest common denominator of the planetary environment.

The force of gravity acts as an environmental factor in many ways. It separates substances of differing density into a layered arrangement, the least dense being at the top and the most dense at the bottom. Air, liquid water, and rock are arranged in order of density because of their response to gravity. As a result, the life layer is defined as an interface between atmosphere and ocean and between atmosphere and the solid land surface.

Trees, animals, cliffs of rock, and Man-made structures must have the strength to withstand the force of gravity, which tends to collapse and crush them. Under a weaker planetary gravity, such structures could rise higher or be constructed of weaker materials to attain a given height. Gravity supplies power for important physical systems of the life layer, particularly streams and glaciers that erode the land. To appreciate the importance of gravity as an environmental factor, we need only speculate as to what would happen if its influence were canceled out and a condition of weightlessness were to take its place. Environmental destruction would be total within a short time!

There are very small, systematic differences in the value of gravity from place to place over the earth. The value at the equator is slightly less than at the two poles; there is also a slight decrease in gravity as we rise to higher elevations above sea level. But, for practical purposes, gravity can be taken as a constant the world over.

The constancy of gravity over the earth's surface might be used in an experiment to prove the earth's sphericity. If we first assume that Newton's law of gravitation is valid, it follows that a given object should register the same weight at all places over the earth's surface. Using a spring balance as the scales, we might travel widely over the earth, repeatedly weighing a small mass of iron and recording the values. If they proved to be unvarying, we could conclude that we have taken our measurements at points all equidistant from the earth's center of mass and thus we are on a spherical surface. Actually, this same experiment, carried out with great precision and using highly refined instruments, has shown that the earth's true figure departs slightly from that of a true sphere.

The Earth as an Oblate Ellipsoid

In 1671 a French astronomer, Jean Richer, was sent by Louis XIV to the Island of Cayenne, French Guiana, to make certain astronomical observations. His clock had been so adjusted that its pendulum, 1 m (39.4 in.) long, beat the exact seconds in Paris (when a pendulum is made shorter, it beats faster; made longer, it beats slower). Upon arriving in Cayenne, which is near the equator, Richer found the clock to be losing about 2½ minutes per day. As soon as Newton's laws of gravitation and motion were published (1687), it became possible to attribute the slowing of the clock at Cayenne to a somewhat reduced value of gravity near the equator. It was soon realized that the smaller value of gravity could be accounted for by supposing that the equatorial region of the earth's surface

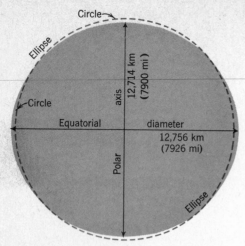

Figure 1.4 **The earth's equatorial and polar diameters, according to the Geodetic Reference System of 1967.**

lies farther from the earth's center than do more northerly regions.

Refined measurements have since revealed that the true form of the earth resembles a sphere that has been compressed along the polar axis and made to bulge slightly around the equator (Figure 1.4). This form is known as an **oblate ellipsoid**. A cross section through the poles gives an **ellipse** rather than a circle. The equator remains a circle and is the largest possible circumference on the ellipsoid. The earth's oblateness is caused by the centrifugal force of the earth's rotation, which deforms the somewhat plastic earth into a form in equilibrium with respect to the forces of gravity and rotation.

Rounding off the earth's dimensions to the nearest whole kilometer, the equatorial diameter is 12,756 km (7926 mi), whereas the length of the polar axis is 12,714 km (7900 mi). The difference is just over 43 km (slightly less than 27 mi). The **oblateness** of the earth ellipsoid, or "flattening of the poles," is the ratio of this difference to the equatorial diameter, or roughly 43/12,756, which reduces to the approximate fraction 1/300. Thus we can say that the earth's polar axis is about 1/300 shorter than the equatorial diameter. Using these figures, the earth's equatorial circumference is about 40,075 km (24,900 mi). For rough calculations, 40,000 km (25,000 mi) is close enough.

Great and Small Circles

For many topics in physical geography, the earth can be treated as if it were a true sphere. For example, flattening of the poles can be disregarded to simplify an understanding of the important concept of the earth as an object turning under the sun's rays.

When a sphere is divided exactly in half by a plane passed through the center, the intersection of the plane with the sphere is the largest circle that can be drawn on the sphere and is known as a **great circle** (Figure 1.5). Circles produced by planes passing through a sphere anywhere except through the center are smaller than great circles and are designated **small circles.**

Become thoroughly familiar with the following properties of great circles because they enter into such global subjects as meridians and parallels, seasons, and map projections:

1. A great circle always results when a plane passes through the center of a sphere, regardless of the orientation of the plane.
2. A great circle is the largest possible circle that can be drawn on the surface of a sphere.
3. An infinite number of great circles can be drawn on a sphere.
4. One and only one great circle can be found that will pass through two given points on the surface of the sphere (unless the two points are exactly opposite, in which case an infinite number of great circles can be drawn through them).
5. An arc of a great circle is the shortest surface distance between any two points on a sphere.
6. Intersecting great circles always bisect each other.

One important use of great circles is in navigation. Whenever ships must travel over vast expanses of open ocean between distant ports, or whenever planes must make long flights, it is desirable in the interests of saving fuel and time to follow the great-circle course between the two points provided, of course, that there are no obstacles or other deterring factors preventing the use of the great circle. Navigators use special types of maps that have the property of always showing great-circle arcs as straight lines. To plot the shortest course between any two points, the navigator simply draws a straight line between the two points on the chart.

Great-circle courses are easily found on a small globe by using only a piece of thin string or a rubber band (Figure 1.6). The string can be held in such a way that it is stretched tightly against the surface of the globe between the two thumbnails, each of which is held on one of the points between which the great-circle course is needed. If a rubber band is used, a complete great circle can be shown; it is of special value for points on opposite sides of the globe. Many globes show great-circle routes between distant ports on the Pacific, Atlantic, or Indian oceans. These may be checked by the stretched piece of string.

Figure 1.5 **A great circle and a small circle.**

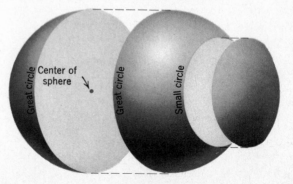

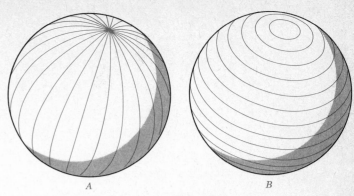

Figure 1.7 **Meridians (A) and parallels (B).**

Figure 1.6 **Forming a great-circle course on a globe, using a length of string. (Charles Phelps Cushing, from H. A. Roberts.)**

Meridians and Parallels

The spinning of the earth on its axis provides two natural points — the poles — upon which to base the **geographic grid,** a network of intersecting lines inscribed on the globe for the purpose of fixing the locations of surface features. The grid consists of a set of north-south lines connecting the poles — the meridians — and a set of east-west lines running parallel with the equator — the parallels (Figure 1.7).

Meridians are halves of great circles, whose ends coincide with the earth's north and south poles. Though it is true that opposite meridians taken together comprise a complete great circle, note that a single meridian is only half of a great circle and contains 180° of arc. Additional characteristics of meridians are as follows:

1. All meridians run in a true north-south direction.
2. Meridians are spaced farthest apart at the equator and converge to a point at each pole.
3. An infinite number of meridians may be drawn on a globe. Thus a meridian exists for any point selected on the globe. For representation on maps and globes, however, selected meridians are spaced at equal distances apart.

Parallels are entire small circles, produced by passing planes through the earth parallel to the plane of the equator. They possess the following characteristics:

1. Parallels are always parallel to one another. Although they are circular lines, any two parallels always remain equal distances apart.
2. All parallels represent true east-west lines.
3. Parallels intersect meridians at right angles. This relationship holds true for any place on the globe,

except the two poles, even though the parallels are strongly curved near the poles.
4. All parallels except the equator are small circles; the equator is unique in being a complete great circle.
5. An infinite number of parallels may be drawn on the globe. Therefore every point on the globe, except the north or south pole, lies on a parallel.

Longitude

The location of points on the earth's surface follows a system in which lengths of arc are measured along meridians and parallels (Figure 1.8). Taking a selected meridian, or prime meridian, as a reference line, arcs are measured eastward or westward to the desired points. Taking the equator as the starting line, arcs are measured north or south to the desired points.

The **longitude** of a place is the arc, measured in degrees, of a parallel between that place and the **prime meridian** (Figure 1.8). The prime meridian is almost universally accepted as the meridian that passes through the old Royal Observatory at Greenwich, near London, England, and it is often referred to as the **Greenwich meridian.** This meridian has the value 0° longitude. The longitude of any given point on the globe is measured eastward or westward from this meridian, whichever is the shorter arc. Longitude may thus range from 0 to 180°, either east or west. It is commonly written in the following form: long. 77° 03′ 41″W, which may be read "longitude 77 degrees, 3 minutes, 41 seconds west of Greenwich."

If only the longitude of a point is stated, we cannot tell its precise location because the same arc of measure applies to an entire meridian. For this reason, a meridian might be defined as a line representing all points having the same longitude. This definition explains why we often use the expression "a meridian of longitude." You may at first be confused by the statement that longitude is measured along a parallel of latitude, but this becomes clear when you realize that to measure the arc between a point and the prime meridian, it is necessary to follow a parallel eastward or westward (Figure 1.8).

The actual length, in kilometers or miles, of a degree of

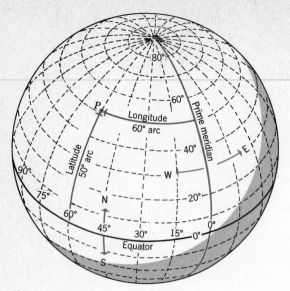

Figure 1.8 Longitude and latitude are measured along arcs of parallels and meridians, respectively. The point *P* has a longitude of 60°W, a latitude of 50°N.

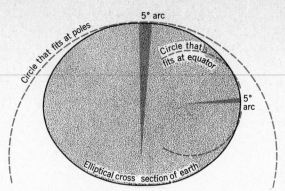

Figure 1.9 The length of a degree of latitude is slightly greater at the poles than at the equator.

longitude will depend on where it is measured. At the equator the approximate length of one degree is computed by dividing the earth's circumference by 360°:

$$\frac{40,075 \text{ km}}{360°} = 111 \text{ km}$$

$$\frac{24,900 \text{ mi}}{360°} = 69 \text{ mi}$$

Because of the rapid convergence of the meridians northward or southward, this equivalent applies close to the equator only. It is also useful to know that the length of 1° of longitude is reduced to about one-half as much along the 60° parallels, or about 55½ km (34½ mi).

Latitude

The **latitude** of a place is the arc, measured in degrees, of a meridian between that place and the equator (Figure 1.8). Latitude thus ranges from 0° at the equator to 90° north or south at the poles. The latitude of a place, written as lat. 34° 10′ 31″N, may be read "latitude 34 degrees, 10 minutes, 31 seconds north." When both the latitude and longitude of a place are given, it is accurately and precisely located with respect to the geographic grid.

If the earth were a perfect sphere, the length of 1° of latitude (a one-degree arc of a meridian) would be a constant value everywhere on the earth. This length is almost the same as the length of a degree of longitude at the equator, so that the value of 111 km (69 mi) per degree may be used for ordinary purposes.

To be precise, and to take into account the oblateness of the earth, we must recognize that a degree of latitude changes slightly in length from equator to poles. The length of 1° of latitude at the equator is 110.6 km (68.7 mi); at the poles it is 111.7 km (69.4 mi), or 1.1 km (0.7 mi)

longer. One degree at the poles is 1 percent longer than at the equator. The difference is by no means trivial and must be taken into account in the precise construction of large-scale maps. Figure 1.9 shows why the length of a degree of latitude is longer at the poles than at the equator. A circle, fitted to the curvature of the ellipse at the pole, is larger in radius than a circle fitted to the ellipse at the equator. It is obvious from the drawing that a 5° arc of the larger circle is longer than a 5° arc of the smaller circle. Thus the length of a degree of latitude changes continuously from a minimum value at the equator to a maximum value at the pole. Table 1.1 gives approximate values of length of degrees of latitude and longitude at 15° intervals from equator to pole.

The Nautical Mile

Both marine and air navigation use the nautical mile as the unit of length or distance. Meteorology (weather science) of the upper atmosphere has also used as the unit of wind speed the mariner's **knot,** which is a speed of one nautical mile per hour. It is therefore worthwhile for a geographer to understand the nautical mile.

The **nautical mile** can be quite simply defined as the length of one minute of arc of the earth's equator. Because measurements of the length of the equator have been refined many times during the past century, precise values

Table 1.1 / Lengths of Degrees of Latitude and Longitude

Latitude, degrees	Length of 1° of latitude		Length of 1° of longitude	
	km	(mi)	km	(mi)
0	110.57	(68.70)	111.32	(69.17)
15	110.64	(68.75)	107.55	(66.83)
30	110.85	(68.88)	96.49	(59.96)
45	111.13	(69.05)	78.85	(48.99)
60	111.42	(69.23)	55.80	(34.67)
75	111.62	(69.36)	28.90	(17.96)
90	111.70	(69.40)	00.00	(00.00)

of the nautical mile have also been revised. As adopted in 1954 by the U.S. Department of Defense, the equivalents in both kilometers and statute miles (English miles) are as follows:

$$1 \text{ nautical mi} = 1.852 \text{ km} = 1.1508 \text{ statute mi}$$

For ordinary calculations, the nautical mile can be considered equal to 1.85 km (1.15 statute mi).

Map Projections

A **map projection** is an orderly system of parallels and meridians used as a basis for drawing a map on a flat surface. The fundamental problem of the mapmaker is to transfer the geographic grid from its actual spherical form to a flat surface in such a way as to present the earth's surface or some part of it in the most advantageous way possible for the purposes desired.

We can avoid the map-projection problem only by using a globe. Unfortunately, a globe has shortcomings. First, we can see only one side of a globe at a time. Second, a globe is on too small a scale for many purposes. On globes ranging from a few centimeters to a meter in diameter, only the barest essentials of geography can be shown. A few large globes in existence, those 3 to 5 m in diameter, show considerable detail. But they serve also to accentuate a third shortcoming of globes — their lack of portability. Flat maps printed on paper can be folded compactly and carried in a pocket, whereas even the smallest globe is a cumbersome and delicate object. Maps can be reproduced easily, whereas making a quality globe requires not only that a map be printed but also that the map be trimmed and carefully pasted in sections on a spherical shell.

The problem of map projection must be faced squarely. You will find it essential to know which types of networks of parallels and meridians are best suited to the illustration of various portions and properties of the earth's surface. Keep in mind that no map projection will ever substitute fully for a globe to show general world relationships. We recommend the use of a globe in conjunction with flat maps.

Developable Geometric Surfaces

Certain geometric surfaces are said to be developable because, by cutting along certain lines, they can be made to unroll or unfold to make a flat sheet. Two such forms are the cone and the cylinder (Figure 1.10). Were the earth conical or cylindrical, the map-projection problem would be solved once and for all by using the developed surface. No distortion of surface shapes or areas would occur, although it is true that the surface would be cut apart along certain lines. The earth, being spherical, belongs to a group of geometric forms said to be undevelopable because, no matter how they are cut, they cannot be unrolled or unfolded to lie flat. It is possible to draw a true straight line in one or more directions on the surface of a developable solid, but this cannot be done anywhere on

an undevelopable form, such as a spherical surface. To make the parts of a spherical surface lie perfectly flat, the surface must be stretched — more in some places than in others. So it is impossible to make a perfect map projection.

When a map is made of a very small part of the earth's surface, for example, an area 10 km across, the map-projection problem can be ignored. If the meridians and parallels are drawn as straight lines, intersecting at right angles and spaced apart correctly, the actual error is probably so small as to fall within the width of the lines drawn and is not worth correcting. As the area included on the map is increased, however, the problem gains in importance. When we attempt to show the whole globe, serious trouble develops. Only by some compromise can the distortion be reduced to a reasonable degree over important parts of the earth's surface.

Scale of Globes and Maps

All globes and maps depict the earths features in much smaller size than the true features they represent. Globes are intended in principle to be perfect models of the earth itself, differing from the earth only in size, but not in shape. The **scale** of a globe is therefore the ratio between the size of the globe and the size of the earth, where size is expressed by some measure of length or distance (but not area or volume). Take, for example, a globe 20 cm in diameter representing the earth, whose diameter is about 13,000 km. The scale of the globe is therefore the ratio between 20 cm and 13,000 km. Dividing 13,000 by 20, this ratio reduces to a scale stated as follows: 1 cm represents 650 km. This relationship holds true for distances between any two points on the globe.

Scale is more usefully stated as a simple fraction, termed the **fractional scale,** or **representative fraction (RF),** which can be obtained by reducing both earth and globe distances to the same unit of measure, which in this case is centimeters:

$$\frac{1 \text{ cm on globe}}{650 \text{ km on earth}} = \frac{1 \text{ cm}}{650 \times 100,000} = \frac{1}{65,000,000}$$

This fraction may be written as 1:65,000,000 for

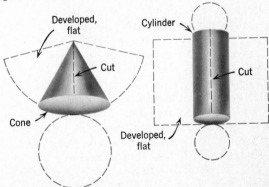

Figure 1.10 **The cone and the cylinder are developable geometric forms.**

convenience in printing. The advantage of the representative fraction is that it is entirely free of any specified units of measure, such as the foot, mile, meter, or kilometer. Persons of any nationality can understand the fraction, regardless of their language or units of measure.

A globe is a true-scale model of the earth in that the same scale of kilometers or miles applies to any distance on the globe, regardless of the latitude or longitude and regardless of the compass direction of the line whose distance is being considered (Figure 1.11*A*). Thus the scale remains constant over the entire globe. Map projections, however, cannot have the uniform-scale property of a globe, no matter how cleverly devised. In flattening the curved surface of the sphere to conform to a plane surface, all map projections stretch the earth's surface in a nonuniform manner so that the map scale changes from place to place. Thus we cannot say about a map of the world, "the scale of this map is 1:65,000,000" because the statement is false for any form of projection.

It is quite possible, however, to have the scale of a flat map remain true, or constant, in certain specified directions. For example, one type of projection preserves constant scale along all parallels, but not along the meridians. This condition is illustrated in Figure 1.11*B*. Another type of projection keeps a constant scale along all merid-

ians, but not along parallels, as shown in Figure 1.11*C*. Still other projections have changing scale along both meridians and parallels, as illustrated in Figure 1.11*D*.

Preserving Areas on Map Projections

Because a globe is a true-scale model of the earth, given areas of the earth's surface are shown to correct relative size everywhere over its surface. If we should take a small wire ring, say 2 cm in diameter, and place it anywhere on the surface of the 20-cm globe, the area enclosed by the ring will represent the same area of earth's surface. But a similar procedure would not enclose constant areas on all parts of most map projections, only on one having the special property of being an **equal-area projection.**

At this point a good question arises. If, as we have stated, no projection preserves a true, or constant, scale of distances in all directions over the projection, how can circles of equal diameter placed on the map enclose equal amounts of earth area? The answer is suggested in Figure 1.12. The square, 1 km on a side, encloses 1 sq km between two meridians and two parallels. The square can be deformed into rectangles of different shapes, but if the dimensions are changed in an inverse manner, each will still enclose 1 sq km. The scale has been changed in one direction to compensate for change in another in just the right way to preserve equal areas of map between corresponding parts of intersecting meridians and parallels. On such a map, any small square or circle moved about over the map surface will enclose a piece of the map representing a constant quantity of area of the earth's surface. Projections shown in Figures 1.21 and 1.22 have the equal-area property, but it is also obvious that these networks have strong distortions of shape, particularly near the outer edges of the map.

Preserving Shapes on Map Projections

A projection is said to be a **conformal projection** when any small piece of the earth's surface has the same shape on the map as it does on a globe. Thus the appearance of small islands or countries is faithfully preserved by a conformal map. One characteristic of a conformal projection is that parallels and meridians cross each other at right angles everywhere on the map, just as they do on

Figure 1.11 **A, Scale true in all directions on a globe. B, Scale true along all parallels but not along all meridians. C, Scale true along all meridians but not along all parallels. D, Scale changes along both parallels and meridians.**

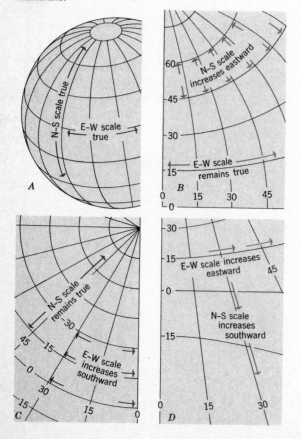

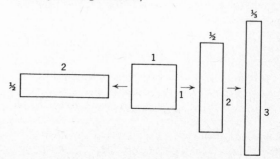

Figure 1.12 **Areas can be preserved even though scales and shapes change radically.**

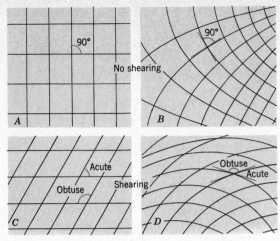

Figure 1.13   **Shearing of areas on map projections.**

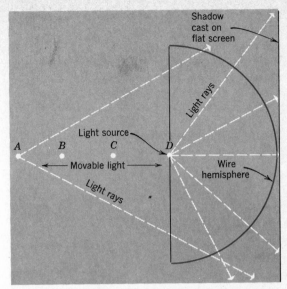

Figure 1.14   **Principle of the zenithal projections.**

the globe. (Not all projections whose parallels and meridians cross at right angles are conformal.)

Another way of saying that parallels and meridians intersect at right angles is that shearing of areas does not occur. Figure 1.13 illustrates the meaning of **shearing.** For projections consisting of straight parallels and meridians, shearing gives parallelograms formed of acute and obtuse angles. For projections with curved meridians and parallels, straight lines are drawn tangent to the curves at the point of intersection. If these tangent lines cross at right angles, the projection is not sheared; but if the tangents form obtuse and acute angles, shearing is present. Although conformal maps are without shearing, not all maps without shearing are conformal. A conformal map cannot also have equal-area properties, so some areas are greatly enlarged at the expense of others. Generally speaking, areas near the margin of a conformal map have a much larger scale than areas near the center.

Whether we should select a conformal projection or an equal-area projection depends on what is to be shown. When the surface extent of something, such as grain crops or forest-covered lands, is to be shown, an equal-area projection is needed. For most general purposes, a conformal type is preferable because physical features most nearly resemble their true shapes on the globe. Many map projections are neither perfectly conformal nor equal area, but represent a compromise between the two. This compromise may achieve a map of more all-around usefulness. In some cases, the projection has some special property that makes it essential for a specific use, such as for navigation.

Classification of Map Projections

Three important classes of map projections are the following: (1) zenithal, (2) conic, and (3) cylindric. Many unique types, not falling into these groups, have also been invented.

Zenithal projections are centered about a point and have a radial, or wheellike, symmetry. Some zenithal projections can actually be demonstrated by use of a wire replica of the earth, in which the wires represent parallels and meridians (Figure 1.14). A tiny light source, such as a flashlight bulb or an arc light, is placed at the center of the wire globe, or at any one of several prescribed positions, shown by letters A, B, C, D, in Figure 1.14. In a darkened room the shadow of the wire globe is cast upon a screen, a wall, or the ceiling. This shadow is a true geometric projection.

All zenithal projections are characterized by the following properties (Figure 1.15):

1. A line drawn from the center point of the map to any other point gives the true compass direction taken by a great circle as it leaves the center point, headed for the outer point.

Figure 1.15   **Properties of zenithal projections.**

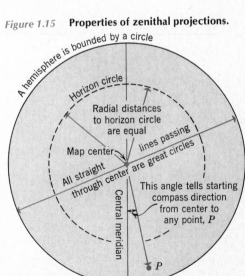

2. When a complete globe or hemisphere is shown, the map is circular in outline. (Because any map can be trimmed down to have a circular outline, this feature is not a reliable criterion of the zenithal class.)

3. The map possesses radial symmetry, in which all its properties are arrayed in wheellike fashion around a center point. All changes of scale and distortion of shapes occur uniformly (concentrically) outward from this center.

4. All points equidistant from the center lie on a circle, known as a **horizon circle.** When the entire globe is shown by a zenithal map, the circular edge of the map represents the opposite point, or **antipode,** on the globe. When a hemisphere is shown, the outer edge of the map represents a great circle, everywhere equidistant from the point on which the projection is centered.

5. All great circles that pass through the center point of the projection appear as straight lines on the map. Likewise, all straight lines drawn through the center point of the map are true great circles.

Zenithal projections appear in three positions, or orientations: (1) polar, (2) equatorial, and (3) oblique, or tilted, as illustrated in Figure 1.18. In the polar position, the center of the projection coincides with the north or south pole; in the equatorial position, the center is somewhere on the equator; and in the oblique position, the center is at some point intermediate between the equator and poles. Although the equatorial and oblique types may not look radially symmetrical, they possess, just as truly as the polar type, the five characteristics of all zenithal projections.

Conic projections are based on the principle of transferring the geographic grid from a globe to a cone, then developing the cone to a flat map. This principle, too, can be demonstrated with the wire globe and a point source of light (Figure 1.16). Instead of a vertical flat screen, however, a clear plastic cone is seated on the wire globe, much as a lampshade is seated on a lamp. The shadow of the wires cast upon the conical shade gives a conic projection. If this shadow were traced in pencil and the cone unrolled, a true conic projection would result.

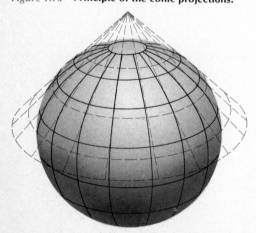

Figure 1.16   **Principle of the conic projections.**

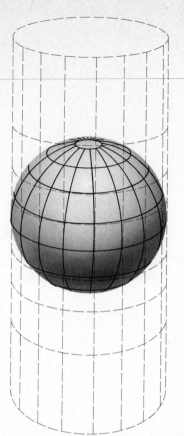

Figure 1.17   **Principle of the cylindric projections.**

Conic projections possess the following features (see Figure 1.19):

1. All meridians are straight lines, converging to a common point at the north (or south) pole.
2. All parallels are arcs of concentric circles, whose common center lies at the north (or south) pole.
3. A complete conic projection is a sector of a circle, never a complete circle.
4. A conic projection cannot show the whole globe and usually shows little more than the northern (or southern) hemisphere.

Cylindric projections are based on the principle of transferring the geographic grid first to a cylinder wrapped about the earth, then unrolling the cylinder to make a flat map (Figure 1.17). Simple cylindric projections are easy to draw because they consist of intersecting horizontal and vertical lines (see Figure 1.21). The completed map is rectangular in outline, and the whole circumference of the globe can be shown. When the cylinder is tangent to the equator (as in Figure 1.17), meridians are equally spaced vertical lines. Parallels are spaced in various ways, according to the particular projection desired.

Many other kinds of map projections have been invented, each based on some unique principle. We have selected several important map projections to illustrate useful types spanning a wide range of properties. Included are both conformal and equal-area types.

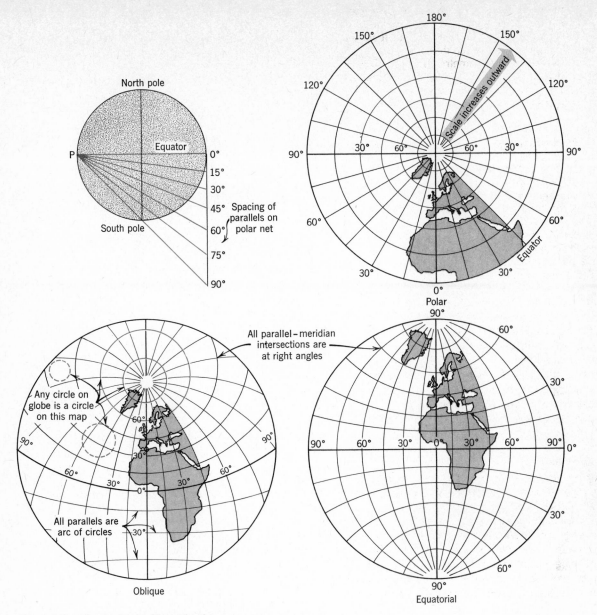

Figure 1.18   **The stereographic projection.**

Stereographic Projection

The **stereographic projection** is a conformal projection belonging to the zenithal class. Figure 1.18 shows how an accurate stereographic net can be constructed by using a compass, a straightedge, and a protractor. The construction lines, or rays, emanate from a point diametrically opposite to the point where the tangent plane touches the globe. The stereographic net can show much more than one hemisphere, although it cannot show the whole globe. Parallels and meridians show closer spacing near the map center and increasingly wider spacing toward the outer margins. On any stereographic projection the parallels and meridians are either straight lines or true arcs

of circles. Because the stereographic projection is truly conformal, all lines that are circles on the globe are shown as circles on the map. The map scale, however, always increases from the map center toward the periphery.

With the enormous growth in importance of polar regions in the age of scientific discovery and long-range missile and aircraft operation, the polar stereographic projection has assumed great importance.

World Aeronautical Charts issued by the U.S. National Ocean Survey are based on a polar stereographic projection for lat. 80° to 90°. The data of scientific observations of the Arctic Ocean and Antarctic continent are usually shown on this projection. The National Weather Service weather maps use a polar stereographic projection.

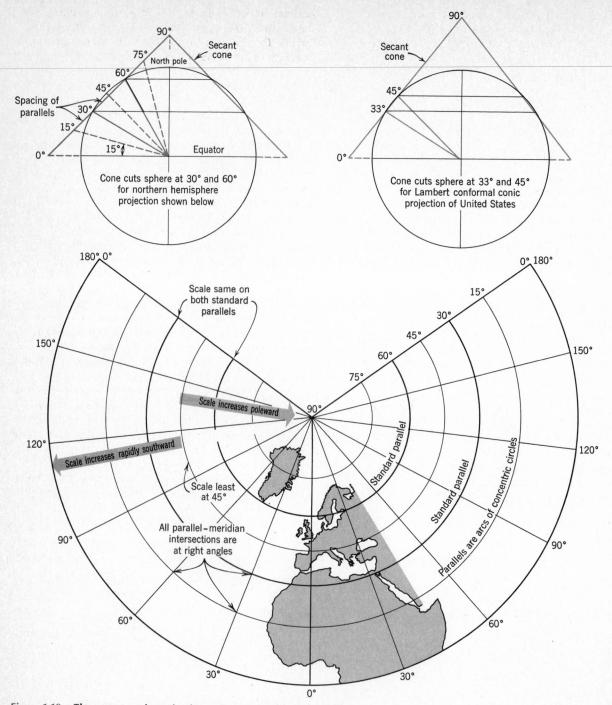

Figure 1.19 **The secant conic projection, using two standard parallels.**

The Secant Conic Projection

Construction principles for the **secant conic projection** are shown in Figure 1.19. Instead of using a tangent cone, as suggested in Figure 1.16, a **secant cone** is selected to pass through the sphere, intersecting the spherical surface on two **standard parallels** of latitude. This method gives a map with much smaller scale distortion in the midlatitude region than when the tangent cone is used. The resulting conic projection consists of straight meridians radiating from the pole, while parallels are concentric circles. No

shearing is present, an important asset to any projection.

The conic projection can easily be made into a true conformal conic projection by spacing the parallels according to a mathematical formula. This adjustment yields the **Lambert conformal conic projection,** an extremely important grid for many categories of U.S. maps. The Lambert projection, as adapted for the United States, uses the 33rd and 45th parallels as standard parallels, reducing the average scale error to the minimum for the 48 contiguous United States. Because a straight line on this map closely approximates a great circle, the map is useful in air navigation. The National Ocean Survey uses the Lambert projection as the base for aeronautical charts of the United States on the scale of 1 : 500,000, and for world aeronautical charts on the scale of 1 : 1,000,000.

Mercator Projection

Perhaps the best known of all map projections is the **Mercator projection,** which was devised by Gerardus Mercator in 1569 (Figure 1.20). It is based on a mathematical formula. The principle can, however, be explained without mathematics. On any cylindrical projection in which the meridians are straight vertical lines, equidistantly spaced, the meridians have had to be spread apart (see right side of Figure 1.20). Only along the equator are meridians the same distance apart as on a globe of the same equatorial scale. To maintain them as parallel lines, the normally converging meridians have had to be spread apart in a greater and greater ratio as the poles are approached. At lat. 60°N and S, the meridians are spread apart twice as far as originally because at that place a degree of longitude is only half what it is at the equator. At the poles, the spreading is infinitely greater because the poles themselves are infinitely tiny points. Now, to maintain the map as a truly conformal map, we must space the parallels increasingly far apart toward the poles, using the same ratio of increase that resulted when the meridians were spread to make vertical lines. For example, near the 60th parallel north, parallels must be spread twice as far apart as on the globe because, as already explained, the meridians here are also spread twice as far apart. Thus, at lat. 60°N, the map scale is double that at the equator. At lat. 80° the map scale is

enlarged almost six times. Near the poles the spacing of parallels increases enormously and rapidly approaches infinity. Therefore the Mercator map is usually trimmed off at about lat. 80° or 85°N and S. The poles can never be shown.

The Mercator chart is a true conformal projection. Any small island or country is shown in its true shape. The scale of the map, however, becomes enormously greater toward the poles.

The really important, unique feature of an equatorial Mercator projection is that a straight line drawn anywhere on the map, in any direction desired, is a line of constant compass bearing. Such a line is known to navigators as a **rhumb line,** or **loxodrome** (Figure 1.20). If this line is followed, the ship's (or plane's) compass will show that the course is always at a constant angle with respect to geographic north. Once the proper compass bearing is determined, the ship is kept on the same bearing throughout the voyage, if the rhumb line is to be followed. The Mercator chart is the only one of all known projections on which all rhumb lines are true straight lines, and vice versa. A protractor can be used with reference to any meridian on the map, and the compass bearing of any straight line can be measured off directly.

The relation of great-circle routes to rhumb lines is also shown in Figure 1.20. The great-circle course is actually the shortest surface distance between two selected points, although on the Mercator map the rhumb line usually appears to be shorter. Along the equator and all meridians (but only on these lines), rhumb lines and great circles are identical and are straight lines on both charts.

Although indispensable for navigational uses, the Mercator projection has serious shortcomings for use as a world map to show geographic information dealing with

Figure 1.20 **The Mercator projection.**

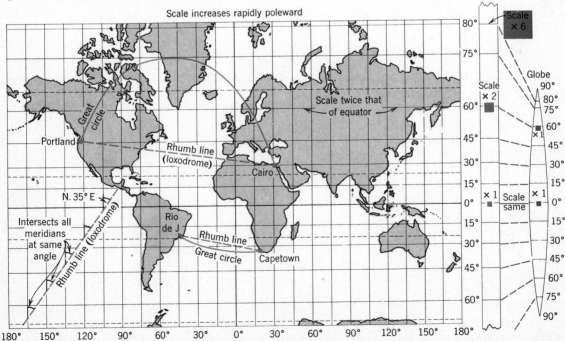

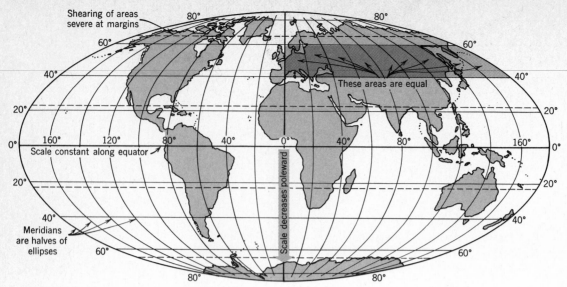

Figure 1.21   The homolographic projection.

areas of distribution. Except for equatorial regions (for which it provides an excellent grid), distortions of scale are very serious. Because of infinite stretching toward the poles, this map fails completely to show how the land areas of North America, Asia, and Europe are grouped around the polar sea. In the mind of an inexperienced user, it may enhance a false sense of isolation between these lands.

On the other hand, certain forms of geographic information are best shown on the Mercator projection. Because of its accurate depiction of the compass directions of lines, the Mercator net is preferred for maps of direction of flow of ocean currents and winds, direction of pointing of the compass needle, or lines of equal value

of air pressure and air temperature. Examples of such uses of the Mercator projection will be seen in later chapters. The Mercator projection is often supplemented by two stereographic projections, one for each polar region.

Homolographic Projection

A projection rather widely used by geographers to show the entire globe is the **homolographic projection** (Figure 1.21). "Homolographic" is a word often used to mean "equal area," a property this projection has. One hemisphere is outlined by a circle; the other hemisphere, divided into two parts, is included within an ellipse. All meridians except the straight central meridian and the

Figure 1.22   The sinusoidal projection.

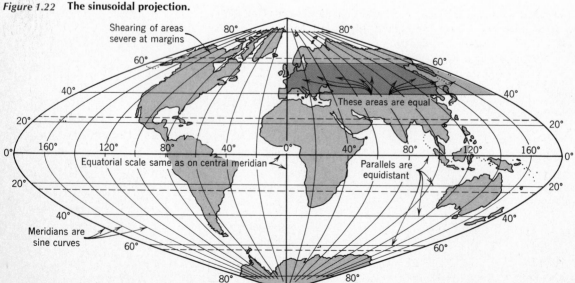

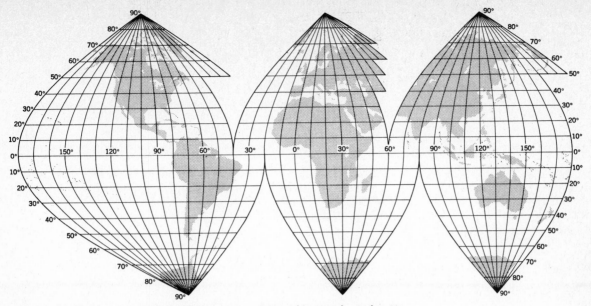

Figure 1.23 **The interrupted sinusoidal projection, designed in 1970 by Arthur N. Strahler. Interruption of Eurasia on the 60th meridian east minimizes distortion in eastern Asia. (From A. N. Strahler, 1971, *The Earth Sciences,* 2nd ed., Harper and Row, New York.)**

hemisphere circle, are halves of ellipses. The equator is twice as long as the central meridian, which is also true on a globe. Parallels are straight, horizontal lines, becoming more closely spaced toward the poles.

The homolographic projection has distinct advantages and disadvantages. Its equal-area property makes it valuable for showing the global distribution of geographic properties that cover areas, such as crops, soil types, or political units. It is an excellent grid for low-latitude regions. For example, it is often used for maps of Africa. Severe distortion in the polar regions, however, has hindered its wider use.

Sinusoidal Projection

In some ways the **sinusoidal projection** is similar to the homolographic projection. It is an equal-area projection with straight central meridian and horizontal straight parallels (Figure 1.22). The difference lies in the type of curve used in meridians. Whereas the homolographic net uses ellipses, the sinusoidal net uses **sine curves.** The parallels are uniformly spaced apart (equidistant). For this reason, more space is available for the high-latitude zones than on the homolographic projecton. Even so, shearing at high latitudes is severe.

Figure 1.24 **Goode's interrupted homolosine projection. (Based on Goode Base Map. Copyright by the University of Chicago. Used by permission of the Geography Department.)**

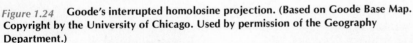

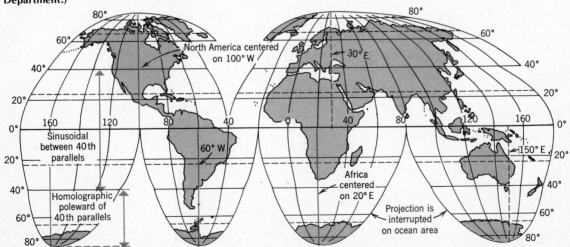

To reduce shearing at high latitudes, the sinusoidal projection can be interrupted (split) along selected meridians, as shown in Figure 1.23. Each sector (gore) of this interrupted projection is centered on a different meridian, minimizing the shearing in that sector. Some overlap of sector boundaries can be shown. The **interrupted sinusoidal projection** is an excellent grid on which to show the areal extent of a geographic property. Examples in a later chapter are world maps of solar radiation and longwave radiation (Figure 4.12).

Homolosine Projection

The **homolosine projection,** invented by Dr. J. Paul Goode in 1923, is a combination of the homolographic and sinusoidal types. The sinusoidal projection is used between lat. 40°N and 40°S, the homolographic for the remaining poleward parts. The interrupted form of the homolosine projection, Figure 1.24, shows North America, Eurasia, South America, Africa, and Australia, each based on the best-suited meridian. In another variation of this map, Eurasia is split down the meridian of 60°E, greatly improving the presentation of eastern Asia.

Finally, as a means of showing information about land areas to greatest possible advantage on the limited space of a book page, the interrupted homolosine projection can be compressed by carving out intervening ocean areas and pulling the continents together. See, for example, our world map of climate, Plate D2.

Selecting a Map Projection

Select a map projection best suited to the class of information it must show. Use an equal-area projection to show the distribution of geographic entities that occupy surface area — soils, vegetation types, climates, or political units. Always choose a conformal projection to show systems of lines whose compass directions are important and should not be distorted — lines of equal temperature or pressure, winds, ocean currents. Use a polar projection for high latitudes.

Certain environmental properties of special interest in physical geography change systematically from the equator toward either pole. These properties are strongly dependent on latitude. Examples are world temperatures, climates, soils, and vegetation types. For such subjects, always choose a projection with horizontal, straight parallels so the eye can easily follow a given latitude zone across the entire map.

Before you begin to analyze data and draw conclusions from a world map, be sure you know the qualities of the projection, whether equal-area, conformal, or neither. If in doubt, compare two or more projections of the same regions and see what possible erroneous concepts each might give. Consult a good globe whenever points or areas are to be related in distance and direction. There is no fully satisfactory substitute for a true-scale model of the earth's surface features.

References for Further Study

Steers, J. A. (1962), *An introduction to the study of map projections,* thirteenth edition, Univ. of London Press.
Robinson, A. H., and R. D. Sale (1969), *Elements of cartography,* third edition, John Wiley and Sons, New York.
Strahler, A. N. (1971), *The earth sciences,* second edition, Harper and Row, New York. See Chapter 10.
Richardus, P., and R. K. Adler (1972), *Map projections: an introduction,* American Elsevier Publishing Co., New York.
Strahler, A. N. (1975), *Physical geography,* fourth edition, John Wiley and Sons, New York. See Chapters 1 and 2.

Review Questions

1. Offer four simple proofs of the earth's approximate sphericity. How did Eratosthenes measure the earth's circumference?
2. Distinguish between gravitation and the earth's gravity. What is the environmental importance of gravity?
3. How did Richer contribute evidence that the earth is an oblate ellipsoid rather than a true sphere?
4. What geometric form has a cross section of the earth cutting through the poles? What is the oblateness of the earth? What fraction approximately expresses the oblateness?
5. What is a great circle? How is it formed? What is a small circle? List six properties of great circles. Of what practical importance are great circles?
6. What is a meridian? How are meridians formed on a globe? List the characteristics of meridians.
7. What is a parallel? How are parallels formed? List the characteristics of parallels.
8. Define longitude. How is longitude written? Give an example. What is a prime meridian? Where is the Greenwich meridian? How long is a degree of longitude at the equator? at lat. 60°? at the poles?
9. Define and explain latitude. How is latitude written? Give an example. How long is a degree of latitude from equator to poles? How much? Why?
10. What is a nautical mile? A knot? Give the approximate equivalent of the nautical mile in terms of kilometers and statute (English) miles.
11. What is a map projection? What is the basic problem of map projection? For what purposes is a globe inferior to a flat map? For what purposes is a globe preferable to a flat map?
12. How do the concepts of developable and undevelopable geometric surfaces apply to the problem of map projections?
13. Explain the concept of globe and map scale. What is the representative fraction? Do any map projections preserve constant scale in all places and directions over the map?
14. What is an equal-area map projection? Name two equal-area types.
15. What is a conformal map projection? Name two conformal types.
16. List the properties of zenithal map projections. In what three positions can a zenithal projection be oriented?

17. What characteristic has a conic projection? a cylindric projection?
18. Why is the stereographic projection favored for polar regions? Why is it a good base for aeronautical charts?
19. How is a secant cone used in constructing a useful conic projection? What advantages has the Lambert conformal conic projection?
20. Explain the principle of the Mercator projection. Is this an equal-area or conformal projection? What shape has a complete Mercator map showing the entire globe? How does the map scale at lat. 60° compare with the equatorial scale?
21. What is a rhumb line? a loxodrome? What projection shows all rhumb lines as straight lines? What use is made of rhumb lines and great circles in navigation?
22. Compare the homolographic and sinusoidal projections. Are these projections equal-area or conformal?
23. What advantages has Goode's interrupted homolosine projection over the homolographic and sinusoidal projections?

Seasons and Time

2

The varied environments of life on our planet depend to a large degree on the way the sun's rays strike a spherical earth. The angle of attack of the solar energy beam, varying greatly with latitude and with time of year, determines many commonplace phenomena — the daily path of the sun across the sky, the changing lengths of day and night, and the annual rhythm of the seasons. These daily and seasonal rhythms in turn act as fundamental controls of air temperatures, winds, ocean circulation, precipitation, and storms — all of which taken together make up the earth's varied climates.

To understand earth-sun relationships requires thinking in three dimensions. You will need to visualize a spherical earth spinning like a top on its axis, but at the same time moving in a circular path around the sun. Superimposed on this simple system of motion is a tilt of the earth's axis with respect to the plane in which it travels. We can look at this earth-sun system from the purely imaginary viewpoint of an observer far out in space; but then the same motions must be recast into the real viewpoint of an observer on the earth's surface, turning with the earth and oriented by the earth's gravity.

Rotation of the Earth

The spinning of the earth on its polar axis is termed **earth rotation.** In the study of earth-sun relationships, we use a period of rotation, the **mean solar day,** consisting of 24 mean solar hours. This day is the average time required for the earth to make one complete turn with respect to the sun.

Direction of earth rotation can be determined by using one of the following rules: (1) Imagine yourself to be looking down on the north pole of the earth; the direction of turning is counterclockwise. (2) Place your finger on a point on a globe near the equator and push eastward. You will cause the globe to rotate in the correct direction (Figure 2.1); this explains the common expression "eastward rotation of the earth." (3) The direction of earth rotation is opposite that of the apparent motion of the sun, moon, and stars. Because these bodies appear to travel westward across the sky, the earth must be turning in an eastward direction.

The speed of travel of a point on the earth's surface in a circular path, because of rotation alone, is roughly computed by dividing the length of parallel at the latitude of the point by 24. Thus at the equator, where the circumference is about 40,000 km (25,000 mi), the eastward velocity of an object on the surface is about 1700 km (1050 mi) per hour. At the 60th parallel the velocity is half this amount, or about 850 km (525 mi) per hour. At the poles it is, of course, zero. We are unaware of this motion because the earth's rotation is constant.

Proof of the Earth's Rotation

Finding a satisfactory proof that the earth rotates on its axis was a frustrating problem of astronomy for several centuries. Those astronomers who believed that a fixed earth was the center of the universe — adherents to the Ptolemaic system — held a powerful grip on science until well into the fifteenth century, when the Polish astronomer Nicolaus Copernicus argued that the earth was one of several planets moving in orbits about the sun and experiencing axial rotation. The Copernican theory became generally accepted through the efforts of another leading astronomer, Johannes Kepler (1571-1630), and was put on a sound physical basis by use of Isaac Newton's laws of motion and gravitation (1687). Newton predicted that a rotating earth would be deformed into a oblate

Figure 2.1 **Direction of rotation is counterclockwise at the north pole (A) and eastward at the equator (B).**

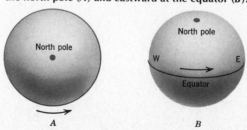

Figure 2.2 **This handsome Foucault pendulum knocks over pins in succession to show that its direction of motion is changing. (Franklin Institute, Philadelphia, Pa.)**

ellipsoid (see Chapter 1). By 1740 geodetic surveys had confirmed that the earth is, indeed, an oblate ellipsoid.

It remained for a French physicist, Jean Bernard Léon Foucault, to devise a public demonstration so forceful that even the layperson could be convinced of the reality of earth rotation. Known as the **Foucault pendulum,** the type of apparatus he used can be seen in action today in the United Nations Building in New York City and in many museums and universities (Figure 2.2). In 1851 Foucault suspended a cannonball from the dome of the Pantheon in

Figure 2.3 **Principle of the Foucault pendulum.**

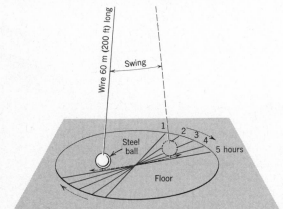

Paris, using a slender wire about 60 m (200 ft) long. Once set in a swinging motion, the ball underwent a steady change in direction of swing. It was obvious that the earth's surface was slowly turning, but that this motion was not being transmitted to the cannonball. Free of mechanical coupling with the earth, the massive ball was maintaining its space path in conformance with Newton's first law of motion. (Every body remains in a state of rest or of uniform motion in a straight line unless compelled to change that state by an external force acting upon it.)

If a Foucault pendulum were set in operation at the north pole, the change of the pendulum swing would be clockwise, a shown in Figure 2.3, and would complete a 360° change of direction in 24 hours. There the change of direction is 15° per hour. At the equator, the swing path does not change direction at all. At intermediate latitudes, the rate of change in swing path ranges between 0° (equator) and 360° per 24 hours (pole). For example, at New York City, about lat. 40°N, the hourly change of direction is about 10°, and 37 hours are required for 360° of change.* Table 2.1 gives values for 15° intervals of latitude.

Table 2.1 / The Foucault Pendulum at Various Latitudes

Latitude	Hourly Change in Pendulum Direction, Degrees	Total Time for 360° Change in Direction, Hours
0°	None	None
15	3.9	93
30	7.5	48
45	10.6	34
60	13.0	28
75	14.5	25
90	15.0	24

Environmental Effects of Earth Rotation

The physical and biological effects of earth rotation are truly profound in terms of the environmental processes of the life layer. First, and perhaps most obvious, is that rotation imposes a daily, or diurnal, rhythm on many phenomena to which plants and animals respond. These phenomena include light, heat, air humidity, and air motion. Plants respond to the daily rhythm by storing energy during the day and releasing it at night. Animals adjust their activities to the daily rhythm, some preferring the day, others the night for food-gathering activities. The daily cycle of input of solar energy and a corresponding cycle of air temperature will be important topics for analysis in Chapter 4 and 5.

Second, as we shall find in the study of the earth's systems of winds and ocean currents, earth rotation causes the flow paths of both air and water to be consistently turned in compass direction: toward the right in the northern hemisphere and toward the left in the southern

*The following formula applies: Rate of direction change, degrees per hour = 15° × sine of latitude.

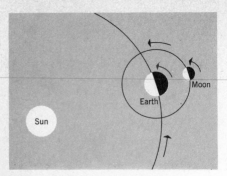

Figure 2.4 **Direction of earth rotation and revolution.**

hemisphere. This phenomenon goes by the name of the Coriolis effect; we shall investigate it in Chapter 6.

A third physical effect of the earth's rotation is of environmental importance. Because the moon exerts its gravitational attraction on the earth, while at the same time the earth is turning with respect to the moon, tidal forces are generated. Tidal forces induce a rhythmic rise and fall of the ocean surface — the ocean tide. These motions in turn cause water currents of alternating direction to flow in shallow waters of the coastal zone. To a grain farmer in Kansas, the ocean tide may have no significance; but, for the clam digger and charter-boat captain on Cape Cod, the tidal cycle is a clock that regulates daily activities. For many kinds of plants and animals of saltwater estuaries, tidal currents are essential to maintain a suitable life environment. The tide and its currents are discussed in Chapter 22.

Earth Revolution

The motion of the earth in its travel path, or **orbit,** around the sun is termed **revolution.** The period of revolution, or **year,** is the time required for the earth to complete one circuit around the sun. The year is, however, defined by astronomers in several ways. For example, the time required for the earth to return to a given point in its orbit

with reference to the fixed stars is called the sidereal year.

For earth-sun relationships we use the **tropical year,** which is the period of time from one vernal equinox to the next. The tropical year has a length of approximately 365¼ days. Every four years the extra one-fourth day difference between the tropical year and the calendar year of 365 days accumulates to nearly one whole day. By inserting a 29th day in February every leap year, we are able to correct the calendar with respect to the tropical year. Further minor corrections are necessary to perfect this system.

In its orbit, the earth moves counterclockwise when viewed from a point in space above the north pole (Figure 2.4). This is the same direction of turning as the earth's rotation.

Perihelion and Aphelion

The average distance between earth and sun is about 150 million km (93 million mi). Because the earth's orbit is an ellipse, rather than a circle, the distance may be 2½ million km (1½ million mi) greater or less than the average value (Figure 2.5). The distance is least, or about 147½ million km (91½ million mi) on about January 3, at which time the earth is said to be in **perihelion.** This word comes from the Greek *peri,* around or near, and *helios,* the sun). On about July 4 the earth is at its farthest point from the sun, or in **aphelion** (Greek *ap,* away from, and *helios*), at a distance of 152½ million km (94½ million mi).

These differences in distance cause minor differences in the amount of solar energy received by the earth, but they are not the cause of summer and winter seasons. This is obvious because perihelion, when the earth receives more heat, falls at the coldest time of the year in the northern hemisphere. Moreover, opposite seasons exist simultaneously in the northern and southern hemispheres, proving that another cause exists. Instead, the seasons are the result of the tilt of the axis of rotation.

In theory, however, summers and winters should be slightly intensified in the southern hemisphere and slightly moderated in the northern hemisphere as a result of the coincidence of perihelion and aphelion with summer and winter season.

Tilt of the Earth's Axis

Imagine the earth's axis to be exactly perpendicular with the plane in which the earth revolves about the sun. Astronomers call the plane containing the earth's orbit the **plane of the ecliptic.** Under these imagined conditions, the earth's equator would lie exactly in the plane of the ecliptic. The sun's rays, which furnish all energy for life processes on earth, would always strike the earth most directly at a point on the equator. In fact, at the equator, the rays would be exactly perpendicular to the earth's surface at noon. The rays would always just graze the north and south poles. Conditions on any given day would be exactly the same as conditions on every other day of the year (assuming the orbit to be circular). In other words, there would be no seasons.

In reality, the earth's axis is not perpendicular to the

Figure 2.5 **Earth's orbit and seasons.**

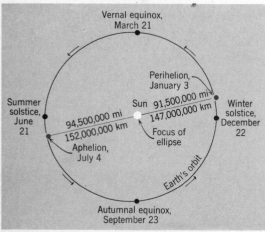

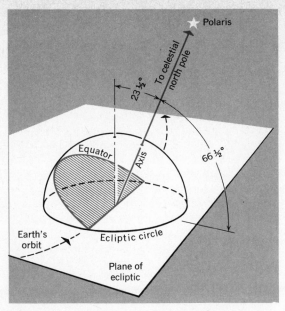

Figure 2.6 **At all times the earth's axis maintains an angle of 66½° with the plane of the orbit (plane of the ecliptic).**

fixed orientation with respect to the stars. The north end of the earth's axis points constantly toward Polaris, the North Star. To visualize this movement, hold a globe so as to keep the axis tilted at 66½° with horizontal. Move the globe in a small horizontal circle, representing the orbit, at the same time keeping the axis pointed at the same point on the ceiling.

Solstice and Equinox

What is the consequence of the facts that (1) the earth's axis keeps a fixed angle with the plane of the ecliptic and (2) the axis always points to the same place among the stars? You will find that at one point in its orbit, the north end of the earth's axis is tilted toward the sun; at an opposite point in the orbit, it is tilted away from the sun. At the two intermediate points, the axis is not tilted with respect to the sun's rays (Figure 2.7). Next, consider the four critical positions in detail.

On June 21 or 22 the earth is so located in its orbit that the north polar end of its axis leans at the maximum angle 23½° toward the sun. The northern hemisphere is tipped toward the sun. This event is named the **summer solstice.** Six months later, on December 21 or 22, the earth is in an equivalent position on the opposite point in its orbit. At this time, known as the **winter solstice,** the axis again is at a maximum inclination with respect to a line drawn to the sun, but now it is the southern hemisphere that is tipped toward the sun.

Midway between the dates of the solstices occur the **equinoxes,** at which time the earth's axis makes a 90° angle with a line drawn to the sun, and neither the north nor the south pole has any inclination toward the sun. The **vernal equinox** occurs on March 20 or 21; the **autumnal equinox** occurs on September 22 or 23. Conditions are

plane of the ecliptic; it is inclined by a substantial angle of tilt, measuring almost exactly 23½° from the perpendicular. Figure 2.6 shows this axial tilt in a three-dimensional perspective drawing. The angle between axis and ecliptic plane is then 66½° (90° − 23½° = 66½°).

To proceed we must couple the fact of axial tilt with a second fact: The earth's axis, while always holding the angle 66½° with the plane of the ecliptic, maintains a

Figure 2.7 **The seasons occur because the tilted earth's axis keeps a constant orientation in space as the earth revolves about the sun.**

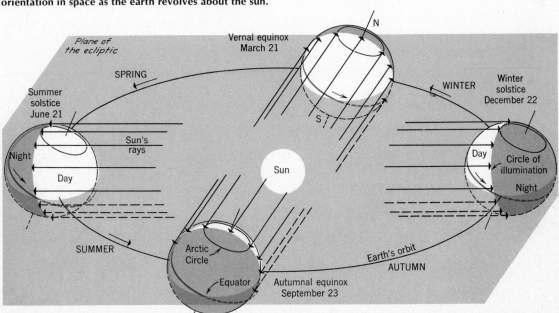

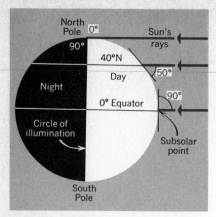

Figure 2.8 Equinox conditions. From this viewpoint the earth's axis appears to have no inclination.

identical on the two equinoxes as far as earth-sun relationships are concerned, whereas on the two solstices the conditions of one are the exact reverse of the other.

The Equinoxes

Consider first the conditions at the equinoxes because this is the simplest case. Figure 2.8 shows that the earth is at all times divided into two hemispheres with respect to the

sun's rays. One hemisphere is lighted by the sun; the other lies in darkness. Separating the hemispheres is a circle, the **circle of illumination;** it divides day from night. Notice that at either equinox the circle of illumination cuts precisely through the north and south poles. Looking at the earth from the side, as in Figure 2.8, conditions at equinox are essentially as follows: The point at which the sun's noon rays are perpendicular to the earth, the **subsolar point,** is located exactly at the equator. Here the angle between the sun's rays and the earth's surface is 90°. At both poles, the sun's rays graze the surface. As the earth turns, the equator receives the maximum intensity of solar energy; the poles receive none. At an intermediate latitude, such as lat. 40°N, the rays of the sun at noon make an acute angle (50°) with the earth's surface. The circle of illumination at the equinoxes passes through the poles and hence coincides with the meridians as the earth turns.

Further details of equinox conditions are shown in Figure 2.9. The parallels are divided into equal halves by the circle of illumination. Consequently, day and night are of exactly equal length, 12 hours each, at all latitudes. This fact explains the word equinox, from the latin *aequus,* equal, and *nox,* night. (We are not taking into account twilight, which extends the period of daylight before sunrise and after sunset.) Conditions are the same for both the northern and southern hemispheres. Sunrise occurs at 6:00 A.M. (local solar time) and sunset occurs at 6:00 P.M. at all places on the globe, except at the poles, where

Figure 2.9 Details of earth-sun relations at equinox. The sun's altitude above the horizon is shown for various latitudes. (From A. N. Strahler, 1971, *The Earth Sciences,* 2nd ed., Harper and Row, New York.)

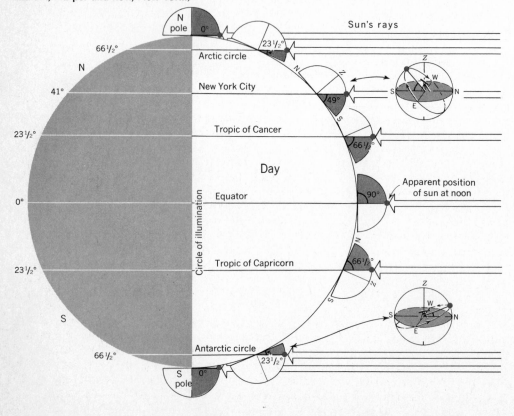

special conditions prevail.

Solar noon occurs when the sun reaches its highest point in the sky. Noon occurs simultaneously at all points having the same longitude, that is, all points lying on a single meridian. The vertical angle of the sun above the horizon at noon, designated the **sun's noon altitude,** may be determined from Figure 2.9 by measuring the angle between a ray from the sun and a line tangent to the globe at a selected latitude. Sun's noon altitude in degrees is given for various latitudes on the right side of Figure 2.9. Although the earth's surface is curved, the apparent world in which we, as tiny individuals, live is a flat world. Within the limits of vision the horizon appears to make a circle on a flat plane. This explains why a straight tangent line may be used for measuring the altitude angle on the diagram.

A few measurements of the sun's noon altitude should reveal that the altitude is always equal to the **colatitude,** or 90° minus the latitude. Thus, on the equinoxes, but not for any other time of year, the sun's noon altitude may be computed by a single simple subtraction, if only the latitude is given. The altitude is the same for similar latitudes both north and south of the equator, but keep in mind that the angle is measured from the southern horizon in the northern hemisphere, from the northern horizon in the southern hemisphere.

The path of the sun in the sky at the equinoxes is illustrated in Figure 2.10. To an earthbound observer, the earth's surface seems to be a flat, horizontal disk. The horizon lies at the circumference of the disk. The sun, moon, and stars seem to travel on the inner surface of a hemispherical dome. (This situation is simulated in the design of a large planetarium, which uses a hemispheric ceiling.) At equinox, the sun rises at a point due east on the horizon and sets at a point due west on the horizon. This fact holds true at all latitudes except at the two poles.

Figure 2.11 is a set of three-dimensional drawings, similar to Figure 2.10, covering all the latitudes for which the sun's noon altitude is shown in Figure 2.9. Imaginary solar paths are shown for the nighttime hours, when the sun is below the horizon. Whereas the conditions shown for New York City are most familiar to persons living in midlatitudes, the sun's path in remote parts of the earth will seem strange, particularly so as you travel into the southern hemisphere.

A good plan will be to start with New York City (C) and travel north. When the arctic circle is reached (B), the sun's path at equinox is low in the northern sky. This path becomes even lower as the north pole is approached. Finally, at the north pole itself (A), the sun seems to follow the horizon through the entire 24 hours. Because the earth turns counterclockwise, the sun will seem to move clockwise (toward the observer's right).

As you travel toward the equator, the sun's path at equinox will seem to rise higher in the sky until, at the equator (E), the path is perpendicular to the horizon plane. Here the sun at noon reaches the zenith position. Then, crossing the equator into southern latitudes, the path inclines toward the northern horizon (F). Approaching the south pole, the noon sun will be seen at a point low in the sky, due north on the horizon (G). At the south pole (H), the sun travels counterclockwise (toward the observer's

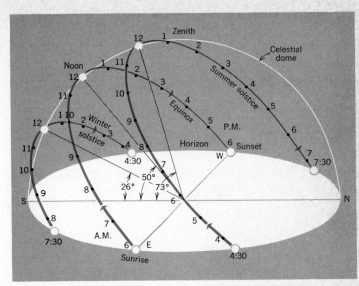

Figure 2.10 **The sun's path in the sky at equinox and both solstices at lat. 40°N.**

left) on the horizon, the direction here being reversed from that at the north pole.

Returning now to the equator (E), we note that the sun has an altitude of 90° at noon. Because the path is in a plane perpendicular to the horizon plane, the sun changes altitude 15° per hour throughout the day. Furthermore, at the equator on these dates the shadow of a vertical rod will point due west from 6:00 A.M. until noon, will disappear precisely at noon, and will point due east from noon until 6:00 P.M.

The Solstices

Conditions at winter (December) solstice are shown in three dimensions in Figure 2.12. Because the maximum inclination of the axis is away from the sun, the entire area lying inside the **arctic circle,** lat. 66½°N, is on the dark side of the circle of illumination. Even though the earth rotates through a full circle during one day, this area poleward of the arctic circle remains in darkness. Conditions in the southern hemisphere during December solstice are shown in Figure 2.13. All of the area lying south of the **antarctic circle,** lat. 66½°S, is under the sun's rays and enjoys 24 hours of day. The subsolar point has shifted to a position on the **tropic of capricorn,** lat. 23½°S.

At summer (June) solstice, conditions are exactly reversed from those of winter solstice. As Figure 2.13 shows, the subsolar point is now on the **tropic of cancer,** lat. 23½°N. Now the region poleward of the arctic circle has a 24-hour day (Figure 2.14). The region poleward of the antarctic circle has a 24-hour night. From one solstice to the next, the subsolar point has shifted over a latitude range of 47°. The progression of changes from equinox to solstice to equinox and back to solstice initiates the **astronomical seasons:** spring, summer, autumn, and winter. These are labeled on Figure 2.7.

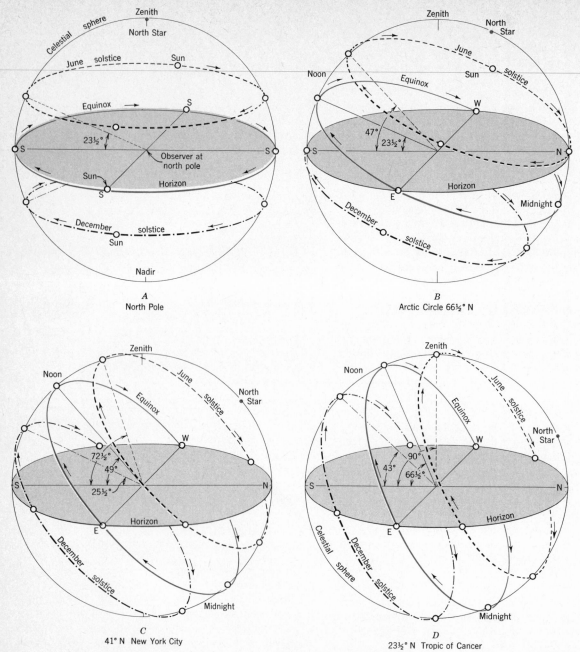

Figure 2.11 **The sun's path in the sky at equinoxes and solstices at the same latitudes shown in Figures 2.9 and 2.15.**

Conditions at Winter Solstice

Details of conditions at winter (December) solstice are shown in Figure 2.15. Keep in mind that "winter" applies only to the northern hemisphere and that the southern hemisphere is experiencing its summer season. At this time, the circle of illumination divides all parallels of latitude that it crosses (except the equator) into unequal parts. The circle of illumination is now tangent to the arctic circle (lat. 66½°N) and the antarctic circle (lat. 66½°S), explaining why these two parallels are given special designation. The circle of illumination bisects the equator in accordance with the law that any two intersecting great circles bisect each other (Chapter 1).

At winter solstice, day and night are unequal in length over most of the globe. This inequality may be estimated from Figure 2.15 by noting what proportions of a given parallel lie to the left and to the right of the circle of illumination. The following facts are evident:

1. Night is longer than day in the northern hemisphere.
2. Day is longer than night in the southern hemisphere.
3. The inequality between day and night increases from the equator poleward.

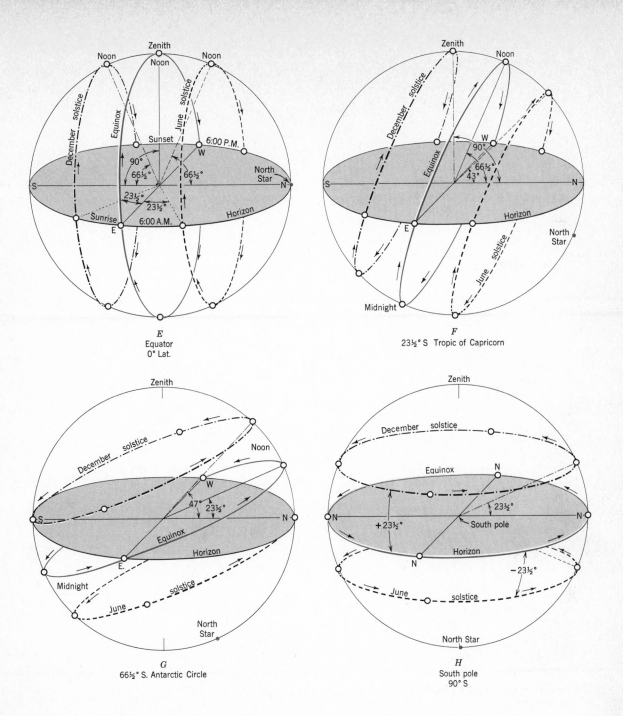

E
Equator
0° Lat.

F
23½° S Tropic of Capricorn

G
66½° S. Antarctic Circle

H
South pole
90° S

4. At corresponding latitudes north and south of the equator, the relative lengths of day and night are in exact opposite relation.
5. Between the arctic circle, lat 66½°N, and the north pole, night lasts the entire 24 hours. (This statement does not take into account twilight, which provides considerable light near the arctic circle.)
6. Between the antarctic circle, lat 66½°S, and the south pole, day lasts the entire 24 hours.

At lat. 23½°S, the tropic of capricorn, the sun's rays at noon strike the earth at an angle of 90° above the horizon. Thus the noon sun is exactly in the zenith point. At the arctic circle, lat. 66½°N, the sun at noon is exactly on the horizon. At the south pole, the noon sun has an altitude of 23½° above the horizon and maintains this angle throughout the full 24 hours. The path of the sun in the sky at the winter solstice is illustrated for various latitudes in Figure 2.11, corresponding with Figure 2.15.

Conditions at Summer Solstice

In every way, conditions at summer solstice, June 21 or 22, are the exact reverse of conditions at winter solstice. Now, the northern hemisphere enjoys the same conditions of increased sunshine that the southern hemisphere had during the winter solstice (Figure 2.13). To see the relation

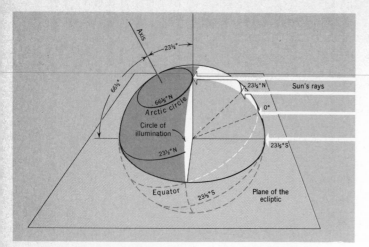

Figure 2.12 At winter solstice, the entire area between the arctic circle and the north pole is in darkness throughout the 24-hour period. (From A. N. Strahler, 1971, *The Earth Sciences,* 2nd ed., Harper and Row, New York.)

of the sun's rays to the earth, turn Figure 2.15 upside down, exchanging "north" for "south," "arctic" for "antarctic," and "tropic of cancer" for "tropic of capricorn." The various statements made in preceding paragraphs concerning circle of illumination, length of day and night, and the sun's noon altitude at the winter solstice may be reread, with suitable wording changes to fit the reversed conditions of the summer solstice.

The path of the sun in the sky on June 21 is shown in Figure 2.11. For all latitudes between the arctic and antarctic circles, the sun rises on the northeastern horizon and sets on the northwestern horizon.

The Seasonal Cycle of Sun's Declination

We have found that the subsolar point travels annually over a 47° latitude range, from lat. 23½°S at winter solstice to lat. 23½°N at summer solstice, crossing the equator twice yearly at the equinoxes. The latitude of the subsolar point at a given instant may be called the **sun's declination.** Figure 2.16 is a graph showing the sun's

declination throughout the year. Notice that, at either solstice, the declination changes very slowly when it is undergoing a reversal in direction. For several days near solstice, the change of declination is too small to be detected without refined telescopic instruments. To the ancient Greeks and Romans, this was a time when the sun "stood still," which explains the derivation of the word *solstice* — from the Latin *sol,* "sun," and *stare,* "to stand."

In contrast, the rate of change of declination is rapid at the two periods of equinox. In each of the months before and after equinox, the declination change is nearly 12 degrees. This fact explains the rapid shortening of days near the autumnal equinox and the rapid lengthening of days near vernal equinox. Plants and animals respond strongly in various forms of activities during these two yearly periods of rapid declination change.

The climate graphs in Chapter 9 reproduce the declination graph in simplified form, so you can use this annual cycle as a reference against which to judge the climatic seasons. A striking fact that will emerge is that the climatic seasons are not in phase with the astronomical seasons.

Time

Global time relationships are important in this age of instantaneous communication and high-speed travel. Before the coming of the telegraph, problems of place-to-place time differences were of little concern to people who lived most of their lives in one community. Even travelers were caused only the inconvenience of resetting their watches to the time used by local communities. The amount of time consumed in getting from one place to another was so much more than the

Figure 2.14 The midnight sun seen in late July from Sunrise Point in Smith Sound, near Etah, Greenland, about lat. 78°N. Eight exposures were taken at intervals of 20 minutes, four before and four after midnight. (American Museum of Natural History.)

Figure 2.13 **Solstice conditions.**

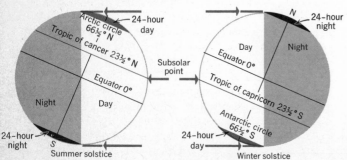

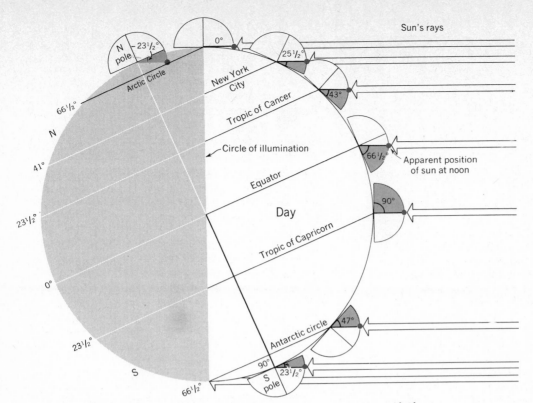

Figure 2.15 **Winter solstice. (From A. N. Strahler, 1971, *The Earth Sciences*, 2nd ed., Harper and Row, New York.)**

difference in watch time between the two places that the time difference was of little consequence.

When it became possible to transmit messages instantaneously by telegraph, differences in local time resulting from differences in longitudinal position were immediately apparent. With the development of rapid travel, it became important to correct schedules for the gain or loss of time incurred by travel across the meridians. Jet aircraft can now fly fast enough to "keep pace with the sun" in midlatitudes. For example, a plane leaving New York at noon, eastern standard time, traveling about 1300 km per hr (800 mph), can arrive in San Francisco at noon, Pacific standard time.

Figure 2.16 **The sun's declination throughout the year.**

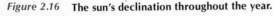
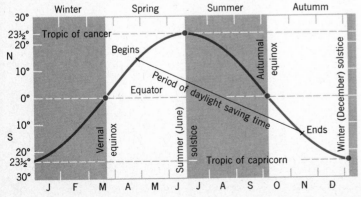

Longitude and Time

To grasp global time relationships, we can return to the days of Claudius Ptolemy (about 250 B.C.) and suppose, as he did, that the sun revolves about the earth. We can think of the earth standing still as the sun completes one circuit about the earth every 24 hours. Imagine that a meridian sweeps westward around the globe at just the right speed to be located always where the sun's rays strike the earth's surface at the highest possible angle. We shall call this line the **noon meridian** (Figure 2.17). Directly opposite the noon meridian, on the other side of the globe, is the **midnight meridian.** It, too, sweeps westward over the globe and remains constantly 180° of longitude apart from the noon meridian. The noon meridian separates the forenoon and afternoon of the same calendar day, while the midnight meridian is the dividing line between one calendar day and the next.

Because the noon meridian sweeps over 360° of longitude every 24 hours, it must cover 15° of longitude every hour, or 1° of longitude every 4 minutes. We therefore find it convenient to state that one hour of time is the equivalent of 15° of longitude. This equality forms the basis for all calculations concerning time belts of the globe. For example, if the noon meridian reaches one place on the globe 4 hours after it leaves another place, the two places are separated by 60° of longitude.

Enlarging this concept of time meridians still further, imagine that in addition to the noon and midnight meridians there are 22 **hour circles,** each one a half great

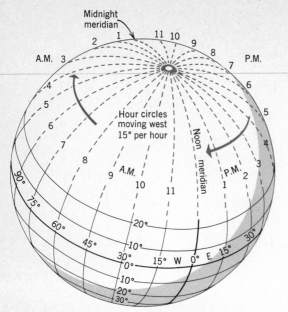

Figure 2.17 **Hour circles, imagined to be moving westward around the globe, provide a model of global time.**

circle, 15° of longitude apart from its neighbor. The hour circles are equidistantly spaced between the noon and midnight meridians (Figure 2.17). Each hour circle will then represent a given hour of the day and can be labeled with a specific hour number, which it keeps permanently. Together with the noon and midnight meridians, the hour circles can be imagined to form a birdcagelike net enclosing the globe and attached only at the north and south poles. As a further convenience in analyzing global time relations, use a globe with meridians drawn for every 15°. Wherever the noon meridian of the time net coincides with a terrestrial meridian, all other hour circles will coincide with meridians printed on the globe.

A working model of global time relations is illustrated in Figure 2.18. Two disks of different radii are attached at their centers in such a way that one disk can be turned while the other remains still. On the inner disk, radii are drawn to represent 15° meridians of a globe seen from a point above the north pole. On the outer disk, similar radii are marked in hours to represent the time net. As a further refinement, the inner disk may be a hemispheric world map. As set in Figure 2.18, it is noon on the Greenwich meridian (long. 0°), while it is midnight on the 180° meridian in mid-Pacific.

Some persons become confused when trying to decide whether the watch time of places to the east (or west) of them is ahead of or behind their own watch time. The problem may be to calculate when a radio or television program will be on the air in different parts of the country or to decide whether to set one's watch ahead or behind one hour when traveling from one standard time zone to another. To avoid confusion, visualize the time meridians moving westward around the globe. Then consider, for example, that you are in New York City and the time there is noon. The noon meridian, which is at New York, left

Greenwich, England, five hours earlier. Therefore, in England, five hours must have elapsed since noon and it must be 5:00 P.M. in that country. The rule is that places located to the east of you have a later hour. Again, consider that the noon meridian is at New York City. Because that meridian will require about three hours to travel westward to reach San Francisco, it must be 9:00 A.M. in that city. A counterpart of the rule just stated is that places located to the west of you have an earlier hour. (Both rules are subject to qualifications where the International Date Line lies between the places.)

Local Time

A century or more ago, the means of establishing a time system for a small community was to take the meridian of longitude passing through some central point in the town or city, for example, the courthouse or a cathedral. All clocks of the community were set to read 12:00 M. (noon) when the sun was directly over that meridian. The time system so derived is called **local time,** defined as mean solar time based on the local meridian. All places located on the same meridian, regardless of how far apart they may be, have the same local time; all places located on different meridians have unlike local times, differing by four minutes for every degree of longitude between them.

Standard Time

As transportation and communication improved through the spread of railroads and the telegraph in the middle 1800s, local time systems had to be replaced. American railroads, about 1870, introduced a standardized system covering large belts of territory; but this system was developed by the railroad companies for their own convenience. Consequently, where several railroads met or passed in a single town, the citizens had to contend with several different kinds of railroad time in addition to

Figure 2.18 **A working model of global time relationships.**

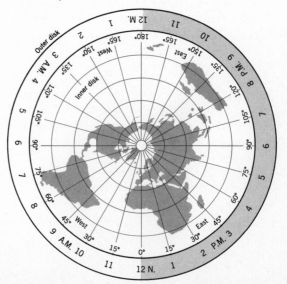

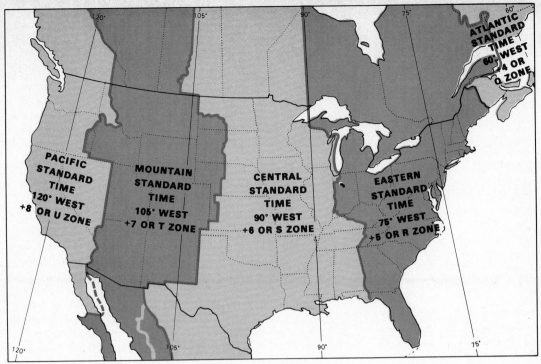

Figure 2.19 **Time-zone map of the United States and southern Canada.**

their own local time. It is said that before 1883 as many as five different time systems were used in a single town and that, altogether, the railroads of the United States followed 53 different systems of time.

The obvious solution to such problems is **standard time,** based on a **standard meridian.** In this system the local time of the standard meridian is arbitrarily given to wide strips of country on both sides. All clocks within the belt are set to a single time. By selecting standard meridians 15° apart, adjacent zones have standard times differing by exactly one hour. Furthermore, if these meridians represent longitudes that are multiples of 15 (e.g., 60°, 75°, 90°, or 105°), each successive standard time zone will differ from the standard time of Greenwich, England, by whole hour units.

Standard Time in the United States

The present system of standard time in the United States was put in operation on November 18, 1883, but it was not until March 19, 1918, that Congress passed legislation directing the Interstate Commerce Commission to determine time-zone boundaries. The standard meridians and boundaries are shown on Figure 2.19. The six standard time zones and their meridians are as follows:

Eastern	75°
Central	90°
Mountain	105°
Pacific	120°
Alaska-Hawaii	150°
Bering	165°

Had it been carried out precisely, the system would have resulted in belts extending exactly 7½° east and west of each standard meridian; but a glance at the map shows that great liberties have been taken in locating the boundaries. Wherever the time-zone boundary could conveniently be located along some already existing and widely recognized line, this was done. Natural boundaries have been used. For example, the eastern time–central time boundary line follows Lake Michigan down its center, and the mountain time–Pacific time boundary follows a ridge-crest line also used by the Idaho-Montana state boundary. Most frequently, the time-zone boundary follows state and county boundaries. For example, the eastern time–central time boundary follows the Alabama-Georgia state line, so as to place all of Georgia in the eastern standard zone.

The time belts are by no means equally distributed on both sides of the standard meridians, as a glance at Figure 2.19 will show. An extreme example is found in western Texas, where the central time–mountain time boundary follows the western boundary of Texas, even crossing west of the standard meridian (long. 105°W) of the mountain time zone. Such deviations of the boundaries permit entire states to operate under one kind of time.

Daylight Saving Time

Because many human activities, especially in urban areas, start well after sunrise but continue long after sunset, it is desirable to set forward the hours of daylight to utilize them to best advantage. A considerable saving in electric power can be made in summer when the early morning daylight period, wasted while schools, offices, and factories are closed, is transferred to the early evening when the majority of persons are awake and busy. The adjusted time system is known as **daylight saving time** and is obtained by setting ahead all timepieces by one hour. Thus, when the sun is over the standard meridian (i.e., noon by the sun), all clocks in that time zone read 1:00 P.M. Sunrise and sunset at the equinoxes or equator, instead of occurring at 6:00 A.M. and 6:00 P.M., would occur at 7:00 A.M. and 7:00 P.M., respectively.

Daylight saving time was adopted in the United States during World War I. After that, it was used locally throughout the United States where authorized by local legislation. During World War II, daylight saving time was used nationally throughout the entire period of February 1942 to October 1945 and was known as war time. During the same war period, England used a double daylight saving time in which clocks were running two hours ahead of Greenwich civil time. This practice was desirable because of the unusually long summer days that England enjoys as a result of her relatively far northerly latitude. Many European countries normally use daylight saving time during part or all of the year. Nations whose time is advanced by an hour throughout the entire year are Spain, France, the Netherlands, Belgium, and the Soviet Union.

In April 1966, the U.S. Congress passed the Uniform Time Act, which requires that daylight saving time be applied uniformly throughout each state, unless the legislature of that state votes to remain on standard time instead. In the latter case, standard time must be applied uniformly throughout the entire state.

Daylight saving time goes into effect at 2:00 A.M. of the last Sunday of April and continues until 2:00 A.M. of the last Sunday of October. Compare these dates with the dates of the vernal and autumnal equinoxes — March 21 and September 22, respectively (Figure 2.16). It is obvious that the period of daylight time was established by Congress to fit the warm season of the year, when outdoor recreation and vacation activity profit most from the added hour of evening daylight. At lat. 40°N (New York City), the sun sets at about 7:50 P.M. standard time on the last Sunday in April, whereas sunset occurs at about 5:05 P.M. standard time on the last Sunday in October. To conserve daylight uniformly according to time of sunset, daylight saving time should be terminated on the third Sunday in August, when the sun's path in the sky is about the same as when daylight time begins at the end of April. On the other hand, if the October date of ending is retained, daylight time should start on the third Sunday in January.

Figure 2.20   Time-zone map of the world. (U.S. Navy Oceanographic Office.)

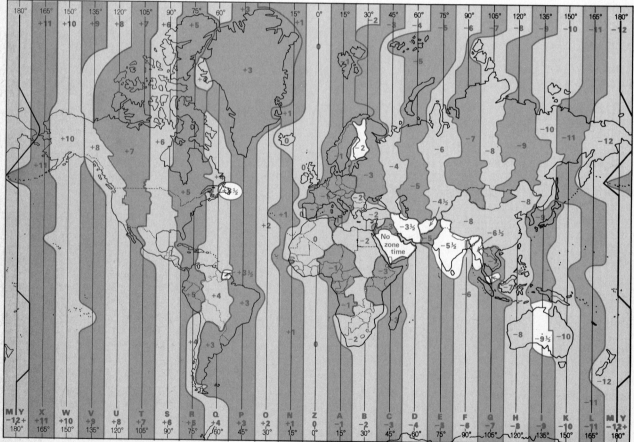

World Time Zones

In 1884 an international congress was held in Washington, D.C. to consider the subject of world standard time. The result was that standard times of countries throughout the world are based on standard meridians, which are multiples of the unit 15° and thus differ from one another by whole hourly amounts. In all global time calculations, the prime meridian of Greenwich, England, is taken as the reference meridian. All time zones of the globe are described in terms of the number of hours' difference between the standard meridian of that zone and the Greenwich meridian. To distinguish whether the time zones lie east or west of the Greenwich meridian, the time is designated *fast* for all places east of Greenwich (east longitude) and *slow* for all places west of Greenwich (west longitude). U.S. eastern standard time, for example, is said to be "five hours slow." An alternative system of designating world time zones uses letters of the alphabet, as shown in Figures 2.19 and 2.20.

Figure 2.20 is a world map on which the 24 principal standard time zones of the world are shown. Fifteen-degree meridians are in black lines; 7½° meridians, which form large elements of the zone boundaries, are in color. Within each time zone is shown the number of hours' difference between the zone time and Greenwich time. Some countries or islands lie about midway between 15° meridians. For these, a standard meridian is chosen halfway between the two and is thus a multiple of 7½°. The standard time of the country is therefore fast or slow by some multiple of a half hour. Iran (3½ hours fast) and Surinam (3½ hours slow) illustrate this point. India, for a large country, is unusual in having the time 5½ hours fast. The country having the greatest east-west extent is the Soviet Union, with eleven standard time zones; but these are all advanced by one hour with respect to the standard time meridians in each zone, to give a perpetual daylight saving time. Canada occupies six time zones. Notice that Newfoundland uses the time 3½ hours slow.

The International Date Line

When we take a world map or globe with 15° meridians and count them in an eastward direction, starting with the Greenwich meridian as 0, we find that 180th meridian is number 12 and that the time of this meridian is, therefore, 12 hours fast. Counting in a similar manner westward from the Greenwich meridian, we find that the 180th meridian is again number 12, but that the time is 12 hours slow. Both results are, of course, correct; and the explanation becomes obvious when we note that the difference in time between 12 hours fast and 12 hours slow is 24 hours, or a full day. At the precise instant when the noon meridian coincides with the Greenwich meridian, the 180th meridian coincides with the midnight hour meridian. At this instant, and only at this instant, the same calendar day exists on both sides of the meridian. At all other times the calendar day on the west (Asiatic) side of the 180th meridian is one day ahead of that on the east (American) side. For example, if it is Monday on the Asiatic side of the

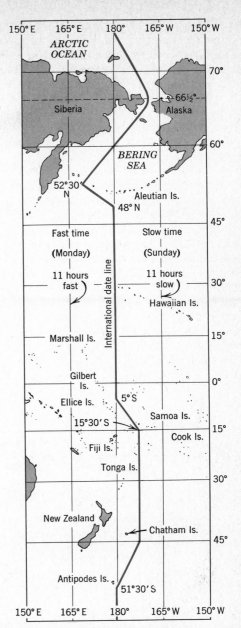

Figure 2.21   The International Date Line.

180th meridian, it is Sunday on the American side.

In the days of slow trans-Pacific travel by sailing vessel and low-powered steamship, an entire calendar day was merely omitted on a westbound voyage and an entire day was repeated on an eastbound voyage. The change was made at any convenient place and time in midocean and usually planned so as to have neither two Sundays in one week nor one week without a Sunday.

Because of their failure to advance the calendar by a whole day, the crew of Magellan's only surviving ship, which reached Spain after circumnavigating the globe in a westward direction, found that in Spain it was September 8, 1522, whereas by their own reckoning it was only September 7 of that year.

When traveling across the Pacific in today's high-speed jet aircraft, the 24-hour correction is made at the time the 180th meridian is crossed. Suppose that the plane is traveling eastward toward North America and crosses the meridian at 4:00 P.M. (standard time, 12 hours fast) on a Tuesday. At the instant of crossing, the time becomes 4:00 P.M. Monday. When traveling westward to the Orient, the time moves ahead by a full day. For example, if the meridian is crossed at 9:30 A.M. Wednesday, the time becomes 9:30 A.M. Thursday.

Because of these peculiar properties, the 180th meridian was designated the **International Date Line** by the International Meridian Conference held in Washington, D.C., in 1884 (Figure 2.21). It is one of the fortuitous occurrences of modern civilization that, after the Greenwich meridian had come into widespread use in English-speaking countries as the international basis for the reckoning of longitude, the 180th meridian should have been found to fall in an almost ideal location: squarely in the middle of the world's largest expanse of ocean. Even so, the International Date Line must deviate both eastward and westward to permit certain land areas

and groups of islands to have the same calendar day (Figure 2.21). Because of an eastward bulge passing through the Bering Strait, the easternmost part of Siberia is included in the Asiatic side, and a westward deflection of the line allows the Aleutian Islands to be included with the Alaskan peninsula. A few degrees south of the equator the date line is shifted eastward 7½° and thus avoids cutting through the Ellice, Wallis, Fiji, and Tonga island groups, which have the same day as New Zealand.

References for Further Study

Strahler, A.N. (1971), *The earth sciences,* second edition, Harper and Row, New York. See Chapters 1–4.

Strahler, A. N. (1975), *Physical geography,* fourth edition, John Wiley and Sons, New York. See Chapters 4, 5.

Royal Astronomical Society of Canada, *The observer's handbook,* issued annually, Univ. of Toronto Press.

U.S. Naval Observatory, *The air almanac,* issued annually, Government Printing Office, Washington, D.C.

World Journal Tribune, *The world almanac and book of facts,* issued annually, Newspaper Enterprise Association, New York. See Astronomical Data.

Review Questions

1. Why are earth-sun relationships important in the study of Man's environment?
2. What is meant by rotation of the earth? What is the period of rotation with reference to the sun? What is the direction of rotation of the earth?
3. Describe and explain the Foucault pendulum experiment. What does it prove? Could you determine your latitude by means of a Foucault pendulum? How?
4. Describe the important environmental effects of earth rotation.
5. What is meant by revolution of the earth? In what two ways can the length and starting point of a year be reckoned? Which one is used in our calendar system? How long is this type of year? How must it be periodically corrected to fit our calendar? What is the direction of the earth's revolution?
6. What form has the earth's orbit? What is perihelion? What is aphelion? On what dates do they occur? What distances separate earth and sun at perihelion and at aphelion? What effect does this have on the seasons?
7. Describe the way the earth's axis is tilted with respect to the plane of the earth's orbit. What is the angle of tilt? Does this angle change throughout the year?
8. Describe each of the two solstice and two equinox positions of the earth with respect to the sun, giving the dates of each in correct sequence.
9. What is the circle of illumination? Where is the circle of illumination located on the date of equinox? On December solstice? On June solstice? What is the subsolar point? Where is it located on the date of equinox?
10. Describe the conditions of global illumination, length of day and night, and path of sun in the sky at equinox. Give details for the equator and poles. How is the sun's noon altitude related to latitude on the date of equinox?

11. Describe the conditions of global illumination, length of day and night, and path of sun in the sky at June solstice. Give details for the equator, tropics of cancer and capricorn, arctic and antarctic circles, and poles.
12. Describe the path of the sun in the sky throughout the year at the north pole. How do conditions differ at the south pole?
13. Describe in general terms how the compass direction of sunrise and sunset varies with season of year and with latitude in the northern hemisphere.
14. Explain how longitude is related to time. How many degrees of longitude are the equivalent of one hour of time? How many hour circles are there on the globe? In which direction do they travel?
15. Do places located east of you have a time that is earlier or later than your time?
16. What is local time? On what is it based? Can places that differ in longitude have the same local time? Can places that differ in latitude have the same local time?
17. What is standard time? Explain the system of standard time used in the United States. Name the time zones. Which meridians are used?
18. How are boundaries between standard time zones determined? Give examples of various kinds of boundaries used in the United States.
19. What is daylight saving time? Why is it used? When is it used?
20. Explain the system of world time zones now in general use. On what prime meridian is it based? What is fast time? Slow time? What advantages are there in conforming to the world time zone system?
21. What is the International Date Line? Describe its location and form. Explain the difference in calendar days on the two sides of this line. Which side has the earlier day? Is it possible for the same day to exist simultaneously on both sides?

The Earth's Atmosphere and Oceans

3

Man lives at the bottom of an ocean of air. Humans are air-breathers dependent on favorable conditions of pressure, temperature, and chemical composition of the atmosphere that surrounds them. Humans also live on the solid outer surface of the earth, which they depend on for food, clothing, shelter, and means of movement from place to place. But the air and the land are not two entirely separate realms; they constitute an interface across which there is a continual flux of matter and energy. Man's surface environment, the life layer, is a shallow but highly complex zone in which atmospheric conditions exert control on the land surface, while at the same time the surface of the land exerts an influence on the properties of the adjacent atmosphere.

Essentially the same statements apply to the surface of the oceans and the atmospheric layer above it. Man utilizes the surface of the sea as a source of food and a means of transportation. There is a continual flow of energy and matter between the sea surface and the lower layer of the atmosphere. Here, again, we find an interface of vital concern to Man. The sea influences the atmosphere above it, while the atmosphere influences the sea beneath it.

Our objective in these early chapters is to examine the atmosphere and oceans with particular reference to the air-land and air-sea interfaces that are so vital to Man. To geographers, the distributions of physical properties of the ocean and atmosphere are matters of special interest, concerned as they are with spatial relationships on a global scale. Physical geographers describe and explain the ways in which the environmental ingredients of weather and climate change with latitude and season and with geographic position in relation to oceans and continents. They seek out the broad patterns of similar regions and attempt to define their boundaries and organize them into systems of classes. More important, geographers try to evaluate the environmental qualities of each region, emphasizing the opportunities as well as the limitations of each for future development of natural resources such as food, water, energy, and minerals.

States of Matter

One of the basic science concepts in physical geography is that of states of matter and their changes. We are concerned with three common states: gas, liquid, and solid.

A **gas** is a substance that expands easily to fill any small empty container, and it is readily compressible. A gas is usually much less dense than liquids and solids of the same chemical composition. While the atmosphere is largely in the gaseous state, it also contains varying amounts of substances in the liquid and solid states.

A **liquid** is a substance that flows freely in response to unbalanced forces but maintains a free upper surface. Liquids are compressed in volume only slightly under strong pressures. Liquids have densities closely comparable with solids of the same composition. Although the ocean is composed largely of water in the liquid state, it also contains substances in the gaseous and solid states. Both gases and liquids are classed as fluids. Layers of fluids tend to assume positions of equilibrium at rest, in which a less dense fluid overlies a more dense fluid.

A **solid** is a substance that resists changes of shape and volume. Solids are typically capable of withstanding large unbalanced forces without yielding. When yielding does occur, it is usually by sudden breakage. Although the earth's crust is largely in the solid state, it also contains substances in gaseous and liquid states.

A change of state occurs frequently in the world of nature. Most important and widespread is the change of state of water from water vapor (a gas) to liquid water and vice versa, and from liquid water to ice (solid state) and vice versa (Chapter 7). Changes of state require either an input of heat energy or the disposal of heat energy, depending on the direction of the change.

Composition of the Atmosphere

The principles we shall review in this chapter belong to two areas of natural science: **meteorology**, the science of the atmosphere, and **physical oceanography**, the

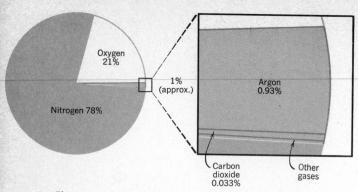

Figure 3.1 Component gases of the lower atmosphere. The figures tell approximate percentage by volume.

marily a neutral substance. Very small amounts of nitrogen are extracted by soil bacteria and made available for use by plants. In contrast to nitrogen, oxygen is highly active chemically and combines readily with other elements in the process of oxidation. The combustion of fuels represents a rapid form of oxidation, whereas certain forms of rock decay (weathering) represent very slow forms of oxidation. Animals require oxygen to convert foods into energy.

The remaining 1 percent of the air is mostly argon, an inactive gas of little importance in natural processes. Part of that 1 percent consists of a very small amount of carbon dioxide, about 0.033 percent. This gas is of great importance in atmospheric processes because of its ability to absorb radiant heat and so allow the lower atmosphere to be warmed by heat rays coming from the sun and from the earth's surface. Green plants, in the process of photosynthesis, utilize carbon dioxide from the atmosphere, converting it with water into solid carbohydrate.

Also present, but in extremely small amounts, are the following gases: neon, helium, krypton, xenon, hydrogen, methane, and nitrous oxide. All the component gases of the lower atmosphere are perfectly diffused among one another so as to give the pure, dry air a definite set of physical properties, just as if it were a single gas.

Atmospheric Pressure

Although we are not constantly aware of it, air is a tangible material substance, exerting **atmospheric pressure** on any solid or liquid surface exposed to it. This pressure is about 1 kg/sq cm (about 15 lb/sq in.). Because atmospheric pressure is exactly counterbalanced by the pressure of air within liquids, hollow objects, or porous substances, its ever-present weight goes unnoticed. The pressure on 1 sq

physical-science aspect of the oceans.

The earth's **atmosphere** consists of a mixture of various gases surrounding the earth to a height of many kilometers. Held to the earth by gravitational attraction, this envelope of air is densest at sea level and thins rapidly upward. Although almost all of the atmosphere (about 97%) lies within 30 km (18 mi) of the earth's surface, the upper limit of the atmosphere can be drawn approximately at a height of 10,000 km (6000 mi), a distance approaching the diameter of the earth itself. From the earth's surface upward to an altitude of about 80 km (50 mi), the chemical composition of the atmosphere is highly uniform throughout in terms of the proportions of its component gases.

Pure, dry air consists largely of nitrogen, about 78 percent by volume, and oxygen, about 21 percent (Figure 3.1). Nitrogen does not enter easily into chemical union with other substances and can be thought of as pri-

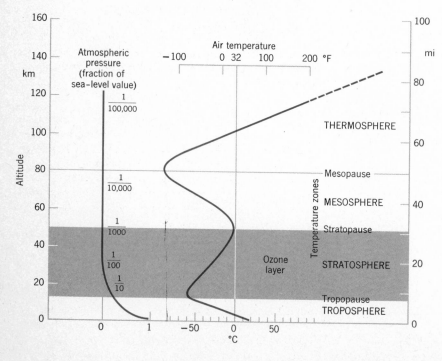

Figure 3.2 Temperature structure of the atmosphere.

cm of surface can be thought of as the actual weight of a column of air 1 cm in cross section extending upward to the outer limits of the atmosphere. Air is readily compressible. That which lies lowest is most greatly compressed and is, therefore, densest. In an upward direction, both density and pressure of the air fall off rapidly. Figure 3.2 shows the upward decrease in pressure in terms of the fraction of the sea-level value.

Meteorologists have used another method of stating atmospheric pressure. It is based on a classic experiment of physics first performed by Evangelista Torricelli in 1643. A glass tube about 1 m (3 ft) long, sealed at one end, is completely filled with mercury. The open end is temporarily held closed. Then the tube is inverted and the end is immersed into a dish of mercury. When the opening is uncovered, the mercury in the tube falls a few centimeters, but then remains fixed at a level about 76 cm (30 in.) above the surface of the mercury in the dish (Figure 3.3). Atmospheric pressure now balances the weight of the mercury column. When the air pressure increases or decreases, the mercury level rises or falls correspondingly. Here, then, is a device for measuring air pressure and its variations.

Any instrument that measures atmospheric pressure is a **barometer.** The type devised by Torricelli is known as the **mercurial barometer.** With various refinements over the original simple device, it has become the standard instrument. Pressure is read in centimeters or inches of mercury, the true measure of the height of the mercury column. Standard sea-level pressure is 76.0 cm (29.92 in.) on this scale.

Another pressure unit is the working standard of meteorologists. This unit is the **millibar** (**mb**). One cm of mercury is equivalent to 13.3 mb (1 in. equals about 34 mb). Standard sea-level pressure is 1013.2 mb.

Another type of barometer is the **aneroid barometer**

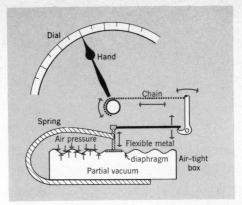

Figure 3.4   **Principle of the aneroid barometer.**

Figure 3.5   **This aneroid barometer is calibrated in both millibars (outer scale) and inches. (Taylor Instrument Co., Rochester, N.Y.)**

(Figures 3.4 and 3.5). It consists of a hollow metal chamber partly emptied of air and sealed. The upper wall of the chamber is a flexible diaphragm that moves up and down as the outside air pressure varies. These movements operate a hand, read against a calibrated circular dial. The aneroid is compact and sturdy. When calibrated in terms of altitude, the aneroid barometer becomes an altimeter, carried in all aircraft.

Vertical Distribution of Atmospheric Pressure

Figure 3.6 shows in detail the rate of decrease of atmospheric pressure (barometric pressure) with altitude up to a height of over 30 km (20 mi). For every 275 m (900 ft) of rise in altitude, pressure is diminished by $1/30$ of itself.

Figure 3.3   **Principle of the mercurial barometer.**

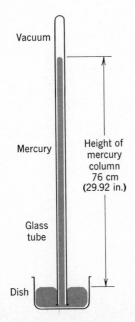

Vacuum

Mercury

Height of mercury column 76 cm (29.92 in.)

Glass tube

Dish

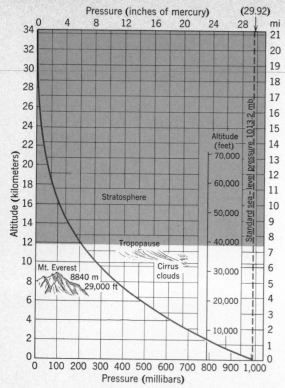

Figure 3.6 **Decrease of atmospheric pressure with altitude.**

Table 3.1 / Decrease of Boiling Point with Altitude

Altitude		Pressure		Boiling point	
m	(ft)	mb	(in.)	°C	(°F)
0	(0)	1,013	(29.9)	100	(212)
1,000	(3,300)	900	(27)	97	(207)
2,000	(6,600)	790	(24)	93	(200)
3,000	(10,000)	700	(21)	90	(194)
4,000	(13,000)	620	(19)	87	(190)
5,000	(16,000)	550	(17)	84	(183)

Temperature Structure of the Lower Atmosphere

The atmosphere has been subdivided into layers according to temperatures and zones of temperature change, shown in Figure 3.2. Of greatest importance to Man and other life forms is the lowermost layer, the **troposphere.** If we sent up a sounding balloon carrying a recording thermometer and repeated this operation many times, we would obtain an average or representative profile of temperature. We would find that air temperature falls quite steadily with increasing altitude. The average rate of temperature decrease is about 6.4C°/1000 m of ascent (3½ F°/1000 ft). This rate is known as the **environmental temperature lapse rate.** When used repeatedly, we shorten this term to "lapse rate".

Figure 3.7 shows a typical air sounding into the troposphere at midlatitudes (lat. 45°N) on a summer day. Altitude is plotted on the vertical axis, temperature on the horizontal axis. The resulting curve is a sloping line. Temperature drops uniformly with altitude to a height of about 13 km (8 mi). Don't be surprised when the captain of your jet aircraft, flying at 12 km (40,000 ft) altitude, announces that the outside temperature is −50°C (−60°F).

The lapse rate changes abruptly at an altitude of about 14 km (9 mi). Instead of continuing to fall, temperature here remains constant with increasing altitude. This level of change is the **tropopause;** it marks a transition into the

Steepening of the curve as it rises shows that the rate of decrease in pressure is rapid at first, but becomes less with increasing altitude.

The physiological effects of a pressure decrease on humans are well known from the experiences of flying and mountain climbing. Lowered pressure decreases the amount of oxygen entering the blood through the lungs. At altitudes of 3000 to 4500 m (10,000 to 15,000 ft) mountain sickness (altitude sickness), characterized by weakness, headache, nosebleed, or nausea, can occur. Persons who remain at these altitudes for a day or two normally adjust to the conditions, but physical exertion is always accompanied by shortness of breath.

Supplementary oxygen is needed for full physical activity above 5400 m (18,000 ft) in unpressurized aircraft. Lightweight oxygen containers enable climbers to reach the summit of Mount Everest (8800 m; 29,000 ft), a feat first accomplished in 1953. The cabins of commercial subsonic jet aircraft are pressurized to about 850 mb, equivalent to an altitude of about 1500 m (4500 ft), whereas the outside pressure at cruising altitude may be only 200 to 300 mb.

As atmospheric pressure becomes less, the boiling point of water becomes lower. Table 3.1 gives data on altitude, pressure, and boiling point. As the boiling point decreases, the time required to cook foods by boiling increases. As many experienced campers know, boiling potatoes is a slow process at altitudes above 2000 m (6600 ft). A pressure cooker is most useful at high altitudes.

Figure 3.7 **A typical environmental temperature lapse-rate curve for a summer day in midlatitudes.**

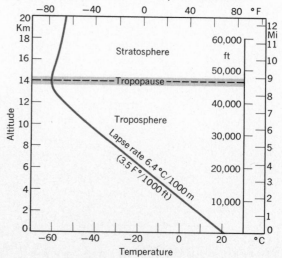

next higher temperature zone, known as the **stratosphere.**

The altitude of the tropopause is highest over the equator (17 km; 10 mi), and lowest over the poles (9–10 km; 5.5–6 mi); see Figure 3.8. There are important seasonal changes in altitude of the tropopause in middle and high latitudes, as shown in Figure 3.8. For example, at lat. 45° the average altitude in January is 12.5 km (8 mi), but rises to 15 km (9 mi) in July. Temperatures at the tropopause are much lower at the equator than at the poles, as shown in Figure 3.8. At first glance, this relationship may seem strange, accustomed as we are to considering the equatorial region as a warm zone and the polar regions as cold. On the other hand, because the lapse rate is quite uniform the world over, the thicker the troposphere, the colder the temperatures at the tropopause will be.

Above the tropopause, temperatures in the stratosphere gradually rise until a value of about 0°C (32°F) is reached at about 50 km (30 mi). Here, at the **stratopause,** a temperature decrease sets in (Figure 3.2). Temperature decreases through the overlying **mesosphere,** a layer extending upward to about 80 km (50 mi), where a low point of −80°C (−120°F) is reached. This level of temperature minimum and reversal is called the **mesopause.** With further increase in altitude, a steep climb in temperature takes place within the **thermosphere.** At this very high altitude, the air is extremely rarefied; the gas molecules are widely separated. This thin air holds very little heat, even though its temperature is high.

The Troposphere and Man

The lowermost atmospheric layer, the troposphere, is of most direct importance to Man. Almost all phenomena of weather and climate that physically affect Man take place within the troposphere (see Plate C.3).

Besides the gases of pure dry air, the troposphere contains **water vapor,** a colorless, odorless gaseous form of water that mixes perfectly with the other gases of the air. The quantity of water vapor present in the atmosphere is of primary importance in weather phenomena. Water vapor can condense into clouds and fog. When condensation is rapid, rain, snow, hail, or sleet — collectively termed precipitation — are produced and fall to earth. Where water vapor is present only in small proportions, extremely dry deserts result. In addition, a most important function is performed by water vapor. Like carbon dioxide, it is a gas capable of absorbing heat in the radiant form coming from the sun and from the earth's surface. Water vapor gives to the troposphere the qualities of an insulating blanket, which inhibits the escape of heat from the earth's surface.

The troposphere contains myriads of tiny dust particles, so small and light that the slightest movements of the air keep them aloft. They have been swept into the air from dry desert plains, lake beds, and beaches, or injected by explosive volcanoes. Strong winds blowing over the ocean lift droplets of spray into the air. These may dry out, leaving as residues extremely minute crystals of salt that are carried aloft. Forest and brush fires are another

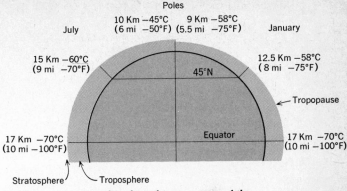

Figure 3.8 Average altitude and temperature of the tropopause for July (left) and January (right). (From A. N. Strahler, 1971, *The Earth Sciences,* 2nd ed., Harper and Row, New York.)

important source of atmospheric dust particles. Countless meteors, vaporizing from the heat of friction as they enter the upper layers of air, have contributed dust particles. Industrial processes involving combustion of fuels are also a major source of atmospheric dust.

Dust in the troposphere contributes to the occurrence of twilight and the red colors of sunrise and sunset, but the most important function of dust particles cannot be seen and is rarely appreciated. Certain types of dust particles serve as nuclei, or centers, around which water vapor condenses to produce cloud particles. In contrast, the stratosphere is almost entirely free of water vapor and dust. Clouds are rare in the stratosphere, but there are high-speed winds in narrow zones.

The Ozone Layer — a Shield to Life

Of vital concern to Man and all other life forms on earth is the presence of an **ozone layer** within the stratosphere. This layer sets in an altitude of about 15 km (9 mi) and extends upward to about 55 km (35 mi). The ozone layer is a region of concentration of the form of oxygen molecule known as **ozone** (O_3), in which three oxygen atoms are combined instead of the usual two atoms (O_2). Ozone is produced by the action of solar radiation on ordinary oxygen atoms.

The ozone layer serves as a shield, protecting the troposphere and earth's surface by absorbing most of the ultraviolet radiation found in the sun's rays. If these ultraviolet rays were to reach the earth's surface in full intensity, all exposed bacteria would be destroyed and animal tissues damaged severely. In this protective role, the presence of the ozone layer is an essential factor in Man's environment.

This brings us to two current environmental issues involving the ozone layer. The first concerns the debate over approval of government appropriations to develop a supersonic transport (SST) jet aircraft in the United States. The project was not approved, but a British-French SST aircraft, the Concorde, is presently in operation. The Concorde flies at an altitude of about 18 km (60,000 ft); the proposed American SST aircraft would operate at about 22 km (70,000 ft), so both levels are well within the stratosphere. Much concern has been expressed by atmospheric scientists that exhaust emissions from these aircraft could, in time, alter the composition of the stratosphere. Should the effect be to reduce the ozone content, increased penetration of ultraviolet radiation would result, with possible undesirable effects to life and with side effects in the form of climate changes in the troposphere. That SST aircraft operation can lead to a reduction of stratospheric ozone has been recognized as a distinct possibility by an investigative panel of the National Academy of Sciences.

An even more serious threat to the ozone layer is posed by the release of aerosol propellants into the atmosphere. Many aerosol spray cans used in the household are charged with a mixture of Freons, synthetic compounds containing carbon, fluorine, and chlorine atoms. Compounds of this class are often called *chloro-fluorocarbons*; they are also widely used as refrigerants. Molecules of chlorofluorocarbons drift upward through the troposphere and eventually reach the stratosphere. As these compounds absorb ultraviolet radiation, they are decomposed and chlorine is released. The chlorine in turn attacks molecules of ozone, converting them in large numbers by a chain reaction into ordinary oxygen molecules. In this way the ozone concentration within the stratospheric ozone layer can be reduced, and the intensity of ultraviolet radiation reaching the earth's surface can be increased. An increase in the incidence of skin cancer in humans is one of the predicted effects. Acting on a report by a special investigative panel of the National Academy of Sciences, the Environmental Protection Agency in 1976 recommended steps to eliminate chlorofluorocarbons from spray cans used in the home.

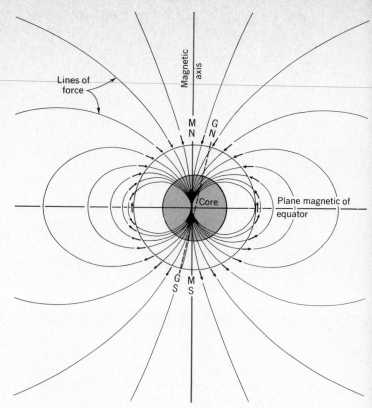

Figure 3.9 **Lines of force of the earth's magnetic field are shown diagrammatically in this cross section drawn through the magnetic and geographic poles. The small arrows show the inclination of lines of force at surface points over the globe. (From A. N. Strahler, 1971,** *The Earth Sciences,* **2nd ed., Harper and Row, New York.)**

The Earth's Magnetic Atmosphere

The earth can be thought of as a simple bar magnet, the axis of which approximately coincides with the earth's geographic axis (Figure 3.9). Magnetism is generated within the earth's metallic core, a central spherical body about half the earth's diameter. The earth's magnetic axis is inclined several degrees with respect to the geographic axis. Consequently, the north and south magnetic poles do not coincide with the geographic north and south poles, nor does the magnetic equator coincide with the earth's geographic equator.

Lines of force of the earth's magnetic field, shown in Figure 3.9, pass outward through the earth's surface and into surrounding space. A magnetic compass needle, which is nothing more than a delicately balanced bar magnet, orients itself in a position of rest parallel with the lines of force. Lines of force extending out into space comprise the earth's **external magnetic field.** This field can be thought of as a magnetic atmosphere. If we assume, for purposes of comparison, that the earth's gaseous atmosphere extends outward to a distance equal to twice its own radius, or 13,000 km (8000 mi), it is evident that the magnetic atmosphere extends far beyond the outer-

most limits of the gaseous atmosphere. All of the region within the limit of the magnetic field is described as the **magnetosphere;** its outer boundary has been named the **magnetopause** (Figure 3.10).

The simplest geometric model for the shape of the magnetosphere would be a doughnut-shaped ring surrounding the earth. The plane of the ring would lie in the plane of the magnetic equator, while the earth would occupy the opening in the center of the doughnut. Actually, this ideal shape does not exist because of the action of the **solar wind,** a more or less continual flow of electrons and protons emitted by the sun. Pressure of the solar wind acts to press the magnetopause close to the earth on the side nearest the sun (Figure 3.10). Lines of force in this region are crowded together, and the magnetic field is intensified. On the opposite side of the earth, in a line pointing away from the sun, the magneto-pause is drawn far out from the earth and the force lines are greatly attenuated. The extent of this "tail" is not known, but the entire shape of the magnetosphere has been described as resembling a comet.

An important environmental role of the magneto-sphere is to shield the inner atmosphere and surface of the earth from an extremely powerful form of energy called

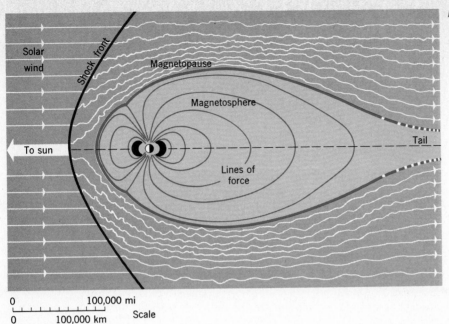

Figure 3.10 Magnetosphere and magnetopause. The Van Allen radiation belt is shown as a black area on either side of earth.

ionizing radiation. This is the same form of dangerous radiation given off by such radioactive elements as uranium, used as a nuclear fuel. Ionizing radiation reaches the earth from the sun in the form of fast-moving electrons and protons, comprising the solar wind (Figure 3.10). Upon encountering the magnetosphere, these particles are trapped and retained within the force lines of the magnetic field. Here they tend to become greatly concentrated at times into a doughnut-shaped ring, called the **Van Allen radiation belt** (Figure 3.10). Trapped particles are also continually escaping from the tail of the magnetosphere. Without the protection of the magnetosphere, ionizing radiation would destroy all life exposed at the earth's surface.

Man and the Oceans

The importance of the oceans to Man is felt in a wide range of dimensions and scales. One environmental role played by the oceans is climatic. The huge water mass of the oceans stores a large quantity of heat. This heat is gained or lost very slowly. As we shall see, the oceans effectively moderate the seasonal extremes of temperature over much of the earth's surface. The oceans supply water vapor to the atmosphere and are the basic source of all rain that falls on the lands. This rainfall, the source of Man's vital freshwater supplies, originates from the ocean surface by a natural process of distillation of salt water.

The oceans sustain a vast and complex assemblage of marine life forms, both plant and animal. This organic production provides humans with a modest but important share of their food. Throughout history the oceans have served as trackless surfaces of transport of people and the commodities that sustain their civilizations. Winds, waves, currents, sea ice, and fog are environmental factors — sometimes favorable and sometimes hazardous — that

the oceans impose on humans and their ships at sea.

The zone of contact between oceans and lands is a unique environment for Man. In Chapter 22 we shall investigate the processes by which ocean waves and currents shape coastal landscape features. Man uses and modifies the coastal zone in various ways, ranging from ports to recreational facilities.

The World Ocean

We shall use the term **world ocean** to refer to the combined ocean bodies and seas of the globe. Let us consider some statistics that emphasize the enormous extent and bulk of this great saltwater layer. The world ocean covers about 71 percent of the global surface (Figure 3.11); its average depth is about 3800 m (12,500 ft), when shallow seas are included with the deep main ocean basins. For major portions of the Atlantic, Pacific, and Indian oceans, the average depth is about 4000 m (13,000 ft).

The total volume of the world ocean is about 1.4 billion cu km (317 million cu mi), a quantity just over 97

Figure 3.11 **Northern and southern hemispheres — a contrast in land-ocean distributions.**

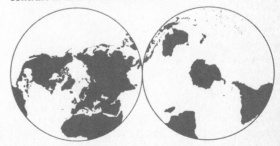

percent of the world's free water. Of the small remaining volume, about 2 percent is locked up in the ice sheets of Antarctica and Greenland, and about 1 percent is represented by fresh water stored on the lands. These figures show the extent of the **hydrosphere,** a general term for the total free water of the earth, in all three of its states — gas, liquid, and solid. The hydrosphere is largely represented by the world ocean. To place the masses of the atmosphere and oceans in their proper planetary perspective, compare the following figures. (The unit of mass used here is 10^{21} kg.)

Entire earth	6000
World ocean	1.4
Atmosphere	0.005

What are the basic differences between the world ocean and the atmosphere in terms of properties and behavior? How do atmosphere and ocean interact in the region of their interface? The answers to these questions are vital to an understanding of environmental processes because marine life depends on the exchanges of matter and energy across the atmosphere-ocean interface. It is also significant that the earliest life forms originated and developed in the shallow layer of water immediately beneath the interface.

The atmosphere, being composed of a gas that is easily compressed, has no distinct upper boundary; it becomes progressively denser toward its base under the load of the overlying gas. The world ocean has a sharply defined upper surface in contact with the densest layer of the overlying atmosphere.

Whereas the most active region of the atmosphere is the lowermost layer, the troposphere, the most active region of the ocean is its uppermost layer. At great ocean depths, water moves extremely slowly and maintains a uniformly low temperature. One reason for intense physical and biological activity in the uppermost ocean layer is that the input of energy from the overlying atmosphere drives water motions in the form of waves and currents. Another reason is that oxygen and carbon dioxide, gases vital to animal and plant growth, enter the ocean from the atmosphere. The atmosphere is also the source layer of heat and of condensed fresh water initially evaporated from the ocean. But the ocean surface also returns heat and water (in the vapor form) to the lower atmosphere, and this phenomenon is of primary importance in driving atmospheric motions. Interaction between atmosphere and ocean surface is a topic we shall explore in later chapters.

The oceans are formed into compartments by intervening continental masses. The land barriers inhibit the free global interchange of ocean waters, whereas the atmosphere is free to move globally. Another difference in the two bodies is that the atmosphere has little ability to resist unbalanced forces. The air therefore moves easily and rapidly, changing its velocity quickly from place to place. In contrast, ocean water can move only sluggishly and is very slow to respond to the changes in force applied by winds.

Composition of Seawater

Seawater is a solution of salts — a brine — whose ingredients have maintained approximately fixed proportions over a considerable span of geologic time. Besides their importance in the chemical environment of marine life, these salts constitute a vast reservoir of mineral matter from which certain constituents may be extracted for industrial uses. One way to describe the composition of seawater is to state the principal ingredients that would be required to make an artificial brine approximately like seawater. These are listed in Table 3.2. Of the various elements combined in these salts, chlorine alone makes up 55 percent by weight of all the dissolved matter, and sodium makes up 31 percent. Magnesium, calcium, sulfur, and potassium are the other four major elements in these salts. Important, but less abundant than elements of the five salts listed in Table 3.2 are bromine, carbon, strontium, boron, silicon, and fluorine. At least some trace of half of the known elements can be found in seawater. Seawater also holds in solution small amounts of all the gases of the atmosphere. Of these, oxygen and carbon dioxide are vital to organic processes in marine animals and plants.

Layered Structure of the Oceans

As with the atmosphere, the ocean has a layered structure. Ocean layers are recognized in terms of both temperature and oxygen content. In the troposphere, air temperatures are generally highest at ground level and diminish upward. In the oceans, temperatures are generally highest at the sea surface and decline with depth. This trend is to be expected because the source of heat is from the sun's rays and from heat supplied by the overlying atmosphere.

With respect to temperature, the ocean presents a three-layered structure in cross section, as shown in the left-hand diagram of Figure 3.12. At low latitudes throughout the year and in midlatitudes in the summer, a warm surface layer develops. Here wave action mixes heated surface water with the water below it to give a warm layer that may be as thick as 500 m (1600 ft), with a temperature of 20° to 25°C (70° to 80°F) in oceans of the equatorial belt. Below the warm layer temperatures drop rapidly, making a second layer known as the **thermocline.** Below the thermocline is a third layer of very cold water

---

Table 3.2 / Composition of Seawater

Name of Salt	Chemical Formula	Grams of Salt per 1000 g of Water
Sodium chloride	NaCl	23.0
Magnesium chloride	$MgCl_2$	5.0
Sodium sulfate	Na_2SO_4	4.0
Calcium chloride	$CaCl_2$	1.0
Potassium chloride	KCl	0.7
With other minor ingredients to total		34.5

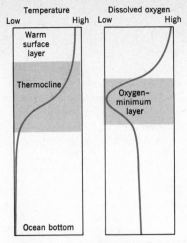

Figure 3.12 *This schematic diagram shows the changes of temperature and oxygen content with depth typical of broad expanses of the oceans in lower latitudes.*

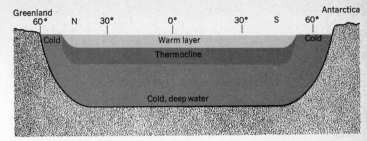

Figure 3.13 **A schematic north-south cross section of the world ocean shows that the warm surface water layer disappears in arctic latitudes, where very cold water lies at the surface.**

extending to the deep ocean floor. Temperatures near the base of the deep layer range from 0° to 5°C (32° to 40°F). In arctic and antarctic regions, the three-layer system is replaced by a single layer of cold water, as Figure 3.13 shows. Temperature is a prime environmental factor controlling the abundance and variety of marine life, the bulk of which thrives in the shallow upper layer.

The content of free oxygen dissolved in seawater shows an oxygen-rich surface layer, accounted for by the availability of atmospheric oxygen and the activity of oxygen-releasing plant life in the sea. As shown in the right-hand diagram of Figure 3.12, oxygen content falls rapidly with depth; over large ocean areas, there is a distinct minimum zone of low oxygen content. Here oxygen has been consumed by biological activity. In deep water the oxygen content holds to a uniform and moderate value down to the ocean floor.

Atmosphere and Oceans in Review

The broad overview of atmosphere and oceans given in this chapter shows us the major elements of physical structure and chemical composition of those two great global layers so vital to life on earth. Most of the information has been about static conditions we would encounter when probing upward into the atmosphere and downward into the oceans.

In the next several chapters we turn to the great systems of flow of matter and energy that continually involve the atmosphere and oceans, making them dynamic rather than static bodies. First, we shall trace the course of radiant energy from the sun as it passes through the atmosphere, reaches the earth, and is returned to outer space. The solar radiation system controls the thermal environment of the life layer, or biosphere. Solar radiation supplies the life layer with the energy needed for

biological processes. We shall then follow with accounts of the vast systems of transport of the atmosphere and oceans. These great flows of air and water redistribute heat over the globe to provide more moderate and more favorable life conditions than would otherwise exist on our planet. Accompanying these circulation systems are intense disturbances of the air and sea — storms that are often severe and hazardous environmental stresses.

Looking over the other planets of the solar system, the uniqueness of Planet Earth as a life environment is most striking. Only Earth has both a great world ocean and a comparatively dense oxygen-rich atmosphere combined with a favorable temperature range. Mars, our nearest planet, has practically no free water in any form and only a very rarefied atmosphere with little oxygen. Venus, matching us closely in size, has a much denser atmosphere than Earth. But water in any form is almost totally lacking on Venus, and surface temperatures there are much higher than on Earth. Little Mercury, with no atmosphere and no water, roasts in the sun's rays. The greater outer planets — Jupiter, Saturn, Uranus, and Neptune — probably have large quantities of water, but it is frozen solid. Their atmospheres, composed in part of ammonia and methane, would be lethal to life such as ours even if the surface temperatures were not impossibly cold. The moon's surface has no free water or atmosphere to offer. So there is really no place for humans to live in large numbers other than Planet Earth.

References for Further Study

Nelson, J. H., L. Hurwitz, and D. G. Knapp (1962), *Magnetism of the earth,* Coast and Geodetic Survey, Publication 40–1, Government Printing Office, Washington, D.C.

Weyl, Peter K. (1970), *Oceanography,* John Wiley and Sons, New York.

Strahler, Arthur N. (1971), *The earth sciences,* second edition, Harper and Row, New York. See Chapters 7, 12.

Riehl, Herbert (1972), *Introduction to the atmosphere,* second edition, McGraw-Hill Book Co., New York. See Chapter 1.

Byers, Horace R. (1974), *General meteorology,* fourth edition, McGraw-Hill Book Co., New York. See Chapter 4.

Van Allen, James (1975), Interplanetary particles and fields, *Scientific American,* vol. 233, no. 3, pp. 161–73.

Review Questions

1. Describe Man's environment on earth in terms of the realms of land, sea, and air. In what way is physical geography concerned with sciences that treat the atmosphere and oceans?
2. Describe the gaseous, liquid, and solid states of matter. Do these states of matter comprise three complete separate and distinct realms of Planet Earth? How are energy changes related to changes of state?
3. List the component gases of the lower atmosphere in order of decreasing proportion by volume, giving approximate percentages for each. Evaluate each gas in terms of its importance to plant and animal life on earth.
4. What causes atmospheric pressure? How is pressure measured? Describe Torricelli's experiment and explain the principle involved. Describe a mercurial barometer.
5. What value is given to standard sea-level pressure in centimeters and inches of mercury? What is the equivalent value in millibars?
6. Explain the principle of the aneroid barometer. What advantages has this type over the mercurial barometer?
7. At what rate does atmospheric pressure diminish with increasing altitude? Is the rate constant? Does the pressure-altitude curve reveal the position of the tropopause?
8. What physiological effects has the upward decrease of air pressure on humans? How does the boiling point of water change with increasing altitude?
9. Describe the temperature structure of the atmosphere, naming the layers in upward order.
10. Describe the change in air temperature upward through the troposphere to the tropopause. What is the amount of the environmental temperature lapse rate? At what altitude is the tropopause encountered over the equator? over the poles? over midlatitudes? What changes in altitude of the tropo-pause take place from summer to winter?
11. Why is the troposphere of special importance to Man's environment? What are the constituents of the troposphere, in addition to gases of the pure, dry air? What importance has each constituent?
12. What is the ozone layer? At what altitude is it found? How is ozone produced? How does the ozone layer protect the environment of plant and animal life at the earth's surface? How does operation of the SST aircraft pose a threat to the ozone layer? How can spray-can propellants (chloro-fluorocarbons) deplete the ozone layer?
13. Describe the earth's magnetic field. Define the magnetic poles and magnetic equator.
14. Describe the magnetosphere. What is the magnetopause? How is the shape of the magnetosphere influenced by the solar wind? How does the magnetosphere protect the earth's surface from ionizing radiation? What is the Van Allen radiation belt?
15. Describe the world ocean in terms of surface extent, average depth, and volume. Compare the mass of the world ocean with that of earth and atmosphere.
16. Compare the atmosphere and oceans with respect to their density distribution, temperature structure, and freedom of motion.
17. What are the principal chemical components of seawater? Of what value are sea salts to Man?
18. Describe the layered structure of the oceans with respect to temperature and oxygen content. How do these layers relate to distribution of organic life and to organic processes?
19. Contrast the surface environment of Earth with those of Mars, Venus, Mercury, the moon, and the outer planets. In what respects is Earth's surface environment uniquely favorable to life?

The Earth's Radiation Balance

4

All life processes on Planet Earth are supported by radiant energy coming from the sun. The planetary circulation systems of atmosphere and oceans are driven by solar energy. Exchanges of water vapor and liquid water from place to place over the globe depend on this single energy source. It is true that some heat flows upward through the earth's crust to reach the surface, but the amount is trivial in comparison with the amount of energy that the earth intercepts from the sun's rays.

The flow of energy from sun to earth and then out into space is a complex system. It involves not only energy transmission by a form of radiation, but also energy storage and transport. Both storage and transport of heat occur in the gaseous, liquid, and solid matter of the atmosphere, hydrosphere, and lithosphere. We can simplify the study of this total system by first examining each of its parts. We shall start with the radiation process itself and develop the concept of a **radiation balance,** in which energy absorbed by our planet is matched by the planetary output of energy into outer space.

Solar energy is intercepted by our spherical planet, and the level of heat energy tends to be raised. At the same time, our planet radiates energy into outer space, a process that tends to diminish the level of heat energy. Incoming and outgoing radiation processes are simultaneously in action (Figure 4.1). In one place and time more energy is being gained than lost; in another place and time more energy is being lost than gained.

The equatorial region receives much more energy through solar radiation than is lost directly to space. In contrast, polar regions lose much more energy by radiation into space than is received. So mechanisms of energy transfer must be included in the energy system. These mechanisms must be adequate to export energy from the region of surplus and to carry that energy into the regions of deficiency. On our planet, motions of the atmosphere and oceans act as heat-transfer mechanisms. A study of the earth's energy balance will not be complete until the global patterns of air and water circulation are described and explained in Chapter 6.

Thinking further, we realize that the movement of water through atmosphere and oceans and on the lands comprises a global system of transport of matter. This system is equal in environmental importance to the flow of energy. We also realize that the activities of these two systems are closely intermeshed. The concept of a water balance will be developed in Chapter 10 and will take its place beside the energy balance. Together, these two great flow systems of energy and matter form a single, grand planetary system and permit us to relate and explain many of the environmental phenomena of our earth within a unified framework.

A systematic approach to the earth's energy balance begins with an examination of the input, or source, of energy from solar radiation. We shall trace this radiation as it penetrates the earth's atmosphere and is absorbed or transformed. We then turn to the mechanism of output of energy by the earth as a secondary radiator.

Electromagnetic Radiation

Our sun, a star of about average size and temperature compared with the overall range of stars, has a surface temperature of about 6000° C (11,000° F). The highly heated, incandescent gas that comprises the sun's surface

Figure 4.1 **A schematic diagram of the global energy balance.**

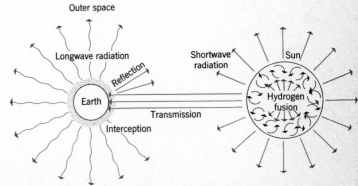

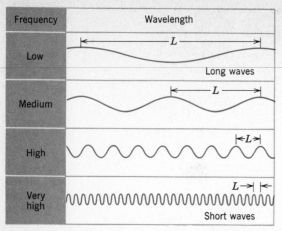

Frequency	Wavelength
Low	$\longleftarrow L \longrightarrow$ Long waves
Medium	$\longleftarrow L \longrightarrow$
High	$\vert\leftarrow L \rightarrow\vert$
Very high	$L \rightarrow\vert\ \vert\leftarrow$ Short waves

Figure 4.2 Relationship of wavelength to frequency. (From A. N. Strahler, 1971, *The Earth Sciences,* 2nd ed., Harper and Row, New York.)

emits a form of energy known as **electromagnetic radiation,** traveling at the uniform speed of 300,000 km (186,000 mi) per second. The energy travels in straight lines radially outward from the sun and requires about 8⅓ minutes to travel the 150 million km (93 million mi) from sun to earth. Although the solar radiation travels through space without energy loss, the rays are diverging as they move away from the sun. Consequently, the intensity of radiation within a beam of given cross section (such as 1 sq cm) decreases inversely as the square of the distance from the sun. The earth intercepts only about one two-billionth of the sun's total energy output.

For many practical purposes, including the study of the earth's radiation balance, electromagnetic radiation can be treated as a wave form of energy transport. The waves can be visualized as resembling the bow waves generated by a ship moving through a perfectly calm sea. Waves of this type are known as sine waves.

Figure 4.2 illustrates how electromagnetic waves differ in dimensions throughout their entire range, or **spectrum.** Waves are described in terms of either wavelength or frequency. **Wavelength,** L, is the actual distance between two successive wave crests (or two successive wave troughs). Metric units of length, for example, centimeters or meters, are always used. Next, we know that all waves of the electromagnetic spectrum travel at the same speed. The number of waves passing a fixed point in a unit of time (one second) is known as the **wave frequency.** Frequency depends on the wavelength. Long waves have low frequency; short waves have high frequency. Frequency is stated in terms of cycles per second; the unit of measure is the **hertz.** One hertz is a frequency of one cycle per second; one **megahertz** is a frequency of one million cycles per second. Knowing the speed of light to be 300,000 km per sec, it is easy to calculate that a frequency of one megahertz is associated with a wavelength of 300 m.

Equivalents between wavelength and frequency can be read directly on the scales of Figure 4.3, which shows the divisions of the electromagnetic spectrum — ranging from the shortest waves at the left to the longest waves at the right. The scale is a logarithmic scale (constant-ratio scale). Frequency in megahertz is given in powers of 10; each scale unit represents a tenfold increase from right to left. For example, 10^3 megahertz is "one thousand megahertz"; 10^9 is "one billion megahertz."

At the short end of the spectrum are **gamma rays** of extremely high frequency and short wavelength. It is customary to describe these short wavelengths in units of **angstroms.** One angstrom is equal to 0.000,000,01 cm (10^{-8} cm). Gamma rays are shorter than 0.03 angstroms. Next come the **X rays.** The shorter X rays, described as "hard," fall in the range 0.03 to about 0.6 angstroms. The longer X rays, described as "soft," range from about 0.6 to about 100 angstroms; they grade into **ultraviolet rays,** which extend to about 4000 angstroms.

For the visible light spectrum it is customary to switch to a longer unit of length, the **micron.** One micron equals 0.0001 cm (10^{-4} cm); and micron thus equals 10,000 angstroms. The **visible light** portion of the spectrum begins with violet at 0.4 microns. Colors then grade successively through blue, green, yellow, orange, and red, reaching the end of the visible spectrum at about 0.7 microns. In Figure 4.3, the visible light spectrum has been expanded to show the limits of the various colors.

Next in the spectrum comes the **infrared region,** consisting of wavelengths from about 0.7 to about 300 microns. These rays are not visible, but can sometimes be felt as "heat rays" from a hot object.

Longer wavelength regions of the spectrum, including microwaves, radar, and radio waves, are described in later paragraphs of this chapter, under the subject of remote sensing. They are not important in the global energy balance.

The Solar Constant

The energy source for solar electromagnetic radiation is in the sun's interior where, under enormous confining pressure and high temperature, hydrogen is converted to helium. In this nuclear fusion process, a vast quantity of heat is generated and finds its way by convection and conduction to the sun's surface. Because the rate of internal production of energy is constant, the energy output of the sun varies only slightly. Consequently, at the average distance from the sun, the solar energy that is received on a unit area of surface held at right angles to the sun's rays is also nearly constant. Known as the **solar constant,** this radiation rate has a value of 2 gram calories per square centimeter per minute (2 cal/sq cm/min). We are assuming, of course, that the solar radiation is measured beyond the limits of the earth's atmosphere so that none has been lost.

One gram calorie per square centimeter (1 cal/sq cm) constitutes a unit measure of heat energy termed the **langley (ly).** Therefore we can say that the solar constant is equal to 2 langleys per minute (2 ly/min). (In English heat units, the solar constant is equivalent to 430 Btu per square foot per hour.)

Orbiting space satellites equipped with suitable

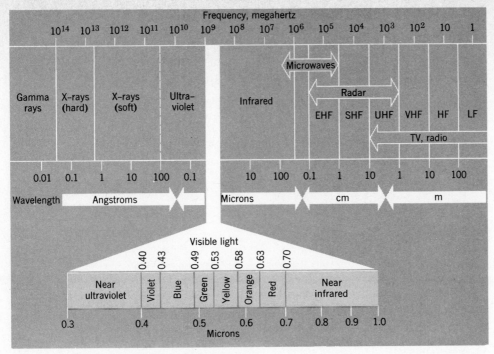

Figure 4.3 **The electromagnetic radiation spectrum.**

instruments for measuring electromagnetic radiation intensity have provided precise data on the solar constant. Minor fluctuations in radiation intensity are observed, and these are attributed to variations in output of ultraviolet rays accompanying disturbances of the sun's surface. For the purpose of evaluating the earth's radiation balance, we can disregard these minor energy fluctuations.

Energy Spectra of Sun and Earth

To understand the basic differences in the kind and intensity of electromagnetic energy emitted by the sun and the earth, you need to know the underlying physical laws governing the radiation of energy in relation to the temperature of a radiation-emitting object. Any substance whose temperature is above absolute zero (−273°C) emits electromagnetic radiation. The total energy emitted for each unit area of surface increases with increased temperature. Thus one square centimeter of the sun's surface at the extremely high temperature of several thousand degrees will emit a vastly greater quantity of energy each second than a square centimeter of the surface of the distant planet Pluto, which has a temperature close to absolute zero. Moreover, the radiation spectrum of emission of a cold surface differs greatly from that of a hot surface. For the cold surface, most of the energy emitted is in the long wavelengths; for the hot surface, most is concentrated in the short (visible and ultraviolet) wavelengths. Comparing the radiation properties of the sun with those of the earth shows how this principle applies.

Figure 4.4 is a graph comparing the intensity of emission of energy of the sun and earth. A logarithmic scale of wavelengths is used, as in Figure 4.3. The vertical scale, which is also logarithmic, shows intensity of energy emission (ly/min) within each micron width of spectrum. Ideal energy curves are shown in Figure 4.4 for both sun (left) and earth (right) by black, smoothly arched lines. These represent the curves for ideal bodies which are perfect radiators of energy; they are called black bodies by physicists.

A **black body** will not only absorb all radiation falling on it, but it will also radiate energy in a manner solely dependent on its surface temperature. The sun closely resembles a black body with a temperature of about 6000°K.* This ideal body has a peak intensity of energy output in the visible light region. The actual radiation intensity curve of the sun is shown by the solid color line; it has minor irregularities and reduced intensity in the ultraviolet region.

Energy within the sun's spectrum is distributed about as follows: ultraviolet and shorter wavelengths, 9 percent; visible light, 41 percent; infrared, 50 percent.

For the earth, the ideal black-body radiation curve is for a planet with an average surface temperature of 300°K (27°C, 80°F). This curve lies entirely within the infrared region and peaks at about 10 microns. Notice that the peak is much lower than for the sun (about one-fifth as intense). The area under the earth curve is much smaller, meaning that the total energy emitted by a square centimeter of area is very much less than from a corresponding area of the sun's surface. This relationship is stated in strict terms by the **Stefan-Boltzmann law:** The

*K stands for *absolute temperature* in *degrees Kelvin*. The degrees are of the same unit value as in the Celsius scale, but the zero point of the Kelvin scale is at −273°C, or absolute zero.

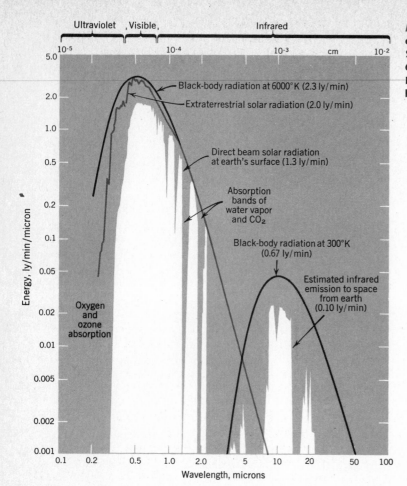

Figure 4.4 **Spectra of solar and earth radiation. (From W. D. Sellers, 1965,** *Physical Climatology,* **Univ. of Chicago Press, p. 20, Figure 6. © 1965 by the University of Chicago.)**

total energy radiated by each unit of surface per unit of time varies as the fourth power of the absolute temperature (°K).

Referring again to Figure 4.4, we find beneath the arched black-body curves of both sun and earth deeply notched curves showing the intensity of energy measured after passage through the earth's atmosphere. The left-hand curve shows the solar spectrum as depleted by passage of the solar beam through the earth's atmosphere. Practically all ultraviolet rays have been removed, while certain bands of the infrared region have been cut out through absorption by atmospheric water vapor and carbon dioxide. The right-hand curve shows the earth's outgoing infrared radiation at the outer limits of the atmosphere. Much of the outgoing infrared energy has been absorbed by water vapor and carbon dioxide in the atmosphere. But in certain wavelength bands, known as **windows,** energy escapes to outer space. The major window, in terms of total energy escape, is between 8 and 14 microns. There are lesser windows in the range of 4 to 6 microns and 17 to 21 microns.

In atmospheric science, the entire solar spectrum is referred to as **shortwave radiation** because the peak of intensity lies in the visible region. In contrast, the earth's outgoing radiation spectrum is referred to as **longwave radiation.** We shall use these terms in describing the earth's energy budget.

Insolation over the Globe

Referring back to Chapter 2 and Figure 2.8, showing equinox conditions, recall that at only one point, the subsolar point, does the earth's spherical surface present itself at right angles to the sun's rays. In all directions away from the subsolar point, the earth's curved surface becomes turned at a decreasing angle with respect to the rays until the circle of illumination is reached. Along that circle the rays are parallel with the surface.

Let us now assume that the earth is a perfectly uniform sphere with no atmosphere. Only at the subsolar point will solar energy be intercepted at the full value of the solar constant, 2 ly/min. We will now use the term **insolation** to mean the interception of solar shortwave energy by an exposed surface. At any particular place on the earth, insolation received in one day will depend on two factors: (1) the angle at which the sun's rays strike the earth and (2) the length of time of exposure to the rays. These factors are varied by latitude and by the seasonal changes in the path of the sun in the sky.

Figure 4.5 shows that intensity of insolation is greatest where the sun's rays strike vertically, as they do at noon at some parallel between the tropic of cancer and the tropic of capricorn. With diminishing angle, the same amount of solar energy spreads over a greater area of ground surface. So, on the average, the polar regions

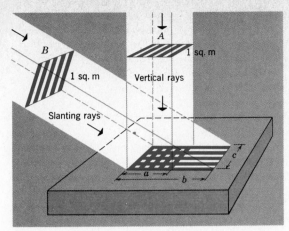

Figure 4.5 *The angle of the sun's rays determines the intensity of insolation on the ground. The energy of vertical rays A is concentrated in square a, but the same energy in the slanting rays B is spread over a rectangle, b.*

receive the least insolation over a year's time.

Consider, now, a hypothetical situation in which the earth's axis is perpendicular to the plane of the ecliptic as the earth revolves around the sun. In other words, equinox conditions would prevail year-round. Under such conditions, the poles would not receive any insolation, regardless of time of year, whereas the equator would receive an unvarying maximum. But we know that the earth's axis is not perpendicular to the plane of the orbit.

As we found in Chapter 2, because of the tilt of the earth's axis, the subsolar point ranges yearly over a total latitude span of 47°, from the tropic of cancer (lat. 23½°N) at June solstice, to the tropic of capricorn (lat. 23½°S) at December solstice. This cycle does not make the yearly total of insolation for the entire globe different from an imaginary situation in which the earth's axis is not inclined; but it does cause a great difference in the

Figure 4.6 **Meridional profile of annual total insolation (solid line).**

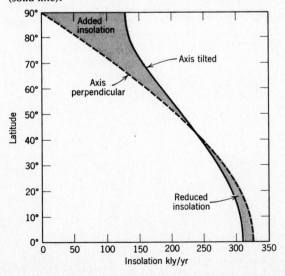

quantities received at various latitudes.

The total annual insolation from equator to poles in thousands of langleys (kilolangleys) per year (kly/yr) is shown in Figure 4.6 by a solid line. A dashed line shows the insolation that would result if the earth's axis had no tilt. Notice how much insolation the polar regions actually receive — over 40 percent of the equatorial value. The added insolation at high latitudes is of major environmental importance because it brings heat in summer to allow forests to flourish and crops to be grown over a much larger portion of the earth than would otherwise be the case.

A second effect of the axial tilt is to produce seasonal differences in insolation at any given latitude. These differences increase toward the poles, where the ultimate in opposites (six months of day, six of night) is reached. Along with the variation in angle of the sun's rays another factor operates — the duration of daylight. At the season when the sun's path is highest in the sky, the length of time it is above the horizon is correspondingly greater. (This effect is clearly shown in Figure 2.11.) The two factors thus work hand in hand to intensify the contrast between amounts of insolation at opposite solstices.

A three-dimensional diagram (Figure 4.7) shows how insolation varies with latitude and season of year. Figure 4.8 shows graphs of insolation at selected latitudes from the equator to the north pole. (These diagrams show insolation at the outer limits of the atmosphere.) Notice that the equator receives two maximum periods (corresponding with the equinoxes, when the sun is overhead at the equator) and two minimum periods (corresponding to the solstices, when the sun's declination is farthest north and south of the equator). At the arctic circle, lat 66½°N, insolation is reduced to nothing on the day of the winter solstice, and with increasing latitude poleward this period of no insolation becomes longer. All latitudes between the tropics of cancer and capricorn have two maxima and two minima, but one maximum becomes dominant as the tropic is approached. From lat. 23½° to 66½°N and S, there is a single continuous insolation cycle with maximum at one solstice, minimum at the other.

Figure 4.7 **Insolation at various latitudes and seasons. (After W. M. Davis.)**

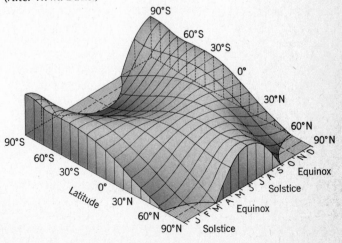

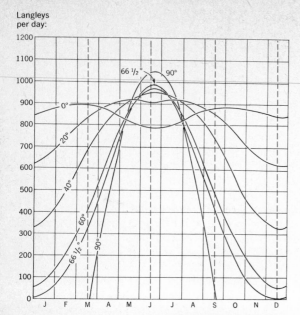

Langleys per day:

Figure 4.8 **Insolation at various latitudes in the northern hemisphere.** (From A. N. Strahler, 1971, *The Earth Sciences,* 2nd ed., Harper and Row, New York.)

World Latitude Zones

The angle of attack of the sun's rays determines the flow of solar energy reaching a given unit area of the earth's surface and so governs the thermal environment of the life layer. This concept provides a basis for dividing the globe into latitude zones (Figure 4.9). The specified zone limits should not be taken as absolute and binding. Instead, the system should be considered as a convenient terminology for identifying world geographic belts throughout this book.

The **equatorial zone** lies astride the equator and covers the latitude belt roughly 10°N to 10°S. Within this zone, the sun throughout the year provides intense insolation, while day and night are of roughly equal duration. Astride the tropics of cancer and capricorn are the **tropical zones,** spanning the latitude belts 10° to 25°N and S. In these zones, the sun takes a path close to the zenith at one solstice and is appreciably lower at the opposite solstice. Thus a marked seasonal cycle exists, but it is combined with a large total annual insolation.

Literary usage and that of many scientific works differ from what is described here; the terms "tropical zone" and "tropics" have been widely used to denote the entire belt of 47° of latitude between the tropics of cancer and capricorn. That is the definition of "tropics" you will find in most dictionaries. Even though correct, that definition is not well suited to a study of Man's physical environment, because it combines belts of very unlike climatic properties; it lumps the greatest of the world's deserts with the wettest of climates.

Immediately poleward of the tropical zones are transitional regions that have become widely accepted in geographers' usage as the **subtropical zones.** For convenience, we have assigned these zones the latitude

belts 25° to 35°N and S, but it is understood that the adjective "subtropical" may be extended a few degrees farther poleward or equatorward of these parallels.

The **midlatitude zones,** lying between lat. 35° and 55°N and S, are belts in which the sun's angle of attack shifts through a relatively large range, so that seasonal contrasts in insolation are strong. Strong seasonal differences in lengths of day and night exist as compared with the tropical zones.

Bordering the midlatitude zones on the poleward side are the **subarctic** and **subantarctic zones,** lat. 55° to 60°N and S. Astride the arctic and antarctic circles, lat. 66½°N and S, lie the **arctic zone** and the **antarctic zone.** These zones have an extremely large yearly variation in lengths of day and night, yielding enormous contrasts in insolation from solstice to solstice. Notice that dictionaries define "arctic" as the entire area from arctic circle to north pole, and "antarctic" in a corresponding sense for the southern hemisphere.

The **polar zones,** north and south, are circular areas between about lat. 75° and the poles. Here the polar regime of a six-month day and six-month night is predominant. These zones experience the ultimate in seasonal contrasts of insolation.

Insolation Losses in the Atmosphere

As the sun's radiation penetrates the earth's atmosphere, its energy is absorbed or diverted in various ways. At an altitude of 150 km (95 mi), the radiation spectrum possesses almost 100 percent of its original energy but, by the time rays have penetrated to an altitude of 88 km (55

Figure 4.9 **A geographic system of latitude zones.**

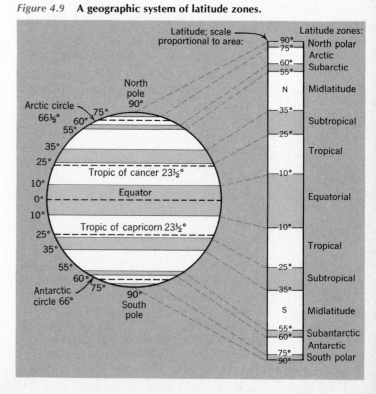

mi), absorption of X rays is almost complete and some of the ultraviolet radiation has been absorbed as well.

As solar rays penetrate more deeply into the atmosphere, they reach the stratosphere. Here, as we explained in Chapter 3, ozone absorbs ultraviolet radiation within the ozone layer. The environmental shielding provided by this layer has already been stressed.

As radiation penetrates into deeper and denser atmospheric layers, gas molecules cause the visible light rays to be turned aside in all possible directions, a process known as **scattering.** Dust and cloud particles in the troposphere cause further scattering, described as **diffuse reflection.** Scattering and diffuse reflection send some energy into outer space and some down to the earth's surface.

As a result of all forms of shortwave scattering, about 5 percent of the total insolation is returned to space and forever lost, as shown in Figure 4.10. Scattered shortwave energy directed earthward is referred to as **down-scatter.**

Another form of energy loss, **absorption,** takes place as the sun's rays penetrate the atmosphere. Both carbon dioxide and water vapor are capable of directly absorbing infrared radiation. Absorption results in a rise of temperature of the air. In this way some direct heating of the lower atmosphere takes place during incoming solar radiation. Although carbon dioxide is a nearly constant quantity in the air, the water vapor content varies greatly from place to place. Absorption correspondingly varies from one global environment to another.

All forms of direct energy absorption — by air molecules, including molecules of water vapor and carbon dioxide, and by dust — are estimated to total as little as 10 percent for conditions of clear, dry air to as high as 30 percent when a cloud cover exists. A global average of 5 percent for absorption generally is shown in Figure 4.10. When skies are clear, reflection and absorption combined may total about 20 percent, leaving as much as 80 percent to reach the ground.

Yet another form of energy loss must be brought into the picture. The upper surfaces of clouds are extremely good reflectors of shortwave radiation. Air travelers are well aware of how painfully brilliant the sunlit upper surface of a cloud deck can be when seen from above. **Cloud reflection** can account for a direct turning back into space of from 30 to 60 percent of total incoming radiation (Figure 4.10). Clouds also absorb radiation. So we see that, under conditions of a heavy cloud layer, the combined reflection and absorption from clouds alone can account for a loss of 35 to 80 percent of the incoming radiation and allow from 45 to 0 percent to reach the ground. A world average value for reflection from clouds to space is about 21 percent of the total insolation. Average absorption by clouds is much less — about 3 percent.

The surfaces of the land and ocean reflect some shortwave radiation directly back into the atmosphere. This small quantity, about 6 percent as a world average, may be combined with cloud reflection in evaluating total reflection losses. Table 4.1 lists the percentages given so far for the energy losses by reflection and absorption. Altogether the losses to space by reflection total 32

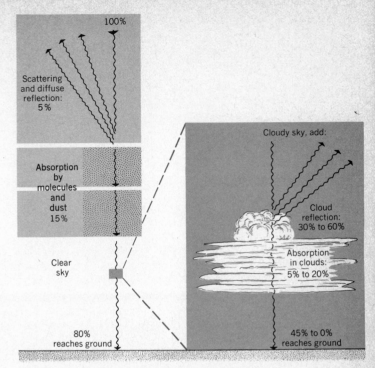

Figure 4.10 **Losses of incoming solar energy by scattering, reflection, and absorption. (From A. N. Strahler, 1971, *The Earth Sciences*, 2nd ed., Harper and Row, New York.)**

percent of the total insolation. Figure 4.11 (upper half) shows the same data in the form of arrows, whose widths are proportionate to the percentages.

The percentage of radiant energy reflected back by a surface is the **albedo.** Orbiting satellites, suitably equipped with instruments to measure the energy levels of shortwave and infrared radiation, both incoming from the sun and outgoing from the atmosphere and earth's surface below, have provided data for estimating the earth's average albedo. Values between 29 and 34 percent have been obtained. The value of 32 percent given in Figure 4.11 lies between these limits.

The earth's albedo lies in a range intermediate between low values of the moon and inner planets (Mercury, 6%; Mars, 16%; moon, 7%) and high values of Venus (76%) and the great outer planets (73 to 94%). Here, again, we find another unique element of the environment of Planet Earth as compared with that of the other planets and the moon.

Figure 4.11 summarizes energy losses by absorption. A global figure of 18 percent is given for the combined losses through absorption by molecules, dust, water vapor, carbon dioxide, and clouds. Fifty percent of the original solar energy remains, and this amount is absorbed by the earth's land and water surfaces. We have now accounted for all the energy of insolation. Our next step is to deal with outgoing energy so as to balance the global energy budget.

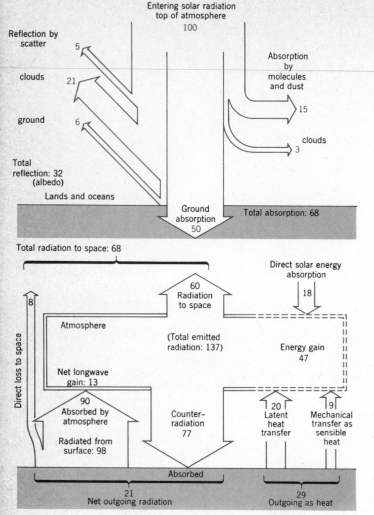

Figure 4.11 **Schematic diagram of the global radiation balance. Figures correspond with those in Table 4.1.**

Table 4.1 / The Global Radiation Budget

Incoming Solar Radiation	Percentage
Total at top of atmosphere	100
Diffuse reflection to space by scattering	5
Reflection from clouds to space	21
Direct reflection from earth's surface	6
Total reflection loss to space by earth-atmosphere system (earth's albedo)	32
Absorbed by molecules, dust, water vapor, CO_2, clouds	18
Absorbed by earth's surface	50
Total absorbed by earth-atmosphere system	68
Sum of absorption and reflection	100

Outgoing Longwave Radiation	Percentage
Radiation balance of earth's surface	
Loss directly to space	8
Loss to atmosphere	90
Total radiation from surface	98
Gain by counterradiation from atmosphere	77
Net outgoing radiation from surface	21
Radiation balance of atmosphere	
Gain from surface radiation	90
Loss by counterradiation	77
Net gain from surface	13
Gain from direct shortwave absorption	18
Gain from latent heat transfer	20
Gain from mechanical heat transfer	9
Total net gain	60
Radiated to space from atmosphere	60
Radiated directly to space from surface	8
Total radiation from planet to space	68

Data source: W. D. Sellers (1965), *Physical climatology,* Univ. of Chicago Press, Chicago and London, Tables 6 and 9.
Note: One hundred percent represents a value of 263 kilolangleys per year.

Longwave Radiation

The earth's ground and ocean surfaces possess heat derived from absorption of insolation. This form of energy is referred to as sensible heat (heat that can be measured by a thermometer); it represents energy in temporary storage. This stored heat is being continually lost by longwave radiation into the overlying atmosphere, a process we can call **ground radiation.** (The term can include radiation from water surfaces, as well.)

The atmosphere also radiates longwave energy both downward to the ground and outward into space, where it is lost. Be sure to understand that longwave radiation is quite different from reflection, in which the rays are turned back directly without being absorbed. Longwave radiation from both ground and atmosphere continues during the night, when no solar radiation is being received.

As Figure 4.4 shows, the intensity of longwave energy leaving our planet is only a small fraction of the intensity of incoming shortwave solar energy. Keep in mind that longwave radiation is constantly emitted from the entire

spherical surface of the earth, whereas insolation falls on only one hemisphere. A single hemisphere presents the equivalent of only a cross section of the sphere to full insolation. Also, do not forget that about 32 percent of the incoming solar radiation is reflected directly back into space. Only the remaining 68 percent that is absorbed must be disposed of by longwave radiation. On the average, year in and year out, for the planet as a whole, the quantity of absorbed solar energy is balanced by an equal quantity of longwave emission to space.

Part of the ground radiation absorbed by the atmosphere is radiated back toward the earth's surface, a process called **counterradiation.** For this reason the lower atmosphere, with its water vapor and carbon dioxide, acts as a blanket that returns heat to the earth. This mechanism helps to keep surface temperatures from dropping excessively during the night or in winter at middle and high latitudes. Somewhat the same principle is used in greenhouses and in homes using direct solar heating; the glass windows permit entry of shortwave energy. Accumulated heat cannot escape by mixing with cooler air outside. Meteorologists use the expression **greenhouse effect** to describe this atmospheric heating principle.

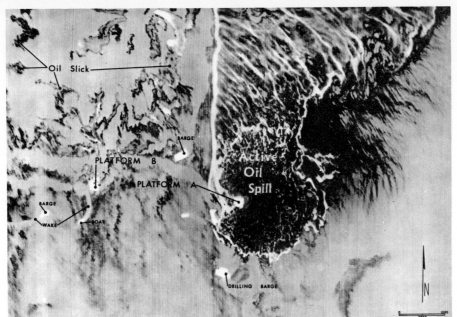

Figure 1 Thermal infrared image of the source area of the Santa Barbara Oil Spill, taken in early February, 1969. This image shows the patterns of emitted thermal infrared energy from the sea surface. Imagery such as this can be acquired day or night. (Photo by courtesy of North American Rockwell Corporation; labels by Geography Remote Sensing Unit, University of California, Santa Barbara.)

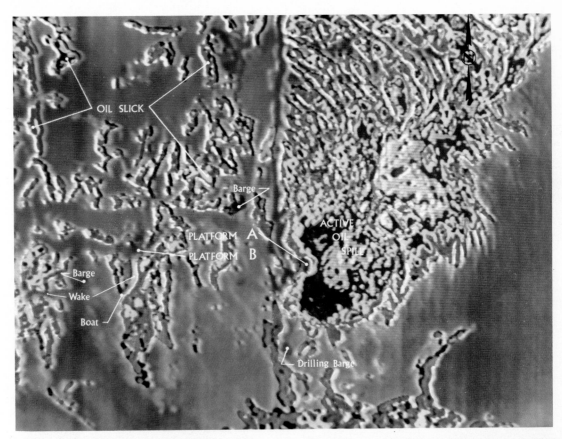

Figure 2 **An electronic enhancement of Figure 1.** Image enhancement systems, such as the one which generated this image, expand the image interpreter's ability to study the meaning of subtle variations in an image's optical density. This effect is achieved by assigning colors to specific gray levels on the original image. In this instance researchers were attempting to ascertain whether or not the patterns of emitted thermal energy emanating from the scene were related to variations in thickness of the oil film on the seawater. (Photo by courtesy of Spatial Data Systems Inc.; labels by Geography Remote Sensing Unit, University of California, Santa Barbara.)

This color brochure illustrating applications of remote sensing was compiled and annotated by John E. Estes and Leslie W. Senger. © 1975 by John Wiley & Sons, Inc., New York.

Remote sensing   **Plate A.1**

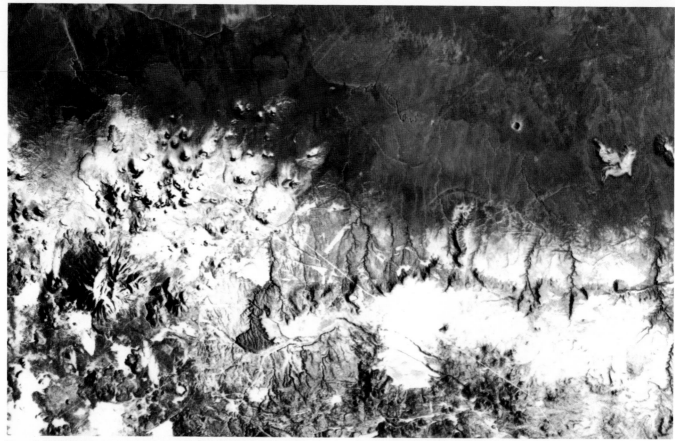

Figure 3 This Skylab 4 color infrared photo of the Flagstaff, Arizona, area was imaged by the Earth Resources Experiment Package (EREP) S-190B (13-cm, 5-in., terrain camera). Humphrey's Peak, the Sunset Crater volcanic field, and Flagstaff are at the left of the photo. Meteor Crater, seen in the right center of the photo, may be used for scale comparison; the crater is approximately 1.6km (1 mi) in diameter. In this photo living vegetation appears in red, snow in white, water in blue. Photos such as this are used for snow mapping in an attempt to gain a better understanding of an area's total water resource capability. (Photo courtesy National Aeronautics and Space Administration.)

Figure 4 Portion of a color infrared vertical photo of the eastern coast of Sicily taken during the Skylab 3 mission with the Earth Resources Experiment Package (EREP) S-190B (13-cm, 5-in., terrain camera). Mt. Etna, the highest volcano in Europe, is still active, as evident by the thin plume of smoke emanating from its crest. On Etna's flanks recent lava flows appear black, while older lava flows and volcanic debris, which now support vegetation, appear in shades of red. Photos such as this can be used not only to study the localized patterns of geologic phenomena, such as volcanic activity or faulting, but regional or global patterns and distributions as well. (Photo by courtesy of National Aeronautics and Space Administration.)

Plate A.2   Remote sensing

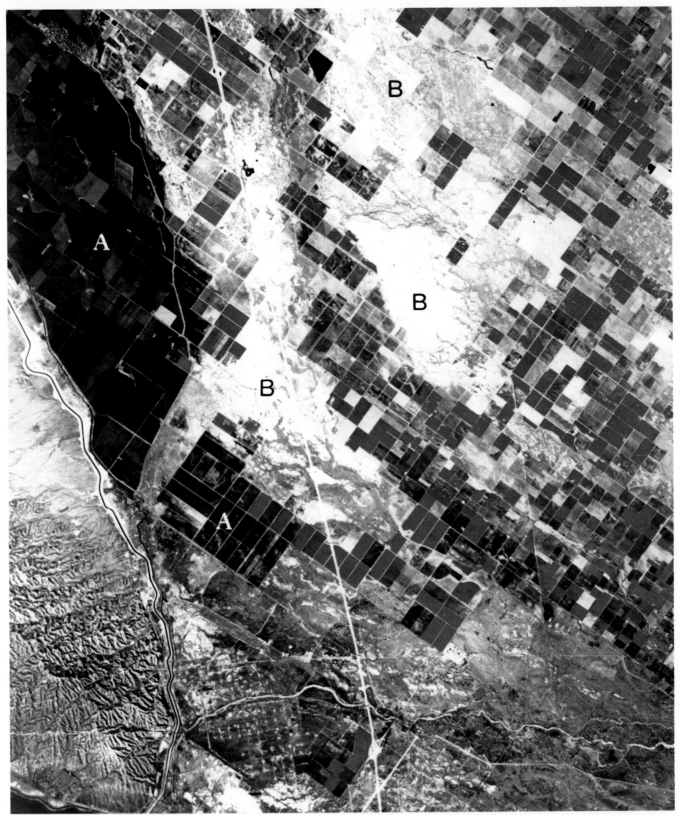

Figure 5 **High-altitude color infrared aerial photograph of an area near Bakersfield, California, in the Southern San Joaquin Valley, taken by a National Aeronautics and Space Administration U-2 aircraft flying at approximately 18km (60,000 ft). The original photo scale was about 1:120,000. Photos such as this are used to study problems associated with agriculture. In this area perched ground water areas seen at (A) and areas of high soil salinity (B) create problems which can affect crop yields. The various red tones in the fields indicate that various types of crops are being grown. By knowing which crops are typically grown in this area and their growing** cycle, determinations can be made of the amount and areal extent of individual crops being cultivated. Estimates of areal, regional, or global yield may be made by adding this type of information to a stratified statistical model with appropriate sampling at the local, regional, or global level and which has as an integral part the monitoring of the growing cycle of the crop or crops of interest and with the addition of information concerning significant meteorological, hydrological, and pedological variables. (Photo by courtesy of National Aeronautics and Space Administration.)

Figure 6 A high-altitude color infrared photograph of Goleta, California, taken in April 1971, by a National Aeronautics and Space Administration RG-57 aircraft from approximately 18 km (60,000 ft) on an original scale of 1:120,000. 1973 was a normal rainfall year for this area and, at this time, grassy natural vegetation (A) appears light pink while chaparral slopes (B) appear darker red. (Photo by courtesy of National Aeronautics and Space Administration.)

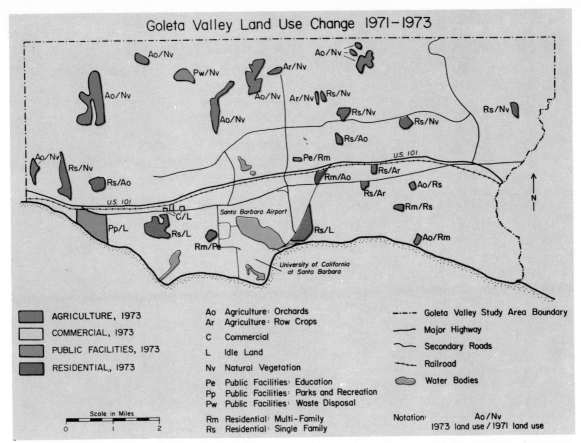

Goleta Valley Land Use Change 1971–1973

	Agriculture: Orchards		
Ao	Agriculture: Orchards		
Ar	Agriculture: Row Crops		
C	Commercial		
L	Idle Land		
Nv	Natural Vegetation		
Pe	Public Facilities: Education		
Pp	Public Facilities: Parks and Recreation		
Pw	Public Facilities: Waste Disposal		
Rm	Residential: Multi-Family		
Rs	Residential: Single Family		

AGRICULTURE, 1973
COMMERCIAL, 1973
PUBLIC FACILITIES, 1973
RESIDENTIAL, 1973

Scale in Miles
0   1   2

–·—·– Goleta Valley Study Area Boundary
——— Major Highway
⌒ Secondary Roads
—+—+— Railroad
Water Bodies

Notation: Ao/Nv
1973 land use / 1971 land use

Figure 8 Land use change map of the Goleta Valley, California, compiled from information interpreted from Figures 6 and 7 on the facing page. Many types of environmental information are perishable. Remotely sensed images, such as those seen here, provide a data bank of information concerning an area at a point in time, which can be used by many types of environmental specialists. By acquiring imagery of an area through time, users can gain insights into the types, rates, and amounts of changes which may be occurring within a given area. Information obtained in this manner may then be analyzed; the data thus extracted can be utilized for a variety of applications. These can range from the testing of theoretical models to the making of policy decisions. Information contained in the photographs of Figures 6 and 7 could be used by rangeland managers in their decision-making process. Decisions concerning the readiness of the area's range grasses could be made so that the animals could be moved on and off the range at the optimum time; thus, neither underuse or overuse of the resource occurs. Land use planners and policy makers can use the information shown here to monitor the changes occurring, and whether or not those changes conform to general planning objectives and requirements. As a final example, it may be possible to relate the land use data, which can be extracted from images such as these, to the development of water consumption equations for the land use categories depicted and the modeling of water demand in the area shown. (Map by courtesy of Geography Remote Sensing Unit, University of California, Santa Barbara.)

Figure 7 (opposite) A second National Aeronautics and Space Administration color infrared photograph of the Goleta Valley imaged in January 1973 by a U-2 aircraft from about 18 km (again with an original contact scale of 1:120,000). This image, taken after a fall period of heavier than normal rainfall, shows grassy vegetation (A) in bright red with the chaparral slopes (B) again slightly darker. (Photo by courtesy of National Aeronautics and Space Administration.)

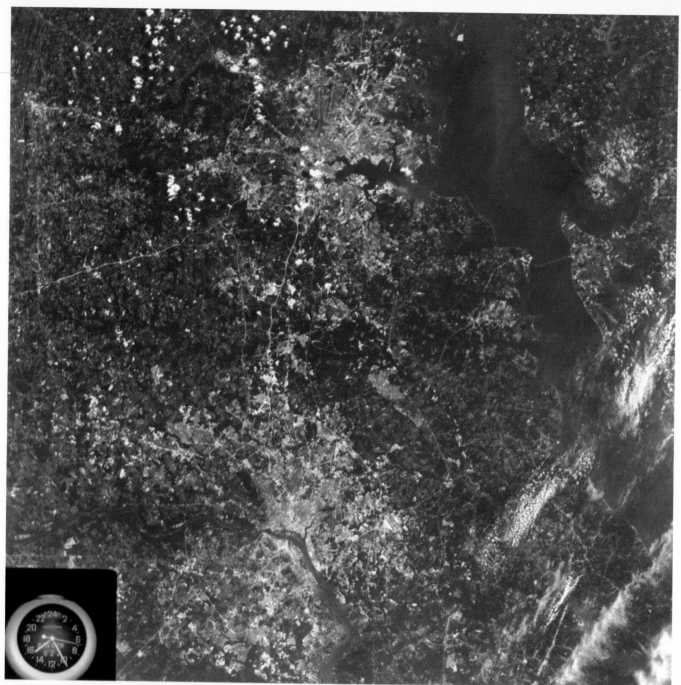

Figure 9 **Color photograph of the Washington, D.C. and Baltimore, Maryland, area imaged from the Skylab 3 Earth Resources Experimental Package (EREP) S-1908 (13-cm, 5-in., terrain camera). From the regional perspective provided by photos such as this, the extent of urbanization is readily apparent, with suburban housing showing up as a pink-gray tone. Beltways** around the cities, Interstate 95 between Baltimore and Washington, Interstate 70N leading west (left) from Baltimore, harbor facilities, bridges across the Potomac, and airfields are easily identifiable. Sedimentation patterns are detectable as lighter tones in the water. (Photo by courtesy of National Aeronautics and Space Administration and Hughes Aircraft Company.)

Figure 10 (opposite) **Portion of an Earth Resources Technology Satellite (Landsat) color infrared composite (Multispectral Scanner Bands 4, 5 and 7) image of essentially the same area seen in Figure 9. Urban patterns and transportation networks can be readily interpreted. Vegetated areas stand out more clearly than other features in this photography because of their high reflectance in the infrared portion of the electromagnetic (E-M)** spectrum and because of the increased haze penetrating ability of this type of photography, which images longer wavelength energy than conventional color photography. Both Figures 9 and 10 have applications for land use analysis, sediment studies, and general resource surveys. (Photo by courtesy of National Aeronautics and Space Administration and Hughes Aircraft Company.)

Plate A.6   Remote sensing

Figure 11 The upper frame is a portion of a false color Earth Resources Technology Satellite (Landsat) photograph of the Los Angeles basin. A standard Landsat photograph covers an area 185 x 185 km (100 x 100) nautical miles. Multispectral Scanner Bands 5, 6, and 7 have been color-combined to show vegetation as yellow and commercial concentrations as a bluish gray (purplish) color. The lower frame is an enlargement of a portion of an image of the Los Angeles area, which has been processed from Landsat computer compatible digital tapes. The image has been electronically enhanced by having a computer use statistical techniques to correlate and assign colors to combinations of gray levels from several Landsat spectral bands. This technique generates a thematic map showing heavy commercial areas as yellow. Enhancement systems, such as the one whose product is shown here, are designed to aid image interpreters in extracting the greatest amount of information possible from each image. (Photos by courtesy of National Aeronautics and Space Administration, Hughes Aircraft Company, and General Electric Company.)

Plate A.8   Remote sensing

Cloud layers are even more important in producing a blanketing effect to retain heat in the lower atmosphere because they are excellent absorbers and emitters of longwave radiation.

The Global Radiation Balance

The lower halves of Table 4.1 and Figure 4.11 show the components of outgoing (longwave) radiation for the earth's surface, the atmosphere, and the planet as a whole, using the same percentage units as for the incoming radiation. The total longwave radiation leaving the earth's land and ocean surface is equivalent to 98 percentage units. Of this, 8 units are lost to space, while 90 units are absorbed by the atmosphere. In turn, the atmosphere emits longwave radiation. As shown in the box representing the atmosphere, the total of this emitted radiation is equivalent to 137 percentage units of the insolation. This figure may seem absurd until we note that this radiation is divided into two parts, one of which goes out into space (60 units) and the other of which is absorbed by the earth's surface as counterradiation (77 units).

We must now introduce two new terms into the energy budget. Much of the heat energy passing from the ground to the atmosphere is carried upward by two mechanisms other than longwave radiation. First, sensible heat is conducted upward by turbulent motions of the air. As Figure 4.11 shows (lower right), this heat transport amounts to about 9 percentage units. Second, heat in latent form is conducted upward in water vapor that has evaporated from land and ocean surfaces. Latent heat transport accounts for about 20 percentage units. (Latent heat is explained in Chapter 7.)

Summarizing the information of the previous paragraph, energy leaving the earth's surface (both land and water surfaces) is divided up as follows:

	Percentage Units
Longwave radiation (net value)	21
Mechanical transport as sensible heat	9
Transport as latent heat	20
Total	50

Recall that 50 percent is the amount of insolation (shortwave energy) absorbed by the earth's surface. We have now balanced the annual energy budget so far as the surface is concerned.

The atmosphere in turn disposes of its heat by longwave radiation into outer space. Together with direct longwave emission from the ground, the total emission to space is 68 percent (Table 4.1). Add this figure to 32 percent loss by reflection (the earth's albedo), and the total is 100 percent.

Latitude and the Radiation Balance

Earlier in this chapter, in relating latitude to insolation, we showed that the tilt of the earth's axis causes a poleward redistribution of insolation, as compared with conditions that would apply if the axis were perpendicular to the orbital plane (Figure 4.6). Let us now look deeper into the wide range in rates of incoming and outgoing energy from low latitudes to the polar regions. We shall deal first with three components in the radiation balance: albedo (reflection), insolation, and outgoing longwave radiation.

Albedo

The percentage of shortwave radiation reflected by land and water surfaces is an important property because it determines the proportion of insolation that is absorbed at the surface and converted into sensible heat. Albedo of a water surface is very low (2%) for nearly vertical rays but high for low-angle rays. It is also extremely high for snow-covered land surfaces and ice-covered water surfaces (45 to 90%). Albedos of fields, forests, and bare ground are of intermediate values, ranging from as low as 5 percent to as high as 30 percent.

For the earth as a whole planet, albedo measured from an orbiting satellite depends on both surface reflection and cloud reflection because, as we noted before, clouds are excellent reflectors of shortwave radiation. Mean planetary albedo is shown in the first of a series of four world maps (Map A) in Figure 4.12. Figure 4.13 is a meridional profile of the earth's mean albedo, averaged by latitude belts 10° wide. Notice that the equatorial zone is a belt of low albedo, mostly in the 15–25 percent range. Low albedo here is explained by extensive ocean surfaces and forests and by the lack of persistent blanketlike cloud layers. Across midlatitudes, albedo increases steadily poleward. Maximum values are over the snow-covered surface of Antarctica.

Insolation

Shortwave radiation received at the earth's surface is shown in Map B of Figure 4.12. The average annual total, in kilolangleys per year, is high in the equatorial belt, with values ranging from 100 to 160 kly/yr. Flanking this belt are tropical zones of exceptionally large values over the continents — 180 to 220 kly/yr — representing the great tropical deserts, over which cloud cover is absent much of the time. Values decline rapidly poleward through midlatitudes and are low in the arctic and polar regions.

Figure 4.14, lower curve, is a meridional profile showing shortwave energy absorbed by the earth's surface. The data differ from Map B in that shortwave energy reflected by the surface has been subtracted. The upper curve shows insolation at the top of the atmosphere; it is essentially the same information as in Figure 4.6. The zone between the upper and lower curves represents energy losses through absorption and reflection processes as the solar rays penetrate the atmosphere.

Longwave Radiation

Outgoing energy in the form of longwave radiation from the entire earth-atmosphere system is shown in Map C of Figure 4.12. These data were obtained by orbiting satellite and have been averaged over a long period. Although the units used — langleys per minute (ly/min) — are not the

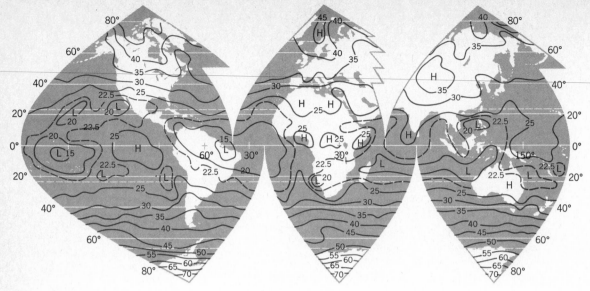

Sinusoidal projection

A. Mean planetary albedo compiled from satellite data. Values are in percent. Letters H and L designate centers of higher and lower values, respectively. (Data from T. H. Vonder Haar and V. E. Suomi, 1969, *Science,* vol. 163, p. 668, Figure 2*a*.)

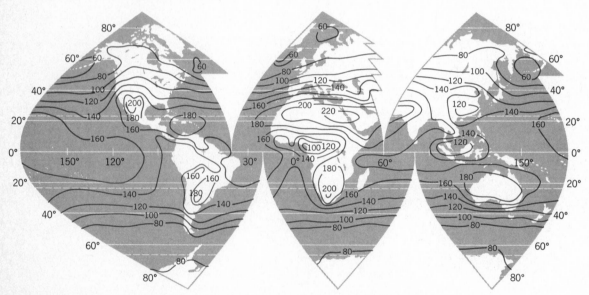

B. Mean annual solar radiation received at the earth's surface. Units are kilolangleys per year. (Based on data from M. I. Budyko, 1963, *Atlas Teplovogo Balansa,* Moscow, USSR, Gidrometeorologischeskoe Izdatel'stvo.)

Figure 4.12 World maps of (*A*) albedo, (*B*) insolation, (*C*) longwave radiation, and (*D*) net radiation. The interrupted sinusoidal projection, an equal-area grid, is explained in Chapter 1 (see Figure 1.23). (From A. N. Strahler, 1971, *The Earth Sciences,* 2nd ed., Harper and Row, New York.)

same as for insolation (Map *B*), the relative amounts can be easily compared. Values are high in low latitudes and decline rapidly toward both polar zones. These data are summarized in a meridional profile in Figure 4.15. Notice maxima occurring in the subtropical zones where the

great deserts are situated. Here ground surface temperatures are high and cloud cover is small, favoring intense outgoing longwave radiation. Over the equatorial belt, where cloud cover on the average is more frequent, radiation is somewhat reduced.

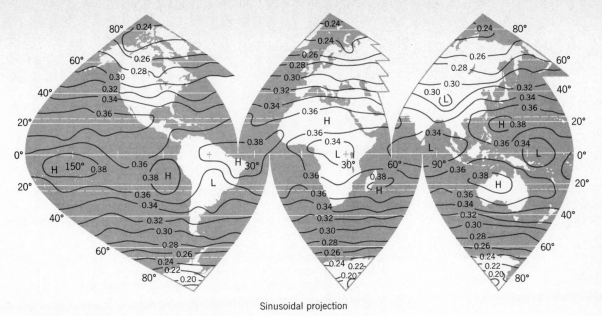

Sinusoidal projection

C. Mean outgoing infrared radiation, based on satellite data. Units are langleys per minute. (Data from T. H. Vonder Haar and V. E. Suomi, 1969, *Science*, vol. 163, p. 668, Figure 2*b*.)

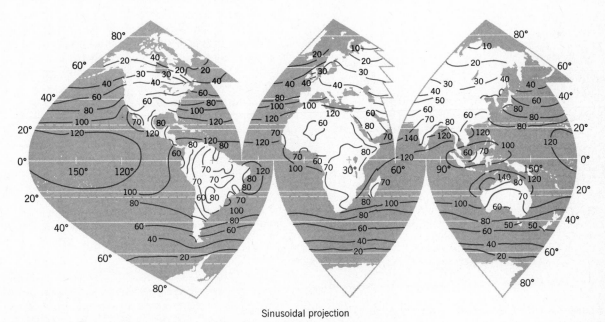

Sinusoidal projection

D. Mean annual net radiation. Units are kilolangleys per year. (Same data source as Map *B*.)

Net Radiation

We now examine **net radiation,** which is the difference between all incoming energy and all outgoing energy carried by both shortwave and longwave radiation. Our analysis of the radiation balance has already shown that for the entire globe as a unit, the net radiation is zero on an annual basis. In some places, however, energy is coming in faster than it is going out, and the energy balance is a

positive quantity, or **energy surplus.** In other places energy is going out faster than it is coming in, and the energy balance is a negative quantity, or **energy deficit.**

Consider first the global distribution of net radiation at the earth's surface shown in Map *D* of Figure 4.12. In Figure 4.16, the uppermost meridional profile averages the same global data by latitude belts. There is a large surplus in low latitudes, with values well over 100 kly/yr generally for the region between lat. 20°N and 20°S. Net radiation

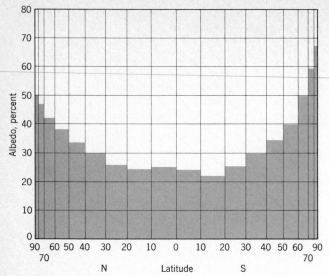

Figure 4.13 Meridional profile of earth's mean albedo based on satellite data. Latitude is scaled proportionately to earth's surface areas between successive 10-degree parallels of latitude. (Data from T. H. Vonder Haar and V. E. Suomi, 1969, *Science,* vol. 163, p. 667, Figure 1. Figure from A. N. Strahler, 1971, *The Earth Sciences,* 2nd ed., Harper and Row, New York.)

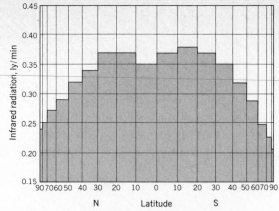

Figure 4.15 Meridional profile of mean longwave radiation from the earth, as determined from satellite data. (Data from T. H. Vonder Haar and V. E. Suomi, 1969, *Science,* vol. 163, p. 667. Figure from A. N. Strahler, 1971, *The Earth Sciences,* 2nd ed., Harper and Row, New York.)

declines rapidly through midlatitudes, reaching zero in the vicinity of 70° latitude in both hemispheres. (High-latitude data are missing from the world map, but show on the profile.) Negative values occur in two small polar zones, showing a radiation deficit.

For the atmosphere, net radiation shows a different global picture than that for the surface. The meridional profile of net radiation for the atmosphere is shown near

the bottom of Figure 4.16. Notice that all values are negative and the deficit always exceeds −60 kly/yr. Next, we combine the data of the earth's surface with those of the atmosphere. In each latitude zone, the deficit quantity (negative value) is subtracted from the surplus quantity (positive value). This calculation produces net radiation values for the combined earth surface-atmosphere system, shown in the middle portion of Figure 4.16. There is a large region of surplus radiation from about lat. 40°N to 30°S and two high-latitude regions of deficit. On the diagram, the areas labeled "deficit" are together equal to the area labeled "surplus," as the radiation balance requires.

It is obvious from Figure 4.16 that the earth's energy

Figure 4.14 Meridional profiles of mean annual values of entering solar radiation at top of atmosphere and radiation absorbed by earth's surface. (Data from W. D. Sellers, 1956, *Physical Climatology.* Figure from A. N. Strahler, 1971, *The Earth Sciences,* 2nd ed., Harper and Row, New York.)

Figure 4.16 Meridional profiles of mean net radiation at earth's surface, from atmosphere, and from combined earth-atmosphere system. (Data from W. D. Sellers, 1965, *Physical Climatology.* Figure from A. N. Strahler, 1971, *The Earth Sciences,* 2nd ed., Harper and Row, New York.)

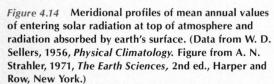

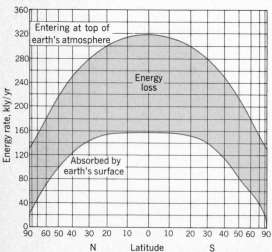

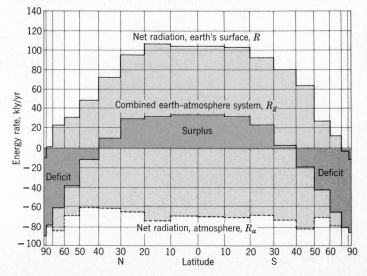

Table 4.2 / Annual Meridional Heat Transport

Latitude (°N)	Heat Transport kcal/yr × 10¹⁹
90	0.00
80	0.35
70	1.25
60	2.40
50	3.40
40	3.91
30	3.56
20	2.54
10	1.21
0	−0.26

Data source: W. D. Sellers (1965), *Physical climatology,* Univ. of Chicago Press, Chicago and London, Table 12.

balance can be maintained only if heat is transported from the low-latitude belt of surplus to the two high-latitude regions of deficit. This poleward movement of heat is described as **meridional transport,** moving north or south along the meridians of longitude. We should expect the rate of meridional heat transport to be greatest in mid-latitudes, and this fact is shown by the figures in Table 4.2. The unit used is the kilocalorie (1000 calories) multiplied by 10 raised to the nineteenth power (10^{19}).

The meridional flow of heat is carried out by the circulation of the atmosphere and oceans. In the atmosphere, heat is transported both as sensible heat and as latent heat. These topics are explained in Chapter 6 and 7.

Remote Sensing in Geography

In various branches of geography, as in other sectors of the earth sciences, a new technical discipline called **remote sensing** has expanded rapidly within the past decade and is adding greatly to our ability to perceive and analyze the physical, chemical, biological, and cultural character of the earth's surface. In its broadest sense, remote sensing is the measurement of some property of an object by using means other than direct contact through the observer's senses of touch, taste, or smell. Hearing and seeing are remote sensing activities of organisms and depend on reception of wave forms of energy transmitted from the object to the observer. Use of an ordinary camera represents a form of remote sensing in which a detecting instrument senses and records the light waves reflected from objects. In its more restricted operational meaning, remote sensing refers to gathering information from great distances and over broad areas, usually through instruments mounted on aircraft or orbiting space vehicles.

All substances, whether naturally occurring or synthesized by Man, are capable of reflecting, absorbing, and emitting energy in forms that can be detected by instruments known collectively as **remote sensors.**

Microwaves

Remote sensors are designed to receive various parts of the electromagnetic spectrum, including the ultraviolet, visible light, and infrared bands, described in detail earlier in this chapter. Also important are longer wavelength bands, illustrated in Figure 4.3.

Infrared rays grade into yet longer wavelengths referred to as **microwaves.** Shifting to centimeters, we can place the microwave region as between about 0.03 and about 1 cm. Most persons are familiar with microwaves as the form of energy used in the microwave oven for the rapid heating or cooking of foods. Microwaves are also used in direct-line transmission of messages from one tower to another across country.

Within the microwave region is the **radar** region beginning at about 0.1 cm and extending through to about 100 cm (1 m). Radar systems are actually microwave sensor systems. Frequencies at which radar systems operate grade into television and radio frequencies, the latter extending into wavelengths exceeding 300 m.

As shown in Figure 4.3, the radar-TV-radio spectrum is divided into equal logarithmic units according to wavelength. The abbreviations shown in Figure 4.3 have the following meanings:

EHF	Extremely high frequency	0.1 to 1 cm
SHF	Superhigh frequency	1 to 10 cm
UHF	Ultrahigh frequency	10 to 100 cm (1 m)
VHF	Very high frequency	1 to 10 m
HF	High frequency	10 to 100 m
LF	Low frequency	Longer than 100 m

Remote sensing makes extensive use of the ultraviolet, visible, infrared, microwave, and radar portions of the spectrum. The longer wavelengths are used largely in TV and radio communications systems.

Sensing Systems

Two classes of electromagnetic sensor systems are recognized: passive systems and active systems. **Passive systems** measure radiant energy reflected or emitted by an object. Mostly, this energy falls in the visible light region (reflected) and the infrared and microwave regions (emitted). The most familiar instrument of this class is the camera, using film sensitive to reflected energy at wavelengths in the visible range.

Active systems use a Man-made beam of wave energy as a source. This type of sensor system sends a beam of energy to an object. A part of the energy is reflected back to the source, where it is recorded by a detector. A simple analogy would be the use of a spotlight on a dark night to illuminate a target and reflect light back to the eye. Active systems used in remote sensing operate mostly in the radar region.

Shortwave Radiation Sensors

Of the passive sensor systems, camera photography in the visual portion of the spectrum is most familiar to the average person. Black-and-white aerial photographs, taken by cameras from aircraft, have been in wide use by geographers and other environmental scientists since before World War II. Commonly, the field of one photograph overlaps the next along the plane's flight path, so that the photographs can be viewed stereoscopically for three-dimensional effect. Color film can be used to increase the level of information on the aerial photographs. Because of its high resolution (degree of sharpness), aerial photography remains one of the most valuable of the older remote-sensing techniques. Photography has been extended to greater distances through the use of cameras operated by astronauts in orbiting space vehicles. Most persons are familiar with the striking color photographs obtained during the early 1960s by the several Gemini missions.

Photography using reflected electromagnetic radiation also extends into the ultraviolet and infrared wavelengths. Conventional cameras equipped with suitable filters and film can be used in the near-ultraviolet region between 3000 and 4000 angstroms (0.3–0.4 microns); see Figure 4.3.

There is a small region of the infrared spectrum, the **near-infrared region** immediately adjacent to the visible red region (Figure 4.3). In this region reflected rays can be recorded by cameras with suitable film and filter combinations. Direct infrared photographs, taken from high-flying aircraft, are extremely sharp and render a great deal of information on vegetation, soil surface conditions, and land use (Plate A, Figure 5).

Scanning Systems

Besides conventional cameras using films, images in the reflected radiation region can be generated by **scanning systems.** The television camera and receiver illustrate one type of scanning system. Scanning is the process of receiving information instantaneously from only a very small area of the frame under surveillance. A receiving beam runs rapidly across the field of view, shifting position after each traverse so as to obtain information along a set of closely spaced parallel lines. After the entire frame is scanned, the process can be immediately repeated. Reflected light is scanned by means of a rapidly rotating scanning mirror that deflects the incoming rays to form a set of lines across a detecting field. This field may be the face of an electron tube, on which the energy of the beam is transformed into luminescence and thus generates an image of the field being scanned. Then an electron beam within the electron tube scans the face of the tube and converts the luminous image into a succession of signals of varying intensities, and these in turn can be stored on tape. Taped data can be reconverted into **imagery,** a graphic form resembling a conventional photograph. Other forms of detectors besides the electron tube can receive the energy of the scanning beam and convert it into electronic signals.

Multiband Spectral Photography

Photography using films or plates is greatly improved as a remote-sensing tool by use of selected narrow frequency bands. A filter placed over the camera lens allows only the desired wavelengths to pass through to the film. For example, one filter passes rays of the blue-green portion of the visible spectrum; a second, the visible red rays; and a third, the near-infrared rays. Because some surfaces reflect one portion of the spectrum much better than another, a comparison of the photographs under different wavelengths allows the type of surface to be identified. The term **multiband spectral photography** applies to this technique; it requires simultaneous photography by several cameras or by a multiple-lens camera.

One of the most recently developed and sophisticated devices for remote sensing from orbiting satellites in the visible and near-infrared regions is the Multispectral Scanning System (MSS). It simultaneously gathers data from four spectral bands in the range from about 0.4 to 1.1 microns (Figure 4.17). A scanner feeds the radiation to detectors and the information is digitized for continuous transmission to a ground receiving station. As the satellite travels, the MSS continuously scans a belt of the earth's surface 185 km (115 mi) wide. On the ground, the data are transformed into square-framed images 185 km on a side. There is a 10 percent forward lap between frames.

As shown in Figure 4.17, various types of earth surfaces have varying degrees of reflectivity over the spectral range covered by the MSS. Water surfaces have their highest reflectivity in the blue-green region. Green vegetation reflects much more strongly in the red and near-infrared region than in the blue-green region. Consequently, a comparison of data from the four bands allows the nature of the surface to be identified.

As a further refinement of the multiband scanning system, each one of three bands of the MSS imagery is reproduced in a different color and the three colors are superimposed in a single print. Although at first glance this print may seem to be a color photograph of the type we get from an ordinary camera, it is by no means the same thing. The colors derived in the MSS imagery are "false" colors. The colors blue, green, and red are selected for use in reproducing bands 1, 2, and 3 (or bands 1, 3, and 4) respectively, of the MSS imagery. Thus the resulting color

Figure 4.17 **Schematic diagram of sensitivities of four filters in the MSS on Landsat, compared with reflectivity of water and green vegetation surfaces.**

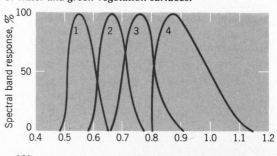

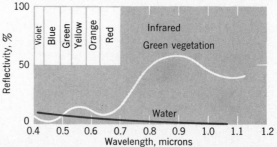

imagery combines the intensities of reflected energy in the three bands at a given point within the frame. Typically, urbanized areas show as blue-gray colors. In contrast, healthy, dense, actively growing vegetation shows as deep red. Various agricultural crops appear as color shades ranging from pink to orange-red to deep red. Mature crops and dried vegetation (dormant grasses) appear yellow or brown. Shallow water areas appear blue; deep water appears dark blue to blue-black. Plate A, Figures 10 and 11 are examples of multiband spectral imagery useful in geographic interpretation and land-use planning.

Thermal Infrared Sensors

We turn next to the longer infrared wavelengths emitted by objects because they possess sensible heat. This radiation, in the range approximately from 1 to 20 microns, is described as **thermal infrared** to distinguish it from reflected radiation in the near-infrared region. Thermal-infrared radiation is sampled by a scanning system, usually of the rotating-mirror type already described. But the detector consists of a surface coated with a material sensitive to infrared wavelengths, rather than to light.

Infrared radiation can be recorded during day or night because it is emitted, rather than reflected, energy. The rays pass readily through haze and smoke — atmospheric contaminants that tend to obscure radiation in the visual range. Because a warmer object emits greater energy than a cooler object, the latter appears lighter in tone on infrared imagery.

Besides absolute temperature, the intensity of infrared emission depends on an intrinsic property of the object, known as **infrared emissivity.** Whereas the black body, mentioned earlier in this chapter, is an ideal perfect radiator of energy, most substances are imperfect radiators, or gray bodies. Infrared emissivity is the ratio of emission of a gray body to that of a black body at the same absolute temperature; it ranges from zero for a body with no emission to unity (100 %) for a black body. For most natural terrestrial surfaces, the infrared emissivity is comparatively high — in the range from 85 to 99 percent. Differences in emissivity can be important in determining the patterns of the infrared image.

Two examples of infrared imagery acquired at night are shown in Figure 4.18. Keep in mind while examining these images that the differences in tone, by means of which objects are delineated, are caused by differences in the level of emission of infrared energy. Differences in temperature are the most important variable in causing the differences in tone, although emissivity plays an important role in the emissions from certain Man-made materials. In Figure 4.18A, an image taken during the early morning hours (about 2:00 to 4:00 A.M.), pavements and water emit stronger infrared radiation and appear light in tone. Trees lining many of the roadways also appear light in tone. Agricultural areas, moist soil surfaces, and buildings are "cooler" and appear darker. The image in Figure 4.18B shows the shore of the Salton Sea in California. Currents in the shallow water containing varying sediment loads appear as swirling bands of lighter and darker tones, allowing the dynamics of water motion to be studied.

Radar Sensing Systems

We turn next to the active mode of remote sensing within the radar portion of the electromagnetic spectrum (0.1 to 100 cm). The active sending system uses pulses of energy emitted by sources mounted on aircraft or space vehicles. A beam of such pulses is directed at the ground, and a portion of the energy is returned as an echo signal. A scanner directs the beam back and forth across the target surface. Radar pulses pass readily through all but the densest of clouds and precipitation.

Radar-sensing systems are used effectively on aircraft. The types most often used to produce imagery are known as **side-looking airborne radar (SLAR)** systems. These systems send their impulses toward either side of the aircraft from a rectangular antenna. A scanning beam is used.

SLAR images show terrain features with remarkable sharpness and contrast (Figure 4.19). Surfaces oriented most nearly at right angles to the radar beam appear lightest in tone; those turned away from the beam are darkest. The effect is to produce an image resembling a relief map using oblique illumination (see Appendix AIII.2). Various types of surfaces, such as forest land, rangeland, and agricultural land, can also be identified by variations in image tone and pattern.

Orbiting Earth Satellites

It is only since the advant of orbiting earth satellites that remote sensing has burgeoned into a major branch of geographic research, going far beyond the limitations of conventional aerial photography.

To cover the entire globe, an earth satellite that orbits over the earth's polar region is needed. An orbit exactly perpendicular to the earth's equator, passing over the poles, would remain fixed in position with respect to the stars. Because remote sensing relies on reflected solar radiation, however, such a truly polar orbit would be undesirable. The sun shifts its position continually with respect to the stars. What is needed is a **sun-synchronous orbit** in which the satellite's plane remains fixed with respect to the sun. Only in this way can the solar illumination over the satellite's earth track be held constant in direction and intensity.

A sun-synchronous orbit is illustrated in Figure 4.20. The plane of this orbit makes an angle of 80° with the plane of the earth's equator, and the satellite's earth track makes a tangent contact with the 80th parallels of latitude north and south. With this orbit, the torque exerted by the earth's equatorial bulge upon the satellite's motion is just sufficient to shift the orbit westward at a rate matching the shift in angle between the sun and stars. In the example illustrated in Figure 4.20, the satellite completes its orbit in a period such that the point of equatorial crossing shifts westward 28½° per orbit. This shift is at the rate of four minutes of time per degree of longitude, or 15° per hour. As explained in Chapter 2, this rate is the same as the rate of earth rotation with respect to the sun. The period of orbit is adjusted so that the satellite can scan the entire globe two or three times daily.

A major step forward in remote sensing of the environment was taken by the National Aeronautics and

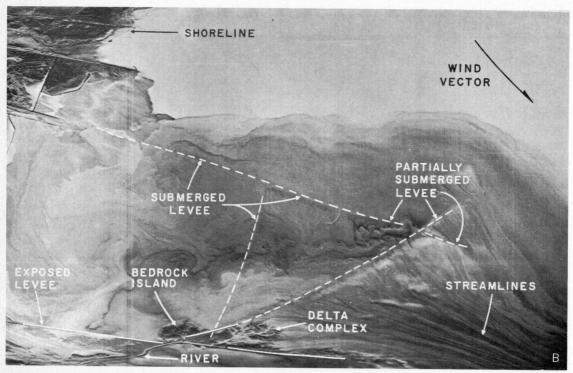

Figure 4.18 **Examples of infrared imagery. *A.* Imagery of Brawley, California, a farm community within the Imperial Valley farming region of southernmost California. *B.* A portion of the shoreline of the Salton Sea, California. Water appears lighter than land, indicating that this is a nighttime image. (Courtesy of Environmental Analysis Department HRB — Singer, Inc.)**

Space Administration (NASA) in July 1972 with the launching of the first Earth Resources Technology Satellite (ERTS-1). The name has since been changed to Landsat. It is the first of two orbiting observatories designed to monitor the natural resources of the planet. Landsat is.a 900-kg (1-ton) satellite orbiting at a height of 918 km (570 mi). It

completes one orbit each 103 minutes, crossing the equator at about 9:30 A.M. local time on its north-south leg. This timing is associated with the best conditions of sunlight and shadow for terrain observation. Each day Landsat completes 14 orbits. It photographs a strip 185 km (115 mi) wide; three such strips cover North America daily. There is an overlap

Figure 4.19 Side-looking radar imagery of a desert landscape in southern Arizona. The flight line parallels the main highway between Phoenix and Tucson. North is toward the lower right. At the left are barren mountain masses surrounded by gentle slopes. Washes (dry stream channels) appear as braided white lines. The "shadows" on northeast slopes show that the radar beam was directed from the southwest. At the right are cultivated lands, laid off into sections and quarter-sections of the Land Office Grid. The larger dark squares are 1.6 km (1 mi) on a side. Built-up areas along the highway are brightly reflective. (Courtesy of Goodyear Aerospace Corporation and Aero Service Corporation.)

of 14 percent in photograph strips near the equator, more overlap near the poles. At any given point on the earth's surface, the satellite passes overhead at the same position once every 18 days, giving an opportunity to match conditions of illumination exactly at 18-day intervals. In this way changes that occur with time can be most readily detected.

In addition to the MSS scanner system described earlier, Landsat carries an imaging system called Return Beam Vidicon (RBV), using three individual high-resolution television cameras. Each camera is equipped with a different spectral filter. One senses the visible blue-green region, another the visible red, the third the near-infrared. As the cameras are simultaneously shuttered, the image is received and stored on the face plates of vidicon tubes. Scanners then read out the information on each face plate by electron beam, transmitting the data to earth by microwave. The cameras are reshuttered every 25 seconds, during which time the satellite has traveled some 165 km (102 mi). Thus sequential images overlap by about 10 percent.

Landsat also carries a data collection system (DCS) that picks up and relays data sent by radio transmission from about 100 ground stations situated in remote locations in North America. These ground stations may monitor such diverse environmental parameters as water quality, rainfall, snow depth, seismic activity, animal migrations, and other earth phenomena.

Imagery obtained by MSS and RBV is transmitted directly to ground stations in the United States and Canada as the satellite moves over North America. Imagery

gathered over other parts of the world is stored in videotape recorders and replayed when the satellite passes over the United States at night. Other ground receiving stations are being constructed in Europe and Australia. Whereas all

Figure 4.20 Earth-track of a sun-synchronous satellite. In the period from March 1 to May 1 the orbit has shifted eastward about 60° with respect to space coordinates. (From A. N. Strahler, 1971, *The Earth Sciences*, 2nd ed., Harper and Row, New York.)

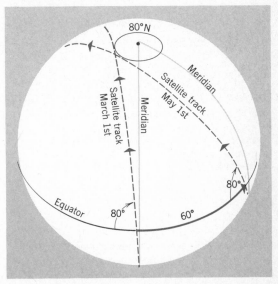

possible scenes are photographed as the satellite passes over North America, photography over other world areas is selective, depending on reports of favorable observing conditions. In the first 10 months of operation, Landsat generated imagery for more than 1½ million photographs and mapped more than 75 percent of the earth's land areas. Some investigators claim that they can identify on the photographs features as small as 90 m (300 ft) in diameter and linear features, such as roads, as narrow as 15 m (50 ft). The use of false-color photographic processing techniques, described earlier, greatly increases the amount of information available from the imagery. A second Landsat was launched in 1976.

Skylab

A new dimension was added to remote sensing by the launching in May 1973 of Skylab, a manned orbiting space laboratory. Skylab traveled in an orbit oblique to the plane of the earth's equator and covered the region between lat. 50°N and 50°S. Traveling at a height of 430 km (270 mi) Skylab orbited the earth once every 93 minutes. It was occupied by three-man crews for work periods of from four to eight weeks.

Observation of earth resources was a major objective of the Skylab missions. Multispectral imagery was obtained by a package of six instruments housed in the Earth Resources Experimental Package (EREP). In addition, cameras allowed photographs of high resolution to be taken at will and the films returned to earth. A variety of terrestrial phenomena covering atmosphere, oceans, and land surfaces was observed visually by Skylab crews and photographed for detailed study. During the second Skylab mission, lasting over 59 days, astronauts brought back 17 miles of magnetic tape bearing stored data of the multispectral EREP cameras, as well as nearly 17,000 frames of film of the earth's surface. Most of the United States was covered by these high-resolution photographs. Subjects included drainage patterns, crustal fracture patterns, and mineral deposits. The third and fourth Skylab missions, carried out in 1973 and 1974, greatly augmented the store of EREP data.

Man's Impact on the Earth's Radiation Balance

Man has profoundly altered the earth's land surfaces by deforestation, crop cultivation, and urbanization. By means of remote sensing imagery, the extent of these changes can be measured and evaluated on a scope never before possible. From our analysis of the earth's radiation balance, it should be obvious that the global radiation balance is a sensitive one, involving as it does a number of variable factors that determine how energy is transmitted and absorbed. We know that agriculture and urbanization have significantly changed the surface albedo and the capacity of the ground to absorb shortwave radiation and to emit longwave radiation. We know that combustion of fuels has already altered the content of the atmosphere through the release of carbon dioxide and dust particles. The nature of these changes and their possible impacts on urban climate and global climate are subjects we will investigate in the next chapter, which deals with heating and cooling of the earth's atmosphere, lands, and oceans.

References for Further Study

Gates, D. M. (1962), *Energy exchange in the biosphere*, Harper and Row, New York.

Sellers, W. D. (1965), *Physical climatology*, University of Chicago Press.

Miller, David H. (1968), *A survey course: the energy and mass budget at the surface of the earth*, Commission on College Geography, Publication No. 7, Association of American Geographers, Washington, D.C.

Halacy, D. S., Jr. (1973), *The coming age of solar energy*, revised edition, Harper and Row, New York.

Estes, J. E., and L. W. Senger, eds. (1974), *Remote sensing techniques for environmental analysis*, Hamilton Publishing Co., Santa Barbara, Calif. (John Wiley and Sons, New York.)

Short, N. M., P. D. Lowman, Jr., S. C. Freden, and W. A. Finch, Jr. (1976), *Mission to earth; Landsat views the world*, NASA SP-360, National Aeronautics and Space Administration, Washington, D. C.

Barrett, E. C., and L. F. Curtis (1977), *Introduction to environmental remote sensing*, Halstead Press, John Wiley and Sons, New York.

Review Questions

1. Explain the concept of the global radiation balance. How is the biochemical energy cycle tied in with the physical energy cycle?
2. Describe electromagnetic radiation. With what speed does it travel through space? In what units are wavelength and frequency stated? Describe the various wavelength bands of the electromagnetic spectrum.
3. What is the solar constant? What unit is used to measure solar radiation? Explain how the frequency and intensity of electromagnetic radiation vary with temperature of the emitting body.
4. How is the sun's radiation spectrum apportioned in terms of frequency and energy? How does longwave radiation escape through the atmosphere? Distinguish between shortwave and longwave radiation.
5. How does the intensity of insolation vary with latitude? What effect does tilt of the earth's axis have on the distribution of insolation over the globe? How does insolation vary with season?
6. Describe the geographic system of latitude zones. Name the zones and give their approximate latitude limits.
7. Describe the various forms of energy losses affecting solar radiation as it penetrates the atmosphere. Explain scattering, diffuse reflection, absorption, and cloud reflection. What is albedo? What is the albedo of the earth as a whole planet?
8. Describe longwave radiation from the earth's surface. Explain why the intensity of longwave radiation from the planet is so much lower than the intensity of incoming shortwave radiation. What is the greenhouse effect and how is it related to counterradiation?

9. Describe the components in the global radiation balance, giving rough percentage figures for the disposal of both incoming shortwave radiation and outgoing longwave radiation. What role do mechanical heat transfer and latent heat transfer play in the global energy balance?

10. Describe the global variation with latitude of albedo, insolation, and longwave radiation. How do these quantities vary in a meridional profile from equator to pole?

11. What is net radiation? In what way does net radiation indicate whether the energy balance shows a surplus or a deficit? How is net radiation distributed over the globe in terms of latitude? Compare the net radiation of the earth's surface with that of the atmosphere. How is the net radiation of the combined surface-atmosphere system distributed with latitude? How is zero global net radiation achieved?

12. What is remote sensing? What class of remote sensors uses radiometers? Describe the microwave and radar bands of the electromagnetic spectrum. Distinguish between passive and active sensing systems.

13. How is direct photography in the near-infrared region used to advantage in remote sensing? What are scanning systems? How do they operate? Explain multiband spectral photography and its uses.

14. How are thermal infrared sensors used to yield information on physical characteristics of the surface? Describe side-looking airborne radar systems. What type of imagery do they produce?

15. Describe the role of orbiting satellites in remote sensing. Why is a sun-synchronous orbit necessary? Describe the programs carried out by Landsat and Skylab.

Heat and Cold at the Earth's Surface

5

Temperature of the air and the soil layer is of major interest to the physical geographer. Temperature is a measure of heat energy available in the air and the soil. Organisms respond directly to heating and cooling of the environmental substance that surrounds them. We can refer to this influence as the **thermal environment.**

Our study of the radiation balance has shown that temperature changes result from the gain or loss of energy by the absorption or emission of radiant energy. When a substance absorbs radiant energy, the surface temperature of that substance is raised. This process represents a transformation of radiant energy into the energy of **sensible heat,** which is the physical property measured by the thermometer. Heat can also enter or leave a substance by conduction or can be lost as latent heat during evaporation.

Many of the biochemical processes taking place within organisms as well as many common inorganic chemical reactions are intensified by an increase in temperature of the solutions in which these reactions are occurring. Severe cold, which is simply the lack of heat energy within matter, may greatly reduce or even completely stop biochemical and inorganic reactions. This is why the vital environmental ingredient of heat — heat in the air, water, and soil — needs to be thoroughly understood. The thermal environment is surely a dominant part of Man's total physical environment.

We are all familiar with natural cycles of temperature change. There is a daily rhythm of rise and fall of air temperature as well as a seasonal rhythm. There are also systematic changes in average air temperatures from equatorial to polar latitudes and from oceanic to continental surfaces. These temperature changes require that the lower atmosphere and the surfaces of the lands and oceans receive and give up heat in daily and seasonal cycles. There must also be great differences in the quantities of heat received and given up annually in low latitudes as compared to high latitudes. Temperature cycles — daily and seasonal — and the influence of latitude on temperature will be dominant themes of this chapter.

Measurement of Air Temperatures

Air temperature is one of the most familiar bits of weather information we hear and read daily through the news media. In the United States, this information comes from National Weather Service observing stations and is taken following a carefully standardized procedure. Thermometers are mounted in an instrument shelter, shown in Figure 5.1. The shelter shades the instruments from sunlight, but louvers allow air to circulate freely past the thermometers. The instruments are mounted from 1.2 to 1.8 m (4 to 6 ft) above ground level, at a height easy to read.

At most stations, only the highest and lowest temperatures of the day are recorded. To save the observer's time, the **maximum-minimum thermometer** is used. This instrument uses two thermometers; one to show the

Figure 5.1 **A standard thermometer shelter. Rain gauge at left. (National Weather Service.)**

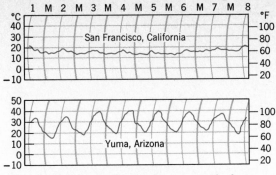

Figure 5.2 **The recording thermometer made these continuous records of the rise and fall of air temperature over a period of one week.**

highest temperature since it was last reset, the other to show the lowest. Another labor-saving device is the recording thermometer (thermograph), which draws a continuous record (a thermogram) on graph paper attached to a slowly moving drum. Figure 5.2 shows typical traces for several days of record.

When the maximum and minimum temperatures of a given day are added together and divided by two, we obtain the **mean daily temperature.** The mean daily temperatures of an entire month can be averaged to give the **mean monthly temperature.** Averaging the daily means (or monthly means) for a whole year gives the **mean annual temperature.** Usually such averages are compiled for records of many years' duration at a given observing station. These averages are used in describing the climate of the station and its surrounding area.

The **Celsius scale (C)** of temperature, in which 0°C is the freezing point and 100°C is the boiling point, is now used officially in all countries of the world except the United States; it is the scale used in physics, chemistry, and meteorology. The **Fahrenheit scale (F)** of temperature is the scale used by the general public in the United States. The freezing temperature is 32°F; the boiling point is 212°F on this scale (Figure 5.3). We give Fahrenheit degrees in parentheses so you can relate temperatures to sensations of bodily comfort or discomfort.

Daily Cycles of Insolation, Net Radiation, and Air Temperature

Because the earth turns on its axis, insolation and net radiation undergo daily cycles of change. These cycles, in turn, produce the daily (diurnal) cycle of rising and falling

air temperature with which we are all familiar. We now examine each of these three daily cycles to determine how they are linked together. The three graphs in Figure 5.4 show average curves of daily insolation, net radiation, and air temperature, greatly generalized for a typical observing station at lat. 40 to 45°N in the interior United States.

Graph A shows insolation; units are langleys per minute (ly/min). At equinox, insolation begins about sunrise (6 A.M. by local time), rises to a peak value at noon, and declines to zero at sunset (6 P.M.). At June solstice, insolation begins about 2 hours earlier (4 A.M.) and ends 2 hours later (8 P.M.). The June peak is much greater than at equinox, and the total insolation for the day is also much greater. At December solstice, insolation begins about 2 hours later (8 A.M.) and ends 2 hours earlier (4 P.M.). Both the peak intensity and daily total insolation are greatly reduced in the winter solstice season.

Insolation, including both the direct solar beam and indirect sky shortwave radiation, can be measured continuously by an instrument known as a **pyranometer** (Figure 5.5). A sensing cell enclosed in a glass bulb receives radiation from the entire hemisphere of the sky. The pyranometer is the standard instrument used at radiation observing stations.

Graph B of Figure 5.4 shows net radiation in the same units as insolation. Net radiation shows a positive value — a surplus — shortly after sunrise and rises sharply to a peak at noon. The afternoon decline reaches zero shortly before sunset and becomes a negative quantity, or deficit. The deficit continues throughout the hours of darkness with a more-or-less constant value. At June solstice, a radiation surplus begins earlier and ceases later than at equinox, generating a much larger daily surplus. At December solstice, the surplus period is greatly shortened and the surplus is small. Now, because the nocturnal deficit period is long, the total deficit exceeds the surplus, so the net daily total is a small negative quantity. (We shall investigate the annual cycle of net radiation in later paragraphs.)

Graph C of Figure 5.4 shows the typical, or average, daily temperature cycle. The minimum daily temperature usually occurs just after sunrise, corresponding with the onset of a radiation surplus. Heat now begins to flow upward from the ground surface to warm the lower air layer. Temperature rises sharply in the morning hours and continues to rise long after the noon peak of net radiation. We should expect the air temperature to rise as long as a radiation surplus is in effect and, in theory, this should produce a temperature maximum just before sunset. Another atmospheric process, however, begins to take

Figure 5.3 **A comparison of the Celsius and Fahrenheit temperature scales.**

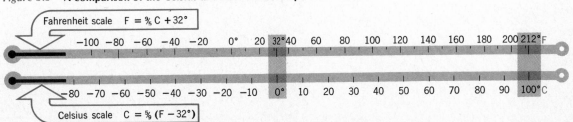

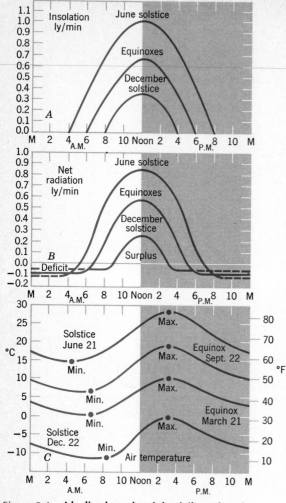

Figure 5.5 **Pyranometer for measurement of the direct solar beam and sky radiation. (Courtesy of WeatherMeasure Corporation, Sacramento, Calif.)**

Figure 5.4 **Idealized graphs of the daily cycles of insolation, net radiation, and air temperature for a midlatitude station in the interior United States.**

effect in the early afternoon. Mixing of the lower air by eddies distributes sensible heat upward, offsetting the temperature rise and setting back the temperature peak to about 3 P.M. (The time of daily maximum varies usually between 2 and 4 P.M., according to local climatic conditions.) By sunset, air temperature is falling rapidly and continues to fall, but at decreasing rate, throughout the entire night.

In midlatitudes and high latitudes, there is a strong seasonal change in the air temperature curve. Upward displacement of the curve to warmer temperatures in summer and downward displacement to colder temperatures in winter constitutes the annual temperature cycle, which we will examine later. At this point, you should note that at summer solstice, the time of minimum daily temperature occurs about 2 hours earlier than at equinox; at winter solstice, the minimum occurs about 2 hours later than at equinox. The time of daily maximum temperature is not greatly affected by the seasonal cycle. For simplicity, in Graph C of Figure 5.4 the maximum is shown as 3 P.M. throughout the year.

Daily Cycle of Soil Temperature

The daily cycle of air temperature can be strongly intensified close to the ground surface, as Figure 5.6 shows. Under the direct sun's rays, a bare soil surface or pavement heats to much higher temperatures than one reads in the thermometer shelter. At night, this surface layer cools to temperatures much lower than in the shelter. These effects are quite weak under a forest floor cover where the ground is shaded and moist. On the dry desert floor and on the pavements of city streets and parking lots, the surface temperature extremes are large.

Soil temperatures play a major role in the seasonal rhythms of plant physiology and in biological activity generally within the soil. Soil scientists regard soil temperature as a major factor in determining many soil properties. As Figure 5.6 suggests, the daily cycle of soil temperature will show its greatest range at the surface, while the daily cycle gradually dies out with increasing depth.

Soil temperatures are studied by placing a series of thermometers in the soil, spacing them at a series of depths, each about twice as deep as the one above it. Figure 5.7 shows temperature data for Pavlovsk, USSR near Leningrad, at lat. 59½°N. The observation depths run in a progression thus: 1, 2, 5, 10, 20, 40, 80, and 160 cm. Curves of temperature are averages for one month, those months being May and January. Looking first at the May graph, we notice immediately that the daily range of temperatures is greatest — about 18C° (32F°)—at the shallowest depth and that this range falls off rapidly with increasing depth.

Notice, too, that the time of maximum temperature of the day is advanced with increasing depth. We can imagine a wave of heating advancing into the soil each day. As this wave passes down by conduction, heat is lost by absorption and the wave crest diminishes in height. There is also a time lag, so the wave crest does not arrive at the 40-cm (16-in.) depth until after midnight. At a

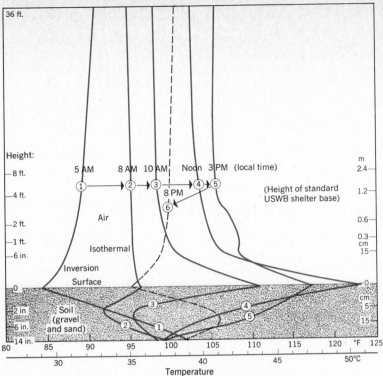

Figure 5.6 **Air and soil temperatures at various levels above and below the ground surface throughout the day and night. Data are averages for July and August during a single summer. Height follows a square-root scale. (Data from Quartermaster Research and Development Branch, U.S. Army.)**

depth of 80 cm (31.5 in.) the wave has practically disappeared, so no daily cyclic changes are perceptible below this depth.

Looking at the curve for January (Graph *B* in Figure 5.7), we find a very different situation. Close to the surface, there is only a slight daily warming in the midday period. At this latitude—about 60°N—the sun is very low in the sky in the winter and daylight hours are few. The weak wave of warming does not penetrate very far and is imperceptible at a depth of 40 cm (16 in.).

Temperature Inversion and Frost

During the night, when the sky is clear and the air is calm, the ground surface rapidly radiates longwave energy into the atmosphere above it. As we have just explained, the soil surface temperature drops rapidly and the overlying air layer becomes colder. When temperature is plotted against altitude, as in Figure 5.8, the straight, slanting line of the normal environmental lapse rate becomes bent to the left in a "J" hook. In the case shown, the air tempera-

Figure 5.7 **Daily cycle of soil temperature in May and January at Pavlovsk, USSR. (Used by permission of the publishers from Rudolph Geiger,** *The Climate Near the Ground,* **rev. ed., Harvard Univ. Press, Cambridge, Mass. Copyright 1950, 1957, 1965 by the President and Fellows of Harvard College.)**

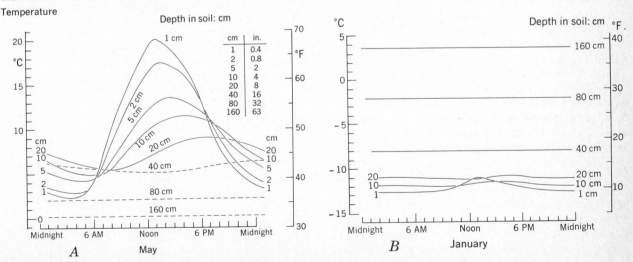

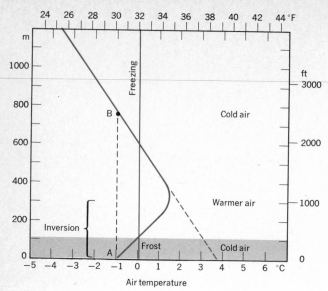

Figure 5.8 **A low-level temperature inversion.**

ture at the surface, point A, has dropped to −1°C (30°F). This value is the same as at point B, some 750 m (2500 ft) aloft. As we move up from ground level, temperatures become warmer up to about 300 m (1000 ft). Here the curve reverses itself and the normal environmental lapse rate takes over.

The lower, reversed portion of the lapse rate curve is called a **low-level temperature inversion.** In the case shown, temperature of the lowermost air has fallen below the freezing point, 0°C (32°F). For sensitive plants, this condition is a **killing frost** when it occurs during the growing season. Killing frost can be prevented in citrus groves by setting up an air circulation that mixes the cold basal air with warmer air above. One method is to use oil-burning heaters; another is to operate powerful motor-driven propellors to circulate the air.

Low-level temperature inversion is also prevalent over snow-covered surfaces in winter. Inversions of this type are intense and often extend a few thousand meters into the air.

Urbanization and the Heat Island

Applying principles of temperatures of the soil and lower air layer, we can anticipate the impact of Man as cities spread, replacing a richly vegetated countryside with blacktop and concrete. In the urban environment the absorption of solar radiation causes higher ground temperatures for two reasons. First, foliage of plants is absent, so the full quantity of solar energy falls on the bare ground. Absence of foliage also means absence of transpiration (evaporation from leaves), which elsewhere removes heat from the lower air layer. Second, roofs and pavements of concrete and asphalt hold no moisture, so evaporative cooling cannot occur as it would from a moist soil.

The thermal effect is that of converting the city into a hot desert. The summer temperature cycle close to the pavement of a city may be almost as extreme as that of the

desert floor. This surface heat is conducted into the ground and stored there. The thermal effects within a city are actually more intense than on a sandy desert floor because the capacity of solid concrete, stone, or asphalt to conduct and hold heat is greater than that of loose, sandy soil. An additional thermal factor is that vertical masonry surfaces absorb insolation or reflect it to the ground and to other vertical surfaces. The absorbed heat is then radiated back into the air between buildings.

As a result of these changes in the energy balance, the central region of a city typically shows summer air temperatures several degrees higher than for the surrounding suburbs and countryside. Figure 5.9 is a map of the Washington, D.C., area, showing air temperatures for a typical August afternoon. The lines of equal air temperature delineate a **heat island.** The heat island persists throughout the night because of the availability of a large quantity of heat stored in the ground during the daytime hours. In winter additional heat is radiated by walls and roofs of buildings, which conduct heat from the inside. Even in summer, Man adds to the city heat output through the use of air conditioners, which expend enormous amounts of energy at a time when the outside air is at its warmest.

In Chapter 7 we shall make further observations about the impact of urbanization on the climate of cities. Other weather phenomena, such as wind speeds, fog, clouds, and precipitation, show measurable changes as a result of heat output and the injection of pollutants into the air.

Annual Cycle of Air Temperature

As the earth revolves about the sun, the tilt of the earth's axis causes an annual cycle in the intensity of insolation, described in Chapter 4. This cycle is also felt in an annual cycle of net radiation that, in turn, generates an annual

Figure 5.9 **A heat island over Washington, D.C. Air temperatures were taken at 10:00 P.M. on a day in early August. (Data from H. E. Landsberg.)**

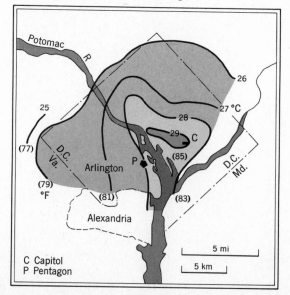

cycle in the mean daily and monthly air temperatures. In this way the climatic seasons are produced.

Figure 5.10 shows the yearly cycle of net radiation for four stations, ranging in latitude from the equator almost to the arctic circle. Figure 5.11 shows mean monthly air temperatures for these same stations. Starting with Manaus, a city on the Amazon River in Brazil, let us compare the net radiation graph with the air temperature graph. At Manaus, almost on the equator, the net radiation shows a large surplus in every month. The average surplus is about 200 ly/day, but there are two minor maxima, approximately coinciding with the equinox, when the sun is nearly straight overhead. A look at the temperature graph of Manaus shows monotonously uniform air temperatures, averaging about 27°C (81°F) for the year. The **annual temperature range,** or difference between the highest mean monthly temperature and the lowest mean monthly temperature, is only 1.7C° (3F°). In other words, near the equator one month is much like the next, thermally. There are no temperature seasons.

We go next to Aswan, Arab Republic of Egypt, on the Nile River at lat. 24°N. Here we are in a very dry desert. The net radiation curve has a strong annual cycle, and the surplus is large in every month. The surplus rises to more than 250 ly/day in June and July, but falls to less than 100 ly/day in December and January. The temperature graph shows a corresponding annual cycle, with an annual range of about 17C° (30F°). June, July, and August are extremely hot, averaging over 32°C (90°F).

Moving farther north, we come to Hamburg, Germany, lat. 54°N. The net radiation cycle is strongly developed. The surplus lasts for nine months, and there is a deficit for three winter months. The temperature cycle reflects the reduced total insolation at this latitude. Summer months reach a maximum of just over 16°C (60°F); winter months reach a minimum of just about freezing, 0°C (32°F). The annual range is 17C° (30F°).

Finally, we travel to Yakutsk, Siberia, lat. 62°N. During the long, dark winters, there is an energy deficit; it lasts about six months. During this time air temperatures drop to extremely low levels. For three of the winter months, monthly mean temperatures are between −35 and −45°C (−30 and −50°F). Yakutsk is one of the coldest places on earth. In summer, when daylight lasts most of the 24 hours, the energy surplus rises to a strong peak, reaching 300 ly/day—a value higher than for any of the other three stations. As a result, air temperatures show a phenomenal spring rise to summer-month values of over 13°C (55°F). In July the temperature is about the same as for Hamburg. The annual range at Yakutsk is enormous—over 61C° (110F°). No other region on earth, even the south pole, has so great an annual range.

Annual Cycle of Soil Temperature

Annual temperature range and mean annual temperature of the soil are important factors in the formation of various soil types. The annual cycle of heating and cooling of the soil is illustrated by data for an observing station on Long Island, New York, lat. 41°N, Figure 5.12. The soil here is sandy and porous. At the shallowest depth, 0.75 m (2.5 ft)

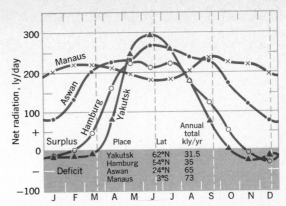

Figure 5.10 **Net radiation for four stations ranging from the equatorial zone to the arctic zone. (Data courtesy of David H. Miller.)**

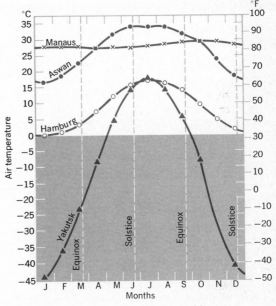

Figure 5.11 **The annual cycles of monthly mean temperatures for the same stations shown in Figure 5.10.**

the annual cycle is strong, with a range of about 25C° (45F°). The annual wave of warming decreases in amplitude with depth, while the peak of temperature comes progressively later with depth. At the 6-m (20-ft) depth, the warmest time is in November. The many minor irregularities in temperature at the shallowest depth reflect the effects of rainstorms, which cool the soil, and the short periods of warmer or cooler air temperature that are normal at all seasons for this climate.

You might anticipate that for a given climate, there is a depth below which the soil or rock temperature is unchanging the year around, and that this constant temperature will be close to the average air temperature near the ground surface. The facts bear out this supposition. For example, the air temperature in Endless Caverns, Virginia, which is constant at 13°C (56°F) throughout the year, is only a little higher than the annual average air

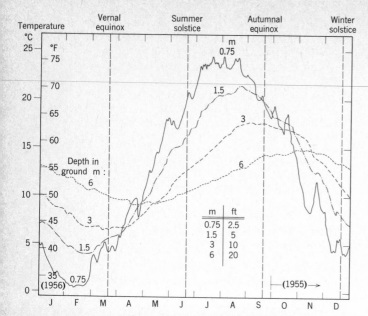

Figure 5.12 Soil temperatures recorded during one year at North Station, Brookhaven, Long Island, New York. (Data from I. A. Singer and R. M. Brown, 1956.)

temperature of 11°C (52°F). The exceptional constancy of subterranean temperatures makes an unusual thermal environment for animals that live in the total darkness of limestone caverns.

Land and Water Temperature Contrasts

The odd distribution of continents and ocean basins gives our planet its great variety of climates. Look back at the two global hemispheres — northern and southern — outlined in Figure 3.12. The northern hemisphere displays a polar sea surrounded by massive continents; the southern hemisphere shows the opposite — a pole-centered continent surrounded by a vast ocean. The Americas form a north-south barrier between two oceans — Atlantic and Pacific. The continents of Eurasia and Africa together form another great north-south barrier. Oceans and continents have quite different properties when it comes to absorbing and radiating energy.

Land surfaces behave quite differently from water surfaces. The important principle is this: The surface of any extensive deep body of water heats more slowly and cools more slowly than the surface of a large body of land when both are subject to the same intensity of insolation.

The slower rise of water-surface temperature can be attributed to four causes (Figure 5.13): (1) Solar radiation penetrates water, distributing the absorbed heat throughout a substantial water layer. (2) The specific heat of water is large (a gram of water heats up much more slowly than a gram of rock). (3) Water is mixed through eddy motions, which carry the heat to lower depths. (4) Evaporation cools the water surface.

In contrast, the more rapid rise of land surface temperature can be attributed to these causes: (1) Soil or rock is opaque, concentrating the heat in a shallow layer; there is little transmission of heat downward. (2) Specific heat of mineral matter is much lower than water. (3) Soil, if it is dry, is a poor conductor of heat. (4) No mixing occurs in soil and rock.

The effect of land and water contrasts is seen in two sets of daily air temperature curves, shown in Figure 5.14. El Paso, Texas, exemplifies the temperature environment of an interior desert in midlatitudes. Soil moisture content is low, vegetation sparse, and cloud cover generally light. Responding to intense heating and cooling of the ground surface, air temperatures show an average daily range of 11 to 14C° (20 to 25F°). North Head, Washington, is a coastal station strongly influenced by air brought from the adjacent Pacific Ocean by prevailing westerly winds. Consequently, North Head exemplifies a maritime temperature environment. The average daily range at North Head is a mere 3C° (5F°) or less. Persistent fogs and cloud cover also contribute to the small daily range. Refer also to Figure 5.2, which shows the same environmental contrast when the record of Yuma is compared with that of San Francisco.

The principle of contrasts in the heating and cooling of water and land surfaces also explains the differences in the annual temperature cycles of coastal and interior stations. Figure 5.15 shows the annual cycle of mean daily temperatures for two places at approximately the same latitude, where insolation is about the same for both. Winnipeg, Manitoba, Canada, is located in the heart of the North American continent. The Scilly Islands, England, are exposed to the Atlantic Ocean.

Although insolation reaches the maximum at summer solstice, the hottest part of the year for inland regions is about a month later because a net radiation surplus persists and heat energy continues to flow into the ground well into August. The air temperature maximum, closely corresponding with maximum ground output of longwave radiation, is correspondingly delayed. (Bear in mind that this cycle applies to middle and high latitudes, but not to the region between the tropics of cancer and capricorn.) Similarly, the coldest time of year for large land areas is January, about a month after winter solstice, because a net radiation deficit persists and the ground continues to lose heat even after insolation begins to rise.

Over the oceans there are two differences: (1) Maximum and minimum temperatures are reached about a

Figure 5.13 Contrasts in heating of land and water.

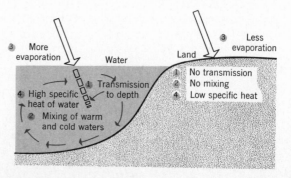

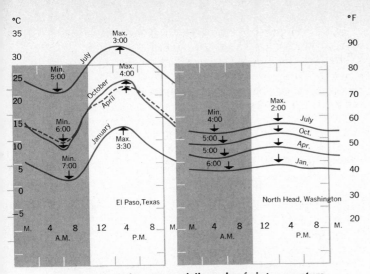

Figure 5.14 **The average daily cycle of air temperature for four different months. Daily ranges and seasonal changes are great at El Paso, a station in the continental interior, but only weakly developed at North Head, close to the Pacific Ocean.**

month later than on land — in August and February, respectively — because water bodies heat or cool much more slowly than land areas; and (2) the yearly range is less than over land, following the law of temperature differences between land and water surfaces. Coastal regions are usually influenced by the oceans to the extent that maximum and minimum temperatures occur later than in the interior. This principle shows nicely for monthly temperatures of the Scilly Islands. February is slightly colder than January (Figure 5.15).

Figure 5.14 reinforces the evidence of Figure 5.15. The annual temperature range at El Paso is about 20C° (35F°), while that at North Head is only about 8C° (15F°).

Air Temperature Maps

The distribution of air temperatures over large areas can best be shown by a map composed of **isotherms,** lines drawn to connect all points having the same temperature. Figure 5.16 is a map on which the observed air temperatures have been recorded in the correct places. These may represent single readings taken at the same time everywhere, or they may represent the averages of many years of records for a particular day or month of a year, depending on the purposes of the map.

Usually, isotherms representing 5 or 10° differences are chosen, but isotherms can be drawn for any selected temperatures. The value of isothermal maps is that they make clearly visible the important features of the prevailing temperatures. Centers of high or low temperature are clearly outlined.

World Patterns of Air Temperature

World maps of air temperature (Plate B.1, B.2) allow us to compare thermal conditions in the two months of greatest extremes over large land areas — January and July. Patterns taken by the isotherms are largely explained by

three controls: (1) latitude, (2) continent-ocean contrasts, and (3) altitude. (Ocean currents play an important, but secondary, role.) Important features to identify when interpreting these maps are as follows:

1. The trend of isotherms is east-west, with temperatures decreasing from equatorial zone to polar zones. This feature shows best in the southern hemisphere because the great Southern Ocean girdles the globe as a uniform expanse of water, while the continent of Antarctica is squarely centered on the south pole. The parallel pattern of isotherms is explained, of course, by the general decrease in intensities of insolation and net radiation from equator to poles.
2. Large landmasses located in the subarctic and arctic zones develop centers of extremely low temperatures in winter. The two large landmasses we have in mind are North America and Eurasia. These centers are conspicuous on the January map.
3. Isotherms change very little in position from January to July over the equatorial zone, particularly over the oceans. This feature reflects the uniformity of insolation throughout the year near the equator.
4. Isotherms make a large north-south shift from January to July over continents in the midlatitude and subarctic zones. Figure 5.17 shows this principle. North America shows this effect well. In January the 15°C isotherm lies over central Florida; in July this same isotherm has moved far north, cutting the southern shore of the Hudson Bay and then looping far up into northwestern Canada. The 15°C isotherm on the Eurasian continent also shows the effect. We can explain the large north-south shift in position over these continents by the law of land and water contrasts.
5. Highlands are always colder than surrounding lowlands. An example is the Andes range, running along the western side of South America. The isotherms loop equatorward in long fingers over this lofty mountain chain.
6. Areas of perpetual ice and snow are always intensely cold. Greenland and Antarctic are the two great ice sheets. Notice how they stand out as cold centers in

Figure 5.15 **Annual cycles of monthly mean air temperature for two stations at lat. 50°N: Winnipeg, Manitoba, Canada, and Scilly Islands, England.**

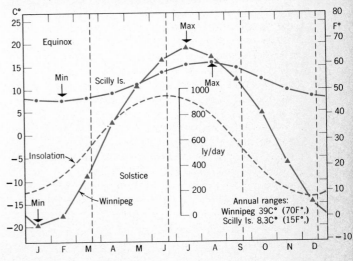

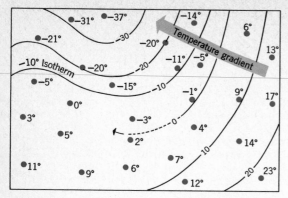

Figure 5.16 **Isotherms are used to make temperature maps. Each line connects those points having the same temperature.**

both January and July. Not only do these ice sheets have high surfaces, rising to over 3000 m (10,000 ft) in their centers, but the white snow surfaces have a high albedo and reflect away much of the insolation. The Arctic Ocean, bearing a cover of floating ice, also maintains its cold throughout the year; but the cold is much less intense in July than over the Greenland Ice Sheet.

The Annual Range of Air Temperatures

Figure 5.18 is a world map showing the annual range of air temperatures. The lines resembling isotherms are *corange lines;* they tell the difference between the January and July monthly means. This map serves as a summary of concepts we have previously emphasized. Note these features:

1. The annual range is extremely great in the subarctic and arctic zones of both Asia and North America. This effect is shown well in the annual temperature graph for Yakutsk (Figure 5.11).
2. The annual range is moderately large on land areas in the tropical zone, near the tropics of cancer and capricorn. North Africa, southern Africa, and Australia are examples. Here the annual ranges are substantially greater than over adjacent oceans.
3. The annual range is very small over oceans in the equatorial zone. At all latitudes, the annual range over all oceans is generally less than over land areas.

Figure 5.17 **Seasonal shift of an isotherm.**

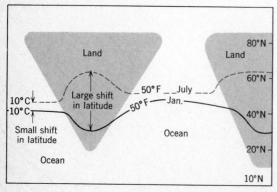

This review of air temperatures over the globe will form the groundwork for an understanding of climates, developed in later chapters.

Solar Radiation and Air Temperatures at High Altitudes

The rapid decrease in density of the atmosphere with increasing altitude brings with it major changes in the environments of radiation and heat at the ground surface. Recall from Chapter 3 and Figure 3.6 that barometric pressure decreases by about one-thirtieth of itself for every 275 m (900 ft) of altitude increase. Thus, at 4600 m (15,000 ft), an altitude repesentative of mountain summits in the higher ranges of the western United States, pressure is only about 570 mb, or a little over half the sea-level value. Air density is correspondingly reduced, with the result that the overlying atmosphere reflects and absorbs a much smaller portion of the incoming solar radiation than at sea level.

Measurements of solar radiation at altitude 3600 to 4300 m (12,000 to 14,000 ft), taken near the summit of a mountain range in the desert region along the border between California and Nevada, showed noon peak values approaching 2.0 ly/min under clear-sky conditions. Compare this value with a June maximum of about 0.7 ly/min, typical for a sea-level station in midlatitudes in a humid climate. At the mountain location, the peak of net all-wave radiation reached almost 1.6 ly/min, a value about triple that at a sea-level station. Of course, the sea-level data include days with cloud cover as well as clear days, whereas the mountain measurements are for clear skies only. Nevertheless, the great daytime intensity of incoming and outgoing radiation at the high-altitude position is truly remarkable.

Increasing intensity of incoming solar radiation with higher altitude has a profound influence on air and soil temperatures. Surfaces exposed to sunlight heat rapidly and intensely; shaded surfaces are quickly and severely cooled. This results in rapid air heating during the day and rapid cooling at night at high-mountain locations. This altitude effect is shown by the increasing spread between high and low daily air temperature readings from left to right in Figure 5.19.

The contrast between exposed and shaded surfaces is particularly noteworthy at high altitudes. It has been found that temperatures of objects in the sun and in the shade differ by as much as 22 to 28C° (40 to 50F°).

Figure 5.19 also shows how monthly mean air temperatures decrease with increasing altitude. The same 15-day period in July is used for all stations. For the total ascent, from sea level to an altitude of 4380 m (14,360 ft), the mean air temperature decreased by nearly 17C° (30F°).

Carbon Dioxide, Dust, and Global Climate Change

Atmospheric changes induced by Man fall into four categories with respect to basic causes: (1) changes in concentrations of the natural component gases of the

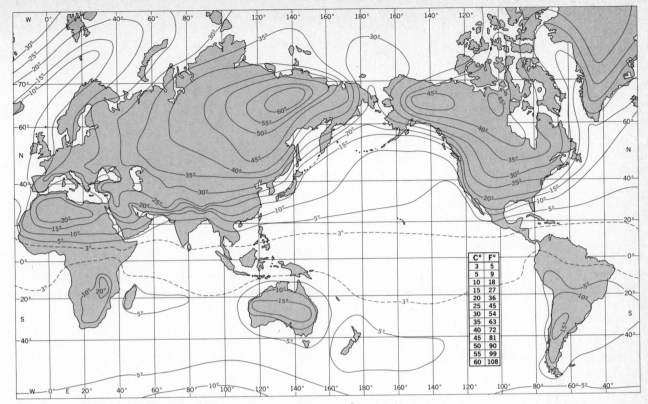

Figure 5.18 **Annual range of air temperatures. Data show differences between January and July means. (Same data sources as Plate B.)**

C°	F°
3	5
5	9
10	18
15	27
20	36
25	45
30	54
35	63
40	72
45	81
50	90
55	99
60	108

lower atmosphere; (2) changes in the water-vapor content of the troposphere and stratosphere; (3) alteration of surface characteristics of the lands and oceans in such a way as to change the interaction between the atmosphere and those surfaces; and (4) introduction of finely divided solid substances into the lower atmosphere, along with gases not normally found in substantial amounts in the unpolluted atmosphere.

Under preindustrial conditions, the atmospheric content of carbon dioxide (CO_2) was fixed at a level of roughly 0.029 percent by volume (290 parts per million). The problem is, of course, that Man has recently

Figure 5.19 **Daily maximum and minimum temperatures for mountain stations in Peru, lat. 15°S. The data cover the same 15-day observation period in July. (Data from Mark Jefferson.)**

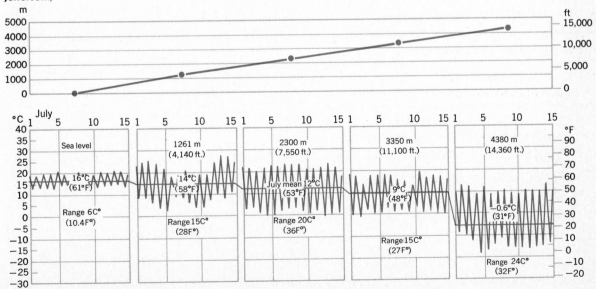

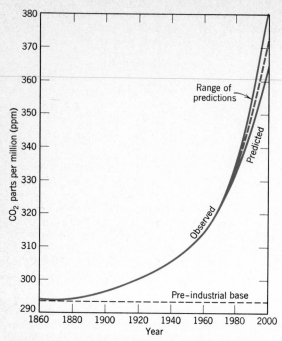

Figure 5.20 **Increase in atmospheric carbon dioxide, observed to 1975 and predicted to the year 2000.**

begun to extract and burn hydrocarbon fuels at a rapid rate, releasing into the atmosphere the combustion products (principally water and CO_2) and also a great deal of heat. How does this activity affect our environment?

During the past 110 years, atmospheric CO_2 has increased about 10 percent. As Figure 5.20 shows, the rate of increase in this period, while slow at first, has become much more rapid toward the end of the period. Projection of the present curve of increase into the future shows a CO_2 value of about 370 parts per million in the year 2000. At that time, if the predictions are correct, the content of CO_2 will have increased by about 25 percent over the 1860 value.

Consider now the environmental effects to be anticipated from an increase of atmospheric CO_2. Because CO_2 is an absorber and emitter of longwave radiation, its presence in larger proportions would raise the level of absorption of both incoming and outgoing radiation, changing the energy balance. The result would be to raise the average level of sensible heat in the atmosphere. In other words, a general air temperature rise is the anticipated result. We shall look next at available temperature data to see if such an increase has occurred.

Figure 5.21 shows the northern hemisphere change in mean air temperature for approximately the past century. From 1880 to 1940, when fuel consumption was rising rapidly, the average temperature increased by about 0.4C° (0.6F°). This relationship follows the predicted pattern. Since 1940, however, the graph shows a drop in temperature, despite the rising rate of fuel combustion. Assuming that atmospheric warming because of increased CO_2 is a valid effect in principle, some other factor, working in the opposite direction, has entered the picture, and its cooling effect has outweighed that of warming by CO_2.

Since the data in Figure 5.21 were made public, atmospheric scientists have shown that average air temperatures in the southern hemisphere have changed in an opposite manner in recent decades. Since about 1960, southern hemisphere temperatures have risen on the order of 0.6C° (1F°) in the latitude range from 60°S to the south pole. This new information suggests that the cause of cooling documented in northern latitudes is confined to that hemisphere and is not global in scope. Thus we must turn to consider a cause of temperature change that can be largely confined to one hemisphere or the other.

Downturn in average northern-hemisphere air temperature since 1940 may be the result of a greatly increased input of dust into the upper atmosphere by a number of major volcanic eruptions, the first of which was the 1947 eruption of the Icelandic volcano Hekla. An increase in volcanic dust at high altitudes increases the losses of insolation to space through diffuse reflection; the result is that less energy arrives at lower atmosphere levels. A reduction in level of sensible heat of the lower atmosphere follows.

It is also suspected by some research workers that Man-made dusts from industrial sources have increased sufficiently to cause, or at least contribute to, the temperature downturn. Increased fallout of dust from the lower troposphere is well documented and is related to industrial activities. At this time, it is not possible to decide the extent of Man's activity in increasing atmospheric dust content at high levels, but the climatic impact may prove to be important eventually.

The Thermal Environment in Review

In this chapter we have covered a vital environmental factor in the life layer. Heat, as recorded by the thermometer, is an essential ingredient of climate near the ground. All organisms respond to changes in temperatures of the medium that surrounds them, whether air, soil, or water. Geographers study temperature records carefully, and they are interested in the mean values based on records of long periods of observation.

Cycles of air-temperature change are particularly important in the thermal environment. Both daily and seasonal changes in air temperature can be explained by daily and seasonal cycles of insolation and net radiation.

Figure 5.21 **Changes in the mean annual air temperature in the zone between the equator and lat. 80°N. (Generalized from data of J. M. Mitchell, Jr., and NOAA.)**

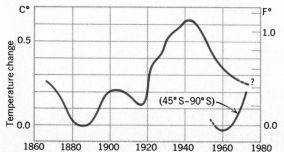

Superimposed on these astronomical rhythms is the powerful effect of latitude. A zoning of thermal environments from the equatorial zone to the polar zones is one of the most striking features of the global climate. But equally important is the way large land areas, especially the continents of North America and Eurasia, subvert the simple latitude zones to cause great seasonal extremes of temperature. In contrast, the southern hemisphere is strongly dominated by the simple effects of latitude.

Besides examining thermal environments on a global scale, we have investigated the small-scale temperature variations close to the ground. Wide swings of temperature near and within the soil surface have profound effects on organisms. Important to city dwellers is the way Man has altered the energy balance and brought about problems of excessive warming.

From the long-range standpoint, Man's combustion of fuels has some serious implications, and these must be studied with care. Sustained research combined with intensified monitoring of the environment deserves high priority.

References for Further Study

Sellers, W. D. (1965), *Physical climatology,* University of Chicago Press.
Chang, J.-H. (1968), *Climate and agriculture,* Aldine Publishing Co., Chicago. See Chapters 8–10.
Strahler, A. N. (1971), *The earth sciences,* second edition, Harper and Row, New York. See Chapter 14.
Frisken, W. R. (1973), *The atmospheric environment,* The Johns Hopkins University Press, Baltimore, Md.
U.S. National Committee for the Global Atmospheric Research Program (1975), *Understanding climatic change,* National Academy of Sciences, Washington, D.C.

Review Questions

1. In what ways is the thermal environment important to physical processes and organisms?
2. Describe the standard procedures and instruments used in measuring air temperature. What temperature statistics are derived from these measurements? Compare the Celsius and Fahrenheit temperature scales.
3. Describe and explain the daily cycles of insolation, net radiation, and air temperature observed near the ground level. How does season of year influence these cycles? What instrument is used to record insolation?
4. Compare the daily cycle of soil temperature with that of the air above it. What changes in the soil-temperature cycle are observed with increasing depth?
5. How does a low-level temperature inversion occur? Describe the curve of air temperature associated with such an inversion. How is inversion related to killing frost?
6. Explain how urbanization alters the radiation and heat balances. What is a heat island?
7. Describe and explain the annual cycle of air temperature. Relate this cycle to net radiation. How does latitude influence net radiation and air temperature? Describe the typical annual cycle of soil temperatures in midlatitudes.
8. Explain the basic differences of land and water surfaces in terms of their properties for absorbing and transmitting insolation. How do these differences influence air temperatures over continents, as contrasted with ocean areas?
9. How does the annual temperature curve for a midlatitude station with an inland location differ from that of a station located on a seacoast? Is there a difference in the dates at which maximum and minimum temperatures normally occur in these two locations? Why?
10. What are isotherms? How are they drawn? In what way are isothermal maps useful?
11. What three basic controls operate to determine the global patterns of isotherms? Describe and explain the important features of those patterns. What influence has the changing sun's declination upon the isotherms? Is the latitudinal shift of isotherms greater over land or water areas? Explain.
12. What effect do the landmasses of North America and Asia have on the isotherms for January and July? Where is the earth's greatest annual temperature range experienced? Where would you expect to find a minimum annual range of temperature?
13. What is the effect of an altitude increase on insolation? On the daily air temperature cycle? On the mean monthly temperature?
14. What are the four categories of Man-induced atmosphere change? How has the carbon dioxide content of the atmosphere changed through the past century? What further change is predicted? How does this change tend to influence global air temperatures? How can the dust of volcanic eruptions influence hemispheric air temperatures?

Winds and the Global Circulation

6

Man's physical environment at the earth's surface depends for its quality as much on atmospheric motions as it does on the flow of heat energy by radiation. In the form of very strong winds — those in hurricanes and tornadoes — air in motion is a severe environmental hazard. Winds also transfer energy to the surface of the sea, as wind-driven waves. Wave energy, in turn, travels to the shores of continents, where it is transformed into vigorous surf and coastal currents capable of reshaping the coastline.

But air in motion has another, more basic, role to play in the planetary environment. Large-scale air circulation transports heat, both as sensible heat and as latent heat present in water vapor. Because of the global energy imbalance — a surplus in low latitudes and a deficit in high latitudes — atmospheric circulation must transport heat across the parallels of latitude from the region of surplus to the regions of deficit. Figure 6.1 shows this meridional transport in schematic form. Notice that circulation of ocean waters is also involved in the transport of sensible heat. But this oceanic circulation is a secondary mechanism, driven largely by surface winds.

Equally important to Man's environment are the rising and sinking motions of air. We shall find in Chapter 7 that precipitation, the source of all fresh water on the lands, requires massive lifting of large bodies of air heavily charged with water vapor. In reverse, large-scale sinking motions in the atmosphere lead to aridity and the occurrence of deserts. In this way the earth's surface becomes differentiated into regions of ample fresh water and regions of water scarcity.

With these broad concepts in mind, we examine the forces that set the atmosphere in motion and drive the ceaseless global circuits of air circulation.

Winds and the Pressure Gradient Force

Wind is air motion with respect to the earth's surface, and it is dominantly horizontal. (Dominantly vertical air motions are referred to by other terms, such as updrafts or downdrafts.) To explain winds, we must first consider barometric pressure and its variations from place to place.

In Chapter 2 we found that barometric pressure falls with increasing altitude above the earth's surface. For an atmosphere at rest, the barometric pressure will be the same within a given horizontal surface at any chosen altitude above sea level. In that case, the surfaces of equal barometric pressure, called **isobaric surfaces,** are horizontal. In a cross section of a small portion of the atmosphere at rest, isobaric surfaces appear as horizontal lines, as shown in Figure 6.2A. (For convenience, we have shown surface pressure as 1000 mb.)

Suppose, now, that the rate of upward pressure decrease is more rapid in one place than another, as shown in Figure 6.2B. As we proceed from left to right across the diagram, the upward rate of pressure decrease is more rapid. The isobaric surfaces now slope down toward the right. At a selected altitude, say 1000 m (horizontal line), barometric pressure declines from left to right. Figure 6.2C is a type of map; it shows that the 1000-m horizontal surface cuts across successive pressure surfaces. The trace of each pressure surface is a line on the map; the line is known as an **isobar.** The isobar is thus a

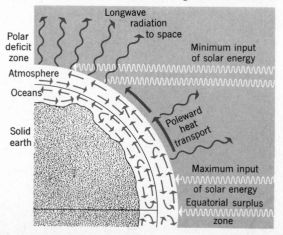

Figure 6.1 **Circulation systems of the atmosphere and oceans are necessary to maintain the global heat balance.**

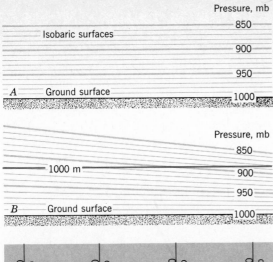

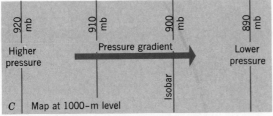

Figure 6.2 **Isobaric surfaces and the pressure gradient. Diagrams A and B are vertical cross sections through the atmosphere. Diagram C is a map.**

line showing the location on a map of all points having the same barometric pressure.

The change in barometric pressure across the horizontal surface of a map constitutes a **pressure gradient;** its direction is indicated by the broad arrow in Figure 6.2C. The gradient is in the direction from higher pressure (at the left) to lower pressure (at the right). You can think of the sloping pressure surface as a sloping hillside; the downward slope of the ground surface is analogous to the pressure gradient.

Where a pressure gradient exists, air molecules tend to drift in the same direction as that gradient. This tendency for mass movement of the air is referred to as the **pressure gradient force.** The magnitude of the force is directly proportional to the steepness of the gradient; that is, a steep gradient is associated with a strong force. Wind is the horizontal motion of air in response to the pressure gradient force.

Sea and Land Breezes

Perhaps the simplest example of the relationship of wind to the pressure gradient force is a common phenomenon of coasts: the sea breeze and land breeze, illustrated in Figure 6.3. Diagram A shows the initial situation in which no pressure gradient exists. During the daytime, more rapid heating of the lower air layer over the land than over the ocean causes a pressure gradient from sea to land (Diagram B). Air moving landward in response to this gradient from higher to lower pressure constitutes the **sea breeze.** At higher levels, a reverse flow sets in. Together

with weak rising and sinking air motions, a complete flow circuit is formed. During the night, when radiational cooling of the land is rapid, the lower air becomes colder over the land than over the water. Higher pressure now develops over land and the barometric gradient is reversed. Air now moves from land to sea as a **land breeze** (Diagram C).

This illustration shows that a pressure gradient can be developed through unequal heating or cooling of a layer of the atmosphere. Air that is warmed also expands and becomes less dense. Air that is cooled contracts and becomes denser. The upward change in barometric pressure is then more rapid within the cooler air layer than within the warmer layer.

Measurement of Surface Winds

A description of winds requires measurement of two quantities: direction and speed. Direction is easily determined by a **wind vane,** a common weather instrument. Wind direction is stated in terms of the direction from which the wind is coming (Figure 6.4). Thus an east wind comes from the east, but the direction of air movement is toward the west. The direction of movement of low clouds is an excellent indicator of wind direction and can be observed without the aid of instruments.

Speed of wind is measured by an **anemometer.** There are several types. The commonest one seen at weather stations is the cup anemometer. It consists of three hemispherical cups mounted as if at the ends of spokes of a horizontal wheel (Figure 6.5). The cups travel with a speed proportional to that of the wind. One type of anemometer turns a small electric generator, and the current it produces can be transmitted to a meter calibrated in units of wind speed. Units are meters per second or miles per hour.

For wind speeds at higher levels, a small hydrogen-filled balloon is released into the air and observed through a telescope. The rate of climb of the

Figure 6.3 **Sea breeze and land breeze.**

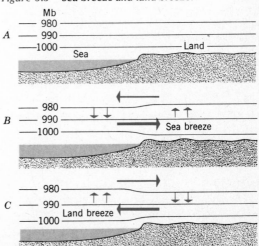

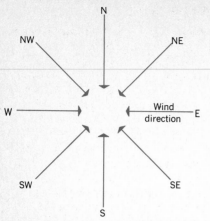

Figure 6.4 Winds are designated according to the compass point from which the wind comes. An east wind comes from the east, but the air is moving westward.

balloon is known in advance. Knowing the balloon's vertical position by measuring the elapsed time, an observer can calculate the horizontal drift of the balloon downwind. For upper-air measurements of wind speed and direction, the balloon carries a target that reflects radar waves and can be followed when the sky is overcast.

The Coriolis Effect and Winds

If the earth did not rotate on its axis, winds would follow the direction of pressure gradient. As mentioned in Chapter 1, earth rotation produces the **Coriolis effect,** which tends to turn the flow of air.* The direction of action of this turning effect is stated in Ferrel's law: Any object or

---

*In strict terms of physics, the Coriolis effect is not a true force.

fluid moving horizontally in the northern hemisphere tends to be deflected to the right of its path of motion, regardless of the compass direction of the path. In the southern hemisphere, a similar deflection is toward the left of the path of motion. The Coriolis effect is absent at the equator but increases in strength toward the poles.

On Figure 6.6 the small arrows show how an initial straight line of motion is modified by the Coriolis effect. Note especially that the compass direction is not of any consequence. If we face down the direction of motion, turning will always be toward the right in the northern hemisphere. Because the deflective effect is very weak, its action is conspicuous only in freely moving fluids such as air or water. Consequently, both winds and ocean current patterns are greatly affected by the Coriolis effect.

Our next step is to apply the Coriolis principle to winds close to the ground surface. Figure 6.7 shows isobars running east-west, forming a ridge of high pressure in each hemisphere. From each ridge, pressure decreases both to the north and south toward belts of low pressure. Broad arrows show the pressure gradient. The Coriolis effect turns the wind so that it crosses the isobars at an angle. For surface winds, the angle of turning is limited by the force of friction of the air with the ground. The diagram shows the wind making an angle of 45° with the isobars. In nature the angle is subject to some variation, depending on the character of the ground surface.

Looking first at the northern hemisphere case, the deflection is to the right. The northward pressure gradient gives a southwest wind. The southward gradient gives a northeast wind. In the southern hemisphere, winds are deflected to the left and the pattern is the mirror image of that in the northern hemisphere.

A general rule for the relationship of wind to pressure

Figure 6.6 Deflective effect of the earth's rotation.

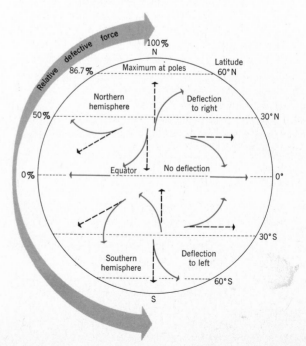

Figure 6.5 A cup-type anemometer. The dial at the base records the number of revolutions made by the shaft in a given period of time. (National Weather Service.)

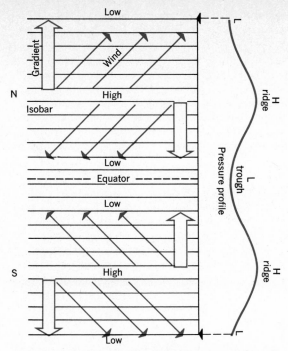

Figure 6.7 Surface winds cross isobars at an angle as the air moves from higher to lower pressure. Turn the figure sideways to view the pressure profile.

in the northern hemisphere is known as Ballot's law, which states: Stand with your back to the wind and the low pressure will be toward your left, the high toward your right.

Where isobars are curved, as shown on the simple weather map in Figure 6.8, the direction of pressure gradient follows a curving trajectory, always cutting the isobars at a right angle. The map shows that where isobars are widely spaced, the gradient is weak; where isobars are closely spaced, the gradient is strong. Short arrows show wind directions. Wind speed is slow where the gradient is weak, fast where gradient is strong.

Figure 6.8 This small portion of a surface weather map shows curving isobars, with both weak and strong pressure gradients. The short arrows show the surface wind directions.

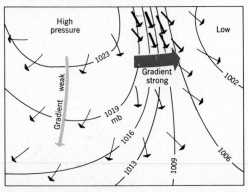

Cyclones and Anticyclones

In the language of meteorology, a center of low pressure is called a **cyclone;** a center of high pressure is an **anticyclone.** Cyclones and anticyclones may be of a stationary type, or they may be rapidly moving pressure centers such as those that create the weather disturbances described in Chapter 8. Isobars form closed, circular patterns around both cyclones and anticyclones.

For surface winds, which move obliquely across the isobars, the systems for cyclones and anticyclones in both hemispheres are shown in Figure 6.9. Winds in a cyclone in the northern hemisphere show an anticlockwise inspiral. In an anticyclone, there is a clockwise outspiral. Note the reversal between the labels "anticlockwise" and "clockwise" in the southern hemisphere.

In both hemispheres the surface winds spiral inward on the center of the cyclone, so the air is converging on the center and must also rise to be disposed of at higher levels. For the anticyclone, by contrast, surface winds spiral out from the center. This motion represents a diverging of airflow and must be accompanied by a sinking (subsidence) of air in the center of the anticyclone to replace the outmoving air.

Global Distribution of Surface Pressure Systems

To understand the earth's surface wind systems, we must first study the global system of barometric pressure distribution. Once we grasp the patterns of isobars and pressure gradients, the prevailing or average winds can be predicted.

World isobaric maps are usually constructed to show average pressures for the two months of seasonal temperature extremes over large landmasses — January and July (Plates B.3, B.4). Because observing stations lie at various altitudes above sea level, their barometric readings must

Figure 6.9 Surface winds in cyclones and anticyclones.

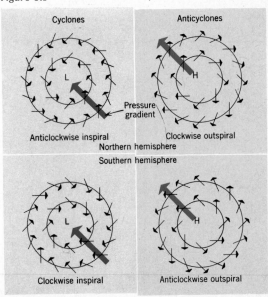

be reduced to sea-level equivalents, using the standard rate of pressure change with altitude (explained in Chapter 3). When this has been done and the daily readings are averaged over long periods of time, small but distinct pressure differences remain.

A reading of 1013 mb is taken as standard sea-level pressure. Readings higher than this will frequently be observed in midlatitudes, occasionally up to 1040 mb or higher. These pressures are designated as "high." Pressures ranging down to 982 mb or below are "low."

Over the equatorial zone is a belt of somewhat lower than normal pressure, between 1011 and 1008 mb, which is known as the **equatorial trough.** Lower pressure is conspicuous by contrast with belts of higher pressure lying to the north and south and centered at about lat. 30°N and S. These are the **subtropical belts of high pressure,** in which pressures exceed 1020 mb. In the southern hemisphere this belt is clearly defined but contains centers of high pressure, called **pressure cells.**

In the southern hemisphere, south of the subtropical high-pressure belt, is a broad belt of low pressure, extending roughly from the midlatitude zone to the arctic zone. The axis of low pressure is centered at about lat. 65°S. This pressure trough is called the **subantarctic low-pressure belt.** Lying over the continuous expanse of Southern Ocean, this trough has average pressures as low as 984 mb. Over the continent of Antarctica is a permanent center of high pressure known as the **polar high.** It contrasts strongly with the encircling subantarctic low.

The pressure belts shift annually through several degrees of latitude, along with the isotherm belts. These annual pressure-belt migrations are important in causing seasonal climate changes. We shall have several occasions to refer to these effects in analyzing world climates.

Northern Hemisphere Pressure Centers

The vast continents of North America and Eurasia and the intervening North Atlantic and North Pacific oceans exert a powerful control over pressure conditions in the northern hemisphere. As a result, the belted arrangement typical of the southern hemisphere is absent.

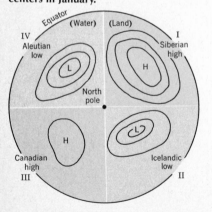

Figure 6.10 **Northern hemisphere pressure centers in January.**

In winter the large, very cold land areas develop high-pressure centers. At the same time, intense low-pressure centers form over the warmer oceans. Over north central Asia in winter we find the **Siberian high,** with pressure exceeding 1030 mb. Over central North America is a clearly defined, but much less intense, ridge of high pressure, called the **Canadian high.** Over the oceans are the **Aleutian low** and the **Icelandic low,** named after the localities over which they are centered. These two low-pressure areas have much cloudy, stormy weather in winter. Figure 6.10 shows these pressure centers as they appear grouped around the north pole. Highs and lows occupy opposite quadrants.

In summer, pressure conditions are exactly the opposite of winter conditions. Land areas develop low-pressure centers because at this season land-surface temperatures rise sharply above temperatures over the adjoining oceans. At the same time, the ocean areas develop strong centers of high pressure. This system of pressure opposites is striking on both the January and July isobaric maps (Plate B.3, B.4), The low in Asia is intense: It is centered over Afghanistan. Over the Atlantic and Pacific oceans are two large, strong cells of the subtropical belt of high pressure. They are shifted northward of their winter position and are considerably expanded. They are called the **Azores high** (or Bermuda high) and the **Hawaiian high.**

The Global Pattern of Surface Winds

Prevailing surface winds during January and July are shown by arrows on the pressure maps of Plate B. The basic wind patterns are stressed in Figure 6.11, which is a highly diagrammatic representation showing the earth as if no land areas existed to modify the belted arrangement of pressure zones.

Let us begin with winds of the tropical zones. From the two subtropical high-pressure belts the pressure gradient is equatorward, leading down to the equatorial trough of low pressure. Following the simple model shown in Figure 6.11, air moving from high to low pressure is deflected by the Coriolis effect. As a result, two belts of **trade winds,** or **trades,** are produced. These are labeled in Figure 6.11 as the northeast trades and southeast trades. The trades are highly persistent winds, deviating very little from a single compass direction. Sailing vessels traveling westward once made good use of the trades.

The pattern of the trades suggests that they must converge somewhere near the equator. Meeting of the trades takes place within a narrow zone called the **intertropical convergence zone,** usually abbreviated to **ITC.** The position of the ITC is marked on the January and July maps (Plate B.3). Converging winds require a rise of air to dispose of the incoming volume of air. This rise takes the form of stalklike flow columns carrying the air toward the top of the troposphere.

Along parts of the equatorial trough of low pressure at certain times of year the trades do not come together in convergence. Instead, a belt of calms and variable winds, called the **doldrums,** forms. Mariners on sailing ships knew that crossing the doldrums was hazardous because of the likelihood of lying becalmed for long periods.

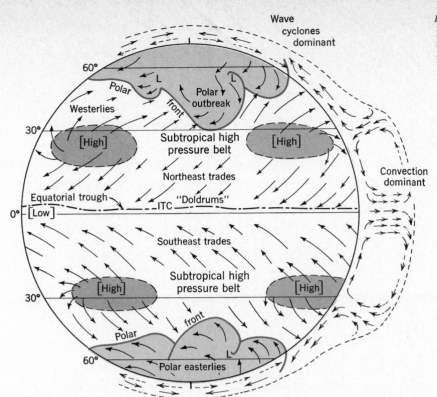

The trades, doldrums, and ITC all shift seasonally north and south, along with the shifting of pressure belts and isotherms. The ITC migrates north and south only a few degrees of latitude over the Pacific and Atlantic oceans, but covers as much as 20° to 30° of latitude over South America, Africa, and the large region of Southeast Asia and the Indian Ocean. Important seasonal changes in winds, cloudiness, and rainfall accompany these migrations of the ITC and the trades.

We now return to the subtropical high-pressure belt, which ranges between lat. 25° and 40°N and S. Here we encounter large, stagnant, high-pressure cells (anti-cyclones). In the centers of the cells, winds are weak and are distributed around a wide range of compass directions; calms prevail as much as a quarter of the time. Because of the high frequency of calms, mariners named this belt the **horse latitudes.** It is said that the name originated in colonial times from the experiences of New England traders, carrying cargoes of horses to the West Indies. When their ships were becalmed for long periods of time, the freshwater supplies ran low and the horses had to be thrown overboard.

Figure 6.12 is a schematic map of large anticyclonic cells centered over the oceans in two hemispheres. Winds make an outspiraling pattern that feeds into the converging trades. On the western sides of the cells, air flows poleward; on the eastern sides, the flow is equatorward. These flows have a strong influence on the climates of adjacent continental margins. Dryness of climate is a dominant general characteristic of the subtropical high-pressure belt and its cells, and we shall emphasize this feature again.

Between lat. 35° and 60°N and S is the belt of **prevailing westerly winds,** or **westerlies.** These surface winds are shown on Figure 6.11 as blowing from a southwesterly quarter in the northern hemisphere and from a northwesterly quarter in the southern hemisphere. This generalization is somewhat misleading, however, because winds from polar directions are frequent and strong. It is more nearly accurate to say that within the westerlies, winds blow from all directions of the compass, but that the westerly components are definitely predominant. Rapidly moving cyclonic storms are common in this belt.

Figure 6.12 **Over the oceans, surface winds spiral outward from the dominant cells of high pressure, feeding the trades and the westerlies.**

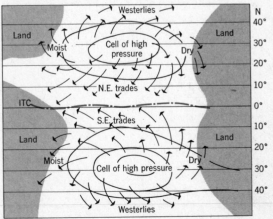

In the northern hemisphere, landmasses disrupt the westerlies but, in the southern hemisphere between lat. 40° and 60°S, there is an almost unbroken belt of ocean. Here the westerlies gather great strength and persistence. Sailors in the great clipper ships called these latitudes "roaring forties," "furious fifties," and "screaming sixties." This belt was used extensively by sailing vessels traveling eastward from the South Atlantic Ocean to Australia, Tasmania, New Zealand, and the southern Pacific Islands. From these places it was then easier to continue eastward around the world to return to European ports. Rounding Cape Horn was relatively easy on an eastward voyage but, in the opposite direction, in the face of prevailing stormy westerly winds, it was a most dangerous operation.

A wind system called the **polar easterlies** has been said to be characteristic of the arctic and polar zones (Figure 6.11). The concept is at best greatly oversimplified and is certainly misleading when applied to the northern hemisphere. Winds in these high-latitude zones take a variety of directions, dictated by local weather disturbances. On the other hand, in the southern hemisphere, Antarctica is an ice-capped landmass resting squarely on the pole and surrounded by a vast oceanic expanse. Here the outward spiraling flow of polar easterlies seems to be a dominant feature of the circulation. The polar maps in Plate B.4 show these easterly winds.

Monsoon Winds of Southeast Asia

The powerful control exerted by the great landmass of Asia on air temperatures and pressures extends to the surface wind systems as well. In summer southern Asia develops a cyclone into which there is a strong flow of air (July map, Figure 6.13). From the Indian Ocean and the southwestern Pacific, warm, humid air moves north-

ward and northwestward into Asia, passing over India, Indochina, and China. This airflow constitutes the **summer monsoon** and is accompanied by heavy rainfall in southeastern Asia.

In winter, Asia is dominated by a strong center of high pressure from which there is an outward flow of air reversing that of the summer monsoon (January map, Figure 6.13). Blowing southward and southeastward toward the equatorial oceans, this **winter monsoon** brings dry weather for a period of several months.

North America does not have the remarkable extremes of monsoon winds experienced by southeastern Asia but, even so, there is a distinct alternation of average temperature and pressure conditions between winter and summer. Wind records show that in summer there is a prevailing tendency for air originating in the Gulf of Mexico to move northward across the central and eastern part of the United States, whereas in winter there is a prevailing tendency for air to move southward from high-pressure sources in Canada. Wind arrows in Plate B.3 show this seasonal alternation in the airflow pattern.

Local Winds

In certain localities, **local winds** are generated by immediate influences of the surrounding terrain rather than by the large-scale pressure systems that produce global winds and large traveling storms. Local winds are of environmental importance in various ways. They can exert a powerful stress on animals and plants when the air is dry and extremely hot or cold. Local winds are also important in affecting the movement of atmospheric pollutants.

One class of local winds — sea breezes and land breezes — was explained earlier in this chapter. The cooling sea breeze (or lake breeze) of summer is an important environmental resource of coastal communities because it adds to the attraction of the shore zone as a recreation site.

Mountain winds and **valley winds** are local winds following a daily alternation of direction in a manner similar to the land and sea breezes. During the day, air moves from the valleys, upward over rising mountain slopes, toward the summits. At this time hillslopes are intensely heated by the sun. At night the air then moves valleyward, down the hillslopes, which have been cooled at night by radiation of heat from ground to air. These winds are responding to local pressure gradients set up by heating or cooling of the lower air.

Still another group of local winds are known as **drainage winds,** or **katabatic winds,** in which cold air flows under the influence of gravity from higher to lower regions. Such cold, dense air may accumulate in winter over a high plateau or high interior valley. When general weather conditions are favorable, some of this cold air spills over low divides or through passes to flow out on adjacent lowlands as a strong, cold wind. Drainage winds occur in many mountainous regions of the world and go by various local names. The **mistral** of the Rhone valley in southern France is a well-known example; it is a cold, dry local wind. On the ice sheets of Greenland and

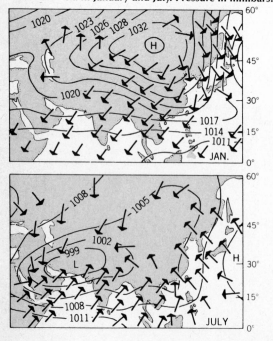

Figure 6.13 **Surface maps of pressure and winds over Southeast Asia in January and July. Pressure in millibars.**

Antarctica, powerful drainage winds move down the gradient of the ice surface and are funneled through coastal valleys to produce powerful blizzards lasting for days at a time.

Another type of local wind occurs when the outward flow of dry air from a strong high-pressure center (an anticyclone) is combined with the local effects of mountainous terrain. An example is the **Santa Ana,** a hot, dry easterly wind that, on occasion, blows from the interior desert region of southern California across coastal mountain ranges to reach the Pacific coast. Locally, this wind is funneled through narrow mountain gaps or canyon floors, where it gains great force. At times the Santa Ana wind carries large amounts of dust. It is greatly feared for its ability to fan intense brush fires out of control. The **bora** of the Adriatic coast of Yugoslavia is another wind of this class. It is a cold winter wind produced by the pressure gradient of a strong anticyclone located over the land. Descending from the coastal mountains to the sea, it can produce gusts up to 160 km (100 mi) per hour.

Still another type of local winds, bearing such names as *foehn* and *chinook,* results when strong regional winds passing over a mountain range are forced to descend on the lee side with the result that the air is heated and dried. These winds are explained in Chapter 8.

Winds Aloft

The global surface wind systems we have examined represent only a shallow basal air layer a few hundred meters deep, whereas the troposphere is several

Figure 6.14   **Wind follows isobars at high levels.**

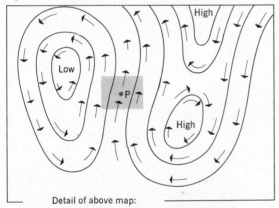

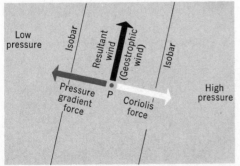

kilometers deep. How does air move at these higher levels? Large, slowly moving high-pressure and low-pressure systems are found aloft, but these are generally simple in pattern, with smoothly curved isobars.

Winds high above the earth's surface are not affected by friction with the ground or water over which they move. The Coriolis effect turns the flow of air until it becomes parallel with the isobars, as Figure 6.14 shows. (The Coriolis effect is shown as if it were a force.) In this position the pressure gradient force and Coriolis force are exactly opposed and exactly balanced.

The ideal wind in this state of balance with respect to the two forces is termed the **geostrophic wind** for cases in which the isobars are straight. Where isobars are curved, centrifugal force must also be taken into account; but, in general, airflow at high altitudes parallels the isobars.

The upper part of Figure 6.14 is a simplified map of pressure and winds high in the troposphere. Notice how the wind arrows run parallel with the isobars, forming elliptic flow patterns around lows and highs. Our rules for winds in cyclones and anticyclones need to be slightly modified, as compared with surface winds. For an airline pilot flying in the northern hemisphere and wishing to keep a tailwind at all times, the rule would be "keep the highs on your right and the lows on your left."

In describing pressure belts and winds at the earth's surface, we gave no explanation of their basic cause in terms of earth rotation and the Coriolis principle. To understand the surface winds, we need to investigate the workings of the entire troposphere, particularly at the higher levels, where surface friction is absent and the airflow closely follows the isobars.

Winds on a Nonrotating Planet

Consider, first, an imaginary nonrotating planet that is heated uniformly around the equatorial belt (where a radiation surplus exists) but cooled severely at both polar regions (where radiation deficits exist). Figure 6.15 shows this ideal case. Air heated at the equator will expand and become less dense because all gases expand in volume when heated. Because the expanded lower air layer is now less dense, the atmospheric pressure at the earth's surface will be lower than average (L, Figure 6.15). The heated air will rise and upon reaching high levels in the atmosphere will spread horizontally, moving poleword. Chilled air over the poles will increase in density, creating high pressure at the earth's surface (H). The chilled air will sink and then spread horizontally, traveling toward the equatorial belt of low pressure. Once the airflow is established, a system of **meridional winds** results, as shown in Figure 6.15. The global winds will now be formed into two circulation cells, one in each hemisphere. As long as heat continues to be supplied to the equatorial belt, the wind system will remain in operation. We have here a **heat engine,** that is, a mechanical system driven by an input of heat energy.

Our model of a nonrotating earth serves to explain one very real feature of the earth's atmospheric circulation: the equatorial belt of low pressure, or equatorial trough, within which heated air rises to high levels. What actually

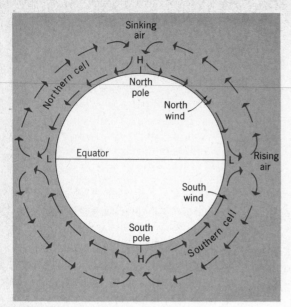

Figure 6.15 An imaginary atmospheric circulation system on a nonrotating planet.

happens to this rising air as it begins to move poleward in the upper atmosphere can be understood only by bringing into account the Coriolis effect.

The Hadley Cell Circulation

Introducing the Coriolis effect into our simple model of meridional circulation, consider what would happen to air beginning to move poleward at high levels from the equatorial belt. As Figure 6.16 shows, a parcel of air starting at point *A* and beginning to move north would be deflected by the Coriolis effect a short distance north of the equator. Its path would then turn eastward in traveling from *A* to *B*. Upon reaching point *C*, the flow would run due east along the parallels of latitude. (In the southern

hemisphere, deflection toward the left would also result in eastward flow.)

With no further poleward progress possible, the eastward-moving air tends to accumulate in the subtropical zone at about lat. 20° to 30°. The piling up of air causes a sinking motion, or **subsidence,** to develop in this zone, which in turn creates a belt of increased pressure. This feature is the subtropical high-pressure belt. Air reaching low levels is now required to move away from the downsinking zone. Part of this air moves poleward, but much of it moves equatorward, as shown in the right-hand diagram of Figure 6.16. Equatorward motion results in deflection to the west, setting up an easterly wind system. At upper levels, these winds are known as the **tropical easterlies;** near the surface, they are the trade winds. The easterlies form a broad, deep airstream over the entire equatorial belt, where they are referred to as the **equatorial easterlies.** In the lower troposphere, the westward-moving air is also slowly converging over the equator in the intertropical convergence zone (ITC). This converging air is rising slowly to complete the entire circuit.

Taking into account only the north-to-south, south-to-north, and vertical components of air movement — in other words, only the meridional flow — we find a cell of atmospheric circulation dominating the tropical and equatorial zones. This system has been named the **Hadley cell** in honor of George Hadley, who postulated its existence in 1735. Ideally there are two matching Hadley cells, one for each hemisphere.

The Upper-Air Westerlies and Rossby Waves

Air spreading poleward from the subtropical high-pressure belt is deflected to the right by the Coriolis effect in the northern hemisphere. Within a short distance the motion is turned to follow the parallels of latitude, moving from west to east and forming the **upper-air westerlies** (Figure 6.16). In the southern hemisphere deflection to the left as

Figure 6.16 The Hadley cell circulation. Air parcels are imagined to be released from initial starting points over the equator (left) and beneath the subtropical highs (right). (From A. N. Strahler, 1971, *The Earth Sciences,* 2nd ed., Harper and Row, New York.)

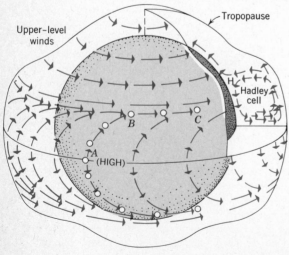
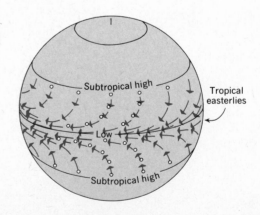

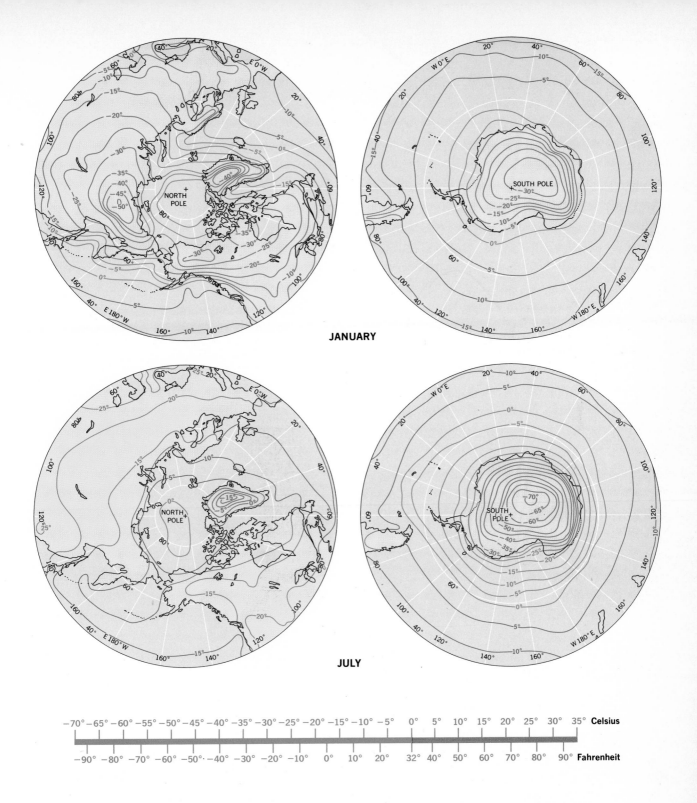

JANUARY

JULY

−70° −65° −60° −55° −50° −45° −40° −35° −30° −25° −20° −15° −10° −5° 0° 5° 10° 15° 20° 25° 30° 35° **Celsius**

−90° −80° −70° −60° −50° −40° −30° −20° −10° 0° 10° 20° 32° 40° 50° 60° 70° 80° 90° **Fahrenheit**

Mean monthly air temperatures for January and July, north and south polar regions. Stereographic polar projection.
(Same data source as maps on following page.)

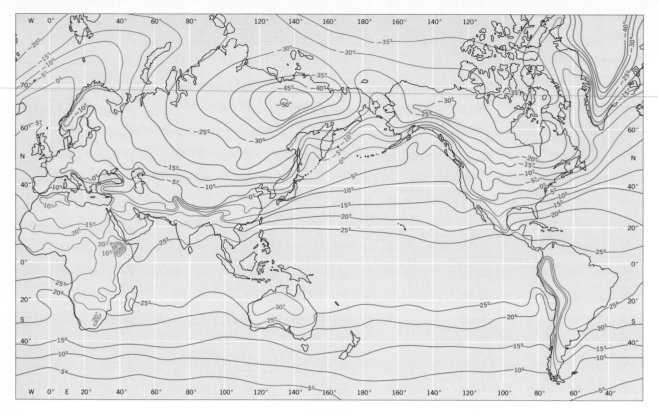

JANUARY

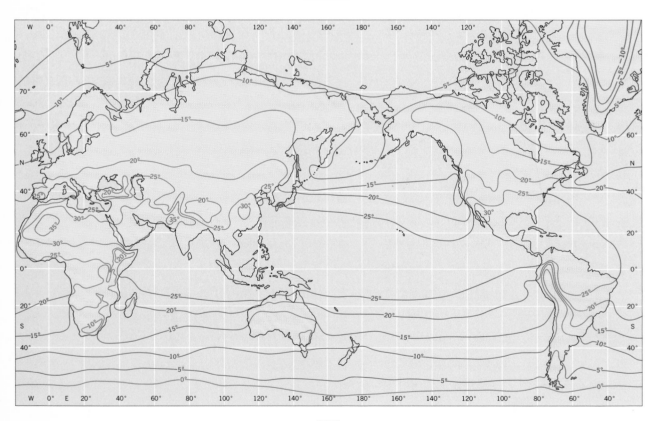

JULY

Mean monthly air temperatures for January and July. Mercator
projection. (Compiled by John E. Oliver from station data by World
Climatology Branch, Meteorological Office, <u>Tables of Temperature</u>, 1958,
Her Majesty's Stationery Office, London; U.S. Navy, 1955, <u>Marine Climatic
Atlas</u>, Washington, D.C.; and P. C. Dalrymple, 1966, American
Geophysical Union.

Plate B.2   World temperatures

Mean monthly atmospheric pressure and prevailing surface winds for
January and July. Pressure units are millibars reduced to sea level. Many
of the wind arrows are inferred from isobars. Mercator projection.
(Compiled by John E. Oliver from published data by Y. Mintz, G. Dean, R.
Geiger, and J. Blüthagen.)

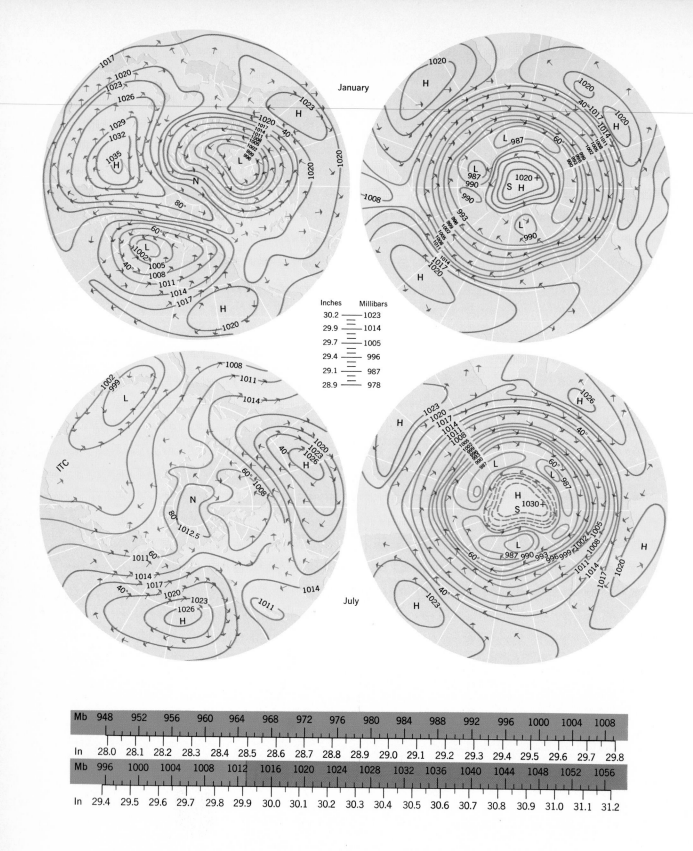

January

Inches	Millibars
30.2	1023
29.9	1014
29.7	1005
29.4	996
29.1	987
28.9	978

July

Mb	948	952	956	960	964	968	972	976	980	984	988	992	996	1000	1004	1008
In	28.0	28.1	28.2	28.3	28.4	28.5	28.6	28.7	28.8	28.9	29.0	29.1	29.2	29.3	29.4	29.5	29.6	29.7	29.8

Mb	996	1000	1004	1008	1012	1016	1020	1024	1028	1032	1036	1040	1044	1048	1052	1056
In	29.4	29.5	29.6	29.7	29.8	29.9	30.0	30.1	30.2	30.3	30.4	30.5	30.6	30.7	30.8	30.9	31.0	31.1	31.2

Mean monthly atmospheric pressure and prevailing winds for January and July, north and south polar regions. Polar stereographic projection. (Same data source as maps on preceding page.)

Plate B.4   World pressures and winds

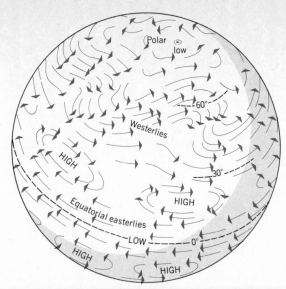

Figure 6.17 **Schematic representation of circulation in the upper part of the troposphere, 6 to 12 km (20,000 to 40,000 ft).**

air moves south also forms upper-air westerlies in that hemisphere.

Figure 6.17 shows in a simplified way the westerly winds in relation to the subtropical high-pressure belt and the tropical easterlies. Notice that a number of high-pressure cells comprise the high-pressure belt. The upper-air westerlies persist into the polar region, where

they form a great vortex. Atmospheric pressure falls rapidly toward the polar region, where the **polar low** is situated.

Westerly winds around the polar lows involve the entire depth of the troposphere. Recall that the troposphere is much thinner at high latitudes than at low latitudes (see Figure 3.8). The uniform flow of the upper-air westerlies is frequently disturbed by the formation of large undulations, called **Rossby waves** (Figure 6.18). These waves develop along a narrow zone of contact between the large body of cold **polar air,** forming the troposphere on the poleward side, and warm **tropical air,** which surrounds the globe on the equatorward side. This contact zone is called the **polar front;** it is an unstable zone, along which severe atmospheric disturbances are generated.

Diagram A of Figure 6.18 shows the polar front in a fairly smooth, stable condition. Minor undulations are, however, beginning to form along the front. These undulations deepen greatly (Diagrams B and C), becoming Rossby waves. Cold polar air pushes into lower latitudes in deep embayments, which form troughs of low pressure at high levels. Two such troughs are shown in Diagram C, their axes marked by dashed lines. Correspondingly, warm tropical air moves into higher latitudes in embayments between the troughs. Here, high-pressure ridges are formed in the troposphere. As we shall find in Chapter 8, this phase of wave development is associated with the outbreak of storms (cyclonic storms) near the surface.

Diagram D shows that the waves have deepened to

Figure 6.18 **Development of upper-air waves in the westerlies. (Modified from J. Namias, NOAA, National Weather Service; Figure from A. N. Strahler, 1971, *The Earth Sciences,* 2nd ed., Harper and Row, New York.)**

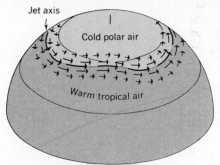

A. Jet stream begins to undulate.

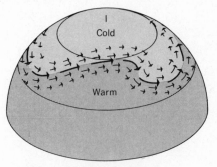

B. Rossby waves begin to form.

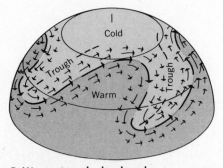

C. Waves strongly developed.

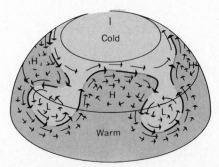

D. Cells of cold and warm air are formed.

the extent that the low-pressure troughs are constricting and becoming detached. When this happens, the detached mass of cold air becomes a **cut-off low,** which is now a cyclone in the upper atmosphere. A warm-air embayment may also become detached to form a **cut-off high,** which is an upper-air anticyclone. The importance of this cutoff process lies in the transport of cold air into lower latitudes and of warm air into high latitudes. Through this **advection process,** or horizontal mixing, the meridional transport of surplus heat into high latitudes is accomplished — a vital process in the maintenance of the global heat balance.

Rossby waves develop slowly. The cutoff process we have described may require many days to complete. The waves may weaken and dissolve without completing the cutoff cycle; they may remain stationary for long periods or may drift slowly in the east-west direction.

The Polar Front Jet Stream

Associated with Rossby waves is a narrow zone of very high wind speed called the **jet stream,** formed along the line of contact of cold air and warm air. The position of the jet stream is shown in Figure 6.18 by a line of heavy arrows. The jet stream is a pulselike flow of air. It resembles a stream of water moving in a hose. Velocity is highest in the center line, or **core,** of the jet stream, which is surrounded by slower-moving zones. Figure 6.19 shows an actual jet stream in cross section, using lines of equal wind speed. Maximum speed in the core was about 300 km/hr (185 mph). Altitude of the core was about 11 km (36,000 ft). Figure 6.20 is a weather map showing that this jet was passing over the western United States, in a path curving southeast toward the Gulf of Mexico, then recurving to the northeast over Florida and outlining a

Rossby wave. Position of the jet core is shown by a solid line. Lines of equal wind speed (isotachs) show that the pulse of highest speed was located over Wyoming.

The jet stream we have described forms at the level of the tropopause, along the polar front, the surface of contact between cold polar air and warm tropical air. For this reason, it is given the name **polar front jet stream.** The existence of the jet is fully explained by principles we have already developed. Figure 6.21 shows the position of the jet core and the polar front at the tropopause in mid-latitudes. The tropopause drops sharply in altitude at the polar front, being much lower over the cold troposphere layer than over the warm troposphere. Atmospheric pressure also changes abruptly at the polar front. Isobaric surfaces, shown by lines on the diagram, drop steeply at the polar front, the steepest drop being at the level of the jet-stream core. The very steep pressure gradient at this point causes the high-speed flow of air. The polar-front jet stream is extremely important through its control of surface weather patterns over midlatitudes. This is a subject we will develop in Chapter 8.

The polar-front jet stream is a factor in the operation of jet aircraft in the range of their normal cruising altitudes in midlatitudes. In addition to strongly increasing or decreasing the ground speed of the aircraft the jet stream carries a form of air turbulence that at times reaches hazardous levels. This is clear air turbulence (CAT); it is avoided when known to be severe.

A second jet stream of major importance in the global circulation forms in the subtropical latitude zone. Called the **subtropical jet stream,** it occupies a position at the tropopause just above the Hadley cell. Here westerly wind speeds reach maximum values of 345 to 385 km/hr (215 to 240 mph). Figure 6.22, a northern-hemisphere map, shows the average position of both the polar-front jet stream and the subtropical jet stream in the winter season.

Figure 6.19 **Cross-sectional diagram through a jet stream over the western United States. See Figure 6.20 for line of section and accompanying map data. (After H. Riehl, 1962, *Jet Streams of the Atmosphere,* Colorado State Univ., Fort Collins.)**

Figure 6.20 **Isotachs (knots) and wind arrows at the 300-mb (9 km) level on April 22, 1958. See Figure 8.10 for explanation of symbols. Solid arrows mark jet stream axis. Cross section along the line X-Y is shown in Figure 6.19. (After H. Riehl, 1962, *Jet Streams of the Atmosphere,* Colorado State Univ., Fort Collins.)**

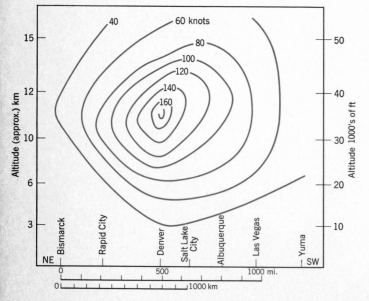

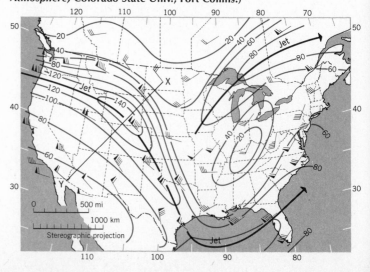

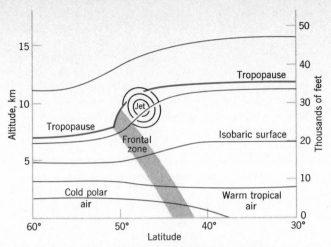

Figure 6.21 **This schematic cross section through the polar front shows the position of the jet-stream core in relation to isobaric surfaces. (After E. R. Reiter, 1967, *Jet streams*, Doubleday and Co., Garden City, N.Y.)**

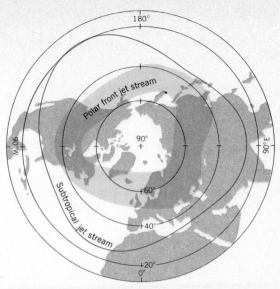

Figure 6.22 **Mean winter position of the axis of the subtropical jet stream and the belt of principal winter activity of the polar-front jet stream. (After H. Riehl, 1962, *Jet Streams of the Atmosphere*, Colorado State Univ., Fort Collins.)**

Figure 6.23 **A schematic diagram of wind directions and jet streams along an average meridian from pole to pole. (From A. N. Strahler, 1971, *The Earth Sciences*, second edition, Harper and Row, New York.)**

A third jet stream system is found at even lower latitudes. Known as the **tropical easterly jet stream**, it runs from east to west — opposite in direction to that of the polar-front and subtropical jet streams. The tropical easterly jet stream occurs only in the summer (high-sun) season and is limited to a northern-hemisphere location over Southeast Asia, India, and Africa. This jet stream is very high in altitude, about 15 km (50,000 ft), and has wind speeds over 180 km (115 mi) per hour. This tropical easterly jet stream is considered by some atmospheric scientists to play a major role in the rainy summer monsoon of Southeast Asia.

Figure 6.23 is a meridional cross section of the atmosphere showing in a schematic way the position of the three jet-stream systems. The diagram shows only the easterly and westerly components of the atmospheric circulation, disregarding the meridional components.

Ocean Currents

An **ocean current** is any persistent, dominantly horizontal flow of ocean water. Ocean currents are important regulators of thermal environments of the earth's surface. On a global scale, the vast current systems aid in exchange of heat between low and high latitudes and are essential in sustaining the global heat balance. On a local scale, warm water currents bring a moderating influence to coasts in arctic latitudes; cool currents greatly alleviate the heat of tropical deserts along narrow coastal belts.

Practically all the important surface currents of the oceans are set in motion by prevailing surface winds. Energy is transferred from wind to water by the frictional drag of the air blowing over the water surface. Because of the Coriolis effect, the water drift is impelled toward the right of its path of motion (northern hemisphere); therefore the current at the water surface is in a direction about 45° to the right of the wind direction. Under the influence of winds, currents may tend to bank up the water close to the

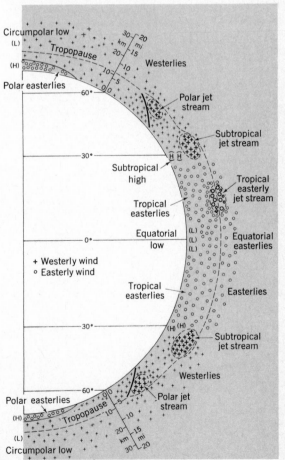

coast of a continent, in which case the force of gravity, tending to equalize the water level, will cause other currents to be set up.

Density differences may also cause flow of ocean water. Such differences arise from greater heating by insolation, or greater cooling by radiation, in one place than in another. Thus surface water chilled in the arctic and polar seas will sink to the ocean floor, spreading equatorward and displacing upward the less dense, warmer water.

Still another controlling influence on water movements is the configuration of the ocean basins and coasts. Currents initially caused by winds impinge on a coast and are locally deflected.

Generalized Scheme of Ocean Currents

To illustrate global surface water circulation, we can refer to an idealized ocean extending across the equator to latitudes of 60° or 70° on either side (Figure 6.24). Perhaps the most outstanding features are the circular movements, called **gyres,** around the subtropical highs, centered about 25° to 30°N and S. An **equatorial current** with westward flow marks the belt of the trades. Although the trades blow to the southwest and northwest, obliquely across the parallels of latitude, the water movement follows the parallels.

A slow, eastward movement of water over the zone of the westerly winds is named the **west-wind drift.** It covers a broad belt between 35° and 45° in the northern hemisphere, and between 30° or 35° and 70° in the southern, where open ocean exists in the higher latitudes. The

equatorial currents are separated by an **equatorial countercurrent.** This is well developed in the Pacific, Atlantic, and Indian oceans (Figure 6.25).

Along the west sides of the oceans in low latitudes the equatorial current turns poleward, forming a warm current paralleling the coast. Examples are the Gulf Stream (Florida or Caribbean stream), the Japan current (Kuroshio) and the Brazil current, which bring higher than average temperatures along these coasts.

The west-wind drift, upon approaching the east side of the ocean, is deflected both south and north along the coast. The equatorward flow is a cool current, produced by the upwelling of colder water from greater depths. It is well illustrated by the Humboldt current (Peru current) off the coast of Chile and Peru; by the Benguela current off the southwest African coast; by the California current off the west coast of the United States; and by the Canaries current off the Spanish and North African coast.

In the northeastern Atlantic Ocean, the west-wind drift is deflected poleward as a relatively warm current. This is the North Atlantic drift that spreads around the British Isles, into the North Sea, and along the Norwegian coast. The port of Murmansk, on the arctic circle, has year-round navigability because of this warm current.

In the northern hemisphere, where the polar sea is largely landlocked, cold water flows equatorward along the west side of the large straits connecting the Arctic Ocean with the Atlantic basin. Three principal cold currents are the Kamchatka current, flowing southward along the Kamchatka Peninsula and Kurile Islands; the Greenland current, flowing south along the east Greenland coast through the Denmark Strait; and the Labrador current, moving south from the Baffin Bay area through Davis Strait to reach the coasts of Newfoundland, Nova Scotia, and New England.

In both the north Atlantic and Pacific oceans, the Icelandic and Aleutian lows in a rough way coincide with two centers of counterclockwise circulation involving the cold arctic currents and the west-wind drifts.

The antarctic region has a relatively simple current scheme consisting of a single **antarctic circumpolar current** moving clockwise around the antarctic continent in latitudes 50° to 65°S, where a continuous expanse of open ocean occurs.

Oceanographers today recognize that oceanic circulation involves the complex motions of water masses of different temperature and salinity characteristics. Sinking and upwelling are both important motions in certain areas of the oceans.

The Global Circulation and Man's Environment

We have outlined the major mechanical circuits within which sensible heat is transported from low latitudes to high latitudes. In low latitudes, the Hadley cell operates like a simple heat engine to transport heat from the equatorial zone to the subtropical zone. Upper-air waves take up the transport and move warm air poleward in exchange for cold air. Ocean currents perform a similar function through the turning of the great gyres.

Figure 6.24   **Schematic map of system of ocean currents.**

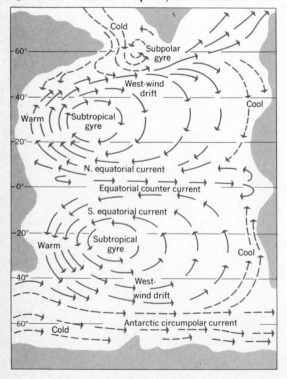

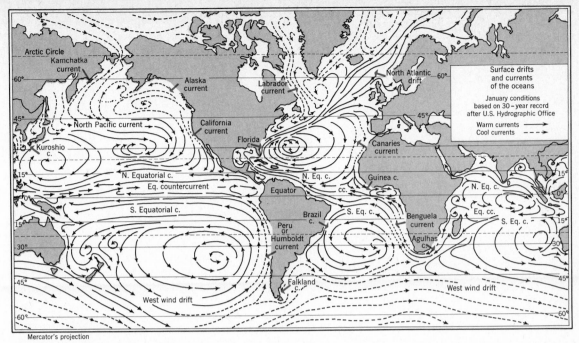

Figure 6.25   **Surface drifts and currents of the oceans in January. (U.S. Navy Oceanographic Office.)**

The global atmospheric circulation also transports heat in the latent form held by water vapor. This heat is released by condensation, a process we shall examine in the next chapter. The movement of water vapor also represents a mass transport of water and is a part of the world water balance.

Wind systems of the lower troposphere have direct environmental significance. Arriving at a mountainous coastal zone after a long travel path over a great ocean, these winds carry a large amount of water vapor, which is deposited as precipitation on the coast. In this way the distribution of our water resources is partly determined by the atmospheric circulation patterns. Winds also transport atmospheric pollutants, carrying them tens and hundreds of kilometers from the sources of pollution. These are environmental topics related to winds; we will investigate them in the next two chapters.

References for Further Study

Reiter, E. R. (1967), *Jet streams,* Doubleday and Co., New York.
Petterssen, S. (1969), *Introduction to meteorology,* third edition, McGraw-Hill Book Co., New York. See Chapters 9–11.
Weyl, P. K. (1970), *Oceanography,* John Wiley and Sons, New York. See Chapter 12.
Strahler, A. N. (1971), *The earth sciences,* second edition, Harper and Row, New York. See Chapters 15, 16.
Riehl, H. (1972), *Introduction to the atmosphere,* second edition, McGraw-Hill Book Co., New York. See Chapter 6.

Review Questions

1. What environmental role is played by atmospheric motions?
2. What is the pressure gradient force? In what direction does it act? Explain in terms of isobars and isobaric surfaces.
3. Explain the workings of the sea breeze and the land breeze.
4. With what instruments are winds measured? How is the direction of a wind designated? How can wind direction and speed in the upper air levels be determined?
5. What is the Coriolis effect? What causes this effect? In what direction does it act? How does it vary with latitude and hemisphere?
6. How are surface winds related to isobars and the pressure gradient? State Ballot's law. How does strength of the pressure gradient influence the speed of surface winds?
7. Describe the patterns of surface winds in cyclones and anticyclones in both hemispheres. What is the direction of spiraling flow in each case? Where do convergence and divergence take place?

8. Describe the principal pressure belts of the globe, giving the latitude and approximate pressures for each. Why do the pressure belts shift in latitude throughout the year? Which pressure belt intensifies in summer?
9. Why do the landmasses of North America and Asia disrupt the belted pressure pattern of the globe? What pressures occur on these land areas in winter? In the summer?
10. What pressure centers dominate the North Atlantic and North Pacific oceans during the summer months? During the winter months?
11. What conditions of winds and calms prevail in the equatorial trough? Why are steady, prevailing winds absent here? What is the intertropical convergence zone (ITC)? How is it formed?
12. Describe the trade winds. In which direction do they blow? Explain the direction of the trade winds. How were the trade winds used by mariners?

13. What conditions of winds and calms prevail in the subtropical high-pressure belt? Describe the cells of high pressure of which this belt is composed. How do they shift in latitude seasonally?

14. Describe the westerly wind belts. How do the westerlies compare with the trades for constancy of direction and strength? How do the westerlies influence transoceanic sailing and flying?

15. Describe the monsoon wind systems of Southeast Asia. What general type of weather condition is associated with the summer monsoon? With the winter monsoon?

16. What are local winds? Explain the mountain and valley winds. Where are drainage (katabatic) winds found? Give examples. What is distinctive about the Santa Ana wind?

17. What is the relationship between isobars and wind direction at high altitudes? What is the geostrophic wind?

18. In what way can winds on an imaginary nonrotating earth contribute to an understanding of the global circulation? How does this system illustrate a heat engine?

19. Describe and explain the motions of air in the Hadley cell. What causes air subsidence in the subtropical zone?

20. Describe the global circulation of the troposphere in middle and high latitudes. What are Rossby waves? Describe their evaluation. What is the significance of cutoff of upper-air lows and highs in terms of the global heat balance?

21. What causes the polar-front jet stream? How is this jet stream related to the polar front? Explain in terms of air temperatures and pressure gradients. Where does the subtropical jet stream occur? Where is the tropical easterly jet stream located? At what season does it occur?

22. What causes surface ocean currents? Show how the Coriolis effect modifies the direction of flow.

23. Sketch a hypothetical ocean modeled after the Pacific or Atlantic and indicate the general scheme of surface currents. Label fully.

24. List three warm currents and three cool currents, giving the location and direction of flow of each. What effect have these currents on air temperatures?

25. What current systems conduct very cold water into the northern Atlantic Ocean? What current prevails in the Southern Ocean, in the region surrounding the continent of Antarctica?

Atmospheric Moisture and Precipitation

7

We have stressed that heat and water are vital ingredients of the environment of the biosphere, or life layer. Plant and animal life of the lands, on which Man depends for food, require fresh water. Humans use fresh water in many ways. The only basic source of fresh water is from the atmosphere through condensation of water vapor. In this chapter we are concerned mostly with water in the vapor state in the atmosphere and the processes by which it passes into the liquid or solid state and ultimately arrives at the surface of the ocean and the lands through the process of precipitation.

Water also leaves the land and ocean surfaces by evaporation and so returns to the atmosphere. Evidently, the global pathways of movement of water form a complex network. There is a global water balance, just as there is an energy balance; the water balance deals with flow of matter and so complements the energy balance. Let us first review basic terms and processes involved in an understanding of the water balance.

Water States and Heat

Water occurs in three states: (1) **solid state,** frozen as crystalline ice; (2) **liquid state,** as water; (3) **gaseous state,** as water vapor. From the gaseous state, molecules may pass into the liquid state by **condensation;** or, if temperatures are below the freezing point, they can pass by **sublimation** directly into the solid state to form ice crystals. By **evaporation,** molecules can leave a water surface to become gas molecules in water vapor. The analogous change from ice directly into water vapor is also designated sublimation. Then, of course, water may pass from liquid to solid state by **freezing,** and from solid state to liquid state by **melting.** These changes can be represented by a triangle in which the three states of water form the corners (Figure 7.1). Arrows show the six possible changes of state.

Of prime importance in meteorology are the exchanges of heat energy accompanying changes of state.

For example, when water evaporates, sensible heat, which we can feel and measure by thermometer, passes into a hidden form held by the water vapor and known as the **latent heat of vaporization.** This change results in a drop in temperature of the remaining liquid. The cooling effect produced by evaporation of perspiration from the skin is perhaps the most obvious example. For every gram of water that is evaporated, about 600 calories of sensible heat pass into the latent form. In the reverse process of condensation, an equal amount of energy is released to become sensible heat and the temperature rises correspondingly.

Similarly, the freezing process releases heat energy in the amount of about 80 calories per gram of water, whereas melting absorbs an equal quantity of heat. This heat is referred to as the **latent heat of fusion.** When sublimation occurs, the heat absorbed by vaporization, or released by crystallization, is approximately equal to the sum of the latent heats of vaporization and fusion.

Figure 7.1 **Three states of water. Changes of state involve absorption or release of heat.**

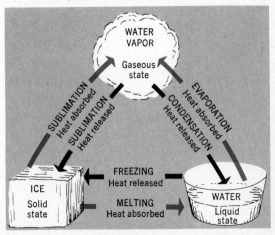

Humidity

The amount of water vapor that may be present in the air at a given time varies widely from place to place. It ranges from almost nothing in the cold, dry air of arctic regions in winter to as much as 4 or 5 percent of a given volume of the atmosphere in the humid equatorial zone.

The general term **humidity** refers to the amount of water vapor present in the air. At any specified temperature, the quantity of moisture that can be held by the air has a definite limit. This limit is the saturation quantity. The proportion of the amount of water vapor present at a given temperature relative to the maximum quantity that could be present is the **relative humidity;** it is expressed as a percentage. For **saturated air,** the relative humidity is 100 percent; when half the total possible quantity of vapor is present, relative humidity is 50 percent, and so on.

A change in relative humidity of the atmosphere can be caused in one of two ways. If an exposed water surface is present, the humidity can be increased by evaporation. This process is slow because it requires that the water vapor diffuse upward through the air. The second way is through a change of temperature. Even though no water vapor is added, a lowering of temperature results in a rise of relative humidity. This change is automatic because the capacity of the air to hold water vapor is lowered by cooling. After cooling, the existing amount of vapor represents a higher percentage of the total capacity of the air. Similarly, a rise of air temperature results in decreased relative humidity, even though no water vapor has been taken away.

The principle of relative-humidity change caused by temperature change is illustrated by a graph of these two properties throughout the day (Figure 7.2). As air temperature rises, relative humidity falls, and vice versa.

A simple example, shown in Figure 7.3, may help to illustrate these principles. At 10 A.M., the air temperature is 16°C (60°F) and the relative humidity is 50 percent. By 3 P.M., the air has become warmed by the sun to 32°C (90°F). The relative humidity has automatically dropped to 20

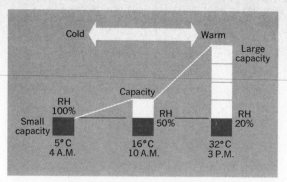

Figure 7.3 **Relative humidity changes with temperature because capacity of warm air is greater than of cold air.**

percent, which is very dry air. Next, the air becomes chilled during the night and, by 4 A.M., its temperature has fallen to 5°C (40°F). Now the relative humidity has automatically risen to 100 percent and the air is saturated. Any further cooling will cause condensation of the excess vapor into liquid or solid form. As the air temperature continues to fall, the humidity remains at 100 percent, but condensation continues and this may take the form of minute droplets of dew or fog. If the temperature falls below freezing, condensation occurs as frost on exposed surfaces.

Dew point is the critical temperature at which air becomes saturated during cooling. Below the dew point, condensation usually sets in, producing minute water droplets. An excellent illustration of condensation caused by cooling is seen in the summer, when beads of moisture form on the outside surface of a pitcher or glass filled with ice water. Air immediately adjacent to the cold glass or metal surface is chilled enough to fall below the dew-point temperature, causing moisture to condense on the surface of the glass.

How Relative Humidity is Measured

Humidity of the air can be measured in two ways. An instrument known as a **hygrometer** indicates relative humidity on a calibrated dial. One simple type uses a strand of human hair that lengthens and shortens according to the relative humidity and, in so doing, activates the dial (Figure 7.4). A continuous record of humidity can be obtained by means of a **hygrograph.** Using the same basic mechanism as the hygrometer, a continuous, automatic record is drawn by a pen on a sheet of paper attached to a rotating drum.

A different principle is applied in the **sling psychrometer.** This instrument is simply a pair of thermometers mounted side by side (Figure 7.5). One is the ordinary type; the other has a piece of wet cloth around the bulb. When the air is fully saturated (relative humidity 100 percent), there is no evaporation from the wet cloth and both thermometers read the same. When the air is not fully saturated, evaporation occurs, cooling the cloth-covered thermometer below the temperature shown on the ordinary thermometer. Because the rate of evaporation depends on dryness of the air, the difference in

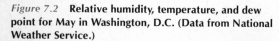

Figure 7.2 **Relative humidity, temperature, and dew point for May in Washington, D.C. (Data from National Weather Service.)**

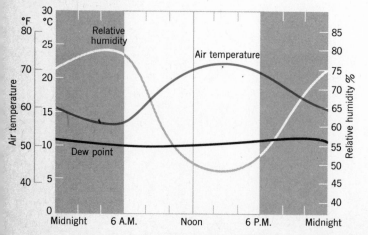

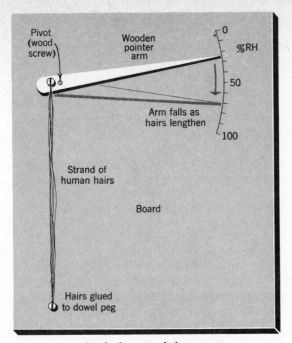

Figure 7.4   A simple, homemade hygrometer.

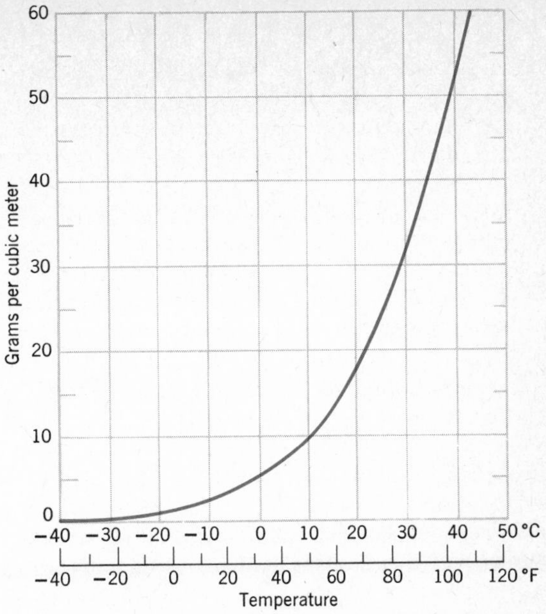

Figure 7.6   **Maximum absolute humidity for a wide range of temperatures.**

temperature shown by the two thermometers will increase as relative humidity decreases.

Standard tables are available to show the relative humidity for a given combination of wet- and dry-bulb temperatures. To be sure that maximum possible evaporation is taking place, the two thermometers are attached by a swivel joint to a handle by which the thermometers can be swung around in a circle by hand. Other types of psychrometers use a fan to blow air past the wet thermometer bulb.

Specific Humidity

Although relative humidity is an important indicator of the state of water vapor in the air, it is a statement only of the relative quantity present compared to a saturation quantity. The actual quantity of moisture present is denoted by **specific humidity,** defined as the mass of water

Figure 7.5   **A sling psychrometer uses paired thermometers. The cloth-covered wet-bulb thermometer projects beyond the dry-bulb thermometer. The handle is used to swing the thermometers in the free air. (National Weather Service.)**

vapor in grams contained in a kilogram of air. For any specified air temperature, there is a maximum mass of water vapor that a kilogram of air can hold (the saturation quantity). Figure 7.6 is a graph showing this maximum moisture content of air for a wide range of temperatures.

Specific humidity is often used to describe the moisture characteristics of a large mass of air. For example, extremely cold, dry air over arctic regions in winter may have a specific humidity of as low as 0.2 g/kg, whereas extremely warm moist air of equatorial regions often holds as much as 18 g/kg. The total natural range on a worldwide basis is such that the largest values of specific humidity are from 100 to 200 times as great as the least.

Figure 7.7 is a graph showing how relative humidity and specific humidity vary with latitude. Notice that the relative humidity curve has two saddles, one over each of the subtropical high-pressure belts where the world's tropical deserts are found. Humidity is high in equatorial and arctic zones. In contrast, the specific humidity curve has a single peak, near the equator, and declines toward high latitudes.

In a real sense, specific humidity is a geographer's yardstick of a basic natural resource — water — to be applied from equatorial to polar regions. It is a measure of the quantity of water that can be extracted from the atmosphere as precipitation. Cold air can supply only a small quantity of rain or snow; warm air is capable of supplying very large quantities.

Condensation and the Adiabatic Process

Falling rain, snow, sleet, or hail are referred to collectively as **precipitation.** Only where large masses of air are experiencing a steady drop in temperature below the dew

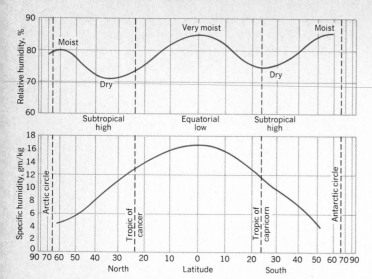

Figure 7.7 Pole-to-pole profiles of average relative humidity (above) and of specific humidity (below). (Data from Haurwitz and Austin.)

point can precipitation occur in appreciable amounts. Precipitation cannot be brought about by the simple process of chilling of the air through loss of heat by longwave radiation during the night. Precipitation requires that a large mass of air be rising to higher altitudes.

An important law of meteorology is that rising air experiences a drop in temperature, even though no heat energy is lost to the outside. The drop of temperature is a result of the decrease in air pressure at higher altitudes, permitting the rising air to expand. Because individual molecules of the gas are more widely diffused and do not move so fast, the sensible temperature of the expanding gas is lowered. This temperature drop is described as an **adiabatic process,** which simply means "occurring without any gain or loss of heat." In the adiabatic process, heat energy as well as matter remains within the system. The process is thus completely reversible. Expansion always results in cooling; compression always results in warming.

Within a rising body of air the rate of drop of temperature, termed the **dry adiabatic lapse rate,** is about 10C° per 1000 m of vertical rise. In English units the rate is 5½F° per 1000 ft. The dry rate applies only when no condensation is taking place. The dew point also declines gradually with rise of air: The rate is 2C° per 1000 m (1F° per 1000 ft).

Adiabatic cooling rate should not be confused with the environmental temperature lapse rate, explained in Chapter 3; that lapse rate applies only to nonrising air, whose temperature is measured at successively higher levels.

Lapse rates can be shown on a simple graph in which altitude is plotted on the vertical scale and temperature on the horizontal scale (Figure 7.8). The chain of circles connected by arrows represents air rising as if it were a

bubble. Suppose that a body of air near the ground has a temperature of 20°C (68°F) and that its dew point temperature is 12°C (54°F); these are the conditions shown in Figure 7.8. If, now, a bubble of air undergoes a steady ascent, its air temperature will decrease much faster than its dew-point temperature. Consequently, the two lines on the graph are rapidly converging. At an altitude of 1000 m (3300 ft), the air temperature, now 10°C (50°F), will have met the dew point temperature. The air bubble has now reached the saturation point. Further rise results in condensation of water vapor into minute liquid particles, and so a cloud is produced. The flat base of the cloud is a common visual indicator of the level of condensation.

As the bubble of saturated air continues to rise, further condensation takes place; but now a new principle comes into effect. When water vapor condenses, heat in the latent form is transformed into sensible heat, which is added to the existing heat content of the air. Recall that latent heat of vaporization amounts to about 600 calories for each gram of water. Originally, this stored energy was obtained during the process of evaporation at the time the water vapor entered the atmosphere.

As condensation continues within a rising mass of air, the latent heat liberated by that condensation partly offsets the temperature drop by adiabatic cooling. As a result, the adiabatic rate is substantially reduced. The reduced rate, which ranges between 3 and 6C° per 1000 m (2 and 3F° per 1000 ft), is termed the **wet adiabatic lapse rate.** On the graph, this reduced rate is expressed by the more steeply inclined section of line above the level of condensation. The lower range of the lapse rate (close to 3C°/1000 m) applies when condensation is very rapid and is typical of comparatively warm rain clouds at low altitudes. As air ascends to high altitudes and becomes very cold, the rate of condensation decreases greatly and the wet adiabatic lapse rate increases to high values (up to 6°C/1000 m and greater), gradually approaching the dry adiabatic lapse rate of 10°C/1000 m.

Figure 7.8 Adiabatic decrease of temperature in a rising air mass.

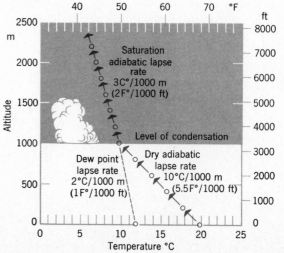

Cloud Particles

A **cloud** is a dense mass of suspended water or ice particles in the diameter range of 20 to 50 microns. Each cloud particle has formed on a **nucleus** of solid matter, which is originally on the order of 0.1 to 1.0 micron in diameter. Nuclei of condensation must be present in large numbers; they must be **hygroscopic,** that is, of such a composition as to attract water vapor molecules. In Chapter 3 we referred to these minute suspended particles collectively as atmospheric dust and noted that one source is the surface of the sea. Droplets of spray from the crests of waves are carried rapidly upward in turbulent air. Evaporation of the water leaves a solid residue of crystalline salt, which is strongly hygroscopic. You are all familiar with the way ordinary table salt becomes moist when exposed to warm humid air.

Although "clean air" is an environmental goal, the term is only relative because all air of the troposphere is charged with dust. As a result, there is no lack of suitable condensation nuclei. As we shall find in the discussion of air pollution, the heavy load of dust carried by polluted air over cities substantially aids in condensation and the formation of clouds and fog.

We are accustomed to finding that liquid water turns to ice when the surrounding temperature falls to the freezing point, 0°C (32°F), or below. Water in such minute particles as those comprising clouds remains in the liquid state at temperatures far below freezing, however. Such water is described as **supercooled.** Clouds consist entirely of water droplets at temperatures down to about −12°C (10°F). Between −12 to −30°C (10 and −20°F), the cloud is a mixture of water droplets and ice crystals. Below −30°C (−20°F), the cloud consists predominantly of ice crystals;

below −40°C (−40°F), all the cloud particles are ice crystals. Very high, thin clouds, formed at altitudes of 6 to 12 km (20,000 to 40,000 ft) are composed of ice particles.

Cloud Forms

Clouds are classified according to altitude and form (Figure 7.9). On the basis of form there are two major classes of clouds: **stratiform,** or layered clouds, and **cumuliform,** or globular clouds.

The stratiform clouds are blanketlike and cover large areas. The important point about stratiform clouds is that they represent air layers being forced to rise gradually over stable underlying air layers of greater density. As forced rise continues, the rising layer of air is adiabatically cooled, and condensation is sustained over a large area. Stratiform clouds can yield substantial amounts of rain or snow. The cumuliform clouds are globular masses representing bubblelike bodies of warmer air spontaneously rising because they are less dense than the surrounding air. Precipitation yielded by a cumuliform cloud is concentrated within a small area.

Figure 7.9 shows that four **cloud families** are recognized. The first three families are defined according to altitude range — high, middle, low. The fourth family consists of the cumuliform types, which have a vertical development and may extend through a great altitude range. Plate C.1 shows several of the cloud types.

The high cloud family includes cirrus and its related forms, cirrostratus and cirrocumulus, within the altitude range of 6 to 12 km (20,000 to 40,000 ft). They are composed of ice crystals. **Cirrus** is a delicate, wispy cloud, often forming streaks or stringers across the sky. It does not interfere with the passage of sunlight or moonlight and

Figure 7.9 **Cloud types are grouped into families according to altitude range and form. (Drawn by A. N. Strahler.)**

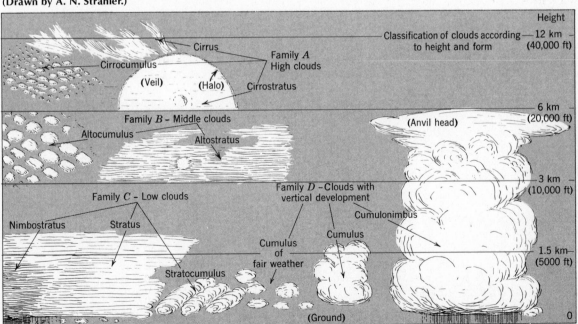

appears to the ground observer to be moving very slowly, if at all. Actually, cirrus clouds may be moving at high speed, as they often indicate the presence of the jet stream aloft. The observer on the ground can estimate direction of upper airflow by means of fibrous cirrus formations.

Cirrostratus is a more complete layer of ice cloud, producing a halo about the sun or moon. Where the layer consists of closely packed globular pieces of cloud, arranged in groups or lines, the name **cirrocumulus** is given. This is the "mackerel sky" of popular description.

The middle cloud family, in the height range 2 to 6 km (6500 to 20,000 ft), includes altostratus and altocumulus. **Altostratus** is a blanket layer, often smoothly distributed over the entire sky. It is grayish in appearance, usually has a smooth underside, and will often show the sun as a bright spot in the cloud. **Altocumulus** is a layer of individual cloud masses, fitted closely together in geometric pattern. The masses appear white, or somewhat gray on the shaded sides, and blue sky can be seen between individual patches or rows. Altostratus is commonly associated with the approach of bad weather, whereas altocumulus is characteristic of generally fair conditions.

In the low cloud family, from ground level to 2 km (6500 ft), are stratus, nimbostratus, and stratocumulus. **Stratus** is a dense, low-lying, dark-gray layer. When rain or snow is falling from this cloud, it is termed **nimbostratus;** the prefix *nimbo* is from the Latin word *nimbus,* for rainstorm. **Stratocumulus** is a low-lying cloud layer consisting of distinct grayish masses of cloud between which is open sky. The individual masses often take on the form of long rolls of cloud, oriented at right angles to the direction of wind and cloud motion. Stratocumulus is generally associated with fair or clearing weather, but sometimes rain or snow flurries issue from individual cloud masses.

Clouds of vertical development — the cumuliform clouds — tend to display a height as great as, or greater than, their horizontal dimensions. **Cumulus** is a white, woolpack cloud mass. Small cumulus clouds are

Figure 7.10 **An isolated cumulonimbus cloud with rain falling from the base. (U.S. Navy.)**

associated with fair weather. Enlarged cumulus clouds (congested cumulus) show a flat base and a bumpy upper surface somewhat resembling a head of cauliflower. These clouds look pure white on the side illuminated by the sun, but may be gray or black on the shaded or underneath side.

Under different conditions, explained later in this chapter, individual cumulus masses grow into **cumulonimbus,** the thunderstorm cloud. This massive, towering cloud yields heavy rainfall, thunder and lightning, and gusty winds (Figure 7.10). A large cumulonimbus cloud may extend from a height of 500 m (1600 ft) at the base up to 9 to 12 km (30,000 or 40,000 ft) at the top. When seen from a great distance, the top of the cumulonimbus cloud is pure white; but to observers beneath, the sky may be darkened to almost nighttime blackness (Plate C.2). Processes occurring within the cumulonimbus cloud will be explained in connection with thunderstorms.

Fog

Fog is simply a cloud layer in contact with the land or sea surface or lying very close to the surface. Fog is a major environmental hazard to Man in an industrialized world. Dense fog on high-speed highways is a cause of terrifying chain-reaction accidents, sometimes involving dozens of vehicles and often taking a heavy toll in injuries and deaths. Landing delays and shutdowns at major airports caused by fog bring economic losses to airlines and inconvenience to thousands of travelers in a single day. For centuries fog at sea has been a navigational hazard. Today, with huge supertankers carrying oil, fog adds to the probability of ship collisions capable of creating enormous oil spills. Polluted fogs are a health hazard to urban dwellers and have at times taken a heavy toll in lives.

One type of fog, known as a **radiation fog,** is formed at night when temperature of the stagnant basal air falls below the dew point. This kind of fog is associated with a low-level temperature inversion (Figure 5.8). Another fog type, **advection fog,** can result from the movement of warm, moist air over a cold or snow-covered ground surface. Losing heat to the ground, the air layer undergoes a drop of temperature below the dew point and condensation sets in. A similar type of advection fog is formed over oceans where air from over a warm current blows across the cold surface of an adjacent cold current. Fogs of the Grand Banks off Newfoundland are largely of this origin because here the cold Labrador current comes in contact with warm water of Gulf Stream origin.

Precipitation Forms

During the rapid ascent of a mass of air in the saturated state, cloud particles grow rapidly and attain a diameter of 50 to 100 microns. They then coalesce through collisions and grow quickly into droplets of about 500 microns diameter (about 1/50 in.). Droplets of this size reaching the ground constitute a drizzle, one of the recognized forms of precipitation. Further coalescence increases drop size and yields **rain.** Average raindrops have diameters of about 1000 to 2000 microns (1 to 2 mm; 1/25 to 1/10 in.), but they can reach a maximum diameter of about 7 mm (1/4 in.). Above this value they become unstable and break into smaller drops while falling. One kind of rain forms directly by liquid condensation and droplet coalescence in warm clouds of the equatorial and tropical zones. Rain of middle and high latitudes, however, is largely the product of the melting of snow as it makes its way to lower, warmer levels.

Snow is produced in clouds that are a mixture of ice crystals and supercooled water droplets. The falling crystals serve as nuclei to intercept water droplets. As these adhere, the water film freezes and is added to the crystalline structure (Figure 7.11). The crystals clot together readily to form larger snowflakes, and these fall more rapidly from the cloud. When the underlying air layer is below the freezing temperature, snow reaches the ground as a solid form of precipitation; otherwise, it will melt and arrive as rain. A reverse process, the fall of raindrops through a cold air layer, results in the freezing of rain and produces pellets or grains of ice. These are commonly referred to in North America as **sleet** but, among the British, sleet refers to a mixture of snow and rain.

Hail, another form of precipitation, consists of large pellets or spheres of ice. The formation of hail will be explained in our discussion of the thunderstorm.

When rain falls on a frozen ground surface that is covered by an air layer of below-freezing temperature, the water freezes into clear ice after striking the ground or other surfaces such as trees, houses, or wires (Figure 7.12). The coating of ice that results is called a **glaze,** and an **ice storm** is said to have occurred. Actually, no ice falls, so that ice glaze is not a form of precipitation. Ice storms cause great damage, especially to telephone and power wires and to tree limbs. Roads and sidewalks are made extremely hazardous.

How Precipitation Is Measured

Precipitation is measured in units of depth of fall per unit of time: for example, centimeters or inches per hour or per day. One cm of rainfall is a quantity sufficient to cover the ground to a depth of 1 cm, provided that none is lost by runoff, evaporation, or sinking into the ground. A simple form of rain gauge can be operated merely by setting out a straight-sided, flat-bottomed pan and measuring the depth to which water accumulates during a particular period. Unless this period is short, however, evaporation seriously upsets the results.

A very small amount of rainfall, such as 2 mm (0.1 in.), makes too thin a layer to be accurately measured. To avoid this difficulty as well as to reduce evaporation loss, the **rain gauge** is made in the form of a cylinder whose base is a funnel leading into a narrow tube (Figure 7.13). A small amount of rainfall will fill the narrow pipe to a considerable height, thus making it easy to read accurately once a simple scale has been provided for the pipe. This gauge requires frequent emptying unless it is equipped with automatic devices for this purpose.

Figure 7.11 These individual snow crystals, greatly magnified, were selected for their variety and beauty. They were photograhed by W. A. Bentley, a Vermont farmer who devoted his life to snowflake photography. (National Weather Service.)

Snowfall is measured by melting a sample column of snow and reducing it to an equivalent in water. In this way rainfall and snowfall records may be combined for purposes of comparison. Ordinarily, a 10-cm layer of snow is assumed to be equivalent to 1 cm of rainfall, but this ratio may range from 30 to 1 in very loose snow to 2 to 1 in old, partly melted snow.

Figure 7.12 Heavily coated wires and branches caused heavy damage in eastern New York State in January 1943 as a result of this ice storm. (National Weather Service and New York Power and Light Co., Albany, N.Y.)

Figure 7.13 **Standard 20-cm (8-in.) rain gauge used by the National Weather Service (left). A plastic 10-cm (4 in.) gauge showing funnel and graduated tube inside (right). (Science Associates, Inc., Princeton, N.J.)**

How Precipitation Is Produced

Precipitation in substantial quantities is induced by two basic types of mechanisms. One is the spontaneous rise of moist air; the other is the forced rise of moist air.

Spontaneous rise of moist air is associated with **convection,** a form of atmospheric motion consisting of strong updrafts taking place within a **convection cell.** Air rises in the cell because it is less dense than the surrounding air. Perhaps a fair analogy is the updraft of heated air in a chimney but, unlike the steady airflow in a chimney, air rise in a convection cell takes place in pulses as bubblelike masses of air ascend in succession.

To illustrate the convection process, suppose that on a clear, warm summer morning the sun is shining on a landscape consisting of patches of open fields and woodlands. Certain of these types of surfaces, such as the bare ground, heat more rapidly and transmit heat by longwave radiation to the overlying air. Air over a warmer patch becomes warmed more than adjacent air and begins to rise as a bubble, much as a hot-air balloon rises after being released. Vertical movements of this type are often called "thermals" by sailplane pilots, who use them to obtain lift.

As the air rises, it is cooled adiabatically so that eventually it is cooled below the dew point. Condensation begins at once, and the rising air column appears as a cumulus cloud. The flat base shows the critical level above which condensation is occurring (Figure 7.14). The bulging "cauliflower" top of the cloud represents the top of the rising warm air column, pushing into higher levels of the atmosphere. When the bubble of air is sufficiently cooled by the adiabatic process, it ceases to rise and condensation stops. Then the small cumulus cloud

dissolves after drifting some distance downwind. Under a different set of atmospheric conditions, convection continues to develop unchecked and the cloud grows into a dense cumulonimbus mass, or thunderstorm, yielding heavy rain.

Why does spontaneous cloud growth sometimes take place and continue beyond the initial cumulus stage, long after the original input of heat energy is gone? Actually, the unequal heating of the lower air layer served only as a trigger effect to release a spontaneous updraft, fed by latent heat energy liberated from the condensing water vapor. Recall that for every gram of water formed by condensation, 600 calories of heat are released. This heat acts like fuel in a bonfire.

Air capable of rising spontaneously during condensation is described as **unstable air.** In such air the updraft tends to increase in intensity as time goes on, much as a bonfire blazes with increasing ferocity as the updraft draws in greater supplies of oxygen. Of course, at very high altitudes, the bulk of the water vapor has condensed and fallen as precipitation, so the energy source is gone. When this happens the convection cell weakens and air rise finally ceases.

Unstable air, given to spontaneous convection in the form of heavy showers and thunderstorms, is most likely to be found in warm, humid areas, such as the equatorial and tropical zones throughout the year and the midlatitude regions during the summer season.

Figure 7.15 will help explain why spontaneous rise of air to produce intense convection can take place only when air properties are favorable. Diagram *A* is a plot of altitude against air temperature. The small circles represent a small parcel of air being forced to rise steadily higher, following the dry adiabatic rate of cooling shown. To the right of this line is a solid line showing the temperature of the undisturbed surrounding air; it decreases upward at the normal environmental lapse rate.

Suppose that the air parcel is forcibly lifted from the ground, where its temperature is 30°C. After the air parcel has been carried up 500 m, its temperature has fallen about 5C° and is now 25°C, whereas the surrounding air (environment) is warmer by 2C° and has a temperature of 27°C. The air parcel would thus be cooler than the environment at 500 m and denser than the surrounding air. If no longer forcibly carried upward, it would tend to

Figure 7.14 **Rise of a bubble of heated air to form a cumulus cloud. (From A. N. Strahler, 1971, *The Earth Sciences,* 2nd ed., Harper and Row, New York.)**

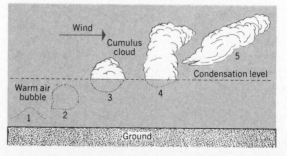

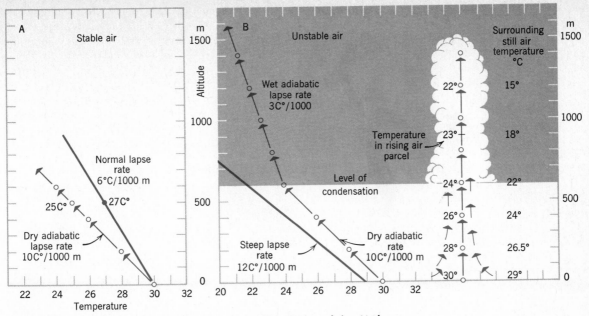

Figure 7.15 Under stable air conditions (A), air would resist forced rise. Under unstable conditions (B), rise is spontaneous.

sink back toward the ground. These conditions represent **stable air,** not able to produce convection cells because the air would resist lifting.

When the lower air layer is excessively heated, the environmental lapse rate is increased and the air becomes unstable. Diagram B of Figure 7.15 shows a steep lapse rate of 12C°/1000 m. (As the graph is constructed, the steeper lapse rate is represented by a line of lower inclination.) An air parcel at ground level is shown to be 1C° warmer than the surrounding air. It begins to rise spontaneously because it is less dense than air over adjacent, less intensely heated ground areas. Although cooled adiabatically while rising, the air parcel at 600 m has a temperature of 24°C, but this is well above the temperature of the surrounding still air.

The air parcel, therefore, is always lighter than the surrounding air and continues its rise. At 600 m, the dew point is reached and condensation sets in. Now the rising air parcel is cooled at a reduced wet adiabatic rate, shown as 3C°/1000 m, because the latent heat liberated in condensation offsets the rate of drop due to expansion. At 1000 m, the rising air parcel is still several degrees warmer than the environment and therefore continues its spontaneous rise. Generally, steep lapse rates in air holding a large quantity of water vapor are associated with instability.

Thunderstorms

Convection manifests itself in the **thunderstorm,** an intense local storm associated with a tall, dense cumulonimbus cloud in which there are very strong updrafts of air. Thunder and lightning normally accompany the storm, and rainfall is heavy, often of cloudburst intensity, for a short period. Violent surface winds may occur at the onset of the storm.

A single thunderstorm consists of individual cells. Air rises within each cell as a succession of bubblelike air bodies (Figure 7.16). As each bubble rises, air in its wake is brought in from the surrounding region, a process called **entrainment.** Rising air in the thunderstorm cell can reach vertical speeds up to 60 km (40 mi) per hour. Precipitation is in the form of rain in the lower levels, mixed water and snow at intermediate levels, and snow at high levels.

Upon reaching high levels, which may be 6 to 12 km (20,000 to 40,000 ft) or even higher, the rising rate diminishes and the cloud top is dragged downwind to form an **anvil top.** Ice particles falling from the cloud top

Figure 7.16 Schematic diagram of interior of a thunderstorm cell.

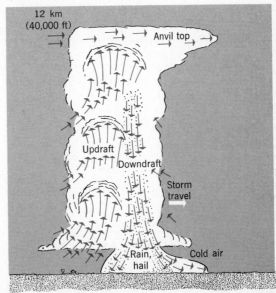

act as nuclei for condensation at lower levels, a process called **cloud seeding.** The rapid fall of raindrops adjacent to the rising air bubbles exerts a frictional drag on the air and sets in motion a downdraft. Striking the ground where precipitation is heaviest, this downdraft forms a local squall wind, which is sometimes strong enough to fell trees and do severe structural damage to buildings.

Recent years have seen a sharp increase in Man's efforts to modify weather phenomena. One goal is to increase precipitation in areas experiencing drought; another is to lessen the severity of storms. One method of inducing convectional precipitation is by artificial cloud seeding, the introduction of minute particles into dense cumulus clouds. The particles, which may be of silver iodide smoke, serve as nuclei for added intensity of condensation and can increase the size of cumulonimbus clouds to generate heavier rainfall. Some measure of success by cloud seeding was obtained in 1970 and 1971 over drought-ridden southern and central Florida.

Figure 7.18　**A severe hailstorm devastated this corn crop. (NCAR photograph.)**

Hail and Lightning — Environmental Hazards

In addition to powerful wind gusts and torrential falls of rain, important environmental hazards connected with the thunderstorm are hail, lightning, and tornadoes. Hailstones are formed by the accumulation of ice layers on ice pellets suspended in the strong thunderstorm updrafts. The phenomenon is much like that of the icing of aircraft flying through a cloud of supercooled water droplets. After the hailstones have grown to diameters that often reach 3 to 5 cm (1 to 2 in.), they escape from the updraft and fall to the ground (Figure 7.17).

An important aspect of planned weather modification is that of reducing the severity of hailstorms. Annual losses from crop destruction by hailstorms are estimated to approach $300 million (Figure 7.18). Damage to wheat crops is particularly severe in a north-south belt of the High Plains, running through Nebraska, Kansas, and Oklahoma. A much larger region of somewhat less hailstorm frequency extends eastward generally from the Rockies to the Ohio Valley. Corn is a major crop over much of this area. Scientists are studying the isolated cumulonimbus clouds that produce hail. Research will

lead to the development of cloud-seeding techniques by means of which the severity of hailstorms can be reduced.

Another effect of convection cell activity is to generate lightning, one of the environmental hazards that results annually in the death of many persons and livestock and in the setting of many forest and building fires. Lightning is a great electric arc — a gigantic spark — passing between cloud and ground, or between parts of a cloud mass. During lightning discharge, a current of as much as 60,000 to 100,000 amperes may develop. Rapid heating and expansion of the air in the path of the lightning stroke sends out intense sound waves, which we recognize as a thunderclap. In the United States, lightning causes a yearly average of about 150 human deaths and property damage of about $100 million, including loss by fires set by lightning.

Orographic Precipitation

Forced ascent of large masses of air occurs under two quite different sets of controlling conditions. When prevailing winds encounter a mountain range, the air layer as a whole is forced to rise to surmount the barrier. Precipitation produced in this way is described as **orographic precipitation,** meaning "related to mountains." A layer of cold air may act in much the same way as a mountain barrier. Warm air in motion often encounters a cold air layer. The cold air is denser than the warm and will remain close to the ground, acting as a barrier to the progress of the warm air. The warm air is then forced to rise over the barrier. In a related type of mechanism, the cold air layer is in motion and forces the warm air to rise over it. Precipitation resulting from these activities is explained in Chapter 8.

Figure 7.19 shows the steps associated with the production of orographic precipitation. Moist air arrives at the coast after passing over a large ocean surface. As the air rises on the windward side of the range, it is cooled at

Figure 7.17　**These hailstones, larger than hens' eggs (arrow), fell at Girard, Illinois. (National Weather Service.)**

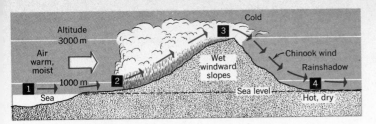

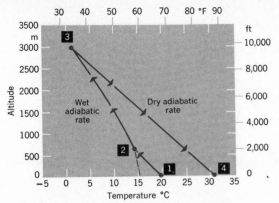

Figure 7.19 **Forced ascent of oceanic air masses produces precipitation and a rainshadow desert.** (From A. N. Strahler, 1971, *The Earth Sciences,* 2nd ed., Harper and Row, New York.)

great desert zone, covering a strip of eastern California and all of Nevada.

Much orographic rainfall at low latitudes is actually of the convectional type in that it takes the form of heavy showers and thunderstorms. The convectional storms are set off by the forced ascent of unstable air as it passes over the mountain barrier. The torrential monsoon rains of the Asiatic and East Indian mountain ranges are largely of this type. For example, Cherrapunji, a hill station facing the summer monsoon air drift in northeast India, averages 1082 cm (426 in.) of rainfall annually.

Latent Heat and the Global Balances of Energy and Water

An understanding of the processes of condensation and precipitation allows us now to take another look at the global energy balance and to include the mechanism of latent heat transport. Figure 7.21 is a schematic diagram to show how evaporation and condensation are involved in

the adiabatic rate. When cooling reaches the dew point, precipitation sets in. After passing over the mountain summit, the air begins to descend the lee side of the range. Now it undergoes compressional warming through the same adiabatic process and, having no source from which to draw up moisture, becomes very dry. Upon reaching sea level, it is much warmer than at the start. A belt of dry climate, often called a **rainshadow desert,** exists on the lee side of the range. Several of the important deserts of the earth are of this type.

Dry, warm **chinook winds** often occur on the lee side of a mountain range in the western United States. These winds cause extremely rapid evaporation of snow or soil moisture. Chinook winds result from the turbulent mixing of lower and upper air in the lee of the range. The upper air, which has little moisture to begin with, is greatly dried and heated when swept down to low levels. A similar type of wind, the **foehn** (föhn) occurs on the north side of the Austrian Alps.

An excellent illustration of orographic precipitation and rainshadow occurs in the far west of the United States. A map of California, Figure 7.20, shows mean annual precipitation using lines of equal precipitation, called **isohyets.** Prevailing westerly winds bring moist air from the Pacific Ocean over the Coast Ranges of central and northern California and the great Sierra Nevada, whose summits rise to 4000 m (14,000 ft) above sea level. Heavy rainfall on the windward slopes of these ranges nourishes rich forests. Passing down the steep eastern face of the Sierra Nevada, air must descend nearly to sea level, even below sea level in Death Valley. The resulting adiabatic heating lowers the humidity, producing part of America's

Figure 7.20 **The effect of mountain topography on rainfall is shown well by the state of California. Isohyets in centimeters.**

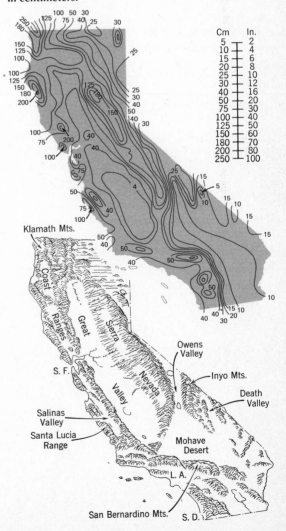

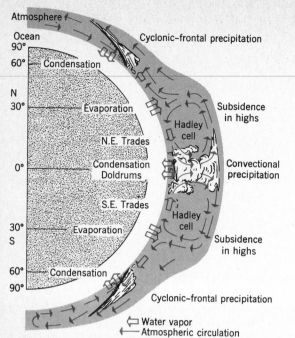

Figure 7.21 Schematic diagram of exchange of water vapor along a meridional profile.

energy exchange in each of the major latitude zones. We can now also summarize the pattern of transport of water vapor across the parallels of latitude; this is a transport of matter and is part of the global water balance. Figure 7.22 shows the average annual rate of vapor transport, in units of 10^{15} kg of water per year. Let us review the energy

Figure 7.22 Graph of mean annual meridional transport of water vapor. (Simplified from data from W. D. Sellers, 1965, *Physical Climatology,* Univ. of Chicago Press, p. 94, Figure 29.)

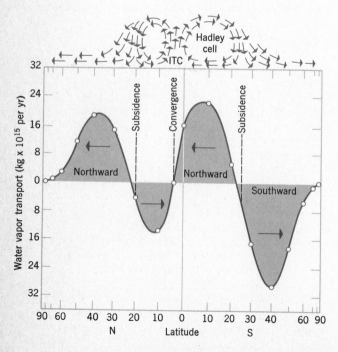

exchanges and water transports for each latitude zone.

The equatorial zone is characterized by a rise of moist, warm air in innumerable convectional cells reaching to the upper limits of the troposphere. As condensation and precipitation occur here, enormous amounts of latent heat energy are liberated. This zone has been called the "fire box" of the globe, in recognition of this intense production of sensible heat by condensation. Moving equatorward to replace the rising air are the tropical easterlies, or trades, within the Hadley cell circulation. Water evaporates from the ocean surface over the subtropical highs and is carried equatorward in vapor form. This transport shows in Figure 7.22 by peaks in the transport curve at about 10°N and 10°S lat. (Where the line of the graph lies above the center line, the movement is northward; where below, movement is southward.)

Next we move into midlatitudes, where upper-air waves constantly form and dissolve in the westerlies. Here, by advection, cyclones and anticyclones exchange cold, polar air for warm, tropical air across the parallels of latitude. Water vapor is also transported poleward and reaches a peak rate of flow at about 40°N and 40°S lat., as Figure 7.22 shows. Condensation in cyclonic storms removes water vapor from the troposphere in increasing quantities into high latitudes, so that the movement of water vapor declines and reaches zero at the poles.

Air Pollution

We can made good use of an understanding of the processes of condensation and precipitation when probing into Man-made changes in the atmospheric environment. These are inadvertent changes, largely the result of urbanization and the growth of industrial processes, which have enormously increased the rate of combustion of hydrocarbon (fossil) fuels in the past half century or so. We have already considered the thermal effects of Man's injecting carbon dioxide into the atmosphere. Now we turn to consider the kinds of foreign matter, or **pollutants,** injected by Man into the lower atmosphere, and the effects of that pollution on air quality and urban climates.

We recognize two classes of atmospheric pollutants. First, there are solid and liquid particles, which are designated collectively as **particulate matter.** Dusts found in the smoke of combustion, as well as droplets naturally occurring as cloud and fog, fall into the category of particles. Second, there are compounds in the gaseous state, included under the general term **chemical pollutants** in that they are not normally present in measurable quantities in clean air remote from densely populated, industrialized regions. (Excess carbon dioxide produced by combustion is not usually classed as a pollutant.)

One group of chemical pollutants of industrial and urban areas is of primary origin, for example, produced directly from a source on the ground. Gases included in this group are carbon monoxide (CO), sulfur dioxide (SO_2), oxides of nitrogen (NO, NO_2, NO_3), and hydrocarbon compounds. These chemical pollutants cannot be treated separately from particulate matter, however, because they are often combined within a single suspended particle. We have already seen that certain dusts, particularly the sea salts, are hygroscopic and easily take on a covering water film. The water film in turn absorbs the chemical pollutants.

When particles and chemical pollutants are present in considerable density over an urban area, the resultant mixture is known as **smog.** Almost everyone living in large cities is familiar with smog because of its irritating effects on the eyes and respiratory system and its ability to obscure distant objects. When concentrations of suspended matter are less dense, obscuring visibility of very distant objects but not otherwise objectionable, the atmospheric condition is referred to as **haze.** Atmospheric haze builds up quite naturally in stagnant air masses as a result of the infusion of various surface materials. Haze is normally present whenever air reaches high relative humidity because water films grow on suspended hygroscopic nuclei. Nuclei of natural atmospheric haze particles consist of mineral dusts from the soil, crystals of salt blown from the sea surface, hydrocarbon compounds (pollens and terpenes) exuded by plant foliage, and smoke from forest and grass fires. Dusts from volcanoes may, on occasion, add to atmospheric haze.

It is evident at this point that what we are calling atmospheric pollutants are of both natural and Man-made origin and that Man's activities can supplement the quantities of natural pollutants present. Table 7.1 illustrates this complexity by listing the primary pollutants according to sources.

Not all Man-made pollution comes from the cities. Isolated industrial activities can produce pollutants far from urban areas. Particularly important are smelters and manufacturing plants in small towns and rural areas. Sulfide ores (metals in combination with sulfur compounds) are processed by heating in smelters close to the mine. There sulfur compounds are sent into the air in enormous concentrations from smokestacks. Fallout over the surrounding area is destructive to vegetation.

Mining and quarrying operations send mineral dusts into the air. For example, asbestos mines (together with asbestos processing and manufacturing plants) send into the air countless threadlike mineral particles, some of which are so small they can be seen only with the electron microscope. These particles travel widely and are inhaled by humans, lodging permanently in the lung tissue. Nuclear test explosions inject into the atmosphere a wide range of particles, including many radioactive substances capable of traveling thousands of miles in the atmospheric circulation.

Man-induced forest and grass fires add greatly to smoke palls in certain seasons of the year. Plowing, grazing, and vehicular traffic raise large amounts of mineral dusts from dry soil surfaces. Bacteria and viruses, which we have not as yet mentioned, are borne aloft in air turbulence when winds blow over contaminated surfaces such as farmlands, grazing lands, city streets, and waste disposal sites.

Figure 7.23 shows major atmospheric pollutants in percentage form as to both component matter and source in recent years. Much of the carbon monoxide, half the hydrocarbons, and about a third of the oxides of nitrogen

Table 7.1 / Sources of Primary Atmospheric Pollutants

Pollutants from Natural Sources	*Sources of Man-made Pollutants*
Volcanic dusts	Fuel combustion (CO_2, SO_2, lead)
Sea salts from breaking waves	Chemical processes
Pollens and terpenes from palnts	Nuclear fusion and fission
(Aggravated by Man's activities:)	Smelting and refining of ores
Smoke of forest and grass fires	Mining, quarrying
Blowing dust	Farming
Bacteria, viruses	

Data source: Association of American Geographers (1968), *Air Pollution*, Commission on College Geography, Resource Paper No. 2, Figure 3, p. 9.

come from exhausts of gasoline and diesel engines in vehicular traffic. Generation of electricity and various industrial processes contribute most of the sulfur oxides because the coal and lower-grade fuel oil used for these purposes are comparatively rich in sulfur. These same sources also supply most of the particulate matter. Fly ash consists of the coarser grades of soot particles emitted from smokestacks of generating plants. These particles settle out quite quickly within close range of the source. Combustion used for heating buildings is a comparatively minor contributor to pollution because the higher grades of fuel oil are low in sulfur and are usually burned efficiently. Finely divided carbon comprises much of the smoke of combustion and is capable of remaining in suspension almost indefinitely because of its colloidal size. Forest fires comprise a secondary contributor of particles. The burning of refuse is a minor contributor in all categories of pollutants.

In the smog of cities there are, in addition to the ingredients already mentioned, certain chemical elements contained in particles contributed by automobile and truck exhausts. Included are particles that contain lead, chlorine, and bromine.

Primary pollutants are conducted upward from the emission sources by rising air currents that are part of the normal convectional process. The larger particles settle under gravity and return to the surface as **fallout.** Particles too small to settle out are later swept down to earth by precipitation, a process called **washout.** By a combination of fallout and washout, the atmosphere tends to be cleaned of pollutants. In the long run a balance is achieved between input and output of pollutants, but there are large fluctuations in the quantities stored in the air at a given time. Pollutants are also eliminated from the air over their source areas by winds that disperse the particles into large volumes of cleaner air in the downwind direction. Strong winds can quickly sweep away most pollutants from an urban area but, during periods when a stagnant anticyclone is present, the concentrations rise to high values.

In polluted air certain chemical reactions take place among the components injected into the atmosphere, generating a secondary group of pollutants. For example,

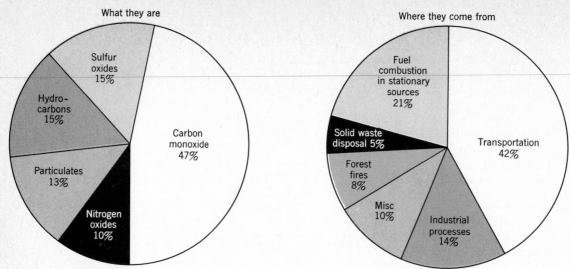

What they are

Sulfur oxides 15%

Hydro-carbons 15%

Carbon monoxide 47%

Particulates 13%

Nitrogen oxides 10%

Where they come from

Fuel combustion in stationary sources 21%

Solid waste disposal 5%

Forest fires 8%

Misc 10%

Transportation 42%

Industrial processes 14%

Figure 7.23 **Air pollution emissions in the United States in percentage by weight. (Data from National Air Pollution Control Administration, U. S. Department of Health, Education, and Welfare.)**

sulfur dioxide (SO_2) may combine with oxygen to produce sulfur trioxide (SO_3), which in turn reacts with water of suspended droplets to yield sulfuric acid (H_2SO_4). This acid is both irritating to organic tissues and corrosive to many inorganic materials. In other typical reactions, the action of sunlight on nitrogen oxides and organic compounds produces ozone (O_3), a toxic and destructive gas. Reactions brought about by the presence of sunlight are described as **photochemical reactions.** One toxic product of photochemical action is ethylene, produced from hydrocarbon compounds.

Low-Level Inversions

The concentration of pollutants over a source area rises to its highest levels when the vertical mixing (convection) of the air is inhibited by a stable configuration of the vertical temperature profile of the air. The principles of stable and unstable air conditions were covered earlier in this chapter, and we shall now apply those principles to the problem of air pollution over cities.

When the normal environmental lapse rate of 6 C°/1000 m (3.5 F°/1000 ft) is present, there is resistance to mixing by vertical movements, as we have already discussed (Figure 7.24, left). Consider next that the environmental lapse rate is steepened by heating of an air layer near the ground because of excess heat radiated and conducted from hot pavements and rooftops (Figure 7.24, right). When the temperature gradient (lapse rate) of the heated air becomes greater than the dry adiabatic rate of 10 C°/1000 m (5.5 F°/1000 ft), a condition of instability exists and a bubble of warm air can begin to rise, like a helium-filled balloon. Assume that the lapse rate lessens with increased altitude, as shown by the curved solid line in Figure 7.24. Cooled at the dry adiabatic rate, the temperature of the rising bubble then falls faster than does the temperature of the surrounding air. When the bubble has reached an altitude at which its temperature (and

therefore also its density) matches that of the surrounding air, it can rise no further and convection ceases.

Now suppose that instead of the bubble of warm air we substitute the hot air from a smokestack (Figure 7.24). The rise of heated air follows essentially the same pattern, although initially faster and in the form of a vertical jet. Carrying up with it the pollutants of combustion, the rising hot air gradually cools and reaches a level of stability, where it spreads laterally. Cooling by longwave radiation and mixing with the surrounding air will reinforce the adiabatic cooling because a truly adiabatic system would not be realistic in nature.

Recall that at night, when the air is calm and the sky is clear, rapid cooling of the ground surface typically produces a low-level temperature inversion, illustrated in Figure 5.8. In cold air, the reversal of the temperature gradient may extend hundreds of meters into the air. A low-level temperature inversion represents an unusually stable air structure. When this type of inversion develops over an urban area, conditions are particularly favorable for the entrapment of pollutants to the degree that heavy smog or highly toxic fog can develop, as shown in Figure 7.25. The upper limit of the inversion layer coincides with

Figure 7.24 **Relation of dry adiabatic lapse rate (circles) to various environmental lapse rates.**

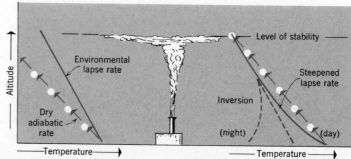

Altitude

Environmental lapse rate

Level of stability

Dry adiabatic rate

Inversion

Steepened lapse rate

(night)

(day)

Temperature

Temperature

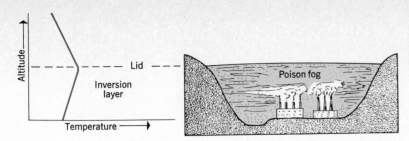

Figure 7.25 **A low-level inversion (left) was the predisposing condition for poison fog accumulation at Donora, Pennsylvania, in October 1948. (From A. N. Strahler, 1972, *Planet Earth,* Harper and Row, New York.)**

the cap, or **inversion lid,** below which pollutants are held. The lid may be situated at a height of perhaps 150 to 300 m (500 to 1000 ft) above the ground.

Although situations dangerous to health during prolonged low-level inversion have occurred a number of times over European cities since the Industrial Revolution began, the first major tragedy of this kind in the United States occurred at Donora, Pennsylvania, in late October 1948. The city occupies a valley floor hemmed in by steeply rising valley walls which prevent the free mixing of the lower air layer with that of the surrounding region (Figure 7.25). Industrial smoke and gases from factories poured into the inversion layer for five days, increasing the pollution level. Because humidity was high, a poisonous fog formed and began to take its toll. In all, 20 persons died and several thousand persons were made ill before a change in weather pattern dispersed the smog layer.

Upper-Level Inversions

Related to the low-level inversion, but caused in a somewhat different manner, is the **upper-level inversion,** illustrated in Figure 7.26. Recall that anticyclones are cells of subsiding air that diverges at low levels. Within the center of the cell, winds are calm or very gentle. As the air subsides, it is adiabatically warmed, so the normal temperature lapse rate is displaced to the right in the temperature-altitude graph, as shown in Figure 7.26 by the diagonal arrows. Below the level at which subsidence is occurring, the air layer remains stagnant. The temperature curve consequently develops a kink in which a part of the curve shows an inversion. The layer of inverted temperature structure strongly resists mixing and acts as a lid to prevent the continued upward movement and dispersal of pollutants. Upper-level inversions develop occasionally over various parts of the United States when an anticyclone stagnates for several days at a stretch.

For the Los Angeles Basin of southern California, and to a lesser degree the San Francisco Bay Area and over other west coasts in these latitudes generally, special climatic conditions produce prolonged upper-air inversions favorable to persistent smog accumulation. The Los Angeles Basin is a low, sloping plain, lying between the Pacific Ocean and a massive mountain barrier on the north side. Cool air is carried inland over the basin on weak winds from the south and southwest, but cannot move farther inland because of the mountain barrier. It is a characteristic of these latitudes that the air on the eastern sides of the subtropical high pressure cells is continually subsiding, creating a more or less permanent upper-air

inversion that dominates the western coasts of the continents and extends far out to sea (Figure 7.27). The effect is particularly marked in the summer season, when the Azores and Bermuda highs are at their largest and strongest. The subsiding air over the Los Angeles Basin is warmed adiabatically as well as heated by direct absorption of solar radiation during the day, so it is markedly warmer than the cool stagnant air below the inversion lid, which lies at an altitude of about 600 m (2000 ft). Pollutants accumulate in the cool air layer, producing the characteristic smog that first became noticeable for its irritating qualities in the early 1940s. Because water vapor content is generally low, Los Angeles smog is better described as a dense haze than as a fog, for it does not shut out the sun or cause a reduction in visibility to a degree that interferes with vehicular operation or the landing and takeoff of aircraft. Close to the coast, true fogs are frequent; but this condition is normally prevalent on dry west coasts adjacent to cold currents. The upper limit of the smog stands out sharply in

Figure 7.26 **An upper-air inversion caused by subsidence. (From A. N. Strahler, 1972, *Planet Earth,* Harper and Row, New York.)**

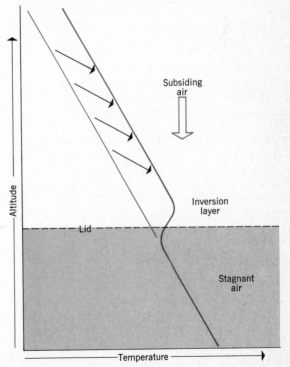

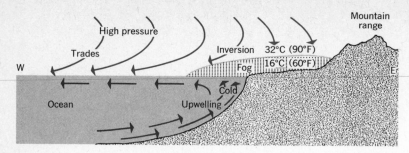

Figure 7.27 **Subsiding air over a western continental coast produces a persistent upper-air temperature inversion, trapping cool air and fog in a surface layer near the shore. (From A. N. Strahler, 1971,** *The Earth Sciences,* **2nd ed., Harper and Row, New York.)**

contrast to the clear air above it, filling the basin like a lake and extending into valleys in the bordering mountains (Plate C.2).

Effects of Urban Air Pollution

One important physical effect of urban air pollution is that it reduces visibility and illumination. A smog layer can cut illumination by 10 percent in summer and 20 percent in winter. Ultraviolet radiation is absorbed by smog, which at times completely prevents these wavelengths from reaching the ground. Reduced ultraviolet radiation may prove to be of importance in permitting increased bacterial activity at ground level. City smog cuts horizontal visibility to some one-fifth to one-tenth of the distance normal for clean air. Where atmospheric moisture is sufficient, the hygroscopic particles acquire water films and can cause formation of true fog with near-zero visiblity. Over cities, winter fogs are much more frequent than over the surrounding countryside. Coastal airports, such as those in New York City, Newark, and Boston suffer severely from a high incidence of fogs augmented by urban air pollution.

A related effect of the urban heat island (Chapter 5) is the general increase in cloudiness and precipitation over a city, as compared with the surrounding countryside. This increase results from intensified convection generated by heating of the lower air. For example, it has been found that thunderstorms over the city of London produce 30 percent more rainfall over the city than over the surrounding country. Increased precipitation over an urban area is estimated to average from 5 to 10 percent over the normal for the region in which it lies.

Table 7.2 summarizes the main climatic differences between a city and the surrounding countryside. Keep in mind that this summary is a generalization applied to highly industrialized nations in midlatitudes, and there are differences among cities with respect to magnitude of a given effect.

As we found in Chapter 5, a large city produces a heat island. Within this warm air layer, pollutants are trapped beneath an inversion lid. The layer of polluted air takes the form of a broad **pollution dome** centered over the city when winds are very light or near calm (Figure 7.28). When there is general air movement in response to a pressure gradient, the pollutants are carried far downwind to form a **pollution plume.** Figure 7.29 has two maps showing plumes from the major cities of the Atlantic seaboard. The color bands show zones of fallout beneath

the plumes and the color dot at the end of each band shows the distance traveled by the air in one day from the source. The left-hand map shows the effects of weak southerly winds on a day in June. In this situation pollution from one city affects another and generally the pollutants remain over the land, contaminating suburban and rural areas over a wide zone. The right-hand map, for a day in February, shows the effect of strong westerly winds, causing the pollution to be carried directly to sea.

Harmful Effects of Atmospheric Pollution

A list of the harmful effects of atmospheric pollutants on plant and animal life and on inorganic substances would be a long one if developed fully. We can only suggest some of these effects. For humans in cities, both sulfur

Table 7.2 / Climate of a City as Compared with that of the Surrounding Countryside

Radiation	
Total insolation	15 to 20 percent less
Ultraviolet (winter)	30 percent less
Ultraviolet (summer)	5 percent less
Sunshine duration	5 to 15 percent less
Temperature	
Annual mean	0.5 to 1.0 C° higher (0.9 to 1.8 F°)
Winter minimum	1.0 to 2.0 C° higher (1.8 to 3.6 F°)
Relative humidity	2 to 3 percent less
Cloudiness	
Cloud cover	5 to 10 percent more
Fog in winter	100 percent more
Fog in summer	30 percent more
Precipitation	
Total quantity	5 to 10 percent more
Snowfall	5 percent less
Particulate matter	10 times more
Gaseous pollutants	5 to 25 times more
Wind speed	
Annual mean	20 to 30 percent lower
Extreme gusts	10 to 20 percent lower
Calms	5 to 20 percent more frequent

Data source: H. E. Landsberg (1970), *Meteorological Monographs,* Vol. 11, p. 91, Table 1.

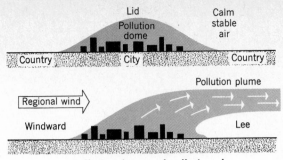

Figure 7.28   **Pollution dome and pollution plume.**

dioxide and hydrocarbon compounds, altered by photochemical reaction to produce sulfuric acid and ethylene, respectively, are irritants to the eyes and the respiratory system. Nitrogen dioxide is also an eye and lung irritant when present in sufficient quantities. Photochemical alteration of nitrogen oxides leads to the production of ozone, which acts as an irritant in smog and would be lethal if it occurred in large concentrations.

For persons suffering from respiratory ailments, such as bronchitis and emphysema, the breathing of heavily polluted air can bring on disability and even death, as statistics clearly show. During the London fog of December 1952, the death rate approximately doubled, and an elevated death rate persisted for weeks after the event. In a more recent London fog, that of December 1962, more than 300 deaths were attributed to the breathing of polluted air. Particularly hard hit were the very young and old. There is also a suspected linkage between the breathing of atmospheric pollutants and lung cancer because the incidence of that disease is higher in cities than in other areas. In addition, the accumulation of atmospheric hydrocarbon compounds on lung tissues may predispose to the onset of lung cancer.

Carbon monoxide is a cause of death when inhaled in sufficient quantities. Everyone knows that the carbon monoxide from automobile exhaust will kill in a short time when breathed in a closed garage. Carbon monoxide levels are a general indicator of the degree of air pollution from vehicular exhausts, but concentrations rarely reach sufficient levels in the open air to be a threat to life. Nevertheless, the long-continued inhalation of small amounts of carbon monoxide is suspected of harmful effects, not as yet evaluated.

Ozone in urban smog has a most deleterious effect on plant tissues, and in some cases has caused the death or severe damage of ornamental trees and shrubs. Sulfur dioxide is injurious to certain plants and is a cause of loss of productivity in truck gardens and orchards in polluted air. Atmospheric sulfuric acid in cities has in some places largely wiped out lichen growth.

Although secondary in the sense that the loss is in dollars, rather than in lives and the health of animals and plants, the deterioration of various materials subjected to polluted air forms an important category of harmful effects. Building stones and masonry are susceptible to the corrosive action of sulfuric acid derived from the atmosphere. Metals, fabrics, leather, rubber, and paint deteriorate and discolor under the impact of exposure to urban air pollutants. In particular, natural rubber is vulnerable to ozone, which causes rubber to harden and crack. The sulfuric acid produced from sulfur dioxide corrodes exposed metals, particularly steel and copper. Not the least of the economic losses from pollution are those from the soilage of clothing, automobiles, furniture, and interior floors, walls, and ceilings. The cleaning bill totaled for a large city is truly staggering when calculated to include the labor and cleaning agents expended by householders.

Lead and other toxic metals in the polluted atmosphere are a particular source of concern for human health in the future. The lead-bearing particles from auto exhausts tend to concentrate in the grass, leaves, and soil near major highways. There is good reason to suppose that humans ingest lead particles directly from the air and that these may prove to be a health hazard. Although lead poisoning from atmospheric sources has not yet been

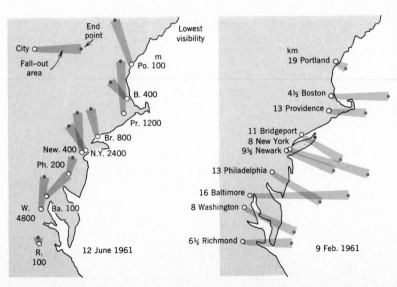

Figure 7.29   **Pollution plumes from cities on the eastern seaboard under conditions of weak southerly winds (left) and strong westerly winds (right). The dot at the end of each plume represents the approximate distance traveled by pollutants at the end of a 24-hour period. (After H. E. Landsberg, 1962, in** *Symposium — Air Over Cities,* **Sanitary Engineering Center Technical Report A62-5, Cincinnati, Ohio.)**

documented in humans, there is now evidence that it has caused the deaths of animals in city zoos. Tests have shown high levels of lead in the tissues of the dead animals, and no source other than atmospheric particles has been found for the ingested lead. Animals in outdoor cages showed higher lead levels than animals kept indoors. These findings are ominous in tone and tend to reinforce the conclusion of the Air Pollution Control Office of the Environmental Protection Agency that atmospheric lead pollution is a possible health hazard. Elimination of lead additives to gasoline has been a major corrective step.

Radioactive substances in the atmosphere are a special form of environmental hazard because of the genetic damage that is done to plant and animal tissues exposed to dangerous radiation.

Late in 1970, the federal Clean Air Act was signed into law, giving the Environmental Protection Agency authority to set national standards for tolerable limits of pollutants in the air. Included in the act is provision for an emergency alert system that will signal the need to reduce fuel combustion and curtail the use of automobiles when the danger point is reached in a given city.

References for Further Study

Ludlum, F. H., and R. S. Scorer (1957), *Cloud study; a pictorial guide,* John Murray, London.

Battan, L. J. (1962), *Cloud physics and cloud seeding,* Doubleday and Co., New York.

Petterssen, S. (1969), *Introduction to meteorology,* third edition, McGraw-Hill Book Co., New York. See Chapters 4–8.

Bach, W. (1972), *Atmospheric pollution,* McGraw-Hill Book Co., New York.

Riehl, H. (1972), *Introduction to the atmosphere,* second edition, McGraw-Hill Book Co., New York. See Chapters 3–5, 11.

Frisken, W. R. (1973), *The atmospheric environment,* Johns Hopkins Univ. Press, Baltimore, Md.

Review Questions

1. In what ways is the presence of water vapor in the air important in Man's environment? What are the sources of atmospheric moisture?
2. Describe the three states in which water may exist. How, and in what amounts, is heat liberated or absorbed as water passes from one state to another?
3. Define relative humidity. What effect has a change of air temperature on the relative humidity of the air? What is the dew-point temperature? Explain the phenomenon of condensation of water vapor.
4. How is humidity of the air measured? Explain the principle of the hygrometer and the psychrometer.
5. What is specific humidity? What use has this measure of water vapor? What range of specific humidity may be expected from equatorial to polar regions?
6. Explain how condensation and precipitation occur as a result of the lifting of a mass of air. What is the dry adiabatic lapse rate? How does this differ from the environmental lapse rate? What is the wet adiabatic lapse rate? Why is it less than the dry rate?
7. Of what kinds of particles are clouds composed? How are clouds classified? What are stratiform and cumuliform clouds? Name the common cloud types, give their height ranges, and describe their forms.
8. In what ways can fog be formed?
9. Name and describe the various forms of precipitation. Is an ice storm a form of precipitation? How does hail form?
10. In what units is rainfall measured? Describe a rain gauge. How is snowfall measured? How does 1 cm of snow compare with 1 cm of rainfall for relative amounts of water?
11. By what two basic mechanisms is precipitation produced? Describe how a convection cell operates to produce rainfall. What keeps the updraft in operation during a convectional storm? Compare the properties of stable air and unstable air.
12. Describe the weather processes within a thunderstorm. What size and height do these storms commonly attain? How does cloud seeding aid the precipitation process? How is hail formed in a thunderstorm? What keeps a thunderstorm active? What is the source of energy?
13. Explain how orographic precipitation occurs. Why is there a rainshadow on the lee side of a mountain chain? Is the air that has reached a rainshadow area warmer or cooler than it was at the same altitude over the coastal zone from which it came? Explain. What is a chinook wind?
14. Describe the transport of water vapor across the parallels of latitude from one zone to another. What changes of state occur in the Hadley cell circulation? In middle and high latitudes?
15. What classes of atmospheric pollutants are recognized? Name the component substances and particles in each class. What are photochemical reactions?
16. To what extent are air pollutants Man-made and to what extent natural? Name the natural sources.
17. Describe fallout, washout, and dispersal of air pollutants. What effect do winds have on urban air pollution?
18. Explain how inversions are conducive to smog formation and persistence. What two basic types of inversions are important in causing a buildup of urban air pollution? What is a pollution dome? A pollution plume?
19. Describe the harmful effects of air pollution.

Air Masses and Cyclonic Storms

8

The atmosphere exerts stress, often severe, on Man and other life forms through weather disturbances involving extremes of wind speeds, cold, and precipitation. Under the category of storms, these phenomena generate environmental hazards directly by the impact of winds and precipitation and indirectly through their attendant phenomena—storm waves and storm surges on the seas and river floods, mudflows, and landslides on the lands. Weather disturbances of lesser magnitude are among the beneficial environmental phenomena because they bring precipitation to the land surfaces and so recharge the vital supplies of fresh water on which Man and all other terrestrial life forms depend.

An understanding of weather disturbances of all intensities enables Man to predict their times and places of occurrence and therefore give warnings and allow protective measures to be taken. This function of the atmospheric scientist can be placed under the heading of environmental protection. It is also possible to a limited degree for Man to modify atmospheric processes in a deliberate way to lessen the wind speeds of storms. These activities come under the heading of planned weather modification. Here, again, we find that the relationship of Man with the atmosphere is one of interaction.

Traveling Cyclones

Much of the unsettled, cloudy weather experienced in middle and high latitudes is associated with traveling cyclones. The convergence of masses of air toward the cyclone centers is accompanied by lift of air and adiabatic cooling, which in turn produce cloudiness and precipitation. By contrast, much of the fair, sunny weather in these latitudes of westerly winds is associated with traveling anticyclones, in which the air subsides and spreads outward, causing adiabatic warming. This process produces stable air, unfavorable to the development of clouds and precipitation.

Many cyclones are mild in intensity, passing with little more than a period of cloud cover and light rain or snow.

On the other hand, when pressure gradients are strong, winds ranging in strength from moderate to gale force accompany the cyclone. In such a case, the disturbance is called a **cyclonic storm.**

Traveling cyclones fall into three general classes: (1) The wave cyclone of midlatitude and arctic zones ranges in severity from a weak disturbance to a powerful storm. (2) The tropical cyclone of tropical and subtropical zones over ocean areas ranges from a mild disturbance to the highly destructive hurricane, or typhoon. (3) Although a very small storm, the tornado is an intense cyclonic vortex of enormously powerful winds; it is on a very much smaller scale of size than other types of cyclones and is related to severe convectional activity.

Air Masses

Cyclones of middle and high latitudes depend for their development on the coming together of large bodies of air of contrasting physical properties. A body of air in which the upward gradients of temperature and moisture are fairly uniform over a large area is known as an **air mass.** In horizontal extent, a single air mass may be as large as part of a continent; in vertical dimension, it may extend through the troposphere. A given air mass is characterized by a distinctive combination of temperature, environmental lapse rate, and specific humidity. Air masses differ widely in temperature from very warm to very cold and in moisture content from very dry to very moist.

A given air mass usually has a sharply defined boundary between itself and a neighboring air mass. This discontinuity is termed a **front.** We found an example of a front in the contact between polar and tropical air masses below the axis of the polar-front jet stream in upper-air waves, as shown in Figure 6.18. This feature we called the polar front; it represents the highest degree of global generalization. Fronts may be nearly vertical, as in the case of air masses having little motion relative to one another. Fronts may be inclined at an angle not far from the horizontal in cases where one air mass is sliding over another. A front may be almost stationary with respect to the earth's surface, but, nevertheless, the adjacent air masses may be moving rapidly with respect to each other along the front.

The properties of an air mass are derived partly from the regions over which it passes. Because the entire troposphere is in more or less continuous motion, the particular air-mass properties at a given place may reflect the composite influence of a travel path covering thousands of kilometers and passing alternately over oceans and continents. This complexity of influences is particularly important in middle and high latitudes in the northern hemisphere, within the flow of the global westerlies.

Over vast tropical and equatorial areas, however, an air mass reflects quite simply the properties of an ocean or a land surface above which it moves slowly or tends to stagnate. Over a warm equatorial ocean surface, the lower levels of the overlying air mass develop a high water-vapor content. Over a large tropical desert, slowly subsiding air forms a warm air mass with low relative humidities. Over cold, snow-covered land surfaces in the arctic zone in winter, the lower layer of the air mass remains very cold with a very low water vapor content. Meteorologists have designated as **source regions** those land or ocean surfaces that strongly impress their temperature and moisture characteristics on overlying air masses.

Air masses move from one region to another, following the patterns of barometric pressure. During these migrations, lower levels of the air mass undergo gradual modification, taking up or losing heat to the surface beneath, and perhaps taking up or losing water vapor as well.

Air masses are classified according to two categories of source regions: (1) the latitudinal position on the globe, which primarily determines thermal properties; and (2) the underlying surface, whether continent or ocean, determining the moisture content. With respect to latitudinal position, five types of air masses are as follows:

Air Mass	Symbol	Source Region
Arctic	A	Arctic ocean and fringing lands
Antarctic	AA	Antarctica
Polar	P	Continents and oceans, lat. 50–60°N and S
Tropical	T	Continents and oceans, lat. 20–35°N and S
Equatorial	E	Oceans close to equator

With respect to type of underlying surface, two further subdivisions are imposed on the preceding types:

Air Mass	Symbol	Source Region
Maritime	m	Oceans
Continental	c	Continents

By combining types based on latitudinal position with those based on underlying surface, a list of six important air masses results (Table 8.1). Figure 8.1 shows the global

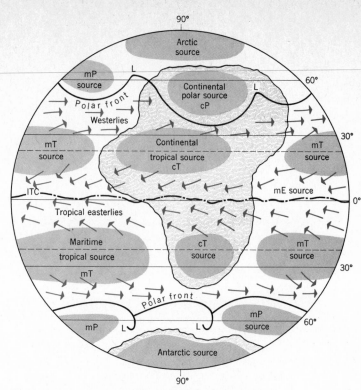

Figure 8.1   **A schematic global diagram showing the source regions of air masses in relation to the polar front and the intertropical convergence zone.**

distribution of source regions of these air masses. Table 8.1 also gives some typical values of temperature and specific humidity at the surface, although a wide range in these properties may be expected, depending on season.

Note that the polar air masses (mP, cP) originate in the subarctic latitude zone, not in the polar latitude zone. The meteorological definition of the word "polar" for air masses has long been in use and has international acceptance; we cannot change this usage to conform with the geographic system of latitude zones defined in Chapter 4.

The maritime equatorial air mass (mE) holds about 200 times as much water vapor as the extremely cold and dry continental arctic and antarctic air masses, cA and cAA. The maritime tropical air mass (mT) and maritime equatorial air mass are quite similar in temperature and water vapor content. With very high values of specific humidity, both are capable of very heavy yields of precipitation. The continental tropical air mass (cT) has its source region over subtropical deserts of the continents. Although it may have a substantial water-vapor content, it tends to be stable and has low relative humidity when highly heated during the daytime. The maritime polar air mass (mP) originates over midlatitude oceans. Although the quantity of water vapor it holds is not large compared with the tropical air masses, the mP air mass can yield heavy precipitation. Much of this precipitation is of the orographic type, over mountain ranges on the western coasts of continents. The continental polar air mass (cP) originates over North America and Eurasia in the subarctic zone. It has low specific humidity and is very cold in winter.

Table 8.1 / Properties of Typical Air Masses

Air Mass	Symbol	Properties	Temperature °C	(°F)	Specific Humidity g/kg
Continental arctic (and continental antarctic)	cA (cAA)	Very cold, very dry (winter)	−46°	(−50°)	0.1
Continental polar	cP	Cold, dry (winter)	−11°	(12°)	1.4
Maritime polar	mP	Cool, moist (winter)	4°	(39°)	4.4
Continental tropical	cT	Warm, dry	24°	(75°)	11.0
Maritime tropical	mT	Warm, moist	24°	(75°)	17.0
Maritime equatorial	mE	Warm, very moist	27°	(80°)	19.0

North American Air Masses

The continental polar air mass of North America originates over north-central Canada (Figure 8.2). This air mass forms tongues of cold, dry air, which periodically extend south and east from the source region to produce anticyclones accompanied, in winter, by low temperatures and clear skies. Over the Arctic Ocean and its bordering lands of the arctic zone, the arctic air mass, which is extremely cold and stable, develops. When this air mass invades the United States, it produces a severe cold wave.

The maritime polar air mass originates over the North Pacific and Bering Strait, in the region of the persistent Aleutian low-pressure center. With ample opportunity to absorb moisture both over the source region and throughout its travel southeastward to the west coast of North America, this air mass is characteristically cool and moist, with a tendency in winter to become unstable, giving heavy precipitation over coastal ranges. Another

maritime polar air mass of the North American region originates over the North Atlantic Ocean. It, too, is cool and moist.

Of the tropical air masses, the commonest visitor to the central and eastern states is the maritime tropical air mass from the Gulf of Mexico. It moves northward, bringing warm, moist, unstable air over the eastern part of the country. In the summer, particularly, this air mass brings hot, sultry weather to the central and eastern United States. This air mass produces many thunderstorms. Closely related is a maritime tropical air mass from the Atlantic Ocean east of Florida, over the Bahamas.

A hot, dry continental tropical air mass originates over northern Mexico, western Texas, New Mexico, and Arizona during the summer. This air mass does not travel widely, but governs weather conditions over the source region.

Over the Pacific Ocean, in the cell of high pressure located to the southwest of Lower California, is a source region of another maritime tropical air mass. Occasionally, in summer, this moist, unstable air mass penetrates the southwestern desert region, bringing severe thunderstorms to southern California and southern Arizona.

Cold and Warm Fronts

Figure 8.3 shows the structure of a front in which cold air is invading the warm-air zone. A front of this type is called a **cold front.** The colder air mass, being the denser, re-

Figure 8.2 **North American air-mass source regions and trajectories. (Data from U.S. Department of Commerce.)**

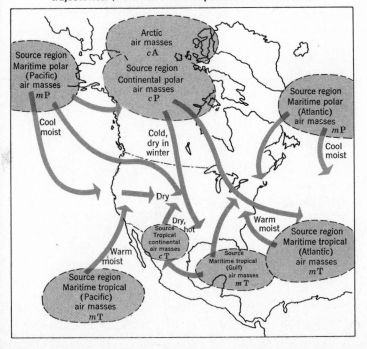

Figure 8.3 **A cold front. (Drawn by A. N. Strahler.)**

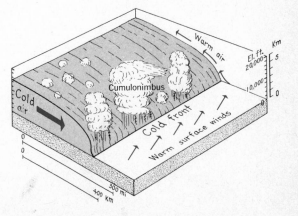

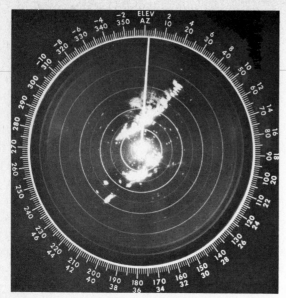

Figure 8.4 Photograph of a radar screen on which lines of thunderstorms show as bright patches. The heavy circles are spaced 80 km (50 nautical miles) apart. (National Weather Service.)

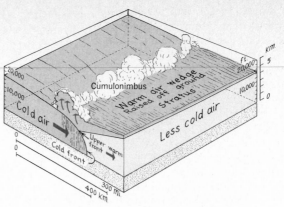

Figure 8.6 An occluded warm front. (Drawn by A. N. Strahler.)

mains in contact with the ground and forces the warmer air mass to rise over it. The slope of the cold-front surface is greatly exaggerated in the figure, being actually of the order of slope of 1 in 40 (meaning that the slope rises 1 km vertically for every 40 km of horizontal distance). Cold fronts are associated with strong atmospheric disturbances. As the unstable warm air is lifted, it may break out in severe thunderstorms. Thunderstorms can be seen on the radar screen (Figure 8.4).

Figure 8.5 illustrates a **warm front,** in which warm air is moving into a region of colder air. Here, again, the cold air mass remains in contact with the ground and the warm air mass is forced to rise, as if it were ascending a long ramp. Warm fronts have lower slopes than cold fronts — on the order of 1 in 80, to as low as 1 in 200. Moreover, warm fronts commonly represent stable atmospheric conditions and lack the turbulent air motions of the cold front. Of course, if the warm air is unstable, it will develop

convection cells and there will be heavy showers or thunderstorms.

Cold fronts normally move along the ground at a faster rate than warm fronts. So, when both types are in the same neighborhood, the cold front overtakes the warm front. An **occluded front** then results (Figure 8.6). The colder air of the fast-moving cold front remains next to the ground, forcing both the warm air and the less cold air to rise over it. The warm air mass is lifted completely free of the ground.

Wave Cyclones

The dominant type of weather disturbance of middle and high latitudes is the **wave cyclone,** a vortex that repeatedly forms, intensifies, and dissolves along the polar front between cold and warm air masses. At the time of World

Figure 8.7 The trough between two high-pressure regions is a likely zone for the development of a wave cyclone.

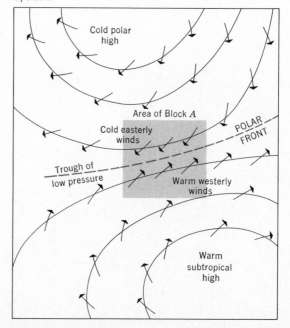

Figure 8.5 A warm front. (Drawn by A. N. Strahler.)

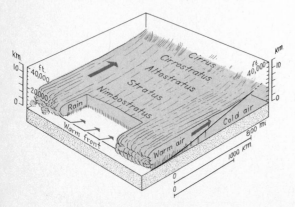

War I, the Norwegian meteorologist Jakob Bjerknes recognized the existence of atmospheric fronts and developed his **wave theory** of cyclones.

The term "front," used by Bjerknes, was particularly apt because of the resemblance of this feature to the fighting fronts in western Europe that were then active. Just as vast armies met along a sharply defined front that moved back and forth, so masses of cold polar air meet in conflict with warm, moist tropical air. Instead of mixing freely, these unlike air masses remain clearly defined; but they interact along the polar front in great spiraling whorls.

A stituation favorable to the formation of a wave cyclone is shown by a surface weather map (Figure 8.7). A trough of low pressure lies between two large anticyclones (highs); one is made up of a cold, dry polar air mass, the other of a warm, moist maritime air mass. Airflow is converging from opposite directions on the two sides of the front, setting up an unstable situation.

A series of individual blocks (Figure 8.8) shows the sequence of stages in the life history of a wave cyclone. At the start of the cycle, the polar front is a smooth boundary along which air is moving in opposite directions. In

Block *A,* the polar front shows a wave beginning to form. Cold air is turned in a southerly direction and warm air in a northerly direction, so each invades the domain of the other.

In Block *B,* the wave disturbance along the polar front has deepened and intensified. Cold air is now actively pushing southward along a cold front; warm air is actively moving northeastward along a warm front. The zone of precipitation is now extensive, but wider along the warm front than along the cold front. In Block *C,* the cold front has overtaken the warm front, reducing the zone of warm air to a narrow sector and producing an occluded front. Eventually, the warm air is forced off the ground (Block *D*), isolating it from the parent region of warm air to the south. The source of moisture and energy thus cut off, the cyclonic storm gradually dies out and the polar front is reestablished as originally (Block *D*).

Cyclones on the Daily Weather Map

Further details of a wave cyclone are illustrated by surface weather maps for two successive days (Figure 8.9). Standard map symbols for wind directions and speeds are

Figure 8.8 **Stages in the development of a wave cyclone. (Drawn by A. N. Strahler.)**

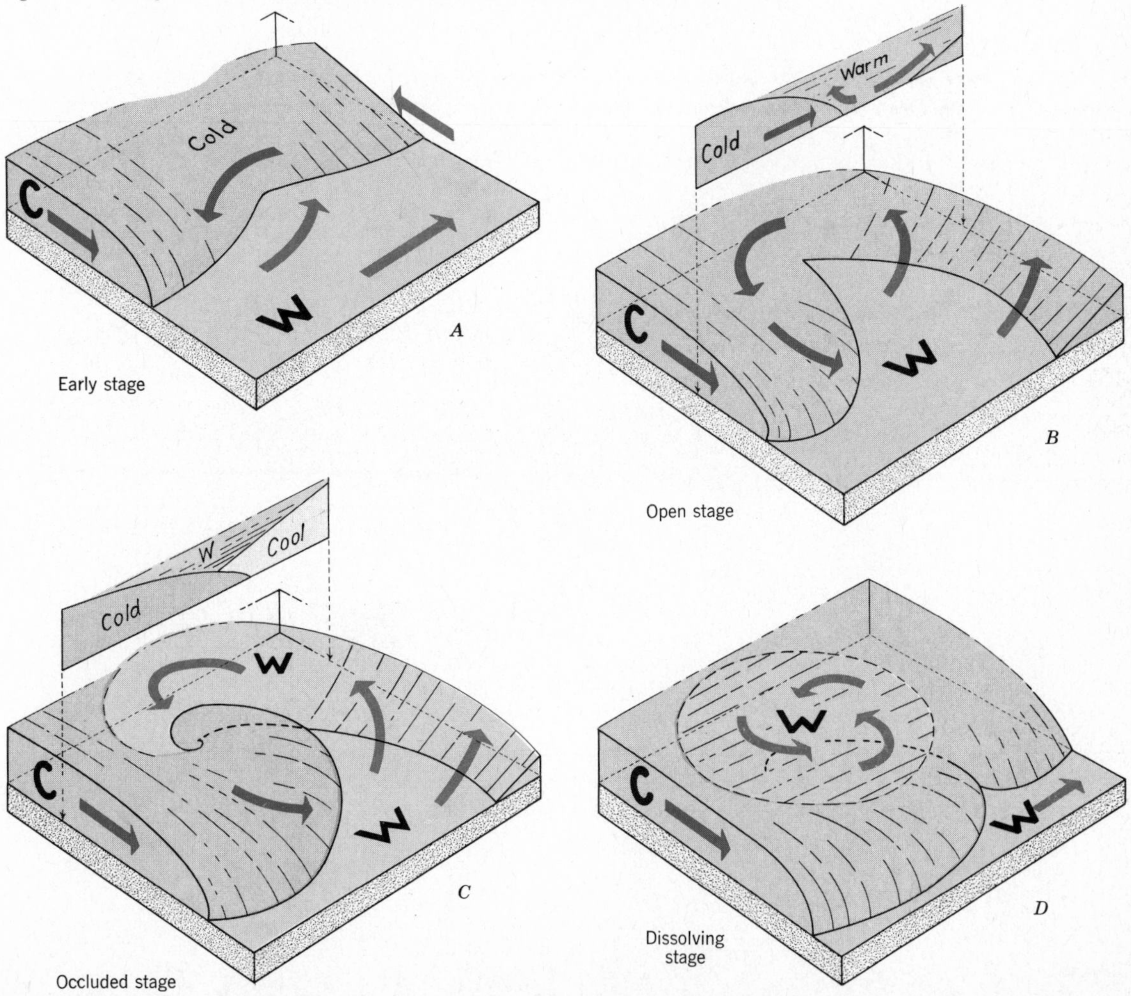

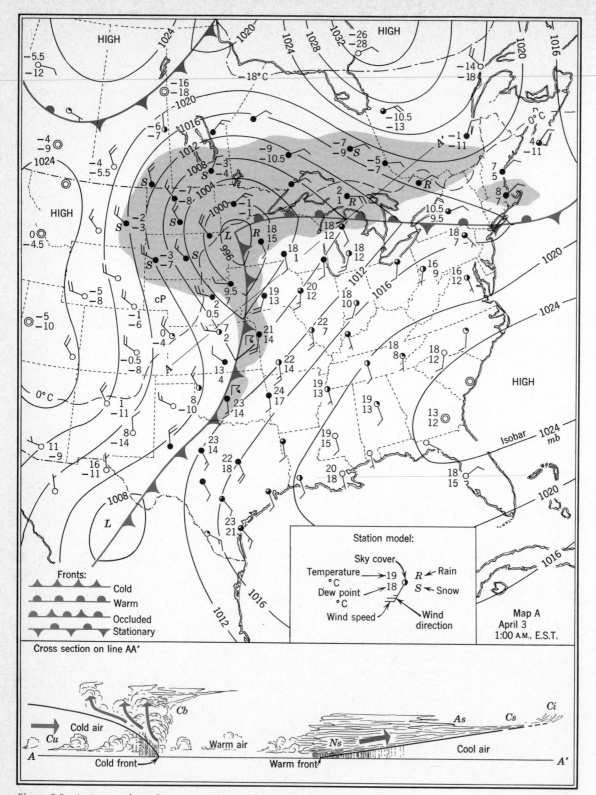

Figure 8.9 A wave cyclone shown on surface weather maps for successive days. Pressures are given in millibars; temperatures are given in Celsius degrees. Areas

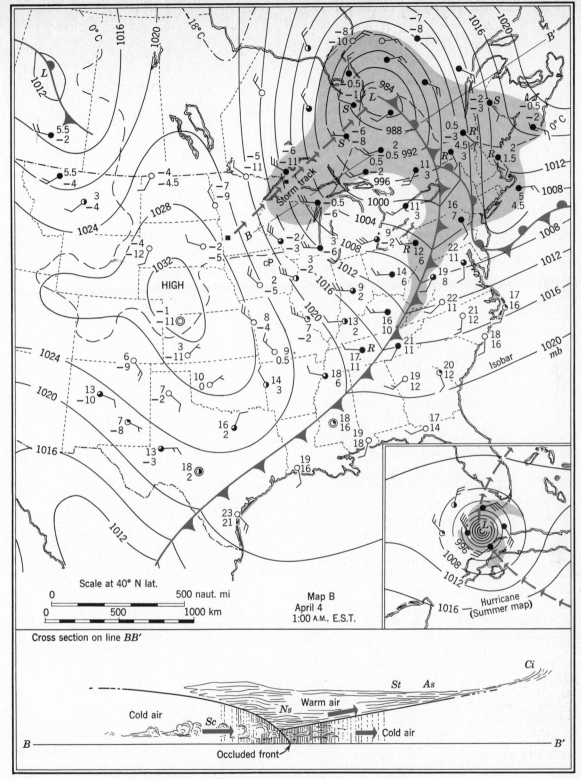

Cross section on line *BB'*

B ——————— Occluded front ——————— B'

Cold air   Sc   Ns   Warm air   Cold air   St   As   Ci

Scale at 40° N lat.

0   500 naut. mi
0   500   1000 km

Map B
April 4
1:00 A.M., E.S.T.

Hurricane
(Summer map)

experiencing precipitation are shaded. See Figure 8.10 for explanation of wind symbols.
(Modified and simplified from daily weather map of the National Weather Service.)

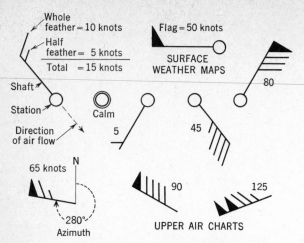

Figure 8.10 Standard weather-map symbols for winds. One knot is equal to 1.85 km/hr (1.15 mi/hr).

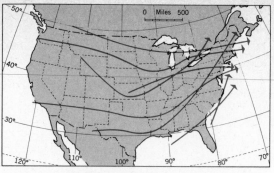

Figure 8.11 Common tracks taken by wave cyclones passing across the United States and southern Canada. (Data from Bowie and Weightman, U.S. Weather Bureau.)

shown in Figure 8.10. Isobars are in millibars; the interval is 4 mb.

Map A of Figure 8.9 shows a cyclone in a stage approximately equivalent to Block B of Figure 8.8. The storm is centered over western Minnesota and is moving northeastward. Note the following features on Map A: (1) Isobars of the low are closed to form an oval-shaped pattern. (2) Isobars make a sharp V where crossing cold and warm fronts. (3) Wind directions, indicated by arrows, are at an angle to the trend of the isobars and form a pattern of counterclockwise inspiraling. (4) In the warm-air sector, warm, moist tropical air flows northward toward the warm front. (5) A sudden shift of wind direction accompanies the passage of the cold front. This wind shift is indicated by the widely different wind directions at stations close to the cold front, but on opposite sides. (6) A severe drop in temperature accompanies the passage of the cold front, as shown by differences in temperature readings at stations on opposite sides of the cold front. (7) Precipitation, shown by color shading, is occurring over a broad zone near the warm front and in the central area of the cyclone, but extends as a thin band down the length of the cold front. (8) Cloudiness, shown by the proportion of blackened area in station circles, extends over the entire cyclone. (9) The low is followed on the west by a high (anticyclone), in which low temperatures and clear skies prevail. (10) The 0°C (32°F) isotherm crosses the cyclone diagonally from northeast to southwest, showing that the southeastern part is warmer than the northwestern part.

A cross section through Map A of Figure 8.9 along the line AA′ shows how the fronts and clouds are related. Along the warm front is a broad area of stratiform clouds. These take the form of a wedge with a thin leading edge of cirrus (Ci) and cirrostratus (Cs). Westward this thickens to altostratus (As), then to stratus (St), and finally to nimbostratus (Ns) with steady rain. Within the warm air mass sector, the sky may partially clear with scattered cumulus (Cu). Along the cold front are cumulonimbus (Cb) clouds with violent thunderstorms and heavy rains, but this activity is limited to a narrow belt and passes quickly.

The second weather map, Map B of Figure 8.9, shows conditions 24 hours later. The cyclone has moved rapidly northeastward into Canada, its path shown by the line labeled "storm track." The center has moved about 1300 km (800 mi) in 24 hours, a speed of just over 65 km (40 mi) per hour. The cyclone has occluded. An occluded front replaces the separate warm and cold fronts in the central part of the disturbance. The high-pressure area, or tongue of cold polar air, has moved in to the west and south of the cyclone. The cold front is now passing over the eastern and Gulf Coast states. With closed isobars around the anticyclone, the skies are clear and the winds are weak. In another day the entire storm will have passed out to sea, leaving the eastern region in the grip of cold but clear weather. A cross section below the map shows conditions along the line BB′, cutting through the occluded part of the storm. Note that the warm air mass is being lifted higher off the ground and is giving heavy precipitation.

Wave-cyclone Tracks

Observations, made over many decades, of the movements of cyclones and anticyclones have shown that certain tracks are most commonly followed. Figure 8.11 is a map of the United States and southern Canada showing these common paths. Notice that some cyclonic storms travel across the entire continent from places of origin in the North Pacific, such as the Aleutian low. Other cyclones originate in the Rocky Mountain region, the

Figure 8.12 Principal tracks of wave cyclones are shown by solid arrows; those of tropical cyclones are shown by dashed arrows. (Data from S. Petterssen.)

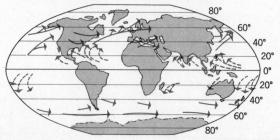

Fibrous cirrus, or mares' tails, above; cumulus of fair weather, below. (Arthur N. Strahler)

Cirrus in broad, parallel bands, indicating a jet stream aloft. (Arthur N. Strahler)

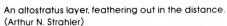

Cirrocumulus forming a mackerel sky. (Arthur N. Strahler)

Altocumulus. (Arthur N. Strahler)

An altostratus layer, feathering out in the distance. (Arthur N. Strahler)

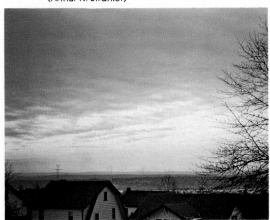

Massive stratocumulus following the passage of a cold front. (Arthur N. Strahler)

Clouds   **Plate C.1**

Active thunderstorm cells, producing cumulonimbus clouds over the Southern Rocky Mountains. (Arthur N. Strahler)

Torrential rain beneath the flat base of a cumulonimbus cloud, North Rim of Grand Canyon, Arizona. (Arthur N. Strahler)

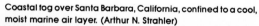

Rain showers falling from a weakening thunderstorm, South Dakota. (Arthur N. Strahler)

A succession of lightning strokes recorded by time exposure. (Arthur N. Strahler)

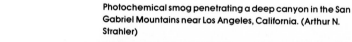

Photochemical smog penetrating a deep canyon in the San Gabriel Mountains near Los Angeles, California. (Arthur N. Strahler)

Coastal fog over Santa Barbara, California, confined to a cool, moist marine air layer. (Arthur N. Strahler)

Plate C.2   Thunderstorms, fog, smog

Viewed from the Apollo 9 spacecraft, the tops of great cumulonimbus clouds over the Amazon jungle of South America show concentric patterns. As each thunderstorm cell rises and attains the top of the troposphere the ice cloud it produces spreads horizontally. Storms of this form are nearly stationary. (NASA AS9-19-3026)

Long rows of cumulus clouds parallel the regional air flow in this Gemini spacecraft view over Alabama, Georgia, and South Carolina. Above the distant horizon, the rarified upper atmosphere produces a blue airglow band that merges into the black of outer space. (NASA S-65-45697)

Hurricane Gladys, photographed from the Apollo 7 spacecraft on October 17, 1968, at an altitude of about 180 km (110 mi). Bands of cumulonimbus clouds spiral around the central eye. The large, smooth cloud patches are high cirrostratus layers derived from the tops of the rising convection cells. (NASA AS7-7-1877)

The atmosphere from space   **Plate C.3**

Planet Earth photographed in 1972 by astronauts of the Apollo 17 mission on their way to the moon. On this date the earth was near the December solstice, with its south polar region tilted toward the sun to receive maximum insolation. Africa is clearly outlined in the central and upper part of the disk, its reddish-brown northern and southern desert belts conspicuously free of clouds. The intervening equatorial zone of the African continent reveals green vegetation between patches of clouds. Cyclonic storms form a white chain over the blue of the Southern Ocean. Antarctica, with its blanket of snow and clouds, is solid white. (NASA 72-HC-928)

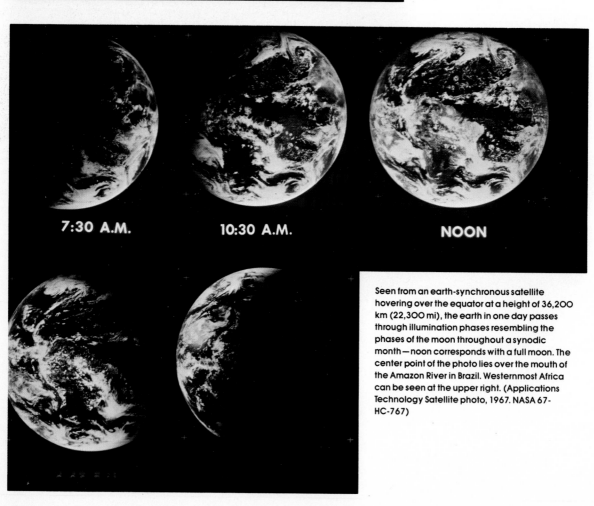

7:30 A.M.   **10:30 A.M.**   **NOON**

Seen from an earth-synchronous satellite hovering over the equator at a height of 36,200 km (22,300 mi), the earth in one day passes through illumination phases resembling the phases of the moon throughout a synodic month—noon corresponds with a full moon. The center point of the photo lies over the mouth of the Amazon River in Brazil. Westernmost Africa can be seen at the upper right. (Applications Technology Satellite photo, 1967. NASA 67-HC-767)

Plate C.4   The earth in space

Figure 8.13   **A daily weather map of the world for a given day during July or August might look like this map, which is a composite of typical weather conditions. (Data from M. A. Garbell.)**

central states, or the Gulf Coast. Most of the tracks converge toward the northeastern states and the St. Lawrence lowland. These cyclones pass out into the North Atlantic, where they tend to concentrate in the region of the Icelandic low.

General world distribution of paths of wave cyclones is shown in Figure 8.12. Notice the heavy concentration in the neighborhood of the Aleutian and Icelandic lows.

Wave cyclones commonly form in succession to travel in a chain across the north Atlantic and north Pacific oceans. Figure 8.13, a world weather map, shows several such **cyclone families.** As each cyclone moves northeastward it deepens and occludes, becoming an intense upper-air vortex. For this reason, cyclones arriving at the western coasts of North America and Europe are typically occluded.

In the southern hemisphere, storm tracks are more nearly along a single lane, following the parallels of latitude. This simplicity of pattern is the result of uniform ocean surface throughout midlatitudes. Only the southern tip of South America breaks the monotonous oceanic expanse. The polar-centered ice sheet of Antarctica provides a centralized source of polar air.

Wave Cyclones and Upper-Air Waves

How are wave cyclones and their accompanying anticyclones related physically to the Rossby waves described in Chapter 6? The wave cyclone with its fronts is a low-level phenomenon in the troposphere and gives way at high levels to the smooth westerly flow of air within the Rossby waves. Figure 8.14 shows the two systems superimposed. The upper diagram is a map; the lower diagram is a cross section along the line x–y of the map.

The key to a linkage of upper and lower systems lies in the streamlines of upper-air flow. Entering the west side (left) of the Rossby wave, the flow lines are coming

together horizontally in a convergence pattern. The converging air is forced to descend to lower levels along the jet-stream axis, or core. As the air descends, it develops an anticyclonic outspiral and produces a low-level anticyclone (high), from which the air diverges horizontally. As the upper-air flow leaves the Rossby wave toward the east (right), the flow lines separate in a divergence pattern. Here, air must rise along the jet core. Air lifted from lower levels develops a cyclonic inspiral and is replaced

Figure 8.14   **Schematic diagram of the coupling of upper-air flow paths with surface cyclones and anticyclones. (Based on a drawing in H. Riehl, 1965,** *Introduction to the Atmosphere,* **McGraw-Hill Book Co., New York, p. 154, Figure 7.7.)**

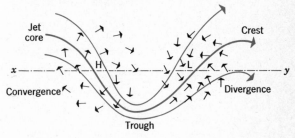

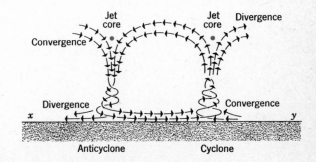

Figure 8.15   **This funnel cloud was photographed by William L. Males at Amarillo, Oklahoma, on May 4, 1961. The tornado was less than 2 km from the observer when the picture was taken.**

by air moving into the cyclone near the surface. As soon as cyclonic flow begins, fronts are formed and a typical wave cyclone comes into existence. The cyclone is now "dragged along" beneath the jet core and moves northeastward. Thus we find that wave cyclones are both initiated and steered by the upper-level flow. Knowledge of upper-air conditions is of great value in forecasting surface weather.

Recall from Chapter 6 that, in an advanced stage of development, the Rossby wave can form a cutoff upper-air low with a cold mass of air circulating within it. At ground level, the cut-off low is represented by a strongly occluded cyclone, so the low-pressure center extends through the entire troposphere. Deep lows of this kind may persist for many days at a stretch.

Tornadoes

The smallest but most intense of all known storms is the **tornado.** It seems to be a typically American storm because it is most frequent and violent in the United States. Tornadoes also occur in Australia in substantial numbers and are reported occasionally in other places in midlatitudes.

The tornado is a small, intense cyclone in which the air is spiraling at tremendous speed. It appears as a dark **funnel cloud** (Figure 8.15) hanging from a cumulonimbus cloud. At its lower end, the funnel may be 100 to 500 m (300 to 1500 ft) in diameter. The funnel appears dark because of the density of condensing moisture, dust, and debris swept up by the wind.

Wind speeds in a tornado exceed anything known in other storms. Estimates of wind speed run as high as 400 km (250 mi) per hour. As the tornado moves across the

country, the funnel writhes and twists. The end of the funnel cloud may alternately sweep the ground, causing complete destruction of anything in its path, and rise in the air to leave the ground below unharmed. Tornado destruction occurs both from the great wind stress and from the sudden reduction of air pressure in the vortex of the cyclonic spiral. Closed houses literally explode. Even corks will pop out of empty bottles, so great is the drop in air pressure as the funnel passes overhead.

Tornadoes occur as parts of dense cumulonimbus clouds traveling in advance of a cold front. They seem to originate where turbulence is greatest. They are most common in the spring and summer, but can occur in any month (Figure 8.16). Where maritime polar air lifts warm, moist tropical air on a cold front, conditions may become favorable for tornadoes. They occur in greatest numbers in the central and southeastern states and are rare over mountainous and forested regions. They are almost unknown from the Rocky Mountains westward and are relatively fewer on the eastern seaboard (Figure 8.17).

Devastation from a tornado is complete within the narrow limits of its path (Figure 8.18). Only the strongest buildings, constructed of concrete and steel, can resist major structural damage. A tornado can often be seen or heard approaching, but those that come during the night may give no warning. The National Weather Service maintains a tornado forecasting and warning system. Whenever weather conditions conspire to favor tornado development, the danger area is alerted and systems for observing and reporting a tornado are set in readiness.

Figure 8.16   **Average number of tornadoes reported in each month in the United States for the period 1916-1960. (From National Weather Service.)**

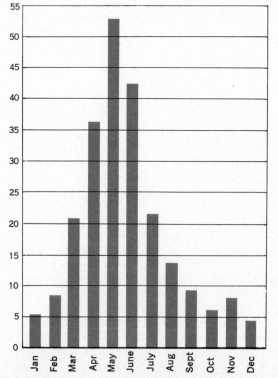

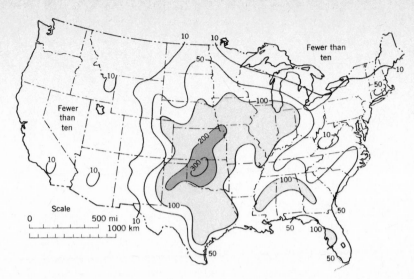

Figure 8.17 **Distribution of tornadoes in the United States from 1955 to 1967. The figures tell the number of tornadoes observed within two-degree squares of latitude and longitude. (Data from M. E. Pautz, 1969, ESSA Tech. Memo WBTM FCST 12, Office of Meteorological Operations, Silver Spring, Md.)**

Waterspouts are similar in structure to tornadoes but form at sea under cumulonimbus clouds. They are smaller and less powerful than tornadoes. Seawater may be lifted 3 m (10 ft) above the sea surface. Rapid condensation of water vapor makes a visible column reaching to the overlying cloud base. Waterspouts are commonly found in subtropical waters of the Gulf of Mexico and off the southeastern coast of the United States. They result from air turbulence along a cold front moving into lower latitudes.

Tropical and Equatorial Weather Disturbances

Weather systems of the tropical and equatorial zones show some basic differences from those of midlatitudes. The Coriolis effect is weak close to the equator, and there is a lack of strong contrast between air masses. Consequently, clearly defined fronts and large, intense wave cyclones are missing. On the other hand, there is intense atmospheric activity in the form of convection cells because of the high moisture content of the maritime air masses in these latitudes. In other words, there is an enormous reservoir of energy in the latent form. This same energy source powers the most formidable of all storms — the tropical cyclone.

The world weather map (Figure 8.13) shows a typical set of weather conditions in the subtropical and equatorial zones. Fundamentally, the arrangement consists of two rows of high-pressure cells, one or two cells to each land or ocean body. The northern row lies approximately along the tropic of cancer; the southern row lies along the tropic of capricorn. Between the subtropical highs lies the equatorial trough of low pressure. Toward this trough the northeast and southeast trades converge along the intertropical convergence zone (ITC). At higher levels in the troposphere, the airflow is almost directly from east to west in the form of persistent tropical easterlies, described in Chapter 6.

One of the simplest forms of weather disturbance is an **easterly wave,** a slowly moving trough of low pressure within the belt of tropical easterlies (trades). These waves occur in lat. 5° to 30°N and S over oceans, but not over the

Figure 8.18 **This wide swath of destruction through Xenia, Ohio, was cut by a tornado on the afternoon of April 4, 1974. The storm was one of 148 tornadoes that struck in 12 central states on two successive days, killing altogether more than 300 persons and injuring over 5000 — the largest single tornado outbreak in the world's history. (Wide World Photos.)**

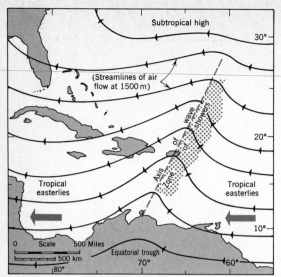

Figure 8.19 An easterly wave passing over the West Indies. (Data from H. Riehl, 1954, *Tropical Meteorology,* McGraw-Hill Book Co., New York, p. 213, Figure 9.3.)

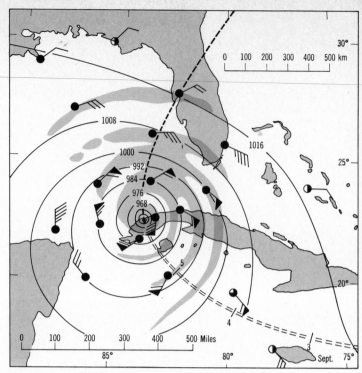

Figure 8.20 Surface weather map of a typical hurricane passing over the western tip of Cuba. See Figure 8.10 for interpretation of wind arrows.

equator itself. Figure 8.19 is a simplified weather map of an easterly wave showing isobars, winds, and the zone of rain. The wave is simply a series of indentations in the isobars to form a shallow pressure trough. The wave travels westward at a rate of 325 to 500 km (200 to 300 mi) per day. Air near the surface converges on the eastern, or rear, side of the wave axis. Moist air is lifted and produces scattered showers and thunderstorms. The rainy period may last for a day or two.

Another related disturbance is the **weak equatorial low,** which forms near the center of the equatorial trough. Moist equatorial air masses converge on the center of the low, causing rainfall from many individual convectional storms. Several such weak lows are shown on the world weather map (Figure 8.13), lying along the ITC. Because the map is for a day in July or August, the ITC is shifted well north of the equator. At this season, the rainy monsoon is in progress in Southeast Asia.

Another distinctive feature of tropical-zone weather is the occasional penetration of powerful tongues of cold polar air from the midlatitudes into very low latitudes. These tongues are known as **polar outbreaks;** they bring unusually cool, clear weather with strong, steady winds moving behind a cold front with squalls. The polar outbreak is best developed in the Americas. Outbreaks that move southward from the United States over the Caribbean Sea and Central America are called "northers" or "nortes"; those that move north from Patagonia into tropical South America are called "pamperos." One such outbreak is shown over South America on the world weather map (Figure 8.13).

Tropical Cyclones

One of the most powerful and destructive types of cyclonic storms is the **tropical cyclone,** otherwise known as the **hurricane** or **typhoon.** The storm develops over

oceans in latitudes 8° to 15°N and S, but not close to the equator, where the Coriolis effect is extremely weak. In many cases an easterly wave simply deepens and intensifies, growing into a deep, circular low. High sea-surface temperatures that are over 27°C (80°F) in these latitudes are important in the environment of storm origin because warming of air at low level creates instability. Once formed, the storm moves westward through the trade-wind belt. It may then curve poleward and penetrate

Figure 8.21 Trace sheet from a barograph at Galveston, Texas, during the hurricane of July 27, 1943. (National Weather Service.)

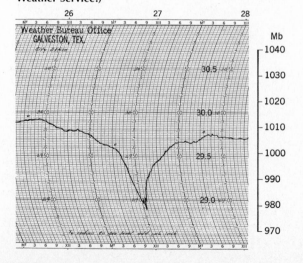

120 Air Masses and Cyclonic Storms

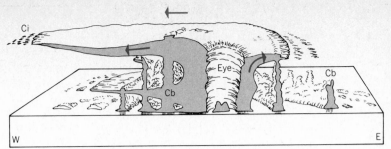

Figure 8.22 Schematic diagram of a hurricane. Cumulonimbus clouds in concentric rings rise through dense stratiform clouds in which the air spirals upward. Width of the diagram represents about 1000 km (600 mi). Highest clouds are at elevations often over 9 km (30,000 ft). (Redrawn from the National Oceanic and Atmospheric Administration, National Weather Service, R. C. Gentry, 1964. *Weatherwise,* vol. 17, p. 182.)

well into the belt of westerly winds.

The tropical cyclone is an almost circular storm center of extremely low pressure into which winds are spiraling at high speed, accompanied by very heavy rainfall (Figure 8.20 and Plate C.3). Storm diameter may be 150 to 500 km (100 to 300 mi). Wind speeds range from 120 to 200 km (75 to 125 mi) per hour, sometimes much higher. Barometric pressure in the storm center commonly falls to 950 mb or lower (Figure 8.21).

A characteristic feature of the tropical cyclone is its **central eye,** in which calm prevails (Figure 8.22). The eye is a cloud-free vortex produced by the intense spiraling of the storm. In the eye, air descends from high altitude and is adiabatically warmed. Passage of the central eye may take about half an hour, after which the storm strikes with renewed ferocity, but with winds in the opposite direction.

World distribution of tropical cyclones is limited to six regions, all of them over tropical and subtropical oceans (Figure 8.23): (1) West Indies, Gulf of Mexico, and Caribbean Sea; (2) western North Pacific, including the Philippine Islands, China Sea, and Japanese Islands; (3) Arabian Sea and Bay of Bengal; (4) eastern Pacific coastal region off Mexico and Central America; (5) south Indian Ocean, off Madagascar; and (6) western South Pacific, in the region of Samoa and the Fiji Islands and the east coast of Australia. Curiously, these storms are unknown in the South Atlantic. Tropical cyclones never originate over land, although they often penetrate well into the margins of continents.

Tracks of tropical cyclones of the North Atlantic are shown in Figure 8.24. Most of the storms originate at lat. 10° to 20°, travel westward and northwestward through

the trades, then turn northeast at about lat. 30° to 35° into the zone of the westerlies. Here the intensity lessens and the storms change into typical midlatitude wave cyclones. In the trade-wind belt, the cyclones travel 10 to 20 km (6 to 12 mi) per hour.

The occurrence of tropical cyclones is restricted to certain seasons of the year, and these vary according to the global location of the storm region. For hurricanes of the North Atlantic, the season runs from May through November, with maximum frequency in late summer or early autumn. The general rule is that tropical cyclones of the northern hemisphere occur in the season during which the ITC has moved north; those of the southern hemisphere occur when it has moved south.

The environmental importance of tropical cyclones lies in their tremendously destructive effect on inhabited islands and coasts (Figure 8.25). Wholesale destruction of cities and their inhabitants has been reported on several occasions. A terrible hurricane that struck Barbados in the West Indies in 1780 is reported to have torn stone buildings from their foundations, destroyed forts, and carried cannons more than 30 m (100 ft) from their locations. Trees were torn up and stripped of their bark, and more than 6000 persons perished.

Coastal destruction by storm waves and greatly raised sea level is perhaps the most serious effect of tropical cyclones. Where water level is raised by strong wind pressure, great storm surf attacks ground ordinarily far inland of the limits of wave action. A sudden rise of water level, known as **storm surge,** may take place as the hurricane moves over a coastline. Ships are lifted and carried inland to become stranded. If high tide accompanies the

Figure 8.23 Typical paths of tropical cyclones in relation to sea surface temperatures (°C) in summer of the respective hemisphere. (Data from Palmén, 1948.)

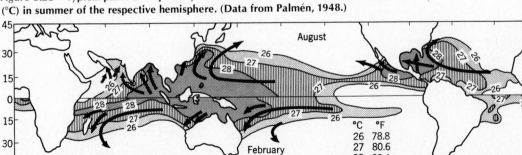

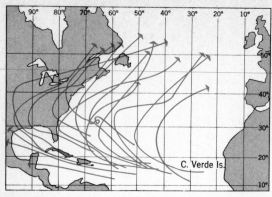

Figure 8.24 **Tracks of some typical hurricanes occurring during August. (After U.S. Navy Oceanographic Office.)**

storm, the limits reached by inundation are even higher. The terrible hurricane disaster at Galveston, Texas, in 1900 was wrought largely by a sudden storm surge, inundating the low coastal city and drowning about 6000 persons. At the mouth of the Hooghly River on the Bay of Bengal, 300,000 persons died as a result of inundation by a 12-m (40-ft) storm surge that accompanied a severe tropical cyclone in 1737. Low-lying coral atolls of the western Pacific may be entirely swept over by wind-driven seawater, washing away palm trees and houses and drowning the inhabitants.

Important, too, is the large quantity of rainfall produced by tropical cyclones. A considerable part of the summer rainfall of some coastal regions can be traced to a few such storms. Although this rainfall is a valuable water resource, it may prove a menace as a producer of unwanted river floods and, in areas of steep mountain

Figure 8.25 **This devastation along the south coast of Haiti was caused by Hurricane Flora on October 3, 1963. (Miami News photo.)**

slopes, gives rise to disastrous mudslides and landslides.

An attempt to reduce the severity of a hurricane by cloud-seeding methods was made in 1969 by scientists of the National Oceanic and Atmospheric Administration (NOAA). The experiment, known as Project Stormfury, apparently produced significant results. A 30 percent reduction in wind speed is claimed to have been achieved for a brief period. In 1976 the research program was shifted to the eastern North Pacific Ocean, in hopes that a larger number of suitable storms would become available for experimental seeding.

Remote Sensing of Weather Phenomena

Remote sensing methods, explained in Chapter 4, have been intensively developed for weather observation through the use of specialized orbiting satellites. The first of these, Nimbus-1, was launched into a sun-synchronous polar orbit in 1964. It transmitted TV photographs as well as high-resolution infrared images, the latter taken in darkness. Later Nimbus satellites and a subsequent NOAA satellite series, all polar orbiters, have greatly improved data-collecting in various categories. Besides transmitting TV and infrared images, these weather satellites record the vertical profile of temperatures in the atmosphere. They also record global distributions of ozone, water vapor, albedo, cloud cover, precipitation, and sea ice.

Figure 8.26 shows two satellite images of an occluded wave cyclone over the eastern Pacific Ocean. These images were prepared in such a way as to form a stereoscopic pair. When viewed simultaneously, one frame with each eye, a strong three-dimensional effect results. A simple lens stereoscope (used in viewing air photos) can be placed over the page to achieve this 3-D effect.

Another class of weather satellites uses a **geostationary orbit,** in which the satellite holds a fixed position over a selected point on the earth's equator. These earth-synchronous satellites are in an equatorial orbit at a height of 36,200 km (22,300 mi); here the satellite speed exactly matches the speed of eastward rotation of the globe. The first of these satellites, called the Synchronous Meteorological Satellite (SMS), was launched in 1974 and has been followed by several more of the same type. The scanning system of the SMS sweeps over the globe from north to south, yielding an image equal to about one hemisphere (Plate C.4). Tropical storms, as well as wave cyclones and fronts of midlatitudes, can be continuously monitored by the SMS system, supplementing imagery returned by the polar-orbiting satellites.

Cyclonic Activity and the Global Environment

Reviewing the major concepts covered in this chapter, we can recognize first the important role that cyclonic storms and their fronts play in the global balances of heat and water. Together with convectional activity in isolated thunderstorms, cyclonic storms liberate enormous quantities of latent heat held in air masses as water vapor. Moist maritime air masses migrate poleward from the low-

Figure 8.26 **An occluded cyclone over the eastern Pacific Ocean shows on this satellite image as a tight cloud spiral. The cold front makes a dense, narrow cloud band sweeping to the south and southwest of the cyclone center. View these frames with a stereoscope for strong three-dimensional effect. (NOAA-2 satellite image, courtesy of National Environmental Satellite Service.)**

latitude belt of surplus energy. These air masses carry heat as well as water, and this heat is released in higher latitudes.

Cyclonic storms supply precipitation vital to native and cultivated plants of the midlatitude and tropical zones. But these same storms also produce environmental hazards to life. As yet, Man has not been able to exert far-reaching or lasting control over the enormous releases of energy in dangerous cyclonic storms. Looking into the future, however, we can anticipate much more effective methods of achieving weather modification than those now being tried. With such success may come side effects that could be unwanted and undesirable, and these will require close watching.

References for Further Study

Flora, S.D. (1954), *Tornadoes of the United States,* Univ. of Oklahoma Press, Norman.

Dunn, G. E., and B. I. Miller (1964), *Atlantic hurricanes,* Louisiana State Univ. Press, Baton Rouge.

Helm, T. (1967), *Hurricanes: weather at its worst,* Dodd, Mead and Co., New York.

Riehl, H. (1972), *Introduction to the atmosphere,* second edition, McGraw-Hill Book Co., New York. See Chapters 5, 7, 8.

Sewell, W. R. D., et al. (1973), *Modifying the weather; A social assessment,* Western Georgaphical Series, vol. 9, Department of Geography, Univ. of Victoria, British Columbia, Canada.

Byers, R. H. (1974), *General meteorology,* fourth edition, McGraw-Hill Book Co., New York. See Chapters 10, 11, 13.

Review Questions

1. How do traveling cyclones influence weather in middle and high latitudes? Name the three classes of traveling cyclones.
2. What is an air mass? What meteorological variables determine the characteristics of an air mass?
3. How are air masses classified and designated? Describe the air masses of North America, giving their source regions, paths of movement, and generally associated weather characteristics. What air masses are found over polar regions? Over the equatorial belt?
4. Describe the structure of a cold front, a warm front, and an occluded front. What types of clouds and precipitation are characteristic of each of these fronts?
5. Explain the wave theory of cyclones as originated by Jakob Bjerknes. What is the polar front?
6. How does a wave cyclone develop and change along the polar front? How are cold and warm fronts involved in the development of the wave cyclone?
7. Describe the changes in wind direction and strength, cloudiness and precipitation, and temperature that an observer experiences when a wave cyclone passes with its center north of the observer. Do the same for a cyclone in which the center passes south of the observer.
8. Describe the weather conditions associated with an anticyclone, or high, such as might normally follow a wave cyclone.
9. What is the prevailing direction of movement of cyclonic storms in the United States? Do storm tracks show any tendency to concentrate at particular places in the northern hemisphere? If so, where? How do concentrations of storm tracks influence the average pressure conditions in these areas, as shown on the world pressure maps?
10. How are wave cyclones and traveling anticyclones linked with Rossby waves and the polar-front jet stream? What importance has this relationship in weather forecasting?
11. Describe the tornado. Where does it occur in relation to air masses and fronts? In what region are tornadoes commonest? Why are tornadoes in the United States most numerous in the spring? What is a waterspout?
12. What is an easterly wave? What weather does it bring? Describe the weak equatorial low. What weather does it bring?
13. Describe polar outbreaks of tropical regions. What names do these outbreaks have locally in Central America? in South America?
14. Describe the weather conditions in a tropical cyclone, or hurricane. Where do these storms originate? What paths do they normally follow? Explain the calm central eye.
15. In what regions of the world do tropical cyclones occur? What are the seasons of occurrence? How is the season of storms related to position of the equatorial trough?
16. Describe some of the destructive effects of tropical cyclones along low-lying coasts. How is ocean level affected by these storms? What is a storm surge?
17. Describe the two basic classes of orbiting satellites used in observing global weather. What kinds of information do these satellites obtain?
18. Explain how cyclonic activity contributes to maintenance of the global heat balance.

Climate and Climate Classification

9

A goal of physical geography is to recognize and describe environmental regions of significance to Man and to all life forms generally. Which are the vital components of the environment of the life layer with respect to Man in particular and to the biosphere in general? Keep in mind that the basic food supply of the biosphere is organic matter synthesized by plants. Plants are the primary producers of organic matter, which sustains them as well as other forms of life.

Because Man is dependent on the primary producers, it seems logical to emphasize those components of the environment that are vital to plant growth. Plants occupy both marine and terrestrial environments. Because plant life of the lands is most directly exposed to the atmosphere for exchanges of energy and matter, and because most of Man's food is derived from terrestrial plants, we will want to concentrate on the land-atmosphere interface.

Two vital, atmospherically derived components vary greatly from place to place and from season to season — energy and water. Both must be in forms available to plants. Plants use the energy of solar radiation to carry out photosynthesis. (Photosynthesis is the process by which carbohydrate molecules are synthesized.) Plants also require carbon dioxide and water to synthesize carbohydrate compounds. Carbon dioxide, however, is largely uniform in its atmospheric concentration over the globe at all seasons, so we can omit it as a variable factor in the environment.

Plants require the presence of sensible heat as measured by air and soil temperatures within specified limits. Plants also require water in the root zone of the soil. Although land animals are consumers they, too, require fresh water and a tolerable range of surrounding temperatures. Plants also release energy and water to the atmosphere through the process of respiration, in which the carbohydrate molecules are broken down into carbon dioxide and water. (Photosynthesis and respiration are discussed in Chapter 12.)

Besides energy, carbon dioxide, and water, plants need nutrients, which they derive from the soil. The nutrients are released by the plants when the plant tissue dies. In this way the nutrients are returned, or recycled, to the soil. (Nutrient cycling is discussed in Chapter 12.)

To summarize, land plants require light energy, carbon dioxide, water, and nutrients. The input of these ingredients comes from two basic sources: (1) the adjacent atmosphere and (2) the soil. Input of light energy, heat energy, and water from the atmosphere is encompassed by the concept of climate. In short, plants depend on climate and soil. Of these two sources of energy and matter, the soil is the more strongly affected by the plants it serves through the recycling of matter. Climate, when broadly defined as a source of energy and water, is an independent agent of control. Climate is determined by latitude and by large-scale air motions and air-mass interactions within the troposphere. If we are to establish a pyramid of priorities and interactions, climate occupies the apex as the independent control (Figure 9.1). Below it and forming the base of the pyramid are (1) the organic process of plants and (2) the soil process. Plants and soil interact with one another at the basal level.

Figure 9.1 **The role of climate in environmental processes of the life layer.**

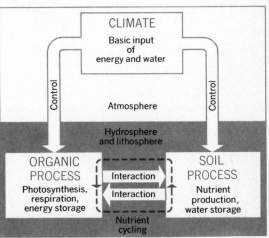

125

Climate and Climate Classification

Climate has always been a keystone in physical geography and has formed the basis for defining physical regions of the globe. Let us first examine the content of traditional **climatology,** the science of climate. In the broadest sense, **climate** is the characteristic condition of the atmosphere near the earth's surface at a given place or over a given region. Components that enter into the description of climate are mostly the same as weather components used to describe the state of the atmosphere at a given instant. If weather information deals with the specific event, climate represents a generalization of weather. A statement of the climate of a given observing station, or of a designated region, is described through the medium of weather observations accumulated over many years' time. Not only are mean, or average, values examined, but the departures from those means and the probabilities that such departures will occur are also taken into account.

Earlier chapters contained much information falling within the definition of climate. For example, world maps showing average January and July barometric pressures, winds, and air temperatures are expressions of climate. The physical components of climate are many; they include measurable quantities such as radiation, sensible heat, barometric pressure, winds, relative and specific humidity, dew point, cloud cover and type, fog, precipitation type and intensity, evaporation and transpiration, incidence of cyclones and anticyclones, and frequency of frontal passages. Which of these components are really significant in our analysis of climate? Approaching climatology as geographers, we shall take only what we need.

Our problem in devising a meaningful system of climate classes (i.e., a classification of climates) is to select categories of available information that correlate closely with the needs of life on the lands. We then use those categories to define and delineate regional classes of climates. If we have done our work well, the regional units within the climate system will reflect strongly the control of the atmosphere on terrestrial life. This same system will give some indication of the opportunities and constraints that the atmospheric environment imposes on humans as they seek to increase their food supplies and water resources at the same time that they extend the areas of urban and industrial land use. The climate classification we seek must have utility in guiding land-use planning and population growth as well as describing the natural environment. This role is particularly crucial in guiding the progress of the developing nations as they seek to augment inadequate food resources.

Let us now select those categories of information that can produce a practical and meaningful climate system.

Net Radiation as a Basis of Climate Classification

Among geographers, the proposal to use net radiation as the basis of a global classification of climates is a recent event. A preliminary attempt to set up a classification system using insolation was made by Werner H. Terjung in 1970; it was followed in the same year by his preliminary system of climate classification using net radiation.

Recall from Chapter 4 that net radiation is an expression of the difference between incoming and outgoing energy in both shortwave and longwave forms. Referring back to Figure 5.10, we see that the annual cycle of net radiation shows striking differences from equatorial to subarctic zones. Terjung analyzed the annual curves of net radiation with respect to the maximum values, the annual range of values, and the form of the curve. Combining these parameters, he derived a system of climate types and a world map showing their distribution.

Net radiation is usually regarded as the best indicator of energy available for plant growth. In combination with a measure of available water in the form of soil moisture, net radiation may eventually succeed as the basis of classifying world climates. At present, however, more conventional systems using other climatic variables are dominant in physical geography; it is to these we turn next.

Air Temperature as a Basis for Climate Classification

All species of organisms, whether plants or animals, are subject to limiting temperatures of the surrounding air, water, or soil; survival is not possible above or below these limits. Few growing plants can survive temperatures over 50°C (120°F) for more than a few minutes. Many species of plants native to the tropical and equatorial latitude zones die if they experience even a short period of below-freezing temperature (0°C; 32°F). Freezing of water in the plant tissues causes physical disruption in plants not adapted to such conditions. Alternate freezing and thawing of the soil can have a disruptive effect on plant roots and is an important factor in limiting plant growth in the arctic and alpine zones.

Apart from extreme limits of temperature tolerance, plants react to an increase in air and water temperature through an increase in the intensity of physical and chemical activity. The optimum rates at which both photosynthesis and respiration take place increase with rising temperature in the range from near-freezing to 20° to 25°C (70° to 80°F), after which the rate of photosynthesis begins to decline. Respiration, however, continues to increase in rate to considerably higher temperatures. Knowing that plants are strongly influenced by the factor of sensible air and soil temperatures that surround them, we include air temperature as one of the essential climate factors.

Besides being an important environmental factor in plant physiology and reproduction, air temperature enters into many activities of animal life (e.g., hibernation and migration). For Man, air temperature is an important physiological factor and relates directly to the quantity of energy expended in space heating and air conditioning of buildings. Nevertheless, air temperature alone does not define meaningful climate classes because the ingredient of water availability is missing.

Temperature of the lower air layer, as measured in the standard thermometer shelter (Chapter 5), has long provided one essential variable quantity in leading systems of climate classification. Monthly data based on daily readings of the maximum-minimum thermometer have been accumulated for many decades at thousands of observing stations the world over. Consequently, availability has been an important factor in favoring the use of air temperature in climate classification.

Thermal Regimes

In Chapter 5 we studied the annual air temperature cycle and its relationship to the annual energy balance cycle. Figure 5.11 showed annual temperature cycles for four stations, ranging from the equatorial zone to the subarctic zone. Let us now carry this investigation of annual temperature cycles a step further to explore the global range of annual cycles.

A comparison of annual air temperature cycles for many observing stations over the globe allows us to recognize a number of distinctive types, which can be called **thermal regimes.** A few of these are shown in Figures 9.2 and 9.3. Each regime has been labeled

according to the latitude zone in which it lies; e.g. equatorial, tropical, midlatitude, and subarctic. Some labels also include words descriptive of the location in terms of position with respect to landmass. "Continental" refers to a continental interior location; "west coast" and "marine" refer to a location close to the ocean, usually on the western side of the continent.

The equatorial regime is uniformly very warm; temperatures are close to 27°C (80°F) year around, and there are no temperature seasons. The tropical continental regime ranges from very hot when the sun is high, near one solstice, to mild at the opposite solstice. Close to the ocean, however, we find the tropical west coast regime with only a weak annual cycle and no extreme heat. The same weak annual cycle can be traced into the midlatitude west coast regime, and it persists even much farther poleward. In the continental interiors, the strong annual cycle prevails through the midlatitude continental regime into the subarctic continental regime, where the annual range is enormous. The ice sheet regime of Greenland is in a class by itself, with severe cold all year. Other regimes can be identified and, because they grade into one another, the list could be expanded indefinitely.

Two important concepts are illustrated in Figures 9.2

Figure 9.2 **Some important thermal regimes, represented by annual cycles of air temperature. (Based on Goode Base Map.)**

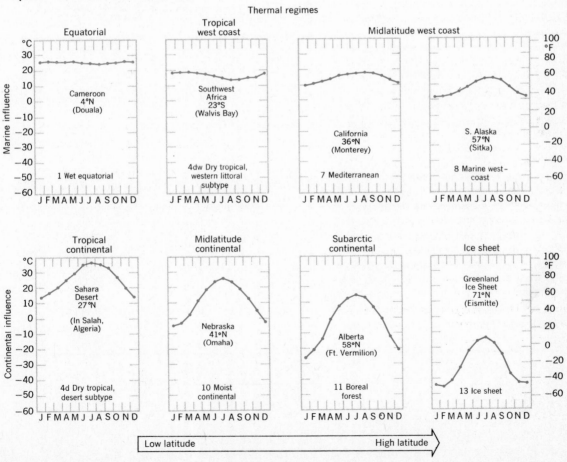

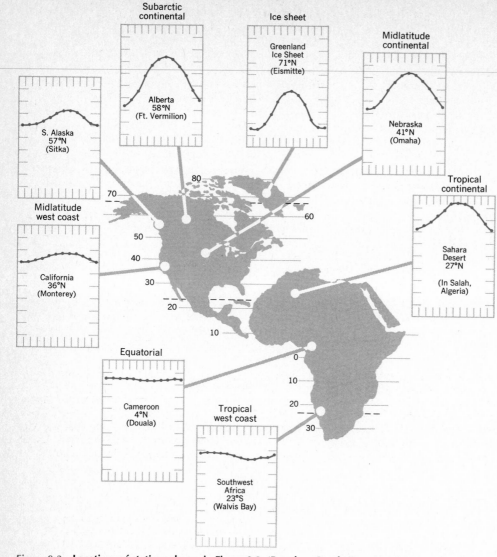

Figure 9.3 Locations of stations shown in Figure 9.2. (Based on Goode Base Map.)

and 9.3. One concept is that of **continentality,** the tendency of a large landmass to impose a large annual range on the annual cycle. Continentality becomes stronger with higher latitude because the insolation cycle shows stronger seasonal extremes with increasing latitude. Another concept is that of the marine influence, which tends to weaken the annual cycle and hold temperatures in a moderate range. This effect is, of course, due to the capacity of the oceans to hold a large amount of heat in storage and to take up and give off that heat very slowly, as compared with land areas.

Precipitation as a Basis for Climate Classification

Precipitation data, obtained by the simple rain gauge (Chapter 7), are abundantly available for long periods of record for thousands of observing stations widely distributed over the globe. Small wonder, then, that monthly and annual precipitation data form the cornerstone of most of the widely used climate classifications.

Average annual precipitation is shown on a world map, Plate D.3. This map uses isohyets, lines drawn through all points having the same annual quantity. For regions where all or most of the precipitation is rain, we use the word "rainfall"; for regions where snow is an important part of the annual total, we use the word "precipitation."

Global patterns of precipitation are related to air-mass source regions and prevailing movements of air masses. Seven precipitation regions can be recognized in terms of annual total in combination with location. Details are given in Table 9.1. Figure 9.4 is a schematic diagram of the global distribution of precipitation, simplifying the picture as much as possible in terms of an imaginary continent. Notice that the adjectives "wet, "humid," "subhumid," "semiarid," and "arid" have been applied to each of five levels of annual precipitation, which range from 200 cm (80 in.) down to zero.

Table 9.1 / World Precipitation Regions

Name	Latitude Range	Continental Location	Prevailing Air Mass	Annual Precipitation Centimeters	Inches
1. Wet equatorial belt	10°N to 10°S	Interiors, coasts	mE	Over 200	Over 80
2. Trade-wind coasts (windward tropical coasts)	5–30° N and S	Narrow coastal zones	mT	Over 150	Over 60
3. Tropical deserts	10–35° N and S	Interiors, west coasts	cT	Under 25	Under 10
4. Midlatitude deserts and steppes	30–50° N and S	Interiors	cT, cP	10–50	4–20
5. Moist subtropical regions	25–45° N and S	Interiors, coasts	mT (summer)	100–150	40–60
6. Midlatitude west coasts	35–65° N and S	West coasts	mP	Over 100	Over 40
7. Arctic and polar deserts	60–90° N and S	Interiors, coasts	cP, cA	Under 30	Under 12

1. The wet equatorial belt of heavy rainfall, over 200 cm (80 in.) annually, straddles the equator and includes the Amazon River basin in South America, the Congo River basin of equatorial Africa, much of the African coast from Nigeria west to Guinea, and the East Indies. Here the prevailingly warm temperatures and high moisture content of the maritime equatorial (mE) air mass favor abundant convectional rainfall. Thunderstorms are frequent year around.

2. Narrow coastal belts of high rainfall, 150 to 200 cm (60 to 80 in.), and locally even more, extend from near the equator to latitudes of about 25° to 30°N and S on the eastern sides of every continent or large island. For example, see the eastern coasts of Brazil, Central

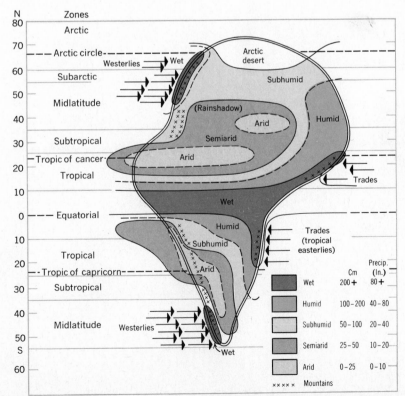

Figure 9.4 A schematic diagram of the distribution of annual precipitation over an idealized continent and adjoining oceans.

America, Madagascar, and northeastern Australia. These are the trade-wind coasts, where the moist maritime tropical (mT) air mass from warm oceans is brought over the land by the trades (tropical easterlies). Encountering coastal hills and mountains, this air mass produces heavy orographic rainfall.

3. In striking contrast to the wet equatorial belt astride the equator are the two zones of vast tropical deserts lying approximately on the tropics of cancer and capricorn. These are hot, barren deserts, with less than 25 cm (10 in.) of rainfall annually and in many places with less than 5 cm (2 in.). They are located under and are caused by the subtropical cells of high pressure, where the subsiding continental tropical (cT) air mass is adiabatically warmed and dried. Such little rain as these areas experience is largely convectional and extremely unreliable. Note that the tropical deserts extend off the west coasts of the lands and out over the oceans.

4. Farther northward, in the interiors of Asia and North America between lat. 30° and 50°, are great continental midlatitude deserts and expanses of semiarid grasslands known as steppes. Annual precipitation ranges from less than 10 cm (4 in.) in the driest areas to 50 cm (20 in.) in the moister steppes. Dryness here results from remoteness from ocean sources of moisture. Located in a region of prevailing westerly

winds, these arid lands occupy the position of rainshadows in the lee of coastal mountains and highlands. For example, the Cordilleran Ranges of Oregon, Washington, British Columbia, and Alaska shield the interior of North America from the moist maritime polar (mP) air mass originating in the Pacific. While descending into the intermontane basins and interior plains, the mP air mass is warmed and dried.

Similarly, mountains of Europe and the Scandinavian peninsula obstruct the flow of moist mP air masses from the North Atlantic into western Asia. The great southern Asiatic ranges likewise prevent the entry of moist mT and mE air masses from the Indian Ocean.

The southern hemisphere has too little land in the midlatitudes to produce a broad continental desert, but the dry north-south strip of Patagonia, lying on the lee side of the Andean chain, is roughly the counterpart of the North American deserts and steppes of Oregon and northern Nevada.

5. On the southeastern sides of the continents of North America and Asia, in lat. 25° to 45°N, are the moist subtropical regions, with 100 to 150 cm (40 to 60 in.) of precipitation annually. Smaller areas of the same kind are found in the southern hemisphere in Uruguay, Argentina, and southeastern Australia. These regions lie on the moist western sides of the subtropical high-pressure centers in such a position that the humid mT

Figure 9.5 **Eight precipitation types selected to show various seasonal patterns.**

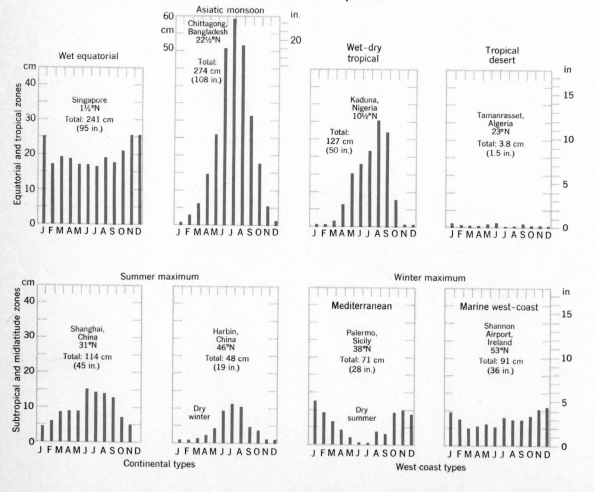

air mass from the tropical ocean is carried poleward over the adjoining land. Commonly, too, these areas receive very heavy rains from tropical cyclones.

6. Another distinctive wet location is on midlatitude west coasts of all continents and large islands lying between about 35° and 65° in the region of prevailing westerly winds. These zones were mentioned in Chapter 7 as good examples of coasts on which abundant orographic precipitation falls as a result of forced lift of mP air masses. Where the coasts are mountainous, as in Alaska and British Columbia, southern Chile, Scotland, Norway, and South Island of New Zealand, the annual precipitation is over 200 cm (80 in.). These rugged coasts formerly supported great valley glaciers that carved the deep bays (fiords) so typically a part of their scenery.

7. A seventh precipitation region is formed by the arctic and polar deserts. Northward of the 60th parallel, annual precipitation is largely under 30 cm (12 in.), except for the west-coast belts. Cold continental polar (cP) and continental arctic (cA) air masses cannot hold much moisture; consequently, they do not yield large amounts of precipitation. At the same time, however, evaporation rates are low.

As you would expect, zones of gradation lie between these precipitation regions. Moreover, this list does not recognize the fact that seasonal patterns of precipitation differ from region to region.

Seasonal Patterns of Precipitation

Although the annual total precipitation is a useful quantity in establishing the characteristics of a climate type, it can be a misleading statistic because there may be strong seasonal cycles of precipitation. It makes a great deal of difference in terms of native plants and agricultural crops if there are alternately dry and wet seasons instead of a uniform distribution of precipitation throughout the year. It also makes a great deal of difference whether the wet season coincides with a season of higher temperatures or with a season of lower temperatures because plants need both water and heat.

The seasonal aspects of precipitation can be largely covered by three major patterns: (1) uniformly distributed precipitation; (2) a precipitation maximum during the summer (or season of high sun), when insolation is at its peak; and (3) a precipitation maximum during the winter or cooler season, when insolation is least. The first pattern can include a range of possibilities from little or no precipitation in any month to abundant precipitation in all months.

A study of monthly means of precipitation throughout the year at many observing stations over the globe shows a number of outstanding precipitation types. Some of these are illustrated in Figures 9.5 and 9.6. For each, the average monthly precipitation is shown by the height of the bar in

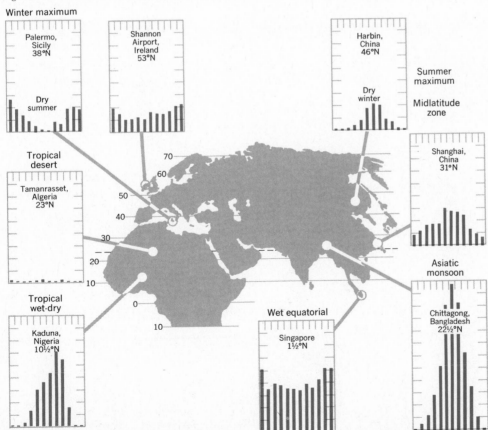

Figure 9.6 Locations of stations shown in Figure 9.5. (Based on Goode Base Map.)

the small graphs.

Dominant in very low latitudes is an equatorial rainy type, illustrated by Singapore, near the equator. Rain is abundant in all months, but some months have considerably more than others. The tropical desert type may have so little rain in any month that it cannot be shown on the graph. The wet-dry tropical type has a very wet season at the time of high sun (summer solstice) and a very dry season at the time of low sun (winter solstice). This seasonal alternation is carried to its greatest extreme in the Asiatic monsoon type, with an extremely wet high-sun monsoon season and a very dry period at low sun.

The summer precipitation maximum is carried into higher latitudes on the eastern sides of continents in a subtropical moist type. This same feature persists into the midlatitude continental type, which has a long, dry winter but a marked summer rain period.

A cycle with a winter precipitation maximum is typical of the west sides of continents in midlatitudes. A Mediterranean type, named for its prevalence in the Mediterranean region, has a very dry summer but a moist winter. This same cycle is carried into higher midlatitudes along narrow strips of west coasts. There is precipitation throughout the year in the marine west-coast type, but with a distinct maximum in winter and a distinct minimum in summer.

Each of these types will appear in more detail in later descriptions of individual climates. The reasons for annual precipitation cycles will be developed further in terms of shifting belts of pressure and winds and the seasonal changes in domination by different air masses.

A Climate System Based on Air Masses and Frontal Zones

Recall from Chapter 8 that air masses are classified according to the general latitude of their source regions and the surface qualities of the source region, whether land or ocean. This air-mass classification has built into it a recognition of the factors of temperature and precipitation because (1) air-mass temperature decreases poleward and (2) precipitation capability of an air mass is related to sources of moisture. Source regions of air masses change seasonally along with shifting belts of air temperature, pressure, and winds. Frontal zones also migrate seasonally. Thus seasonal cycles of both air temperature and precipitation can be described in terms of changing air-mass activity.

The climate system we shall develop uses the concepts of thermal regimes and precipitation types, but explains them in terms of air-mass activity. This explanatory system makes use of all the global climatic information developed systematically in previous chapters. The system allows that information is to be used to interpret a given climate and to appreciate the reasons for its occurrence in a particular location.

Our climate system is based on location of air-mass source regions and the nature and movement of air masses, fronts, and cyclonic storms. A schematic global diagram (Figure 9.7) shows the principles of the classification. Notice that this figure is basically the same as Figure 8.1.

Figure 9.7 **The three major climate groups are related to air-mass source regions and frontal zones.**

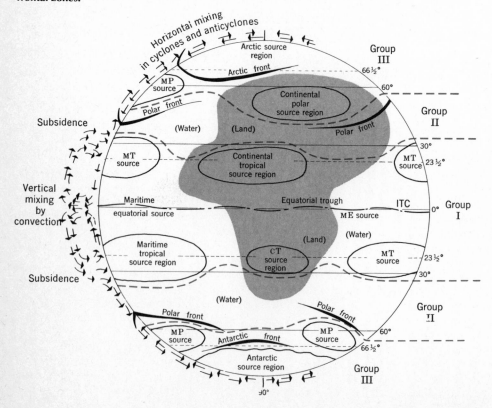

Group I Low-Latitude Climates

Group I includes the tropical air-mass source regions and the intertropical convergence zone (ITC) that lies between. Climates of Group I are controlled by the subtropical high-pressure cells, or anticyclones, which are regions of air subsidence and are basically dry, and by the great equatorial trough of convergence that lies between them. Although air of polar origin occasionally invades the tropical and equatorial zones, the climates of Group I are almost wholly dominated by the tropical and equatorial air masses. Tropical cyclones are important in this climate group.

Group II Midlatitude Climates

Climates of Group II are in a zone of intense interaction between unlike air masses: the **polar front zone.** Tropical air masses moving poleward and polar air masses moving equatorward are in conflict in this zone, which contains a procession of eastward-moving wave cyclones. Locally and seasonally, either tropical or polar air masses may dominate in these regions, but neither has exclusive control.

Group III High-Latitude Climates

Climates of Group III are dominated by polar and arctic (including antarctic) air masses. The two polar continental air-mass source regions of northern Canada and Siberia fall into this group, but there is no southern hemisphere counterpart to these continental centers. In the arctic belt of the 60th to 70th parallels, air masses of arctic origin meet polar continental air masses along an **arctic front zone,** creating a series of eastward-moving wave cyclones.

Climate Types

Within each climate group are a number of **climate types** (or simply, *climates*): four in Group I, six in Group II, and three in Group III, for a total of thirteen climate types. For ease in identification, the climate types are numbered; but the numbers have no significance as a code. Ordering of climates within each group is in many cases quite arbitrary, particularly for the climates in Group III. You need use only the climate names because these are descriptive of climate quality and global location.

In presenting the 13 climate types, we shall make use of a graphic device called a **climograph** (Figure 9.8). It shows simultaneously the cycles of monthly mean air temperature and monthly mean precipitation, combining the temperature and precipitation graphs used in Figures 9.2 and 9.5. The climograph allows you to integrate at a glance the important basic features of the climate.

Another useful graphic device is the generalized diagram of the climate groups, climate types, and climate subtypes shown in Plate D.1. Boundaries are drawn in smooth, simple lines on an imaginary supercontinent, as we did for precipitation (Figure 9.4). Coastal highlands are shown, because they have strong local effects in producing orographic recipitation belts and rain-shadows. The diagram shows typical positioning of each climate type and the common boundaries between climates. Boundaries between groups are labeled.

A world map of climates, Plate D.2, shows the actual distribution of climate types and subtypes on the continents. This map is based on available data of a large number of observing stations. It is, however, a simplified map; the climate boundaries are uncertain in many areas where observing stations are thinly distributed. Precise definitions of the climates and their boundaries are presented in Chapter 10. In the remainder of this chapter, we will analyze the 13 climates in purely descriptive terms, using actual figures of temperature and precipitation very broadly and only for purposes of example.

Dry and Moist Climates

All but two of the thirteen climate types are classified as either dry climates or moist climates. **Dry climates** are those in which evaporative water losses from the soil and from plant cover exceed the incoming precipitation by a wide margin. Generally speaking, the dry climates do not support permanent streams and the land surface is clothed with sparse plant cover — either grasses or shrubs — or lacks a plant cover. **Moist climates** are those with sufficient rainfall to maintain the soil in a moist condition through much of the year and to sustain the year-round flow of the larger streams. Moist climates support forests or prairies of tall grasses. A precise distinction between dry climates and moist climates is given in connection with the Köppen climate system (discussed later in this chapter) and the newer climate system based on the soil-water balance (Chapter 10).

Within the dry climates is a wide range of degree of aridity, ranging from very dry deserts through transitional levels of aridity to arrive at adjacent moist climates. We will refer to three dry climate subtypes: semiarid (or steppe), semidesert, and desert. The *semiarid subtype* (steppe subtype), designated by the letter *s,* has enough precipitation to support sparse grasslands of a variety referred to by geographers as *steppes,* or *steppe grasslands.* The steppe climate subtype will be found adjacent to moist climates. The *desert subtype,* letter *d,* is extremely dry and has so little precipitation that only scattered hardy plants can grow. The *semidesert subtype,* letters *sd,* is transitional between semiarid and desert subtypes. (Precise definitions of these subtypes and their boundaries are stated in Chapter 10.)

Within the moist climates is a wide range of degree of wetness. On the one hand is a *subhumid subtype,* designated by the letters *sh,* in which the evaporative losses from soil and plant cover approximately balance the precipitation on an annual average basis. Where precipitation is great enough to produce stream flow through most of the year and to support forests, the *humid subtype* prevails, designated by the letter *h.* Where precipitation is very heavy, with copious stream flow, the *perhumid subtype* prevails, letter *p.* (Precise definitions and boundaries of these subtypes are given in Chapter 10.)

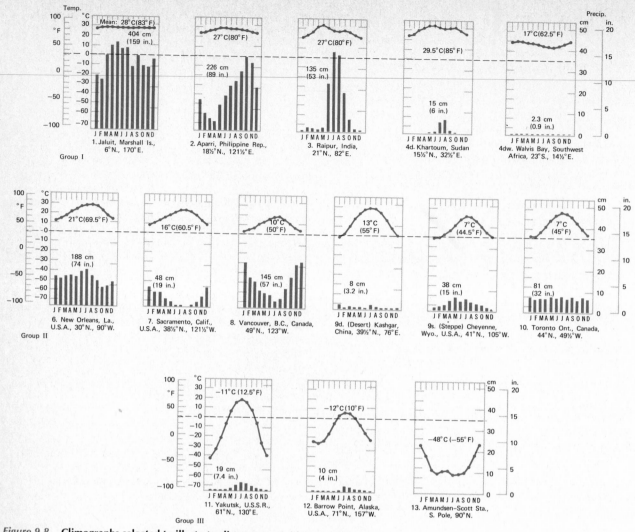

Figure 9.8 Climographs selected to illustrate climate types within each climate group.

DESCRIPTIONS OF CLIMATES

GROUP I / LOW-LATITUDE CLIMATES
(controlled by equatorial and tropical air masses)

1. Wet equatorial climate Latitude range: 10°N to 10°S

Climate of the intertropical convergence zone (ITC), is dominated by warm, moist maritime equatorial (mE) and maritime tropical (mT) air masses yielding heavy convectional rainfall. Rainfall is copious in all months, annual total often being over 250 cm (100 in.). Strong seasonal differences in monthly rainfall are typical and can be attributed to changes in position of the ITC and to local orographic effects. Remarkably uniform temperatures prevail throughout the year. Both monthly mean and mean annual temperatures are typically close to 27°C (80°F).

Major regions of occurrence: Amazon lowland of South America; Congo Basin of equatorial Africa; and East Indies, from Sumatra to New Guinea.

Example. Figure 9.9 is a climograph for Iquitos, Peru, a typical wet equatorial station located close to the equator in the broad, low basin of the upper Amazon River. Note that the annual range in temperature is only 2.2C° (4°F) and that the annual rainfall total is just over 250 cm (100 in.). In all but one month, the mean monthly rainfall is more than 15 cm (6 in.).

2. Monsoon and trade-wind littoral climate Latitude range: 5° to 25°N and S

Trade winds (tropical easterlies) bring maritime tropical (mT) air masses from moist western sides of oceanic high-pressure cells to give heavy orographic rainfall along narrow eastern littorals. (*Littoral* means a "narrow coastal zone.") Shower activity is intensified by arrivals of easterly waves. Rainfall shows a strong annual cycle, peaking in the season of high sun, when the ITC is close by. A marked

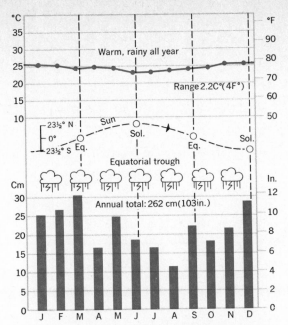

Figure 9.9 Iquitos, Peru, is a rainforest station near the equator. Temperatures differ very little from month to month, and there is copious rainfall throughout the year.

short season of reduced rainfall occurs at time of low sun. In Southeast Asia, the rainy period represents the summer monsoon season. Temperatures are very warm throughout the year, but with a marked annual cycle. Minimum temperatures occur at time of low sun.

Major regions of occurrence: Trade-wind littorals occur along the east sides of Central and South America, the

Caribbean Islands, Madagascar (Malagasy), Indochina, the Philippines, and northeast Australia. Monsoon west coasts are found in western India (Malabar Coast) and Burma. Large inland extensions of the monsoon climate occur in Bangladesh and Assam, central and western Africa, and southern Brazil.

Example. Figure 9.10 is a climograph for Belize City, Belize, on the Central American coast at lat. 17°N, exposed to the tropical easterlies. Rainfall is copious from June through November, when the ITC is in this latitude zone. Easterly waves are common in this season, and an occasional tropical cyclone strikes the coast, bringing torrential rainfall. Following the winter solstice (end of December), rainfall is greatly reduced, with minimum values in March and April. At this time, the ITC lies farthest away and the climate is dominated by the subtropical high-pressure cell. Air temperatures show an annual range of 5°C (9°F), with maximum in the high-sun months.

Example. Figure 9.11 is a climograph for Cochin, India, lat. 10°N. Located on the west coast of lower peninsular India, Cochin occupies a windward location with respect to the southwest winds of the rainy (summer) monsoon season. In the rainy season, monthly rainfall is extremely high — over 64 cm (25 in.) per month in June and July. A strongly pronounced season of low rainfall occurs at time of low sun — December through March. Air temperatures show only a very weak annual cycle, cooling a bit during the rains; but the annual range is small at this low latitude.

Figure 9.11 Cochin, India, on a windward coast at lat. 10°N, shows an extreme peak of rainfall during the rainy monsoon, contrasting with a short dry season at time of low sun.

Figure 9.10 This climograph for Belize, a Central American east-coast city at lat. 17°N, shows a marked season of low rainfall following the period of low sun.

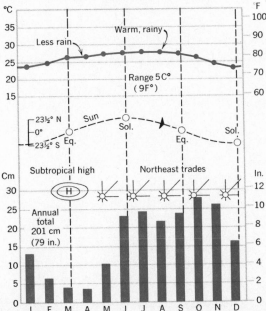

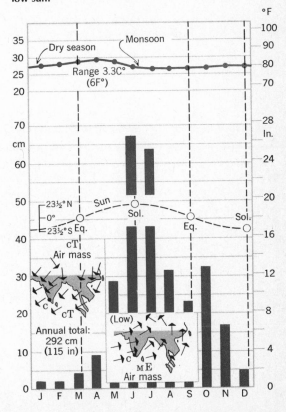

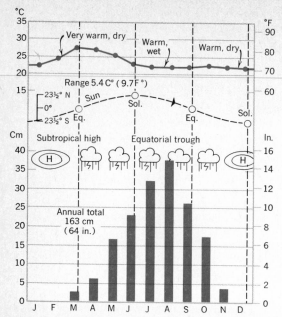

Figure 9.12 **Timbo, Guinea, at lat. 10½°N, is in the wet-dry tropical climate of West Africa. A long wet season at time of high sun alternates with an almost rainless dry season at time of low sun.**

3. Wet-dry tropical climate Latitude range: 5° to 20°N and S (Asia: 10° to 30°N)

Seasonal alternation occurs between dominance of moist mT or mE air masses along the ITC and dry continental tropical (cT) air masses of the subtropical high-pressure belt. As a result, there is a very wet season at time of high sun and a very dry season at time of low sun. Cooler temperatures accompany the dry season, but give way to a very hot period before the rains begin.

Major regions of occurrence: India; Indochina; West Africa; southern Africa; South America, both north and south of the Amazon lowland; and north coast of Australia.

Example. Figure 9.12 is a climograph for Timbo, Guinea, at lat. 10½°N in West Africa. The rainy season begins just after vernal equinox, reaching a peak in July and August, or about two months following summer solstice. At this time, the ITC has migrated to its most northerly position and large infusions of moist mE air masses occur from the ocean lying to the south. Rainfall then declines as the low-sun season arrives. Three months — December through February — are practically rainless. At this season the subtropical high-pressure cell dominates the climate, with a stable subsiding cT air mass over the area. The temperature cycle is linked closely with precipitation. In February and March air temperature rises sharply, bringing on a brief hot season. As soon as the rains set in, the effect of cloud cover and evaporation of rain causes the temperature to decline. By July temperatures have resumed an even level at about 22°C (72°F).

Example. Figure 9.13 is a climograph for Allahabad, India, showing the wet-dry tropical climate with a strong Asiatic monsoon influence. Notice that this city lies at a much

higher latitude (25°N) than Timbo, Guinea (10½°N), or about 1450 km (900 mi) farther north. The onset of the rainy season at Allahabad is much more sudden than at Timbo. The annual range of temperature at Allahabad, about 18°C (32°F), is much greater than at Timbo. The hot season at Allahabad reaches extreme values — over 32°C (90°F) — and there is a marked cool season at time of low sun, when monthly mean temperature falls to 18°C (65°F).

4. Dry tropical climate Latitude range: 15° to 25°N and S

The tropical dry climate occupies source regions of the cT air mass in high-pressure cells centered over the tropics of cancer and capricorn. This subsiding air mass is stable and dry, becoming highly heated at the surface under intense insolation. A strong annual temperature cycle follows the changing declination of the sun. The high-sun period brings extreme heat; the low-sun season, a comparatively cool season. Extremely dry areas, recognized as a desert subtype (4d), are over the tropics of cancer and capricorn. This dry zone grades on the equatorward side through a narrow zone of semidesert subtype (4sd), into a semiarid, or steppe climate subtype (4s). In this transition zone, a short rainy season occurs, grading into the long rainy season of the wet-dry tropical climate (3). A narrow western littoral zone (4dw) has a cool, uniform thermal regime because of the presence of a cool marine air layer.

Major regions of occurrence: Sahara-Arabia-Iran-Thar desert belt of North Africa and southern Asia; a large part of Australia; small areas in Central America; South America; and South Africa. Important areas of the steppe subtype (4s) are found in India and Thailand, with many small, scattered dry areas to the lee of highlands in the trade-wind belt.

Figure 9.13 **Allahabad, India, lat. 25°N, lies in the rich agricultural lowland of the Ganges River. Here the monsoon rains are usually ample, but there is a long dry season. Note the very high temperatures in May and June, just before the onset of rains.**

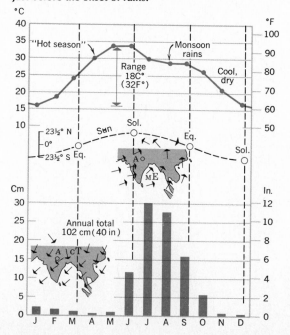

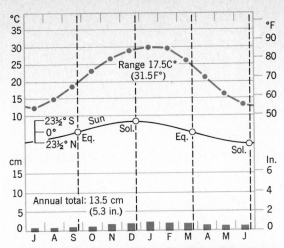

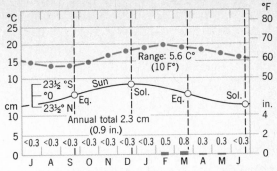

Figure 9.16 Walvis Bay, Namibia, is a desert station on the west coast of southwest Africa at lat. 23°S. Air temperatures are cool and remarkably uniform throughout the year.

Figure 9.14 Charlotte Waters, in the heart of the Great Australian Desert, lat. 26°S, represents the tropical desert climate. Note that the yearly cycle begins with July and that the temperature peak occurs in January.

Example. Figure 9.14 is a climograph of a tropical desert (4d) station in the interior of Australia. Charlotte Waters, located at lat. 26°S, shows a strong annual temperature cycle with a hot season in time of high sun and a comparatively cool season at low sun. (Note that the graph begins with July, so that the peak of insolation — December solstice — is placed in the center of the graph.) The annual temperature range is 17.5C° (31.5 F°). Annual precipitation total is a mere 13.5 cm (5.3 in.). There is a marked annual precipitation cycle, with maximum at time of high sun, but no monthly average is more than 0.2 cm (0.8 in.).

Figure 9.15 Kayes, Mali, at lat. 14½°N, lies in the Sahel of West Africa. There is a short wet season in normal years. The dry season is long, with a succession of rainless months.

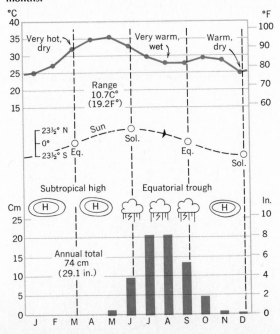

Example. Figure 9.15 is a climograph for Kayes, Mali, located at lat. 14½°N in West Africa. This station lies in the tropical semiarid or steppe subtype (4s) and is in the transition zone between the dry Sahara Desert to the north and the wet-dry tropical climate (3) immediately to the south (see climograph of Timbo, Figure 9.12). The short wet season brings a total annual rainfall of 75 cm (29 in.), mostly in four months. The long dry season is nearly rainless for four consecutive months. Extremely high temperatures characterize the hot season that precedes the rains.

Example. Figure 9.16 shows the desert climate typical of narrow western littorals (4dw). The station is Walvis Bay, a port city on the west coast of Southwest Africa (Namibia), at lat. 23°S. (The yearly cycle begins with July because this is a southern hemisphere station.) For a location nearly on the tropic of capricorn (23½°S), the monthly temperatures are remarkably cool; the warmest month mean is only 19°C (67°F). The coolest month is 14°C (57°F), making an annual range of only 5C° (10 F°). The cool Benguela Current, with upwelling of cold, deep water, chills the lower air layer. This condition explains both coolness and uniformity of the air temperatures. Coastal fog is a persistent feature of this littoral climate.

GROUP II / MIDLATITUDE CLIMATES
(controlled by both tropical and polar air masses)

5. Dry subtropical climate Latitude range: 25° to 35°N and S

Caused by the same air-mass patterns, the dry subtropical climate is simply a poleward extension of the dry tropical climate. The annual temperature range is greater than that for the dry tropical climate. There is a distinct cool season in the lower latitude portions and a cold season in the higher latitude portions. The cold season, occurring at time of low sun, is due in part to invasions of continental polar (cP) air masses from higher latitudes, and in this way the climate shows the influence of polar air masses. Precipitation that occurs in the low-sun season is produced by midlatitude cyclones that make incursions into the subtropical zone. As in the case of the dry tropical climate, subtypes include steppe subtype (5s), semidesert subtype (5sd), and desert subtype (5d).

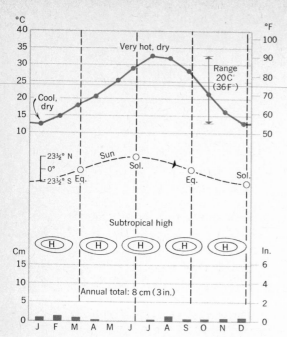

Figure 9.17 **Yuma, Arizona, lat. 33°N, is a North American example of the subtropical desert climate.**

Major regions of occurrence: North Africa; Near East (Jordan, Syria, Iraq); southwestern United States and northern Mexico; southern part of Australia; Argentina (Pampa and Patagonia); and South Africa.

Example. Figure 9.17 is a climograph for Yuma, Arizona, a city close to the Mexican border at lat. 33°N. The annual temperatures show a strong seasonal cycle with hot summers. A cold season brings monthly means as low as

13°C (55°F). The annual range is 20C° (36 F°). Precipitation, which totals about 8 cm (3 in.), is small in all months, but with two maxima. The August maximum is caused by invasions of mT air masses, which result in thunderstorm activity. Higher rainfalls from December through March are produced by midlatitude cyclones following a southerly path. Two months, May and June, are nearly rainless.

6. Moist subtropical climate Latitude range: 20° to 35°N and S

Subtropical eastern continental margins are dominated by the mT air mass, flowing out from the moist western sides of the oceanic high-pressure cells. This air mass brings copious summer rainfall, much of it convectional. An occasional tropical storm brings heavy rainfall. Winter precipitation is also copious, produced in midlatitude cyclones. Invasions of the cP air mass are frequent in winter, bringing spells of subfreezing weather. Summers are warm, with persistent high humidity. No winter month has a mean temperature lower than 0°C (32°F). Subhumid, humid, and perhumid (very wet) subtypes are identified. This climate is characterized in Southeast Asia by a strong monsoon effect, with increased summer rainfall.

Major regions of occurrence: Southeastern United States, southern China; Formosa, southernmost Japan, Uruguay and adjoining parts of Brazil and Argentina, and eastern coast of Australia.

Example. Figure 9.18 is a climograph for Charleston, South Carolina, located on the eastern seaboard at lat. 33°N. In this region, a marked summer maximum of precipitation is typical. Total annual rainfall is copious (120 cm, 47 in.); at least 5 cm (2 in.) of precipitation falls in every month. The annual temperature cycle is strongly developed, with an annual range of about 17C° (31F°).

Figure 9.18. **Charleston, South Carolina, lat. 33°N, has a mild winter and a warm summer. There is ample precipitation in all months, but a definite summer maximum.**

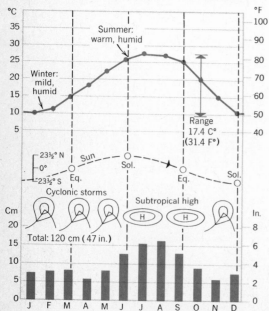

Figure 9.19 **Nagasaki, in southern Japan, lat. 33°N, shows the effect of the monsoon by the strong rainfall peak in summer.**

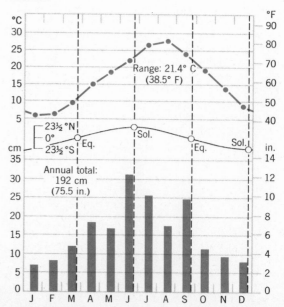

Winters are mild, with the January mean temperature a full 10C° (18F°) above the freezing mark.

Example. Figure 9.19 is a climograph for Nagasaki, a west-coast city at lat. 33°N on Kyushu, the southernmost island of Japan. The climate falls in the perhumid subtype, with total annual precipitation over 190 cm (75 in.). Rainfall peaks strongly in summer. Both June and July receive over 25 cm (10 in.), reflecting the Asiatic rainy monsoon. In contrast, precipitation is low in the winter months, when a dry air mass from interior Asia dominates the weather. The air temperature cycle at Nagasaki shows a greater range than at Charleston, although summer temperatures are almost the same in both places. The winter months in Nagasaki are colder than at Charleston; even so, the coldest month, January, has an average well above the freezing point.

7. Mediterranean climate Latitude range: 30° to 45°N and S

This wet-winter, dry-summer climate results from a seasonal alternation of the same conditions that cause the dry subtropical climate (5), which lies at lower latitudes, and the moist marine west-coast climate (8), which lies on the poleward side. The moist maritime polar (mP) air mass invades in winter with cyclonic storms, generating ample rainfall. In summer, subsiding cT and mT air masses are dominant, with extreme drought of several months' duration. In terms of total annual rainfall, the Mediterranean climate spans a wide range from arid to humid, depending on location. Temperature range is moderate, with warm to hot summers and mild winters. Coastal zones between 30° and 35° show a smaller annual range, with very mild winters.

Major regions of occurrence: Central and southern California; coastal zones bordering the Mediterranean Sea; coastal western Australia, South Australia, and Victoria; Chilean coast; and Cape Town region of South Africa.

Example. Figure 9.20 is a climograph for Monterey, California, a coastal city at lat. 36½°N. The annual temperature cycle is very weak, with a range of only about 7C° (12 F°), reflecting the strong control of the cold California Current and its cool marine air layer. Coolness of the summer, with monthly means not exceeding 17°C (62°F) is typical of the narrow western littorals. The minimum-month winter temperature of 10.5°C (51°F) is very high compared with inland locations at this latitude. Rainfall drops to nearly zero for four consecutive summer months, but rises to more than 7.5 cm (3 in.) per month in the rainy winter season. The total annual rainfall is only about 42 cm (17 in.), and the climate as a whole can be classed as semiarid.

Example. Figure 9.21 is a climograph for Naples, Italy, located at lat. 40½°N on the west coast of the Italian peninsula. Compared with Monterey, the annual temperatue range is much greater — 16C° (28.5C°) — and summer temperatures are very much warmer. No cold coastal ocean current is present to subdue the temperature cycle. Total annual rainfall (86 cm, 34 in.) is much greater than at Monterey, and there is a small amount of rain, on the average, in every summer month.

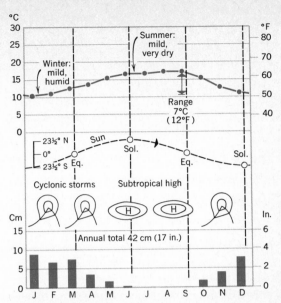

Figure 9.20 **Monterey, California, lat. 36½°N, has a very weak annual temperature cycle because of its closeness to the Pacific Ocean. The summer is very dry.**

8. Marine west-coast climate Latitude range: 35° to 60°N and S

Midlatitude west coasts, windward to the prevailing westerlies, receive frequent cyclonic storms involving the cool, moist mP air mass. In this humid climate, precipitation is copious in all months, but with a distinct winter maximum. Where the coast is mountainous, the orographic effect causes a very large annual precipitation,

Figure 9.21 **Naples, Italy, lat. 40½°N, has a much warmer summer than Monterey, but winter temperatures are much the same.**

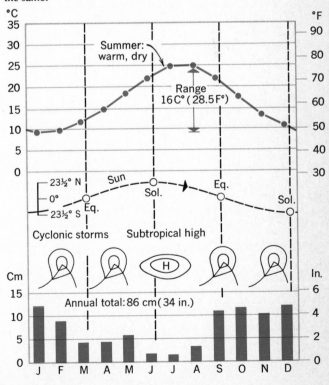

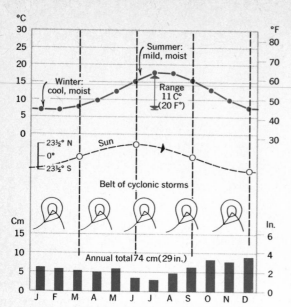

Figure 9.22 Brest, France, lat. 49°N, has a mild summer and a small annual temperature range. Precipitation shows a marked decline in the summer months.

or perhumid subtype. The annual temperature range is comparatively small for midlatitudes. Winter temperatures are very mild compared with inland locations at equivalent latitudes.

Major regions of occurrence: Western coast of North America, spanning Oregon, Washington, and British Columbia; western Europe and the British Isles; Victoria and Tasmania; New Zealand; and Chile, south of 35°S.

Example. Figure 9.22 is a climograph for Brest, France, a port city at lat. 49°N, on the Brittany coast. For a station this far north, the annual temperature cycle is weak, with annual range only 11C° (20F°). Nearness of the cold North Atlantic Ocean keeps summer temperatures cool, but prevents winter-month temperatures from falling below 7°C (45°F). Precipitation is moderate in every month, but with a summer period of reduced rainfall and a winter maximum. The total precipitation — 74 cm (29 in.) — establishes this as a humid climate.

9. Dry midlatitude climate Latitude range: 35° to 55°N

Limited almost exclusively to North America and Eurasia, this continental dry climate occupies a rainshadow position with respect to mountain ranges on the west or south. Maritime air masses are blocked effectively much of the time, so the continental polar cP air mass dominates the climate in winter. In summer, a dry continental air mass of local origin is dominant. Summer rainfall is caused by sporadic invasions of maritime air masses. Steppe and semidesert subtypes (9s, 9sd) are extensive; true desert (9d) occurs only in basins of interior Asia. The annual temperature cycle is strongly developed, with a large annual range. Summers are warm to hot, but winters are very cold.

Major regions of occurrence: Western North America (Great Basin, Columbia Plateau, Great Plains); Eurasian interior, from steppes of eastern Europe to Gobi Desert and North China; and a small area in southern Patagonia.

Example. Figure 9.23 is a climograph for Pueblo, Colorado, located at 38°N, just east of the Rocky Mountains. The climate is of the semiarid (steppe) subtype (9s), with a total annual precipitation of 31 cm (12 in.). Most of this precipitation is in the form of summer rainfall, when the moist mT air mass invades from the Gulf of Mexico and causes thunderstorms. In winter, because the mP air mass is effectively dried by passage over mountain ranges, snowfall is light and yields only small monthly precipitation averages. The temperature cycle has a large annual range — 24C° (43 F°) — with warm summers and cold winters. January, the coldest winter month, has a mean below 0°C (32° F).

Example. Figure 9.24 is a climograph for Kazalinsk, Russia, a city at lat. 46°N, located just east of the Aral Sea in Kazakhstan. The climate is semidesert (9sd), with an annual precipitation total of only 13 cm (5 in.), or less than half that of Pueblo. The annual temperature range is extremely great — 37C° (67 F°) — and winters are bitterly cold. Winter-month temperatures average below freezing (0°C, 32°F) for five consecutive months.

10. Moist continental climate Latitude range: 30° to 55°N (Europe: 45° to 60°)

Located in central and eastern parts of North America and Eurasia in the midlatitude zone, this climate is in the polar front zone — the battleground of polar and tropical air masses. Seasonal temperature contrasts are strong, while day-to-day weather is highly variable. Ample precipitation throughout the year is increased in summer by the invading mT air mass. Eastern maritime locations are perhumid. Cold winters are dominated by cP and continental arctic (cA) air masses from subarctic source regions. In eastern Asia (China, Korea, Japan), the monsoon effect is strongly evident in a summer rainfall

Figure 9.23 Pueblo, Colorado, lat. 38°N, lies in the semiarid (steppe) continental climate, just east of the Rockies. Rainfall peaks strongly in the summer months.

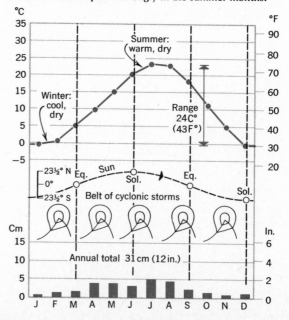

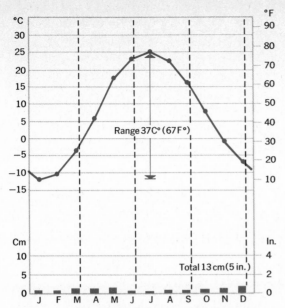

Figure 9.24 At Kazalinsk, Russia, lat. 46°N, winters are bitterly cold and summers are warm. Little precipitation falls in any month in this interior desert.

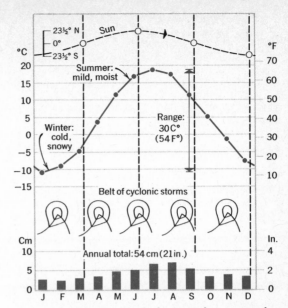

Figure 9.26 Moscow, Russia, lat. 56°N, has an annual range about the same as Madison; but summers in Moscow are not as warm.

maximum and a relatively dry winter. In Europe, the moist continental climate lies in a higher latitude belt (45° to 60°N) and receives precipitation from the mP air mass coming from the North Atlantic.

Major regions of occurrence: Eastern parts of the United States and southern Canada, northern China, Korea, Japan, and central and eastern Europe.

Example. Figure 9.25 is a climograph for Madison, Wisconsin, lat. 43°N, in the American Midwest. The annual temperature range is very large (31C°, 56 F°). Summers are warm but winters are cold, with three

consecutive monthly means well below freezing (0°C, 32°F). Precipitation is ample in all months and totals 81 cm (32 in.), giving a humid climate. There is a well-developed summer maximum of precipitation when the mT air mass invades and thunderstorms are formed along moving cold fronts and squall lines. Much of the winter precipitation is in the form of snow, which remains on the ground for long periods.

Example. Figure 9.26 is a climograph for Moscow, Russia, at lat. 56°N. This location is over 600 km (1000 mi) farther north than Madison, so Moscow summers are not as

Figure 9.25 Madison, Wisconsin, lat. 43°N, has cold winters and warm summers, making the annual temperature range very large.

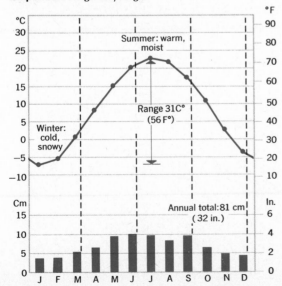

Figure 9.27 Inchon, Korea, lat. 37½°N, has two very rainy months during the summer monsoon, but very little precipitation in winter.

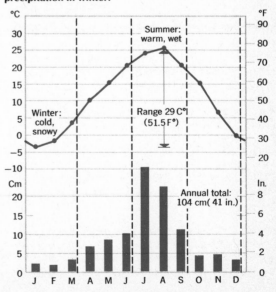

warm and winters are colder. Annual total precipitation at Moscow is less (54 cm, 21 in.) than at Madison, with smaller monthly amounts throughout the year. Winter precipitation, largely snow, remains on the ground for many months.

Example. Figure 9.27 is a climograph for Inchon, Korea, at lat. 37½°N. Compared with Madison, Inchon has somewhat warmer summers; but winters are not as cold. The major difference in the two stations is in the precipitation cycle. Inchon shows the Asiatic monsoon effect with two very rainy summer months and greatly diminished precipitation in winter. Because of the summer monsoon, with influx of a maritime air mass from the western Pacific Ocean, Inchon has a larger total annual precipitation (104 cm, 41 in.) than either Madison or Moscow.

GROUP III / HIGH-LATITUDE CLIMATES
(controlled by polar and arctic air masses)

11. Boreal forest climate Latitude range: 50° to 70°N

This is a continental climate with long, bitterly cold winters and short, cool summers. This climate occupies the source region of the cP air mass, which is cold, dry, and stable in the winter. Invasions of the very cold cA air mass are common. The annual range of temperature is greater than that for any other climate, attaining a value of 60C° (110 F°) in Siberia. Precipitation is substantially increased in summer, when maritime air masses penetrate the continent with traveling cyclones; but the total annual precipitation is small. Although much of the boreal forest climate is classed as humid, with precipitation of 50 to 100 cm (20 to 40 in.), large areas in western Canada and Siberia have annual precipitation totals of less than 40 cm (16 in.) and are classed as subhumid (11sh) or semiarid (11s) subtypes.

Major regions of occurrence: central and western Alaska; Canada, from Yukon Territory to Labrador; southernmost Greenland; Iceland; and Eurasia, from northern Europe across all of Siberia to the Pacific Coast.

Example. Figure 9.28 is a climograph for Fort Vermilion in Alberta, Canada, at lat. 58°N. The very great annual temperature range (41C°, 74 F°) is typical for North America. Monthly mean air temperatures are below freezing (0°C, 32°F) for seven consecutive months. The summers are short and cool. Precipitation shows a marked annual cycle with a strong summer maximum; but the total annual precipitation is only 31 cm (12 in.), and the climate can be characterized as subhumid (11sh). Although precipitation in winter is small, a snow cover remains over solidly frozen ground throughout the entire winter. On the same climograph, temperature data are shown for Yakutsk, U.S.S.R., a Siberian city at lat. 62°N. The enormous annual range is evident, also the extremely low winter-month means. January reaches a mean of about −45°C (−50°F), making this region the coldest on earth except for the ice-sheet interiors of Antarctica and Greenland. Precipitation is not shown for Yakutsk, but the annual total is only about 18 cm (7 in.) — much less than for Fort Vermilion — and indicates a semiarid climate subtype (11s).

12. Tundra climate Latitude range: 60° to 75°N and S

The tundra climate occupies the arctic coastal fringes, dominated by cP, mP, and cA air masses, and with frequent cyclonic storms. Winters are long and severe. There is a very short mild season, which many climatologists do not recognize as a true summer. A moderating influence of the nearby ocean water prevents winter temperatures from falling to the extreme lows found in the continental interior. The tundra climate ranges from a humid subtype (12h) bordering the Atlantic Ocean to subhumid (12sh) and semiarid (12s) subtypes bordering the Arctic Ocean.

Figure 9.28 **Extreme winter cold and a very great annual range in temperature characterize the boreal forest climate, illustrated by these climographs for Fort Vermilion in Alberta, Canada, and Yakutsk, Russia.**

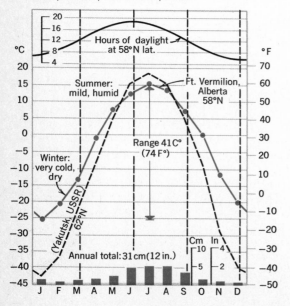

Figure 9.29 **Upernivik, Greenland, lat. 73°N, is a typical arctic tundra station.**

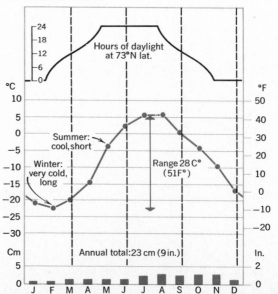

Major regions of occurrence: Arctic slope of North America, Hudson Bay region and Baffin Island, Greenland coast, northern Siberia bordering the Arctic Ocean, and Antarctic Peninsula.

Example. Figure 9.29 is a climograph for Upernivik, located on the west coast of Greenland at lat. 73°N. A short milder season, equivalent to a summer season in lower latitudes, has monthly means barely over 5°C (41°F). The long, very cold winters bring monthly means as low as −20°C (−9°F). The annual temperature range (28C°, 51F°) is not as large as for the Boreal forest climate to the south. Total annual precipitation is small — 23 cm (9 in.) — representing the subhumid subtype (12 sh). Increased precipitation beginning in July is explained by the melting of the sea-ice cover and a warming of ocean water temperatures, increasing the moisture content of the local air mass.

13. Ice-sheet climate Latitude range: 65° to 90°N and S

Source regions of arctic (A) and antarctic (AA) air masses are situated on vast, high ice sheets, and over polar sea ice of the Arctic Ocean. Mean annual temperature is much lower than that of any other climate, with no above-freezing monthly mean. Strong temperature inversions develop over the ice sheets. The strong net radiation deficit in winter at high surface altitudes intensifies the cold. Strong cyclones with blizzard winds are frequent. Precipitation, almost all occurring as snow, is very small, but accumulates because of the continuous cold.

Example. Figure 9.30 shows temperature graphs for five representative ice-sheet stations. The graph for Eismitte, Greenland, shows the northern-hemisphere temperature cycle, whereas the other four examples are all from Antarctica. Temperatures in the interior of Antarctica have proved to be far lower than at any other place on earth. The Russian meteorological station Vostok, located about

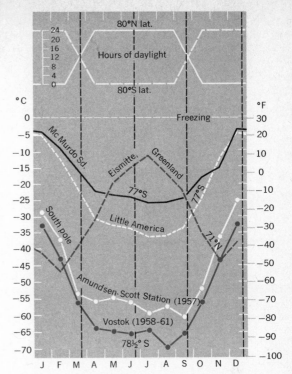

Figure 9.30 **Temperature graphs for five ice-sheet stations.**

1300 km (800 mi) from the south pole at an altitude of about 3500 m (11,400 ft) may be the world's coldest spot. Here a record low of −88.3°C (−127°F) was observed. At the pole itself (Amundsen-Scott Station), July, August, and September of 1957 had averages of about −60°C (−76°F). Temperatures run roughly 22C° (40 F°) higher, month for month, at Little America because it is located close to the Ross Sea and is at low altitude.

Highland Climates

Highland climates shown by a distinctive pattern on the world climate map, are cool to cold, usually moist climates occupying mountains and high plateaus. Many occur as narrow belts, with strong climatic gradients, becoming steadily colder with increasing altitude. The climate of a given highland area is usually closely related to the climate of the surrounding lowland in seasonal characteristics, particularly the form of the annual temperature cycle and the times of occurrence of wet and dry seasons. In lowlands of arid climate, narrow mountain ranges tend to be islands of moist climate. Highland climates are not usually included in the broad schemes of climate classification. Designation of highland areas on most climate maps is somewhat arbitrary. Many small highland areas are simply not shown on a world map.

The upward decrease in monthly mean temperatures was explained in Chapter 5 and illustrated with a graph for several stations in the Andes of South America (Figure 5.19). In Figure 9.31, an example of the altitude effect in the tropical zone is shown by climographs for two stations

in close geographic proximity. New Delhi, the capital city of India, lies in the Ganges lowland. Simla, a mountain refuge from the hot weather, is located about 2000 m (6500 ft) in altitude, in the foothills of the Himalayas. When hot-season averages are over 32°C (90°F), Simla is enjoying a pleasant 18°C (65°F), which is a full 14C° (25F°) cooler. Notice, however, that the two temperature cycles are quite similar in phase, with the minimum month being January for both.

The general effect of increased altitude is to bring an increase in precipitation, at least for the first few kilometers of altitude increase. This change is due to the production of orographic rainfall, generated by the forced ascent of air masses (see Chapter 7). The altitude effect shows nicely in the monthly rainfalls of New Delhi and Simla (Figure 9.31). New Delhi shows the typical rainfall pattern of the tropical climate of Southeast Asia, with monsoon rains peaking in July and August. Simla has the same pattern, but the amounts are larger in every month and the monsoon peak is very strong. Simla's annual total is well over twice that of New Delhi. Note that Simla also receives rainfall in the low-sun season.

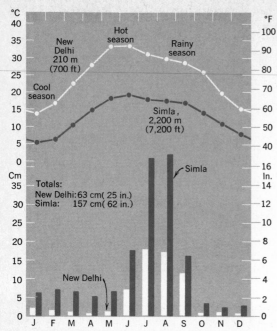

Figure 9.31 **Climographs for New Delhi and Simla, both in northern India. Simla is a welcome refuge from the intense heat of the Gangetic Plain in May and June.**

The Köppen Climate System

Among professional geographers, a widely used climate classification system is that devised originally by Dr. Wladimir Köppen in 1918 and subsequently revised by his students, R. Geiger and W. Pohl. The version we present here was published in 1953 and is sometimes called the Köppen-Geiger-Pohl system. Köppen was both a climatologist and a plant geographer, so his main interest lay in finding climate boundaries that coincided approximately

with boundaries between major vegetation types. Although he was not entirely successful in achieving this goal, his climate system has appealed to geographers because it is strictly empirical and allows no room for subjective decisions.

Under the Köppen system, each climate is defined according to assigned values of temperature and precipitation, computed in terms of annual or monthly values. Any given station can be assigned to its particular climate group and subgroup solely on the basis of the records of temperature and precipitation at that place provided, of course, that the period of record is long enough to yield meaningful averages. To use the Köppen system, it is not necessary to understand how climate controls operate.

The Köppen system features a shorthand code of letters designating major climate groups, subgroups within the major groups, and further subdivisions to distinguish particular seasonal characteristics of temperature and precipitation.

The five major climate groups are designated by capital letters as follows:

A *Tropical rainy climates.* Average temperature of every month is above 18°C (64.4°F). These climates have no winter season. Annual rainfall is large and exceeds annual evaporation.

B *Dry Climates.* Evaporation exceeds precipitation on the average throughout the year. There is no water surplus; hence no permanent streams originate in *B* climate zones.

C *Mild, humid (mesothermal) climates.* Coldest month has an average temperature under 18°C (64.4°F), but above −3°C (26.6°F); at least one month has an average temperature above 10°C (50°F). The C climates thus have both a summer and a winter season.

D *Snowy-forest (microthermal) climates.* Coldest month average temperature is under −3°C (26.6°F).

Figure 9.32 **Highly generalized world map of major climatic regions according to the Köppen classification. Highland areas are in black. (Based on Goode Base Map.)**

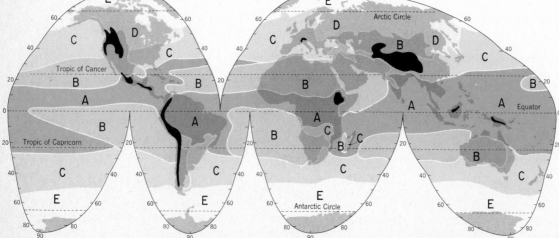

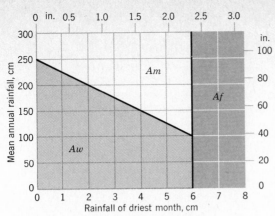

Figure 9.33 **Boundaries of the A climates.**

and no dry season. This modifier is applied to *A, C,* and *D* groups.

w Dry season in winter of the respective hemisphere (low-sun season).

s Dry season in summer of the respective hemisphere (high-sun season).

m Rainforest climate despite short, dry season in monsoon type of precipitation cycle. This applies only to *A* climates.

From combinations of the two letter groups, 12 distinct climates emerge:

Af **Tropical rainforest climate.** Rainfall of the driest month is 6 cm (2.4 in.) or more.

Am **Monsoon variety of Af.** Rainfall of the driest month is less than 6 cm (2.4 in.). The dry season is strongly developed.

Aw **Tropical savanna climate.** At least one month has rainfall less than 6 cm (2.4 in.). The dry season is strongly developed.

Figure 9.33 shows the boundaries betweeh *Af, Am,* and *Aw* climates as determined by both annual rainfall and rainfall of the driest month.

BS **Steppe climate.** A semiarid climate characterized by grasslands. It occupies an intermediate position between the desert climate (*BW*) and the more humid climates of the *A, C,* and *D* groups. Boundaries are determined by formulas given in Figure 9.34.

BW **Desert climate.** An arid climate with annual precipitation usually less than 40 cm (15 in.). The boundary with the adjacent steppe climate (*BS*) is determined by formulas given in Figure 9.34.

Cf **Mild humid climate with no dry season.** Precipitation of the driest month averages more than 3 cm (1.2 in.).

Cw **Mild humid climate with a dry winter.** The wettest month of summer has at least 10 times the precipitation of the driest month of winter. (Alternate definition: Seventy percent or more of the mean annual precipitation falls in the warmer six months.)

Cs **Mild humid climate with a dry summer.** Precipitation

Average temperature of the warmest month is above 10°C (50°F), that isotherm coinciding approximately with the poleward limit of forest growth.

E **Polar climates.** Average temperature of the warmest month is below 10°C (50°F). These climates have no true summer.

Note that four of these five groups (*A, C, D,* and *E*) are defined by temperature averages, whereas one (*B*) is defined by precipitation-to-evaporation ratios. Groups *A, C,* and *D* have sufficient heat and precipitation for the growth of forest and woodland vegetation. Figure 9.32 shows the boundaries of the five major climate groups.

Subgroups within the five major groups are designated by a second letter, according to the following code.

S Semiarid (steppe).
W Arid (desert).

(The capital letters *S* and *W* are applied only to the dry *B* climates.)

f Moist. There is adequate precipitation in all months

Figure 9.34 **Boundaries of the B climates.**

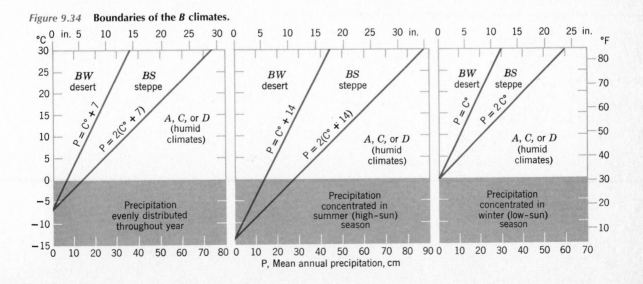

Table 9.2 / Köppen equivalents to climates classified by air masses and front zones

Group I: Low Latitude Climates

1. Wet equatorial climate	Af:	Tropical rainforest climate
2. Monsoon and trade-wind littoral climate	Am:	Tropical rainforest climate, monsoon type; and some areas of
	Af:	Tropical rainforest climate
3. Wet-dry tropical climate	Aw:	Tropical savanna climate; and
	Cw:	Mild climate with dry winter
4. Dry tropical climate	BWh:	Desert climate, hot; and
	BSh:	Steppe climate, hot

Group II: Midlatitude Climates

5. Dry subtropical climate	BWh:	Desert climate, hot; and
	BSh:	Steppe climate, hot
6. Moist subtropical climate	Cfa:	Temperate rainy climate, hot summer
7. Mediterranean climate	Csa:	Temperate rainy climate, dry hot summer; and
	Csb:	same, with dry warm summer
8. Marine west-coast climate	Cfb:	Temperate rainy climate with warm summers; also some parts of
	Csb:	Temperate rainy climate, dry warm summer
9. Dry midlatitude climate	BWk:	Desert climate, cold, and
	BSk:	Steppe climate, cold
10. Moist continental climate	Dfa:	Cold snowy-forest climate, hot summers; and
	Dfb:	same, with warm summers

Group III: High Latitude Climates

11. Boreal forest climate	Dfc:	Cold snowy-forest climate, cool summers; and
	Dw:	Cold snowy-forest climate, dry winters; and
	Cfc:	Temperate rainy climate, cool short summers
12. Tundra climate	ET:	Polar tundra climate
13. Ice-sheet climate	EF:	Polar climate, perpetual frost

of the driest month of summer is less than 3 cm (1.2 in.). Precipitation of the wettest month of winter is at least three times as much as the driest month of summer. (Alternate definition: Seventy percent or more of the mean annual precipitation falls in the six months of winter.)

Df *Snowy-forest climate with a moist winter.* No dry season.

Dw *Snowy-forest climate with a dry winter.*

ET *Tundra climate.* Mean temperature of the warmest month is above 0°C (32°F) but below 10°C (50°F).

EF *Perpetual frost climate.* Ice-sheet climate. Mean monthly temperatures of all months are below 0°C (32°F).

To denote further variations in climate, Köppen added a third letter to the code group. Meanings are as follows:

a With hot summer; warmest month is over 22°C (71.6°F); C and D climates.

b With warm summer; warmest month is below 22°C (71.6°F); C and D climates.

c With cool, short summer; less than four months are over 10°C (50°F); C and D climates.

d With very cold winter; coldest month is below −38°C (−36.4°F); D climates only.

h Dry-hot; mean annual temperature is over 18°C (64.4°F); B climates only.

k Dry-cold; mean annual temperature is under 18°C (64.4°F); B climates only.

As an example of a complete Köppen climate code, BWk refers to a *cool desert climate*, and Dfc refers to a *cold, snowy forest climate with cool, short summer.*

Plate D.4 is a world map of Köppen climates. Table 9.2 gives approximate Köppen equivalents for the 13 climate types based on air masses and frontal zones.

Vegetation, Soils, and Climate

Plant geographers are keenly aware that plants are responsive to differences in climate. A particular combination of climatic factors most favorable to its growth is associated with each plant species, as are certain extremes of heat, cold, or drought beyond which it cannot survive. Plants tend to adapt their physical forms to meet the stresses of climate. We find a wide range of forms taken by assemblages of the dominant plant species. These overall form patterns, or habits, closely reflect climate controls.

Since the late nineteenth century, soil scientists have recognized that the several fundamental classes of mature soils, seen on a worldwide scope, are strongly controlled by climatic elements. But plants also contribute to determining the properties of the very soils on which they depend.

To develop the concept of an interplay among climate controls, vegetation, and soils, we must make a special study of water within the soil layer. Although precipitation is an essential climate ingredient, what really counts in the growth of plants is the amount of water available in the soil zone where the plant roots are situated. Evaporation from soil and plant surfaces can return a large proportion of the incoming precipitation to the atmosphere. Within the soil is a balance between incoming and outgoing water. As in the case of the energy balance, there can be seasonal water surpluses and water deficits. In the next chapter we shall investigate the soil-water balance as an important step in increasing our understanding of the varied environments of the life layer.

References for Further Study

Köppen, Wladimir, and Geiger, R. (1954), *Klima der Erde* (map), Justus Perthes, Darmstadt, Germany; American distributor, A. J. Nystrom and Co., Chicago.

Trewartha, G. T. (1968), *An introduction to climate,* fourth edition, McGraw-Hill Book Co., New York.

Oliver, J. E. (1973), *Climate and man's environment; an introduction to applied climatology,* John Wiley and Sons, New York.

Mather, J. R. (1974), *Climatology: fundamentals and applications,* McGraw-Hill Book Co., New York.

Review Questions

1. What is climatology? How does climatology differ from meteorology? Which of the physical components of climate are most useful in setting up a geographic system of climate classification?
2. Explain how life processes of the biosphere are dependent on the atmospheric environment. What are the vital components of the atmospheric environment with respect to plant life on the lands?
3. How can net radiation be used as the basis of a system of climate classification? What vital component of climate is not included in such a system?
4. Explain how temperature can be used as a basis for a climate classification. Why is this element alone unsatisfactory as a basis for a climate system?
5. What is a thermal regime? Describe several different thermal regimes and explain their characteristics. What is the concept of continentality? Illustrate with an example.
6. How can annual precipitation be used as a basis for a climate classification? Why does the factor of evaporation limit the value of this classification?
7. Name and describe the seven basic precipitation regions. How is each related to prevailing air masses?
8. Why are seasonal patterns of precipitation important in climate description? Describe six basic seasonal patterns and explain each in terms of air-mass controls.
9. Explain how a climate classification system can be based on the air-mass source regions, patterns of air-mass travel, and frontal zones.
10. How may the climates of the globe be classified into three major groups on the basis of air-mass types?
11. What is the distinction between dry and moist climates? What subtypes are recognized in each category?
12. Describe the four climates that comprise a group controlled by equatorial and tropical air masses. How are they related to the system of pressure and wind belts? What Köppen climates are equivalent?
13. Describe the six climates controlled by both tropical and polar air masses. What Köppen climates are equivalent?
14. Describe the three climates controlled by polar and arctic air masses. Explain how the large northern hemisphere landmasses control these climates. What are the equivalent Köppen climates?
15. What changes in temperature and precipitation are typically associated with an increase in surface elevation? Give an example of a tropical-zone climate at a high elevation. Is the seasonal pattern different from that of a nearby station at low elevation?
16. Describe the Köppen climate system. What are the five major climate groups? What code system is used to designate 12 climates? Name these and give the symbols for each. What do each of the following letters mean in the Köppen system: *a, b, c, d, h, k*?

The Soil-Water Balance and World Climates

10

A basic concept of physical geography is that the availability of water to plants is a more important factor in the environment than precipitation itself. Much of the water received as precipitation is lost in a variety of ways and is not usable for plants. Just as in a fiscal budget, when the monthly or yearly loss of moisture exceeds the precipitation, a budgetary deficit results; when precipitation exceeds the losses, a budgetary surplus results. An analysis of the water budget of a given area of the earth's surface is actually arrived at in much the same way as a fiscal budget. The accounting requires only additions and subtractions of amounts within fixed periods of time, such as the calendar month or year.

To understand why such moisture gains and losses occur, we need to study the physical processes affecting water in its vapor, liquid, and solid states—not only in the atmosphere but also in the soil and rock and in the exposed water of streams, lakes, and glaciers. The science of **hydrology** treats such relationships of water as a complex but unified system on the earth. The primary concern of this chapter is the hydrology of the soil zone.

Surface and Subsurface Water

We classify water of the lands according to whether it is **surface water,** flowing exposed or ponded on the land, or **subsurface water,** occupying openings in the soil or rock. Water held in the soil within a meter or two of the surface is the **soil water;** it is the particular concern of the plant ecologist, soil scientist, and agricultural engineer. Water held in the openings of the bedrock is referred to as ground water and is studied by the geologist. Surface water and ground water are the subjects of Chapter 18.

The Hydrologic Cycle and the Global Water Balance

Water of oceans, atmosphere, and lands moves in a great series of continuous interchanges of both geographic position and physical state, known as the **hydrologic cycle** (Figure 10.1). A particular molecule of water might, if we could trace it continuously, travel through any one of a number of possible circuits involving alternately the water vapor state and the liquid or solid state.

The pictorial diagram of the hydrologic cycle given in Figure 10.1 can be quantified for the earth as a whole. Figure 10.2 is a mass-flow diagram relating the principal pathways of the water circuit. We can start with the oceans, which are the basic reservoir of free water. Evaporation from the ocean surfaces totals about 455,000 cu km per year (English equivalents are given in Figure 10.2). At the same time, evaporation from soil, plants, and water surfaces of the continents totals about 62,000 cu km. Thus the total evaporation is 517,000 cu km; it represents the quantity of water that must be returned annually to the liquid or solid state.

Precipitation is unevenly divided between continents

Figure 10.1   **The hydrologic cycle (After Holtzman.)**

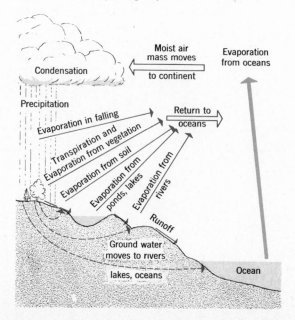

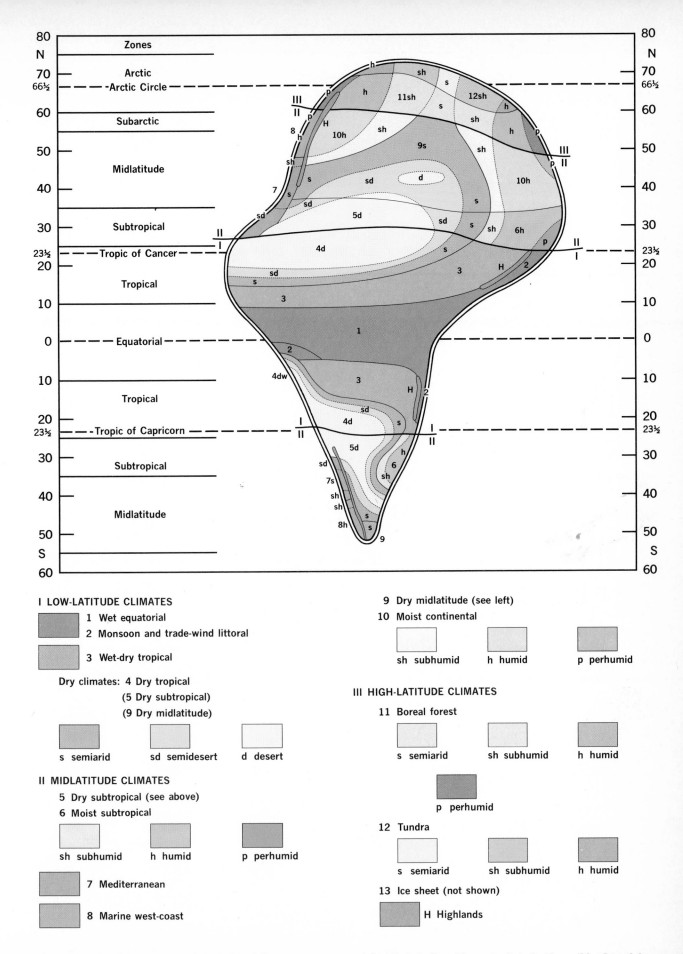

Schematic Diagram of World Climates on an Idealized Continent, by Arthur N. Strahler

I LOW-LATITUDE CLIMATES

1 Wet equatorial

2 Monsoon and trade-wind littoral

3 Wet-dry tropical

Dry climates: 4 Dry tropical
(5 Dry subtropical)
(9 Dry midlatitude)

s semiarid   sd semidesert   d desert

II MIDLATITUDE CLIMATES

5 Dry subtropical (see above)
6 Moist subtropical

sh subhumid   h humid   p perhumid

7 Mediterranean

8 Marine west-coast

9 Dry midlatitude (see left)

10 Moist continental

sh subhumid   h humid   p perhumid

III HIGH-LATITUDE CLIMATES

11 Boreal forest

s semiarid   sh subhumid   h humid

p perhumid

12 Tundra

s semiarid   sh subhumid   h humid

13 Ice sheet (not shown)

H Highlands

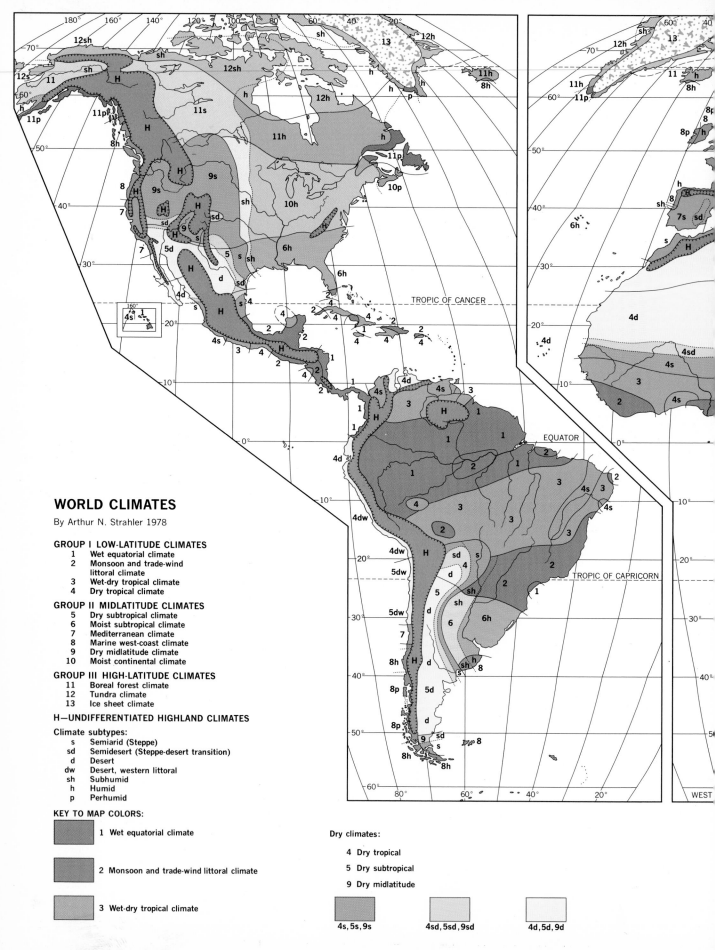

WORLD CLIMATES

By Arthur N. Strahler 1978

GROUP I LOW-LATITUDE CLIMATES
1 Wet equatorial climate
2 Monsoon and trade-wind littoral climate
3 Wet-dry tropical climate
4 Dry tropical climate

GROUP II MIDLATITUDE CLIMATES
5 Dry subtropical climate
6 Moist subtropical climate
7 Mediterranean climate
8 Marine west-coast climate
9 Dry midlatitude climate
10 Moist continental climate

GROUP III HIGH-LATITUDE CLIMATES
11 Boreal forest climate
12 Tundra climate
13 Ice sheet climate

H—UNDIFFERENTIATED HIGHLAND CLIMATES

Climate subtypes:
s   Semiarid (Steppe)
sd Semidesert (Steppe-desert transition)
d   Desert
dw Desert, western littoral
sh Subhumid
h   Humid
p   Perhumid

KEY TO MAP COLORS:

1 Wet equatorial climate

2 Monsoon and trade-wind littoral climate

3 Wet-dry tropical climate

Dry climates:

4 Dry tropical

5 Dry subtropical

9 Dry midlatitude

4s, 5s, 9s

4sd, 5sd, 9sd

4d, 5d, 9d

Plate D.2   World climates

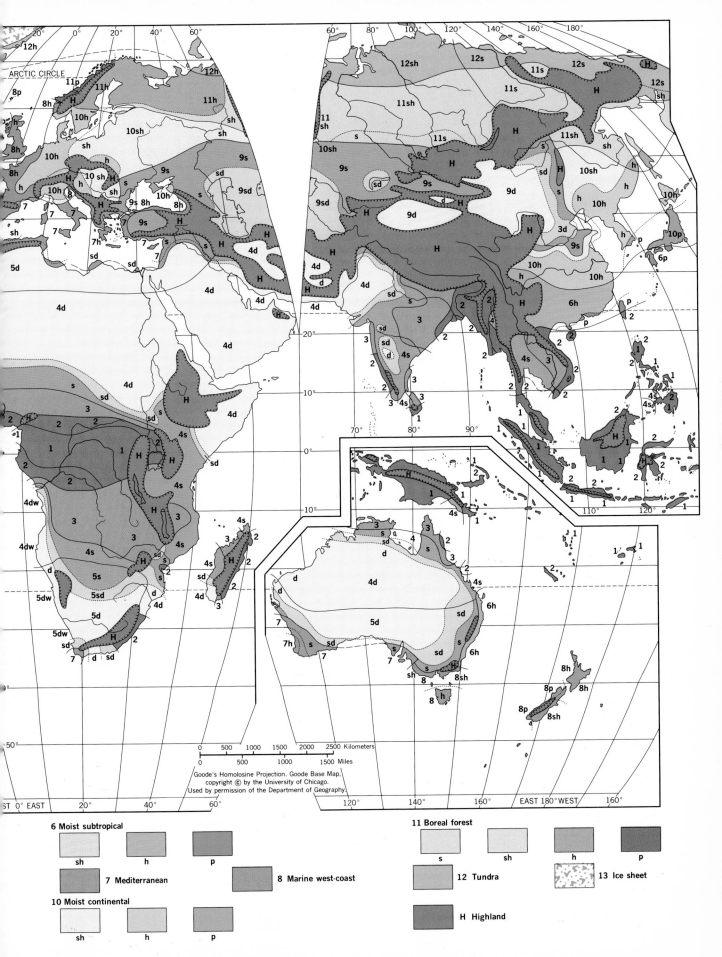

World climates   **Plate D.2**

6 Moist subtropical

sh   h   p

7 Mediterranean

8 Marine west-coast

10 Moist continental

sh   h   p

11 Boreal forest

s   sh   h   p

12 Tundra

13 Ice sheet

H Highland

Goode's Homolosine Projection. Goode Base Map,
copyright © by the University of Chicago.
Used by permission of the Department of Geography

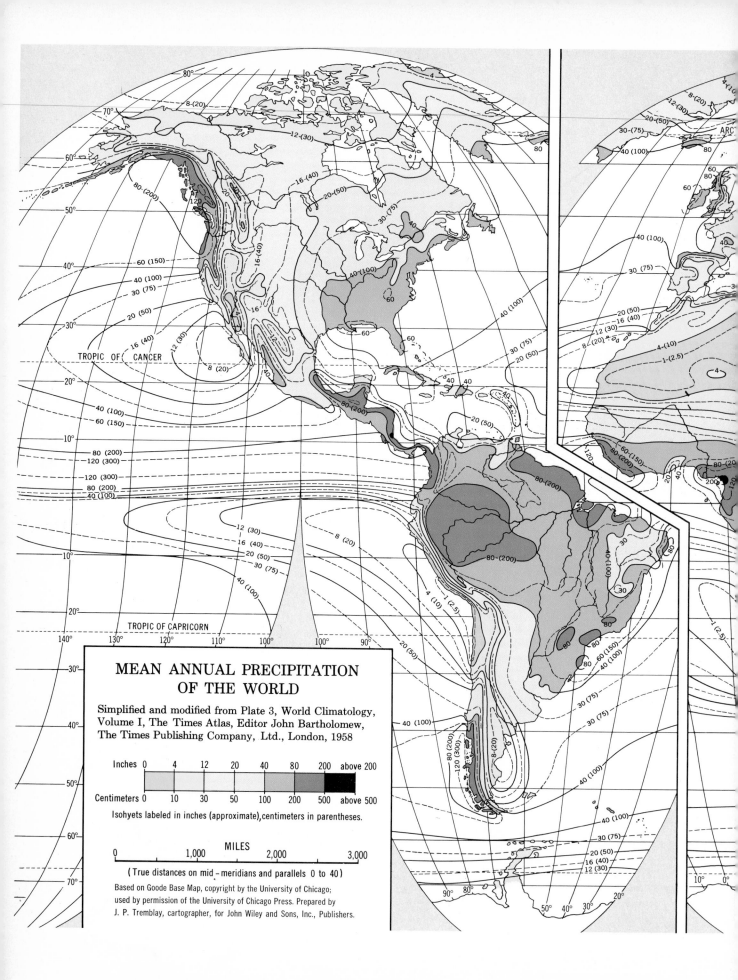

MEAN ANNUAL PRECIPITATION OF THE WORLD

Simplified and modified from Plate 3, World Climatology, Volume I, The Times Atlas, Editor John Bartholomew, The Times Publishing Company, Ltd., London, 1958

Inches	0	4	12	20	40	80	200	above 200
Centimeters	0	10	30	50	100	200	500	above 500

Isohyets labeled in inches (approximate), centimeters in parentheses.

MILES
0   1,000   2,000   3,000

(True distances on mid – meridians and parallels 0 to 40)

Based on Goode Base Map, copyright by the University of Chicago; used by permission of the University of Chicago Press. Prepared by J. P. Tremblay, cartographer, for John Wiley and Sons, Inc., Publishers.

Plate D.3   World precipitation

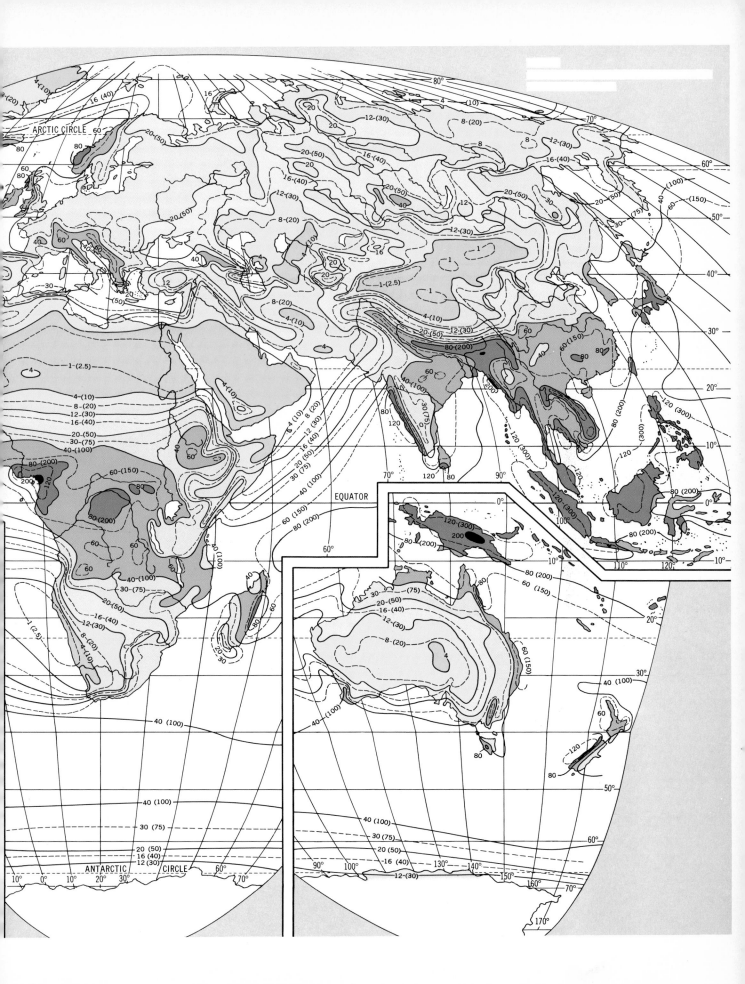

World precipitation   **Plate D.3**

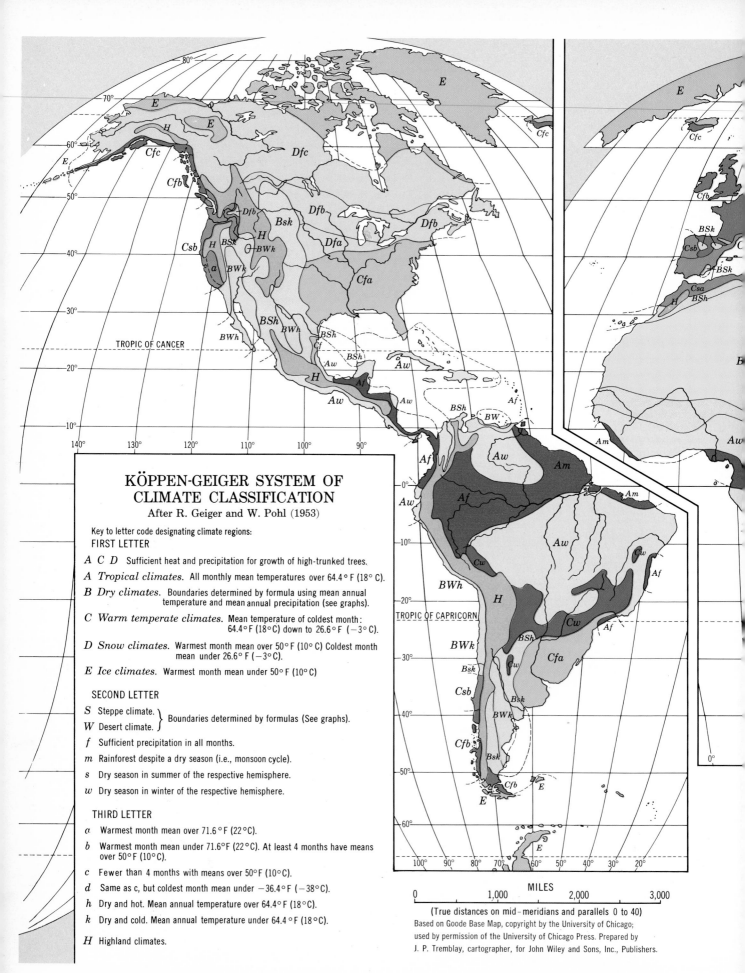

KÖPPEN-GEIGER SYSTEM OF CLIMATE CLASSIFICATION

After R. Geiger and W. Pohl (1953)

Key to letter code designating climate regions:

FIRST LETTER

A C D   Sufficient heat and precipitation for growth of high-trunked trees.

A   *Tropical climates.* All monthly mean temperatures over 64.4 °F (18 °C).

B   *Dry climates.* Boundaries determined by formula using mean annual temperature and mean annual precipitation (see graphs).

C   *Warm temperate climates.* Mean temperature of coldest month: 64.4 °F (18 °C) down to 26.6 °F (−3 °C).

D   *Snow climates.* Warmest month mean over 50 °F (10 °C) Coldest month mean under 26.6 °F (−3 °C).

E   *Ice climates.* Warmest month mean under 50° F (10° C)

SECOND LETTER

S   Steppe climate. } Boundaries determined by formulas (See graphs).

W   Desert climate.

f   Sufficient precipitation in all months.

m   Rainforest despite a dry season (i.e., monsoon cycle).

s   Dry season in summer of the respective hemisphere.

w   Dry season in winter of the respective hemisphere.

THIRD LETTER

a   Warmest month mean over 71.6 °F (22 °C).

b   Warmest month mean under 71.6°F (22 °C). At least 4 months have means over 50°F (10 °C).

c   Fewer than 4 months with means over 50°F (10 °C).

d   Same as c, but coldest month mean under −36.4 °F (−38 °C).

h   Dry and hot. Mean annual temperature over 64.4 °F (18 °C).

k   Dry and cold. Mean annual temperature under 64.4 °F (18 °C).

H   Highland climates.

MILES

0   1,000   2,000   3,000

(True distances on mid-meridians and parallels 0 to 40)

Based on Goode Base Map, copyright by the University of Chicago;
used by permission of the University of Chicago Press. Prepared by
J. P. Tremblay, cartographer, for John Wiley and Sons, Inc., Publishers.

Plate D.4   World Köppen climates

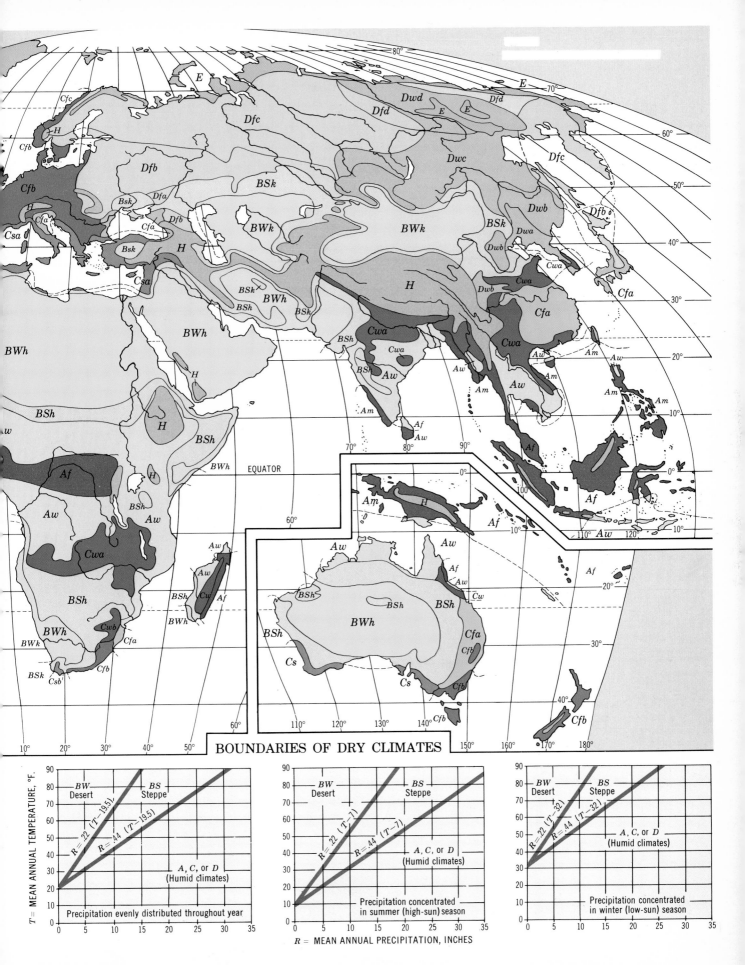

BOUNDARIES OF DRY CLIMATES

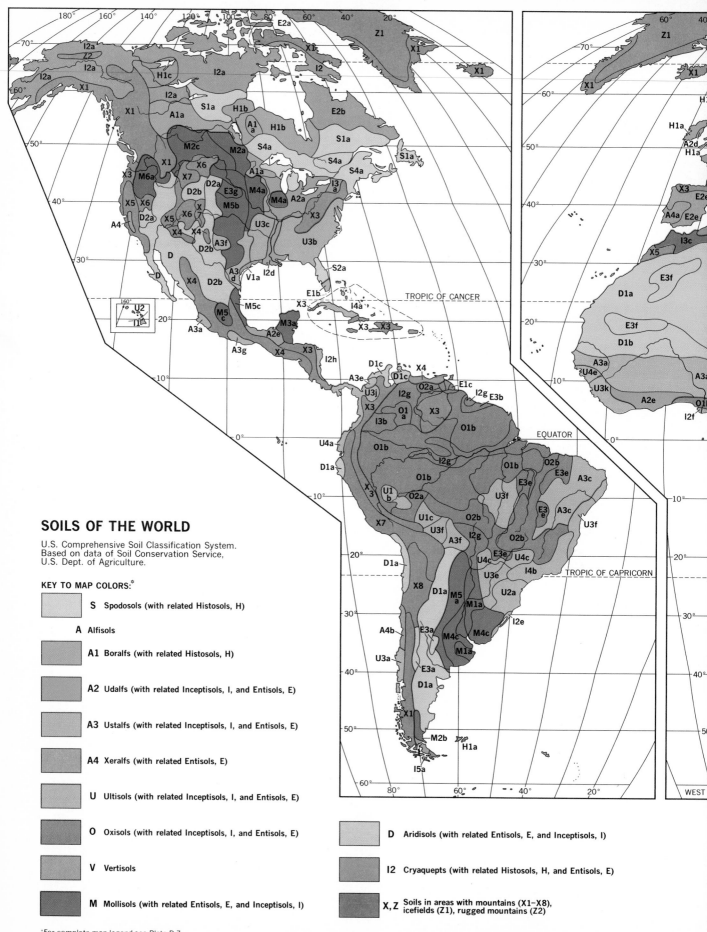

SOILS OF THE WORLD

U.S. Comprehensive Soil Classification System.
Based on data of Soil Conservation Service,
U.S. Dept. of Agriculture.

KEY TO MAP COLORS:[*]

S Spodosols (with related Histosols, H)

A Alfisols

A1 Boralfs (with related Histosols, H)

A2 Udalfs (with related Inceptisols, I, and Entisols, E)

A3 Ustalfs (with related Inceptisols, I, and Entisols, E)

A4 Xeralfs (with related Entisols, E)

U Ultisols (with related Inceptisols, I, and Entisols, E)

O Oxisols (with related Inceptisols, I, and Entisols, E)

V Vertisols

M Mollisols (with related Entisols, E, and Inceptisols, I)

D Aridisols (with related Entisols, E, and Inceptisols, I)

I2 Cryaquepts (with related Histosols, H, and Entisols, E)

X, Z Soils in areas with mountains (X1–X8),
icefields (Z1), rugged mountains (Z2)

*For complete map legend see Plate D.7.

Plate D.5   World soils

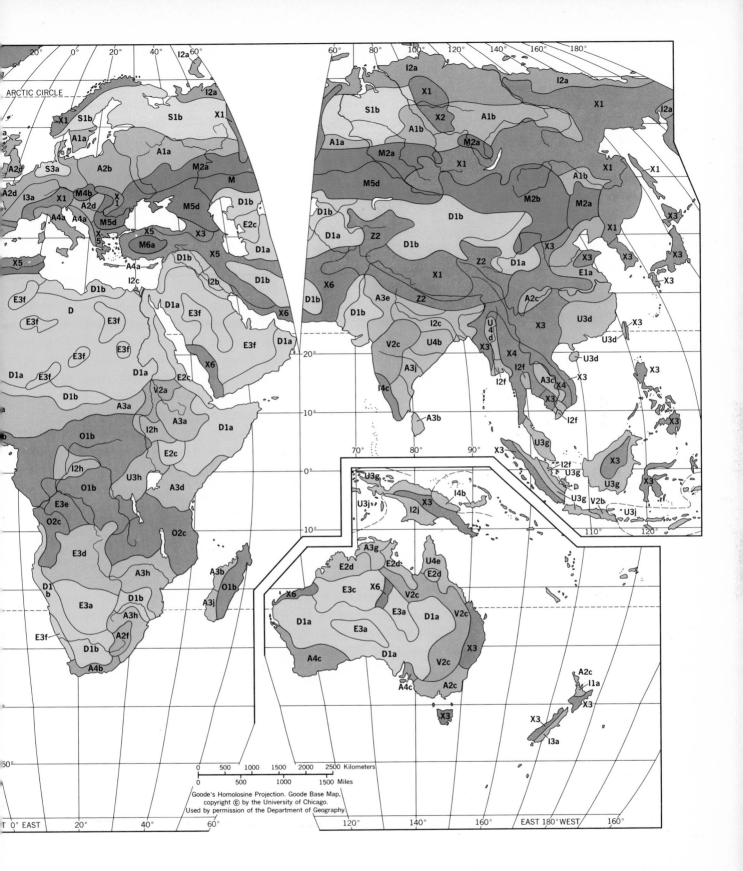

World soils   **Plate D.5**

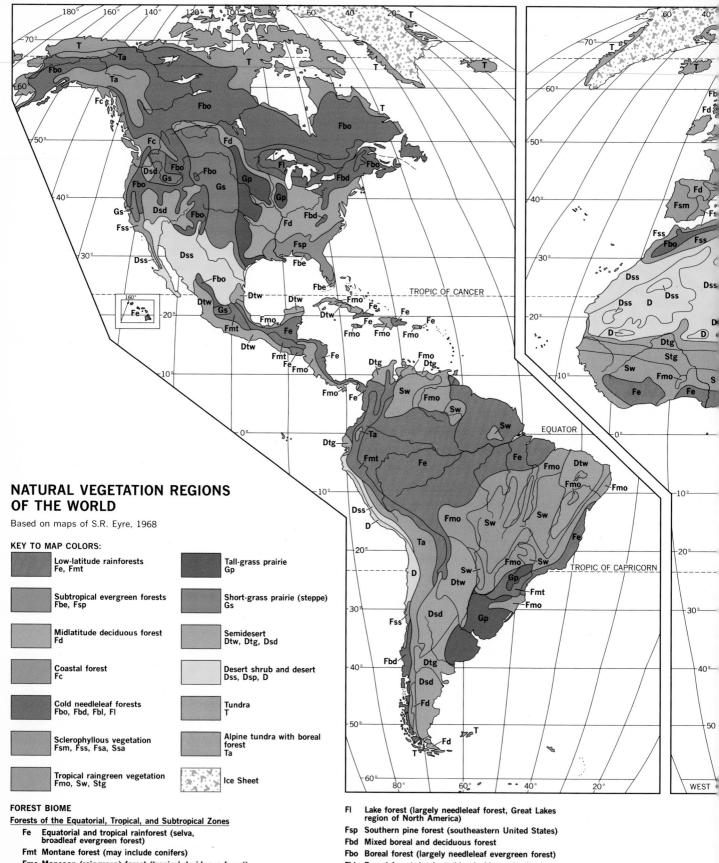

NATURAL VEGETATION REGIONS OF THE WORLD

Based on maps of S.R. Eyre, 1968

KEY TO MAP COLORS:

Low-latitude rainforests
Fe, Fmt

Subtropical evergreen forests
Fbe, Fsp

Midlatitude deciduous forest
Fd

Coastal forest
Fc

Cold needleleaf forests
Fbo, Fbd, Fbl, Fl

Sclerophyllous vegetation
Fsm, Fss, Fsa, Ssa

Tropical raingreen vegetation
Fmo, Sw, Stg

Tall-grass prairie
Gp

Short-grass prairie (steppe)
Gs

Semidesert
Dtw, Dtg, Dsd

Desert shrub and desert
Dss, Dsp, D

Tundra
T

Alpine tundra with boreal forest
Ta

Ice Sheet

FOREST BIOME

Forests of the Equatorial, Tropical, and Subtropical Zones

Fe Equatorial and tropical rainforest (selva, broadleaf evergreen forest)

Fmt Montane forest (may include conifers)

Fmo Monsoon (raingreen) forest (tropical deciduous forest)

Fbe Subtropical broadleaf evergreen forest (laurel forest; may include mixed broadleaf-needleleaf forest)

Forests of the Midlatitude and Subarctic Zones

Fd Midlatitude deciduous (summergreen) forest

Fc Coastal forest (largely needleleaf evergreen forest, west coast of North America)

Fl Lake forest (largely needleleaf forest, Great Lakes region of North America)

Fsp Southern pine forest (southeastern United States)

Fbd Mixed boreal and deciduous forest

Fbo Boreal forest (largely needleleaf evergreen forest)

Fbl Boreal forest dominated by deciduous larch (<u>Larix dahurica</u>)

Sclerophyllous Forests of the Subtropical and Midlatitude Zones

Fsm Mediterranean evergreen mixed forest

Fss Sclerophyllous scrub (dwarf forest, chaparral; may be transitional to desert biome)

Fsa Australian sclerophyll (<u>Eucalyptus</u>) forest

Plate D.6   World natural vegetation

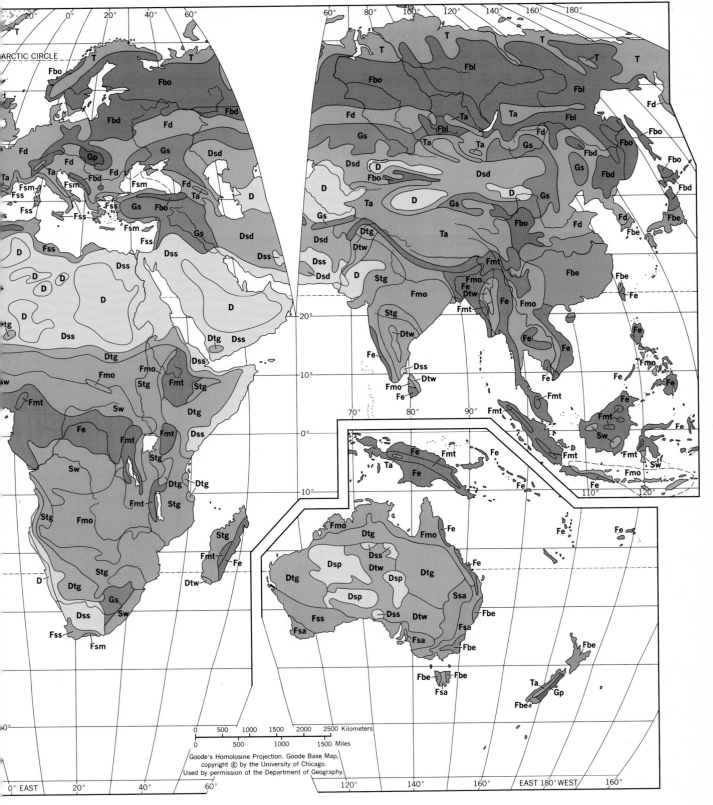

SAVANNA BIOME

Sw   Savanna woodland (broadleaf tree savanna)

Stg Thorntree-tall grass savanna

Ssa Australian sclerophyllous tree savanna

GRASSLAND BIOME

Gp   Tall-grass prairie

Gs   Short-grass prairie (steppe)

DESERT BIOME

Dtw Thorn forest and thorn woodland (may be transitional
     to forest)

Dtg Thorntree-desert grass savanna

Dsd Semidesert scrub and woodland

Dss Semidesert scrub

Dsp Desert alternating with porcupine grass semidesert

D   Desert

TUNDRA BIOME

T   Arctic tundra

Ta   Alpine tundra (includes boreal forest)

Data source: S. R. Eyre, Vegetation and soils; a world picture,
Second Edition, Aldine Publishing Company, copyright ⓒ 1968 by S. R. Eyre.
See Appendix I, Maps 1–10. Map boundaries and classes have been
simplified and modified by the authors with permission of S. R. Eyre,
Edward Arnold (Publishers) Ltd., and The Aldine Publishing Company.

World natural vegetation   **Plate D.6**

A   **Alfisols**
  A1   *Boralfs*
    A1a   with Histosols
    A1b   with Spodosols
  A2   *Udalfs*
    A2a   with Aqualfs
    A2b   with Aquolls
    A2c   with Hapludults
    A2d   with Ochrepts
    A2e   with Troporthents
    A2f   with Udorthents
  A3   *Ustalfs*
    A3a   with Tropepts
    A3b   with Troporthents
    A3c   with Tropustults
    A3d   with Usterts
    A3e   with Ustochrepts
    A3f   with Ustolls
    A3g   with Ustorthents
    A3h   with Ustox
    A3j   Plinthustalfs with
        Ustorthents
  A4   *Xeralfs*
    A4a   with Xerochrepts
    A4b   with Xerorthents
    A4c   with Xerults

D   **Aridisols**
  D1   *Aridisols, undifferentiated*
    D1a   with Orthents
    D1b   with Psamments
    D1c   with Ustalfs
  D2   *Argids*
    D2a   with Fluvents
    D2b   with Torriorthents

E   **Entisols**
  E1   *Aquents*
    E1a   Haplaquents with
        Udifluvents
    E1b   Psammaquents with
        Haplaquents
    E1c   Tropaquents with
        Hydraquents
  E2   *Orthents*
    E2a   Cryorthents
    E2b   Cryorthents with Orthods
    E2c   Torriorthents with
        Aridisols
    E2d   Torriorthents with Ustalfs
    E2e   Xerorthents with Xeralfs
  E3   *Psamments*
    E3a   with Aridisols
    E3b   with Orthox
    E3c   with Torriorthents
    E3d   with Ustalfs
    E3e   with Ustox
    E3f   of shifting sands
    E3g   Ustipsamments with
        Ustolls

H   **Histosols**
  H1   *Histosols, undifferentiated*
    H1a   with Aquods
    H1b   with Boralfs
    H1c   with Cryaquepts

I   **Inceptisols**
  I1   *Andepts*
    I1a   Dystrandepts with
        Ochrepts

  I2   *Aquepts*
    I2a   Cryaquepts with Orthents
    I2b   Halaquepts with Salorthids
    I2c   Haplaquepts with
        Humaquepts
    I2d   Haplaquepts with
        Ochraqualfs
    I2e   Humaquepts with
        Psamments
    I2f   Tropaquents with
        Hydraquents
    I2g   Tropaquepts with
        Plinthaquults
    I2h   Tropaquepts with
        Tropaquents
    I2j   Tropaquepts with
        Tropudults
  I3   *Ochrepts*
    I3a   Dystrochrepts with
        Fragiochrepts
    I3b   Dystrochrepts with Orthox
    I3c   Xerochrepts with Xerolls
  I4   *Tropepts*
    I4a   with Ustalfs
    I4b   with Tropudults
    I4c   with Ustox
  I5   *Umbrepts*
    I5a   with Aqualfs

M   **Mollisols**
  M1   *Albolls*
    M1a   with Aquepts
  M2   *Borolls*
    M2a   with Aquolls
    M2b   with Orthids
    M2c   with Torriorthents
  M3   *Rendolls*
    M3a   with Usterts
  M4   *Udolls*
    M4a   with Aquolls
    M4b   with Eutrochrepts
    M4c   with Humaquepts
  M5   *Ustolls*
    M5a   with Argialbolls
    M5b   with Ustalfs
    M5c   with Usterts
    M5d   with Ustochrepts
  M6   *Xerolls*
    M6a   with Xerorthents

O   **Oxisols**
  O1   *Orthox*
    O1a   with Plinthaquults
    O1b   with Tropudults
  O2   *Ustox*
    O2a   with Plinthaquults
    O2b   with Tropustults
    O2c   with Ustalfs

S   **Spodosols**
  S1   *Spodosols, undifferentiated*
    S1a   cryic regimes, with Boralfs
    S1b   cryic regimes, with
        Histosols
  S2   *Aquods*
    S2a   Haplaquods with
        Quartzipsamments
  S3   *Humods*
    S3a   with Hapludalfs
  S4   *Orthods*
    S4a   Haplorthods with Boralfs

U   **Ultisols**
  U1   *Aquults*
    U1a   Ochraquults with Udults
    U1b   Plinthaquults with Orthox
    U1c   Plinthaquults with
        Plinthaquox
    U1d   Plinthaquults with
        Tropaquepts
  U2   *Humults*
    U2a   with Umbrepts
  U3   *Udults*
    U3a   with Andepts
    U3b   with Dystrochrepts
    U3c   with Udalfs
    U3d   Hapludults with
        Dystrochrepts
    U3e   Rhodudults with Udalfs
    U3f   Tropudults with Aquults
    U3g   Tropudults with
        Hydraquents
    U3h   Tropudults with Orthox
    U3j   Tropudults with Tropepts
    U3k   Tropudults with Tropudalfs
  U4   *Ustults*
    U4a   with Ustochrepts
    U4b   Plinthustults with
        Ustorthents
    U4c   Rhodustults with Ustalfs
    U4d   Tropustults with
        Tropaquepts
    U4e   Tropustults with Ustalfs

V   **Vertisols**
  V1   *Uderts*
    V1a   with Usterts
  V2   *Usterts*
    V2a   with Tropaquepts
    V2b   Tropofluvents
    V2c   with Ustalfs

X   **Soils in areas with mountains**
  X1   Cryic great groups of Entisols,
    Inceptisols, and Spodosols.
  X2   Boralfs and cryic great groups of
    Entisols and Inceptisols.
  X3   Udic great groups of Alfisols,
    Entisols, and Ultisols;
    Inceptisols.
  X4   Ustic great groups of Alfisols,
    Inceptisols, Mollisols, and
    Ultisols.
  X5   Xeric great groups of Alfisols,
    Entisols, Inceptisols, Mollisols,
    and Ultisols.
  X6   Torric great groups of Entisols;
    Aridisols.
  X7   Histic and cryic great groups of
    Alfisols, Entisols, Inceptisols, and
    Mollisols; ustic great groups of
    Ultisols; cryic great groups of
    Spodosols.
  X8   Aridisols; torric and cryic great
    groups of Entisols, and cryic
    great groups of Spodosols and
    Inceptisols.

Z   **Miscellaneous**
  Z1   Ice sheets
  Z2   Rugged mountains, mostly
    devoid of soil (includes glaciers,
    permanent snowfields, and in
    some places, areas of soil.)

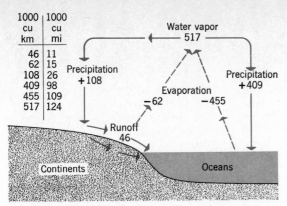

1000 cu km	1000 cu mi
46	11
62	15
108	26
409	98
455	109
517	124

Figure 10.2 **The global water balance. Figures give average annual water flux in and out of world land areas and world oceans. (Data of M. I. Budyko, 1971.)**

and oceans; 108,000 cu km is received by the land surfaces and 409,000 cu km by the ocean surfaces. Notice that the continents receive about 46,000 cu km more water as precipitation than they lose by evaporation. This excess quantity flows over or under the ground surface to reach the sea; it is collectively termed **runoff.**

We can state the **global water balance,** as follows:

$$P = E + G + R$$

where P = precipitation
E = evaporation
G = net gain or loss of water in the system, a storage term
R = runoff (positive when out of the continents, negative when into the oceans)

All terms have the dimensions of volume per unit time (e.g., cubic kilometers per year). When applied over the span of a year, and averaged over many years, the storage

Figure 10.3 **The water balance for 10-degree latitude zones. (Data from several sources compiled by W. D. Sellers, 1965, Figure from A. N. Strahler, 1971,** *The Earth Sciences,* **2nd ed., Harper and Row, New York.)**

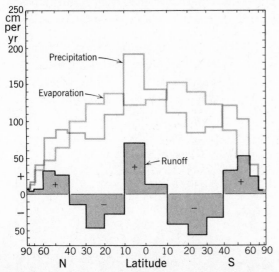

term G can be considered as zero because the global system is essentially closed so far as matter is concerned. The quantities of water in storage in the atmosphere, on the lands, and in the oceans will remain about constant from year to year.

The equation then simplifies to

$$P = E + R$$

Using the figures given for the continents,

$$108,000 = 62,000 + 46,000$$

and for the oceans,

$$409,000 = 455,000 - 46,000$$

For the globe as a whole, combining continents and ocean basins, the runoff terms cancel out:

$$108,000 + 409,000 = 62,000 + 455,000$$
$$517,000 = 517,000$$

The Global Water Balance by Latitude Zones

From a global water balance in terms of water volume per unit of time (cu km/yr; cu mi/yr), we now turn to water balance data in terms of water depth per unit of time (cm/yr; in./yr). Whereas the first set of data tells us the total quantity of water transported into or out of a specified area in one year, the second set will tell us the intensity of water flow, independent of total surface area and total water quantity.

The water balance can be estimated for latitude belts of 10-degree width. This treatment reveals a response to the latitudinal changes in radiation and heat balances from the equatorial zone to the polar zones. Figure 10.3 shows the average annual values of precipitation, evaporation, and runoff. The runoff term R in this case includes import or export of water in or out of the belt by ocean currents as well as by stream flow. Notice the equatorial zone of water surplus with positive values of R, in contrast to the subtropical belts, with an excess of evaporation. The runoff term R, which has a negative sign in the subtropical belts, represents the importation of water by ocean currents to furnish the quantity needed for evaporation. Water surpluses occur poleward of the 40th parallel; but the values of all three terms decline rapidly at high latitudes, going nearly to zero at the poles. This graph should be compared with Figure 7.22, which shows water-vapor transport across parallels of latitude. Notice that the precipitation surpluses are sustained by importation of water vapor and that evaporation surpluses are sustained by export of water vapor.

Global Water in Storage

The total global water resource is stored in the gaseous, liquid, and solid states. Water of the oceans constitutes over 97 percent of the total, as we would expect (Figure 10.4). Next comes water in storage in glaciers, a little over 2 percent. Of the remaining quantity, almost all is ground

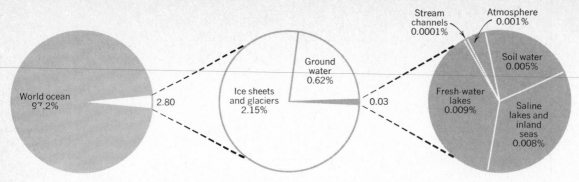

Figure 10.4 The total volume of global water in storage is largely held in the world ocean.

water, so surface water in lakes and streams is a very small quantity indeed. But it is surface water, together with the very small quantity of soil water, that sustains all life of the lands. Some environmentalists consider that fresh surface water will prove to be the limiting factor in the capacity of our planet to support the rapidly expanding human population. The quantity of water vapor held in the atmosphere is also very small—only about 10 times greater than that held in streams; but this atmospheric moisture is the source of all fresh water of the lands.

Global Water Balance and Climate Change

Long-term changes in water-balance quantities are associated with atmospheric environmental changes. For example, atmospheric cooling on a global scale would bring a reduction of atmospheric water-vapor storage. This change would, in turn, reduce precipitation and runoff generally. But at the same time, a greater proportion of that precipitation would be in the form of snow, so the storage in ice accumulations would rise and the storage in ocean waters would fall. These changes describe the changing water balance associated with onset of an ice age, or glaciation—a major environmental change already experienced by our planet at least four times in the past two million years (see Chapter 21).

Infiltration and Runoff

Most soil surfaces in their undisturbed, natural states are capable of absorbing the water from light or moderate rains. This absorption process is known as **infiltration.** Most soils have natural passageways between poorly fitting soil particles, as well as larger openings, such as earth cracks resulting from soil drying, borings of worms and animals, cavities left from decay of plant roots, or openings made by growth and melting of frost crystals. A mat of decaying leaves and stems breaks the force of falling drops and helps keep these openings clear. If rain falls too rapidly to be passed downward through these soil openings, the excess amount flows as a surface-water layer down the direction of ground slope. This surface runoff, called overland flow, is described in Chapter 18.

As stated in Chapter 7, rainfall is measured in units of centimeters or inches per hour. This is the depth to which water will accumulate in each hour if rain is caught in a flat-bottomed, straight-sided container, assuming none to be lost by evaporation or splashing out (Figure 10.5). Similarly, infiltration is stated in centimeters or inches per hour and might be thought of as the rate at which the water level in the same container might drop if the water were leaking through a porous base. Runoff, also stated in cm (in.) per hour, may be thought of as the amount of overflow of the container per hour when rain falls too fast to be disposed of by leaking through the base.

Evaporation and Transpiration

Between periods of rain, water held in the soil is gradually given up by a twofold drying process. First, direct evaporation into the open air occurs at the soil surface and progresses downward. Air also enters the soil freely and may actually be forced alternately in and out of the soil by atmospheric pressure changes. Even if the soil did not "breathe" in this way, there would be a slow diffusion of water vapor surfaceward through the open soil pores. Ordinarily only the upper 30 cm (1 ft) of soil is dried by evaporation in a single dry season. In the prolonged drought of deserts, a dry condition extends to depths of many meters.

Second, plants draw the soil water into their systems through vast networks of tiny rootlets. This water, after being carried upward through the trunk and branches into the leaves, is discharged through leaf pores into the

Figure 10.5 Rainfall, infiltration, and overland flow. (From A. N. Strahler, 1971, *The Earth Sciences,* 2nd ed., Harper and Row, New York.)

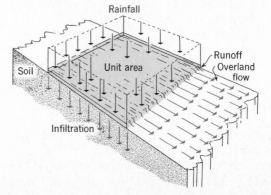

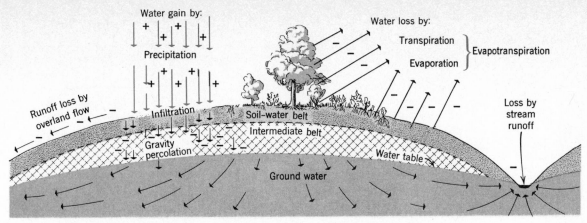

Figure 10.6 **The soil-water belt occupies an important position in the hydrologic cycle.**

atmosphere in the form of water vapor. The process is termed **transpiration.**

In studies of climatology and hydrology, it is convenient to use the term **evapotranspiration** to cover the combined water loss from direct evaporation and the transpiration of plants. The rate of evapotranspiration slows down as soil water supply becomes depleted during a dry summer period because plants use various devices to reduce transpiration. In general, the less water remaining, the slower the loss through evapotranspiration.

Figure 10.6 illustrates the various terms explained up to this point and serves to give a more detailed picture of that part of the hydrologic cycle involving the soil. The soil layer from which plants can draw moisture is the **soil-water belt.** This belt gains water through precipitation and infiltration. As the minus signs show, the soil loses water through transpiration, evaporation, and overland flow. Excess water also leaves the soil by downward **gravity percolation** to the ground water zone below. Between the soil-water belt and the ground water zone is an **intermediate belt.** Here water is held at a depth too great to be returned to the surface by evapotranspiration because it is below the level of plant roots.

Ground Water and Stream Flow

Gravity percolation carries excess water down to the ground water zone, in which all pore spaces are fully saturated (Figure 10.6). Within this zone water moves slowly in deep paths, eventually emerging by seepage into streams, ponds, and lakes.

Excess water leaves the area by stream flow. Streams are fed directly by overland flow in periods of heavy, prolonged rain or rapid snowmelt. Streams that flow throughout the year—perennial streams—derive much of their water from ground water seepage. Streams fed only by overland flow run intermittently, and their channels are dry much of the time between rain periods. We shall investigate ground water flow and stream flow in detail in Chapter 18.

Water in the Soil

When infiltration occurs, the water is drawn downward through the soil pores, wetting successively lower layers. This activity is called **soil-water recharge.** Eventually the soil layer holds the maximum possible quantity of water, although the larger pores remain filled with air. Water movement then continues downward through the underlying intermediate belt.

Suppose now that the rain stops and several days of dry weather follow. Excess soil water continues to drain downward under gravity, but some water clings to the soil particles and effectively resists the pull of gravity because of the force of capillary tension. We are all familiar with the way a water droplet seems to be enclosed in a "skin" of surface molecules, drawing the droplet together into a spherical shape so that it clings to the side of a glass indefinitely without flowing down. Similarly, tiny films of water adhere to the soil grains, particularly at the points of grain contacts, and will stay until disposed of by evaporation or absorption into plant rootlets.

When a soil has first been saturated by water, then allowed to drain under gravity until no more water moves downward, the soil is said to be holding its **storage capacity** of water. (Among soil scientists, storage capacity is referred to as **field capacity.**) Drainage takes no more than two or three days for most soils. Most excess water is drained out within one day. Storage capacity is measured in units of depth, usually centimeters or inches, just as precipitation. For example, a storage capacity of 2 cm in the uppermost 10 cm of the soil means that for a given cube of soil, 10 cm on a side (1 cu decimeter), all the extractable water would form a layer of water 2 cm deep in a 10 × 10 cm container with flat bottom and vertical sides. This depth would be equivalent to complete absorption of a 2-cm rainfall by a completely dry 10-cm layer of soil.

Storage capacity of a given soil depends largely on its texture. Sandy soil has a very small storage capacity; clay soil has a large capacity.

The Soil-Water Cycle

We can turn next to consider the annual water budget of the soil. Figure 10.7 shows the annual cycle of soil water for a single year at an agricultural experiment station in Ohio. This example can be considered generally representative of conditions in moist, midlatitude climates where there is a strong temperature contrast between winter and summer. Let us start with the early spring (March). At this time the evaporation rate is low because of low air temperatures. The abundance of melting snows and rains has restored the soil water to a surplus quantity. For two months the quantity of water percolating through the soil and entering the ground water keeps the soil pores nearly filled with water. This is the time of year when one encounters soft, muddy ground conditions, whether driving on dirt roads or walking across country. This, too, is the season when runoff is copious and major floods may be expected on larger streams and rivers. In terms of the soil-water budget, a **water surplus** exists.

By May the rising air temperatures, increasing evaporation, and full growth of plant foliage bring on intense evapotranspiration. Now the soil water falls below the storage capacity, although it may be restored temporarily by unusually heavy rains in some years. By midsummer a large **water deficit** exists in the water budget. Even the occasional heavy thunderstorm rains of summer cannot restore the water lost by heavy evapotranspiration. Small springs and streams dry up, and the soil becomes firm and dry. By November, however, the soil water again begins to increase. This is because the plants go into a dormant state, sharply reducing transpiration losses. At the same time, falling air temperatures reduce evaporation. By late winter, usually in February at this location, the storage capacity of the soil is again fully restored.

The Soil-Water Balance

From the example of soil-water change in a single year, we move forward to a more generalized concept. The gain, loss, and storage of soil water are accounted for in the **soil-water balance.** Figure 10.8 is a flow diagram that illustrates the components of the balance. **Soil-water storage** (S), the actual quantity of water held in the soil-water zone, is increased by recharge during precipitation (P) but decreased by use through evapotranspiration (E). Any **water surplus** (R) is disposed of by downward percolation to the ground-water zone or by overland flow.

The soil column is assumed to have a unit cross-sectional area, for example, 1 sq cm. We can therefore use flow units of depth per unit time (e.g., cm/month or cm/year).

The water-balance equation is as follows:

$$P = E + G + R$$

where P = precipitation
E = evapotranspiration
G = change in soil-water storage
R = water surplus

To proceed, we must recognize two ways to define evapotranspiration. First is **actual evapotranspiration** (Ea), which is the true or real rate of water-vapor return to the atmosphere from the ground and its plant cover. Second is **potential evapotranspiration** (Ep), representing the water-vapor flux under an ideal set of conditions. One condition is that there be present a complete (or closed) cover of uniform vegetation consisting of fresh green leaves, and no bare ground exposed through that cover. The leaf cover is assumed to have a uniform height above ground—whether the plants be trees, shrubs, or grasses. A second condition is that there be an adequate water

Figure 10.7 **A typical annual cycle of soil-water change in the Midwest shows a short period of surplus in the spring and a long period of deficit in the summer and fall. (Data of Thornthwaite and Mather.)**

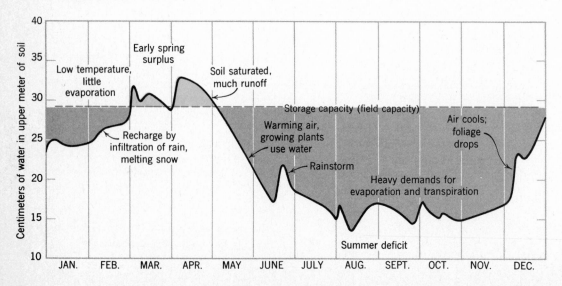

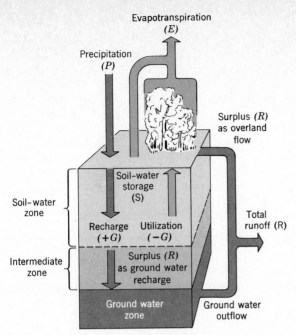

Figure 10.8 **Schematic diagram of the soil-water balance in a soil column.**

supply, such that the storage capacity of the soil is maintained at all times. This condition can be fulfilled naturally by abundant and frequent precipitation or artificially by irrigation. To simplify the ponderous terms we have just defined, they may be transformed as follows:

"actual evapotranspiration" is **water use** (Ea)
"potential evapotranspiration" is **water need** (Ep)

The word "need" signifies the quantity of soil water needed if plant growth is to be maximized for the given conditions of solar radiation and air temperature and the available supply of nutrients.

The difference between water use and water need is the **soil-water shortage** (D). This is the quantity of water that must be furnished by irrigation to achieve maximum crop growth within an agricultural system.

All the terms in the soil-water budget are stated in centimeters of water depth, the same as for precipitation. (We will not give English equivalents on the following pages.) A particular value of storage capacity of the soil is established in advance. In the soil-water budgets we present here, it has been assumed that the soil layer has a storage capacity of 30 cm (about 12 in.).

Several rigorous methods have been devised for estimating the monthly water need (potential evapotranspiration) at any given location on the globe. The method used in this chapter was developed by C. Warren Thornthwaite, a distinguished climatologist and geographer. Thornthwaite's method is based on air temperature, as well as latitude and date. Latitude and season determine intensity and duration of solar radiation received at the ground. Put another way, water need is a

measure of the maximum capability of a land surface to return energy from the surface to the atmosphere by the mecahnism of latent heat flow under the defined conditions of plant cover and water supply.

A Simple Soil-Water Budget

A simplified **soil-water budget** for the year is shown in Figure 10.9, a graph on which monthly means are plotted as points and connected by smooth lines. In this example, precipitation (P) is much the same in all months, with no strong yearly cycle. In contrast, water need (Ep) shows a strong seasonal cycle: low values in winter and a high summer peak. For midlatitudes in a moist climate, this model is about right.

At the start of the year a large water surplus (R) exists, disposed of by runoff. By May conditions have switched over to a water deficit. In this month plants begin to withdraw soil water from storage. **Storage withdrawal** (−G) is represented by the difference between the water-use curve and the precipitation curve (−G is calculated as Ea − P, when Ea exceeds P). As storage withdrawal continues, the plants reduce their water use to less than the optimum quantity, so without irrigation the water-use curve departs from the water-need curve. Storage withdrawal continues throughout the summer. The deficit period lasts through September. The area labeled "soil-water shortage" represents the total quantity of water needed by irrigation to insure maximum growth throughout the deficit period. Soil-water shortage (D), is calculated by subtracting Ea from Ep (D = Ep − Ea).

In October precipitation (P) again exceeds water need (Ep), but the soil must first absorb an amount equal to the summer storage withdrawal (−G). So we next have a period of **storage recharge** (+G); it lasts through November. In December the soil has reached its full storage

Figure 10.9 **A simplified soil-water budget typical of a humid midlatitude climate.**

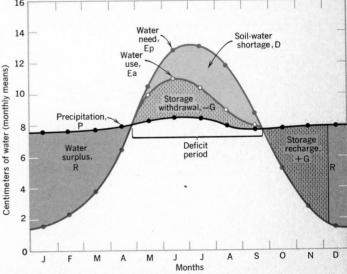

Equation	P	= Ea	+G	+R	Ep	(Ep − Ea) = D
January	11.0	= 1.0		+10.0	1.0	0.0
February	9.0	= 2.0		+ 7.0	2.0	0.0
March	6.0	= 3.5		+ 2.5	3.5	0.0
April	3.0	= 6.0	−3.0		6.0	0.0
May	2.5	= 7.0	−4.5		8.5	1.5
June	2.0	= 6.0	−4.0		9.5	3.5
July	2.5	= 5.0	−2.5		9.0	4.0
August	4.0	= 4.5	−0.5		7.0	2.5
September	7.0	= 4.5	+2.5		4.5	0.0
October	9.0	= 3.0	+6.0		3.0	0.0
November	10.5	= 1.5	+6.0	+ 3.0	1.5	0.0
December	12.0	= 1.5		+10.5	1.5	0.0
Totals	78.5	= 45.5	−14.5	+14.5	+33.0	57.0	11.5
	78.5	= 78.5				

Table 10.1 / Simplified Example of a Soil-Water Budget

capacity (30 cm). Now a water surplus (R) again sets in, lasting throughout the winter.

Thornthwaite was concerned with practical problems of crop irrigation. He developed the calculation of the soil-water budget to place crop irrigation on a precise, accurate basis. The Thornthwaite method was later adopted by soil scientists of the U. S. Department of Agriculture to identify classes of soils associated with various soil-water budgets. Water need (potential evapotranspiration) is a difficult quantity to measure, and several other methods have been developed to estimate its true value.

In a world beset by severe and prolonged food shortages, the Thornthwaite concepts and calculations are of great value in assessing the benefits to be gained by increased irrigation. Only in a few parts of the tropical and midlatitude zones is precipitation ample to fulfill the water need during the growing season. In contrast, the equatorial zone generally has a large water surplus throughout the year.

Calculating a Simple Soil-Water Budget

Table 10.1 gives simplified data for a hypothetical example of a soil-water budget. Columns are arranged so that the terms appear in the same sequence as in the soil-water balance equation. Values are rounded to multiples of 0.5 cm to simplify addition and subtraction. Figure 10.10 shows the same data plotted in graphic form.

By means of simple arithmetic, we have calculated the soil-water budget for each month singly and for the year as a whole. Tallied separately at the right are the monthly differences between Ep and Ea, giving a total soil-water shortage, D, of 11.5 cm for the year. This is the quantity of water that would have to be supplied by irrigation in an average year to sustain the full value of Ep. The importance of the water budget calculation in estimating the need for irrigation of crops should be obvious. We have also been able to calculate the annual surplus (R = 33 cm). This information can be used to estimate the

recharge of ground water and the runoff into streams. In this way an assessment of the water-resource potential of a region can be made—a vital consideration in planning regional economic development and resource management.

The Global Range of Water Need

Figure 10.11 is a world map showing mean annual water need (Ep, potential evapotranspiration) calculated by the Thornthwaite method. Isopleths of water need in centimeters per year are greatly generalized and are deleted from major highland regions. What counts is the general pattern of poleward decrease of Ep from high values in the equatorial and tropical zones to very low values in the arctic zone. A second important feature is the effect of deserts, which produce abnormally high values compared with areas of moist climate in which temperatures are moderated by precipitation and persistent cloud cover.

Figure 10.12 illustrates the effect of latitude on the annual cycle of water need. Both air-temperature cycles and insolation cycles are reflected in these cycles. Each chart has been labeled as to climate, according to the system given in Chapter 9.

In the wet equatorial climate (1), water need is high all year and the total is over 150 cm (60 in.). For the monsoon and trade-wind littoral climate (2) in the tropical latitude zone, there is a pronounced annual cycle, but a large annual total water need: almost 150 cm (60 in.). In the tropical desert climate (4d), the annual cycle is very strongly developed and the total water need remains high: nearly 127 cm (50 in.). In midlatitudes on the west coast, the Mediterranean climate (7) shows a well-developed annual cycle; but monthly values of water need remain high throughout the mild winter. In the humid continental climate (10) winter months have no water need, but the summer peak is high; the total is 75 cm (30 in.). In the marine west-coast climate (8), close to the Pacific Ocean, however, water need remains substantial

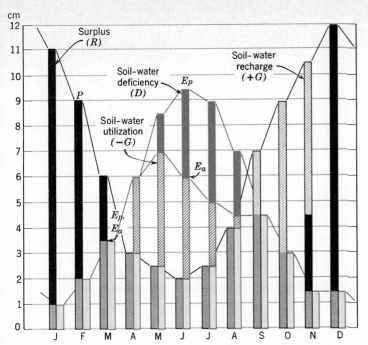

Figure 10.10 **A model soil-water budget. Bars are correctly scaled to agree with the figures in Table 10.1.**

shows nine consecutive winter months of zero water need and a very narrow summer peak. The total water need here is least of all the examples: 20 cm (8 in.).

The details of these examples are not important. What counts is the trends they show. Total annual water need diminishes from a maximum in the equatorial zone to a minimum in the arctic zone. At the same time, the annual cycle becomes stronger and stronger in the poleward direction. As the cold of winter makes its effects more pronounced, the number of months of zero water need increases. Of course, plant growth is essentially zero when soil water is frozen, so the growing season is nicely reflected in the graphs of water need. Water need persists throughout the winter far poleward along narrow western coastal zones (littorals), in contrast to the continental interior regions.

The Soil-Water Balance as a Basis of Climate Classification

The soil-water balance of C. W. Thornthwaite is precise and quantitative in its use of station data of monthly mean precipitation and monthly mean air temperature (used to derive water need). The soil-water balance offers us the opportunity to set up a quantitative climate classification. The Köppen climate system, as we noted, has been held in esteem for many decades by geographers because it uses precise, quantitative definitions of climate types based on monthly mean precipitation and temperature data. But a climate system based on the soil-water balance has certain advantages over the Köppen system. First, the Thornthwaite method gives information of direct value in

throughout the winter because of the mild temperatures. Farther poleward in the boreal forest climate (11), the number of months of zero water need increases to six and the annual total water need diminishes to 40 cm (16 in.). North of the arctic circle, the tundra climate station (12)

Figure 10.11 **World map of annual water need (potential evapotranspiration). (Based on data of C. W. Thornthwaite Associates, Laboratory of Climatology, Centerton, N.J.)**

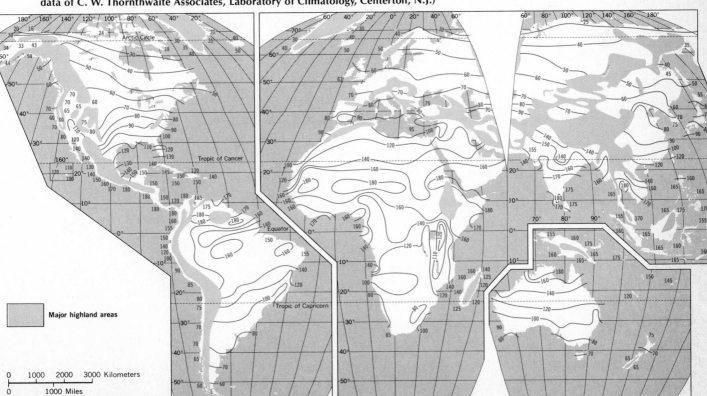

Major highland areas

0   1000   2000   3000 Kilometers

0   1000 Miles

Goode's Homolosine Projection. Goode Base Map,
copyright © by the University of Chicago. Used by
permission of the Department of Geography

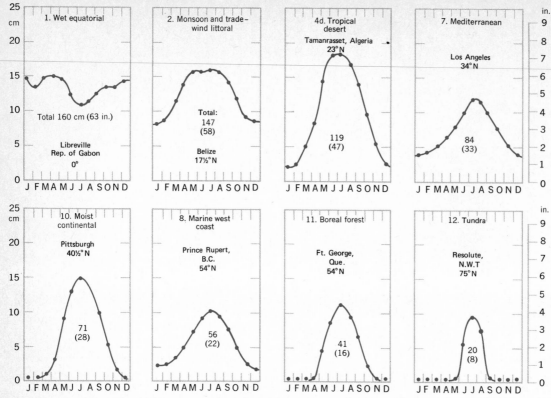

Figure 10.12 **(above and right) Some representative annual cycles of water need (potential evapotranspiration) over a wide latitude range. (Data of C. W. Thornthwaite Associates. Goode Base Map.)**

assessing the conditions favorable or unfavorable to the growth of plants—both native plants and cultivated crops. Second, the soil-water balance has recently been applied by soil scientists to a modern system of soil classification, which we will present in Chapters 13 and 14. Thus a quantitative water-balance system of climatology is a necessity. The Köppen system does not fill this need and is not used in the U.S. comprehensive soil classification system.

We offer here a climate system based on recognition of distinctive soil-water regimes. All that is needed to establish this climate classification is a set of definitions. The classification retains the 13 climate classes recognized in Chapter 9 on the basis of air-mass source regions and frontal zones. The names of the 13 classes remain valid—only the frame of reference is changed. The new quantitative frame of reference simply reinforces the explanatory-descriptive framework already established.

Source of Data

A climate system based on the soil-water balance requires a large base of station data derived in a consistent manner from long records of monthly mean precipitation and monthly mean air temperature. We make use of published water-balance data of the continents prepared by the Laboratory of Climatology in Centerton, New

Jersey, which was originally under the direction of Dr. Thornthwaite. Over a 13-year period, the staff of the Laboratory of Climatology computed the water balances of more than 13,000 stations. Representative station data have been published in a series of 8 volumes, covering the world's inhabited land areas. These data were used in drawing our world map of climates, Plate D.2. The Thornthwaite data have also been used by the Soil Conservation Service of the U.S. Department of Agriculture as the basis for recognizing soil-water regimes associated with the major classes of soils (see Chapters 13 and 14).

Water Need as an Indicator of Available Heat

Our three major climate groups—I. Low-Latitude, II. Midlatitude, and III. High-Latitude—can be put on a firm quantitative basis using the total annual water need, Ep. In a general way, Ep grades from maximum values in low latitudes to minimum values in the arctic zone. This gradation reflects the poleward decrease in both annual net radiation and mean annual air temperature.

The Low-Latitude Climates, Group I, are those climates with total annual Ep greater than 130 cm. This criterion applies rigorously to the dry tropical climate (4) and the wet-dry tropical climate (3); but Ep may be

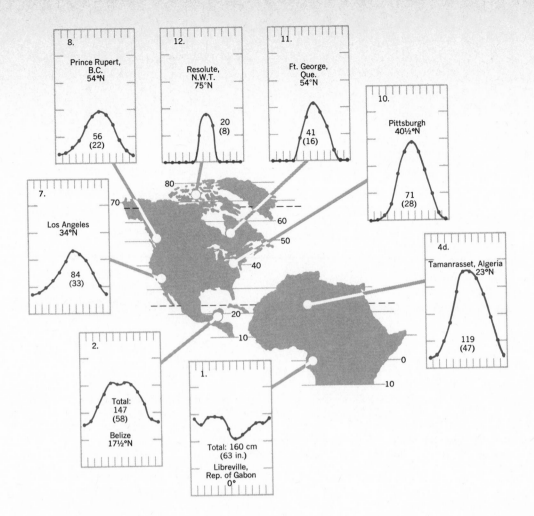

somewhat less (down to 110 cm) in some parts of the wet equatorial climate (1) and in the monsoon and trade-wind littoral climate (2). On our world climate map, Plate D.2, the poleward boundaries of climates 3 and 4 are based on the 130-cm isopleth of Ep. This boundary is particularly important in separating the dry tropical climate (4) from the dry subtropical climate (5) in northern Africa, North America, Australia, South America, and southern Africa.

The Midlatitude Climates, Group II, are those climates with total annual Ep ranging from 130 cm down to 52.5 cm. Thus the poleward boundary of Group II climates follows the 52.5-cm isopleth of Ep. This boundary is found in both North America and Eurasia, separating the moist continental climate (10) from the boreal forest climate (11). It is a line approximately coinciding with the southern limit of the great boreal forests. To the south lie deciduous forests and grasslands of warmer climates.

The High-Latitude Climates, Group III, have a total annual water need, Ep, of less than 52.5 cm. This group includes the boreal forest climate (11), the tundra climate (12), and the ice-sheet climate (13). In the tundra climate (12), Ep is less than 35 cm. Thus the boundary between boreal forest climate (11) and tundra climate (12) is the 35-cm isopleth of Ep. It is a line marking the northern limit of the boreal forests in North America and Eurasia.

Figure 10.13 is a schematic diagram showing the boundaries of the three climate groups in terms of total annual water need, Ep. The horizontal layering in the diagram expresses the availability of heat to plants in the life layer. The second climate ingredient, availability of soil water, is expressed by the vertical columns, defined in terms of soil-water shortage (D), soil-water storage (S), and water surplus (R). We turn next to the definitions of climate types in terms of availability of soil water.

Defining Dry and Moist Climates

An essential step in designing a climate system based on the soil-water balance is to distinguish precisely a dry climate from a moist climate. A **dry climate** is one in which the total annual soil-water shortage, D, is 15 cm (5.9 in.) or larger, while at the same time there is no water surplus, R. A **moist climate** is one in which the soil-water shortage, D, is less than 15 cm. (There is no requirement that a moist climate must show a water surplus.) These definitions enable us to draw a map boundary between dry and moist climates. We will find that this boundary, the 15-cm isopleth of D, coincides in midlatitudes quite well with the boundary between steppe grass-lands (short-grass prairie) and tall-grass prairie.

Table 10.2 / Definitions of Climate Subtypes

Symbol	Name	Definition
s	Semiarid subtype (Steppe subtype)	At least two months in which soil-water storage, S, is equal to or exceeds 6 cm.
sd	Semidesert subtype	Fewer than two months with S greater than 6 cm, but at least one month with S greater than 2 cm.
d	Desert subtype	No month with S greater than 2 cm.
sh	Subhumid subtype	Soil-water shortage, D, is greater than zero but less than 15 cm, when there is no water surplus, R. Otherwise, D is greater than R, when R is not zero.
h	Humid subtype	R is 1 mm or greater but less than 60 cm. (R is always greater than D.)
p	Perhumid subtype	R is 60 cm or greater.

Subdivisions of the Dry Climates

Within the dry climates, three degrees of dryness (aridity) are defined precisely in terms of monthly values of soil-water storage, S. Recall that the Thornthwaite system of calculating the soil-water budget assumes a soil storage capacity (field capacity) of 30 cm (about 12 in.). Values of soil-water storage, S, can thus range from 0 to 30 cm. Values of S are zero, or very close to zero, throughout the year in dry desert locations. In very moist climates S runs at, or close to, 30 cm in all months.

Table 10.2 defines three levels of aridity within the dry climates; these form three climate subtypes described in Chapter 9.

Boundaries between each of the three subtypes are drawn on the world climate map, Plate D.2, in accordance with these definitions. Areas falling within each subtype are assigned a different color in the range from light brown to pale yellow. Three climates fall in the dry category; these are distinguished on the map by their numbers (4, 5, and 9).

Subdivisions of the Moist Climates

The degree of wetness of the moist climates spans a wide range—from climates with no water surplus, R, to those with a very large water surplus. Our system uses three levels of wetness based on total annual water surplus, R, defining precisely the three climate subtypes already described in Chapter 9. Definitions of subtypes are given in Table 10.2.

Boundaries separating each of the three moist subtypes are shown on the climate map, Plate D.2. Each subtype is assigned a different color intensity, from light to dark. Note that these subtypes are recognized only for moist climates in midlatitude and high-latitude groups.

The schematic diagram, Figure 10.13, shows the subtypes of dry and moist climates, arranged from greatest water shortage (left) to greatest water surplus (right).

Climates with Strong Moisture Seasons

Two of the 13 climate types cannot be described as either dry or moist by the definitions we have given. Instead, these two types have a very wet season alternating with a very dry season, with the result that they show both a substantial soil-water shortage, D, and a substantial soil-water surplus, R. In the low-latitude group, the wet-dry tropical climate (3) has D greater than 20 cm and R greater than 10 cm. In the humid subtype of the Mediterranean climate (7), both D and R exceed 15 cm. In the schematic diagram, Figure 10.13, these two wet-dry climates (3 and 7) are shown as alternating between dry and moist extremes.

Figure 10.13 Schematic diagram of world climates in terms of water need and degree of dryness and wetness.

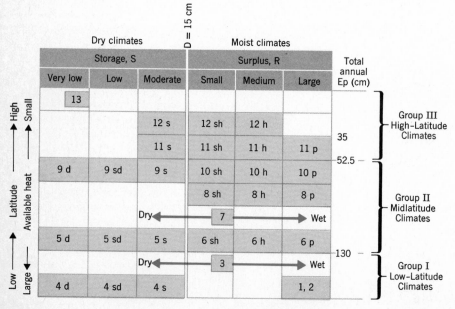

GLOBAL CLIMATE CLASSIFIED BY SOIL-WATER BUDGETS*

GROUP I / LOW-LATITUDE CLIMATES

1. Wet equatorial climate

A warm, very wet climate found in the equatorial zone. Precipitation, P, is heavy throughout the entire year. In nearly all months, P exceeds water need, Ep. As a result, there is a substantial water surplus, R, in all or nearly all months. Soil-water storage, S, is large in all months and is usually over 25 cm in 10 or more months. For many stations, S is 30 cm in all 12 months.

Example. Singapore, Malay Peninsula, lat. 1½°N (Figure 10.14). Precipitation, P, at Singapore is large in every month, the least being 15 cm, the greatest 29 cm. The annual total precipitation is very great: 240 cm. Ep is highly uniform in monthly values, which range from 13 cm to 15 cm. The yearly total Ep is thus very great: 172 cm. P is much greater than Ep in all months but one (July), so 11 months show a water surplus. Total annual R is nearly 70 cm. There is no soil-water shortage, D, in any month. Moreover, soil-water storage, S, stands at the maximum value of 30 cm in every month. It is obvious that optimum conditions for plant growth exist year around at Singapore. Equatorial rainforest is the typical natural vegetation. Stream flow is copious throughout the year.

2. Monsoon and trade-wind littoral climate

A warm, very moist climate showing a strong rainfall peak at one season (usually at time of high sun) and a short period of reduced rainfall at another season (usually at time of low sun). This cycle represents the monsoon system in Southeast Asia. A small soil-water shortage, D, develops in the short dry season. Water need, Ep, exceeds 4 cm in every month or the yearly total Ep is more than 130 cm, or both. A large water surplus, R, is generated in the rainy months. Soil-water storage, S, is greater than 20 cm through 6 to 9 consecutive months.

Example. Aparri, Luzon Island, Philippine Islands, lat. 18°N (Figure 10.15). This trade-wind littoral station shows a strong annual precipitation cycle, peaking strongly following the high-sun period in a wet season, with P greater than 25 cm in three consecutive months.

---

*Climate definitions and boundaries are summarized in Appendix I.

Figure 10.14 **Soil-water budget for Singapore, Malay Peninsula, lat. 1½°N. (Data from C. W. Thornthwaite Associates, Laboratory of Climatology, Centerton, N.J.)**

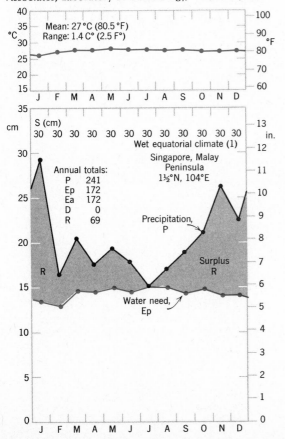

Figure 10.15 **Soil-water budget for Aparri, Luzon Island, Philippine Islands, lat. 18°N. (Data from C. W. Thornthwaite Associates, Laboratory of Climatology, Centerton, N.J.)**

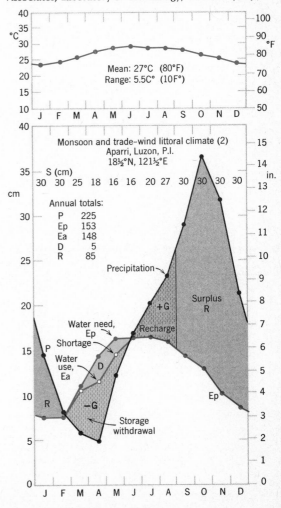

Following the period of low sun, when the ITC is in the southern hemisphere, a period of greatly reduced rainfall develops, with P falling below 10 cm in three consecutive months. The annual cycle of soil-water need, Ep, also has a strong annual cycle; but the maximum values occur from May through August, which is well in advance of the rainfall peak. As a result, a very small soil-water shortage, D, is developed through May, June, and July. When P rises to exceed Ep, recharge begins. By September a large surplus is developed. The total water surplus, R, is very large: 85 cm. Soil-water storage, S, remains at 30 cm for six consecutive months (September through February). The lowest monthly values of S, about 16 cm, occur in May and June, which is the brief water-shortage period. Thus, despite a short dry season, soil water remains ample for plant growth. Rainforest thrives in this environment. Stream flow, which reaches flood proportions in the wet season, diminishes greatly during the brief dry season.

3. Wet-dry tropical climate

A tropical-zone climate characterized by a very wet season alternating with a very dry season. The wet season, or rainy monsoon, occurs in the high-sun period; the dry season, in the low-sun period. Water need, Ep, exceeds 4 cm in every month or the yearly total Ep is more than 130 cm, or both. A substantial water surplus, R, occurs in the wet season; a substantial soil-water shortage, D, occurs in the dry season. R is 10 cm or greater; D is 20 cm or greater. Soil-water storage, S, exceeds 20 cm in five months or fewer.

Example. Raipur, Madhya Pradesh, India, lat. 21°N (Figure 10.16). Dominated by the Asiatic monsoon system, Raipur shows a very wet high-sun season with P over 20 cm in four consecutive months (June through September), which together account for 117 cm, or 87 percent, of the total annual rainfall. During the long, cool dry season, seven consecutive months have P less than 3 cm. The annual cycle of water need, Ep, is strongly developed and the annual total Ep, 152 cm, is very large. But because Ep rises to high values while P is very low in March, April, and May, a large soil-water shortage, D, is accumulated: 43 cm. This shortage period is also the hot season. Storage recharge, +G, takes place in June and July, so by August a substantial water surplus, R, is developed. The annual surplus, R, however, is only 25 cm, an amount about half as much as the shortage, D. This ratio of D to R (about 2 to 1) is typical of the climate. Soil-water storage, S, is greater than 20 cm in five consecutive months (July through November) and attains 30 cm (storage capacity) in August and September. The period of high values of S is the season of rapid plant growth. The long drought period requires that native plants also be adapted to very low soil-water levels. Grasses, deciduous trees, and certain shrubs can survive this cycle of extreme conditions. Stream flow is copious in the wet season, with some severe flooding, but streams shrink greatly in discharge or disappear entirely in the dry season.

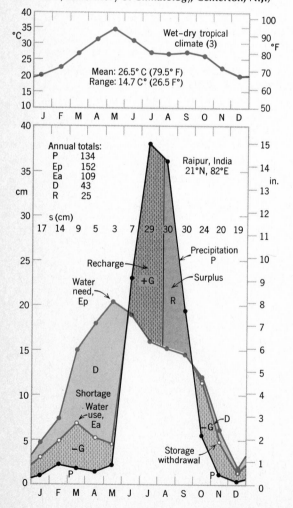

Figure 10.16 **Soil-water budget for Raipur, Madhya Pradesh, India, Lat. 21°N. (Data from C. W. Thornthwaite Associates, Laboratory of Climatology, Centerton, N.J.)**

4. Dry tropical climate

A warm to hot, dry climate, widespread in the tropical zone. Large central areas of true desert (4d), located over the tropics of cancer and capricorn, are bordered on the equatorward side by narrow parallel belts of semidesert (4sd) and semiarid (steppe) (4s) subtypes. Annual total water need, Ep, exceeds 130 cm or all monthly values of Ep exceed 4 cm, or both. Large areas of North Africa between lat. 10 and 20°N have total Ep between 160 and 180 cm, with all monthly values of Ep over 10 cm. A special desert subtype (4dw) is recognized on very narrow western littorals under the influence of a cool marine air layer.

Example. Khartoum, Sudan, lat. 15½°N (Figure 10.17). Khartoum lies in the heart of an enormous area of tropical desert climate (4d) that spans North Africa and the Arabian Peninsula. The total annual soil-water need, Ep, is extremely large: 183 cm. The annual cycle of Ep is strongly developed, while all monthly values of Ep exceed 7 cm. Monthly values in excess of 15 cm occur in 7 successive months, during which time very high air temperatures prevail. Precipitation, P, shows a sharp peak in July and August, with more than 6 cm in each of those months. Total annual P is, however, only 18 cm. Note that water use, Ea, is identical with P in every month. Every month shows a substantial soil-water shortage, D. The annual total of D is very large: 165 cm. Soil-water storage, S, is zero in all months. Few plants can survive in this extremely dry, hot environment; those that can are

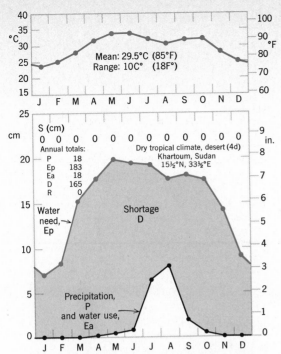

Figure 10.17 Soil-water budget for Khartoum, Sudan, lat. 15½°N. (Data from C. W. Thornthwaite Associates, Laboratory of Climatology, Centerton, N.J.)

adapted to profit from rains in July and August, which briefly moisten the uppermost soil layer.

Example. Fort Lamy, Chad, lat. 12°N (Figure 10.18). Fort Lamy lies in a narrow belt of tropical steppe (4s) climate subtype crossing North Africa from west to east; it lies immediately north of the belt of wet-dry tropical climate (3). Total annual precipitation, P, is 62 cm; it occurs mostly in a short rainy season that peaks sharply in August, with a monthly mean of about 25 cm. The total water need, Ep, is very high (175 cm) and shows an annual cycle similar to that of Khartoum, but with a sharp dip in August, coinciding with the brief period of rain. Because P exceeds Ep in August by a substantial margin, soil-water recharge, +G, occurs in August in the amount of about 11 cm. The remainder of the year sees small monthly withdrawals from soil-water storage. Soil-water shortage, D, is heavy throughout most months, with a large annual total of 113 cm. No water surplus, R, is developed. Soil-water storage, S, exceeds 6 cm in two consecutive months (August and September). This criterion establishes the climate subtype as semiarid, or steppe (4s). Tropical savanna grassland, the native vegetation in this climate, can support limited grazing and even some unirrigated grain farming. Year-to-year variations in precipitation are great, so human occupation is hazardous.

GROUP II / MIDLATITUDE CLIMATES

5. Dry subtropical climate

A dry climate, transitional between the dry tropical climate (4) and the dry midlatitude climate (9). It occupies an intermediate geographic zone between those climates.

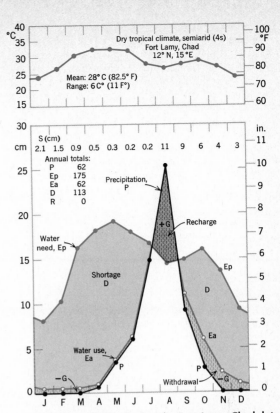

Figure 10.18 Soil-water budget for Fort Lamy, Chad, lat. 12°N. (Data from C. W. Thornthwaite Associates, Laboratory of Climatology, Centerton, N.J.)

Figure 10.19 Soil-water budget for Parker, Arizona, lat. 34°N. (Data from C. W. Thornthwaite Associates, Laboratory of Climatology, Centerton, N.J.)

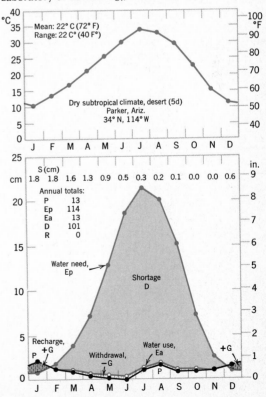

Water need, Ep, of the coolest month is less than 4 cm, but greater than 0.7 cm. Annual total Ep is less than 130 cm. Semiarid (5s), semidesert (5sd), and desert (5d) subtypes are recognized.

Example. Parker, Arizona, lat. 34°N (Figure 10.19). Located on the Colorado River, Parker lies in the desert subtype (5d) of the dry subtropical climate. Certain distinctive differences exist between this climate and the dry tropical desert subtype (4d), illustrated by the graph for Khartoum (Figure 10.17). Total annual water need, Ep, is 114 cm, much less than at Khartoum and well under the boundary value of 130 cm separating the two climates. The annual cycle of Ep is very strongly developed, with a sharp summer peak. Monthly Ep in the cold season falls to low values (1 cm) in December and January—much lower than in the tropical dry climates. Precipitation, P, which totals 13 cm, shows two distinctive maxima: (1) a cold-season maximum when midlatitude cyclones pass across the region; and (2) a summer maximum, when maritime tropical air masses that penetrate from the Gulf of Mexico or the Gulf of California set off thunderstorms. P exceeds Ea by a small margin in December and January, allowing a small amount of storage recharge (+G). This water is utilized in very small monthly amounts during the remainder of the year. The total soil-water shortage, D, is very large, and there is no water surplus, R. As required for the desert subtype, soil-water storage, S, does not exceed 2 cm in any month.

6. Moist subtropical climate

A moist climate characterized by a moderate to large water surplus, R, and a small seasonal soil-water shortage,

D. The annual cycle of Ep is strongly seasonal, with fairly low values in the winter months. Ep is less than 4 cm in at least one month. In all months, Ep is at least 0.8 cm. Thus the coldest month temperature mean is not below 0°C. In the subhumid subtype (6sh), D is greater than zero but less than 15 cm when R is zero; otherwise, D is greater than R when R is not zero. In the humid subtype (6h) R is greater than zero but not over 60 cm, while R is greater than D. In the perhumid subtype (6p), R is greater than 60 cm.

Example. Baton Rouge, Louisiana, lat. 30°N (Figure 10.20). Precipitation, P, is copious in all months, but with a pronounced dip in late summer and autumn. There is a small summer precipitation peak. Otherwise monthly values of P are uniformly about 12 cm and the annual total is large: 143 cm. There is a strong cycle of soil-water need, Ep, with summer-month values exceeding 16 cm and winter values falling to about 2 cm. Because winter-month mean temperatures are well above freezing, however, some plant growth can continue through the mild winter with active transpiration. Storage withdrawal (−G) occurs in 4 summer months, but the shortage, D, is small in every month and totals only 2 cm for the year. In contrast, the water surplus, R, is large (39 cm). Thus the climate qualifies as the humid subtype (6h). Stream flow is copious and continues throughout the year. Forest is the natural plant cover in this humid climate.

7. Mediterranean climate

A subtropical climate distinguished by the alternation of a very dry summer with a mild, rainy winter. Water need, Ep, is 0.8 cm or greater in every month, separating the Mediterranean climate from colder climates (9 and 10) bordering on the interior side. The dry aspect of the Mediterranean climate is expressed in a soil-water shortage, D, always 15 cm or larger. Water surplus, R, may

Figure 10.20 **Soil-water budget for Baton Rouge, Louisiana, lat. 30½°N. (Data from C. W. Thornthwaite Associates, Laboratory of Climatology, Centerton, N.J.)**

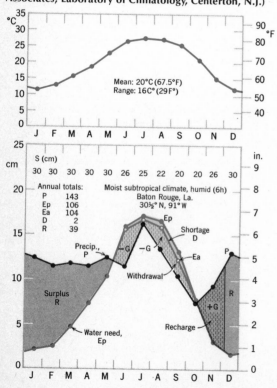

Figure 10.21 **Soil-water budget for Los Angeles, California, lat. 34°N. (Data from C. W. Thornthwaite Associates, Laboratory of Climatology, Centerton, N.J.)**

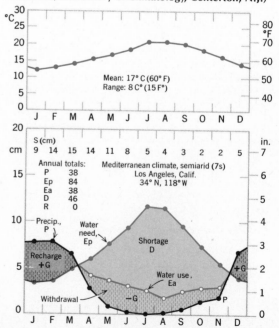

be zero (in the semiarid and semidesert subtypes.) A diagnostic feature of the climate is the large amplitude of the annual cycle of soil-water storage, S. For all subtypes, the *storage index** is 75 percent or greater. The following subtypes are recognized:

7sd semidesert Soil-water storage, S, exceeds 6 cm in fewer than 2 months. S exceeds 2 cm in at least 1 month.

7s semiarid (steppe) S is 6 cm or greater in 2 or more months.

7sh subhumid Water surplus, R, ranges from 0 to 15 cm, so that D is larger than R.

7h humid R exceeds 15 cm.

Both semidesert (7sd) and semiarid (7s) subtypes are true dry climates. No desert subtype is included. Adjacent desert climates showing the winter rainfall maximum fall into the dry subtropical climate (5d) or the dry midlatitude climate (9d). The Mediterranean precipitation cycle also extends into the adjacent marine west-coast climate (8), in which D is always less than 15 cm.

Example. Los Angeles, California, lat. 34°N (Figure 10.21). Los Angeles is typical of the semiarid subtype (7s) of the Mediterranean climate found along the entire California coast from Monterey to Oceanside. The Mediterranean precipitation cycle, with its long dry summer, is a striking feature that contrasts with the annual cycle of water need, Ep, in opposite phase. Thus the summer peak of Ep coincides with the season of almost zero rainfall, greatly accentuating the soil-water shortage, D. Storage withdrawal, −G, sets in early in the year (April) and continues until December. The large accumulated shortage, D, totals 46 cm. Storage recharge, +G, begins in December and continues throughout the winter, but no water surplus, R, is generated. Native plants—mostly grasses and hard-leaved shrubs or trees—are adapted to the long dry summer. Massive amounts of summer irrigation are required to maintain green lawns and garden crops.

Example. Perth, Western Australia, lat. 32°S (Figure 10.22). This coastal city represents the humid subtype (7h) of the Mediterranean climate. (Note that the graph begins with July at the left.) Water surplus, R, is 28 cm, a large quantity. Soil-water shortage, D, is also large: 27 cm. The result is a true wet-dry climate—a subtropical equivalent to the wet-dry tropical climate (3), from which it is far separated in latitude. Compared with Los Angeles, which is of the

---

*The storage index, stated as a percentage, is calculated from maximum and minimum monthly values of S, as follows:

$$\frac{S_{max} - S_{min}}{S_{max}} \times 100$$

An alternative method of identifying the Mediterranean climate uses monthly values of P and Ea. For July (northern hemisphere) or January (southern hemisphere), the ratio of P to Ea is calculated in percent, as follows:

$$\frac{P}{Ea} \times 100$$

Stations with ratios lower than 40 percent are assigned to the Mediterranean climate. For nearly all stations, there is agreement between the two methods.

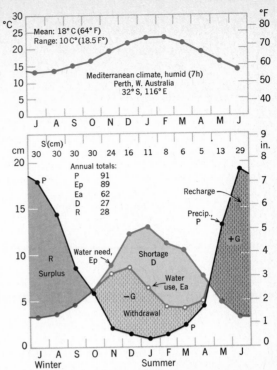

Figure 10.22 **Soil-water budget for Perth, Western Australia, lat. 32°S. (Data from C. W. Thornthwaite Associates, Laboratory of Climatology, Centerton, N.J.)**

semiarid subtype (7s), Perth has much greater winter rainfall and a total annual P over twice as large as at Los Angeles. Perth has small monthly amounts of rainfall throughout the summer, whereas rainfall is nearly zero at Los Angeles in the equivalent period. Perth has a large water surplus, R, whereas Los Angeles has none. The humid climate of the Perth region supports an evergreen forest that is composed largely of species of eucalyptus trees.

8. Marine west-coast climate

A cool, moist climate with soil-water surplus, R, ranging from small to very large. Soil-water shortage, D, ranges from moderate to very small and may be zero. Annual total water need, Ep, is less than 80 cm, largely because summers are exceptionally cool for the higher midlatitudes. Monthly Ep is 0.8 cm or greater in every month, so no month has a mean temperature as low as 0°C.

Example. Cork, Ireland, lat. 52°N (Figure 10.23). Located on the south coast of Ireland and exposed to westerly winds of the North Atlantic, Cork illustrates the humid subtype (8h) of the marine west-coast climate. Precipitation, P, shows a strong annual cycle, with a winter maximum and a summer minimum, reflecting a poleward persistence of the Mediterranean cycle. P is copious in all months, and the annual total is large: 105 cm. Water need, Ep, shows a strong annual cycle, but in all winter months Ep is 1.8 cm or greater. Thus the lowest mean monthly temperature is only about 6°C; growth of many perennial plants can continue slowly throughout the winter months. In the summer months, Ep is not much greater than P,

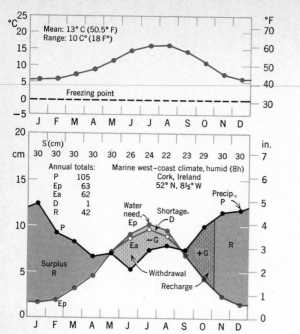

Figure 10.23 **Soil-water budget for Cork, Ireland, lat. 52°N. (Data from C. W. Thornthwaite Associates, Laboratory of Climatology, Centerton, N.J.)**

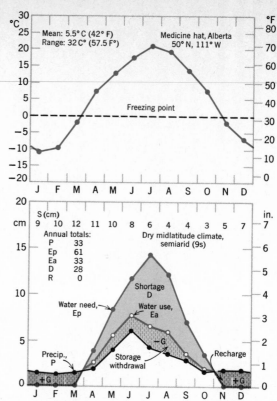

Figure 10.24 **Soil-water budget for Medicine Hat, Alberta, Canada, lat. 50°N. (Data from C. W. Thornthwaite Associates, Laboratory of Climatology, Centerton, N.J.)**

so the water shortage, D, is very small, totaling only 1 cm. The water surplus, R, is large: 42 cm. Even small streams can flow continuously throughout the year. The humid climate supports a native vegetation of forest.

9. Dry midlatitude climate

A dry climate, found almost exclusively in the northern hemisphere, poleward of the dry subtropical climate (5). Monthly Ep is 0.7 cm or less in at least one month. Thus winters are cold; usually at least one month has a mean temperature of 0°C or lower. Near the northern limit, Ep is zero for as many as five consecutive months. The lower limiting value of annual total Ep is 52.5 cm, occurring along the boundary with the boreal forest climate (11). Semiarid (9s), semidesert (9sd), and desert (9d) subtypes are recognized.

Example. Medicine Hat, Alberta, Canada, lat. 50°N (Figure 10.24). Located near the northern limit of the Great Plains, Medicine Hat illustrates the semiarid (steppe) subtype (9s) of the dry midlatitude climate. The annual cycle of water need, Ep, peaks strongly in summer, following 5 consecutive months in which Ep is zero because of severe winter cold. Precipitation shows a distinct annual cycle; the summer months have about double the precipitation of the winter months. A substantial soil-water shortage, D, develops in summer and totals 28 cm. But soil-water storage, S, remains above 3 cm, even in the late summer and early autumn months. Storage recharge, +G, is not sufficient to bring S to levels approaching storage capacity, so no surplus, R, is generated. Recharge accumulates in the frozen state throughout the winter, to be released rapidly in the spring thaw. Short-grass prairie is the native plant cover of this northern steppe climate. Spring wheat, grown here, uses the accumulated soil water of early spring and matures in summer.

10. Moist continental climate

A moist climate with strongly defined winter and summer seasons, found only in the northern hemisphere. Water surplus, R, ranges from small to large, except in a narrow subhumid subtype (10sh) bordering the dry continental climate (9). Total annual water need, Ep, is more than 52.5 cm, while R is greater than D. In the perhumid subtype is typically zero in from one to five winter months. A summer maximum of precipitation, P, is typical, but is not present everywhere. In the subhumid subtype (10sh), D is greater than zero, but less than 15 cm when R is zero; otherwise D is greater than R when R is not zero. In the humid subtype (10h), R is greater than zero but not over 60 cm, while R is greater than D. In the perhumid subtype (10p), R is greater than 60 cm.)

Example. Pittsburgh, Pennsylvania, lat. 40°N (Figure 10.25). Located in the interior eastern region, Pittsburgh illustrates the humid subtype (10h) of the moist continental climate. Precipitation, P, runs rather uniformly throughout the year, though with a slight summer maximum. Total annual P is substantial: 92 cm. The annual cycle of soil-water need, Ep, peaks strongly in summer, after 3 consecutive winter months of zero values, when soil water is frozen and plants are dormant. A small soil-water shortage, D, develops in summer but is small in every month and totals only 4 cm. A substantial water surplus, R, occurs in winter and early spring; the yearly total is 24 cm. Some of this water is held in the frozen state in winter, to be released rapidly in the early spring when thawing

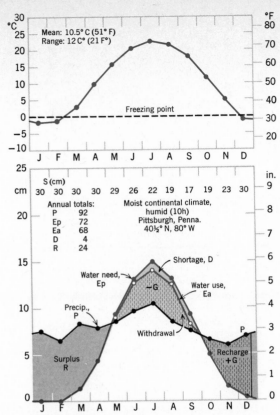

Figure 10.25 **Soil-water budget for Pittsburgh, Pennsylvania, lat. 40°N. (Data from C. W. Thornthwaite Associates, Laboratory of Climatology, Centerton, N.J.)**

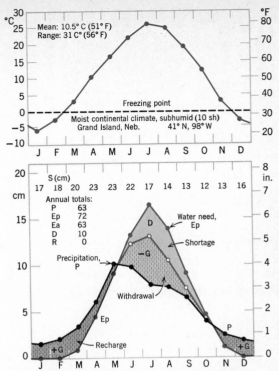

Figure 10.26 **Soil-water budget for Grand Island, Nebraska, lat. 41°N. (Data from C. W. Thornthwaite Associates, Laboratory of Climatology, Centerton, N.J.)**

occurs. Spring floods are highly probable. Larger streams maintain their flow throughout the summer. Forest is the natural plant cover.

Example. Grand Island, Nebraska, lat. 41°N (Figure 10.26). Located almost exactly midway between the Atlantic and Pacific oceans, Grand Island illustrates the subhumid subtype (10sh) of the moist continental climate. Water need, Ep, peaks very sharply in summer, but falls to zero in three winter months. Precipitation shows a strong summer maximum in the warmer months, but the peak months occur in early summer—May and June. A soil-water shortage, D, sets in by June and continues throughout October, with a total value of 10 cm. Storage recharge, +G, in the fall, winter, and early spring is not enough to bring about a surplus. Soil-water storage, S, is high in late spring, then falls off throughout the summer. The native plant cover was prairie grassland, but today wheat and corn farming dominates the landscape. This area lies close to the boundary of dry and moist climates.

GROUP III / HIGH-LATITUDE CLIMATES

11. Boreal forest climate

A climate characterized by long, severely cold winters. The total annual water need, Ep, ranges from 35 to 52.5 cm and is effectively zero for five, six, or seven consecutive winter months. During the short summer, Ep peaks sharply. Across Canada, subtypes are arranged in order from west to east from subhumid (11sh) in the Yukon

and Northwest Territories, through humid (11h), to perhumid (10p) in Labrador and Newfoundland. (see climate 10 for definition of these subtypes.) Across Eurasia, this order is reversed in direction. An area in eastern Siberia qualifies as a dry subtype (11s), with D greater than 15 cm.

Example. Trout Lake, Ontario, Canada, lat. 53½°N (Figure 10.27). Trout Lake lies in central Ontario, in the heart of the Canadian Shield not far south of Hudson Bay. This station illustrates the humid subtype (11h) of the boreal forest climate. The cycle of water need, Ep, rises to a sharp peak following 6 consecutive winter months of zero evapotranspiration. Precipitation, P, also rises sharply during the summer, but there is a brief period of soil-water use, −G. Water shortage, D, is extremely small: 1 cm. Recharge is completed in October, after which time soil water becomes solidly frozen. Throughout the remainder of the winter, precipitation accumulates as snow. Soil-water storage, S, is shown on the graph to exceed 30 cm from November through April. The excess represents accumulated snow; it is released as runoff in May during the spring thaw. The soil retains a high level of water storage throughout the summer. Streams flow copiously during the summer. Needleleaf forest (evergreen forest) is the native plant cover. The growing season for crops is very short.

12. Tundra climate

An arctic-zone climate without a true summer. Total annual water need, Ep, is less than 35 cm. During eight or more consecutive months, soil water is frozen and Ep is

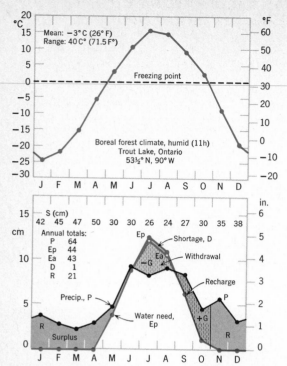

Figure 10.27 Soil-water budget for Trout Lake, Ontario, Canada, lat. 53½°N. (Data from C. W. Thornthwaite Associates, Laboratory of Climatology, Centerton, N.J.)

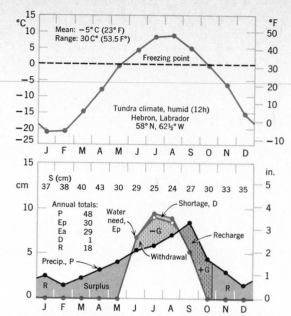

Figure 10.28 Soil-water budget for Hebron, Labrador, Canada, lat. 58°N. (Data from C. W. Thornthwaite Associates, Laboratory of Climatology, Centerton, N.J.)

effectively zero. Some parts of the tundra climate are humid (12h), with a substantial water surplus, R; other areas are subhumid (12sh). (See climate 10 for definition of these subtypes.) In eastern Siberia, the tundra belt qualifies as a dry climate of the semiarid subtype (12s).

Example. Hebron, Labrador, Canada, lat. 58°N (Figure 10.28). Located on the Atlantic coast, Hebron illustrates the humid subtype (12h) of the tundra climate. Because of the great length and severity of the winter, water need, Ep, is effectively zero for eight consecutive months. Total annual Ep is only 30 cm. Ep peaks sharply in the short warm season, when the sun is in the sky for much of the 24-hour day. Total annual precipitation is 48 cm, a substantial quantity. Precipitation shows a strong annual cycle with a summer maximum, peaking in September,

just prior to the onset of winter. Storage withdrawal, −G, occurs during June, July, and August, and the soil-water shortage, D, is extremely small (actually less than 1 cm). Soil-water storage, S, remains very high throughout the summer period. Storage capacity (30 cm) is reached before the end of October. Snow that falls throughout the winter is held for release in May, when the spring thaw occurs. Native vegetation of this region is the treeless arctic tundra, consisting of a few species of hardy plants in a low ground layer. When the surface layer is thawed, the ground is wet and boggy over large areas.

13. Ice-sheet climate

A severely cold climate found on the Greenland and Antarctic ice sheets and over the ice-covered Arctic Ocean. Water need, Ep, is effectively zero throughout the year. No monthly temperature mean is above 0°C. Almost all the scanty precipitation is in the form of snow, which on land accumulates as glacial ice.

Soil-Water Climatology and Man's Food Resources

This brief survey of 13 climate types, defined according to the terms used in the soil-water balance, illustrates the important types encountered over the habitable lands of the globe. We must apply this information to two grave questions facing the human race: (1) Can the developing nations increase their food production fast enough to stave off starvation? Answers given by well-informed specialists span the range from extreme pessimism to extreme optimism. (2) Will there be enough fresh water to supply the rapidly increasing demands of the energy-consuming

industrial nations? Your appreciation of current problems of agriculture and freshwater supplies over the lands of the globe can be greatly increased by application of water-balance climatology.

Perhaps the most important practical lesson of this chapter is that few soil-water budgets provide the full water need of food plants during the growing season, with neither too great a shortage nor too great a surplus. Severe soil-water shortages require irrigation to achieve crop production. On the other hand, large year-round surpluses of water lead to removal of plant nutrients, so costly applications of fertilizers are needed. Irrigation in a dry climate has serious side effects, such as the accumulation of salts in the soil and the rise of ground water to saturate the soil. The soil-water budget sets stringent limitations on Man's expansion of agricultural resources. Intelligent global planning for environmental management depends heavily on an understanding of all phases of the soil-water balance.

Fundamental though it is, the soil-water balance and the global climate system based on it comprise only one of three major topics to be mastered before we can achieve a thorough understanding of global soils. The next chapter deals with the second of these topics: the nature and properties of the parent mineral matter of the soil. The third major topic, fundamental to an understanding of soil-forming processes, is the flow of energy and matter in the biosphere (Chapter 12).

References for Further Study

Thornthwaite, C. W. (1948), An approach toward a rational classification of climate, *Geographical Review,* Vol. 38, pp. 55–94.

Thornthwaite, C. W., and J. R. Mather (1955), *The water balance,* Drexel Institute of Technology, Laboratory of Climatology, *Publications in Climatology,* vol. 8, no. 1, Centerton, N.J.

U.S. Dept. Agriculture (1955), *Water; yearbook of agriculture, 1955,* Government Printing Office, Washington, D.C.

Sellers, W. D. (1965), *Physical climatology,* Univ. of Chicago Press. See Chapter 7.

Chang, Jen-Hu (1968), *Climate and agriculture; an ecological survey,* Aldine Publishing Co., Chicago. See Chapters 12, 13, and 19.

Review Questions

1. Distinguish between surface water and subsurface water and between soil water and ground water.
2. Give a complete general account of the hydrologic cycle. What quantities of water are evaporated annually from the oceans? From the lands? How is evaporation balanced by precipitation and runoff?
3. Where is most of the world's water held in storage? What environmental effects would result from changes in quantity of water held in storage as glacial ice?
4. Describe how infiltration takes place. In what units is infiltration measured? How is infiltration related to runoff?
5. Describe the process of evaporation of soil water. What is transpiration? What is evapotranspiration? Distinguish between the soil-water belt and the intermediate belt. By what flow paths is excess precipitation disposed of?
6. Describe the process of soil-water recharge. What role does capillary force play in the storage of soil water? Define storage capacity.
7. Describe the typical annual soil-water cycle in a midlatitude region of moist climate. When does a water surplus usually occur? When does a water deficit usually occur?
8. Explain the principle of the soil-water balance. Name and define each term used. Distinguish between actual evapotranspiration (water use) and potential evapotranspiration (water need). What is meant by a soil-water shortage? Name the scientist most closely linked to the development of the concept of the soil-water balance.
9. Explain how a soil-water budget is calculated. Define each of these terms: storage withdrawal, soil-water shortage, and storage recharge. What practical value has the soil-water budget?
10. Give a general description of the global pattern of annual water need (potential evapotranspiration). How does latitude influence the annual water need?
11. Compare seasonal cycles of water need in low, middle, and high latitudes. Why is water need effectively zero for several consecutive months in the subarctic zone? How do plants respond to annual totals and seasonal cycles of water need?
12. Explain how the soil-water balance can be used as the basis of a system of climate classification. What advantages has this system over a purely descriptive climate system based on air masses?
13. What source of data is used in the climate system based on the soil-water balance? How does water need serve as an indicator of available heat? Use this concept in defining the three major climate groups.
14. Give rigorous definitions of dry and moist climates and of the subtypes within each, using soil-water shortage and soil-water storage as the basis of each definition. How can climates with strong moisture seasons be recognized in terms of the soil-water budget?
15. For each of the 13 climates described in Chapter 9, supply a set of definitions based on the soil-water budget. Give one example for each climate type and subtype.

Parent Mineral Matter of the Soil

11

The soil, a thin surface layer composed partly of mineral substances, provides nutrient elements required by plants. These nutrient elements are derived from various kinds of rock by geologic processes that break the rock into fine particles and by chemical processes that decompose rock-forming minerals. A knowledge of geologic processes and their products is essential if we are to understand how soils differ from place to place over the lands.

Our plan in this chapter is to examine the basic classes of rocks with emphasis on their mineral composition. Because solid rock is a sterile base, incapable of supporting the higher plants — those with roots, stems, and leaves — we shall emphasize the ways rock is altered to produce the parent matter of the soil. Familiarity with the various kinds of parent matter is essential to an understanding of soil-forming processes.

Elements of the Earth's Crust

From the standpoint of the environment of Man, the most significant zone of the solid earth is the thin outermost layer — the earth's crust. This mineral skin, averaging about 17 km (10 mi) thickness for the globe as a whole, contains the continents and ocean basins and is the source of soil and sediment, salts of the sea, gases of the atmosphere, and all free water of the oceans, atmosphere, and lands.

Figure 11.1 displays in order the eight most abundant elements of the earth's crust in terms of percentage by weight. Oxygen (O), the predominant element, accounts for about half the total weight. It occurs in combination with silicon (Si), the second most abundant element. Oxygen is a major element in organic substances. Silicon is used in minor amounts by plants.

Aluminum (Al) and iron (Fe) are third and fourth on the list. Both are nutrients required by plants, although in smaller quantities. Both metals are of primary importance in Man's industrial civilization, and it is most fortunate that they are comparatively abundant elements. Four metallic elements follow: calcium (Ca), sodium (Na),

potassium (K), and magnesium (Mg). All four are on the same order of abundance — 2 to 4 percent. Calcium, potassium, and magnesium are major plant nutrients; their presence is essential if a soil is to have a high level of fertility.

If we were to extend the list, the ninth-place element would be titanium, followed in order by hydrogen, phosphorus, barium, and strontium. Both hydrogen (H) and phosphorus (P) are essential nutrient elements in plant growth. Hydrogen, combined with oxygen in the form of water (H_2O), is used by plants to form organic molecules.

Rocks and Minerals

The elements of the earth's crust are organized into compounds that we recognize as minerals. A **mineral** is a naturally occurring, inorganic substance, usually possess-

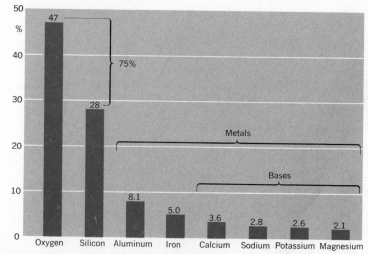

Figure 11.1 The average composition of the earth's crust is given here in terms of the percentage by weight of the eight most abundant elements.

ing a definite chemical composition and a characteristic atomic structure. Vast numbers of mineral varieties exist, together with a great number of their combinations into rocks.

Rock is broadly defined as any aggregate of minerals in the solid state. Rock comes in a wide range of compositions, physical characteristics, and ages. A given rock is usually composed of two or more minerals, and usually many minerals are present; however, a few rock varieties consist almost entirely of one mineral. Most rock of the earth's crust is extremely old in terms of human standards, the time of formation ranging back many millions of years. But rock is also being formed at this very hour as a volcano emits lava that solidifies on contact with the atmosphere.

Rocks of the earth's crust fall into three major classes: (1) **Igneous rocks** are solidified from mineral matter in a high-temperature molten state, that is, from a **magma.** (2) **Sedimentary rocks** are layered accumulations of mineral particles derived in various ways from preexisting rocks. (3) **Metamorphic rocks** are igneous or sedimentary rocks that have been physically and chemically changed by application of heat and high pressures during mountain-making activities. Although we have listed these classes in a conventional sequence, it will become obvious in a later chapter that no one class has first place in terms of origin. Instead, they form a continuous circuit through which the crustal minerals have been recycled during many millions of years of geologic time.

The Silicate Minerals

Igneous rocks make up the vast bulk of the earth's crust. Practically all igneous rock consists of **silicate minerals,** which are all compounds containing silicon atoms in combination with oxygen atoms in a close linkage. In the crystal structure of silicate minerals, one atom of silicon is linked with four atoms of oxygen as the unit building block of the compound. Most of the silicate minerals also contain one, two, or more of the metallic elements listed in Figure 11.1.

Figure 11.2 gives the names and chemical compositions of seven major silicate minerals, or groups of silicate minerals. The large bulk of all igneous rocks consists of two or more of these silicate minerals in varying proportions.

Among the commonest minerals of most rock types is **quartz;** its composition is silicon dioxide (SiO_2). There follow five mineral groups, collectively forming the **aluminosilicates** because all contain aluminum. Two groups of **feldspars** are set apart: the **potash feldspars** contain potassium (K) as the dominant metallic ion, but sodium (Na) is commonly present in various proportions. The **plagioclase feldspars** form a continuous series, beginning with the sodic, or sodium-rich varieties, and grading through with increasing proportions of calcium to the calcic, or calcium-rich varieties.

Belonging to the **mica group,** which is familiar because of its property of splitting into very thin, flexible layers, is biotite, a dark-colored mica with a complex chemical formula. Potassium, magnesium, and iron are present in

biotite, along with some water. The **amphibole group,** of which the common black mineral hornblende is a common representative, is a complex aluminosilicate containing calcium, magnesium, iron and water. Similar in outward appearance and having essentially the same component elements is the **pyroxene group,** with the mineral augite as a representative. Last on the list is **olivine,** a dense greenish mineral that is a silicate of magnesium and iron, but without aluminum.

Density is an important property of a given mineral. **Density** is defined as the mass of substance per unit of volume and is stated in grams per cubic centimeter (gm/cc). (Density of pure water is 1.0 g/cc.)

Looking down the list of densities of the silicate minerals in Figure 11.2, you will notice that there is a progressive increase from the least dense (quartz, 2.6 g/cc) to the most dense (olivine, 3.3 g/cc). This change reflects the decreasing proportion of aluminum and sodium, elements of low atomic weights, and the increasing proportion of calcium and iron, elements of considerably greater atomic weights.

The list as a whole is conveniently divided into two major groups of silicate minerals: **felsic minerals,** consisting of quartz and the feldspars, and **mafic minerals,** consisting of those silicates rich in magnesium and iron. The coined word *felsic* is easily recognized as a combination of "fel," for feldspar, and "si," for silica. The syllable "ma" in *mafic* stands for magnesium, the letter "f" for iron (Fe). The felsic minerals are light in color and of comparatively low density; the mafic minerals are dark in color and of comparatively high density.

Two important mafic minerals that are not silicates occur in many igneous rocks. These are magnetite, an oxide of iron (Fe_3O_4) and ilmenite, an oxide of iron and titanium ($FeTiO_3$). Both of these minerals are black and have high densities — 4½ to 5½ gm/cc.

Silicate Magmas

From the geologic standpoint, the silicate minerals can be viewed as the fundamental materials out of which other rock groups — sedimentary and metamorphic — are created. About 99 percent of the bulk of the igneous rocks of the earth's crust consists of the seven silicate minerals or mineral groups listed in Figure 11.2. The remainder consists of minerals of secondary importance in bulk, although their number is very large. Fortunately, the eight silicate minerals or groups combine to form only a dozen or so igneous rock varieties having widespread occurrence. We shall simplify the list to five representative rock types.

Igneous rocks are derived from **silicate magma** formed at depths of many kilometers in the earth under conditions of very high temperatures and pressures. Here, silicate magmas probably have temperatures in the range from 500 to 1200°C (900 to 2200°F) and are under pressures 6000 to 12,000 times as great as atmospheric pressure at sea level.

As magma cools at or near the surface, crystallization occurs over a certain critical range of temperatures and

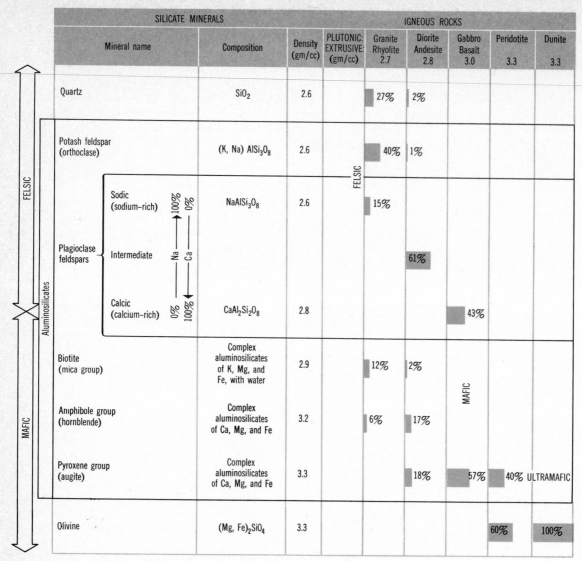

SILICATE MINERALS				IGNEOUS ROCKS				
Mineral name	Composition	Density (gm/cc)	PLUTONIC: EXTRUSIVE: (gm/cc)	Granite Rhyolite 2.7	Diorite Andesite 2.8	Gabbro Basalt 3.0	Peridotite 3.3	Dunite 3.3
Quartz	SiO_2	2.6		27%	2%			
Potash feldspar (orthoclase)	$(K, Na) AlSi_3O_8$	2.6		40%	1%			
Plagioclase feldspars — Sodic (sodium-rich)	$NaAlSi_3O_8$	2.6		15%				
Plagioclase feldspars — Intermediate					61%			
Plagioclase feldspars — Calcic (calcium-rich)	$CaAl_2Si_2O_8$	2.8				43%		
Biotite (mica group)	Complex aluminosilicates of K, Mg, and Fe, with water	2.9		12%	2%			
Amphibole group (hornblende)	Complex aluminosilicates of Ca, Mg, and Fe	3.2		6%	17%			
Pyroxene group (augite)	Complex aluminosilicates of Ca, Mg, and Fe	3.3			18%	57%	40% ULTRAMAFIC	
Olivine	$(Mg, Fe)_2SiO_4$	3.3					60%	100%

Figure 11.2 Simplified chart of common silicate minerals and abundant igneous rocks. (From A. N. Strahler, *Planet Earth*, Harper and Row, New York.)

pressures. Through a complex series of interactions, the eight elements named in Figure 11.1 are gathered into compounds as individual crystals of the various silicate minerals. The character of the igneous rocks that are formed by crystallization varies greatly, depending on both the initial composition of the magma and its cooling history.

Volatiles in Magmas

Paramount in importance in terms of the origin of the atmosphere and hydrosphere is the presence in magmas of substances besides the elements that comprise solidified silicate rock. These substances are known as **volatiles** because they remain in a gaseous or liquid state at much lower temperatures than the silicate compounds. Consequently, the volatiles are separated from the magma as it cools and solidifies. The volatiles escape into the atmosphere from volcanoes and from gas and steam vents in

geothermal localities.

From analyses of gas samples collected from volcanic emissions, we know that water is the major constituent of the volatile group. Table 11.1 lists the volatiles found in gases emanating from magma of active Hawaiian volcanoes. For comparison, the table lists the proportion of these volatiles in the atmosphere and hydrosphere. Notice that the abundances of the various constituents are of the same general order of magnitude in both lists.

The emanation of volatiles from the earth's crust is called **outgassing** and is the source of free water of the earth's hydrosphere as well as of the atmospheric gases carbon dioxide, nitrogen, argon, and hydrogen. Chlorine and sulfur compounds of seawater have derived their chlorine and sulfur from outgassing. We can conclude, then, that silicate magmas, with their enclosed volatiles, have through geologic time supplied almost all the essential components of the atmosphere, hydrosphere, and lithosphere.

Table 11.1 / Volatiles in Gases of Magmas

	Lava Gases from Mauna Loa and Kilauea Volcanoes (Percent by Weight)	Volatiles Free in Earth's Atmosphere and Hydrosphere Percent by Weight)
Water, H_2O	60	93
Carbon, as CO_2 gas	24	5.1
Sulfur, S_2	13	0.13
Nitrogen, N_2	5.7	0.24
Argon, A	0.3	Trace
Chlorine, Cl_2	0.1	1.7
Hydrogen, H_2	0.04	0.07
Flourine, F_2	—	Trace

Data source: W. W. Rubey, 1952, *Geol. Soc. Amer., Bull.,* vol. 62, p. 1137. Figures have been rounded off.

Textures of Igneous Rocks

Igneous rocks are classified not only on the basis of mineral composition, but also on the basis of grades of sizes of the component crystals. The term **rock texture** applies both to crystal size and to the arrangement of crystals of mixed sizes.

Very gradual cooling of a magma enclosed in solid crustal rock results in formation of large mineral crystals and forms a rock with coarse-grained texture (Figure 11.3). Great bodies of coarse-grained igneous rock, solidified at great depth, consist of **plutons.** Magma that reaches the earth's surface, pouring out through gaping cracks to form streamlike masses, is called **lava** (Figure 11.4). The lavas are classed as **extrusive rocks,** in contrast to the **intrusive rocks,** which have solidified beneath the surface. Rapid cooling of lava results in very small mineral crystals, not usually distinguishable with the unaided eye, and gives the rock a fine-grained texture. Very sudden cooling of outpouring magma yields a natural volcanic glass; the black variety is called *obsidian* (Figure 11.5, right). Frothing of magma as its enclosed gases expand gives

Figure 11.4 **A freshly solidified basalt lava flow with ropy surface texture. Craters of the Moon National Monument in Idaho. (George A. Grant, U.S. Department of Interior.)**

a porous, spongelike rock known as *scoria* or *pumice* (Figure 11.5, left).

Volcanoes often emit great volumes of solid particles formed by rapid cooling of emerging magma and impelled by the explosive release of gases under great pressure. These particles fly through the air and land on the ground at varying distances from the vent, depending on their weight and the strength of prevailing winds. The finest particles, which are in the form of minute glass shards, travel long distances and accumulate in layers as **volcanic ash** (Plate E.1). Later we will include volcanic ash as one of the classes of sediment (pyroclastic sediment). Particles the size of gravel or pebbles fall close to the vent. **Tephra** is the collective term for all sizes of solid particles blown from a volcanic vent.

Figure 11.3 **Seen close up this granite, a coarse-grained intrusive rock, proves to be made of tightly interlocking crystals of a few kinds of minerals.**

Figure 11.5 **A frothy, gaseous lava solidifies into a light, porous scoria (left). Rapidly cooled lava may form a dark volcanic glass (right).**

Classification of Igneous Rocks

Using the simplest possible classification of igneous rocks, we recognize five coarse-grained (plutonic) types. These are named on the top line above the columns in Figure 11.2. For the first three rocks listed, important equivalent extrusive types (lavas) are recognized. Bars of varying width, with attached percentages, show the typical mineral compositions of these igneous rocks.

Granite and its extrusive equivalent, **rhyolite,** are rich in quartz and potash feldspar, with lesser amounts of sodic plagioclase, biotite, and amphibole. **Diorite** and its extrusive equivalent, **andesite,** are almost totally lacking in quartz and potash feldspar, but consist dominantly of intermediate plagioclase and lesser amounts of the mafic minerals.

Going progressively in the direction of domination by mafic minerals, we come to **gabbro** and its lava equivalent, **basalt.** In these, plagioclase feldspar is of the calcic type, making up nearly half the rock, while pyroxene makes up the remainder. In a common variety of basalt, olivine is present at the expense of part of the feldspar.

The next rock, *peridotite,* is not abundant in the crust, but probably makes up the bulk of the next lower layer, or mantle. It is composed mostly of pyroxene and olivine. Finally, we list *dunite,* a rare rock composed almost entirely of olivine, as an example of the extreme mafic end of the mineral series.

Granite and diorite, rocks rich in felsic minerals, are collectively described as **felsic rocks,** whereas gabbro and basalt are **mafic rocks;** the extreme mafic types are **ultramafic rocks.**

Densities of the igneous rocks are proportional to the densities of the component minerals. Thus granite has a density of about 2.7 gm/cc; gabbro and basalt, about 3.0; and peridotite and dunite, 3.3.

Rock Weathering

Poets and advertising copywriters assure us that the highly polished granite slab is an everlasting monument in a changing world. But in truth, the surface environment is poorly suited to the preservation of an igneous rock formed under conditions of high pressure and high temperature. Most silicate minerals formed in magmas do not last long, geologically speaking, in the low temperatures and pressures of atmospheric exposure, particularly because free oxygen, carbon dioxide, and water are abundantly available. Rock surfaces are also acted on by physical forces of disintegration, tending to break up the igneous rock into small fragments and to separate the component minerals, grain from grain. Fragmentation is essential for the chemical reactions that follow because it results in a great increase in mineral surface area exposed to attack by chemically active solutions.

Weathering is the general term applied to the combined action of all processes causing rock to be disintegrated physically and decomposed chemically because of exposure at or near the earth's surface. The products of rock weathering tend to accumulate in a soft surface layer, called **regolith** (Figure 11.6). The regolith

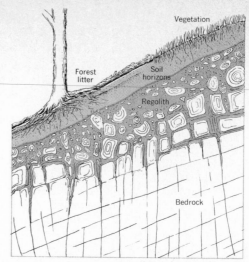

Figure 11.6   In a typical exposure cut into a hillside, the bedrock (broken into blocks by joints) grades upward into the regolith. Immediately beneath the surface is the soil, penetrated by roots of plants.

grades downward into solid, unaltered rock, known simply as **bedrock.** Regolith, in turn, provides the source for **sediment,** consisting of detached mineral particles transported and deposited in a fluid medium, which may be water, air, or glacial ice. Both regolith and sediment comprise parent materials for the formation of the true soil, a surface layer capable of supporting the growth of plants. Our objective here is to explain physical and chemical processes that form regolith and sediment because this information is essential to an understanding of soil-forming processes.

Physical Weathering

Physical weathering processes produce fine particles from massive rock by the exertion of stresses great enough to fracture the rock. One of the most important physical weathering processes is **frost action,** the repeated growth and melting of ice crystals in the pore spaces or natural fractures of bedrock. Almost everywhere, bedrock is broken into blocks by systems of fractures called **joints** (Figure 11.6). Rarely is igneous rock or rock of the other groups free of numerous joints that permit the entry of water. Frost action is, of course, limited to those mid-latitude and high-latitude climates with cold winters and to cold alpine climates at high altitudes. As igneous rock becomes softened by chemical decomposition, water is able to penetrate the contact surfaces between mineral grains; here the water can freeze and exert strong forces to separate the grains.

Closely related to ice-crystal growth as a disruptive process is the growth of salt crystals in fractures and other openings in rock. This process operates extensively in dry climates. During long dry periods, water deep within the rock is drawn to the surface by capillary force. This water carries dissolved mineral salts. As evaporation takes place, minute salt crystals remain behind. The growth force of these crystals is capable of causing grain-by-grain disintegration of the outer rock layer. The same process can be seen in action on building stones and concrete in cities. Deicing salt, spread on streets in winter, is remarkably effective in causing the disintegration of rock and concrete structures close to the ground.

Another possible set of forces of rock disruption is temperature change. Rock-forming minerals expand when heated, but contract when cooled. Where rock surfaces are exposed daily to intense heating by direct solar rays, alternating with intense cooling by longwave radiation at night, the resulting expansion and contraction of mineral grains tends to break them apart. Given sufficient time, in which tens of thousands of daily cycles of expansion and contraction occur, the accumulated effect may be important as an agent of physical weathering. The intense heat of forest and brush fires is known to cause rapid flaking and scaling of exposed rock surfaces.

Plant roots, growing between joint blocks and along minute fractures between mineral grains, exert an expansive force tending to widen those openings. Heaving and cracking of concrete sidewalk slabs by root growth of nearby trees is commonplace evidence of the effectiveness of plants in contributing to physical weathering.

In Chapter 17, we shall investigate the surface features produced by physical weathering. Our purpose here has been to view physical weathering as a process contributing to the formation of regolith and sediment — the parent materials of the soil.

Size Grades of Mineral Particles

Physical weathering, as well as the grinding and crushing processes to which mineral particles are subjected when transported by streams, waves and currents, wind, and ice, continually reduce those particles to smaller and smaller diameters. Large particles, such as pebbles, are reduced in size by **abrasion,** the grinding away of the surface. Abrasion thus produces myriads of very small particles. When a particle is reduced by abrasion to the size of a sand grain, it is quickly pulverized between much larger fragments. Thus the parent matter of soil can consist of physically fragmented particles of a wide range of sizes. To describe this material, a system of particle grade sizes is needed.

The U.S. Department of Agriculture has established size limits and names for various grades of mineral particles in soils. Table 11.2 shows these grades. The three basic categories are **sand** (2.0 to 0.05 mm), **silt** (0.05 to 0.002 mm), and **clay** (finer than 0.002 mm). Sand is subdivided into five subgrades; size limits are given in the table. Although larger particles are commonly present in the soil, they are removed by screening when a soil sample is analyzed. The larger size grades of sediment particles are pebbles (2 to 64 mm), cobbles (64 to 256 mm), and boulders (larger than 256 mm).

Clay particles range downward in diameter from 0.002 mm, or 2 microns, to smaller than 0.001 micron. Clay particles finer than about 0.01 micron are classed as **colloids.** A property of mineral colloids is their ability to

Table 11.2 / Grade Sizes of Sediment Particles, U.S. Department of Agriculture System

Grade name	Diameter limits
	2.0 mm
Very coarse sand	
	1.0 mm
Coarse sand	
	0.5 mm
Medium sand	
	0.25 mm
Fine sand	
	0.10 mm
Very fine sand	
	0.05 mm
Silt	
	0.002 mm (2 microns)
Clay noncolloidal	2 to 0.01 microns
colloidal	Below 0.01 microns

remain in suspension indefinitely in water, once the particles have become dispersed (separated one from another). A colloidal suspension appears clouded or murky.

Grains of sand, as well as pebbles and cobbles, are often well rounded in shape as a result of slow mechanical abrasion during transport by water or wind. Spherical quartz grains of coarse sand size, shown in Figure 11.7, were shaped by wind action in moving dunes. Silt grains and the coarser grades of clay are usually highly angular and may appear under the microscope as if they were particles of crushed glass. Fine clay particles, those of colloidal dimensions, are typically in the form of thin scales and plates (see Figure 11.9).

Decreasing particle size brings a great increase in surface area of particles contained within a given volume; when the colloidal size is reached, the surface area is truly enormous. Table 11.3 illustrates this point by assuming that we start with a single cube, 1 cm on a side (pebble

Figure 11.7 **Rounded quartz grains from an ancient sandstone. The grains average about 1 mm (0.04 in.) in diameter. (Andrew McIntyre, Columbia University, New York.)**

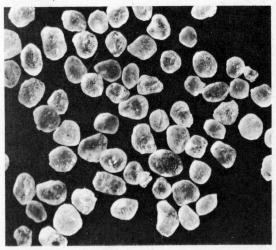

Table 11.3 / Surface Area of Cubes Obtained by Subdividing a One-Centimeter Cube

Cube Dimensions Length of Side (cm)	Grade Equivalent	Number of Particles (per cc)	Total Surface Area (sq cm)
1	Pebble	1	6
0.1	Coarse sand	10^3	60
0.0001	Fine clay	10^{12}	60,000
0.000,01	Colloidal	10^{15}	600,000
(0.1 micron)	clay		(8 m × 8 m)

size); it has a volume of 1 cc and a surface area of 6 sq cm. Consider next that we slice the cube into cubes 0.1 cm on a side (size of coarse sand); this move yields 1000 cubes with a total surface area of 60 sq cm. Subdividing the cube into 10^{12} cubes, each 0.0001 cm on a side (fine clay size) gives a total surface area of 60,000 sq cm. Finally, subdivision of the original cube into 10^{15} cubes, each 0.000,01 cm (0.1 micron) on a side (colloidal clay size), yields a total surface area of 600,000 sq cm; if spread out into a continuous horizontal surface, this would be an area about 8 m by 8 m.

This relationship between surface area and particle size is important because nutrients and water are held on soil particle surfaces. We shall investigate this phenomenon in Chapter 13.

Chemical Weathering

Chemical weathering, also known as **mineral alteration,** consists of a number of chemical reactions; all these reactions change the original silicate minerals of igneous rock, the **primary minerals,** into new compounds, the **secondary minerals,** that are stable in the surface environment. Chemical weathering also affects various types of sedimentary and metamorphic rocks.

Chemical weathering includes several types of chemical reactions, all of which take place more or less simultaneously. Surface water — whether it be raindrops, water of streams and lakes, or water held in the soil — contains gases of the atmosphere. The presence of atmospheric gases dissolved in natural water is a matter of great environmental importance because of the inorganic reactions with mineral matter and because the presence of two gases in particular — oxygen and carbon dioxide — is vital to life processes of plants and animals living in water.

The presence of dissolved oxygen in water in contact with mineral surfaces leads to **oxidation,** which is the chemical union of oxygen atoms with atoms of those metallic elements (calcium, potassium, magnesium, iron) abundant in the silicate minerals. At the same time, carbon dioxide (CO_2) in solution forms a weak acid, **carbonic acid,** capable of reaction with several minerals. In addition, where decaying vegetation is present, soil water containing complex organic acids is capable of interaction with mineral compounds. Some common minerals, such as rock salt (sodium chloride, NaCl), dissolve directly in water; but direct solution is not particularly effective for the silicate minerals.

Water itself combines with certain mineral compounds in a reaction known as **hydrolysis.** This process is not merely a soaking or wetting of the mineral, but a true chemical change producing a different compound and a different mineral. The reaction is not readily reversible under atmospheric conditions, so the products of hydrolysis are stable and long-lasting — as are the products of oxidation. In other words, these changes represent a permanent adjustment of mineral matter to a new environment of pressures and temperatures.

Because water is required for mineral alteration, you might think that the rate of rock decay would be directly proportional to the amount of free water available in the rock and soil and that the dry deserts would be environments of very limited rock decay. To some extent this is a valid conclusion. Polished stone surfaces of the ancient Egyptian monuments have been almost perfectly preserved throughout the centuries in a dry, hot desert climate; these same monuments, taken to midlatitude cities of humid climate and exposed to the atmosphere, undergo rapid disintegration. (Frost action and the action of sulfuric acid derived from polluted air are also factors in rapid disintegration.) Although mineral alteration is perhaps much slower in dry deserts than in humid lands, there is nevertheless enough water present as water vapor and as dew to allow alteration to proceed; we observe the decay products in abundance on igneous rock surfaces in most deserts.

The effect of cold is quite another matter. When soil water is frozen, chemical reactions are greatly slowed. Minerals in arctic regions of perennially frozen soil and rock (permafrost) show little chemical decay because seasonal thaw affects only a thin surface layer. Most chemical reactions take place more rapidly at high temperatures than at low. Consequently, chemical alteration of minerals is most rapid in warm (and moist) climates of low latitudes.

Mineral Products of Hydrolysis and Oxidation

Figure 11.8 shows schematically some important alteration products of the common silicate minerals and mineral groups. Chemical formulas are given for each mineral. Certain of these products are clay minerals. A **clay mineral** is one that has plastic properties when moist because it consists of minute thin flakes lubricated by layers of water molecules. Potash feldspar undergoes hydrolysis to become **kaolinite,** a soft, white clay mineral with the composition $Al_2Si_2O_5(OH)_4$. Kaolinite becomes plastic when moistened. It is an important ceramic mineral used to make chinaware, porcelain, and tile. Kaolinite can also be derived from the plagioclase feldspars.

Bauxite is an important alteration product of feldspars, occurring typically in warm climates of tropical and equatorial zones where rainfall is abundant year around or in a rainy season. Bauxite is actually a mixture of minerals, the dominant constituent being diaspore, with the formula $Al_2O_3 \cdot 2H_2O$. The combination of two atoms of aluminum with three atoms of oxygen is known as **sesquioxide of**

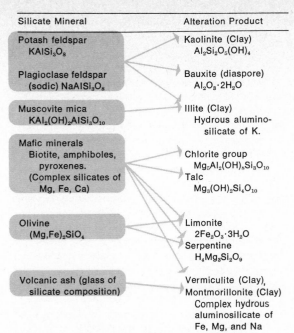

Silicate Mineral	Alteration Product

Potash feldspar
KAlSi$_3$O$_8$

Plagioclase feldspar
(sodic) NaAlSi$_3$O$_8$

Muscovite mica
KAl$_2$(OH)$_2$AlSi$_3$O$_{10}$

Mafic minerals
Biotite, amphiboles,
pyroxenes.
(Complex silicates of
Mg, Fe, Ca)

Olivine
(Mg,Fe)$_2$SiO$_4$

Volcanic ash (glass of
silicate composition)

Kaolinite (Clay)
Al$_2$Si$_2$O$_5$(OH)$_4$

Bauxite (diaspore)
Al$_2$O$_3$·2H$_2$O

Illite (Clay)
Hydrous alumino-
silicate of K.

Chlorite group
Mg$_5$Al$_2$(OH)$_8$Si$_3$O$_{10}$
Talc
Mg$_3$(OH)$_2$Si$_4$O$_{10}$

Limonite
2Fe$_2$O$_3$·3H$_2$O
Serpentine
H$_4$Mg$_3$Si$_2$O$_9$

Vermiculite (Clay),
Montmorillonite (Clay)
Complex hydrous
aluminosilicate of
Fe, Mg, and Na

Figure 11.8 **Common alteration products of silicate minerals.** "Clay" denotes a clay mineral. (From A. N. Strahler, 1971, *The Earth Sciences,* second edition, Harper and Row, New York.)

aluminum. Full oxidation has taken place, yielding an unusually stable compound. Unlike kaolinite, which is a true clay with plastic properties, bauxite forms massive rocklike lumps and layers below the soil surface.

A second clay mineral is **illite,** formed as an alteration product of feldspars and muscovite mica. Illite is a hydrous aluminosilicate of potassium. It occurs as minute thin flakes of colloidal dimensions and is carried long distances in streams (Figure 11.9). Skipping down to the last two minerals listed in Figure 11.8 we find **montmorillonite,** a common clay mineral (more correctly a group of minerals) derived from the alteration of feldspar, or certain of the mafic minerals, or volcanic ash. Fragments of montmorillonite are seen together with illite in Figure 11.9.

Another group of clay minerals important in the soil is **vermiculite,** which is similar in composition to montmorillonite. Whereas montmorillonite expands greatly when it absorbs water, vermiculite does not. Vermiculite is a hydrous aluminosilicate rich in magnesium and iron. It is formed by hydrolysis of mafic silicate minerals, such as biotite mica and hornblende, and is a common product of the chemical weathering of mafic volcanic rocks.

Alteration products of mafic minerals include chlorite (a mineral group) and talc; both are hydrous aluminosilicates of magnesium. They are soft, scaly substances.

Two important alteration products of the mafic minerals are **hematite,** sesquioxide of iron (Fe$_2$O$_3$), and **limonite,** a hydrous iron compound with the formula 2Fe$_2$O$_3$·H$_2$O. Iron sesquioxide in hematite and limonite is a stable form of iron and is found widely distributed in rocks and soils. It is closely associated with bauxite. Hematite supplies the typical reddish to chocolate-brown colors of soils and rocks.

Another common alteration product of mafic minerals, particularly of olivine, is serpentine, a hydrous magnesium silicate with the formula H$_4$Mg$_3$Si$_2$O$_9$. Serpentine has resulted from the alteration of large bodies of peridotite, a rock rich in olivine.

Silicate minerals differ in their susceptibility to alteration. Olivine is most susceptible to alteration, followed by the pyroxenes, amphiboles, biotite, and sodic plagioclase feldspar. Potash feldspars are somewhat less susceptible. Muscovite mica is comparatively resistant to alteration, while quartz is immune to hydrolysis and to further oxidation.

As hydrolysis takes place, **silica** (silicon dioxide, SiO$_2$) is released. It has the same composition as the mineral quartz found in the felsic igneous rocks. (We can refer to quartz of igneous origin as primary quartz.) Grains of primary quartz released from decomposing rock are also subject to being dissolved. Dissolved silica, whether primary or derived from hydrolysis, is commonly redeposited in finely crystalline forms adhering to other minerals. Several kinds of reformed silica are found in the regolith and soil and are recognized as distinct minerals. For example, **chalcedony** is a form of silica with crystals too small to be individually separated. The banded varieties of chalcedony are familiar to us as agate, an ornamental or gem stone.

Important in many kinds of soils is a mineral substance called **allophane,** a product of chemical weathering. Allophane is amorphous, meaning that it lacks organized crystal structure. Instead, allophane takes the form of a gel, which is a solid state taken by colloids when they lose water. Thus allophane forms hard, glassy coatings on surfaces of other minerals. Allophane consists of varying proportions of sesquioxide of aluminum (Al$_2$O$_3$) and silica,

Figure 11.9 **Seen here enlarged about 20,000 times are tiny flakes of the clay minerals illite (sharp outlines) and montmorillonite (fuzzy outlines). These particles have settled from suspension in San Francisco Bay. (Harry Gold; Courtesy of R. B. Krone, San Francisco District Corps of Engineers, U.S. Army.)**

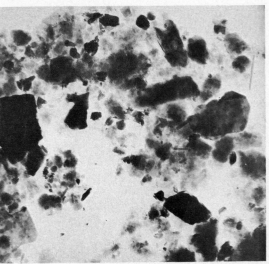

with varying amounts of water. Also present are small amounts of iron sesquioxide. Allophane occurs in important amounts in soils produced by the chemical weathering of volcanic ash.

Clastic Sediment

Sediment deposited in layers after transport by streams, waves and currents, wind, or ice, is an important class of parent material of the soil. Sediment deposits are environmentally important in forming the physical base, or substrate, for many forms of plant and animal life. Water-saturated sands, silts, and clays form the life environments of thousands of species of aquatic organisms under the shallow seas, in tidal estuaries, in lake bottoms and stream beds, and in swamps and marshes.

A principle of paramount importance is that organisms modify the sediment in which they live and feed; they also create sediment through life processes, for example, as shells and skeletons. Consequently, a major class of sediments is organically derived, in contrast to the chemically derived products of rock alteration and the physically derived particles of rock disintegration.

Clastic and nonclastic are the two major divisions of sediments. **Clastic sediments** are those derived directly as particles broken from a parent rock source, in contrast to the **nonclastic sediments,** which are of newly created mineral matter precipitated from chemical solutions or from organic activity. The clastic sediments include solid particles (tephra) that issue directly from volcanoes. These are **pyroclastic sediments.**

Clastic sediments are derived from any one of the rock groups — igneous, sedimentary, and metamorphic — which gives a wide range of parent minerals. One sediment source is from the primary silicate minerals and the secondary alteration products of those minerals. The highly susceptible primary minerals — mostly mafic — are often altered prior to transportation, whereas quartz is immune to such alteration. Consequently, the most important single component of coarse clastic sediment is quartz (see Figure 11.7). Second in abundance are fragments of unaltered fine-grained parent rocks. Feldspar and mica are also commonly present. Clay minerals, particularly kaolinite, illite, and montmorillonite, are major constituents of very fine clastic sediment.

Also important in coarse clastic sediment are the **heavy minerals.** By "heavy" we mean a mineral of high density. Two good examples of heavy minerals are magnetite and ilmenite; both were mentioned earlier in this chapter. The heavy minerals are derived from any one of a variety of preexisting rocks. They are extremely resistant to abrasion and, like quartz, travel long distances in streams and along beaches. Concentrations of these minerals can be seen as black layers in most beach sands and in sand bars on stream beds.

The grade size of particles in a particular body of clastic sediment determines the ease and distance of travel of the particles in transport by water currents. Obviously, the finer the particles, the more easily they are held in suspension in the fluid; the coarser particles tend to settle to the bottom of the fluid layer. In this way a separation of grades, called **sorting,** occurs and determines the texture of the sediment deposit and of the sedimentary rock derived from that sediment. Colloidal clays do not settle out unless they are made to clot together into larger groups. This process, called **flocculation,** takes place when fresh water carrying the sediment mixes with salt water of the ocean.

The concept of sorting of size grades in clastic sediment is extremely important when applied to the parent mineral matter of the soil. On the one hand, when sorting is carried almost to perfection, all the particles in a sample fall into a narrow range of diameters. For example, we may find a sediment consisting almost entirely of fine sand and medium sand, with only small percentages of very fine sand and coarse sand — not even a trace of silt or clay and no pebbles. This sample of extremely good sorting would prove to be taken from a sand dune, in which the winnowing action of the wind has been extremely effective in isolating clastic grains in a narrow range of diameters. In contrast, we may find that a sediment has substantial proportions of clay, silt, sand, and pebbles, all well mixed together. In this sediment, sorting is lacking because the process of transportation did not involve any physical mechanism for separating size grades. An example of unsorted sediment would be the material dragged along beneath a moving glacier. The degree of sorting varies greatly according to the process by which the particles are transported and deposited.

Colluvium and Alluvium

Clastic particles, originating by the weathering of bedrock on upland surfaces and accumulating as regolith, are acted upon by downhill flow of the surface layer of water (overland flow) generated during heavy rains and periods of rapid snowmelt. Erosion of the soil surface takes place, a subject we will describe in detail in Chapter 19. At such times, mineral particles are swept down the hillsides. Reaching the base of a hillside, where the gradient of the surface becomes gentle, deposition takes place; the process forms a layer of sandy sediment known as **colluvium.** This layer is one of the many varieties of the parent matter of soil.

Sediment swept farther down the slope enters the channel of a stream, to be carried downvalley for a long distance. This sediment may then accumulate in layers on the broad, flat floodplain of the stream. (See Chapter 19 for details of floodplain forms.) Sediment deposited by a stream is referred to as **alluvium.** It may consist of layers of gravel, sand, silt, or clay; the texture depends on the conditions present at the time of deposition (Plate E.1). Alluvium is an important class of parent material from which soil is formed. Sorting of grades is usually good within individual layers.

Alluvium deposited by streams flowing across valley floors in arid regions is typically of coarse texture — sand, pebbles, and sometimes boulders. Also found in dry climates are accumulations of muddy debris swept from mountainsides by local torrents and carried out upon plains adjacent to the mountains. These **mudflow deposits** are a mixture of grades, including clay, silt, sand, and larger particles ranging upward in size to boulders (Plate G.4). The material is unsorted, in contrast to the well-sorted alluvial deposits in the beds of streams.

Alluvium deposited on river floodplains in areas of moist climates is usually of fine texture grades — clay, silt, and fine sand. Thus the physical qualities of alluvium as a parent material of soil differ greatly, depending in part on the climate under which deposition has taken place.

Wind-Deposited Sediment

Sediment deposited by wind action is another important class of parent material of the soil. Volcanic ash, already referred to as pyroclastic sediment, is one variety of wind-transported sediment. Because the ash particles are of fine grade sizes, they are rapidly altered into clay minerals by chemical weathering. Volcanic ash is an important class of soil parent matter.

Mineral particles in the size range of silt and clay are carried through the atmosphere for long distances. This material is swept upward from barren surfaces of regolith and alluvium in arid regions. (The process is described in Chapter 23.) The dust finally settles out in an adjacent region, forming a thin layer of surface particles. Over spans of hundreds and thousands of years, a sediment layer known as **loess** accumulates. Loess is one of the most important types of parent matter of soils, it is a well-sorted sediment composed largely of silt grades of crushed mineral fragments. Under a moist climate, the fine particles are rapidly altered into clay minerals. Loess is porous and soft, but has sufficient cohesion to stand in vertical walls (Plate J.4; Figure 23.13). Details of the origin and distribution of loess deposits are given in Chapter 23.

Dune sand is another type of clastic sediment important as parent material of soil. Grains of dune sand are moved close to the ground under the force of strong winds. Dunes occur in many kinds of environments, including the coastal zone and inland regions of dry climates. (Dune types are described in Chapter 23.) As a parent matter of soil, dune sand is characterized by a high degree of uniformity of grain sizes and a large proportion of open pore space between the grains. Water penetrates dune sand rapidly and drains out rapidly as well, so the field capacity for soil water is low. Most dune sands consist largely of quartz, although minor amounts of other silicate minerals and heavy minerals may be present. Clay minerals are almost totally lacking.

Sediments of the Coastal Zone

As parent matter of soil, coarse clastic sediment deposited in the beach zone of ocean and lake shorelines has much in common with dune sand. Beaches are formed largely of sand, which is well sorted and can range in grade from fine sand to coarse sand. Pebbles and cobbles make up some types of beaches. Where sand beaches have been uplifted beyond the reach of further wave and current action, they form a parent matter of soil having much the same qualities we have described for dune sand.

Fine sediment carried in suspension by currents into deep water comes to rest to form yet another class of

parent matter of soil. This marine sediment consists of layers of silt or clay. Sorting is good within individual layers. Should the land experience an uplift, marine-sediment layers may become the soil parent matter over broad, flat coastal plains. Clay minerals are usually abundant in marine sediment. Freshwater lakes also receive accumulations of fine sediment (lacustrine sediment). Should geologic events cause the lake to be permanently drained, the lake floor can become a flat plain underlain by fine clastic sediment rich in clay minerals.

Sediments Deposited by Glaciers

Finally, in this list of sediment groups important as parent matter of the soil, we come to deposits made by glaciers. (Details of erosion, transportation, and deposition by glaciers are given in Chapter 21.) The advancing ice of a valley glacier or a continental ice sheet picks up rock fragments in a great range of sizes from the bedrock surface over which it moves. Boulders the size of houses may be moved by the ice. At the other extreme, vast numbers of particles of silt and clay grades are produced by ice abrasion performed by rock fragments held fast in the moving ice and serving as cutting tools. All these fragments tend to be very angular, as would be expected of hard bedrock crushed and ground under high pressures beneath a thick ice layer. As the glacial ice melts in a stagnant marginal zone, the rock particles it holds are lowered to the solid surface beneath, there they form a layer of debris (Figure 11.10). This **residual till** shows no

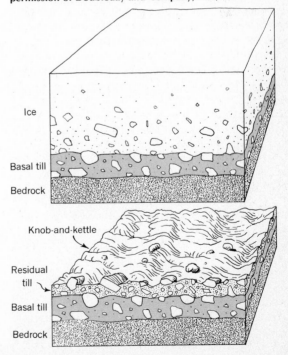

Figure 11.10 **Basal till and residual till. (From *A Geologist's View of Cape Cod* by Arthur N. Strahler. Copyright © 1966 by Arthur N. Strahler. Reproduced by permission of Doubleday and Company, Inc.)**

sorting and often consists of a mixture of sand and silt, with many angular pebbles and boulders. Beneath this residual layer there may be a layer of dense **basal till,** consisting of clay-rich debris previously dragged forward beneath the moving ice.

As parent matter for soil, glacial till has a unique combination of characteristics. It has an abundance of clay particles, which may be in part clay minerals derived from older regolith; or they may be finely crushed minerals, capable of being altered into clay minerals during a long period of chemical weathering following disappearance of the ice. The clay fraction in dense till impedes the movement of soil water, so water drainage is often very slow. Some residual tills of sandy composition, on the other hand, are of loose texture and drain rapidly. The presence of cobbles and boulders, called *erratics,* can make soil tillage difficult or even impossible. Because glacial tills range greatly in age and composition, each occurrence must be evaluated separately in terms of physical and chemical properties. In Chapter 21, the various landforms shaped from glacial till are described.

An entirely different class of sediment derived from glacial ice is glaciofluvial sediment, transported from stagnant ice masses by streams of meltwater and deposited in broad, shallow channels as a form of coarse alluvium. As layer upon layer of sand accumulates near the ice margin, an **outwash deposit** is formed (Figure 11.11). Outwash sands form well-drained upland plains. The sand can hold little soil water and is often lacking in clay minerals. Thus outwash deposits share many physical properties in common with dune sands and beach sands.

In North America and northern Europe, during the most recent of the glacial epochs, numerous temporary lakes existed close to the margins of the great ice sheets

(Chapter 21). These lakes received sediment (glaciola-custrine sediment) derived from nearby melting ice. The sediment commonly shows banding, with alternating dark and light layers, termed **varves** (Figure 11.12). An individual varve consists of a light-colored silt layer beneath a dark layer of fine clay. The silt layers are interpreted as summer deposits from highly turbid lake water; the clay layers, as still-water deposits formed in winter when the lakes were sealed over by an ice cover. Glacio-lacustrine sediments are often rich in primary silicate minerals, capable of becoming altered into clay minerals.

Figure 11.12 **Samples of varved clays from glacial lake beds near New York City. The columns measure about 20 cm (8 in.) from top to bottom of the photograph. (C. A. Reeds, American Museum of Natural History.)**

Figure 11.11 **Thick layers of outwash sands and gravels such as these on the north shore of Long Island were excavated in great quantities for use in highway and building construction. The dark layer at the top is a bed of glacial till left by a glacial advance. Boulders in the foreground are glacial erratics that have rolled down from the till bed. (A. K. Lobeck.)**

The former lake floors are extremely flat, poorly drained, and often consist of marshland.

This list of sediment types capable of serving as parent mineral matter of soil has enormous significance in the geography of soils, as we shall learn in later chapters. Although the character of the parent matter is only one factor in determining the properties of the soil layer, it is a powerful factor because it determines not only the soil texture but also the basic store of nutrient elements that give a soil its initial level of fertility. The importance of parent matter is greatest in young soils — soils that have had only a few tens or hundreds of years in which to develop.

Clastic Sedimentary Rocks

Streams carry sediment to lower levels and to locations where permanent accumulation is possible. (Wind and glacial ice also transport sediment, but not necessarily to lower elevations or to places suitable for accumulation.) Usually these sites of accumulation are shallow seas bordering the continent, but they may also be inland seas and lakes. Here the sediment is reworked and redistributed by wave and current action. Over long spans of time, thick masses of sediments undergo physical or chemical changes, or both, to become compacted and hardened, producing sedimentary rock. The process of compaction and hardening is referred to as **lithification.** (*Diagenesis* is a more general term for complex changes with time, usually involving chemical change and mineral replacement.)

Sandstone is formed of the sand grades of sediment, cemented into a solid rock by silica (SiO_2) or calcium carbonate ($CaCO_3$). The sand grains are commonly of quartz, as shown in Figure 11.7, and sometimes of feldspar; but in other cases they are sand-sized fragments

of fine-grained rock containing several minerals. *Siltstone* and **shale** are lithified layers of silt and clay particles, respectively, while mixtures of silt and clay, often with some sand, produce *mudstone*. (*Mud* is a mixture of silt and clay, with small amounts of sand.)

Shale is the most abundant of the sedimentary rocks. It is formed largely of the clay minerals kaolinite, illite, and montmorillonite. The compaction of the original clay and mud involves a considerable loss of volume as water is driven out.

A characteristic feature of sedimentary rocks is their layered arrangement, the layers being called **strata.** Typically, layers of different textural composition are alternated or interlayered. The planes of separation between layers are known as **bedding planes.** Bedding planes of strata deposited on the ocean floor (marine strata) are usually horizontal in attitude, or nearly so (see Figure 26.7).

Nonclastic Sediments and Sedimentary Rocks

The nonclastic sediments are of particular importance in the environment because they represent enormous storages of carbon, obtained from atmospheric carbon dioxide and changed into carbonate and hydrocarbon compounds by both inorganic and organic processes.

We recognize two major divisions of the nonclastic sediments: (1) **Chemical precipitates** are compounds precipitated directly from water in which the ions are transported; these sediments are described as **hydrogenic.** (2) **Organically derived sediments** are created by the life processes of plants and animals; these are described as **biogenic.**

Chemical precipitates are of major importance in sediments of the seafloor; but a second important environment of deposition is in inland salt lakes of desert regions, where evaporation greatly exceeds precipitation.

The most important hydrogenic and biogenic minerals (hydrocarbon compounds are not included) are listed in Table 11.4, together with their compositions. The first three — calcite, aragonite, and dolomite — are classed as **carbonates. Calcite** is the dominant carbonate mineral and occurs in many forms. *Aragonite* is of the same composition as calcite, but of different crystal structure; it is secreted in the shells of certain invertebrate animals. *Dolomite* contains magnesium as well as calcium. All three carbonate minerals are soft substances, as compared with the silicate minerals.

The carbonate minerals and rocks are highly susceptible to a form of chemcal weathering known as **carbonation.** Atmospheric carbon dioxide (CO_2), dissolved in surface water, forms a weak acid — carbonic acid (H_2CO_3). This acid reacts with calcium carbonate ($CaCO_3$) to produce a soluble product, calcium bicarbonate, which is removed with surplus water and conducted out of the region by runoff. This process is reversible, so calcium-bearing solutions, when undergoing evaporation, deposit calcium carbonate in crystalline form in the soil, regolith, or alluvium. Carbonate deposition is a dominant geologic process in regions of dry climates;

Table 11.4 / Common Hydrogenic and Biogenic Minerals

	Mineral Name	Composition	Density gm/cc
CARBONATES	Calcite	Calcium carbonate $CaCO_3$	2.72
	Aragonite	Calcium carbonate $CaCO_3$	2.9–3.0
	Dolomite	Calcium-magnesium carbonate $CaMg(CO_3)_2$	2.9
EVAPORITES	Anhydrite	Calcium sulfate $CaSO_4$	2.7–3.0
	Gypsum	Hydrous calcium sulfate $CaSO_4 \cdot 2\,H_2O$	2
	Halite	Sodium chloride $NaCl$	2.1–2.3
	Hematite	Sesquioxide of iron Fe_2O_3	4.9–5.3
	Chalcedony (chert, flint)	Silica SiO_2	2.6

it is an important soil-forming process as well.

The second group of minerals listed in Table 11.4 are hydrogenic. They are **evaporites,** typically formed where seawater is evaporated in shallow bays and gulfs. *Anhydrite* and **gypsum** are composed of calcium sulfate, the latter in combination with water. The third evaporite, *halite,* is commonly known as rock salt, with the composition sodium chloride (NaCl). In refined form, it is the table salt we use in cooking and flavoring. Evaporites formed on the floors of shallow, temporary lakes in a desert climate include many varieties of salts of sodium, calcium, magnesium, and potassium.

Hematite, a sesquioxide of iron (Fe_2O_3), is a common hydrogenic mineral in sedimentary rocks; it is a major ore of iron. No less important than the carbonates and evaporites is chalcedony, which we have described as commonly formed during mineral alteration in regolith and soil.

The minerals listed in Table 11.4 can be chemically precipitated from seawater or the water of saline lakes, or they can be secreted by organisms to produce the nonclastic sedimentary rocks. Most common of the carbonate rocks is **limestone,** of which there are many varieties. One source is in reefs built by corals and algae; another form is *chalk,* made up of the skeletons of a marine form of the algae. Other limestones are formed of shell fragments or other broken carbonate matter. Some limestones are densely crystalline. Dolomite, a rock of the same name as the mineral that composes it, may have been derived through the alteration of limestone, as magnesium atoms of seawater gradually replaced calcium atoms.

Chert, composed of chalcedony, is an important siliceous sedimentary rock. It occurs as nodules in limestone and, in some cases, as solid rock layers referred to as *bedded cherts.* Gypsum, anhydrite, and halite are layered rocks of their respective mineral compositions and occur in association with clastic sedimentary rocks.

Hydrocarbon Compounds in Sedimentary Rocks

Hydrocarbon compounds form a second group of biogenic sediments. These organic substances occur both as solids (peat and coal) and as liquids and gases (petroleum and natural gas), but only coal qualifies physically for designation as a rock.

Peat, a soft, fibrous substance of brown to black color, accumulates in a bog environment where the continual presence of water inhibits decay and oxidation of plant remains. One form of peat is of freshwater origin and represents the filling of shallow lakes (see Figure 15.9). Thousands of such peat bogs are found in North America and Europe; they occur in depressions remaining after recession of the great ice sheets of the Pleistocene epoch (Chapter 21). This peat has been used for centuries as a low-grade fuel (Plate E.1). Peat of a different sort is formed in the saltwater environment of tidal marshes (Chapter 22).

At various times and places in the geologic past, conditions were favorable for the large-scale accumulation of plant remains, accompanied by subsidence

of the area and burial of the compacted organic matter under thick layers of inorganic sediments. Thus **coal** seams interbedded with shale, sandstone, and limestone strata came into existence. Individual seams range in thickness from a centimeter to as great as 12 m (50 ft) in the exceptional case (see Figure 26.11). Coal consists mostly of compounds of carbon, hydrogen, and oxygen and usually contains a small proportion of sulfur compounds. Water and hydrocarbon volatiles make up a large part of lower-grade coals, but fixed carbon is the principal constituent of the highest grades of coal.

Petroleum, or crude oil, as the liquid form is correctly called, includes many hydrocarbon compounds. Typically, a sample of crude oil might consist of 82 percent carbon, 15 percent hydrogen, and 3 percent oxygen and nitrogen. **Natural gas,** found in close association with accumulations of petroleum, is a mixture of gases. The principal gas is methane (marsh gas), CH_4, and there are minor amounts of ethane, propane, and butane, all of which are hydrocarbon compounds. Petroleum and natural gas are not classed as minerals, but they originated as organic compounds in sediments. The occurrence of accumulations of petroleum and natural gas in sedimentary strata is explained in Chapter 26.

Metamorphic Rocks

Any of the types of igneous or sedimentary rocks may be altered by the tremendous pressures and high temperatures that accompany mountain-building movements of the earth's crust. The result is a rock so changed in appearance and composition as to be classified as a **metamorphic rock.** Generally speaking, metamorphic rocks are harder and more compact than their original types, except when the latter are igneous rocks. Moreover, the kneading action and baking that metamorphic rocks have undergone produces new structures and even new minerals. We can regard these metamorphic minerals as primary minerals because most are silicates.

Shale, after being squeezed and sheared under mountain-making forces, is altered into **slate.** This gray or brick-red rock splits neatly into thin plates so familiar as roofing shingles and as flagstones of patios and walks.

With continued application of pressure and internal shearing, slate changes into **schist,** the most advanced grade of metamorphic rock. Schist has a structure called *foliation,* consisting of thin but rough and irregularly curved planes of parting in the rock. Schist is set apart from slate by the coarse texture of the mineral grains, the abundance of mica, and the presence of scattered large crystals of new minerals, such as garnet. These crystals have grown during the process of internal shearing of the rock.

The metamorphic equivalent of conglomerate, sandstone, and siltstone is **quartzite,** which is formed by addition of silica to fill completely the interstices between grains. This process is carried out by the slow movement of underground waters carrying the silica into the sandstone, where it is deposited. Pressure and kneading of the rock is not essential in producing a quartzite.

Limestone, after undergoing metamorphism, becomes

Figure 11.13 **Outcrop of banded gneiss of Precambrian age, east coast of Hudson Bay, south of Povungnituk, Quebec, Canada. (Photograph G.S.C. No. 125221 by F. C. Taylor, Geological Survey of Canada, Ottawa.)**

marble, a rock of sugary texture when freshly broken. During the process of internal shearing, the calcite mineral of the limestone has reformed into larger, more uniform crystals than before. Bedding planes are obscured, and masses of mineral impurities are drawn out into swirling streaks and bands.

Finally, the important metamorphic rock **gneiss** may be formed either from intrusive igneous rocks or from clastic sedimentary rocks that have been in close contact with intrusive magmas. A single description will not fit all gneisses because they vary considerably in appearance, mineral composition, and structure. One conspicuous variety of gneiss is strongly banded into light and dark layers, or lenses (Figure 11.13), which may be contorted

into wavy folds. These bands, which have differing mineral compositions, are interpreted as the relics of sedimentary strata, such as shale and sandstone, to which new mineral matter has been added from nearby intrusive rocks.

Chemical weathering of the metamorphic rocks includes hydrolysis and oxidation, affecting the aluminosilicate minerals. Carbonate varieties are susceptible to being dissolved by the same carbonic action that affects limestone and dolomite.

The Cycle of Rock Transformation

We can now bring together the formative processes of rocks into a single unified concept of recycling of matter through geologic time. A schematic diagram, Figure 11.14, distinguishes between a **surface environment** of low pressures and temperatures and a **deep environment** of high pressures and temperatures. The deep environment is the realm of the igneous and metamorphic rocks. The surface environment is one of mineral alteration and sediment deposition.

Seen in its complete form, the total circuit of rock changes in response to environmental stress constitutes the **cycle of rock transformation.** Figure 11.14 emphasizes that mineral matter is continually recycled through the three major rock classes. Igneous rocks are by no means the "original" rocks of the earth's crust. Actually, there is no known record of the rocks that first formed the earth's crust; they were consumed and recycled long ago.

Earth Materials and Soils

In this chapter we have gathered together for study those materials of the lithosphere important in the formation of the soil. We have traced the components of the parent matter of soil from an igneous origin, as primary silicate minerals, through processes of alteration, to form secondary minerals and regolith. We then examined the production of sediment and its occurrence as alluvium and

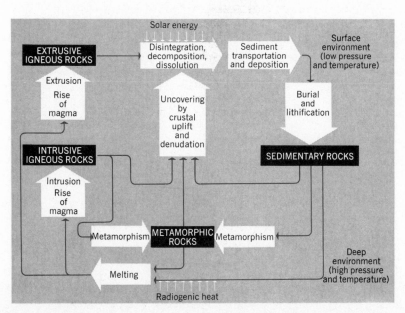

Figure 11.14 **Schematic diagram of the cycle of rock transformations.**

other soil parent materials transported and deposited by the actions of streams, waves and currents, winds, and glacial ice. Also important as soil materials are inorganic compounds produced by hydrogenic and biogenic processes in water bodies and the various hydrocarbon compounds representing altered remains of plant tissues.

All the above processes and products are important in the origin and evolution of the soil. This information will be applied in Chapters 12 and 13 as we investigate soil-forming processes and the classification of soils.

References for Further Study

Spock, L. E. (1962), *Guide to the study of rocks,* second edition, Harper and Row, New York.

Ernst, W. G. (1969), *Earth materials,* Prentice-Hall, Englewood Cliffs, N.J.

Loughnan, F. C. (1969), *Chemical weathering of the silicate minerals,* American Elsevier Publishing Co., New York.

Strahler, A. N. (1971), *The earth sciences,* second edition, Harper and Row, New York. See Chapters 20–22.

Birkeland, P. W. (1974), *Pedology, weathering, and geomorphological research,* Oxford Univ. Press, New York.

Review Questions

1. Name in order the eight most abundant elements in the earth's crust. Give approximate percentages for each. Comment on the importance of each element in terms of use by plants or for their economic value to Man.

2. What is a mineral? A rock? What are the three major classes of rocks? What two elements are always present in silicate minerals? How are they linked together in the crystal structure?

3. Give the names and chemical compositions of the seven major silicate minerals or mineral groups. What range in density is found in this list? Explain how these minerals are grouped as mafic and felsic types. Name two mafic minerals that are oxides.

4. Describe a silicate magma. What volatile substances are contained in a magma? What importance have these volatiles in the atmosphere and hydrosphere?

5. Explain how the texture of igneous rocks is affected by the rate of cooling. How do intrusive and extrusive rocks differ in texture? How are volcanic glass and volcanic ash formed?

6. Name and describe five igneous rocks, including plutonic forms and lavas. Which are felsic? mafic? ultramafic?

7. Why are igneous rocks susceptible to weathering when exposed at the earth's surface? Distinguish between regolith and sediment. Describe the action of several physical weathering processes.

8. What system of grade sizes is used in soil science? Name the grades and give grade limits in millimeters. Distinguish between noncolloidal clay and colloidal clay. How is particle shape related in a general way to size grade? Why is increase of surface area with decreasing particle size important in mineral weathering?

9. What processes are included in chemical weathering (mineral alteration)? Distinguish between primary minerals and secondary minerals. How does oxidation affect minerals? What acids are present in surface water? Describe hydrolysis. Relate the intensity of mineral alteration to climate controls.

10. Name the principal minerals produced by hydrolysis and oxidation. Give chemical compositions. Which products are clay minerals? Which are oxides? What is a sesquioxide? Which of the silicate minerals are most susceptible to alteration? Which are least susceptible?

11. What forms does silica take when reformed in regolith and soil? What is the composition of allophane?

12. Distinguish between clastic and nonclastic sediments. What is the source of clastic sediment? What are heavy minerals? How does sorting of sediment occur during transportation? How does flocculation affect suspended colloids?

13. Distinguish between colluvium and alluvium. Of what sediment grades does alluvium consist and how well is it sorted? How does alluvium in arid regions differ from that in humid regions? Compare mudflow deposits with alluvium in terms of degree of sorting.

14. Describe loess; compare it in terms of grade size and composition with dune sand. What types of sediment are deposited by waves and currents? Describe marine and lacustrine sediments.

15. Compare residual and basal till in terms of origin, grade sizes, and sorting. Describe outwash deposits and the sediments of marginal glacial lakes.

16. What changes occur in sediments when they are transformed into sedimentary rock? Name and describe the clastic sedimentary rocks. What distinctive structure do they show?

17. Distinguish between chemical precipitates and organically derived sediments and between hydrogenic and biogenic sediments. Name the most important hydrogenic and biogenic minerals and give their chemical compositions. Which are carbonates? Which are evaporites? How does carbonation affect the carbonate rocks? How can carbonate minerals be deposited in the soil and regolith?

18. Under what environmental conditions do evaporites form? What sedimentary rock is composed largely of silica?

19. Name the common mineral forms of hydrocarbon compounds. How does peat originate? How is coal formed? What is the approximate element composition of coal? of petroleum? of natural gas? What are the fossil fuels? Are they a renewable energy resource?

20. Describe the metamorphic rocks as a group. Of what mineral groups are they composed? Describe slate, schist, quartzite, marble, and gneiss.

21. Explain the concept of a cycle of rock transformation based on changes in both surface and deep environments.

Energy Flows and Material Cycles in the Biosphere

12

All living organisms of the earth, together with the environments with which they interact, constitute the **biosphere.** Organisms, whether belonging to the plant or animal kingdom, also interact with each other. The study of these interactions — in the form of exchanges of matter, energy, and stimuli of various sorts — between life forms and the environment is the science of **ecology,** very broadly defined. The total assemblage of components entering into the interactions of a group of organisms is known as an ecological system, or more simply, an **ecosystem.** The root "eco" comes from a Greek word connoting a house in the sense of household, which implies that a family lives together and interacts within a functional physical structure.

To the geographer, ecosystems are part of the physical composition of the life layer. A forest, for example, is not only a living community of organisms; it is also a collection of physical objects on the land surface. A forest is physically very different from an expanse of prairie grassland. Plant geographers are aware of these place-to-place differences in the physical aspect of vegetation. They attempt to categorize the varied types; they map the global distribution of vegetation forms.

Geographers also view ecosystems as natural resource systems. Food, fiber, fuel, and structural material are products of ecosystems; they represent organic compounds placed in storage by organisms through the expenditure of energy derived basically from the sun. Geographers are interested in the influence of climate on ecosystem productivity. Our basic understanding of global climates and the global range of soil-water budgets will prove most useful in explaining the global pattern of ecosystems on the lands.

Ecosystems have inputs of matter and energy that are used to build biological structures, reproduce, and maintain necessary internal energy levels. Matter and energy are also exported from an ecosystem. An ecosystem tends to achieve a balance of the various processes and activities within it. For the most part, these balances are quite sensitive and can be easily upset or destroyed.

Physical geography meshes closely with ecology. For example, organisms and their life processes play an important role in shaping the characteristics of the soil layer.

The Ecosystem and the Food Chain

As an example of an ecosystem, consider a salt marsh (Figure 12.1). A variety of organisms are present: algae and aquatic plants, microorganisms, insects, snails, and crayfish, as well as such larger organisms as fishes, birds, shrews, mice, and rats. Inorganic components will be found as well: water, air, clay particles and organic sediment, inorganic nutrients, trace elements, and light energy. Energy transformations in the ecosystem occur by means of a series of steps or levels, referred to as a **food chain.**

The plants and algae in this chain are the **primary producers.** They use light energy to convert carbon dioxide and water into carbohydrates (long chains of sugar molecules) and eventually into other biochemical molecules needed for the support of life. This process of energy conversion is called photosynthesis. Organisms engaged in photosynthesis form the base of the food chain.

At the next level are the **primary consumers** (the snails, insects, and fishes); they live by feeding on the producers. At a still higher level are the **secondary consumers** (the mammals and birds); they feed on the primary consumers. As in many ecosystems, still higher levels of feeding are evident: the marsh hawks and owls. The **decomposers** feed on detritus, or decaying organic matter, derived from all levels. They are mostly microscopic organisms (microorganisms) and bacteria.

The food chain is really an energy flow system, tracing the path of solar energy through the ecosystem. Solar energy is stored by one class of organisms, the primary producers, in the chemical products of photosynthesis. As these organisms are eaten and digested by consumers, chemical energy is released. This chemical energy is used to power new biochemical reactions, which again

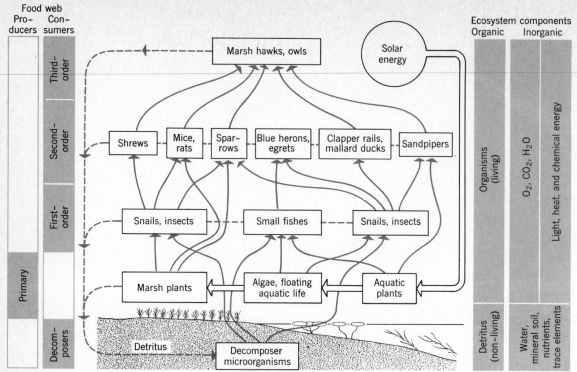

Figure 12.1 Flow diagram of a salt-marsh ecosystem in winter. The arrows show how energy flows from the sun to producers, consumers, and decomposers. (Food chain after R. L. Smith, 1966, *Ecology and Field Biology,* Harper and Row, New York, p. 30, Figure 3.1.)

produce stored chemical energy in the bodies of the consumers.

At each level of energy transformation, energy is lost as waste heat. In addition, much of the energy input to each organism must be used in respiration. Respiration can be thought of as the burning of fuel to keep the organism operating; it will be discussed in more detail in the next section. Energy expended in respiration is used for bodily maintenance and cannot be stored for use by other organisms higher up in the food chain. This means that, generally, both the numbers of organisms and their total amount of living tissue must decrease drastically up the food chain.

Photosynthesis and Respiration

Stated in the simplest possible terms, **photosynthesis** is the production of carbohydrate. **Carbohydrate** is a general term for a class of organic compounds consisting of the elements carbon, hydrogen, and oxygen. Carbohydrate molecules are composed of chains of carbon molecules with hydrogen (H) atoms and hydroxyl (OH) atom-pairs attached to their sides. We can symbolize a single carbon atom with its attached hydrogen atom and hydroxyl molecule as $-CHOH-$. The leading and trailing dashes indicate that the unit is just one portion of a longer chain.

Photosynthesis of carbohydrate requires a series of complex biochemical reactions using water (H_2O) and carbon dioxide (CO_2) as well as light energy. A simplified chemical reaction for photosynthesis can be written as follows:

$$H_2O + CO_2 + \text{light energy} \longrightarrow -CHOH- + O_2$$

Oxygen in the form of gas molecules (O_2) is a by-product of photosynthesis.

Respiration is the process opposite to photosynthesis, in which carbohydrate is broken down and combined with oxygen to yield carbon dioxide and water. The overall reaction is as follows:

$$-CHOH- + O_2 \longrightarrow CO_2 + H_2O + \text{chemical energy}$$

As in the case of photosynthesis, the actual reactions are far from simple. The chemical energy released is stored in many types of energy-carrying molecules and used later to synthesize all the biological molecules necessary to sustain life.

At this point, it is helpful to link photosynthesis and respiration in a continuous cycle involving both the primary producer and the decomposer. Figure 12.2 shows one closed loop for hydrogen (H), one for carbon (C), and two loops for oxygen (O). We are not taking into account that there are two atoms of hydrogen in each molecule of water and carbohydrate, or that there are two atoms of oxygen in each molecule of carbon dioxide and oxygen gas. Only the flow pattern counts in this representation.

A good place to start is the soil, from which water is drawn up into the body of a living plant. In the green leaves of the plant, photosynthesis takes place while light

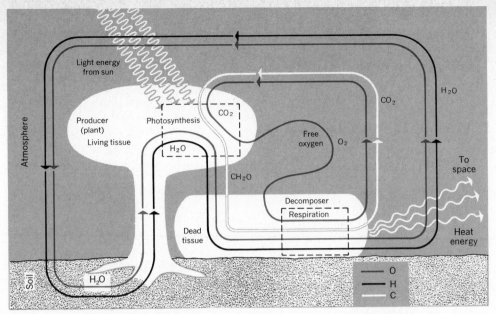

Figure 12.2 A simplified flow diagram of the essential components of photosynthesis and respiration through the biosphere.

energy is absorbed by the leaf cells. Carbon dioxide is being brought in from the atmosphere at this point. Here oxygen is liberated and begins its atmospheric cycle. The plant tissue then dies and falls to the ground, where it is acted on by the decomposer. Through respiration, oxygen is taken out of the atmosphere or soil air and combined with the decomposing carbohydrate. Energy is now liberated. Here both carbon dioxide and water enter the atmosphere as gases.

An important concept emerges from this flow diagram. Energy passes through the system. It comes from the sun and returns eventually to outer space. On the other hand, the material components — hydrogen, oxygen, and carbon — are recycled within the total system. Of course, many other material components are recycled in the same way. These are plant nutrients, essential in the growth of plants. Nutrients are constantly recycled. Because the earth as a planet is a closed system, the material components never leave the total system; but they can be stored

in other ways and forms where they are unavailable for use by plants for prolonged periods of geologic time. We shall develop this concept more fully later in this chapter.

Net Photosynthesis

Because both photosynthesis and respiration go on simultaneously in a plant, the amount of new carbohydrate placed in storage is less than the total carbohydrate being synthesized. We must thus distinguish between gross photosynthesis and net photosynthesis. **Gross photosynthesis** is the total amount of carbohydrate produced by photosynthesis; **net photosynthesis** is the amount of carbohydrate remaining after respiration has broken down sufficient carbohydrate to power the plant. Stated as an equation,

Net photosynthesis = Gross photosynthesis − Respiration

Because both photosynthesis and respiration occur in the same cell, gross photosynthesis cannot be measured readily. Instead, we shall deal with net photosynthesis. In most cases, respiration will be held constant, so use of the net instead of the gross will show the same trends.

The rate of net photosynthesis is strongly dependent on the intensity of light energy available, up to a limit. Figure 12.3 shows this principle. The rate of net photosynthesis is indicated on the vertical axis by the rate at which a plant takes up carbon dioxide. On the horizontal axis, light intensity increases from left to right. At first, net photosynthesis rises rapidly as light intensity increases; but the rate then slows and reaches a maximum value, shown by the plateau in the curve. Above this maximum, the rate falls off because the incoming light is also causing heating. Increased temperature, in turn, increases the rate of respiration, which offsets gross production by photosynthesis.

Figure 12.3 The curve of net photosynthesis shows a steep initial rise, then levels off as light intensity rises.

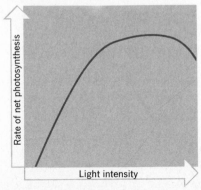

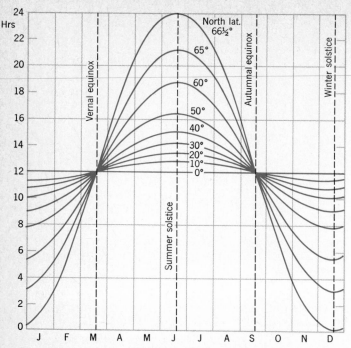

Figure 12.4 Duration of the daylight period at various latitudes throughout the year. The vertical scale gives the number of hours the sun is above the horizon.

Light intensity sufficient to allow maximum net photosynthesis is only 10 to 30 percent of full summer sunlight for most green plants. Additional light energy is simply ineffective. Duration of daylight then becomes the important factor in the rate at which products of photosynthesis accumulate as plant tissues. On this subject, you can draw on your knowledge of the seasons and the changing angle of the sun's rays with latitude. Figure 12.4 shows the duration of the daylight period with changing seasons for a wide range of latitudes in the northern

Figure 12.5 Respiration and gross and net photosynthesis vary with temperature. (Data of Stofelt, 1937, in A. C. Leopold, 1964, *Plant Growth and Development,* McGraw-Hill, New York, p. 31, Figure 2.26.)

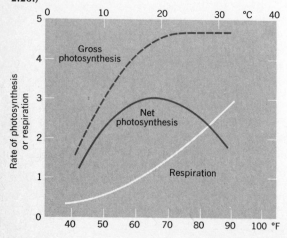

hemisphere. At low latitudes, days are not far from the average 12-hour length throughout the year; at high latitudes, days are short in winter but long in summer. The seasonal contrast in day length increases with latitude. In subarctic latitudes, photosynthesis can obviously go on in summer during most of the 24 hours, and this factor can compensate partly for the shortness of the season.

Photosynthesis also increases in rate as air temperature increases, up to a limit. Figure 12.5 shows the results of a laboratory experiment in which sphagnum moss was grown under constant illumination. Gross photosynthesis increased rapidly to a maximum at about 20°C (70°F), then leveled off. Respiration increased quite steadily to the limit of the experiment. Net photosynthesis, which is the difference between the values in the two curves, peaked at about 18°C (65°F), then fell off rapidly.

Energy Flow Along the Food Chain

The primary producers trap the energy of the sun and make it available to support consuming organisms at higher levels in the food chain. But remember that energy is lost during each upward step in the chain, so the number of steps is limited. In general, anywhere from 10 to 50 percent of the energy stored in organic matter at one level can be passed up the chain to the next level. Four levels of consumers are about the normal limit.

Figure 12.6 is a bar graph showing the percentage of energy passed up the chain when only 10 percent moves from one level to the next. The horizontal scale is in powers of ten. In ecosystems of the lands, the mass of organic matter also decreases with each upward step in the chain; the number of individuals of the consuming animals also decreases with each upward step. In the food chain shown in Figure 12.1, there are only a few marsh hawks and owls in the third level of consumers, whereas there are countless individuals in the primary level. These facts serve as a clear message to Mankind; they tell us that when we raise beef cattle, swine, sheep, and fowl to eat, we are being quite wasteful of the world's total food resource. These animals, in storing food for humans to consume, have already wasted two-thirds or more of the energy they consumed as plant matter from the producing level. If we instead subsisted largely on grains and other food plants, such as legumes, grown on the available farm lands, our food resource could support a much larger population. Although this observation does not apply to sheep and cattle subsisting entirely on grazing lands that are unfit for cereal crops, it is all the same a potent concept.

Net Primary Production

Plant ecologists measure the accumulated net production by photosynthesis in terms of the **biomass,** which is the dry weight of the organic matter. This quantity could, of course, be stated for a single plant or animal; but a more useful statement is made in terms of the biomass per unit of surface area within the ecosystem — the hectare, square meter, or square foot. Of all ecosystems, forests have the greatest biomass; that of grasslands and crop-

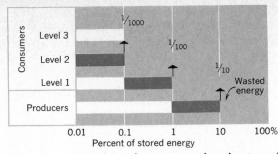

Figure 12.6 **Percentage of energy passed up the steps of the food chain, assuming 90 percent is lost energy at each step.**

Table 12.1 / Net Primary Production for Various Ecosystems

	Grams per Square Meter per Year	
	Average	*Typical Range*
Lands		
Rainforest of the equatorial zone	2000	1000–5000
Freshwater swamps and marshes	2000	800–4000
Midlatitude forest	1300	600–2500
Midlatitude grassland	500	150–1500
Agricultural land	650	100–4000
Lakes and streams	500	100–1500
Extreme desert	3	0–10
Oceans		
Estuaries (tidal)	2000	500–4000
Continental shelf	350	300–600
Open ocean	125	1–400

lands is very small in comparison. For freshwater bodies and the oceans, the biomass is even smaller — on the order of one-hundredth that of the grasslands and croplands.

The annual rate of production of organic matter is the vital information we need, not the biomass itself. Granted that forests have a very large biomass and grasslands very little, what we want to know is, Which ecosystem is the more productive? In other words, which ecosystem produces the greatest annual yield in terms of resources useful to Man? For the answer we turn to figures on the **net primary production** of various ecosystems. Table 12.1 gives this information in units of grams of dry organic matter produced annually from 1 sq m of surface. The figures are rough estimates, but they are nevertheless highly meaningful. Note that the highest values are in two quite unlike environments: forests and wetlands

(estuaries). Agricultural land compares favorably with grassland; but the range is very large in agricultural land, reflecting many factors, such as availability of soil water, soil fertility, and use of fertilizers and machinery.

Productivity of the oceans is generally low. The deep-water oceanic zone is the least productive of the marine ecosystems. At the same time, however, it comprises about 90 percent of the world ocean area. Continental shelf areas are a good deal more productive and, in fact, support much of the world's fishing industry at the present time (Figure 12.7). Except for the shallow-

Figure 12.7 **Distribution of world fisheries. Coastal areas and upwelling areas together supply over 99 percent of world production. (Compiled by the National Science Board, National Science Foundation.)**

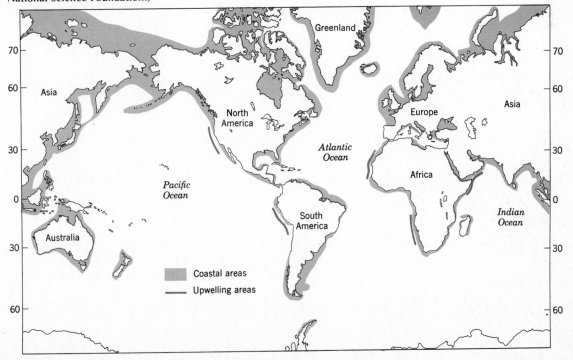

water estuarine and reef ecosystems, the upwelling zones are most productive.

Upwelling of cold water from great depths brings nutrients to the surface and greatly increases the growth of microscopic floating plants (phytoplankton). These, in turn, serve as food sources for marine animals in the food chain. Consequently, zones of upwelling are fisheries of great wealth. An example is the Peru Current off the west coast of South America. Here, countless individuals of a single species of small fish, the anchoveta, provide food for larger fish and for birds. The birds, in turn, excrete their wastes on the mainland coast of Peru. The accumulated deposit, called guano, is a rich source of fertilizer.

Net Production and Climate

The geographer is interested in the climatic factors controlling organic productivity. Besides light intensity and temperature, an obvious factor will be availability of water. Soil-water shortage or surplus might be the best climatic factor to examine, but data are not available. Ecologists have related net annual primary production to mean annual precipitation, as shown in Figure 12.8. The production values are for plant structures above the ground surface. Although the productivity increases rapidly with precipitation in the lower range from desert through semiarid to subhumid climates, it seems to level off in the humid range. Apparently, a large soil-water surplus carries with it some compensating influence, such as removal of plant nutrients by leaching.

Combining the effects of light intensity, temperature, and precipitation, you would guess that the maximum net productivity year around would be in wet equatorial climates where soil water is adequate at all times. Productivity would be low in all desert climates and would decrease generally with increasing latitude. Productivity would be reduced to near zero in the arctic zone, where the combination of a short growing season and low temperatures would act together to slow the growth process.

These deductions are nicely confirmed by a schematic

diagram of net primary productivity on an imagined supercontinent (Figure 12.9). Comparing this diagram with the world climate diagram (Plate D.1), we can assign rough values of productivity to each of the climates (units are grams of carbon per square meter per year):

Highest (over 800)	Wet equatorial (1)
Very high (600–800)	Monsoon and trade-wind littoral (2)
	Wet-dry tropical (3)
High (400–600)	Wet-dry tropical (3) (Southeast Asia)
	Moist subtropical (6)
	Marine west coast (8)
Moderate (200–400)	Mediterranean (7)
	Moist continental (10)
Low (100–200)	Dry tropical, semiarid (4s)
	Dry midlatitude, semiarid (9s)
	Boreal forest (10)
Very low (0–100)	Dry tropical, desert (4d)
	Dry midlatitude, desert (9d)
	Boreal forest (10)
	Tundra (12)

The wet-dry tropical climate (3) is in a narrow transition zone between very high productivity and low productivity, so some areas are in the "very high" rating and others are in the "high" rating. Transition zones such as this are highly sensitive to alternating cycles of drought and excessive rainfall. They pose the most difficult famine problem of all global regions because they carry dense populations, which can be fed adequately only under the best of conditions.

Material Cycles in Ecosystems

We have seen how energy of solar origin flows through ecosystems, passing from one part of the food chain on to the next, until it is ultimately lost from the biosphere as energy radiated to space. Matter also moves through ecosystems, but because gravity keeps surface material earthbound, matter cannot be lost to the global ecosystem. As molecules are formed and reformed by chemical and biochemical reactions within an ecosystem, the atoms that compose them are not changed or lost. Thus matter is conserved within an ecosystem, and atoms and molecules can be used and reused, or cycled, within ecosystems.

Atoms and molecules move through ecosystems under the influence of both physical and biological processes. The pathways of a particular type of matter through the earth's ecosystem comprise a **material cycle** (sometimes referred to as a *biogeochemical cycle,* or *nutrient cycle*).

Ecologists recognize two types of material cycles — gaseous and sedimentary. In the **sedimentary cycle,** the compound or element is released from rock by weathering, then follows the movement of running water either in solution or as sediment to the sea. Eventually, by precipitation and sedimentation, these materials are converted into rock. When the rock is uplifted and exposed to weathering, the cycle is completed.

Figure 12.8 **Net primary production increases rapidly with increasing precipitation, but levels off in the higher values. Observed values fall mostly within the shaded zone. (Data of Whittaker, 1970.)**

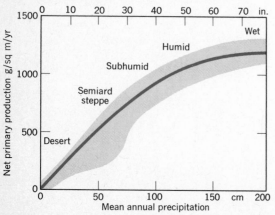

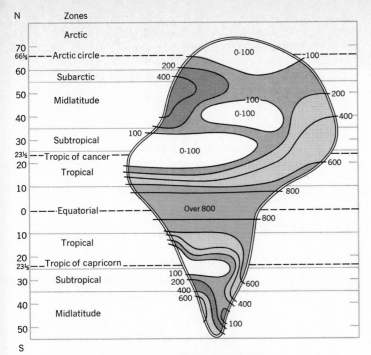

Figure 12.9 **Schematic diagram of net primary production of organic matter on an imagined supercontinent. The units are grams of carbon per square meter per year. The quantities shown are estimates only. Compare this map with the world climate diagram, Plate D.1. (Based on data of D. E. Reichle, 1970.)**

In the **gaseous cycle,** a shortcut is provided — the element or compound can be converted into a gaseous form. The gas diffuses throughout the atmosphere and thus arrives over land or sea, to be reused by the biosphere, in a much shorter time. The primary constituents of living matter — carbon, hydrogen, oxygen, and nitrogen — all move through gaseous cycles.

The major features of a material cycle are diagrammed in Figure 12.10. Any area or location of concentration of a material is a **pool.** There are two types of pools: **active pools,** where materials are in forms and places easily accessible to life processes, and **storage pools,** where materials are more or less inaccessible to life. A system of pathways of material flows connects the various active and storage pools within the cycle. Pathways between active pools are usually controlled by life processes, whereas pathways between storage pools are usually controlled by physical processes.

The magnitudes of the total storage and total active pools can be very different. In many cases, the active pools are much smaller than storage pools, and materials move more rapidly between active pools than between storage pools or in and out of storage. Taking an example from the carbon cycle, photosynthesis and respiration will cycle all the carbon dioxide in the atmosphere (active pool) through plants in about 10 years. But it may be many millions of years before the carbonate sediments (storage pool) now forming as rock will be uplifted and decomposed to release carbon dioxide.

Nutrient Elements in the Biosphere

We will take the 15 most abundant elements in living matter as a whole and designate their total mass as 100 percent. Percentages of each of the elements are given in Table 12.2. The three principal components of carbohydrate — hydrogen, carbon, and oxygen — account for almost all living matter and are called **macronutrients.** The remaining one-half percent is divided among 12 elements. Six of these are also macronutrients: nitrogen, calcium, potassium, magnesium, sulfur, and phosphorus. The macronutrients are all required in substantial quantities for organic life to thrive. The first three macronutrients—hydrogen, carbon, oxygen—are materials whose pathways we have already followed in the photosynthesis-respiration circuits (Figure 12.2). We will undertake a more detailed analysis of the gaseous cycles of carbon and oxygen because of their involvement with atmospheric changes induced by Man through the combustion of hydrocarbon compounds. We will also give special attention to nitrogen, the fourth most abundant element in the composition of living matter; it also moves in a gaseous cycle. Of the remaining macronutrients, three — calcium, potassium, and magnesium — are elements derived from silicate rocks through mineral weathering. Two other macronutrients derived from rock weathering are sulfur and phosphorus.

Quite a number of additional elements, not among the nine macronutrients, are also vital to life processes. Their presence is needed in mere traces and, for this reason, they are called **micronutrients.** The micronutrient list includes iron, copper, zinc, boron, molybdenum, manganese, and chlorine.

Figure 12.10 **General features of a material cycle.**

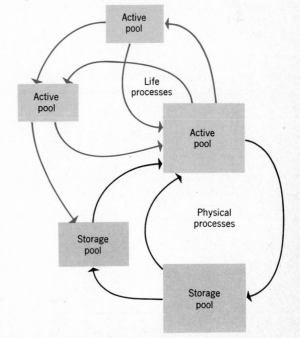

Table 12.2 / Elements Comprising Global Living Matter, Taking 100 Percent as the Total of the 15 Most Abundant Elements.

Basic carbohydrate

Hydrogen (M)	49.74
Carbon (M)	24.90
Oxygen (M)	24.83
	Subtotal 99.47

Other nutrients

Nitrogen (M)	0.272
Calcium (M)	0.072
Potassium (M)	0.044
Silicon	0.033
Magnesium (M)	0.031
Sulfur (M)	0.017
Aluminum	0.016
Phosphorus (M)	0.013
Chlorine	0.011
Sodium	0.006
Iron	0.005
Manganese	0.003
	M - macronutrient

Data source: E. S. Deevey, Jr. (1970), *Scientific American*, Vol. 223.

The Carbon Cycle

The movements of carbon through the life layer are of great importance because all life is composed of carbon compounds of one form or another. Of the total carbon available, most lies in storage pools as carbonate sediments below the earth's surface. Only about two-tenths of a percent are readily available to organisms as CO_2 or as decaying biomass in active pools.

Some details of the **carbon cycle** are shown in a schematic diagram, Figure 12.11. In the gaseous portion of its cycle, carbon moves as carbon dioxide (CO_2) as a free gas in the atmosphere and as a gas dissolved in fresh water of the lands and in salt water of the oceans. In the sedimentary portion of its cycle, carbon resides in carbohydrate molecules in organic matter, as hydrocarbon compounds in rock (petroleum, coal), and as mineral carbonate compounds such as calcium carbonate ($CaCO_3$). The world supply of atmospheric carbon dioxide is represented in Figure 12.11 by a box. It is a small portion of the carbon in active pools, constituting less than 2 percent. This atmospheric pool is supplied by respiration from plant and animal tissues in the oceans and on the lands. Under natural conditions, some new carbon enters the atmosphere each year from volcanoes by outgassing in the form of CO_2 and carbon monoxide (CO). Man's industrial role is to inject substantial amounts of carbon into the atmosphere through combustion of fossil fuels. This Man-made increment and its probable effects on global air temperatures were discussed in Chapter 5.

Carbon dioxide leaves the atmospheric pool to enter the oceans, where it is used in photosynthesis by minute marine plants, the phytoplankton. These organisms are primary producers in the ocean ecosystem and are consumed by marine animals in the food chain. Phytoplankton also build skeletal structures of calcium carbonate. This mineral matter settles to the ocean floor to accumulate as sedimentary strata, an enormous storage

Figure 12.11 **The carbon cycle. (After A. N. Strahler, 1972, *Planet Earth*, Harper and Row, New York.)**

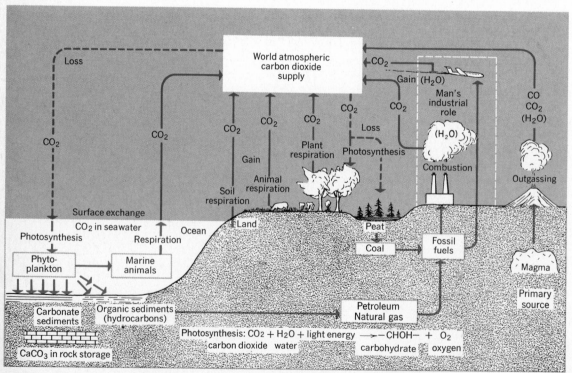

pool not available to organisms until released later by rock weathering. Organic compounds synthesized by phytoplankton also settle to the ocean floor and eventually are transformed into the hydrocarbon compounds making up petroleum and natural gas. On the lands accumulating plant matter has, under geologically favorable circumstances, formed massive layers of peat, which were transformed into coal. Petroleum, natural gas, and coal comprise the fossil fuels, and these represent huge storage pools of carbon.

The Oxygen Cycle

Details of the **oxygen cycle** are shown in schematic form in Figure 12.12. The complete picture of the cycling of oxygen also includes its movements and storages when combined with carbon in carbon dioxide and in organic and inorganic compounds. These we have covered in the carbon cycle.

The world supply of atmospheric free oxygen is shown in Figure 12.12 by a box at the top of the diagram. Oxygen enters this storage pool through release in photosynthesis, both in the oceans and on the lands. Each year a small amount of new oxygen comes from volcanoes through outgassing, principally as CO_2 and H_2O (shown in Figure 12.11). Balancing the input to the atmospheric storage pool is loss through organic respiration and mineral oxidation. Adding to the withdrawal from the atmospheric oxygen pool is Man's industrial activity through the combustion (oxidation) of wood and fossil fuels. Forest fires and grass fires (not shown) are another means of oxygen consumption. The oceans also contain a

small active pool of dissolved gaseous oxygen. Some oxygen is continuously placed in storage in mineral carbonate form in ocean-floor sediments.

Man reduces the amount of oxygen in the air by (1) burning fossil fuels; (2) clearing and draining land, which speeds the oxidation of soils and soil organic matter; and (3) reducing photosynthesis by clearing forests for agriculture and by paving and covering previously productive surfaces. The importance of urbanization can be appreciated by the fact that every six months a land area about the size of Rhode Island is covered by new construction in the United States alone. In fact, calculations have shown that the oxygen consumed in the continental United States in 1966 exceeded the amount produced by nearly 70 percent. Fortunately, the oxygen pool is so large that Man's impact or potential impact is very small, at least at this time.

The Nitrogen Cycle

Nitrogen moves through the biosphere in a gaseous **nitrogen cycle** in which the atmosphere, containing 78 percent nitrogen by volume, is a vast storage pool (Figure 12.13). Nitrogen in the atmosphere, in the form N_2, cannot be assimilated directly by plants or animals. Only certain microoganisms possess the ability to utilize N_2 directly, a process termed **nitrogen fixation.** One class of such microorganisms consists of certain species of free-living soil bacteria. Some blue-green algae can also fix nitrogen.

Another class consists of the symbiotic nitrogen fixers. In a symbiotic relationship, two species of organisms live in close physical contact, each contributing to the life

Figure 12.12 **The oxygen cycle. (After A. N. Strahler, 1972, *Planet Earth*, Harper and Row, New York.)**

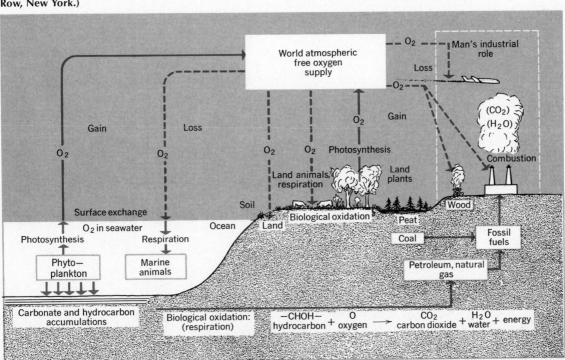

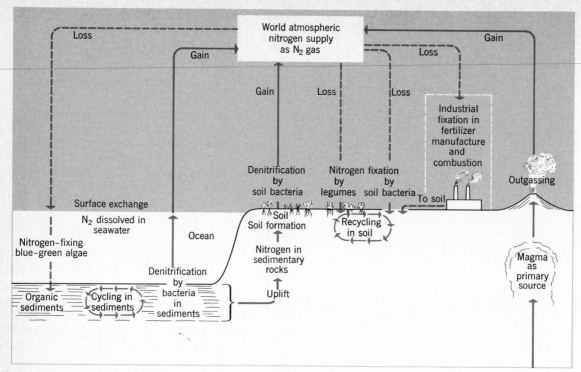

Figure 12.13   **The nitrogen cycle.**

processes or structures of the other. Symbiotic nitrogen fixers are bacteria of the genus *Rhizobium;* they are associated with some 190 species of trees and shrubs as well as almost all members of the legume family. Legumes important as agricultural crops are clover, alfalfa, soybeans, peas, beans, and peanuts. *Rhizobium* bacteria infect the root cells of these plants in root nodules produced jointly by action of the plant and the bacteria. The bacteria supply the nitrogen to the plant through nitrogen fixation, while the plant supplies nutrients and organic compounds needed by the bacteria. Crops of legumes are often planted in seasonal rotation with other food crops to ensure an adequate nitrogen supply in the soil. Both the action of nitrogen-fixing crops and of soil bacteria are shown in the nitrogen cycle diagram (Figure 12.13).

Nitrogen is lost to the biosphere by **denitrification,** a process in which certain soil bacteria convert nitrogen from usable forms back to N_2. This process is also shown in the diagram. Denitrification completes the organic portion of the nitrogen cycle, as nitrogen returns to the atmosphere.

At the present time, nitrogen fixation is far exceeding denitrification, and usable nitrogen is accumulating in the life layer. This excess of fixation is produced almost entirely by human activities. Man fixes nitrogen in the manufacture of nitrogen fertilizers and by oxidizing nitrogen in the combustion of fossil fuels. Widespread cultivation of legumes has also greatly increased worldwide nitrogen fixation. At present rates, Man-induced nitrogen fixation nearly equals all natural biological fixation.

Much of the excess nitrogen fixed by human activities

is carried from the soil into rivers and lakes and ultimately reaches the ocean. Major water pollution problems can arise when nitrogen stimulates the growth of algae and phytoplankton, which can be detrimental to desirable forms of aquatic life. These problems will be accentuated in years to come because industrial fixation of nitrogen in fertilizer manufacture is doubling about every 6 years at present. Just what impact such large amounts of nitrogen reaching the sea will have on the earth's ecosystem remains uncertain.

Sedimentary Cycles

The oxygen, carbon, and nitrogen cycles are all referred to as gaseous cycles because they possess a gaseous phase in which the element involved is present in significant quantities in the atmosphere. Many other elements move in sedimentary cycles, that is, from land to ocean in running water, returning after millions of years as uplifted terrestrial rock. These elements are not present in the atmosphere except in small quantities as blowing dust or condensation nuclei in precipitation.

Figure 12.14 shows how some important macronutrients move in sedimentary cycles. Within the large box representing the lithosphere are smaller compartments representing the parent matter of the soil and the soil itself. In the soil, nutrients are held as ions on the surfaces of soil colloids and are readily available to plants. (This will be explained in the following chapter.)

The nutrient elements are also held in enormous storage pools, where they are unavailable to organisms. These storage pools include seawater (unavailable to land

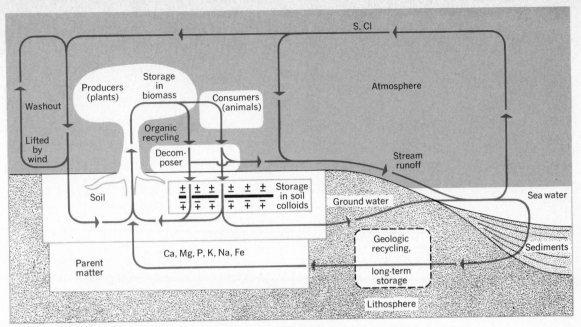

Figure 12.14 **A flow diagram of the sedimentary cycle of materials in and out of the biosphere and within the inorganic realm of the lithosphere, hydrosphere, and atmosphere.**

organisms), sediments on the seafloor, and enormous accumulations of sedimentary rock beneath both lands and oceans. Eventually, elements held in the geologic storage pools are released into the soil by weathering. Soil particles are lifted into the atmosphere by winds and fall back to earth or are washed down by precipitation. Chlorine and sulfur are shown as passing from the ocean into the atmosphere and entering the soil by the same mechanisms of fallout and washout.

The organic realm, or biosphere, is shown in three compartments: producers, consumers, and decomposers. Considerable element recycling occurs between organisms of these three classes and the soil. The elements used in the biosphere, however, are continually escaping to the sea as ions dissolved in stream runoff and groundwater flow.

Agricultural Ecosystems

The principles of energy use and flow in natural ecosystems also apply to **agricultural ecosystems.** Important differences exist, however, between natural ecosystems and those that are highly managed for agriculture. The first major difference is the reliance of agricultural ecosystems on inputs of energy that are ultimately derived from fossil fuels. The most obvious of these inputs is the fuel that runs the machinery used to plant, cultivate, and harvest the crops. Another is the application of fertilizers and pesticides, which have required large expenditures of fuel to extract or synthesize and to transport. Fossil fuels also indirectly power such activities as the breeding of plants that have higher yields and are resistant to disease and the development of new chemicals to combat insect pests. Fuel is expended in

transporting crops to distant sources of consumption, thus enabling large areas of similar climate and soils to be used for the same crop. In these and many other ways, the high yields obtained today are brought about only at the cost of a large energy input derived from fossil fuels.

A schematic flow diagram, Figure 12.15, shows how fuel-energy input enters into both the purchased inputs and the operations performed on the farm to raise and harvest crops. Solar energy of photosynthesis is, of course, a "free" input. But, to be delivered to the humans or animals that consume it, the raw food or feed product of this flow system requires the expenditure of fossil fuel energy.

Agricultural ecosystems, unlike natural ecosystems, are simple in structure and function. They often consist of one genetic strain of one species. Such ecosystems are overly sensitive to attacks by one or two well-adapted insects that can multiply rapidly to take advantage of an abundant food source. Thus pesticides are constantly needed to reduce insect populations. Weeds, too, are a problem, adapted as they are to rapid growth on disturbed soil in sunny environments. Weeds can divert much of the productivity to undesirable forms. Herbicides are often the immediate solution to these problems.

In natural ecosystems, nutrient elements are returned to the soil following the death of the plants that concentrate them. In agricultural ecosystems, this recycling is usually interrupted by harvesting the crop for consumption at a distant location. Continuous removal of nutrients in crop biomass introduces a new pathway in the nutrient cycle. To preserve soil fertility, still another pathway must be introduced — from the fertilizer plant to the farm. Keeping this new nutrient subcycle functioning requires considerable fossil fuel energy input.

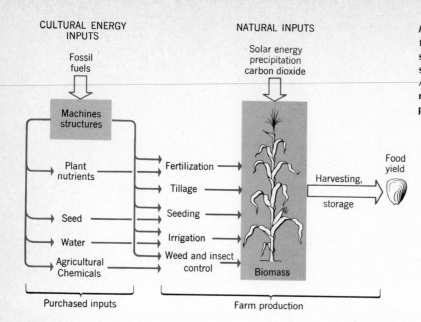

CULTURAL ENERGY
INPUTS

NATURAL INPUTS

Figure 12.15 A schematic diagram of the inputs of cultural energy into various stages in the agricultural production system. (After G. H. Heichel, 1976, Agricultural production and energy resources, *American Scientist,* vol. 64, p. 65.)

Productivity and Efficiency of Agricultural Ecosystems

The inputs of energy that Man adds to managed ecosystems in the form of agricultural chemicals and fertilizers, as well as those in the form of the work of farm machinery, have acted to boost greatly the net primary productivity of the land. Table 12.3 gives some examples of how agricultural yields have responded to these inputs. (Units used in the table are the energy equivalents of the biomass.) The table shows that Man has acted to raise the net primary productivity of agricultural ecosystems more than five times over through the use of energy contained in fossil fuels.

As the United States, in company with many other nations, enters a period of apprehension over the future adequacy of energy resources to meet growing demands, we can do well to evaluate the level of efficiency of our agricultural system in delivering food energy in return for the energy we expend on that production. The total U.S.

energy consumption in 1974 is estimated to have amounted to 18,900 × 10¹² kilocalories per year (kcal/yr). Table 12.4 gives a breakdown of the proportions of the total energy consumed by U.S. agriculture. The total agricultural energy use of 498 × 10¹² kcal/yr represented 2.6 percent of the total U.S. energy consumption. This agricultural energy use is about 10 percent of the amount expended in fuel for transportation and about 14 percent of that expended for the heating of buildings.

Energy consumed in agriculture is subdivided in the table into two major categories: (1) petroleum and electricity expended on farms and (2) energy represented by materials expended and equipment used on farms. Petroleum expended as fuel on the farm amounted to 51

Table 12.3 / Crop Productivity and Efficiency with and without Fossil Fuel Energy Subsidy

Crop	Net Productivity, as Harvested (kcal/m²/day)	Efficiency of Primary Production (percent)
Without fossil fuel energy inputs:		
Grain, Africa, 1936	0.72	0.02
All U.S. farms, 1880	1.28	0.03
With fossil fuel energy inputs:		
Grain, North American average, 1960	5	0.12
Rice, U.S., 1964	10	0.25

Data source: Data from several sources, compiled in H. T. Odum, 1971, *Environment, power and society,* Wiley-Interscience, New York, p. 116, Table 4.1.

Table 12.4 / Energy Consumed by Agriculture in the United States

	10¹² Kcal	Percent of total U.S. energy budget
Farm production		
Crops		
Petroleum	189	1.0
Electricity	3	0.02
Livestock		
Petroleum	63	0.3
Electricity	9	0.05
Purchased inputs:		
Fertilizer	140	0.7
Petroleum	49	
Feeds and additives	25	
Animal and marine oils	9	0.5
Farm machinery	8	
Pesticides	3	
Total	498	2.6

Data source: Economic Research Service, 1974, as presented by G. H. Heichel, 1976, *American Scientist,* vol. 64, p. 64.

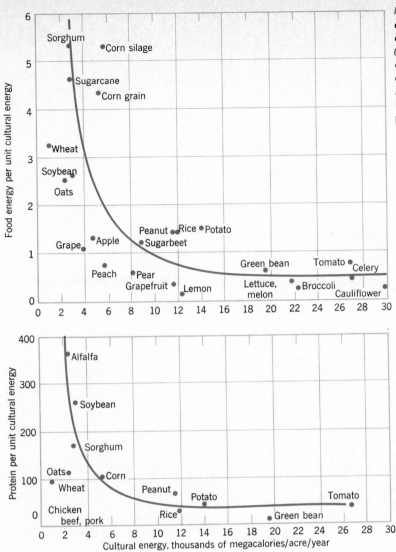

Figure 12.16 (upper graph) Cultural energy used to produce certain food crops in relation to yield of food energy. (lower graph) Protein yield in relation to cultural energy for several kinds of crops. (After G. H. Heichel, 1976, Agricultural production and energy resources, *American Scientist*, vol. 64, p. 66.)

percent of the total agricultural expenditure. (Petroleum consumption listed under "purchased inputs" represents fuel consumed in the manufacture of those input materials.) Fertilizer accounted for about 28 percent of the total agricultural energy used, whereas pesticides used a very small proportion, about 0.6 percent.

Excluding the solar energy of photosynthesis, energy expended on the production of a raw food or feed crop is referred to as **cultural energy.** Let us examine the cultural energy input needed to produce various feed and food crops. The upper graph in Figure 12.16 shows the relative efficiency of food production of 24 crops. The horizontal scale shows cultural energy expended in terms of thousands of megacalories per acre per year (mcal/acre/yr). The vertical scale shows the ratio of food energy produced to input of cultural energy; the higher the ratio, the greater the efficiency (and the lower the cultural input required). Notice that field crops, such as sorghum and corn, which are used largely as animal feeds, have the highest levels of efficiency, whereas foods consumed directly by humans have relatively low efficiencies.

Garden crops require a very high cultural-energy input per unit area of land and also have very low efficiencies.

The lower graph in Figure 12.16 shows, on the vertical scale, the protein derived from each unit of cultural energy. Alfalfa and soybeans rate very high on this scale. (Soybean protein is presently being processed in various forms for human diet as a meat substitute.) Notice that oats, wheat, and corn have intermediate values of protein yield, but that rice ranks very low. (Protein deficiency is a serious health problem for peoples subsisting largely on rice.) A small area of the graph at the lower left is labeled "chicken, beef, pork." This insertion serves to show that protein obtained from meat has an energy efficiency only about one-tenth as great as that from soybeans. Indeed, the energy available from edible meat represents only about 10 percent of the energy expended in animal feed, a fact previously pointed out in our discussion of energy flow in the food chain.

The data thus indicate clearly that our food production system is not efficient in terms of cultural energy expended to furnish food to humans. The most highly coveted part of the American diet, meat and garden vegetables, is extremely wasteful of cultural energy as compared with, say, diets based largely on grain foods (bread, breakfast cereals) and soybean products.

Our discussion of agricultural ecosystems has emphasized their general nature and how they differ from natural ecosystems. We will return to a fuller discussion of natural ecosystems in Chapters 15 and 16, which will emphasize the diversity in structure of primary producers as related to climate and soils. Our closing chapters will return to the theme of agricultural ecosystems and their potential for human use. A complete understanding of agricultural ecosystems, however, will also require a thorough examination of soils and soil-forming processes, and these are the subject of our next chapter.

References for Further Study

Strahler, A. N., and Strahler, A. H. (1974), *Introduction to environmental science,* Hamilton Publishing Co., Santa Barbara, Calif. (John Wiley and Sons, New York) See Chapters 20, 21.

Odum, E. P. (1971), *Fundamentals of ecology,* W. B. Saunders Co., Philadelphia.

Scientific American (1970), The biosphere, W. H. Freeman and Co., San Francisco.

Review Questions

1. Define the science of ecology. What is an ecosystem? Why are geographers interested in ecosystems?
2. How does energy move through the food chain? Describe the levels of energy flow in an ecosystem and use the salt marsh as an example. How does respiration limit energy flow up the food chain? What implications does this have for Man?
3. Compare the processes of photosynthesis and respiration. How are photosynthesis and respiration linked in a cycle with carbon, hydrogen, and oxygen?
4. What is net photosynthesis? How is it dependent on light intensity and temperature?
5. What is net primary production? Compare the productivities of various natural ecosystems as they are influenced by climate.
6. Define a material cycle. What types of material cycles are recognized? Describe the types of pools within a material cycle and how flows in and out of pools are controlled.
7. What is a macronutrient? Which elements are considered macronutrients? Which are micronutrients?
8. Describe the flows and pathways of carbon within the carbon cycle. How does Man impact the carbon cycle?
9. Describe the flows and pathways of oxygen in the oxygen cycle. What human activities reduce the amount of free atmospheric oxygen?
10. Describe the flows and pathways of nitrogen in the nitrogen cycle, including nitrogen fixation and denitrification. How does Man affect the nitrogen cycle?
11. What are the main features of a sedimentary cycle? How does a sedimentary cycle differ from a gaseous material cycle?
12. Contrast natural and agricultural ecosystems. In what ways are agricultural ecosystems dependent on cultural energy? What are the most important energy inputs to U.S. agriculture?
13. Which crops need lesser inputs of cultural energy? Which need more? Which crops yield more protein per unit of cultural energy invested? What are the implications of differing efficiencies for the survival of Man?

Soil-Forming Processes

13

The soil is the very heart of the life layer on the lands. The soil layer is a place in which plant nutrients are produced and held. As we emphasized in Chapter 9, the soil layer also holds water in storage for plants to use. In Chapter 10 we explained that the role of climate is to vary the input of water and heat into the soil. This same heat energy and water are responsible for the breakup and chemical change of rock to produce the parent mineral matter of the soil (Chapter 11). That mineral matter is the source of many plant nutrients.

But climate acting on rock cannot make a soil layer capable of sustaining a rich plant cover. Plants themselves, together with many forms of animal life, play a major role in determining the qualities of the soil layer. Those qualities have evolved through centuries by the inter-action of organic processes with physical and chemical soil processes. The organic processes include the synthesis of organic compounds, and these are eventually added to the body of the soil. Plants use the mineral nutrients to build complex organic molecules. Upon the death of plant tissues, these nutrients are released and reenter the soil, where they are reused by living plants. So the familiar concept of nutrient cycling by plants is one of the keys to understanding the development of the soil layer.

The Dynamic Soil

The soil is a dynamic layer in the sense that many complex physical and chemical activities are going on simul-taneously within it. Because climate and plant cover vary greatly from place to place over the globe, the combined effects of soil-forming activities are expressed differently from place to place. Anyone can observe that the pale gray soil beneath a spruce forest in Maine is quite different in composition and structure from the dark brown soil beneath the prairie farmlands of Iowa. But, in each of these localities, the soil has reached a physical and chemical state related to the climate controls and soil-forming processes prevailing there.

The geographer is keenly interested in the differences

in soils from place to place. The capability of a given soil type to furnish food crops largely determines which areas of the globe support the bulk of the human population. Despite changes in population distribution made possible by technology and industrialization, most of the world's inhabitants still live where the soil furnishes them food. Many of those same humans die prematurely because the soil does not furnish adequate nutrition for all who need it.

The substance of the soil exists in all three states—solid, liquid, and gas. The solid portion consists of both inorganic (mineral) and organic substances. The liquid present in the soil is a complex solution capable of engaging in a multitude of important chemical reactions. Gases present in the open pores of the soil consist not only of the atmospheric gases, but also of gases liberated by biological activity and chemical reactions within the soil. Soil science, often called **pedology,** is obviously a highly complex body of knowledge. We shall do no more than cover a few of the high points of this science.

The Nature of Soil

We have used the word "soil" in previous chapters without giving it a precise definition. **Soil,** as the term is used in soil science, is a natural surface layer containing living matter and supporting or capable of supporting plants. Substance of the soil includes both inorganic (mineral) matter and organic matter, the latter both living and dead. Living matter in the soil consists not only of plant roots, but of many kinds of organisms, including microorganisms. The upper limit of the soil is air or shallow water. Horizontal limits of the soil may be deep water or barren areas of rock or ice.

The lower limit of the soil is often difficult to define in simple and exact terms. Nonsoil below the soil may be bedrock or any one of the soil parent materials described in Chapter 11 (regolith and various kinds of sediment accumulations) devoid of living roots and other signs of biologic activity. In contrast, soil shows evidences of biologic activity.

Soils show soil horizons, which are distinctive horizontal layers set apart from other soil zones or layers by differences in physical and chemical composition, organic content, structure, or a combination of those properties. Soil horizons are developed by the interactions, through time, of climate, living organisms, and the configuration of the land surface (relief).

The word "soil" is used by civil engineers and geologists to mean any surface layer of unconsolidated mineral matter of low strength as compared with strong, hard bedrock. Any of the parent materials we described in Chapter 11, and some nonlithified sedimentary rocks as well, are referred to as "soil" by the civil engineer. Our definition of soil excludes surficial materials that do not support the growth of plants.

Although most classes of soil—those with horizons—require a long span of time in which to develop, a layer capable of supporting plants can come into existence very rapidly. An example would be an accumulation of silt on a river floodplain. Here the parent matter may include organic matter and nutrients formed elsewhere and transported to the new location. Generally, the transformation of raw parent matter into a soil with horizons requires one to two centuries at the least. The time required for a soil to reach an equilibrium state with the environment is usually estimated in thousands of years.

Concept of the Pedon*

Modern soil science makes use of the concept of the **polypedon,** which is the smallest distinctive division of the soil of a given area. A unique single set of properties applies to the polypedon, and this set differs from that applying to adjacent polypedons. The polypedon is conceived in terms of space geometry as being composed of pedons. A **pedon** is a soil column extending down from the surface to reach a lower limit in some form of regolith or bedrock. As Figure 13.1 shows, soil scientists often visualize a pedon as a six-sided (hexagonal) column. The surface area of a single pedon ranges from 1 to 10 sq m (10 to 100 sq ft). The **soil profile** is the display of horizons on one face of the pedon. Obviously, the same soil profile is displayed on all six faces of the pedon. In practice, a soil scientist digs a deep pit, exposing a soil profile in the side of the pit.

Figure 13.1 shows a number of soil horizons. Most horizons are visibly set apart on the basis of color or texture. Mineral soil horizons are designated by a set of capital letters and numeral subscripts, starting with A at the top. In Figure 13.1 we see A, B, and C horizons. An organic horizon, designated by the letter O, lies on the A horizon. (See also Figure 13.6.)

The **soil solum** consists of the A and B horizons of the soil profile; these are the dynamic and distinctive layers of

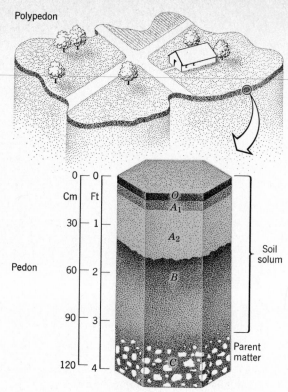

Figure 13.1 **Concept of the pedon and polypedon.**

the soil. The C horizon, by contrast, is the parent material. The soil solum occupies the zone in which living plant roots exert control on the soil horizons; the C horizon lies below that level of root activity.

Soil Color

For persons unfamiliar with soil science, color—easily perceived at a distance—is the most obvious property of a soil. The dark brown to black color of the soil is conspicuous to the traveler in farmlands of Iowa and Nebraska; the pervasive red color of soil in the Piedmont upland of Georgia does not escape notice. Certain color relationships are quite simple. Black color usually indicates the presence of abundant organic matter (humus); red color usually indicates the presence of sesquioxide of iron (hematite). The soil color may in some areas be inherited from the parent matter; but, more generally, it is a property generated by the soil-forming processes.

Description of soil color has been placed on an objective basis through the use of books of standard (Munsell) colors adapted to the needs of soil science. Three measurable variables determine color. One is the hue, or dominant color of the pure spectrum, depending on wavelength (see Figure 4.3). A second variable is value, the degree of darkness or lightness of the color. The third is chroma, the purity or strength of the spectral color. By using standard color books, the observer can express soil color as a letter-numeral code, telling hue, value, and chroma.

*Throughout this chapter and the next, numerous phrases and sentences have been taken verbatim from the following source: Soil Survey Staff (1975), *Soil Taxonomy,* Soil Conservation Service, U.S. Dept. of Agriculture, Agriculture Handbook No. 436, Government Printing Office, Washington, D.C.

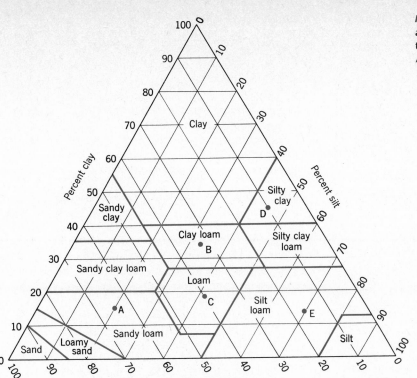

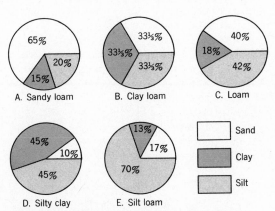

Figure 13.3 **Typical compositions of five soil texture classes. These examples are shown as lettered points on Figure 13.2. (U.S. Department of Agriculture.)**

Soil Texture Classes

A scale of grade sizes of mineral particles was explained in Chapter 11 (see Table 11.2). The same grade scale is used in describing soil texture. **Soil-texture classes** are based on varying proportions of sand, silt, and clay, expressed as percentages. The standard system of classes used by the U.S. Department of Agriculture is shown in a triangular diagram, Figure 13.2. Percentages of all three components are shown simultaneously. The corners of the triangle represent 100 percent of each of the three grades of particles—sand, silt, or clay. **Loam** is a mixture in which no one of the three grades dominates over the other two. Loams therefore appear in the central region of the triangle. A particular soil whose components give it a

position at Point A in the triangle has 65 percent sand, 20 percent silt, and 15 percent clay; it falls into a texture class known as *sandy loam*. Another soil, whose texture is represented by Point B has 33⅓ percent sand, 33⅓ percent silt, and 33⅓ percent clay; it falls into the class of a *clay loam*. Figure 13.3 gives five examples of soil textures; their positions are shown on Figure 13.2.

Texture is important because it largely determines the water retention and transmission properties of the soil. Sand may drain too rapidly; in a clay soil, the individual pore spaces are too small for adequate drainage. Where clay and silt content is high, root penetration may be difficult.

Recall from Chapter 10 that storage capacity (field capacity) of a soil is its capacity to hold water against the pull of gravity. Figure 13.4 shows how storage capacity varies with soil texture. Pure sand holds the least water, while pure clay holds the most. Loams hold intermediate amounts. Sand transmits the water downward most rapidly, clay most slowly. When planning the quantity of irrigation water to be applied, these factors must be taken into account. Sand reaches its full capacity very rapidly, and added water is wasted. Clay-rich loams take up water very slowly and, if irrigation is too rapid, water will be lost by surface runoff. By the same token, sandy soils require more frequent watering than clay-rich soils. The organic content of a soil also strongly affects its water-holding capacity. The intermediate loam textures are generally best as agricultural soils because they drain well, but also have favorable water-retention properties.

Soil texture is largely an inherited feature of a given soil and depends on the composition of the parent matter. Texture of parent matter was stressed in Chapter 11. Some

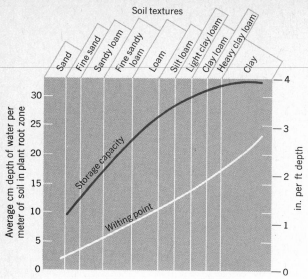

Figure 13.4 Storage capacity and wilting point vary according to soil texture.

kinds of parent matter furnish a large spread in particle sizes; others yield mostly sand or mostly clay.

Agricultural soil scientists also use a measure of soil-water storage termed the **wilting point.** Soil water in an amount less than the value at the wilting point cannot be absorbed by plants rapidly enough to meet their needs. At this point, the foliage of plants not adapted to drought will wilt. As Figure 13.4 shows, the wilting point depends on texture.

Soil Consistence

Soil consistence refers to the quality of stickiness of wet soil and to the plasticity of moist soil, as well as the degree of coherence or hardness of the soil when it holds small amounts of moisture or is in the dry state. Stickiness of a wet soil is evaluated by pressing a quantity of soil between thumb and finger, then separating the digits and observing the extent to which soil adheres to the skin. Plasticity is evaluated by rolling a small amount of wet soil into a rod shape. If plasticity is high, the soil can be rolled into a thin wire. Consistence when a soil is dry is expressed by various levels of hardness, ranging from loose (noncoherent) to extremely hard. Cementation, which usually occurs in a particular horizon, is not affected by wetting and ranges from weakly cemented material, easily broken in the hands, to an indurated condition resembling hard rock. Cementation is caused by the accumulation of mineral substances, such as calcium carbonate, silica, or iron oxides.

Soil Structure

Soil structure refers to the presence of aggregations (lumps or clusters) of soil particles. Each aggregate is separated from adjoining aggregates by natural surfaces of weakness (cracks). In some instances, the aggregates are coated with

surface films of material that help keep them apart. In other cases, aggregates are simply held together by forces of internal cohesion. An individual natural soil aggregate is called a **ped.** (An aggregate caused by breakage during plowing is a *clod.*) Soil structure is described in terms of the shape, size, and durability of peds.

Four primary types of soil structure are recognized: platy, prismatic, blocky, and spheroidal. These are illustrated in Figure 13.5. Platy soil structure consists of plates—thin flat pieces—in a horizontal position. In prismatic structure, peds are formed into vertical columns, often flat-sided, which may be 0.5 to 10 cm (0.2 to 4 in.) across. Blocky structure consists of angular, equi-dimensional peds with flattened surfaces that fit the surfaces of adjacent peds. Spheroidal structure consists of peds more or less rounded in outline with surfaces that do not fit those of adjacent peds. In the granular variety of spheroidal structure, shown in Figure 13.5, the peds are small and the soil is very porous.

A description of soil structure includes not only the form of the peds, but also their sizes (fine, medium, coarse) and their degree of durability (weak, strong). Soil structure is a physical property of great agricultural importance because it influences the ease with which water will penetrate a dry soil, the susceptibility of the soil to erosion, and the ease of cultivation.

Related to soil structure is the presence of thin films or coatings, called **cutans** (skins), on soil peds or individual coarse mineral grains. Cutans of clay (argillans) are clay skins coating the outer surfaces of peds or sand grains (Plate E.4). These coatings consist of clay particles carried down through the soil by infiltrating water. Other cutan types include thin coatings of oxides of iron or manganese on mineral grains, and films of organic matter.

Also important in soil structure is the nature of the void spaces between peds. Size and degree of interconnection of voids are important in determining the ease with which water and air move through the soil.

Soil Horizons

Soil horizons range greatly in thickness and distinctness. In some cases the upper and lower boundaries of a horizon are sharply defined; in others, the boundaries are gradual or diffuse (Plates E.2 and E.3). In general, soil horizons are of two major classes: organic horizons and mineral horizons.

Organic horizons, designated by the capital letter *O,* overlie the mineral horizons and are formed of accumulations of organic matter derived from plants and animals (Figure 13.6). Typically, the uppermost organic horizon, designated as O1, consists of vegetative matter in original forms recognizable with the unaided eye. Beneath the O1 horizon lies the O2 horizon, which consists of altered remains of parts of plants and animals not recognizable with the naked eye. Material of the O2 horizon is referred to as **humus;** it consists largely of plant tissues partly oxidized by consumer organisms (Chapter 12). The process by which the O2 horizon is produced has been called **humification.**

Mineral horizons consist predominantly of inorganic

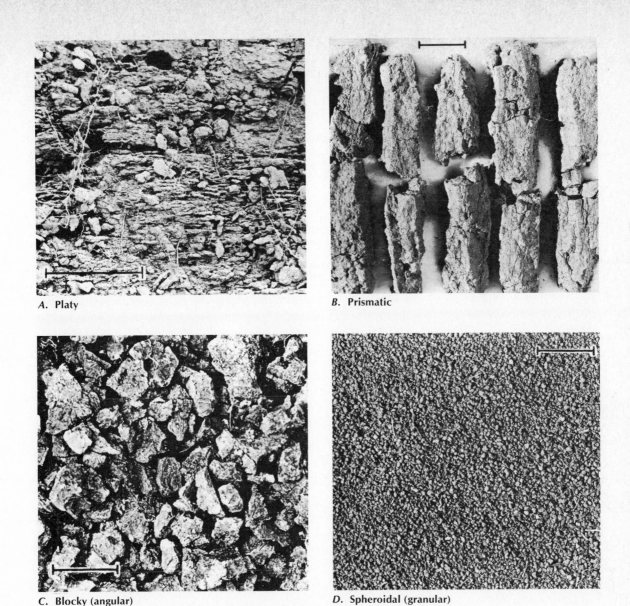

A. Platy

B. Prismatic

C. Blocky (angular)

D. Spheroidal (granular)

Figure 13.5 **Four basic soil structures are illustrated. The black bar on each photograph represents 2.5 cm (1 in.). (Division of Soil Survey, U.S. Dept. of Agriculture.)**

mineral matter, of which two basic groups are recognized: (1) skeletal minerals and (2) clay minerals and related weathering products. The **skeletal minerals,** mostly in the form of particles of sand and silt grades, comprise the bulk of most soils. Skeletal materials may consist of individual grains of a single mineral—quartz, for example—or grains that are aggregates of several minerals. The clay minerals and related alteration products are those described in Chapter 11. They form the mineral soil fraction most important in soil-forming processes, in the development of horizons, and in determining the natural fertility of soils. The clay minerals have special physical and chemical properties because of their colloidal size and because of the platelike shape of the individual clay particles.

Mineral horizons are designated by the letters *A* and *B,* with various subdivisions designated by adding numerals

to the letters. Our emphasis in this description of the properties of the A and B horizons is on soils of moist climates formed under forest cover. Our description would not apply closely to soils of semiarid and arid climates. Mineral horizons have less than 20 percent organic matter when no clay is present, less than 30 percent organic matter when the mineral fraction consists of 50 percent or more clay. The A horizon is often subdivided into two subhorizons: A1 and A2.

The *A1* horizon is usually rich in organic matter and is therefore usually darker than the A2 horizon below. The *A2* horizon is characterized by loss of clay minerals and of oxides of iron and aluminum. A concentration of quartz grains of sand or coarse silt grade usually remains, and the horizon is often pale.

The B horizon typically shows a gain of mineral matter,

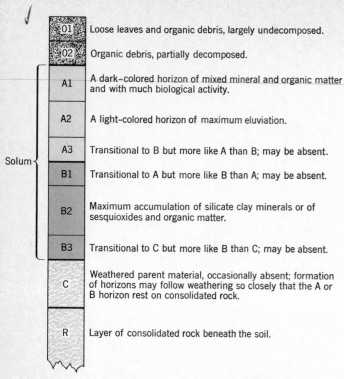

O1	Loose leaves and organic debris, largely undecomposed.
O2	Organic debris, partially decomposed.
A1	A dark-colored horizon of mixed mineral and organic matter and with much biological activity.
A2	A light-colored horizon of maximum eluviation.
A3	Transitional to B but more like A than B; may be absent.
B1	Transitional to A but more like B than A; may be absent.
B2	Maximum accumulation of silicate clay minerals or of sesquioxides and organic matter.
B3	Transitional to C but more like B than C; may be absent.
C	Weathered parent material, occasionally absent; formation of horizons may follow weathering so closely that the A or B horizon rest on consolidated rock.
R	Layer of consolidated rock beneath the soil.

Figure 13.6 **Designations of horizons of a hypothetical soil profile that might represent a forest soil in a cool, moist climate. (Roy W. Simonson, Soil Conservation Service, U. S. Department of Agriculture.)**

which may come from the A1 and A2 horizons above. High concentrations of clay minerals, oxides of iron and aluminum, and organic matter (humus) are often found in the B horizon. Thus the B horizon is typically less friable than the A horizons; it may be dense and tough, and cementation may also occur.

The C horizon, lying beneath the B horizon, is a mineral layer of regolith or sediment (but not bedrock), little affected by biologic activity. The C horizon is not part of the soil solur, and is described simply as the layer of parent material. The C horizon is, however, affected by physical and chemical processes. An example is the accumulation of calcium carbonate in dry climates, which causes cementation in some soils. In these environments, the C horizon may also show accumulations of silica or soluble salts. Bedrock underlying the C horizon (or the B horizon when the C horizon is missing) is designated the R horizon.

To indicate special characteristics of a soil horizon, a number of lowercase letters are used after the capital letters A or B. Here are a few examples:

b   buried horizon
ca   accumulation of carbonate matter
f   soil frozen
h   accumulation of humus
ir   accumulation of iron
hir accumulation of humus and iron
p   horizon disturbed by plowing
si   accumulation of silica
t   accumulation of translocated clay
x   brittle layer (fragipan)

The Soil Solution

Both air and water combine to form the **soil solution,** which comprises the environment for chemical reactions affecting the solid fraction of the soil. The soil atmosphere consists basically of air that enters pore spaces in the soil, diffusing into all interconnected openings. Fluctuations in barometric pressure are believed capable of inducing soil air to move alternately inward and outward, resulting in some degree of circulation.

Three of the atmospheric gases present in soil air play active roles in soil processes: molecular oxygen (O_2), molecular nitrogen (N_2), and carbon dioxide (CO_2). These active roles require that the gases be dissolved in water. As gas molecules dissolved in soil water, neither nitrogen nor oxygen is directly involved in chemical reactions affecting clay minerals and carbonate minerals in the soil. Carbon dioxide, on the other hand, is of major importance in direct reactions because it combines with soil water to form a weak solution of carbonic acid. Complex organic acids, produced during the decomposition of organic matter, are also important reagents in the soil solution. An acid serves as the active agent in attacking the tightly bonded atoms of the crystal structure of clay minerals.

Ions

To understand chemical activity in the soil solution, we must refer to the activity of ions. An **ion** is an atom or group of atoms bearing an electrical charge. When certain compounds are dissolved in water, the atoms become separated as ions. For example, ordinary table salt is the compound sodium chloride, consisting of atoms of sodium (Na) and atoms of chlorine (Cl) in a one-to-one ratio. Thus the chemical formula for sodium chloride is NaCl. In its solid state, sodium chloride is a crystalline substance formed of atoms of sodium bonded weakly to adjacent atoms of chlorine. When placed in water, NaCl dissolves, meaning that the Na and Cl atoms separate and move freely among the water molecules. Separation from the crystal structure results in a sodium atom having a single positive electrical charge; it is now an ion, indicated by the symbol Na^+. A chlorine atom assumes a negative charge, becoming a chlorine ion: Cl^-.

Chemists refer to a positively charged ion as a **cation,** to a negatively charged ion as an **anion.** Some kinds of ions consist of two different kinds of atoms joined

Figure 13.7 **Schematic diagram of a thin, flat colloidal particle with negative surface charges and a layer of positively charged ions (cations) held to the surface.**

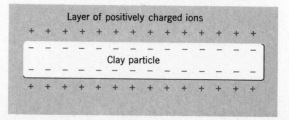

together. For example, the ammonium ion consists of one atom of nitrogen (N) joined with four atoms of hydrogen (H) with the formula NH_4^+; it is a cation. The sulfate ion is another example, with one atom of sulfur (S) joined to four atoms of oxygen (O), with the formula SO_4^{--}; it is an anion. Notice that the sulfate ion has two negative charges. Some kinds of ions have a single charge, others a double charge, and others a triple charge.

Ions important in soils include the following:

Cations

H^+	Hydrogen
Al^{+++}	Aluminum
$Al(OH)^{++}$	Hydroxyl aluminum
Ca^{++}	Calcium
Mg^{++}	Magnesium
K^+	Potassium
Na^+	Sodium
NH_4^+	Ammonium

Anions

Cl^-	Chlorine
SO_4^{--}	Sulfate
OH^-	Hydroxide
HCO_3^-	Bicarbonate
NO_3^-	Nitrate

The soil solution contains several varieties of ions derived from precipitation. Rainwater holds sea salts, suspended mineral particles, and various pollutants. When carried down to earth as rain, sea salts contribute all the ions that are present in seawater. Most of these ions are chlorine (Cl^-) and sodium (Na^+). Ions of magnesium (Mg^{++}), sulfate (SO_4^{--}), calcium (Ca^{++}), and potassium (K^+), are contributed in minor amounts. Mineral dusts lifted from the ground into the atmosphere and carried

upward in turbulent wind account for many of the potassium, calcium, and magnesium ions found in rainwater.

Sulfate ions (SO_4^{--}) are usually present in rainwater. They come largely from sulfate particles and gaseous sulfur compounds injected into the atmosphere by fossil-fuel combustion, forest fires, volcanoes, and biologic activity. Also present are nitrate ions (NO_3^-) and ammonium ions (NH_4^+) produced from gaseous forms of nitrogen introduced into the atmosphere from a variety of sources, including the combustion of fuels, the decay of organic matter, and fertilizers. Phosphate ions (PO_4^{---}) are also present in rainwater, but in much smaller amounts than ammonium and nitrate.

An important point about the pollutant ions is that they may form acids in the soil solution. Sulfate ions form sulfuric acid (H_2SO_4); nitrate ions form nitric acid (HNO_3). In certain regions over which washout of heavily polluted air occurs, these acids cause many undesirable impacts on biologic activity and may to some extent be causing changes in the normal soil-forming processes.

Soil Colloids and Cation Exchange

Clay mineral particles of colloidal dimensions are chemically active in the soil because of their great surface area. A clay particle can be pictured as a very thin, platelike object with flat, parallel upper and lower surfaces (Figures 13.7 and 13.8). (Figure 11.9 is an electron microscope photograph of colloidal clay particles.) The crystalline structure of the clay minerals is such that the atoms are arranged in systematic repeating geometric patterns, called *crystal lattices*. (The same is true of all crystalline minerals.) For the clay minerals, the lattice structure takes the form of flat, parallel **lattice layers** of extreme thinness. For this reason the clay minerals are referred to as **layer silicates.**

The chemical bonds that hold together the atoms within each lattice layer are strong, whereas the bonds between layers are weaker. Because of this structure, water molecules and various free ions can penetrate between the layers of the clay mineral, leading to its chemical alteration and to its physical disruption.

The lattice layer structure of clay minerals is such that oxygen atoms, which are negatively charged, are nearest the upper and lower surfaces. This condition is indicated in Figure 13.7 by minus signs on the clay particle. As a result, positive ions, or cations, will be attracted to the clay particle surface and held there by electrostatic attraction. Cations of hydrogen (H^+), aluminum (Al^{+++}), sodium (Na^+), potassium (K^+), calcium (Ca^{++}), and magnesium (Mg^{++}) are commonly present in soil solutions, and all are found on clay particle surfaces. In many soil reactions, these cations replace one another in the process of **cation exchange.**

Cation exchange is governed by a replacement order, indicating which ion is capable of replacing another. This is a kind of seniority system in which the ion of a given rank can take over the position of ions of lower seniority. The aluminum ion can displace any of the other metallic ions, so it occupies the top position on the list. In order

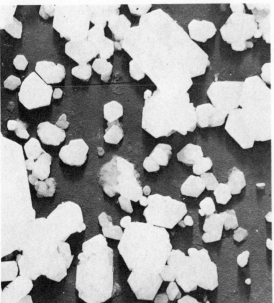

Figure 13.8 **Electron microscope photograph of kaolinite crystals magnified about 20,000 times. (Paul F. Kerr.)**

Table 13.1 / Cation-exchange Capacities of Various Soil Colloids

Material	CEC
Organic matter (humus colloids)	150–500
Clay minerals	
Vermiculite	100–150
Montmorillonite	80–150
Illite	10–40
Kaolinite	3–15
Hydrous sesquioxides of aluminum and iron	4
Feldspar, quartz	1–2

of replacement ability, cations of hydroxyl aluminum, calcium, magnesium, potassium, and sodium follow.

The capacity of a given quantity of soil to hold and exchange cations is called the **cation-exchange capacity (CEC)** and is a general indicator of the degree of chemical activity of a soil. Capacity is indicated in a unit known as the *milliequivalent,* which is a measure of the ratio of weight of ions to weight of soil. The exact definition of this unit is not important here, but its relative magnitude does concern us. The exchange capacities of various soil colloids are given in Table 13.1. Among the clay minerals, both vermiculite and montmorillonite have high CEC. Illite has intermediate CEC, while kaolinite has a low value. Even lower is the CEC of the sesquioxides of iron and aluminum. Particles of feldspar and quartz have nearly zero CEC.

The continued chemical weathering of a soil over a very long period of time tends to bring a shift in clay mineral composition. In earlier stages of weathering, the content of high CEC minerals — vermiculite and montmorillonite — may be comparatively large, so the soil as a whole has a high CEC. In advanced stages, the minerals of high CEC are removed or altered in favor of minerals of low CEC — kaolinite and the sesquioxides of aluminum and iron. As a result, total CEC gradually declines to low values. Generally, when CEC falls below 10 in the B horizon the soil is classed as having low CEC. Soils with high CEC usually have a high capacity to store plant nutrients (base cations) and are potentially fertile soils (if the soil is not strongly acid in chemical balance).

Soil Acidity and Alkalinity

The various soil cations capable of being readily exchanged on colloidal particles belong to two general classes. One class, already mentioned in Chapter 12 as being important plant nutrients, consists of the **base cations** (or simply, **bases**). The base cations most important in soils are the following:

Calcium	Ca^{++}
Magnesium	Mg^{++}
Potassium	K^+
Sodium	Na^+

When base cations comprise the large majority of cations held by soil colloids, the soil is in a condition described as **alkaline.**

The other class consists of **acid-generating cations.** Three acid-generating cations are important in soils. One is the aluminum ion, Al^{+++}; it is associated with extreme acidity. The second is the hydroxyl aluminum ion, $Al(OH)^{++}$, associated with a moderate degree of acidity. The third is the hydrogen ion, H^+, forming about 10 percent of the acid-generating ions in acid soils. These ions must be exchangeable, that is, free to change places with other ions on the surfaces of colloids. Usually present are many Al^{+++} and H^+ ions too tightly bound to be readily exchangeable. A soil is described as **acid** when the total numbers of readily exchangeable acid-generating cations comprise from 5 to 60 percent of the total cation exchange capacity; the larger the percentage, the greater the degree of acidity.

The range of alkalinity or acidity of a soil is measured in terms of a number known as the **pH** of the soil solution.* A pH of 7.0 is neutral in this scale; values below 5 represent a strongly acid soil solution; values above 10 represent a strongly alkaline soil solution. Table 13.2 shows a classification of soils according to acidity and alkalinity. For agricultural soils, this quality is very important because certain crops require near-neutral values of pH and cannot thrive on acid soils. Plants differ considerably in their preference for soil acidity or

---

*pH is a measure of the concentration of hydrogen ions; it is the logarithm to the base 10 of the reciprocal of the weight in grams of hydrogen ions per liter of water. Consequently, the smaller the pH number, the greater the hydrogen ion concentration.

Table 13.2 / Soil Acidity and Alkalinity

pH	4.0	4.5	5.0	5.5	6.0 6.5	6.7	7.0	8.0	9.0	10.0 11.0
Acidity	Very strongly acid	Strongly acid	Moderately acid	Slightly acid	Neutral		Weakly alkaline	Alkaline	Strongly alkaline	Excessively alkaline
Lime requirements	Lime needed except for crops requiring acid soil	Lime needed for all but acid-tolerant crops		Lime generally not required	No lime needed					
Occurrence	Rare	Frequent	Very common in cultivated soils of humid climates				Common in sub-humid and arid climates		Limited areas in deserts	

Data source: C. E. Millar, L. M. Turk, and H. D. Foth (1958), *Fundamentals of soil science,* third edition, John Wiley and Sons, New York. See Chart 4.

alkalinity, and this is an important factor in the distribution of plant types.

As Table 13.2 shows, agricultural soils with pH under about 6 require the application of lime for the successful cultivation of many crops. **Lime,** as the word is used in agriculture, may be either calcium oxide (CaO) or calcium carbonate ($CaCO_3$). Nearly all lime used in agriculture is natural limestone of calcium carbonate composition. It is ground into powder and spread over fields. Lime eliminates acid-producing ions from soil colloids, replacing them with calcium ions. After the soil pH has been raised to the desired level, fertilizers rich in macronutrients (nitrogen, phosphorus, potassium) must also be added to the soil because these nutrients are deficient in most acid soils.

Table 13.3 lists some of the relationships between pH and soil conditions. Figure 13.9 shows this information in the form of graph. In highly acid soils, pH below about 4.5, exchangeable aluminum ions (Al^{+++}) are dominant, with some exchangeable hydrogen ions (H^+) and a substantial proportion of bound hydrogen and aluminum ions (not freely exchangeable). As pH increases, Al^{+++} gives way to exchangeable hydroxyl aluminum ions ($Al(OH)^{++}$). As the neutral region is approached, acidity is caused by H^+ only. Exchangeable bases increase in proportion steadily as pH rises above 4 until, at pH 7, they form about 85 percent of the total CEC. Above pH 8, they

form 100 percent of the total CEC. Then, with increasing pH, the proportion of sodium ions (Na^+) increases. At pH 10, the soil is 15 percent or more saturated with Na^+.

Base Status of Soils

As in many forms of human society, soils are stratified into "status" levels and are classified into major groups on that basis. Status in soils is determined by the **percentage base saturation (PBS),** defined as the percentage of exchangeable base cations with respect to the total cation exchange capacity of the soil. A value of 35 percent has been used by soil scientists as a dividing number separating one class of soils of **high base status** (PBS greater than 35%) from those of another class of **low base status** (PBS less than 35%). Soils of high base status have high natural fertility for food crops; those of low base status are naturally low in fertility and require special treatment and the application of chemicals to correct the deficiency. Base status of soils thus has enormous impact on Man's food resources and on the possibilities for future expansion of agricultural food production into areas not now under cultivation.

Soil Temperature Regimes

Soil temperature, which we introduced in Chapter 5, is an important factor in determining the characteristics of a soil. Temperature acts as a control over biologic activity and influences the intensity of chemical processes affecting the clay minerals. Below the freezing point, 0°C (32°F), there is no biologic activity; chemical processes affecting minerals are inactive. Between 0°C and 5°C (32 and 41°F), root growth of most plants and germination of

Table 13.3 / Relationship between Soil pH and Soil Conditions

pH range	Acidity or alkalinity	Cause, source material, or occurrence
3.5	Extremely acid Al^{+++}	Acid sulfates in coastal-marsh clays, mine spoil. Excess of Al^{+++} may render soil toxic to plants.
4.5	Highly acid H^+ and Al^{+++}	Organic layer (histic epipedon). Sulfur and sufides in salt-marsh clays and mine spoil.
4.0–5.2	Very acid H^+ and Al^{+++}	Organic matter (histosols, histic epipedon).
4.5–5.8	Moderately acid H^+ and $Al(OH)^{++}$	Mineral soils.
5.8–6.5	Slightly acid	Soil 70–90% base saturated.
6.5–8.0	Neutral to weakly alkaline No Al^{+++}	Soil fully base saturated. Some $CaCO_3$ may be present.
8.0–8.5	Alkaline	Exchangeable cations are mostly Ca^{++}, Mg^{++}. Free $CaCO_3$ present.
8.5–10	Strongly alkaline	Large amounts of soluble salts present. Exchangeable Na^+ present.
>10	Excessively alkaline	Soil is sodium saturated. Alkali soil. Sodium ions may be toxic to many plant species.

Data source: S. W. Buol, F. D. Hole, and R. J. McCracken, 1973, *Soil genesis and classification,* Iowa State Univ. Press, Ames, pp. 66–68.

Figure 13.9 **Generalized diagram of the proportions of exchangeable base cations and acid-generating cations in relation to soil pH. Bound hydrogen and aluminum ions are so tightly held to the colloids as not to be readily exchangeable. (Based on a diagram by H. O. Buckman and N. C. Brady, 1969, *The nature and properties of soils,* Macmillan Co. of Canada, Toronto, Fig. 14.1.)**

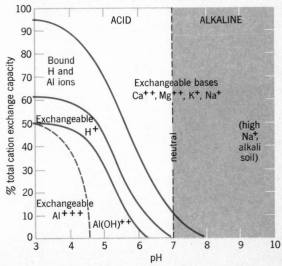

most seeds is impossible; but water can move through the soil and carbonic acid activity may be important. A horizon as cold as 5°C (41°F) acts as a thermal barrier to the roots of most plants. The germination of seeds of many low-latitude plants requires a soil temperature of 24°C (75°F) or higher.

At any moment the temperature within a soil varies from horizon to horizon. The temperature near the surface experiences both a diurnal and an annual cycle (Chapter 5). The range of these cycles may be small or large according to latitude and the degree of continentality in the thermal regime. (See Chapter 9 and Figure 9.2.) The annual cycle will be almost imperceptible in the wet equatorial climate (1), but very large in moist continental (10) and boreal forest (11) climates.

Each pedon has a characteristic **soil-temperature regime** that can be measured and described. For purposes of classifying soils, the temperature regime can be described by the mean annual soil temperature and the average seasonal fluctuations from that mean.

Each pedon has a mean annual temperature that is essentially the same in all horizons at all depths. Some representative examples are as follows:

Location	Mean Annual Soil Temperature	
	°C	(°F)
Irkutsk, USSR	−2	(28)
Bozeman, Montana	6	(43)
Urbana, Illinois	10	(50)
Colombo, Sri Lanka	27	(81)

Table 13.4 gives the names and temperature boundaries of six soil temperature regimes. Two criteria are specified: (1) Mean annual temperature, T, and (2) difference between mean of summer (or warm season) and mean of winter (or cold season), Ts−Tw, at a depth of 50 cm (20 in.). (The notation 0° < T < 8° is read "mean

Table 13.4 / Soil Temperature Regimes

Name of regime	Mean annual soil temperature, °C (T)		Difference between mean temperature (°C) of warm season and of cold season. (Ts − Tw)
Pergelic	T < 0°		—
Cryic	0° < T < 8°		—
Frigid	T < 8°		>5°
Mesic	8° < T < 15°		>5°
Thermic	15° < T < 22°		>5°
Hyperthermic	T > 22°		>5°
Equivalents:	°C	°F	
	0	32	
	8	47	
	15	59	
	22	72	
5°C = 9°F			

Data source: Soil Survey Staff, 1975, *Soil taxonomy*, Agriculture Handbook No. 436, Government Printing Office, Washington, D.C., pp. 62–63.

annual temperature greater than 0°C but less than 8°C.")

The soil temperature regimes are used as a means of classifying certain subclasses of soils. In so doing, a prefix is added to the soil name as follows:

Regime	Prefix used
Cryic	Cry−, Cryo−
Frigid	Bor−
Thermic	Trop−

We will make use of these prefixes in Chapter 14.

Soil-Water Regimes

Soil-water regimes, explained in Chapter 10, are used by American soil scientists in the modern system of soil classification. (In publications of the Soil Survey Staff of the U.S. Department of Agriculture, "soil moisture" is used instead of "soil water.") The Thornthwaite method of calculating the soil-water budget has been adopted, with minor simplifications, for the purpose of estimating the actual soil-water conditions. These estimates do not apply under special conditions. For example, in a dry climate the soil-water budget calls for virtually zero soil-water storage in all months. In some low-lying places in deserts, however, the ground-water table is close to the surface, supplying the roots of plants with adequate water during much of the year.

An important criterion in determining the soil-water regime is the distinction between a dry soil and a moist soil. As a soil loses water by evaporation, the remaining films of capillary water clinging to grain contacts shrink. As they shrink, the films set up increasing resistance to being drawn up into the tiny rootlets of plants. The measure of this resistance is the tension (force) of the water films in units of *bars*. One bar is about equal to the standard pressure of the atmosphere (approximately 1000 millibars). When the drying out of the soil reaches a point that the soil-water tension is 15 bars or greater, the remaining soil water is considered to be unavailable to plants. Thus 15 bars is the dividing point between a dry condition and a moist condition.

The definition of dry and moist conditions is applied to the **soil-water control section.** The upper boundary of this control section is defined as the depth to which 2.5 cm (1 in.) of water, applied to the surface of a dry soil, will moisten the soil. "Moisten" means to cause the tension to fall to a value below 15 bars. The lower boundary of the control section is the depth to which the application of 7.5 cm (3 in.) of water at the soil surface will moisten a dry soil. In soils of coarse loam grade the soil-water control section has a depth of 20 cm (8 in.) at the upper boundary and 60 cm (24 in.) at the lower boundary. The depth is strongly influenced by soil texture class, being shallower for fine textures and deeper for coarse textures.

Five soil-water regimes are recognized and given names under the modern U.S. system of soil classification:

Aquic regime (L. *aqua*, "water"). The soil is saturated with water most of the time because the ground-water table lies at or close to the surface throughout most of the year. The aquic regime is found in bogs, marshes, and swamps. It occurs independently of the various soil-water regimes associated with well-drained upland locations.

Udic regime (L. *udus,* "humid"). The soil is not dry in any part of the control section as long as 90 days per year. The udic regime is found in the moist climates (6h, 8h, 10h, 11h), as we have defined them in Chapter 10. The soil-water budget shows little or no water deficiency (D) in the growing season (summer), and there is a seasonal water surplus that causes water to move through the soil at some part of the year. For soil-water regimes with a water surplus in all months (Climates 1, 2, 6p, 8p, 10p, 11p) the name is modified to *perudic.* Under a perudic regime, even in the dry season moisture tension rarely becomes as great as 1 bar.

Ustic regime (L. *ustus,* "burnt," implying dryness). The soil has a moderate quantity of water in storage during the season when conditions are favorable to plant growth and the soil water is not frozen, but the soil is dry for 90 or more cumulative days in most years. The ustic regime is associated with the semiarid subtype (s) of the dry climates (4s, 5s, 9s) and with the tropical wet-dry climate (3).

Aridic (torric) regime (L. *aridus,* "dry," and *torridus,* "hot and dry"). The two names for this regime are used in different categories of the soil classification system. For warm soils the soil is never moist in some or all parts for as long as 90 consecutive days. The aridic (torric) regime applies to the semidesert (sd) and desert (d) subtypes of the dry climates (4, 5, 9), in a wide latitude range from tropical zone to midlatitudes.

Xeric regime (Gr. *xeros,* "dry"). This regime applies to areas of Mediterranean climate (7s, 7sd, 7sh) with a long, dry summer and a rainy winter. The soil-water control section is dry in all parts for 45 or more consecutive days in the dry summer season, but moist in all parts for 45 or more consecutive days in the moist winter season.

In the modern soil classification system, prefixes taken from the names of the soil-water regimes are used to designate one of the major classes of soils and many of the subclasses. The prefixes are as follows:

Regime name	Prefix
Aquic	Aqu-
Udic	Ud-
Ustic	Ust-
Aridic	Aridi-
Torric	Torr-
Xeric	Xer-

The soil-water regime is a powerful controlling factor in determining the properties of soils and the nature of soil-forming processes. The quantity of soil water in storage determines the rate of primary production by plants and general organic activity within the soil. In the udic regime, water moving through the soil is capable of removing ions (leaching). In the ustic regime, limited water storage is associated with the accumulation of calcium carbonate. In the aridic regime, soluble salts tend to accumulate in the soil. In the aquic regime, water saturation prevents oxygen from entering the soil. It would not be an understatement to say that the combined influence of soil-temperature regime and soil-water

regime shapes almost the entire complex of chemical and biological properties of a soil, assuming only that the parent matter originally contained a wide spectrum of silicate minerals.

Landform and Soil

Configuration of the ground surface as a factor in soil formation can be included in the single word, **landform.** Landform includes the varying steepness of the ground surface, or **slope,** as well as the compass orientation—the **aspect**—of an element of ground surface. Another landform property is the **relief,** or average elevation difference between adjacent high and low points (hilltops versus valley bottoms). Strong relief and steep slopes are combined in many areas of hills and mountains. Small relief and gentle slopes are combined in low plains and flat uplands.

Figure 13.10 shows how landform influences the thickness of A and B horizons of the soil solum. Taking the profile in an undulating upland surface (left) as the normal condition, the flat upland shows thickened horizons because removal of the soil surface by erosion is much slower on very gentle slopes. In the hilly region with steep slopes, erosion removes the upper horizons more rapidly and the profile is reduced in thickness. The soil profile beneath the poorly drained meadow is entirely different in character. The combined A and B horizons are rich in organic matter, and a gray layer (glei horizon) is formed beneath the organic layer, where oxygen is deficient. In the adjacent bog a thick upper layer of peat is usually present.

Aspect of the ground surface influences both the soil temperature and the soil-water regimes. In midlatitudes, where the sun's rays strike at an intermediate angle between the horizon and zenith, slopes facing north (northern hemisphere) receive greatly reduced insolation. Here soil temperatures are cooler than average, and the soil-water regime is pushed in the direction of a moister climate.

Biologic Processes in Soil Formation

The total role of biologic processes in soil formation includes the presence and activities of living plants and animals as well as their nonliving organic products. Living plants contribute to soil formation in two basic ways. First is the production of organic matter—the biomass—both above the soil as stems and leaves and within the soil as roots. This primary production, discussed in Chapter 12, provides the raw material of organic matter in the O horizon and in lower horizons. The decomposer organisms process this raw material, reducing it to humus and ultimately to its initial components—carbon dioxide and water. Second is the recycling of nutrients from the soil to plant structures above ground and their return to the soil in dead plant tissues. Nutrient recycling is a mechanism by which nutrients are prevented from escaping through the leaching action of surplus soil water moving downward through the soil. In our discussion of

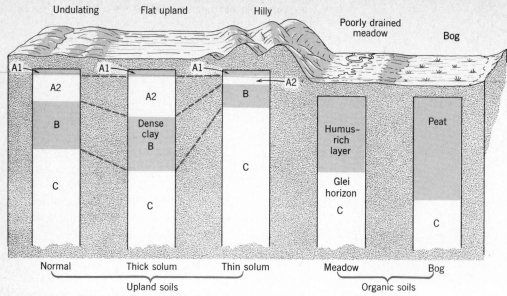

Figure 13.10 **Relief and slope strongly influence the thickness and composition of the soil profile. (After U.S. Dept. of Agriculture, *Yearbook of American Agriculture*, 1938.)**

the basic soil types, Chapter 14, this process will receive special attention.

Animals living in the soil, or entering and leaving the soil by means of excavated passageways, span a wide range in species and individual dimensions. The total roles of animals in biologic processes of soil cannot be overestimated in soils with sufficient heat and moisture to support large animal populations. For example, earthworms continually rework the soil not only by burrowing, but also by passing the soil through their intestinal tracts. They ingest large amounts of decaying leaf matter, carrying it down from the surface and incorporating it into the mineral soil horizons. The granular structure of the darkened A1 horizon derives its quality from this activity. Many forms of insect larvae perform a similar function. Small tubular soil openings are also formed by many species of burrowing insects. Large openings are made by larger animals—moles, gophers, rabbits, badgers, prairie dogs, and many other species. The growth of roots followed by decay leaves tubular openings in the soil.

In moist climates, the evolution of a soil from parent mineral matter is accompanied by increasing plant growth and changes in plant species. We will examine this evolution process—called plant succession—in Chapter 15. Various close relationships between ecosystems and soil characteristics will become evident in descriptions of the widely varied classes of soils.

Man is also a potent agent in influencing the physical and chemical nature of the soil. Large areas of agricultural soils have been tilled and fertilized for centuries. Both structure and composition of these agricultural soils have undergone profound changes and can now be recognized as distinct soil classes of importance equal to natural soils. The new system of soil classification presented in the next chapter includes soil types produced by or greatly modified by human activities.

Review of Basic Pedogenic Processes

Facts and concepts developed in this chapter and in Chapter 11 have prepared the way for a brief review of several basic soil-forming processes, or **pedogenic processes.**

Pedogenic processes can be classified into four groups: (1) addition of material to the soil body, (2) losses from the soil body, (3) translocations of materials within the soil body, and (4) transformations of material within the soil body.

Additions of materials to the soil body are covered by the general term **soil enrichment.** Inorganic enrichment can come as sediment added to the soil surface by running water (colluvium) and by wind (loess or volcanic ash). Another form of enrichment is from the organic litter of plants growing in the soil. This material accumulates in the O horizon and produces finely divided humus that is carried down into the mineral horizons.

Losses of material from the soil body include removal of surface material by soil erosion and by **leaching,** the downward washing out and removal by surplus soil water percolating through the soil.

Translocation of materials within the soil takes place in a number of quite different ways, each with a different cause and often uniquely related to a particular soil-water regime. Two simultaneous processes of downward translocation are eluviation and illuviation, typical of the udic soil-water regime. **Eluviation** consists of the downward transport of fine particles, particularly the colloids (both mineral and organic), carrying them out of an upper soil horizon (usually the lower part of the A horizon); see Figure 13.11. Eluviation leaves behind coarse skeletal mineral grains. In cool moist climates, the A2 horizon formed by eluviation contains largely quartz in sand or coarse silt grade sizes. **Silication** is a term applied to this increase in proportion of silica because it remains

208 Soil-Forming Processes

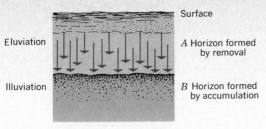

Figure 13.11 Eluviation and illuviation, leading to the formation of A and B horizons.

behind while other materials are removed.

Illuviation is the accumulation of materials in a lower horizon, brought down from a higher horizon (Figure 13.11). Typically, illuviation occurs in the B horizon. The materials that accumulate may be clay particles, organic particles (humus), or sesquioxides of iron and aluminum.

The translocation of calcium carbonate is another important process. Removal of calcium carbonate, or **decalcification,** takes place as carbonic acid reacts with carbonate mineral matter. The soluble products are carried down into a lower horizon. Their accumulation constitutes **calcification;** it may take place in the B horizon or in the C horizon below the soil solum. Precipitation of soluble salts and the reverse process of removal of salts are **salinization** and **desalinization,** respectively.

Transformations within the soil body affect both inorganic and organic materials. Decomposition of primary to secondary minerals is one such transformation, already described in detail in Chapter 11. By synthesis,

new minerals and organic compounds can be formed from the products of decomposition. Decomposition also affects organic materials. Humification, the process of transformation of plant tissues into humus, may be followed by total disappearance of the organic matter as water and carbon dioxide by respiration.

Horizonation, the degree of development of soil horizons, is the result of complex combinations of the pedogenic processes already listed. In the next chapter, as each class and subclass of soil is described, we will have an opportunity to explain how horizons vary from place to place under the complex interaction of the various soil-forming factors—parent material, temperature and soil-water regimes, biologic activity, and time.

References for Further Study

Loughnan, F. C. (1969), *Chemical weathering of the silicate minerals,* American Elsevier Publishing Co., New York. See Chapters 2, 3.

Foth, H. D., and L. M. Turk (1972), *Fundamentals of soil science,* fifth edition, John Wiley and Sons, New York. See Chapters 1–8.

Buol, S. W., F. D. Hole, and R. J. McCracken (1973), *Soil genesis and classification,* Iowa State Univ. Press, Ames. See Chapters 1–12.

Birkeland, P. W. (1974), *Pedology, weathering, and geomorphological research,* Oxford Univ. Press, New York.

Soil Survey Staff (1975), *Soil taxonomy,* Soil Conservation Service, U.S. Department of Agriculture, Agriculture Handbook No. 436, Government Printing Office, Washington, D.C. See Chapters 1–6.

Review Questions

1. In what sense is the soil a dynamic layer? Define the word *soil* as the term is used in pedology. What are the upper, lateral, and lower limits of the soil? What are soil horizons?

2. Explain the concepts of the pedon and polypedon. What is a soil profile? What is the soil solum? What determines the lower boundary of the soil solum?

3. What causes differences in color of the soil from one region to another? What three measurable variables determine soil color, as described objectively in soil science?

4. Describe the system of soil-texture classes used by the U.S. Department of Agriculture. Give examples of several classes. How does soil texture influence soil-water retention and transmission? What is the wilting point? How is it related to soil texture?

5. What physical properties are included in soil consistence? How are stickiness and plasticity measured? What causes cementation?

6. What is meant by soil structure? What is a ped? Name and describe the four primary types of soil structure. What are cutans? How are they formed?

7. Describe the organic horizons of soils. What is the nature and origin of the organic matter? What is humification? What materials comprise the mineral horizons of soils? What two groups of minerals are recognized? Which group is active in soil-forming processes?

8. How are the mineral horizons of a soil designated? What process is typical of the A1 horizon? of the A2 horizon? What

are the characteristic features of the B horizon? What activities occur in the C horizon? Give examples of special characteristics of horizons, as indicated by lowercase letters.

9. What important role is played by the soil solution? How are atmospheric gases involved in the soil solution? What is an ion? Give examples of anions and cations common in the soil solution. What common ions are contributed to the soil by rainwater?

10. Describe the shape of clay mineral particles of colloidal size. How is particle shape influenced by crystal lattice structure? Why do clay particles attract and hold cations?

11. What is cation exchange? Name in ranking order the common cations involved in exchange. What is cation-exchange capacity (CEC)? Which soil colloids have the highest CEC? The least? Why is CEC important in soils?

12. Distinguish between base cations and acid-generating cations. Name the common acid-generating cations. Under what conditions is the soil solution acid? Alkaline? How does pH indicate the degree of acidity or alkalinity of a solution? How can excessive acidity be corrected in soils? In what range of pH are aluminum ions largely the cause of soil acidity? Under what conditions do sodium ions dominate the exchangeable cations?

13. What is meant by percentage base saturation (PBS)? How are high base status and low base status defined? What is the significance of base status in terms of soil fertility?

14. What role does soil temperature play in biologic activity in

soils? How are soil-temperature regimes defined? Name six basic soil-temperature regimes and give the mean annual soil temperature range for each.

15. How is the Thornthwaite system of soil-water budget analysis used in soil science? What is the precise distinction between a dry soil and a moist soil? How is the soil-water control section defined? How does soil texture affect the upper and lower limits of the control section?

16. Name the five principal soil-water regimes. Define each in terms of dry and moist conditions in the control section. For each regime, name the associated climate or climates, using the climate system described in Chapter 10. Explain the importance of the soil-water regime in soil-forming processes.

17. What is meant by landform? How does landform influence the thickness of horizons in the soil solum? How does aspect of a slope influence the soil-temperature and soil-water regimes?

18. Describe the total role of biologic processes in soil formation. How do animals affect the structure of the soil? Give examples. How does human activity influence soil properties?

19. Name at least a dozen important pedogenic processes and explain the action and effects of each. What is horizonation?

Soil Classification and Global Soils

14

From the standpoint of physical geography, the most important aspect of pedology is the classification of soils into major classes and subclasses recognized in terms of their areal distribution over the earth's land surfaces. Geographers are particularly interested in the ways climate, parent materials, and landform are linked with the distribution of types of soils. Geographers are also interested in the kinds of natural vegetation associated with each of the major soil types. The geography of soils is thus an essential ingredient in establishing the quality of environmental regions of the globe — important because soil fertility, along with availability of fresh water, is the basic measure of the potential of an environmental region to produce food for the human race.

Earlier Soil Classifications

The founder of modern theories of soil origin and classification was V. V. Dokuchaiev, a Russian geologist. His studies between 1882 and 1900 led him to the concept that soil is an independent body whose character is determined primarily by climate and vegetation. A Russian follower of Dokuchaiev, K. D. Glinka, expanded the concepts of horizons in the soil profile.

For the development of modern soil science in the United States during the 1920s and 1930s, much of the credit rests with C. F. Marbut, who served for many years as chief of the Soil Survey Division of the U.S. Department of Agriculture (USDA). Marbut became acquainted with Russian pedological views, adapted them to conditions in the United States, and created a comprehensive system of soil classification. Built on his work, but also departing from it, was the system outlined in the *1938 Yearbook of Agriculture, Soils and Men*. The 1938 system was used with a number of modifications for the next 25 years.

The 1938 USDA system recognized the existence of three orders of soils: zonal, intrazonal, and azonal (Table 14.1). Zonal soils, formed under conditions of good soil drainage through the prolonged action of climate and vegetation, are by far the most important and widespread of the three orders. Intrazonal soils are formed under conditions of very poor drainage (such as in bogs, flood-plain meadows, or in lake basins of deserts), on regolith with high calcium carbonate content, or where soluble salts or sodium or both are high.

Azonal soils have no well-developed profile characteristics, either because they have had insufficient time to develop or because they are on slopes too steep to allow profile development. Azonal soils include Lithosols (thin soils on bedrock of the earth's mountain regions), Alluvial soils, and dune sands. Azonal soils cannot be easily classified, whereas the zonal and intrazonal soils have distinctive profile characteristics as the result of long development.

Within the zonal and intrazonal orders are several suborders. Within each suborder are great soil groups, of which 36 are named in Table 14.1. Some of the names of the great soil groups reflect the Russian influence (Podzol, Chernozem). Other names reflect the American experience (Prairie soils, Chestnut soils, Brown soils).

Despite minor improvements in the 1938 USDA system, made in the 1940s, the inadequacy of the system became increasingly apparent to professional soil scientists. New information on soils was being rapidly accumulated from remote areas of the globe. New varieties of soils, particularly soils from the tropical and equatorial zones, could not be accommodated within the classes of the system. It was also realized that the system rested on unsupported assumptions about the role of climate and other pedogenic processes in determining soil characteristics. A fresh viewpoint was needed, one that would bring to bear a high level of objectivity and new, rigorous, quantitative methods of investigation and description.

The Comprehensive Soil Classification System

In the early 1950s, a United States national cooperative effort was launched to develop a completely new scheme of soil classification. Soil scientists of the Soil Conservation Service, the faculties of land-grant universities, and

211

Table 14.1 / U.S. Department of Agriculture 1938 classification of soils.

Zonal Order

Suborders	Great Soil Groups
Soils of the cold zone	1. Tundra soils
Light-colored soils of arid regions	2. Desert soils 3. Red desert soils 4. Sierozem 5. Brown soils 6. Reddish Brown soils
Dark-colored soils of the semiarid, subhumid, and humid grasslands	7. Chestnut soils 8. Reddish Chestnut soils 9. Chernozem soils 10. Prairie soils 11. Reddish Prairie soils
Soils of the forest-grassland transition	12. Degraded Chernozem soils 13. Noncalcic Brown or Shantung Brown soils
Light-colored podzolized soils of forested regions	14. Podzol soils 15. Brown Podzolic soils 16. Gray-Brown Podzolic soils
Lateritic soils of forested subtropical and tropical regions	17. Yellow Podzolic soils 18. Red Podzolic soils (and Terra Rossa) 19. Yellowish-Brown Lateritic soils 20. Reddish-brown Lateritic soils 21. Laterite soils

Intrazonal Order

Suborders	Great Soil Groups
Halomorphic (saline and alkaline) soils of imperfectly drained arid regions and littoral deposits	1. Solonchak or saline soils 2. Solonetz soils 3. Soloth soils
Hydromorphic soils of marshes, swamps, seep areas, and flats	4. Wiesenböden (Meadow soils) 5. Alpine Meadow soils 6. Bog soils 7. Half Bog soils 8. Planosols 9. Ground-Water Podzol soils 10. Ground-Water Laterite soils
Calomorphic soils	11. Brown Forest soils (Braunerde) 12. Rendzina soils

Azonal Soils

	1. Lithosols 2. Alluvial soils 3. Sands (dry)

Data source: U.S. Department of Agriculture (1938), *Soils and men; yearbook of agriculture 1938.* Government Printing Office, Washington, D.C., see pp. 993–995.

many soil scientists of other nations participated in the new developments. After progressing through a succession of stages over a period of several years, the new scheme was ready for presentation by American pedologists to the Seventh International Congress of Soil Science

in 1960. Known at the time as the Seventh Approximation (because it was the seventh in the series of revisions), the new system was prepared by the Soil Survey Staff of the Soil Conservation Service and was published in 1960. Many modifications and refinements were made in the decade that followed. The system was named the **Comprehensive Soil Classification System (CSCS).** It is documented in a 1975 summary treatise titled *Soil Taxonomy, A Basic System of Soil Classification for Making and Interpreting Soil Surveys,* prepared by the Soil Survey Staff of the U.S. Soil Conservation Service.

The CSCS differs in important ways from the 1938 USDA system. The new system defines its classes strictly in terms of the morphology and composition of the soils, that is, in terms of the soil characteristics themselves. Moreover, the definitions are made as nearly quantitative as possible. Every effort was made to use definitions in terms of features that can be observed or inferred so that subjective or arbitrary decisions as to classification of a given soil will be avoided.

The CSCS does not recognize a distinction between zonal soils (mostly well-drained, upland soils) and intrazonal soils (mostly poorly drained soils) in the highest category of classification, which is the soil order. Thus, within a particular soil order, we find soils of wet places grouped with soils of well-drained upland surfaces. No attempt is made in the new classification to assign more importance to one class of soils than another, particularly with respect to the amount of surface area it occupies.

The CSCS recognizes and gives equal importance to classes of soils deriving their characteristics from Man's activities, such as long-continued cultivation and applications of lime and fertilizers and the accumulation of Man-made wastes. Because Man's occupancy and agricultural exploitation of large expanses of soils have gone on for centuries in various parts of the world, recognition of such Man-modified soils is desirable and realistic in a classification system.

The CSCS terminology uses a large number of newly coined words. The new terms give the system a major advantage because the syllables were selected to convey precisely the intended meaning with respect to the properties or genetic factors relating to the soil class.

The classification system of the CSCS is known as the **Soil Taxonomy;** it is based on a hierarchy of six categories, or levels, of classification. They are listed here with the numbers of classes recognized within each category:

Orders	10
Suborders	47
Great groups	185
Subgroups	1,000 (approx.)
Families	5,000 (approx.)
Series	10,000 (approx.)

Numbers given for the lowest three categories refer only to soils of the United States. When soils of all continents are eventually included, the numbers of at least the two lowest categories will be greatly increased. On the other hand, numbers given for the first three categories are expected to remain approximately the same because they are designed to provide for classification of all known soils.

This tree stump, buried in layers of fine-textured alluvium, gives evidence of continued sediment deposition on the Mackenzie River delta, Northwest Territories, Canada. Permafrost ice layers are visible in the lower part of this riverbank exposure. (Mark A. Melton)

Sloping layers of partly decomposed basaltic volcanic ash (tephra) on the lower slopes of extinct volcano Haleakala, Island of Maui, Hawaii. The reddish-brown colors indicate the presence of iron oxides—hematite and limonite. (Arthur N. Strahler)

Well-sorted layers of sand and gravel make up this glaciofluvial deposit, a kame in southern New York State. (Arthur N. Strahler)

Wind-deposited loess, showing vertical partings, exposed in a river bluff near Vermillion, South Dakota. The brown soil is a fertile Mollisol. (Arthur N. Strahler)

Blocks of freshly cut peat have been set aside to dry out; they will then be used as a domestic fuel. The soil exposed on this moor is a Histosol. Island of Skye, Inner Hebrides, Scotland. (Mark A. Melton)

This gray desert soil, an Aridisol, has proved highly productive when cultivated and irrigated. The locality is near Palm Springs, California, in the Coachella Valley. (Ned L. Reglein)

Parent material of the soil   **Plate E.1**

HISTOSOLS

A **Fibrist**
(Minnesota)

OXISOLS

B **Orthox**
(Puerto Rico)

ULTISOLS

C **Udult**
(North Carolina)

VERTISOLS

D **Udert**
(Texas)

ALFISOLS

E **Udalf**
(Minnesota)

ALFISOLS

F **Ustalf**
(Texas)

ALFISOLS

G **Xeralf**
(California)

SPODOSOLS

H **Humod**
(France)

Plate E.2   Soil profiles

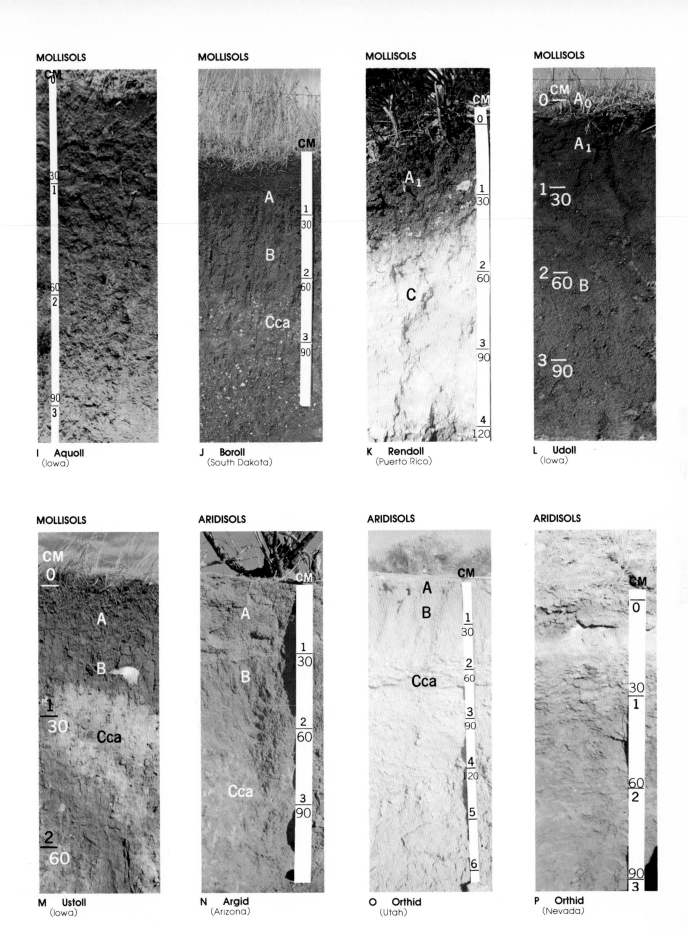

MOLLISOLS

CM
0
30
1
60
2
90
3

I Aquoll
(Iowa)

MOLLISOLS

CM
A
1
30
B
2
60
Cca
3
90

J Boroll
(South Dakota)

MOLLISOLS

CM
0
A₁
1
30
2
60
C
3
90
4
120

K Rendoll
(Puerto Rico)

MOLLISOLS

0 CM A₀
A₁
1
30
2
60 B
3
90

L Udoll
(Iowa)

MOLLISOLS

CM
0
A
B
1
30
Cca
2
60

M Ustoll
(Iowa)

ARIDISOLS

CM
A
1
30
B
2
60
Cca
3
90

N Argid
(Arizona)

ARIDISOLS

CM
A
B
1
30
2
60
Cca
3
90
4
120
5
6

O Orthid
(Utah)

ARIDISOLS

CM
0
30
1
60
2
90
3

P Orthid
(Nevada)

Soil profiles   Plate E.3

This roadcut in Story County, Iowa, exposes the profile of a Udoll (order Mollisols) with a thick mollic epipedon. The corn crop in the adjacent field is typical of this area. (Soil Conservation Service)

A thin layer of wind-deposited silt and dune sand (pale brown layer at top) has buried a Spodosol profile on outer Cape Cod. The pale grayish A2 horizon overlies a reddish B horizon, or spodic horizon. (Arthur N. Strahler)

A salic horizon, appearing as a white layer, lies close to the surface in this Orthid profile (order Aridisols) in the Nevada desert. The scale is marked in feet; 1 ft = 30 cm. (Soil Conservation Service)

Magnified view of clay skins (cutans) coating and connecting grains of quartz sand. The white bar is 2 mm long. (Soil Conservation Service)

Red mottles characterize this plinthite horizon in a Udult profile (order Ultisols) in North Carolina. The area shown is about 50 cm (20 in.) wide. (Soil Conservation Service)

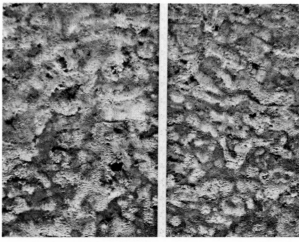

Plate E.4   Soil horizons

We shall concentrate on the uppermost two levels of the soil taxonomy — orders and suborders — for the purpose of understanding the major differences of soils from region to region. We emphasize those units of greatest geographic extent and of greatest importance in understanding the patterns of natural vegetation — forest, grassland, desert — and the ways major soil units reflect the varying intensity of the soil-forming processes we studied in Chapter 13.

Diagnostic Horizons for Classification

To understand how the 10 soil orders are differentiated requires a knowledge of several **diagnostic horizons.** The diagnostic horizons are freshly named and rigorously defined as an integral part of the CSCS; they are not simply relics of older soil classification systems.

The basic definition of a soil horizon is the same as given in Chapter 13, but requires precise restatement when applied to the CSCS. A **soil horizon** is a layer that is approximately parallel with the soil surface and has a set of properties that have been produced by soil-forming processes. The properties of a soil horizon are not like those of the layer above or below it. A soil horizon is differentiated from those adjacent to it partly by characteristics that can be seen or measured in the field, such as color, structure, texture, consistence, and the presence or absence of carbonates. In some cases, however, laboratory measurements are required to supplement field observations.

We consider first the diagnostic horizons of mineral soils — those with only a small proportion (less than 20 percent) of organic carbon by weight. Diagnostic criteria of organic soils are separately considered; they define the kind of organic material present.

Soil horizons fall into two major groups: epipedons and subsurface diagnostic horizons. An **epipedon** (Gr. *epi,* "over" or "upon") is simply a horizon that forms at the surface. It is not, however, identical in meaning with the A horizon, which we defined in Chapter 13. The subsurface horizons originate below the soil surface. In many cases they are identified as parts of the A horizon or as the B horizon. The brief descriptions of selected diagnostic horizons that follow are intended to give a general picture of the horizons rather than to supply rigorous definitions; exact definitions, required in soil classification, are extremely detailed and precise and are often lengthy.

Epipedons

Mollic epipedon (L. *mollis,* "soft"). A relatively thick, dark-colored surface horizon. The dark color is due to presence of organic matter (humus) derived from roots or carried underground by animals. The horizon is usually rich in base cations — calcium, magnesium, and potassium — so the base saturation (PBS) is over 50 percent and often much higher. Structure is usually granular or blocky, with the peds loosely organized when the soil is dry.

Umbric epipedon (L. *umbra,* "shade," hence "dark"). A dark surface horizon resembling the mollic epipedon, but with PBS less than 50 percent.

Histic epipedon (Gr. *histos,* "tissue") A thin surface horizon of peat, usually formed in a wet place. The horizon is saturated with water for 30 consecutive days or more during the year (unless the soil has been artificially drained). Very thick accumulations of peat are associated with organic soils; they are not classed as epipedons.

Ochric epipedon (Gr. *ochros,* "pale"). A surface horizon that is light is color and contains less than 1 percent organic matter. Also included are surface horizons too thin, too dry, or too hard to qualify as any one of the previously listed epipedons.

Plaggen epipedon (Ger. *plaggen,* "sod"). A Man-made surface layer greater than 50 cm (20 in.) thick. It has been produced by long-continued manuring, incorporating sod or other livestock bedding materials into the soil. This horizon is common in western Europe, but rare elsewhere.

Subsurface Diagnostic Horizons

Argillic horizon (L. *argilla,* "white clay"). An illuvial horizon (usually the B horizon) in which layer-lattice clay minerals have accumulated by illuviation (see Chapter 13). Clay cutans, called "argillans," are usually present. The argillic horizon normally forms beneath an eluvial (A1) horizon.

Agric horizon (L. *ager,* "field"). An illuvial horizon formed under cultivation and containing significant amounts of illuvial silt, clay, and humus. Plowing facilitates the down-washing of these materials, which accumulate immediately beneath the plow layer.

Natric horizon (L. *natrium,* "sodium"). Although similar to an argillic horizon, a natric horizon has prismatic structure and a high proportion of Na^+, amounting to 15 percent or more of the CEC.

Calcic horizon A horizon of accumulation of calcium carbonate or magnesium carbonate.

Petrocalcic horizon A hardened calcic horizon that does not break apart when soaked in water.

Gypsic horizon A horizon of accumulation of hydrous calcium sulfate (gypsum).

Plinthite (Gr. *plinthos,* "brick"). Iron-rich concentrations, usually in the form of dark red mottles, present in deeper horizons and capable of hardening into rocklike material with repeated wetting and drying.

Salic horizon Horizon enriched by soluble salts.

Albic horizon (L. *albus,* "white"). A pale, often sandy horizon from which clay and free iron oxides have been removed. Commonly it is the A2 horizon, overlying a spodic horizon.

Spodic horizon (Gr. *spodos,* "wood ashes"). A horizon containing precipitated amorphous materials composed of organic matter and sesquioxides of aluminum, with or without iron. (Allophane, described in Chapter 11, is such a material.) The spodic horizon is found mostly in quartz sands and is formed partly by illuviation. It is designated a Bir or Bhir horizon and normally underlies an A2 horizon.

Cambic horizon (L. *cambiare,* "to exchange"). An altered horizon with texture as fine as or finer than very fine sand that has lost sesquioxides or bases, including carbonates, through leaching. Although considered a B horizon, it has accumulated little clay, as compared with an argillic horizon; it lacks the dark color and organic-matter content of a histic, a mollic, or an umbric epipedon.

Oxic horizon (Fr. *oxide,* "oxide"). A highly weathered horizon at least 30 cm (12 in.) thick, rich in clays and sesquioxides of low CEC (16 or less). There remain few or no primary minerals capable of releasing base cations. Oxic horizons are very old and seldom found outside equatorial, tropical, and subtropical latitude zones.

Other Horizons or Layers of Diagnostic Value

Duripan (L. *durus,* "hard"). A dense, hard subsurface horizon cemented by silica that does not soften during prolonged soaking.

Fragipan (L. *fragilis,* "brittle"). Dense, moderately brittle layer, often mottled in color. Weak cementation may be due to close packing and binding by clay. A fragipan tends to retard downward movement of water.

Diagnostic Materials of Organic Soils

Fibric soil materials (L. *fibra,* "fiber"). Organic matter consisting of fibers readily identifiable as to botanical origin. An example is *Sphagnum* peat from bogs in cold climates.

Hemic soil materials (Gr. *hemi,* "half"). Organic matter of intermediate degree of composition between fibric and sapric materials.

Sapric soil materials (Gr. *sapros,* "rotten"). Highly decomposed organic matter; denser than fibric and hemic materials, with little content of identifiable fibers.

The Soil Orders

Every polypedon falls within one, and only one, of the 10 **soil orders.** Each order has its unique criteria, so selected that the criteria for a given order exclude members of all other orders. Criteria may include (1) gross composition, whether organic or mineral or both (e.g., percent clay or percent organic matter); or (2) degree of development of horizons; or (3) presence or absence of certain diagnostic horizons; or (4) degree of weathering of the soil minerals, expressed as cation exchange capacity (CEC) or as percent base saturation (PBS).

Those soil scientists who devised the Soil Taxonomy were well aware that a polypedon derives its properties from some unique combination of soil-forming processes and influences — climate and living organisms acting on parent materials over time and influenced by landform. A working soil taxonomy cannot, however, be constructed in terms of the soil-forming processes — on a genetic basis, that is — because this information is not always available. Moreover, the origin of a given soil feature is often subject to differences of interpretation in terms of the processes involved. That is why the Soil Taxonomy of the CSCS is based on observable features of the soil profile and additional facts that can be determined by laboratory analysis of soil samples.

As geographers, we cannot rest easily with an alphabetical listing of the 10 soil orders (as is the practice of

the Soil Survey Staff). Instead, a natural grouping and arrangement of orders will prove both meaningful and helpful. The three major classes we use are not, however, recognized by the CSCS. A brief descriptive statement introduces each soil order in the following sequence:

Soils with poorly developed horizons or no horizons and capable of further mineral alteration:

Entisols	Soils lacking horizons.
Inceptisols	Soils having weakly developed horizons and containing weatherable minerals.

Soils with a large proportion of organic matter:

Histosols Soils with a thick upper layer of organic matter.

Soils with well-developed horizons or with fully weathered minerals, resulting from long-continued adjustment to prevailing soil-temperature and soil-water regimes:

Oxisols	Very old, highly weathered soils of low latitudes, with an oxic horizon and low CEC.
Ultisols	Soils of mesic and warmer soil-temperature regimes, with an argillic horizon and low base status (PBS < 35%).
Vertisols	Soils of subtropical and tropical zones with high clay content, developing deep, wide cracks when dry and showing evidence of movement between aggregates.
Alfisols	Soils of humid and subhumid climates, with high base status (PBS > 35%) and an argillic horizon (B horizon).
Spodosols	Soils with a spodic horizon (B horizon), an albic horizon (A2 horizon) with low CEC, and lacking carbonate minerals.
Mollisols	Soils chiefly of midlatitudes, with a mollic epipedon and very high base status, associated with subhumid and semiarid soil-water regimes.
Aridisols	Soils of dry climates, with or without argillic horizons, and with accumulations of carbonates or soluble salts.

The Soil Suborders

Within each soil order are **suborders** ranging in number from as few as two to as many as seven. Suborders are defined in various ways. The basis for defining suborders within one order may be quite different from that used for defining suborders within another order. For example, within some orders the list of suborders begins with soils of wet places and moves in sequence to well-drained and increasingly dry environments. Either soil-temperature regime or soil-water regime, or both, may be used to define suborders within certain orders. In still other orders, suborders are defined in terms of organic content or mineral content.

Names of suborders combine two sets of syllables called **formative elements.** The formative elements of the soil orders are given in Table 14.2 The formative elements used for suborders are given in Table 14.3; these elements

Table 14.2 / Formative Elements in Names of Soil Orders

Name of order	Formative element	Derivation of formative element	Pronunciation of formative element
Entisol	ent	Meaningless syllable	recent
Inceptisol	ept	L. *inceptum*, beginning	inept
Histosol	ist	Gr. *histos*, tissue	histology
Oxisols	ox	Fr. *oxide*, oxide	ox
Ultisol	ult	L. *ultimus*, last	ultimate
Vertisol	ert	L. *verto*, turn	invert
Alfisol	alf	Meaningless syllable	alfalfa
Spodosol	od	Gr. *spodos*, wood ash	odd
Mollisol	oll	L. *mollis*, soft	mollify
Aridisol	id	L. *aridus*, dry	arid

form prefixes, attached to the formative elements of the order names. For example, a *Boralf* is a suborder of the Alfisols. The element *bor* connotes its occurrence in a cold (boreal) climate; *alf* is the formative element in Alfisol.

Throughout the remainder of this chapter, we shall take up each order in turn, giving a description of the order followed by a list and description of the suborders. A world map of soils, Plate D.5, shows the world distribution of soil orders and a number of the most important suborders. Bear in mind that, over large areas (particularly in low latitudes), the map boundaries are not well substantiated by field observations. For some orders, such as Aridisols, Mollisols, Spodosols, Ultisols, and Oxisols, the world pattern strongly reflects climate controls. A single color is assigned to each of these orders. For other orders — Alfisols, Entisols, Inceptisols, and Histosols —

the regions of occurrence of each order are widespread in terms of latitude. Within the Alfisols, each of four suborders is represented by a different color. The colors show that each suborder is strongly related to climate and latitude zone. Entisols and Inceptisols are shown in the same colors as related, contiguous orders with well-developed horizons. This treatment helps to simplify the map and to accentuate the relationship between soils and climate; it carries forward our objective as geographers to recognize important global environmental patterns and their interrelationships.

Figure 14.1 is a map of soil orders and suborders of the United States and southern Canada. In this region, map boundaries are well established by detailed field observations.

Entisols*

Entisols have in common the combination of a mineral soil and the absence of distinct pedogenic horizons that would persist after normal plowing. (The syllable *ent* in Entisol has no root meaning, but can be associated with "recent.")

*Much of the description of soil orders and suborders in the remainder of this chapter is selectively quoted or paraphrased from one of the following two sources: (1) Guy D. Smith, *The soil orders of the classification used in the U.S.A.*, Romanian Geological Institute, Technological and Economic Bulletin, Series C — Pedology, no. 18, pp. 509–31, 1970. Dr. Guy D. Smith is the former Director, Soil Survey Investigations, Soil Conservation Service, U.S. Department of Agriculture. (2) Roy W. Simonson and Darrell L. Gallup, *Soils of the world*, unpublished report, 1972. Dr. Roy W. Simonson is the former Director of Soil Classification and Correlation, Soil Survey, Soil Conservation Service, U.S. Department of Agriculture. This material has been used with permission of the authors.

Table 14.3 / Formative Elements in Names of Soil Suborders

Formative element	Derivation	Memory word (Mnemonicon)	Connotation
Alb	L. *albus*, white	albino	Presence of albic horizon
And	Modified from *andosol*		Resembling Ando soils of Indonesia, formed from volcanic ash
Aqu	L. *aqua*, water	aquarium	Aquic soil-water regime
Ar	L. *arare*, to plow	arable	Horizons mixed by plowing
Arg	L. *argilla*, white clay	argillaceous	Presence of argillic horizon
Bor	Gr. *boreas*, northern	boreal	Formed in cold climate
Ferr	L. *ferrum*, iron	ferrous	Presence of iron
Fibr	L. *fibra*, fiber	fibrous	Least decomposed stage of organic matter
Fluv	L. *fluvius*, river	fluvial	Formed on floodplain deposits
Fol	L. *folia*, leaf	foliage	Layer of forest litter
Hem	Gr. *hemi*, half	hemisphere	Intermediate stage of decomposition of organic matter
Hum	L. *Humus*, earth	humus	Presence of organic matter
Ochr	Gr. *ochros*, pale	ocher	Presence of ochric epipedon
Orth	Gr. *orthos*, true	orthodox	The common suborder
Plagg	Ger. *plaggen*, sod		Presence of plaggen epipedon
Psamm	Gr. *psammos*, sand		Sand texture
Rend	From *Rendzina*, a soil order in 1938 USDA system.		High carbonate content
Sapr	Gr. *sapros*, rotten	saprolite	Most decomposed stage of organic matter
Torr	L. *torridus*, hot and dry	torrid	Torric soil-water regime
Ud	L. *udus*, humid	udder	Udic soil-water regime
Umbr	L. *umbra*, shade	umbrella	Presence of umbric epipedon
Ust	L. *ustus*, burnt	combustion	Ustic soil-water regime
Xer	Gr. *xeros*, dry	xerophyte	Xeric soil-water regime

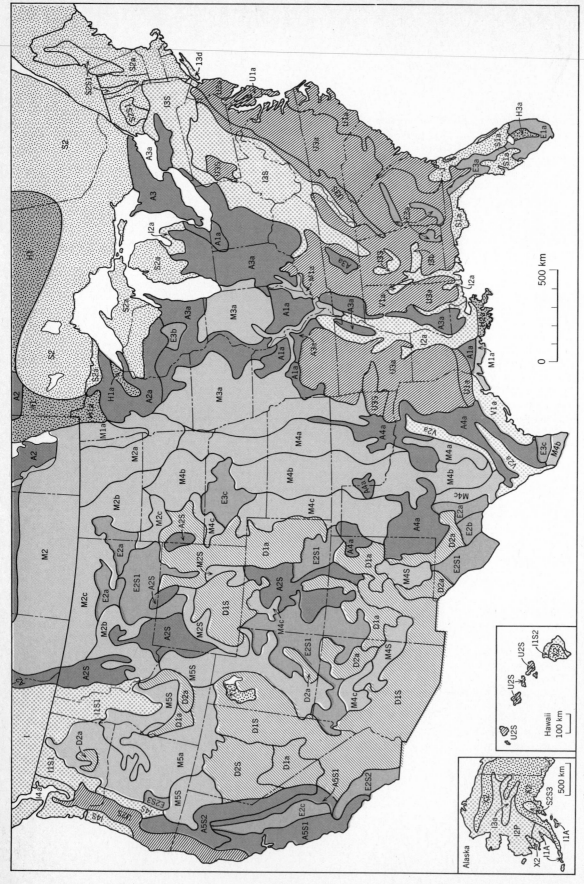

Figure 14.1 **Soil orders and suborders of the United States and southern Canada. Legend on facing page. (Soil Conservation Service, U.S. Department of Agriculture, 1972.)**

ALFISOLS

AQUALFS

A1a—Aqualfs with Udalfs, Haplaquepts, Udolls; gently sloping.

BORALFS

A2a—Boralfs with Udipsamments and Histosols; gently and moderately sloping.

A2S—Cryoboralfs with Borolls, Cryochrepts, Cryorthods, and rock outcrops; steep.

UDALFS

A3a—Udalfs with Aqualfs, Aquolls, Rendolls, Udolls, and Udults; gently or moderately sloping.

USTALFS

A4a—Ustalfs with Ustochrepts, Ustolls, Usterts, Ustipsamments, and Ustorthents; gently or moderately sloping.

XERALFS

A5S1—Xeralfs with Xerolls, Xerorthents, and Xererts; moderately sloping to steep.

A5S2—Ultic and lithic subgroups of Haploxeralfs with Andepts, Xerults, Xerolls, and Xerochrepts; steep.

ARIDISOLS

ARGIDS

D1a—Argids with Orthids, Orthents, Psamments, and Ustolls; gently and moderately sloping.

D1S—Argids with Orthids, gently sloping; and Torriorthents, gently sloping to steep.

ORTHIDS

D2a—Orthids with Argids, Orthents, and Xerolls; gently or moderately sloping.

D2S—Orthids, gently sloping to steep, with Argids, gently sloping; lithic subgroups of Torriorthents and Xerorthents, both steep.

ENTISOLS

AQUENTS

E1a—Aquents with Quartzipsamments, Aquepts, Aquolls, and Aquods; gently sloping.

ORTHENTS

E2a—Torriorthents, steep, with borollic subgroups of Aridisols; Usterts and aridic and vertic subgroups of Borolls; gently or moderately sloping.

E2b—Torriorthents with Torrerts; gently or moderately sloping.

E2c—Xerorthents with Xeralfs, Orthids, and Argids, gently sloping.

E2S1—Torriorthents; steep, and Argids, Torrifluvents, Ustolls, and Borolls; gently sloping.

E2S2—Xerorthents with Xeralfs and Xerolls; steep.

E2S3—Cryorthents with Cryosamments and Cryandepts; gently sloping to steep.

PSAMMENTS

E3a—Quartzipsamments with Aquults and Udults; gently or moderately sloping.

E3b—Udipsamments with Aquolls and Udalfs; gently or moderately sloping.

E3c—Ustipsamments with Ustalfs and Aquolls; gently or moderately sloping.

HISTOSOLS

HISTOSOLS

H1a—Hemists with Psammaquents and Udipsamments; gently sloping.

H2a—Hemists and Saprists with Fluvaquents and Haplaquepts; gently sloping.

H3a—Fibrists, Hemists, and Saprists with Psammaquents; gently sloping.

INCEPTISOLS

ANDEPTS

I1a—Cryandepts with Cryaquepts, Histosols, and rock land; gently or moderately sloping.

I1S1—Cryandepts with Cryochrepts, Cryumbrepts, and Cryorthods; steep.

I1S2—Andepts with Tropepts, Ustolls, and Tropofolists; moderately sloping to steep.

AQUEPTS

I2a—Haplaquepts with Aqualfs, Aquolls, Udalfs, and Fluvaquents; gently sloping.

I2P—Cryaquepts with cryic great groups of Orthents, Histosols, and Ochrepts; gently sloping to steep.

OCHREPTS

I3a—Cryochrepts with cryic great groups of Aquepts, Histosols, and Orthods; gently or moderately sloping.

I3b—Eutrochrepts with Uderts; gently sloping.

I3c—Fragiochrepts with Fragiaquepts, gently or moderately sloping; and Dystrochrepts, steep.

I3d—Dystrochrepts with Udipsamments and Haplorthods; gently sloping.

I3S—Dystrochrepts, steep, with Udalfs and Udults; gently or moderately sloping.

UMBREPTS

I4a—Haplumbrepts with Aquepts and Orthods; gently or moderately sloping.

I4S—Haplumbrepts and Orthods; steep, with Xerolls and Andepts; gently sloping.

MOLLISOLS

AQUOLLS

M1a—Aquolls with Udalfs, Fluvents, Udipsamments, Ustipsamments, Aquepts, Eutrochrepts, and Borolls; gently sloping.

BOROLLS

M2a—Udic subgroups of Borolls with Aquolls and Ustorthents; gently sloping.

M2b—Typic subgroups of Borolls with Ustipsamments, Ustorthents, and Boralfs; gently sloping.

M2c—Aridic subgroups of Borolls with Borollic subgroups of Argids and Orthids, and Torriorthents; gently sloping.

M2S—Borolls with Boralfs, Argids, Torriorthents, and Ustolls; moderately sloping or steep.

UDOLLS

M3a—Udolls, with Aquolls, Udalfs, Aqualfs, Fluvents, Psamments, Ustorthents, Aquepts, and Albolls; gently or moderately sloping.

USTOLLS

M4a—Udic subgroups of Ustolls with Orthents, Ustochrepts, Ustipsamments, Aquents, Fluvents, and Udolls; gently or moderately sloping.

M4b—Typic subgroups of Ustolls with Ustipsamments, Ustorthents, Ustochrepts, Aquolls, and Usterts; gently or moderately sloping.

M4c—Aridic subgroups of Ustolls with Ustalfs, Orthids, Ustipsamments, Ustorthents, Ustochrepts, Torriorthents, Borolls, Ustolls, and Usterts; gently or moderately sloping.

M4S—Ustolls with Argids and Torriorthents; moderately sloping or steep.

XEROLLS

M5a—Xerolls with Argids, Orthids, Fluvents, Cryoborolls, Cryoborolls, and Xerorthents; gently or moderately sloping.

M5S—Xerolls with Cryoboralfs, Xeralfs, Xerorthents, and Xererts; moderately sloping or steep.

SPODOSOLS

AQUODS

S1a—Aquods with Psammaquents, Aquolls, Humods, and Aquults; gently sloping.

ORTHODS

S2a—Orthods with Boralfs, Aquents, Orthents, Psamments, Histosols, Aquepts, Fragiochrepts, and Dystrochrepts; gently or moderately sloping.

S2S1—Orthods with Histosols, Aquents, and Aquepts; moderately sloping or steep.

S2S2—Cryorthods with Histosols; moderately sloping or steep.

S2S3—Cryorthods with Histosols, Andepts and Aquepts; gently sloping to steep.

ULTISOLS

AQUULTS

U1a—Aquults with Aquents, Histosols, Quartzipsamments, and Udults; gently sloping.

HUMULTS

U2S—Humults with Andepts, Tropepts, Xerolls, Ustolls, Orthox, Torrox, and rock land; gently sloping to steep.

UDULTS

U3a—Udults with Udalfs, Fluvents, Aquents, Quartzipsamments, Aquepts, Dystrochrepts, and Aquults; gently or moderately sloping.

U3S—Udults with Dystrochrepts; moderately sloping or steep.

VERTISOLS

UDERTS

V1a—Uderts with Aqualfs, Eutrochrepts, Aquolls, and Ustolls; gently sloping.

USTERTS

V2a—Usterts with Aqualfs, Orthids, Udifluvents, Aquolls, Ustolls, and Torrerts; gently sloping.

Areas with little soil

X1—Salt flats.

X2—Rock land (plus permanent snow fields and glaciers).

Slope classes

Gently sloping—Slopes mainly less than 10 percent, including nearly level.

Moderately sloping—Slopes mainly between 10 and 25 percent.

Steep—Slopes mainly steeper than 25 percent.

Entisols are soils in the sense that they support plants, but they may be found in any climate and under any vegetation. Lack of distinct horizons in the Entisols may be the result of a parent material, such as quartz sand, in which horizons do not readily form; or it may be the result of lack of time for horizons to form in recent deposits of volcanic ash or alluvium, or on actively eroding slopes, or in soils recently disturbed by plowing to a depth of one meter or more.

Entisols occur throughout the full global range from equatorial to arctic latitude zones. One large area identified on the world soils map (Plate D.5) as dominantly Entisols (E2) is the heavily glaciated Ungava region of Quebec and Labrador. Entisols occur as a number of large dune-sand areas (E3) of the North African-Arabian desert, as a large part of the Kalahari Desert of southwestern Africa, and as parts of the central and northern desert of Australia. A very different type of region having Entisols is indicated in central and northern China (E1). This is an expanse of floodplains and deltaic plains of the Yellow and Yangtze rivers.

Entisols of the subarctic zone and tropical deserts (along with arctic areas of Inceptisols) are the poorest of all soils from the standpoint of potential agricultural productivity. In contrast, Entisols and Inceptisols of floodplains and deltaic plains in warm and moist climates are among the most highly productive agricultural soils in the world because of their favorable texture, ample nutrient content, and large soil-water storage. Dense agricultural populations in central China and in the Ganges-Brahmaputra plain of India and Bangladesh bear witness to this fact.

Suborders. To some extent, the following definitions of the five suborders of Entisols allow us to sort out the productive suborders from the nonproductive suborders.

Aquents (E1) are Entisols of wet places (aquic regime). They are dominantly gray throughout and stratified or are saturated with water at all times. Aquents include some soils formerly called *low-humic gley soils* in the 1938 USDA system.

Arents are Entisols with recognizable fragments of pedogenic horizons that have been mixed by deep plowing, spading, or other disturbance by Man.

Fluvents are Entisols formed from recent alluvium, rarely saturated with water, usually stratified, and brownish to reddish throughout — at least in the upper 50 cm (20 in.) — and with loamy or clayey texture in some part of the upper 1 m (40 in.). Fluvents include some soils formerly called *Alluvial soils* in the 1938 USDA system.

Orthents (E2) are Entisols on recent erosional surfaces. They have a shallow depth to consolidated rock or have unconsolidated materials, such as loess, and have skeletal, loamy, or clayey texture in some part of the upper 1 m (40 in.). Orthents include some soils formerly called *Lithosols* in the 1938 USDA system.

Psamments (E3) are Entisols that have sandy texture (sands and loamy sands) throughout the upper 1 m (40 in.). Parent material is usually dune sand or beach sand. Psamments include some soils formerly called *Sands* in the 1938 USDA system.

Inceptisols

Inceptisols are uniquely defined by a combination of the following properties: (1) Soil water is available to plants more than half the year for more than three consecutive months during a warm season. The soil-water regime is mostly udic or perudic. (2) One or more pedogenic horizons formed by alteration or concentration of matter are present, but without accumulation of translocated materials other than carbonate minerals or amorphous (noncrystalline) silica. (3) Soil textures are finer than loamy sand. (4) The soil contains some weatherable minerals. (5) The clay fraction of the soil has a moderate to high CEC.

Inceptisols are found in a wide range of latitudes having the requisite soil-water regimes. One area of widespread occurrence is in the region of tundra climate (12). Compare the subarctic and arctic regions of Inceptisol occurrence on the world soils map (Plate D.5) with the regions of tundra climate shown on the world climate map (Plate D.2). The correspondence is generally quite good. Inceptisols are also found in high mountains associated with alpine tundra climate; such locations are found at all latitudes, ranging down to the equatorial zone.

The Inceptisols are mostly found on relatively young geomorphic surfaces, for example, surfaces shaped by glacial ice of the last great ice advance (Wisconsin glaciation, Chapter 21). Because of their youth, these Inceptisols usually lack distinct horizons. Under the cold tundra climate with its permafrost, chemical decomposition of minerals is much inhibited.

Inceptisols also occur on recently accumulated alluvial sediments of floodplains where sedimentation is no longer active and pedogenic horizons have developed. These occurrences are shown on the world soils map as ranging from equatorial to midlatitude zones. Examples are floodplains of the Mississippi, Amazon, Congo, and Ganges-Brahmaputra rivers and deltaic plains of the Nile, Irrawaddy, and Mekong rivers.

Suborders. Definitions of the five suborders of Inceptisols make clear the point that they can occur over the full geographic range from the arctic zone to the equator.

Andepts commonly have a dark surface horizon, are rich in humus and amorphous silica (allophane), and have a high CEC. Andepts are formed chiefly from recent volcanic ash deposits. An example is the occurrence of Andepts on the island of Hawaii, where volcanic activity is in progress.

Aquepts are Inceptisols of wet places (aquic regime). They have a gray to black mollic or histic epipedon, a gray subsoil, and are seasonally saturated with water. The Aquepts include some soils formerly called *Tundra soils* and some of the *Humic gley soils* and *Low-humic gley soils* in the revised 1938 USDA system.

Ochrepts have a brownish horizon (ochric epipedon) of altered materials at or close to the surface. Ochrepts are found in midlatitude and high-latitude climates. They include some soils formerly called *Brown forest soils* in the revised 1938 USDA system.

Plaggepts are Inceptisols with a plaggen epipedon — a thick surface horizon of materials added by Man through long habitation or long-continued manuring. Most are in Europe and the British Isles.

Tropepts are Inceptisols of low latitudes. They have a brownish or reddish horizon of altered material (ochric epipedon) at the surface. The Tropepts include some soils formerly called *Latosols* in the revised 1938 USDA system.

Umbrepts are Inceptisols with a dark acid surface horizon (umbric or mollic epipedon) more than 25 cm (10 in.) thick. They contain crystalline clay minerals and are brownish in color. They occur in moist midlatitude and high-latitude climates. The Umbrepts include some soils formerly referred to as *Brown forest soils, alpine meadow soils,* and *Tundra soils* in the 1938 USDA system.

Histosols

Histosols are unique in having a very high content of organic matter in the upper 80 cm (32 in.); (see Plate E.3A). More than half this thickness has from 20 to 30 percent of organic matter or more, or the organic horizon rests on bedrock or rock rubble. Most Histosols are peats or mucks consisting of more or less decomposed plant remains that accumulated in water; but some have formed from forest litter or moss or both in a cool, continuously humid climate, but are freely drained.

Histosols of cool climates are commonly acid in reaction and low in plant nutrients. Important areas of such Histosols are shown on the world soils map, Plate D.5, in the Northwest Territories of Canada and in an area lying to the south of Hudson Bay in Manitoba, Ontario, and Quebec. Histosols are found in low latitudes as well, where conditions of poor drainage have favored thick accumulations of plant matter.

Histosols that are mucks (fine black materials of sticky consistency) are agriculturally valuable in midlatitudes, where they occur in the beds of countless former lakes in the glaciated zone. After appropriate drainage and application of lime and fertilizers, these mucks are highly productive for truck-garden vegetables. Peat bogs are extensively used for cultivation of cranberries (cranberry bogs); *Sphagnum* peat is dried and baled for sale as a mulch for use on suburban lawns and shrubbery beds. Dried peat from bogs has been used for centuries in Europe as a low-grade fuel (Plate E.1).

Suborders. Histosols are subdivided into four suborders on the basis of soil-water regime and degree of decomposition of the plant remains.

Fibrists are Histosols composed mostly of *Sphagnum* moss or are Histosols that are saturated with water most of the year or artificially drained and consist mainly of recognizable plant remains so little decomposed that rubbing does not destroy the bulk of the fibers. Fibrists were formerly classed as *peats* in the 1938 USDA system.

Folists are Histosols formed with free drainage and consist of forest litter resting on bedrock or rock rubble.

Hemists are Histosols saturated with water most of the year or artificially drained, with between 10 and 40 percent of the volume remaining as fibers after rubbing. Hemists were formerly classed as *peaty mucks* in the 1938 USDA system.

Saprists are Histosols saturated with water most of the year or artificially drained, and so decomposed that after rubbing less than 10 percent of the volume remains as fibers. Saprists were formerly classed as *mucks* in the 1938 USDA system.

Oxisols

Oxisols are unique through the combination of three properties: (1) extreme weathering of most minerals, other than quartz, to sesquioxides of aluminum and iron and to kaolinite; (2) very low CEC of the clay fraction; and (3) a loamy or clayey texture (sandy loam or finer). An oxic horizon is usually present within 2 m (6 ft) of the surface. Plinthite may be present in the deeper parts of the profile.

Oxisols characteristically develop in equatorial, tropical, and subtropical zones on land surfaces that have been stable over long periods of time. Most of these surfaces are of early Pleistocene age or much older (i.e., older than half a million years). During soil development, the climate must be moist. In some areas, however, the Oxisols now occupy seasonally dry environments (wet-dry tropical climate) because a climate change has occurred since the time of their formation.

Oxisols of low latitudes lack distinct horizons, except for darkened surface layers, even though they are formed in strongly weathered parent material. Red, yellow, and yellowish-brown colors are normal to the well-drained Oxisols because of the presence of iron sesquioxides (Plate E.2B). Supplies of plant nutrients are commonly low. When phosphorus is added as a fertilizer, the soil has the capacity to fix the phosphorus into forms unavailable to plants. The clay fraction consists largely of silicate clay minerals, but with an important proportion of iron and aluminum sesquioxides. Despite the large clay fraction, however, the soils are friable (easily broken apart) and are easily penetrated by water and roots.

The world map of soils (Plate D.5) shows Oxisols to be dominant in vast areas of equatorial and tropical South America and Africa. These are regions of wet equatorial climate (1) and wet-dry tropical climate (3). (See world climate map, Plate D.2),

Oxisols and Ultisols of low latitudes were used under systems of shifting cultivation prior to the advent of modern technology; a large proportion of these soils is still used in that way. The levels of plant nutrients under natural conditions are so low that the yields obtainable under hand tillage are very low, especially after a garden patch has been used for a year or two. Substantial use of lime, fertilizers, and other industrial inputs is necessary for high yields and sustained production.

Suborders. Five suborders of Oxisols are recognized on the basis of their soil-water and soil-temperature regimes, or other properties very closely associated with climate.

Aquox are Oxisols of wet places (aquic regime). They are dominantly gray, or mottled dark red and gray, and are seasonally saturated with water. Aquox include some soils formerly called *Ground-Water Laterite soils* in the 1938 USDA system.

Humox are Oxisols of relatively cool moist regions, with very large accumulations of organic carbon. Humox occur on high plateaus; they include some soils formerly called *Black-red latosols* and some *Humic latosols* in the revised 1938 USDA system.

Orthox are Oxisols of warm, humid regions with short dry seasons or no dry season (perudic regime). These soils are never dry in any horizon for as long as 60 consecutive days. Orthox are widespread in the equatorial zone. They include soils formerly called *Red latosols* and *Red-yellow latosols* in the 1938 USDA system.

Torrox are Oxisols of the aridic or torric soil-water regime. They are dry in all horizons for more than six months of the year, are never continuously moist for as long as three consecutive months, and are not appreciably darkened by humus. Although occurring in dry climates, Torrox may have originated under a formerly moist climate.

Ustox are Oxisols of the ustic soil-water regime. They are moist in some horizon more than six months (cumulative) or more than three consecutive months, but are dry in some horizon for more than 60 consecutive days (roughly equivalent to a three-month dry season). Ustox are closely associated with the wet-dry tropical climate. They include some soils formerly called *low-humic latosols* in the 1938 USDA system.

Ultisols

Ultisols are unique through the combination of three properties: (1) There is an argillic horizon, (2) the supply of exchangeable bases (CEC) is low, particularly in the lower horizons; and (3) the mean annual soil temperature is greater than 8°C (47°F). A point of importance is that the percentage of base saturation (PBS) diminishes rapidly with depth. The highest PBS is within the upper few centimeters, reflecting the recycling of base cations by plants or the additions of fertilizers.

Ultisols characteristically form under forest vegetation in climates with a slight to pronounced seasonal soil-water deficit alternating with a surplus (udic and ustic regimes). Thus, at the season of soil-water surplus, some water passes through the soil into the substratum, allowing leaching to occur. As a result, bases not held in plant tissue are depleted. The B horizon of well-drained Ultisols is characteristically red or yellowish-brown in color due to concentration of sesquioxides of iron (Plate E.2C).

The world soils map (Plate D.5) shows a large area of dominantly Ultisols in the southeastern United States and another in southern China. There are also important areas of Ultisols in Bolivia, southern Brazil, western and central Africa, India, Burma, the East Indies, and northeastern Australia.

Reference to the world climate map, Plate D.2, shows that the climate in areas of Ultisols ranges from the moist subtropical climate (6) in the subtropical zone to the wet-dry tropical climate (3) and monsoon and trade-wind littoral climate (2) in the tropical zone. These climates have large water surpluses in one season, although they cover a large spread in temperature characteristics. The land surfaces in these areas have been subjected to prolonged erosion and weathering, so igneous bedrock beneath is often deeply decayed, forming thick saprolite.

Ultisols of low latitudes were largely used under systems of shifting cultivation prior to the development of modern technology, as they still are where the benefits of modern technology and capital are not available. The soils are low enough in plant nutrients so that growth of cultivated plants is poor without the general use of lime and fertilizers. Given essential industrial products and adequate management skills, however, the Ultisols can be highly productive.

Suborders. The five suborders of Ultisols are defined in terms of their soil-water regimes and organic content.

Aquults (U1) are Ultisols of wet places (aquic regime). They are seasonally saturated with water and are dominantly gray in color throughout. A few have black epipedons on gray subsoils. The Aquults include some soils formerly called *Low-humic gley soils* and *Humic gley soils* in the revised 1938 USDA system.

Humults (U2) are Ultisols with large accumulations of organic matter formed under relatively high, well-distributed rainfall in middle and low latitude zones. Humults include some soils formerly called *Reddish-brown lateritic soils* and *Humic latosols* in the revised 1938 USDA system.

Udults (U3) are Ultisols of the udic and perudic soil-water regimes. They have moderate to small amounts of organic matter, reddish or yellowish *B* horizons, and no periods or only short periods when some part of the soil is dry. Udults include some soils formerly called *Red-yellow podzolic* and *Reddish-brown lateritic soils* in the revised 1938 USDA system.

Ustults (U4) are Ultisols of the ustic soil-water regime, in which the soil is dry during a pronounced dry season. The soil-temperature regime is usually thermic. Ustults are brownish to reddish throughout. The suborder include some soils formerly called *Red-yellow podzolic soils* in the 1938 USDA system.

Xerults are Ultisols of Mediterranean climate (7) with cool moist winters and rainless summers. The xeric soil-water regime prevails. Xerults are brownish to reddish throughout. They include some soils formerly called *Noncalcic Brown soils* and *Red-yellow podzolic soils* in the 1938 USDA system.

Vertisols

Vertisols are uniquely defined through a combination of the following properties: (1) a high content of clay (montmorillonite) that shrinks and swells greatly with changes in soil-water storage; (2) deep wide cracks at some season; and (3) evidences of soil movement in the form of slickensides, gilgae, or structural aggregates tilted between horizontal and vertical. Horizons are normally weakly expressed and may not be apparent (Plate E.2D).

The evidences of movement, just mentioned, require description. **Slickensides** are soil surfaces with grooves or striations (scratch marks) made by movement of one mass of soil against another while the soil is in a moist, plastic state. **Gilgae** are very small relief features that may be knobs and basins or narrow ridges with valleys between. The height of these features is seldom over 1 m (3 ft). Heaving of the soil causes tilting of various objects supported by the soil, including fence posts and sidewalk slabs (Figure 14.2).

Vertisols typically form under grass or savanna vegetation in subtropical and tropical climates with a moderate or pronounced seasonal soil-water deficit. These climates include dry tropical climate, semiarid subtype (4s), and wet-dry tropical climate (3). At some season the soils dry out enough that the soil becomes deeply cracked (Figure 14.3). When rains come, some masses of surface soil drop into the cracks before they close, so the soil slowly "swallows itself." Vertisols the world over have had more names than any other kind of soil. Examples are *black cotton soils, tropical black clays,* and *regur.* In the 1938 USDA system, most Vertisols were called *Rendzina soils* for some years and then renamed *Grumusols.*

Vertisols are found only in the latitude range of the tropical and subtropical zones. The U.S. soils map, Figure 14.1, shows three narrow belts of Vertisols on the coastal plain of Texas. One large tropical area is in the Deccan region of India, where the Vertisols are formed on weathered basalts. Other areas shown on the world map are the Sudan region of east-central Africa and a large north-south belt in eastern Australia.

Figure 14.3 **Soil cracks in a Texas Vertisol, the Houston black clay, during an extremely dry period. The crop is cotton. (Soil Conservation Service, U.S. Department of Agriculture.)**

Figure 14.2 **Sidewalk slabs tilted by movements in Vertisol (Houston black clay), Travis County, Texas. (Soil Conservation Service, U.S. Department of Agriculture.)**

Vertisols are very high in exchangeable bases such as calcium and magnesium; most are near neutral in pH; and most have intermediate amounts of organic matter. The soils retain large amounts of water because of their fine texture, but much of this soil water is held by the montmorillonite clay and is not available to plants. Where soil cultivation depends on human or animal power, agricultural yields are low. Problems of using these soils are especially difficult; for example, the soil becomes highly plastic and difficult to till when wet. Many areas of Vertisols have thus been left in grass or savanna, which may provide grazing for cattle. With the use of modern technology and its energy resources, however, production of food and fiber from Vertisols could be substantial.

Suborders. The four suborders of Vertisols are defined on the basis of soil-water regimes. The suborders of Vertisols have no equivalents in the 1938 USDA system.

Torrerts are Vertisols of the torric regime. They have cracks that remain open throughout the year in most years.
Uderts (V1) are Vertisols of the udic regime. They have cracks that remain open for only short periods in most years, less than 90 cumulative days and less than 60 consecutive days.
Usterts (V2) are Vertisols of the ustic regime with distinct wet and dry seasons. They are widespread in areas of wet-dry tropical climate. The cracks remain open more than 90 cumulative days in most years.
Xererts are Vertisols of the xeric regime in the Mediterranean climate. The soil cracks close in winter but open in summer every year for more than 60 consecutive days.

Alfisols

Alfisols are uniquely defined by the following combination of properties: (1) a gray, brownish, or reddish horizon (ochric epipedon) not darkened by humus close to the surface; (2) a horizon of clay accumulation (argillic horizon); (3) a medium to high percentage base saturation (PBS); and (4) soil water available to plants more than half the year or more than three consecutive months during a warm season. The argillic horizon is a B horizon (Plate E.2E). It is enriched by accumulated silicate clay minerals and is moderately saturated with exchangeable bases, such as calcium and magnesium. The A2 horizon above it is marked by some loss of bases, silicate clay minerals, and sesquioxides.

On the world soils map (Plate D.5), dominant areas of Alfisols are shown in central North America, Europe, central Siberia, north China, and southern Australia. These areas are identified with the moist continental climate (10), the marine west-coast climate (8), and the Mediterranean climate (7); see the world climate map, Plate D.2. Other important areas of Alfisols shown on the map are located in eastern Brazil; western, eastern, and southern Africa; western Madagascar; northern Australia; eastern India; and Indochina. These occurrences are in areas of wet-dry tropical climate (3) and the semiarid subtypes of the tropical and subtropical dry climates (4s, 5s).

On the whole, Alfisols are agriculturally fairly productive under simple management because soil water is usually adequate in one season and bases are not depleted. In western Europe and China, Alfisols, along with the Inceptisols occurring within the same regions, have supported dense populations for centuries.

Suborders. Five suborders of the Alfisols are defined in terms of soil-water and soil-temperature regimes. Four of the five suborders are shown with distinctive colors on the world soils map, Plate D.5.

Aqualfs are Alfisols of wet places (aquic regime). They are dominantly gray throughout and are seasonally saturated with water. Aqualfs were classed as *Low-humic gley soils* and *Planosols* in the revised 1938 USDA system.

Boralfs (A1) are Alfisols of boreal forests or high mountains. They have a gray surface horizon, a brownish subsoil, and are associated with a mean annual soil temperature less than 8°C (47°F). The boralfs were called *Gray wooded soils* in the 1938 USDA system.

Udalfs (A2) are brownish Alfisols of the udic soil-water regime. They occur in regions having no periods or only short periods (fewer than 90 days in most years) when part or all of the soil is dry. A deciduous forest was the typical vegetation on Udalfs, but today these soils are intensively farmed. Udalfs were classed as *Gray-Brown Podzolic soils* in the 1938 USDA system.

Ustalfs (A3) are Alfisols of the ustic soil-water regime, with long periods (more than 90 cumulative days in most years) when the soil is dry. Ustalfs are brownish to reddish throughout (Plate E.2F). They were designated as *Reddish-Brown soils* and *Reddish-Chestnut soils* in the 1938 USDA system.

Xeralfs (A4) are Alfisols of the xeric soil-water regime, under a Mediterranean climate with cool, moist winters and rainless summers. Xeralfs are brownish to reddish throughout. A typical occurrence is in coastal and inland valleys of central and southern California (Figure 14.1). Xeralfs were designated as *Noncalcic Brown soils* in the 1938 USDA system.

Spodosols

Spodosols have a unique property: a spodic horizon (*B* horizon) of accumulation (illuviation) of dark amorphous materials. The amorphous materials contain organic material, compounds of aluminum, and commonly iron, all brought downward by eluviation from an overlying A horizon (Plate E.2H). In undisturbed Spodosols, a bleached gray to white A2 horizon overlies the B horizon. This is an albic horizon. Spodosols are strongly acid, low in plant nutrients, such as the base cations of calcium and magnesium, and generally low in humus. Spodosols usually have sand texture and small water-holding capacities.

Spodosols, with some exceptions, were formed under forest cover in a midlatitude climate that is both cool and moist. In comparing the world distribution of Spodosols with that of climates (compare Plate D.5 with Plate D.2), the correspondence with the boreal forest climate (11) is striking. Spodosols also occupy northerly portions of the moist continental climate (10). Correspondence with needleleaf (boreal) forest is also striking (see Plate D.6).

As is the case of the northern Inceptisols, regions of Spodosols are largely those that experienced the most recent continental glaciation; the soils are therefore very young. Nevertheless, Spodosols can be recognized as occurring in latitudes down to the equatorial zone. For example, soils of much of northern and central Florida are classed as Spodosols.

Spodosols of middle and high latitudes are naturally poor soils in terms of agricultural productivity. Because they are acid, the application of lime is a necessity. Because they are poor in the nutrients required by most crops, the application of fertilizers is required. With proper management and the input of required industrial products, the Spodosols can be highly productive. Another unfavorable factor is shortness of the growing season in the more northerly areas of Spodosols.

Suborders. Four suborders of the Spodosols are defined in terms of soil-water regimes and composition. (Areas of undifferentiated Spodosols are designated S1 on the world soils map.)

Aquods (S2) are Spodosols of wet places (aquic regime), seasonally saturated with water. The surface horizon may be either black or white and lacks free iron oxides. The B horizon is black, brownish, or reddish and may be firm or cemented. Aquods were called *Ground Water Podzols* in the 1938 USDA system.

Ferrods are freely drained Spodosols having B horizons of accumulation of free iron oxides. This suborder is of rare occurrence.

Humods (S3) are freely drained Spodosols in which at least the upper part of the B horizon has accumulated humus and aluminum, but not iron. Humods were formerly classed with the *Podzol soils* in the 1938 USDA system.

Orthods (S4) are freely drained Spodosols with B horizons in which iron, aluminum, and humus have accumulated. Orthods are widespread in North America and northern Europe. Orthods were identified with the *Podzol soils* and *Brown Podzolic soils* in the 1938 USDA system.

Mollisols

Mollisols are uniquely defined through a combination of the following properties: (1) a mollic epipedon, which is a very dark brown to black surface horizon that is more than one-third of the combined thickness of the A and B horizons or is more than 25 cm (10 in.) thick and has a loose structure or soft consistence when dry; (2) a dominance of calcium among the extractable base cations in the A and B horizons; (3) a dominance of crystalline clay minerals of moderate or high cation exchange capacity (CEC); and (4) less than 30 percent of clay in some horizon if the soil has deep wide cracks at some season.

Mollisols typically form under grass in climates with a moderate to pronounced seasonal soil-water deficiency. A few form in marshes or on marls (compacted muds of calcium carbonate composition) in moist climates. Comparison of Mollisol distribution on the world soils map (Plate D.5) with distribution of climates in Plate D.2 shows that the Mollisols are closely associated with the semiarid subtype of the dry midlatitude climate (9s), with the subhumid subtype and adjacent parts of the moist continental climate (10sh, 10h), and with the subhumid subtype of the moist subtropical climate (6sh). In North America, Mollisols dominate the Great Plains region, the Columbia Plateau, and the northern Great Basin. Smaller areas exist along the Gulf coastal plain of Mexico and the Yucatan Peninsula. In South America, a large area of Mollisols covers the Pampas of Argentina and Uruguay. In Eurasia, a great belt of Mollisols stretches from Rumania eastward across the steppes of Russia, Siberia, and Mongolia, into Manchuria.

Because of their well-developed granular texture and high base saturation, Mollisols are among the naturally most fertile soils in the world. They now produce the bulk of the grain that moves in commercial trade channels. Most of these soils have not been widely used for crop production except during the last century. Prior to that time they were used mainly for grazing by nomadic herds. The Mollisols have favorable characteristics for growing cereals in large-scale farming and are relatively easy to manage. Production varies considerably from one year to the next because seasonal rainfall is highly variable and soil-water storage is the limiting factor in productivity.

Suborders. Seven suborders of Mollisols are defined primarily in terms of their soil-water and soil-temperature regimes, and in one suborder by the carbonate content.

Albolls (M1) are Mollisols with a bleached gray to white horizon (albic horizon) above a slowly permeable horizon of clay accumulation, saturated with water in at least the surface horizons at some season. Albolls include some soils called *Planosols* and *solodized solonetz* in the 1938 USDA system.

Aquolls are Mollisols of wet places (aquic regime). They have a nearly black mollic epipedon resting on a mottled grayish B horizon and are seasonally saturated with water (Plate E.3*I*). Aquolls include soils called *humic-gley soils, calcium carbonate Solonchak,* and a few *Solonetz soils* in the 1938 USDA system.

Borolls (M2) are Mollisols of cold-winter semiarid plains (steppes) or high mountains (Plate E.3*J*). They have a mean annual soil temperature lower than 8°C (47°F). Borolls include some soils called *Chernozem soils, Chestnut soils,* and *Brown soils* in the 1938 USDA system.

Rendolls (M3) are Mollisols formed on highly calcareous parent materials and with more than 40 percent calcium carbonate in some horizon within 50 cm (20 in.) of the surface, but without a horizon of accumulation of calcium carbonate (Plate E.3*K*). Rendolls are mostly formed in moist climates under forest vegetation. They were called *Rendzina soils* in the 1938 USDA system.

Udolls (M4) are Mollisols of the udic soil-water regime. They have brownish hues throughout and no horizon of accumulation of calcium carbonate in soft powdery forms (Plate E.3*L*). Udolls include soils called *Prairie soils* and *brunizems* in the 1938 USDA system.

Ustolls (M5) are Mollisols of the ustic soil-water regime, with long periods (more than 90 cumulative days in most years) in which the soil is dry. Most Ustolls have a horizon of accumulation of soft, powdery calcium carbonate starting at a depth of about 50 to 100 cm (20 to 40 in.). (This horizon is labeled *Cca* in Plate E.3*M*.) A petrocalcic horizon may be formed within 1 m (3 ft) of the soil surface. In the southwestern United States, this material is known as *caliche* (Figure 14.4). Ustolls include some soils called *Chernozems, Chestnut soils,* and *Brown soils* in the 1938 USDA system.

Xerolls (M6) are Mollisols of the xeric regime under a Mediterranean climate with cool, moist winters and rainless summers. Xerolls include some soils called *brunizems, Chestnut soils,* and *Brown soils* in the 1938 USDA system.

Aridisols

Aridisols are uniquely defined through a combination of the following properties: (1) a lack of water available for most plants for very extended periods, (2) one or more

Figure 14.4 **A petrocalcic horizon consisting of rocklike slabs and nodules of calcium carbonate, formed in an area of Ustolls on the Pecos Plains of New Mexico. This material is known locally as caliche. (A. N. Strahler.)**

pedogenic horizons, (3) a surface horizon or horizons not significantly darkened by humus, and (4) absence of deep wide cracks. The Aridisols have practically no available soil water most of the time that the soil is warm enough for plant growth, for example, higher than 5°C (41°F). Soil water, in most years, is not available to plants for as long as 90 consecutive days when the soil temperature is above 8°C (47°F). Carbonate accumulations are often large at depth in the profile.

Aridisols form under conditions that preclude entry of much water into the soil — either extremely scanty rainfall or slight rainfall that for one reason or another does not enter the soil. The vegetation consists of scattered plants, ephemeral grasses and forbs, cacti, and shrubs, all of which are well adapted to sustained drought.

The world soils map (Plate D.5) shows that Aridisols occupy regions shown on the climate map (Plate D.2) as semidesert and desert subtypes of the dry tropical, subtropical, and midlatitude climates (4sd, 4d, 5sd, 5d, 9sd, 9d) and the Mediterranean climate (7s, 7sd). These same arid regions include large areas of Entisols, occupying areas of dune sand (Psamments) or rocky desert floor.

Most Aridisols are used, as they have been through the ages, for nomadic grazing. This use is dictated by the limited rainfall, which is inadequate for crops without irrigation. Locally, where water supplies from exotic streams or ground water permit, Aridisols can be highly productive for a wide variety of crops under irrigation (Plate E.1).

Suborders. Two suborders of Aridisols are defined by the presence or absence of an argillic horizon (a B horizon of clay accumulation). (Areas of undifferentiated Aridisols are designated D1 on the world soils map.)

Argids (D2) are Aridisols with an argillic horizon (Plate E.3*N*). Argids have formed largely on surfaces of late Pleistocene (Wisconsin) age or older and, consequently, have developed throughout one or more of the Pleistocene pluvial (moist) periods. Illuviation of the B horizon is thought to have taken place during these moist periods. Argids include some soils called *Desert soils* and *Red Desert soils* and associated *Solonetz soils* in the 1938 USDA system.

Orthids are soils without an argillic horizon but with one or more pedogenic horizons that extend at least 25 cm (10 in.) below the surface, so some horizon persists in part if the soil is plowed (Plate E.3*O*). One variety of Orthids (Salorthids) is characterized by having a salic horizon (horizon of salt accumulation). This horizon appears as a white layer in Plate E.4. Orthids include some soils called *Desert soils* and *Red Desert soils, calcisols,* and *Sierozem soils* in the 1938 USDA system.

Soils and Altitude

As we explained in Chapters 5 and 7, an increase in altitude brings lower air temperatures and increased orographic precipitation. Correspondingly, the soil-temperature and soil-water regimes undergo a change. The altitude effect is particularly conspicuous on the flanks of isolated mountain ranges within the dry mid-latitude climate (9). Figure 14.5 shows a schematic soil profile, changing gradually with increased elevation as an observer traces soil horizons from a semiarid basin floor to the summit of a high mountain range — in this case the Bighorn Mountains in Wyoming. At the lowest elevation (left), the soil profile is that of an Aridisol (an Argid). With increasing elevation the profile becomes that of a Ustoll, which grades into the profile of a Udoll under cool, moist conditions above 2300 m (7500 ft). On the summit uplands, above 2400 m (8000 ft), conditions are like those of the boreal forest climate. Here the soil profile is that of a Spodosol, formed under needleleaf forest.

Notice that a similar series of profile intergradations can be found in a west-to-east traverse across the United States, as shown in Figure 14.6, starting in the northern desert of the Great Basin and crossing the Midwest.

Global Soils in Review

In two chapters devoted to the basic elements of soil science, we have been limited to important concepts of soil formation and to a brief description of the soil orders and suborders. One dominant theme has been the role of climate in shaping the characteristics of the soil. This theme has made good use of a knowledge of the global heat and water balances and, more specifically, of the

Figure 14.5 **This schematic diagram shows the gradation of soils from a semiarid basin (left) to a cool, moist climate (right) as one ascends the west slope of the Bighorn Mountains, Wyoming. (After J. Thorp.)**

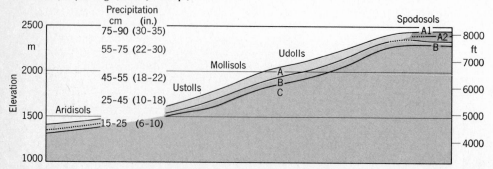

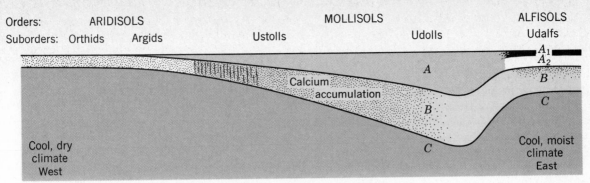

Orders: ARIDISOLS MOLLISOLS ALFISOLS

Suborders: Orthids Argids Ustolls Udolls Udalfs

A_1
A_2
A
B
C
Calcium accumulation
B
C

Cool, dry climate West

Cool, moist climate East

Figure 14.6 **A schematic diagram of the changing soil profile from a cool, dry desert on the west to a cool, moist climate on the east. (After C. E. Millar, L. M. Turk, and H. D. Foth, *Fundamentals of Soil Science,* John Wiley and Sons, New York.)**

soil-water balance. Soils associated with water-surplus budgets are strikingly different in structure, composition, and fertility from those associated with budgets featuring a large soil-water shortage. Surprisingly, the most fertile soils from the standpoint of plant nutrients are those of the semiarid climates — Mollisols, for example. A great surplus of water in a warm climate is not conducive to producing rich agricultural soils because, under such conditions, the nutrient bases are leached from the soil and the clay minerals are altered to varieties with low capacity to hold base cations — as in the Oxisols, for example.

We have stressed too that soil character is also shaped by organisms living in or on the soil. Plant tissues supply humus that profoundly influences the soil profile. Moreover, plants recycle nutrient bases and thus delay or prevent their escape from the soil.

In the next two chapters, we return to a study of the biosphere. First, we develop concepts of biogeography — a geographic approach to ecosystems and their environments. Second, we examine the global distribution of vegetation of the lands, looking for distributional patterns

linked to soil types and controlled by global climate patterns. Finally, in closing chapters of this book, soils, vegetation, and climate will be brought together in a survey of the global terrestrial environments.

References for Further Study

U.S. Department of Agriculture (1938), *Soils and men; yearbook of agriculture 1938,* Government Printing Office, Washington, D.C.

Cruickshank, J. G. (1972), *Soil geography,* David and Charles (Publishers), Newton Abbot, Devon, England.

Foth, H. D., and L. M. Turk (1972), *Fundamentals of soil science,* fifth edition, John Wiley and Sons, New York. See Chapter 10.

Buol, S. W., F. D. Hole, and R. J. McCracken (1973), *Soil genesis and soil classification,* Iowa State Univ. Press, Ames.

Soil Survey Staff (1975), *Soil taxonomy: a basic system of soil classification for making and interpreting soil surveys,* Soil Conservation Service, U.S. Department of Agriculture, Agriculture Handbook No. 436, Government Printing Office, Washington, D.C.

Steila, D. (1976), *The geography of soils,* Prentice-Hall, Englewood Cliffs, N.J.

Review Questions

1. Describe the evolution of pedology as a science. What three soil orders were recognized in the 1938 U.S. Department of Agriculture classification system? For what reasons was this system judged to be inadequate to meet the needs of modern soil science?

2. Give a general description of the Comprehensive Soil Classification System (CSCS). What goals and principles guided the development of the CSCS?

3. Name the six categories or levels of the Soil Taxonomy of the CSCS. Approximately how many classes are included within each level?

4. What are diagnostic horizons? How are diagnostic horizons used in the Soil Taxonomy? What two classes of diagnostic horizons are recognized?

5. What is an epipedon? Name and describe five important kinds of epipedons. Name and describe six subsurface diagnostic horizons. Name and describe at least five other horizons or layers of diagnostic value. What diagnostic

materials are used in the classification of organic soils?

6. What types of criteria are used to define the soil orders? Why is the origin (genesis) of a soil not considered a workable criterion of classification? Name the 10 soil orders, arranged in terms of three larger groups of geographic significance.

7. On what basis are suborders of soils defined? Give examples. How are formative elements used in creating the names of soil suborders? Give several examples.

8. What properties distinguish Entisols from other orders? Give examples of Entisols formed under highly varied combinations of parent material and climate. Illustrate by naming and describing at least four suborders of Entisols.

9. What combination of properties uniquely defines the Inceptisols? How do Inceptisols differ from Entisols? Are Inceptisols typically young soils or old soils? Name and describe four suborders of Inceptisols to illustrate the wide range of environments in which they form.

10. What unique feature distinguishes Histosols from soils of all

other orders? What is the basis for subdividing the Histosols into four suborders? Which of these suborders are important in agriculture?

11. What three properties are uniquely combined in the Oxisols? Is an oxic horizon always present? Explain the characteristic colors of Oxisols. Are Oxisols young soils or old soils? Do they have high or low levels of natural fertility in terms of food crops? Name and describe four suborders of the Oxisols. In what parts of the world do Oxisols occupy large areas? Why?

12. What three properties are uniquely combined in the Ultisols? Under what soil-water regime or regimes are most Ultisols formed? Where do Ultisols occur in large areas? Distinguish between Udults, Ustults, and Xerults.

13. What is unique about Vertisols? What is meant by the expression that the soil "swallows itself"? Is the base status of Vertisols high or low? Explain. Are Vertisols highly productive as agricultural soils the world over? Explain. On what basis are Vertisols separated into suborders?

14. List four properties uniquely combined in the definition of the Alfisols. How does the argillic horizon form? Are the Alfisols soils of low or high base status? Give the important global regions of occurrence of Alfisols. Name the suborders of Alfisols and assign each to a controlling soil-temperature regime or soil-water regime.

15. What single unique property defines the Spodosols? Describe the spodic horizon. How is it formed? Are Spodosols acid or alkaline? Rich or poor in nutrient bases? How must Spodosols be treated to achieve satisfactory levels of agricultural productivity? On what basis are Spodosols divided into suborders?

16. What four properties are uniquely combined in defining the Mollisols? With what climate and soil-water regime are the Mollisols most closely associated? Name the most important global regions where Mollisols occur. How do the Mollisols rank in terms of base status? In what way are Mollisols important to Man? Name and describe each of the seven suborders of Mollisols.

17. What four properties are uniquely combined in the definition of the Aridisols? How does climate control the properties of Aridisols? What natural vegetation is associated with Aridisols? Distinguish between Argids and Orthids.

18. How does the soil profile change with increasing altitude as one ascends a mountain range in western North America?

Concepts of Biogeography

15

A major field within physical geography is **biogeography,** the study of the distribution patterns of plants and animals on the earth's surface and the processes that produce those patterns. We shall devote two chapters to the field of biogeography. This chapter deals with the processes that differentiate the world ecosystem into distinctive assemblages of plants and animals; the next chapter deals with the spatial distribution patterns and unique features of the world's vegetation. Throughout our two chapters, emphasis will be on terrestrial plants. Not only is vegetation the most obvious part of the biotic landscape, but it is also the most important ecologically because the plants provide the primary production on which the animals depend.

In considering how various factors of the physical environment influence plants and animals, two scale ranges can be treated. One is the global scale, which considers such climatic factors as the seasonal and latitudinal patterns of insolation, light and darkness, temperature, precipitation, and prevailing winds. The other scale of consideration is that of variations of the physical environment found within a relatively small area. Thus, within a generally humid region, a few small areas (such as a dune or cliff) that are extremely dry places will be found. We may also find in a large desert some small places (such as a seep or spring) that are extremely wet most of the time.

Air temperatures and soil-water availability are the most important factors governing plant and animal distribution at both global and local scales. This chapter will begin by examining how plants and animals respond to the varying climatic factors of temperature and water availability.

The Biological Role of Water

Water is probably the most important of the factors that determine the global distribution patterns of plants and animals. Water is important because, through evolution, plants and animals have become specialized, or adapted, to excesses or deficiencies in water availability.

The availability of water to terrestrial organisms at a particular point in time or space is determined by the balance among precipitation, evaporation, runoff, and infiltration. This balance is, in turn, affected by organisms — mainly the plant cover. Through transpiration, plants return much of the soil water to the atmosphere. By obstructing overland flow and increasing soil porosity, plants reduce runoff and increase infiltration. Although these movements of water are vital from the viewpoint of organic life, their effects are small compared to those of the physical processes that control the major features of the water cycle. Thus the major pattern of variation in water from place to place is still determined by the overall dynamics of the atmosphere and oceans. Because such a large portion of the earth's surface lies in areas where water shortages exist in at least some portion of the year, our discussion will emphasize the ways in which plants and animals are adapted to dry conditions.

Organisms and Water Needs

Ecologists and biogeographers often classify plants according to their water requirements. Terms associated with the water factor are built on three simple prefixes of Greek roots: *xero-, "dry"; hygro-, "wet"; and meso-, "intermediate."* Plants that grow in dry habitats are **xerophytes;** those that grow in water or in wet habitats are **hygrophytes;** those of habitats of an intermediate degree of wetness and relatively uniform soil-water availability are **mesophytes.**

The xerophytes are highly tolerant of drought and can survive in habitats that dry quickly following rapid drainage of precipitation (e.g., on sand dunes, beaches, and bare rock surfaces). The plants typical of dry climates are also xerophytes; cactus is an example (Plate F.4). The hygrophytes are tolerant of excessive water and may be found in shallow streams, lakes, marshes, swamps, and bogs (Figure 15.1). The mesophytes are found in upland habitats in regions of ample rainfall. Here the drainage of soil water is good and moisture penetrates deeply, where it can later be used by the plants.

Water loss from plant tissues occurs through the

Figure 15.1 **Bog vegetation in Emmet County, Michigan. A bog mat lies at the edge of the lake, and a black spruce forest is in the background. (Pierre Dansereau.)**

process of transpiration, explained in Chapter 10. The rate of transpiration varies greatly according to the type of plant and the prevailing atmospheric conditions. High temperatures, low humidities, and winds favor high rates of transpiration. The plant structure, particularly of the leaf, largely determines the rate of water loss. Plants with large total foliage surfaces composed of broad, thin leaves have higher rates of loss than plants bearing needleleaves, spines, or small thick leaves. Under conditions of low water supply but high rates of evaporation, only those plants that minimize transpiration losses by their special leaf structures and their small size can survive.

The adaptation of plant structures to soil-water budgets with large water deficiences is of particular interest to the plant geographer. Transpiration occurs largely from specialized leaf pores, called **stomata,** which are openings in the outer cell layer, and through the **cuticle,** the outermost protective layer of the leaf. Surrounding the openings of the stomata are **guard cells** that can open and close the openings and thus regulate the flow of water vapor and other gases (Figure 15.2). Although most of the transpiration occurs through the stomata, some water vapor may pass through the cuticle. This latter form of loss is reduced in some plants by thickening of the outer layers of cells or by the deposition of wax or waxlike material on or near the leaf surface. Thus many desert plants have thickened cuticle or wax-coated leaves, stems, or branches.

Other means of reducing transpiration are the development of stomata deeply sunken into the leaf surface to retard outward diffusion of water vapor into dry air and the restriction in location of stomata to the shaded undersurfaces of leaves. A plant may also adapt to a desert environment by greatly reducing the leaf area or by bearing no leaves at all. Thus needleleaves and spines representing leaves greatly reduce loss from transpiration. In cacti, the foliage leaf is not present and transpiration is limited to fleshy stems.

In addition to developing leaf structures that reduce water loss by transpiration, plants in a water-scarce environment improve their means of obtaining water and of storing it. Roots become greatly extended to reach soil moisture at increased depth. In cases where the roots reach to the ground-water table, a steady supply of water is assured. Plants drawing from such a source are termed **phreatophytes** and may be found along dry channels (draws) and alluvial valley floors in desert regions. Other desert plants produce a widespread but shallow root system enabling them to absorb the maximum quantity of water from sporadic desert downpours that saturate only the uppermost soil layer. Stems of desert plants are commonly greatly thickened by a spongy tissue in which much water can be stored. Plants employing this adaptation are termed **succulents.**

A quite different adaptation to extreme aridity is seen in many species of small desert plants that complete a very short cycle of germination, leafing, flowering, fruiting, and seed dispersal immediately following a desert downpour. Because they appear so briefly, these plants are referred to as **ephemeral annuals.**

Certain climates, such as the wet-dry tropical climate and the moist continental climate, have a yearly cycle with one season in which water is unavailable to plants because of lack of precipitation or because the soil water is frozen. This season alternates with one in which there is abundant water. Plants adapted to such regimes are called **tropophytes,** from the Greek word *trophos,* meaning "change," or "turn." Tropophytes meet the impact of the season of unavailable water by dropping their leaves and becoming dormant. When water is again available, they leaf out and grow at a rapid rate. Trees and shrubs that seasonally shed their leaves are **deciduous plants;** in distinction, **evergreen plants** retain most of their leaves in a green state throughout the year.

Figure 15.2 **Cell structure of a foliage leaf. (After W. W. Robbins and T. E. Weier, *Botany,* copyright 1950 by John Wiley & Sons. Reprinted by permission of John Wiley & Sons, Inc.)**

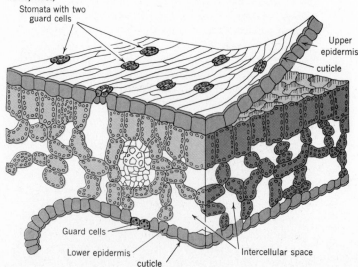

The Mediterranean climate also has a strong seasonal wet-dry alternation, the summers being dry and the winters wet. Plants in this climate adopt the habit of xerophytic plants and characteristically have hard, thick, leathery leaves. An example is the live oak, which holds most of its leaves throughout the dry season. Such hard-leaved evergreen trees and woody shrubs are called **sclerophylls.** The prefix *scler* is from the Greek word for "hard." Plants that hold their leaves throughout a dry or cold season have the advantage of being able to resume photosynthesis immediately when growing conditions become favorable, whereas the deciduous plants must grow a new set of leaves.

To cope with shortage of water, **xeric animals,** those adapted to dry conditions, have evolved methods that are somewhat similar to those used by the plants. Many of the invertebrates exhibit the same pattern as the ephemeral annuals—evading the dry period in dormant stages. When rain falls, they emerge to take advantage of the new and short-lived vegetation. For example, many species of birds regulate their behavior to nest only when the rains occur, the time of most abundant food for their offspring. The tiny brine shrimp of the Great Basin may wait many years in dormancy until normally dry lake beds fill with water, an event that occurs perhaps three or four times a century. The shrimp then emerge and complete their life cycles before the lake evaporates.

The mammals are by nature poorly adapted to desert environments, but many survive there by using a variety of mechanisms to avoid water loss. Just as plants reduce transpiration to conserve water, so many desert mammals do not sweat through skin glands; they rely instead on other methods of cooling. Many of the desert mammals conserve water by excreting highly concentrated urine and relatively dry feces. The desert mammals also evade the heat by nocturnal activity. In this respect, they are joined by most of the rest of the desert fauna, spending their days in cool burrows in the soil and their nights foraging for food.

Organisms and Temperature

The temperature of air and soil, another of the important climatic factors in ecology, acts directly on organisms through its influence on the rates at which the physiological processes take place. For plants, we can say that each species has an optimum temperature associated with each of its functions, such as photosynthesis, flowering, fruiting, and seed germination. Some overall optimum yearly temperature conditions exist for its growth in terms of size and numbers of individual plants. There are also limiting lower and upper temperatures for the individual functions of the plant and for its total survival.

Temperature also acts as an indirect factor. Higher air temperatures increase the water vapor capacity of the air, inducing greater transpiration as well as faster rates of direct evaporation of soil water.

In general, the colder the climate, the fewer are the species that are capable of surviving. A large number of tropical plant species cannot survive below-freezing temperatures. In the severely cold arctic and alpine environments of high latitudes and high altitudes, only a few species can survive. This principle explains why a forest in the equatorial zone has many species of trees, whereas a forest of the subarctic zone may be dominated by just a few. Tolerance to cold is closely tied up with the ability of the plant to withstand the physical disruption that accompanies the freezing of water. If the plant has no means of disposing of the excess water in its tissues, the freezing of that water will damage the cell tissue.

The effects of temperature variations on animals are moderated by their physiology and by their ability to seek sheltered environments. Most animals lack a physiological mechanism for internal temperature regulation. These animals, including reptiles, invertebrates, fish, and amphibians, are **cold-blooded animals;** their body temperatures passively follow the environment. With a few exceptions (notably fish and some social insects), these animals are active only during the warmer parts of the year. They survive the cold weather of the midlatitude zone winter by becoming dormant. Some vertebrates enter a dormant state termed **hibernation,** in which metabolic processes virtually halt and body temperatures closely parallel those of the surroundings. Most hibernators seek out burrows, nests, or other environments where winter temperatures do not reach extremes or fluctuate rapidly. Because the annual range of soil temperatures is greatly reduced below the uppermost layers, soil burrows are particularly suited to hibernation.

Other animals maintain their tissues at a constant temperature by internal metabolism. This group includes the birds and mammals. These **warm-blooded animals** possess a variety of adaptations to maintain a constant internal temperature. Fur, hair, and feathers act as insulation by trapping dead air spaces next to the skin surface, reducing heat loss to the surrounding air or water. A thick layer of fat will also provide excellent insulation. Other adaptations are for cooling; for example, sweating or panting uses the high latent heat of vaporization of water to remove heat. Heat loss is also facilitated by exposing blood-circulating tissues to the cooler surroundings. The seal's flippers and bird's feet serve this function.

Other Climatic Factors

The factor of light is also of importance in determining local plant distribution patterns. The amount of light available to a plant will depend in large part on the plant's position. Tree crowns in the upper layer of a forest receive maximum light, but correspondingly reduce the amount available to lower layers. In extreme cases, forest trees so effectively cut off light that the forest floor is almost free of shrubs and smaller plants. In certain deciduous forests of midlatitudes, the period of early spring, before the trees are in leaf, is one of high light intensity at ground level, permitting the smaller plants to go through a rapid growth cycle. In summer these plants will largely disappear as the leaf canopy is completed. Other low plants in the same habitat require shade and do not appear until later in the summer.

Treated on a global basis, the factor of light available for plant growth varies by latitude. Duration of daylight in

Figure 15.3 **Trees deformed by the effects of cold, dry winds and driving snow. Timberline scene in Arapaho National Forest, Colorado. (U.S. Forest Service.)**

summer increases rapidly with higher latitude and reaches its maximum poleward of the arctic and antarctic circles, where the sun may be above the horizon for 24 hours (see Figure 12.4). Thus, although the growing season for plants is greatly shortened at high latitudes by frost, the rate of plant growth in the short frost-free summer is greatly accelerated by the prolonged daylight.

In midlatitudes, where vegetation is of a deciduous type, the annual rhythm of increasing and decreasing periods of daylight determines the timing of budding, flowering, fruiting, leaf-shedding, and other vegetation activities. As to the importance of light intensity itself, even on overcast days there is sufficient light to permit most plants to carry out photosynthesis at their maximum rates.

Light also influences animal behavior. The day-night cycle controls the activity patterns of many animals. Birds, for example, are generally active during the day, whereas small foraging mammals, such as weasels, skunks, and chipmunks, are more active at night. Light also controls seasonal activity through **photoperiod,** or day length, in midlatitudes. As autumn days grow shorter and shorter, squirrels and other rodents hoard food for the coming winter season. Later, increasing photoperiod will trigger such activities as mating and reproduction in the spring. Although these rhythms of activity are important to the ecologist, they are of lesser interest to the biogeographer.

Wind is also an important environmental factor in the structure of vegetation in highly exposed positions. Close to timberline in high mountains and along the northern limits of tree growth in the arctic zone, trees are deformed by wind in such a way that the branches project from the lee side of the trunk only (flag shape). Some trees may show trunks and branches bent to near-horizontal attitude, facing away from the prevailing wind direction (Figure 15.3). The effect of wind is to cause excessive drying, damaging the exposed side of the plant. The tree limit on mountainsides thus varies in elevation with degree of exposure of the slope to strong prevailing winds and will extend higher on lee slopes and in sheltered pockets.

Bioclimatic Frontiers

Taken separately or together, climatic factors of moisture, temperature, light, and wind can act to limit the distribution of plant and animal species. Biogeographers recognize that there is a critical level of climatic stress beyond which a species cannot survive and that there will be a geographic boundary marking the limits of the potential distribution of a species. Such a boundary is sometimes referred to as a **bioclimatic frontier.** Although the frontier is usually marked by a complex of climatic elements, it is sometimes possible to single out one climatic element related to soil water or temperature that coincides with it.

The distribution of ponderosa pine *(Pinus ponderosa)* in western North America provides an example (Figure 15.4). In this mountainous region, annual rainfall varies sharply with elevation. The 50 cm (20 in.) isohyet of annual total precipitation encloses most of the upland areas having the yellow pine. It is the parallelism of the isohyet with forest boundary, rather than actual degree of coincidence, that is significant. The sugar maple *(Acer saccharum)* is a somewhat more complex case (Figure 15.5). Here the boundaries on the north, west, and south coincide roughly with selected values of annual precipitation, mean annual minimum temperature, and mean annual snowfall.

Although bioclimatic limits must exist for all species, no plant or animal need necessarily be found at its frontier. Many other factors may act to keep the spread of a species in check. A species may be limited by diseases or predators found in adjacent regions. In another example, a species (especially a plant species) may migrate slowly and may still be radiating outward from the location in which it evolved. Or, a species may be dependent on another species and therefore be limited by the latter's distribution.

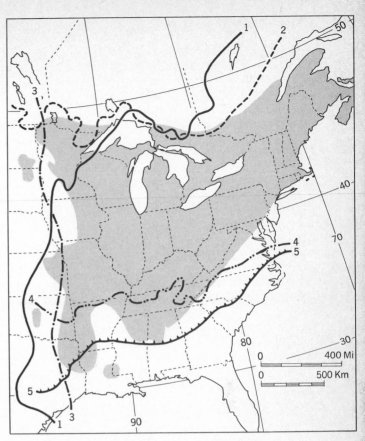

Figure 15.4 **Areas of ponderosa pine *(Pinus ponderosa)* in western North America are shown in solid black. The edge of the shaded area represents the isohyet of 50 cm (20 in.) annual precipitation. (Based on *Biogeography — an ecological perspective* by Pierre Dansereau. Copyright © 1957, The Ronald Press Company, New York.)**

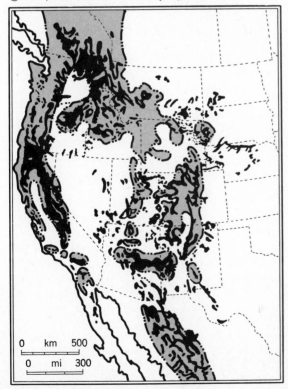

Figure 15.5 **Bioclimatic limits of the sugar maple *(Acer saccharum)* in eastern North America. The shaded area represents the distribution of the sugar maple. Line 1, 76 cm (30 in.) annual precipitation. Line 2, −40°C (−40°F) mean annual minimum temperature. Line 3, eastern limit of yearly boundary between arid and humid climates. Line 4, 25 cm (10 in.) mean annual snowfall. Line 5, −10°C (16°F) mean annual minimum temperature. (Based on *Biogeography — An Ecological Perspective,* by Pierre Dansereau. Copyright © 1957, The Ronald Press Company, New York.)**

Terrestrial Ecosystems — the Biomes

Because plants and animals have become adapted through evolution to varying environments, many different ecosystems exist, each adjusted to a different range of environmental conditions. Ecosystems fall into two major groups: aquatic and terrestrial. The **aquatic ecosystems** include life forms of the marine environments and the freshwater environments of the lands. Marine ecosystems include the open ocean, coastal estuaries, and coral reefs. Freshwater ecosystems include lakes, ponds, streams, marshes, and bogs. The **terrestrial ecosystems** comprise the assemblages of land plants and animals spread widely over the upland surfaces of the continents. The terrestrial ecosystems are determined largely by climate and soil and, in this way, are closely woven into the fabric of physical geography.

Within terrestrial ecosystems, the largest recognizable subdivision is the **biome.** Although the biome includes the total assemblage of plant and animal life interacting within the life layer, the green plants dominate the biome physically because of their enormous biomass, as compared with that of other organisms. Thus biogeographers classify the biomes by the characteristic life form of the green plants within the biome.

The following are the principal biomes, listed in order of availability of soil water and heat.

Forest   Ample soil water and heat.

Savanna   Transitional between forest and grassland.

Grassland Moderate shortage of soil water; adequate heat.

Desert   Extreme shortage of soil water; adequate heat.

Tundra   Insufficient heat.

Biogeographers subdivide the biomes into smaller units, called **formation classes,** based on the size, shape, and structure of the plants. For example, at least four and perhaps as many as six kinds of forests are easily recognizable within the forest biome. At least two kinds of grasslands are easily recognizable. Deserts, too, span a wide range in terms of the abundance and life form of plants. Reviewing these formation classes and their unique features and adjustments to characteristic climates and soils will be our goal in the following chapter.

Communities and Habitats

Although the distribution of the major biomes is dependent on climate, there is much local variation in plants and animals. **Biotic communities** are local associations of plants and animals that are interdependent and often found together. The total biotic cover of a region is actually a mosaic of small community types that recur in different places on the landscape (see Plate A, Figures 9 and 10).

A primary influence on the distribution of biotic communities is the effect of varying landform and soil on vegetation. Landform refers to the configuration of the land surface, including such features as hills, valleys, ridges, or cliffs. Vegetation on an upland — relatively high ground with thick, well-drained soil — is quite different from that on an adjacent valley floor, where water lies near the surface much of the time. Vegetation is often different on rocky ridges and steep cliffs, where water drains away

rapidly and soil is thin or largely absent. Because the animals in an area are dependent on the vegetation for primary production, both plant and animal components of the biotic community will respond to these differences in the physical environment.

Habitat is the word used to refer to a type of physical environment that has a characteristic biotic community. For example, Figure 15.6 shows six habitats within the Canadian forest: upland, bog, bottomland, ridge, cliff, and active dune. In establishing formation classes for generalized maps, biogeographers usually base their classes on the communities of the upland habitat because it is here that a middle range of environmental conditions prevails. Just where each habitat is located and how large an area it occupies depends largely on factors of soil and landform.

Geomorphic Factors

Geomorphic factors (landform factors) influencing ecosystems are essentially the same as landform factors influencing the soil (Chapter 13). They include such elements as slope steepness (the angle that the ground surface makes with respect to the horizontal), slope aspect (the orientation of a sloping ground surface with respect to geographic north), and relief (the difference in elevation of divides and adjacent valley bottoms). In a much broader sense, geomorphic factors include the entire sculpturing of the landforms of a region by processes of erosion, transportation and deposition by streams, waves, wind, and ice, and by forces of volcanism and mountain building. These are topics covered in detail in Chapters 17 through 26.

Slope steepness acts indirectly by influencing the rate at which precipitation is drained from the surface. On steep slopes, surface runoff is rapid and soil-water recharge by infiltration is reduced. On gentle slopes, much of the precipitation can penetrate the soil and be retained. More rapid erosion on steep slopes may result in thin soil, whereas that on gentler slopes is thicker. Slope aspect has a direct influence on plants by increasing or decreasing the exposure to sunlight and prevailing winds. Slopes facing the sun have a warmer, drier environment than slopes facing away from the sun and therefore lying in shade for much longer periods of the day. In midlatitudes, these slope-aspect contrasts may be so strong as to produce quite different biotic communities on north-facing and south-facing slopes (Figure 15.7).

Geomorphic factors are in part responsible for the dryness or wetness of the habitat within a region having

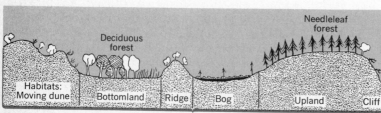

Canadian forest

Figure 15.6 **Habitats within the Canadian forest. (Modified from Pierre Dansereau, 1951, *Ecology,* vol. 32.)**

Figure 15.7 Contrasts in vegetation on opposing valley-side slopes. The heavily wooded slope on the left faces northeast and is in shade during the afternoon. The sparsely wooded slope on the right, facing southwest, receives intense insolation when air temperatures are highest. Long Canyon of the Rio Hondo, Carson National Forest, New Mexico. (U.S. Forest Service.)

essentially the same overall climate. Each community has its own microclimate. On divides, peaks, and ridge crests, the soil tends to dryness because of rapid drainage and because the surfaces are more exposed to sunlight and drying winds. By contrast, the valley floors are wetter because surface runoff over the ground and into streams causes water to converge there. In humid climates, the ground water table in the valley floors may lie close to or at the ground surface to produce marshes and swamps, ponds and bogs.

Edaphic Factors

Edaphic factors are those related to the soil. In Chapters 13 and 14, the principles of soil development (pedogenesis) were taken up systematically. In terms of biogeography, we can look at soils in two perspectives. One views the broad patterns of soil distribution as controlled by the climatic regimes. Patterns of soils and climates are closely correlated with the global patterns of the formation classes. These broad relationships will be taken up in Chapter 16.

A second viewpoint is in terms of habitats — the small-scale mosaic of place-to-place variations of the earth's surface. Among the edaphic factors important in differentiating the habitat are soil texture and structure, humus content; presence or absence of soil horizons; soil alkalinity, acidity, or salinity; and the activity of bacteria and animals in the soil.

Although this book treats the systematic principles of soil science ahead of those of natural ecosystems, a good argument might be made for reversing this order of treatment on the grounds that plants and animals play a leading role in the development of soil characteristics. Given a barren habitat, recently formed by some geologic

event, such as the outpouring of lava or the emergence of a coastal zone from beneath the sea, the gradual evolution of a soil profile goes hand in hand with the occupance of the habitat by a succession of biotic communities. The plants profoundly alter the soil by such processes as contributing organic matter or producing acids that act on the mineral matter. Animal life, feeding on the plant life, also makes its contribution to physical and chemical processes of soil evolution. This change in communities through time forms our next topic.

Ecological Succession

The phenomenon of change in ecosystems through time is familiar to us all. A drive in the country reveals patches of vegetation in many stages of development — from open, cultivated fields through grassy shrublands to forests. Clear lakes, gradually filled with sediment from the rivers that drain into them, become bogs. These kinds of changes, in which biotic communities succeed one another on the way to a stable end point, are referred to as **ecological succession.**

In general, succession leads to formation of the most complex community of organisms possible in an area, given its physical controlling factors of climate, soil, and water. The series of communities that follow one another on the way to the stable stage is called a **sere.** Each of the temporary communities is referred to as a **seral stage.** The stable community, which is the end point of succession, is the **climax.** If succession begins on a newly constructed deposit of mineral sediment, it is termed **primary succession.** If succession occurs on a previously vegetated area that has been recently disturbed by such agents as fire, flood, windstorm, or Man, it is referred to as **secondary succession.**

A new site on which primary succession occurs may have one of several origins: a sand dune, a sand beach, the surface of a new lava flow or freshly fallen layer of volcanic ash, or the deposits of silt on the inside of a river bend that is gradually shifting. Such a site will not have a true soil with horizons; rather, it may perhaps be little more than a deposit of coarse mineral fragments. In other cases, such as that of the floodplain silt deposits, the surface layer may represent redeposited soil endowed with substantial proportions of soil colloids and exchangeable base cations.

The first stage of a succession is a **pioneer stage;** it includes a few plant species unusually well adapted to adverse conditions of rapid water drainage and drying of soil and to excessive exposure to sunlight, wind, and extreme ground and lower-air temperatures. As these plants grow, their roots penetrate the soil; their subsequent decay adds humus to the soil. Fallen leaves and stems add an organic layer to the ground surface. Bacteria and animals begin to live in the soil in large numbers. Grazing mammals feed on the small plants. Birds forage the newly vegetated area for seeds and grubs.

Soon conditions are favorable for other species that invade the area and displace the pioneers. The new arrivals may be larger plant forms providing more extensive cover of foliage over the ground. In this case the

A. Beachgrass is a pioneer on beach dunes and helps stabilize the dune against wind erosion.

B. The low, matlike shrubs in the center of the photo replace dunegrass in the better stabilized areas.

C. This beach thicket of poison ivy, bayberry, and wild cherry paves the way for the development of the forest climax.

D. The climax forest on dunes. Here at Sandy Hook, New Jersey, holly (*Ilex opaca*), on the left, is an important constituent of the climax forest. Note the leaves and organic matter on the forest floor.

Figure 15.8 **Stages in beach-dune succession. (A. H. Strahler.)**

climate near the ground, or **microclimate,** is considerably altered toward one of less extreme air and soil temperatures, higher humidities, and less intense insolation. Now still other species can invade and thrive in the modified environment. When the succession has finally run its course, a climax community of plant and animal species in a more or less stable composition will exist.

The colonization of a sand dune provides an example of primary succession. Growing foredunes bordering the ocean or lakeshore present a sterile habitat. The dune sand — usually largely quartz, feldspar, and other common rock-forming minerals — lacks such important nutrients as nitrogen, calcium, and phosphorus, and its water-holding ability is very low. Under the intense solar radiation of the day, the dune surface is a hot, drying environment. At night, radiation cooling in the absence of moisture produces low surface temperatures. One of the first pioneers of this extreme environment is beachgrass (Figure 15.8A).

This plant reproduces vegetatively by sending out rhizomes (creeping underground stems), and the plant thus slowly spreads over the dune. Beachgrass is well adapted to the eolian environment; it does not die when buried by moving sand, but instead puts up shoots to reach the new surface.

After colonization, the shoots of beachgrass act to form a baffle that suppresses movement of sand, and thus the dune becomes more stable. With increasing stabilization, plants that are adapted to the dry, extreme environment but cannot withstand much burial begin to colonize the dune. Typically these are low, matlike woody shrubs, such as beach wormwood or false heather (Figure 15.8B).

On older beach and dune ridges of the central Atlantic coastal plain, the species that follow matlike shrubs are typically larger woody plants and such trees as beach plum, bayberry, poison ivy, and choke cherry (Figure 15.8C). These species all have one thing in common —

their fruits are berries that are eaten by birds. The seeds
from the berries are excreted as the birds forage among the
low dune shrubs, thereby sowing the next stage of
succession. As the scrubby bushes and small trees spread,
they shade out the matlike shrubs and any remaining
beachgrass. Pines may also enter at this stage.

At this point, the soil begins to accumulate a
significant amount of organic matter. No longer dry and
sterile, it now possesses organic compounds and nutrients,
and has accumulated enough colloids to hold water for
longer intervals. These soil conditions encourage the
growth of such broadleaf species as red maple, hackberry,
holly, and oaks, which shade out the existing shrubs and
small trees (Figure 15.8D). Once the forest is established, it
tends to reproduce itself; the species of which it is
composed are tolerant to shade, and their seeds can
germinate on the organic forest floor. Thus the climax is
reached. The stages through which the ecosystem has developed constitute the sere, progressing from beachgrass
to low shrubs to higher shrubs and small trees to forest.

Although this example has stressed the changes in
plant cover, animal species are also changing as
succession proceeds. Table 15.1 shows how some typical
invertebrates appear and disappear through succession on
the Lake Michigan dunes. Note that the seral stages shown
in the table for these inland dunes are somewhat different
than those described for the coastal environment.

Another example of primary succession is that of **bog
succession** (see Figure 15.1). Extensive land areas of North
America and Europe have innumerable bogs. These are
former glacial lake basins now filled with partially decomposed plant matter that we recognize as **freshwater peat.**

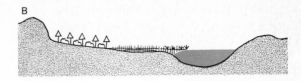

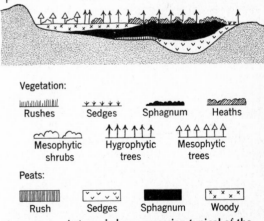

Figure 15.9 **Autogenic bog succession typical of the
Laurentian shield area of Canada. (After Dansereau and
Segadas-Vianna, 1952,** *Canadian Journal of Botany,*
vol. 30.)

Table 15.1 / Invertebrate Succession on the Lake Michigan Dunes

	Successional Stages				
Invertebrate	Beach-grass—Cotton-wood	Jack Pine Forest	Black Oak Dry Forest	Oak and Oak-Hickory Moist Forest	Beech-Maple Forest Climax
White tiger beetle	x				
Sand spider	x				
Long-horn grasshopper	x	x			
Burrowing spider	x	x			
Bronze tiger beetle		x			
Migratory locust		x			
Ant lion			x		
Flatbug			x		
Wireworms			x	x	x
Snail			x	x	x
Green tiger beetle				x	x
Camel cricket				x	x
Sowbugs				x	x
Earthworms				x	x
Woodroaches				x	x
Grouse locust					x

Data source: V. E. Shelford, as presented in E. P. Odum, 1971,
Fundamentals of ecology, W. B. Saunders Co., Philadelphia, p. 259.

The peat accumulates because decay of plant matter is
slow in these cold climates. Plant matter that accumulates
below the water level of a lake is in a continually
saturated condition, with little oxygen available to

promote the activity of decomposers.

Figure 15.9 shows, by a series of diagrams the stages in the filling of lakes during bog succession. At the water's edge is a zone of sedges, followed by rushes. These construct a floating layer that encroaches upon the open water. There follows a zone of sphagnum (peat moss), which eventually completely fills the lake. Now the peat deposit supports hygrophytic trees (largely spruce), which produce a woody peat. The soil is now a Histosol. This community may in turn be replaced by mesophytic trees, marking the climax stage.

Old-Field Succession

Where disturbance alters an existing community, secondary succession can occur. **Old-field succession,** taking place on abandoned farmland, is a good example of secondary succession. In the eastern United States, the first stages of the sere often depend on the last use of the land before abandonment. If row crops were last cultivated, one set of pioneers, usually annuals and biennials, will appear; if small grain crops were cultivated, the pioneers are often perennial herbs and grasses. If pasture is abandoned, those pioneers that were not grazed will have a head start. Where mineral soil was freshly exposed by plowing, pines are often important following the first stages of succession because pine seeds favor disturbed soil and strong sun for germination. Although slower growing than other pioneers, the pines will eventually shade the others out and become dominant. Their dominance is only temporary, however, because their seeds cannot germinate in shade and litter on the forest floor. Seeds of hardwoods such as maples and oaks, however, can germinate under these conditions, and as the pines die, hardwood seedlings grow quickly to fill the holes produced in the canopy. The climax, then, is the hardwood forest, which can reproduce itself. Figure 15.10 is a schematic diagram showing one example of old-field succession.

It is important to note that the changes of the sere result from the action of the plants and animals themselves; one set of inhabitants paves the way for the next. As long as nearby populations provide colonizers, the changes lead in an automatic fashion from old field to forest. This type of succession is often termed **autogenic** (self-producing) **succession.**

In many cases, however, autogenic succession does not run its full course. Environmental disturbances, such as wind, fire, flood, or clearing by Man, may occur often enough to permanently alter or divert the course of succession. In addition, habitat conditions, such as site exposure, unusual bedrock, impeded drainage, and the like can hold back the course of succession so successfully that the climax is never reached; instead, an earlier stage of the sere becomes more or less permanent and is as stable at that site as the climax may be on more favorable sites. Thus a mosaic of biotic communities will be the stable biogeographic pattern in an area with a diversity of habitats.

Man's Impact on Natural Ecosystems

Two concepts of biogeography stand opposed but inseparable, like the two sides of a coin. One is the concept of the **natural ecosystem,** an ecosystem that attains its development without appreciable interference by Man and is subject to natural forces of modification and destruction, such as storms or fires. The other concept is that of the ecosystem sustained in a modified state by Man's activities. Extremes of both cases are found throughout the world.

Natural ecosystems can still be seen over vast areas of the wet equatorial climate where rainforests are as yet scarcely touched by Man. Much of the arctic tundra and the needleleaf forest of the subarctic zones is in a natural state. In contrast, much of the continental surface in midlatitudes is almost totally under Man's control through intensive agriculture, grazing, or urbanization. You can drive across an entire state, such as Ohio or Iowa, without seeing a vestige of the plant cover it bore before the coming of the White Man. Only if you know where to look for them can you find a few small plots of virgin prairie or virgin forest.

Some areas of natural vegetation appear to be

Figure 15.10 **Old-field succession in the Piedmont region of the southeastern United States, following the abandonment of cornfields and cottonfields. This is a pictorial graph of continuously changing composition spanning about 150 years. (After E. P. Odum, 1973,** *Fundamentals of Ecology,* **W. B. Saunders Co., Philadelphia.)**

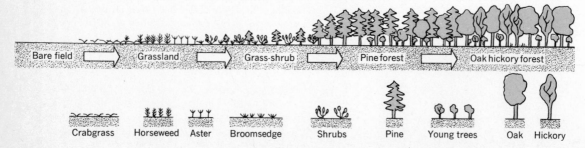

untouched but are dominated by Man in a more subtle manner. Certain of our national parks and national forests have been protected from fire for many decades, setting up an unnatural condition in terms of what might be expected of the natural ecosystem. When lightning starts a forest fire, the fire fighters parachute down and put out the flames as fast as possible. But periodic burning of forests and grasslands is a natural phenomenon, and it performs vital functions in the ecosystem. One such vital function is to release nutrients from storage in the biomass so that the soil can be revitalized. Another is to increase the areas covered by grasses and herbs on which grazing animals are dependent. Already those who manage our parks and forests are experimenting with a hands-off policy in fire control.

Man has influenced ecosystems in yet another way: by moving plant and animal species from their indigenous habitats to foreign lands and foreign environments. The eucalyptus tree is a striking example. The various species of eucalypts have been transplanted from Australia to such far-off places as California, North Africa, and India. Sometimes exported plants thrive like weeds, forcing out natural species and becoming a major nuisance. Transporting the jackrabbit to Australia resulted in a devastating rabbit population explosion, only recently checked by importing rabbit diseases as well. It is said that scarcely one of the kinds of grasses seen clothing the coast ranges of California is a native species, but the casual observer might think that these represent the native vegetation.

Inadvertent introduction of a disease from a foreign continent can cause the extinction of a particular plant or animal species. An example is the chestnut blight, which eliminated the American chestnut from the forests of the northeastern United States. Accidentally imported insects can also wipe out most of the mature individuals of a native plant when there are no native insect predators available to combat the invasion. These are just a few of the ways Man interferes with natural ecosystems.

References for Further Study

Dansereau, Pierre (1957), *Biogeography; an ecological perspective,* Ronald Press, New York.

Daubenmire, R. F. (1959), *Plants and environment: a textbook of plant autecology,* second edition, John Wiley and Sons, New York.

Polunin, Nicholas (1960), *Introduction to plant geography and some related sciences,* McGraw-Hill Book Co., New York.

Odum, E. P. (1973), *Fundamentals of ecology,* W. B. Saunders Co., Philadelphia.

Strahler, A. N., and Strahler, A. H. (1974), *Introduction to environmental science,* Hamilton Publishing Co., Santa Barbara, Calif. (John Wiley and Sons, New York).

Review Questions

1. Define biogeography. Describe some of the processes and problems with which biogeographers are concerned.
2. What two factors are most important in determining the global distribution pattern of plants and animals? What kinds of processes control the two factors?
3. How are plants classified by water need? Give examples. How do plants lose water? Describe the transpiration process and the adaptations by which plants reduce water loss.
4. What plant adaptations are associated with an annual wet-dry cycle? What climates are involved?
5. What adaptations have animals evolved to deal with dry conditions? Give some examples.
6. How does temperature affect plants directly and indirectly? How does mean annual temperature affect the number of species present?
7. How does temperature affect animals? What adaptations do animals use to cope with varying temperatures?
8. How does light act to determine local plant distribution patterns? How does the seasonal rhythm affect plant growth? How does light influence animal behavior?
9. What are the effects of wind on the shape and form of vegetation?
10. What bioclimatological law relates a species to the level of climatic stress imposed on it? Give two examples of bio-climatic frontiers. What factors may keep a species from reaching the limits of its frontier?
11. What two major groups of ecosystems are recognized? What is a biome? Name the principal biomes and the availability of soil water and heat in each. What are formation classes?
12. What is a biotic community? How are biotic communities arranged on the landscape? Define habitat. How does habitat relate to community?
13. Discuss the geomorphic factors that differentiate habitats and explain how each factor affects local communities.
14. What are edaphic factors? Give some examples. What two viewpoints are used in discussing the effects of soils on ecosystems?
15. Describe the process of ecological succession. What is primary succession? Secondary succession? What are the stages involved in succession? Describe dune succession and bog succession as examples.
16. What is old-field succession? Describe the stages of a typical old-field succession.
17. Describe the ways in which Man influences ecosystems.

Distribution of Natural Vegetation

16

The distribution of natural vegetation and its relationship to soils and climate are of primary concern to geographers. To organize and classify the world's vegetation, biogeographers have adopted a number of systems. Most such systems are based on two criteria. First in importance is the structure of the vegetation in a region. Structure includes the growth forms of the dominant plants as well as their organization and arrangement in space. A second criterion is the flora, or list of plants found in a region. A biogeographer may wish to distinguish different vegetation regions, or floristic realms, on the basis of species composition. The world vegetation map described in this chapter is based on both structural and floristic criteria.

Structural Description of Vegetation

The structural description of vegetation is based on the physical properties and outward forms of plants. Six categories of information have been included in a system of structural description set up by Pierre Dansereau.*

---

*Pierre Dansereau (1957), *Biogeography, an ecological perspective,* Ronald Press Co., New York.

These six categories tell us about the growth form of the plants, their size and stratification, the degree to which they cover the ground, their time functions (periodicity), and their leaf forms.

1. Life-form. Plants can be classified according to their **life-form.** The first two forms, trees and shrubs, are erect, woody plants (Figure 16.1). A **tree** is a perennial plant having a single upright main trunk, often with few branches in the lower part, but branching in the upper part to a crown which, in the mature individual, contributes to one of the upper layers of vegetation (Figure 16.2). A **shrub** is a woody plant having several stems branching near the ground, so as to place the foliage in a mass starting close to ground level. Third are woody vines that climb on trees: the **lianas** (Figure 16.3). Although the term "liana" is usually associated with woody vines of tropical forests, the meaning may be broadened to include woody vines of midlatitude forests, such as the poison ivy and Virginia creeper. Thus life-form classes cut across the taxonomic categories of the plant world. For example, palms and tree ferns look much alike outwardly but are entirely different botanically.

Fourth of the life-forms is the plant group known as

Figure 16.1 **Schematic diagram of life forms in a beech-maple-hemlock forest. The tree layer consists of sugar maple (As), ash (Fa), beech (Fg), and hemlock (Tc) and includes a liana (Cs). The shrub layer includes elder (Sp), dogwood (Ca), and a young hemlock (Tc). An epiphyte (Ul) grows on the hemlock. Plants designated Te, Mr, Cb, and Cu form the herb layer. Moss (Du) forms the lowest layer. (From Pierre Dansereau, 1951, *Ecology,* Vol. 32.)**

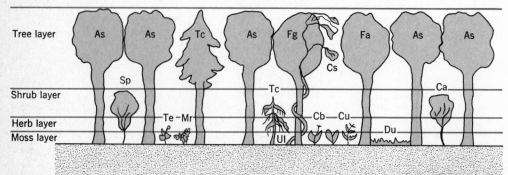

Figure 16.2 **A stand of mature sugar maples, 160 to 200 years old, in the Allegheny National Forest, Pennsylvania, 1939. (U.S. Forest Service.)**

to make precise descriptions of vegetation. Further standardization is achieved by setting height limits for a series of layers, numbered successively from the ground up (Table 16.1).

3. Coverage. The degree to which the foliage of individual plants of a given life-form cover the ground beneath them is the *coverage*. We may use four terms descriptive of the coverage: barren or very sparse, discontinuous, in tufts or groups, and continuous. For example, the trees may be of discontinuous coverage whereas the herb layer is continuous, or vice versa.

4. Periodicity. Of primary importance in the classification of forms of natural vegetation is the response of the plant foliage to the annual climatic cycle. Deciduous plants shed their leaves and become dormant in an unfavorable season, which is either too cold or too dry to permit growth. Evergreen plants retain green foliage year around, although in some cases they become almost dormant in a cold or dry season. Where the climate is equable (i.e., moist and not cold throughout the year), evergreen plants grow continuously. A third class, the **semideciduous plants,** are those that shed their leaves at intervals not in phase with a season. Thus a semideciduous forest would not at any one time have all its individuals devoid

the **herbs.** These are usually small, tender plants lacking woody stems. They occur in a wide variety of forms and leaf structures, including annuals and perennials, broad-leaved plants and grasses. Broad-leaved herbs are termed **forbs** in distinction with grasses. The adjective **herbaceous** is applied to this life-form. The herbaceous layer will normally occupy a low position in a layered or stratified plant community. Smaller, and lying in close contact with the ground or attached to tree trunks, are the **bryoids,** a life-form that includes the mosses and liverworts.

Finally, among the life-forms are the **epiphytes,** plants that use other plants as supporting structures and thus live above ground level, out of contact with soil (Plate F.1). Familiar to all are the tropical orchids that live high on tree limbs and are sometimes referred to as "air plants." Ferns also commonly exist as epiphytes.

Not included in our list of life-forms are lower forms of plants, grouped under the name of *thallophytes*. These include bacteria, algae, molds, and fungi — all of which are plants lacking true roots, stems, and leaves. **Lichens,** plant forms in which algae and fungi live together to form a single plant structure, are included within the life-form classification of bryoids. Lichens occur as tough, leathery coatings or crusts and as leaflike masses attached to rocks and tree trunks (Figure 16.4). In arctic and alpine environments, lichens may grow in profusion and dominate the vegetation in the near absence of more conspicuous plant forms.

2. Size and stratification. Each of the life-forms just described may be classified according to size. The words "tall," "medium," and "low" may be given definite limits for each life-form (Table 16.1). For example, a tree higher than 25 m (82 ft) is "tall"; from 10 to 25 m (33 to 82 ft) is "medium"; 8 to 10 m (26 to 33 ft) is "low." Such standardization of size enables the plant geographer

Figure 16.3 **Lianas of the equatorial rainforest, near Belém, Brazil. (Otto Penner, Instituto Agronómico do Norte.)**

Table 16.1 / Plant Size and Stratification of Vegetation

Plant size

Tall:	Tree	Over 25 m
	Shrub	2 to 8 m
	Herb	Over 2 m
Medium:	Tree	10 to 25 m
	Shrub	0.5 to 2 m
	Herb	0.5 to 2 m
	Bryoid	Over 10 cm
Low:	Tree	8 to 10 m
	Shrub	0.5 m
	Herb	Under 0.5 m
	Bryoid	Under 10 cm

Stratification

Layer number	Range
7	Over 25 m
6	10 to 25 m
5	8 to 10 m
4	2 to 8 m
3	0.5 to 2 m
2	10 cm to 0.5 m
1	0 to 10 cm

Data source: Pierre Dansereau (1958), Botanical Institute of the University of Montreal, *Contributions* No. 72, pp. 31–32.

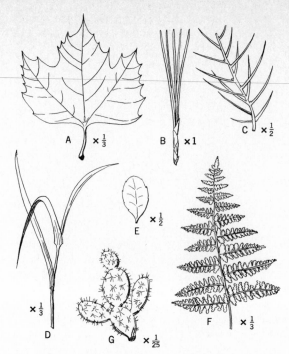

Figure 16.5 **Leaf shapes. A, Large thin leaf (sycamore). B, Needleleaf (pine). C, Spine. D, Graminoid leaf (grass). E, Small leaf, F, Compound leaf (fern). G, Succulent stem (cactus).**

of foliage. As a fourth class we recognize **evergreen-succulent plants,** those with very thick, fleshy leaves that retain their foliage year around, and **evergreen-leafless plants,** those with fleshy stems but no functional leaves, such as the cacti.

5. *Leaf shape and size.* Recognition of the leaf form of a plant is an essential part of the structural description (Figure 16.5). One form is the **broadleaf,** familiar to us in such common trees as oak, maple, beech, and rhododendron. In contrast is the **needleleaf,** also familiar in the pine, spruce, fir, and hemlock. A similar form is the **spine,** which in some plants represents the transformation of

Figure 16.4 **Reindeer moss, the lichen *Cladonia rangifera*. (Larry West.)**

the entire leaf. The slender, tapering leaves of grass are **graminoid** in form. We may also recognize the **small leaf,** as in the birch or holly, and the **compound leaf,** as in the hickory and ash.

6. *Leaf texture.* Leaf textures range widely according to climate and habitat because of the different degrees to which the water loss from the leaf into the air must be controlled. Leaves of average thickness are described as **membranous;** those that are thin and delicate (as in the maidenhair fern) are described as **filmy.** Leaves that are hard, thick, and leathery are **sclerophyllous;** a forest dominated by trees and shrubs having such leaves is termed a **sclerophyll forest.** Very greatly thickened leaves, capable of holding much water in their spongy structure, are described as **succulent.**

The Major Biomes

The biome is a broad major grouping of natural ecosystems that includes animal life as well as plants (Chapter 15). The major biomes, however, are recognized primarily on the basis of their distinctive types of vegetation: forest, savanna, grassland, desert, and tundra.

The **forest biome** includes all regions of **forest** — an assemblage of trees growing close together, forming a layer of foliage that largely shades the ground. Forests often show stratification, with more than one layer. Shading of the ground gives a distinctly different microclimate than that found over open ground. Forests require a relatively large annual precipitation, but it does not need to be uniformly distributed throughout the year. In terms of the soil-water balance, forest is most closely associated with the humid and perhumid subtypes of the moist climates, as defined in Chapter 10. Soil-water storage usually remains high throughout most of the year, although it may be depleted during a dry season. Consequently, the forest biome spans a wide range of climates and latitudes, from the wet equatorial climate (1) to the boreal forest climate (11).

Vegetation of the **savanna biome** consists of a combination of trees and grassland in various proportions. The appearance of the vegetation can be described as parklike, with trees spaced singly or in small groups and surrounded by, or interspersed with, surfaces covered by grasses or by some other plant life form, such as shrubs or annuals in a low layer. Savanna is closely associated with warm climates having alternate wet and dry seasons. The most widespread areas of savanna are identified with the wet-dry tropical climate (3) and with the semiarid subtype (4s) of the dry tropical climate. The latter climate shows a brief, but often intense, rainy season.

Vegetation of the **grassland biome** consists largely or entirely of herbs, which may include grasses, grasslike plants, and forbs (broadleaf herbs). Degree of coverage may range from continuous to discontinuous and there may be stratification. Grassland can include some trees in moister habitats of valley floors and along stream courses where ground water is within reach of tree roots. The grassland biome is closely associated with both the semiarid subtype of the dry midlatitude climate (9s) and the subhumid subtypes of the moist subtropical climate (6sh) and the moist continental climate (10sh). Soil-water deficit ranges from zero to 20 cm or greater, and there is little or no water surplus. Thus soil-water storage is well below the storage capacity throughout most of the year.

The **desert biome,** associated with climates of extreme aridity, has thinly dispersed plants. A large proportion of the surface is bare ground, exposed to direct insolation and to the forces of wind and water erosion or to freeze-thaw action. Although essentially treeless, desert may have scattered woody plants, grasses, and perennial herbs.

Vegetation of the **tundra biome** consists of grasses and grasslike plants and an abundance of flowering herbs, dwarf shrubs, mosses, and lichens. Ground cover can range from complete to spotty. Tundra climate is cold throughout most of the year, with severe frost possible during all the warmer months. Tundra is encountered not only in the arctic zone, but also at high elevations as alpine tundra.

Formation Classes

Biogeographers subdivide the biomes into formation classes, smaller vegetation units based largely on vegetation structure. The formation classes we introduce are major widespread types clearly associated with twelve climate types and their soil-water budgets. Association with major soil types will also be important. In this way we survey the global scope of plant geography as a synthesis of climate with organic and physical processes of the life layer. Table 16.2 lists 16 formation classes and their approximate associations with climate types and soil types. Figure 16.6 is a schematic diagram to show how the

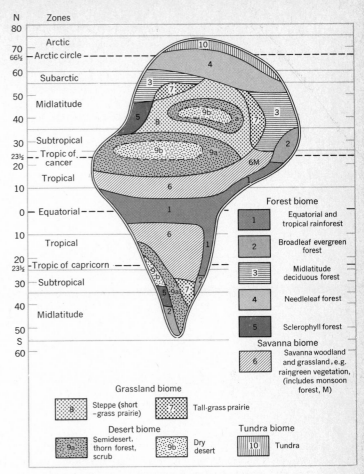

Figure 16.6 **A schematic diagram of the vegetation formation classes on an idealized continent. Compare with the world vegetation map, Plate D.6**

vegetation formation classes would be arranged on an idealized continent, combining Eurasia and Africa into a single supercontinent.

On world vegetation maps, it is possible to recognize many more than 16 formation classes. When further subdivisions are made, floristic composition becomes of equal importance to vegetation structure. Thus we can define lesser units to distinguish between vegetation types that are structurally very similar but consist of greatly differing assemblages of species. Some subordinate units are also necessary to show transitions or mixes between major vegetation types.

Our world vegetation map, Plate D.6, recognizes 28 distinctive vegetation units, using both structural and floristic criteria. It is based on a world system of vegetation types prepared by Professor S. R. Eyre. The Eyre system recognizes 34 vegetation units.* We have combined certain of of Eyre's classes that are closely related but of relatively minor occurrence.

We now turn to a systematic examination of the major formation classes, their composition and structure, and their relationships with climate types and soil orders.

---

*S. R. Eyre (1968), *Vegetation and soils: a world picture,* second edition, Aldine Publishing Co., Chicago. Appendix I. Vegetation Maps of the Continents.

Table 16.2 / The major formation classes

Formation class and map symbol	Associated climate types	Associated soil orders or suborders
Forest Biome		
Equatorial and tropical rainforest (Fe)	(1) Wet equatorial (2) Monsoon and trade-wind littoral	Oxisols Ultisols
Montane forest (Fmt)	Highlands, climates (1) and (2)	Oxisols Ultisols
Monsoon forest (Fmo)	(3) Wet-dry tropical (4s) Dry tropical, semiarid	Ultisols Oxisols Ustalfs Vertisols
Broadleaf evergreen forest (laurel forest) (Fbe)	(6) Moist subtropical (8) Marine west-cost	Udults Udalfs
Midlatitude deciduous forest (Fd)	(8) Marine west-coast (10) Moist continental	Udalfs Boralfs Udolls
Needleleaf forest (Fcl, Fl, Fsp, Fbo, Fbl)	(8) Marine west-coast (North America) (10) Moist continental (North America) (11) Boreal forest	Spodosols Boralfs Histosols Cryaquepts
Sclerophyll forest (Fsm, Fss, Fsa)	(7) Mediterranean	Xeralfs Xerolls Xerults
Savanna Biome		
Savanna woodland (Sw)	(3) Wet-dry tropical	Ustalfs Ultisols Oxisols Vertisols
Thorntree-tall grass savanna (Stg)	(4s, 4sd) Dry tropical, semiarid, semidesert (5s, 5sd) Dry subtropical, semiarid, semidesert	Ustalfs Ultisols Oxisols Vertisols
Grassland Biome		
Prairie (tall-grass) (Gp)	(6) Moist subtropical (10) Moist continental	Udolls
Steppe (short-grass prairie) (Gs)	(9s) Dry midlatitude, semiarid (10sh) Moist continental, subhumid	Borolls Ustolls Xerolls Aridisols
Desert Biome		
Thorntree semidesert (Dtw, Dtg)	(4sd, 4d) Dry tropical, semidesert, desert (5sd, 5d) Dry subtropical, semidesert, desert	Aridisols Ustalfs
Semidesert (Dsd, Dss)	(4sd, 4d) Dry tropical, semidesert, desert (5sd, 5d) Dry subtropical, semidesert, desert (9sd, 9d) Dry midlatitude, semidesert, desert	Aridisols Psamments
Dry desert (D, Dsp)	(4d) Dry tropical, desert (5d) Dry subtropical, desert (9d) Dry midlatitude, desert	Aridisols Psamments
Tundra Biome		
Arctic tundra (T)	(12) Tundra	Cryaquepts Cryorthents
Alpine tundra (Ta)	Highland climate, alpine zone	

Equatorial rainforest (Fe) consists of tall, closely set trees. Their crowns form a continuous canopy of foliage and provide dense shade for the ground and lower layers (Figure 16.7). The trees are characteristically smooth barked and unbranched in the lower two-thirds. Trunks commonly are buttressed at the base by radiating, wall-like roots (Figure 16.8). Tree leaves are large and evergreen; from this characteristic the equatorial rainforest is often described as "broadleaf evergreen forest." Crowns of the trees tend to form into two or three layers, or strata, of which the highest layer consists of scattered emergent crowns rising to 40 m (130 ft) and protruding conspicuously above a second layer, 15 to 30 m (50 to 100 ft), which is continuous (Figure 16.9). A third, lower layer consists of small, slender trees 5 to 15 m (15 to 50 ft) high with narrow crowns.

Typical of the equatorial rainforest are thick, woody lianas, supported by the trunks and branches of tall trees. Some are slender, like ropes; others reach thicknesses of 20 cm (8 in.). They rise to heights of the upper tree levels where light is available and may have profusely branched crowns. Many lianas have tendrils or suckers to cling to a growing tree. Others rise by winding about the tree trunk.

Epiphytes are numerous in the equatorial rainforest. These plants are attached to the trunk, branches, and foliage of trees and lianas, using the "host" solely as a means of physical support. Epiphytes are of many plant classes and include ferns, orchids, mosses, and lichens. Some epiphytes are **stranglers.** These woody vines send down their roots to the soil and may eventually surround the tree, perhaps ultimately replacing it. The strangling fig (*Ficus*) is an example (Plate F.1). Other stranglers begin as lianas.

A particularly important botanical characteristic of the equatorial rainforest is the large number of species of trees that coexist. It is said that as many as 1000 species may be found in a square kilometer. Individuals of a species are often widely separated. Consequently, if a particular tree species is to be extracted from the forest for commercial uses, considerable labor is involved in seeking out the trees and transporting them from their isolated positions. Representative trees of the rainforest of the Amazon valley are the Brazilnut (*Bertholletia excelsa*) and the silk-cotton tree (species of *Bombax*).

The floor of the equatorial rainforest is usually so densely shaded that plant foliage is sparse close to the ground. The forest has an open aspect, making it easy to traverse. The ground surface is covered only by a thin litter of leaves. Rapid consumption of dead plant matter by bacterial action results in the absence of humus on the soil surface and within the soil profile.

Equatorial rainforest is a response to an equable climatic regime that is continuously warm, frost free, and has abundant precipitation in all months of the year (or, at most, only one or two dry months). The equatorial rainforest is thus closely correlated in extent with the wet equatorial climate (1) and the monsoon and trade-wind littoral climate (2). A large water surplus characterizes the

Figure 16.7 **Equatorial rainforest near Belém, Brazil. This forest of tall, broad-leaved evergreen trees and numerous lianas is on the relatively high ground *(terra firme)* of the Amazon lowland. (Otto Penner, Instituto Agronómico do Norte.)**

annual water budget, so soil water is adequate at all times. In the absence of a cold or dry season, plant growth goes on continuously throughout the year. Individual species have their own seasons of leaf shedding, possibly caused by slight changes in photoperiod.

Soil orders of the equatorial rainforest are typically Oxisols and Ultisols on well-drained upland surfaces. Recall from Chapters 13 and 14 that soils of these orders have low base status. The meager supply of nutrient base cations is held in a thin surface layer, where it is effectively recycled by rainforest plants.

Variations in the equatorial rainforest structure are found in specialized habitats and where Man has disturbed the vegetation. Where the forest has been cleared by cutting and burning (as for small-plot agriculture), the returning plant growth is low and dense and may be

Figure 16.8 Buttress roots at the base of a large tree *(Bombacopsis fendleri)* of the rainforest on Barro Colorado Island, Canal Zone. (American Museum of Natural History.)

pine *(Pandanus).* Typical of recently formed coastal deposits are belts of palms, such as the cocoanut palm *(Cocos nucifera),* see Plate F.1.

World distribution of equatorial rainforest is shown on the vegetation map, Plate D.6, by those areas of rainforest (Fe) located close to the equator. The principal world areas are the Amazon lowland of South America; the Congo lowland of Africa and a coastal zone extending westward from Nigeria to Guinea; and the East Indian region, from Sumatra on the west to the islands of the western Pacific on the east. Poleward of these areas, equatorial rainforest regions are transitional into tropical rainforest of higher latitudes.

Tropical rainforest (Fe) is quite similar in structure to the equatorial variety, but extends through the tropical zone (lat. 10°–25°N and S) along coasts windward to the trades. The tradewind littoral climate in which the tropical rainforest thrives has a short dry season, but not intense enough to deplete the soil water. There is also a marked annual temperature cycle resulting from the variations in height of the sun's path in the sky in tropical latitudes. The cooler temperatures, coinciding approximately with the period of reduced rainfall, impose some stress on the plants. As a result, there are fewer species and fewer lianas. Epiphytes are abundant, however, because of continued exposure to humid air and cloudiness of the maritime tropical air masses that impinge on the coastal hill and mountain slopes. Soils of the tropical rainforest are largely Oxisols and Ultisols.

In global distribution, the tropical rainforest is represented on the world map (Plate D.6) by those areas of rainforest, designated Fe, that lie between lat. 10° and 25°. The Caribbean lands represent one important area of tropical rainforest, although the rainforest is predominantly limited to windward locations. The Everglades of southernmost Florida contains small isolated communities

described as **jungle.** Jungle can consist of lianas, bamboo scrub, thorny palms, and thickly branching shrubs. This tangled growth can form an impenetrable barrier to travel, in contrast to the open floor of the climax rainforest.

Coastal vegetation in areas of equatorial rainforest is highly specialized. Coasts that receive suspended sediment (mud) from river mouths, and where water depths are shallow, typically have **mangrove swamp forest,** consisting of stilted trees (Plate F.1). The mangrove prop roots trap sediment from ebb and flood tidal currents, so the land is gradually extended seaward. Mangroves commonly consist of several shoreward belts of the red *(Rhizophora),* the black *(Avicennia),* and the white *(Laguncularia)* mangrove. Another common coastal salt-marsh plant of the humid low-latitude zone is the screw

Figure 16.9 Diagram of the structure of tropical rainforest in Trinidad, British West Indies. A representative species of the tallest trees is *Mora exelsa.* (After J. S. Beard, 1946, *The Natural Vegetation of Trinidad,* Clarendon Press, Oxford.)

This air-plant, or epiphyte, is a bromeliad supported on a tree branch. Everglades National Park, Florida. (Arthur N. Strahler)

Rainforest of the western Amazon lowland, northwest of Iquitos, Peru. An exploration well is being drilled from a barge, floated 3700 km (2300 mi) up the Amazon River and its tributary, seen here, the Cuinico River. (Phillips Petroleum Company)

(left) A coastal mangrove forest growing in muddy water. Everglades National Park, Florida. (Arthur N. Strahler)

(right) This strangler fig has surrounded the trunk of a living oak tree. Everglades National Park, Florida. (Arthur N. Strahler)

A coastal grove of coconut palms, City of Refuge, Kona Coast, Island of Hawaii. (Arthur N. Strahler)

(above) **Mixed deciduous-needleleaf forest of the midlatitude zone is represented by this beech-hemlock stand in Great Smoky Mountains National Park, Tennessee. Lightning has shattered the tree at the left. (Donaldson Koons)**

(above) **Broadleaf evergreen forest of the Gulf Coast region is represented here by Evangeline oaks bearing Spanish "moss." The ground beneath is maintained as a lawn. Evangeline State Park, Bayou Teche, Louisiana. (Arthur N. Strahler)**

(far left) **An open needleleaf forest of western yellow pine (Pinus ponderosa), Kaibab National Forest, Arizona. (Arthur N. Strahler)**

(left) **Coastal needleleaf forest is represented by this grove of great redwood trees (Sequoia sempervirens) in Humboldt Redwoods State Park, California. (Alan H. Strahler)**

(below) **Boreal needleleaf forest, largely red and black spruce, in southern Ontario. (Alan H. Strahler)**

(right) **Lichen woodland, a form of open boreal forest, consists of scattered black spruce, small shrubs, and a carpet of lichen (Cladonia). The locality is near Ft. McKenzie, lat. 57°N, in northern Quebec. (R. N. Drummond)**

Plate F.2   Forest biome in middle latitudes

Sawgrass savanna with scattered pines in the dry season (winter), Everglades National Park, Florida. Clumps of tropical forest (left) are known as hammocks. (Arthur N. Strahler)

Savanna woodland of the Serengeti Plains, Tanzania, East Africa. Acacia trees with flattened crowns remain green, although the coarse grasses have turned to straw in the dry, cool season. (H. Barad/Photo Researchers, Inc.)

Closeup view of virgin tall-grass prairie preserved in Kalsow Prairie, a State Botanical Monument, Iowa. Among the tall leaves of grass are forbs in flower. (Gene Ramsay, Webb Publishing Company.)

(above) Cattle and horses grazing on steppe grasslands, southwestern South Dakota. The soil beneath the sod is a Ustoll (order Mollisols). (Arthur N. Strahler)

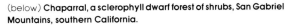

(below) Chaparral, a sclerophyll dwarf forest of shrubs, San Gabriel Mountains, southern California.

(below) A closeup view of the chaparral shown in the view at lower left. The rain gauge, mounted in a tilted position, rests on a patch of bare ground. San Dimas Experimental Forest. (Arthur N. Strahler)

Sagebrush semidesert along the base of the White Cliffs in southern Utah. The classic automobile is an air-coooled Franklin, <u>ca</u> 1934. (Donald L. Babenroth)

Desert scene near Phoenix, Arizona. The tall, columnar plant is saguaro cactus; the delicate wandlike plant is ocotillo. Small clumps of prickly-pear cactus are seen between groups of hardleaved shrubs. (Alan H. Strahler)

(left) Caribou herds grazing on cotton-grass tundra in northern Alaska. (William R. Farrand)

(above) This prickly-pear cactus (<u>Opuntia</u>) grows on the rocky desert floor of Havasu Canyon, a tributary to the Grand Canyon. (Donald L. Babenroth)

(right) Alpine tundra near the summit of the Snowy Range, Wyoming. A broad cirque with a tarn occupies the foreground. In the background is a cirque headwall of quartzite, with talus cones along the base. (Arthur N. Strahler)

Plate F.4   Desert and tundra biomes

of tropical rainforest, termed "hammocks," in which one finds the mahogany tree and strangler fig, along with abundant epiphytes. The coast of the same area has extensive mangrove swamp forest.

In southern and southeastern Asia, tropical rainforest is extensive in coastal zones and highlands that have heavy monsoon rainfall and a very short dry season. The Western Ghats of India and the coastal zone of Burma have tropical rainforest supported by orographic rains of the southwest monsoon. In the zone of combined northeast trades and Asiatic summer monsoon are rainforests of the coasts of Vietnam and the Philippine Islands. In the southern hemisphere, belts of tropical rainforest extend down the eastern Brazilian coast, the Madagascar coast, and the coast of northeastern Australia.

Within the regions of equatorial and tropical rainforest are many islandlike highlands of cooler climate in which precipitation is increased by the orographic effect. Here rainforest extends upward on the rising mountain slopes. Between 1000 and 2000 m (3000 and 6000 ft), the rainforest gradually changes in structure, becoming **montane forest (Fmt).** Several areas of montane forest are shown on the vegetation map — for example, in Central America and along the eastern slopes of the Andes mountains of South America. Montane forest is lower in tree height and less dense in canopy than equatorial and tropical rainforest of adjacent lowlands; with increasing elevation, forest height becomes even lower. Tree ferns and bamboos are numerous; epiphytes are particularly abundant (Figure 16.10). Heavy accumulations of mosses attached to tree branches become conspicuous at higher altitudes. In these **mossy forests** (cloud forests), enshrouded much of the time in clouds, mist and fog are persistent. Relative humidity is high, and condensed moisture keeps the plant surfaces wet. Near its upper limit, in **elfin forest,** the forest trees are dwarfed and densely festooned with mosses. Above the forest limit, which occurs at about 3600 m (12,000 ft) in the equatorial zone, forest gives way to a treeless vegetation that may be alpine tundra, scrub, grassland, or heath, depending on altitude and available moisture.

Monsoon forest (Fmo) presents a more open tree growth than the equatorial and tropical rainforests. Consequently, there is less competition among trees for light but a greater development of vegetation in the lower layers. Maximum tree heights range from 12 to 35 m (40 to 100 ft), which is less than in the equatorial rainforest. Many tree species are present and may number 30 to 40 in a small tract. Tree trunks are massive; the bark is often thick and rough. Branching starts at comparatively low level and produces large round crowns.

Perhaps the most important feature of the monsoon forest is the deciduousness of most of the tree species present. The shedding of leaves results from the stress of a long dry season that occurs at time of low sun and cooler temperatures. Thus the forest in the dry season has somewhat the dormant winter aspect of deciduous forests of midlatitudes (Figure 16.11). A representative example of a monsoon forest tree is the teakwood tree (Tectona grandis).

Lianas and epiphytes are locally abundant in monsoon rainforest but are fewer and smaller than in the equatorial

Figure 16.10 **A tree limb covered by epiphytic ferns. Upland rainforest of Kenya, Africa. (American Museum of Natural History.)**

rainforest. Undergrowth is often a dense shrub thicket. Where second-growth vegetation has formed, it is typically jungle. Clumps of bamboo are an important part of the vegetation in climax teakwood forest.

Monsoon forest is a response to a wet-dry tropical climate in which a rainy season with a water surplus alternates with a dry, rather cool season having a soil-water deficit. These conditions are most strongly developed in the Asiatic monsoon climate, but are not limited to that area. Typical regions of monsoon forest are in India, Burma, Thailand, and Cambodia. Deciduous tropical forest also occurs in Africa, Central and South America, and northern Australia.

Monsoon forest is closely identified with Ultisols and Oxisols in Southeast Asia and South America. These are

Figure 16.11 **Monsoon forest in Chieng Mai Province, northern Thailand. Scale is indicated by a line of porters crossing the clearing. (Robert L. Pendleton, American Geographical Society.)**

soils of low base status and may have developed in periods of moister climate in the Pleistocene epoch. According to our world map of soils, Plate D.5, several areas of monsoon forest coincide with areas of Ustalfs, the ustic-regime suborder of Alfisols associated with a climate having a dry season. Base status of Alfisols is high, in contrast to the Ultisols and Oxisols.

Broadleaf evergreen forest (Fbe), also referred to as *temperate rainforest* or *laurel forest,* differs from the equatorial and tropical rainforests in having relatively few species of trees. Thus there are large populations of individuals of a species. Trees are not as tall as in the low-latitude rainforests; the leaves tend to be smaller and more leathery, the leaf canopy less dense. Among the trees commonly found in the broadleaf evergreen forests of southern Japan and the southeastern United States are evergreen oaks (such as *Quercus virginiana*) and members of the laurel and magnolia families (such as *Magnolia grandiflora*). A quite different evergreen forest flora is found in New Zealand and consists of large tree ferns, large conifers such as the kauri tree (*Agathis australis*), podocarp trees (*Podocarpus*), and small-leaved southern beeches (*Nothofagus*) (see Figure 16.12). Another important type of broadleaf evergreen forest, found in the Azores and Canary island groups, is the Canary laurel forest, which formerly covered Europe in the Miocene geologic epoch.

Broadleaf evergreen forests tend to have a well-developed lower stratum of vegetation that, in different places, may include tree ferns, small palms, bamboos, shrubs, and herbaceous plants. Lianas and epiphytes are abundant. An example of conspicuous epiphyte accumulation is the Spanish "moss" (*Tillandsia usneoides*) that festoons the Evangeline oak and bald cypress trees of the Gulf Coast of the southeastern United States (Plate F.2).

Broadleaf evergreen forest represents a response to a moist climate strongly influenced by maritime air masses. The annual range of temperature is small or moderate, and rainfall is abundant and well distributed throughout the year. Broadleaf evergreen forest of North America and eastern Asia occurs in the moist subtropical climate (6); that of New Zealand and Tasmania is in the marine

Figure 16.12 **This podocarp forest illustrates a variety of evergreen forest in a marine west-coast climate, Hari Hari, western coast of South Island of New Zealand. (Pierre Dansereau.)**

west-coast climate (8).

Broadleaf evergreen forest of the southeastern United States and southern China is closely identified with the Udults, the udic-regime suborder of the Ultisols. Low base status is a characteristic of this soil order. Nutrient bases are concentrated close to the surface, where they are recycled by forest plants. Broadleaf evergreen forest of southeastern coastal Australia coincides with an area of Udalfs, the udic-regime suborder of Alfisols.

Midlatitude deciduous forest (Fd), sometimes called *summergreen deciduous forest,* is familiar to inhabitants of eastern North America and western Europe as a native forest type. It is dominated by tall, broadleaf trees that provide a continuous and dense canopy in summer but shed their leaves completely in the winter (Plate F.2). Lower layers of small trees and shrubs are weakly developed. In spring a luxuriant low layer of herbs develops quickly, but is greatly reduced after the trees have reached full foliage and have shaded the ground.

The deciduous forest is almost entirely limited to the midlatitude landmasses of the northern hemisphere. Figure 16.13, a map of vegetation of the United States and southern Canada, shows forest subdivisions emphasizing floristic composition. Common trees of the deciduous forests of eastern North America, central and eastern Europe, and eastern Asia (all in the moist continental climate) are oak (*Quercus*), beech (*Fagus*), birch (*Betula*), hickory (*Carya*), walnut (*Juglans*), maple (*Acer*), basswood (*Tilia*), elm (*Ulmus*), ash (*Fraxinus*), tulip tree (*Liriodendron*), chestnut (*Castanea*), and hornbeam (*Carpinus*). In western Europe, under a marine west-coast climate, dominant trees are mostly oak and ash, with beech (*Fagus sylvatica*) in cooler and moister areas.

In poorly drained habitats, the deciduous forest consists of such trees as alder, willow, ash, and elm and many hygrophytic shrubs. Where the deciduous forests have been cleared in lumbering, pines readily develop as second-growth vegetation.

Because the regions of midlatitude deciduous forest have for centuries been the areas of dense populations, few remnants of a primeval forest have survived. Instead, most existing forests are modified by practices of tree farming. Large areas are completely and permanently removed from forest growth by farming, urban development, and roadways.

The midlatitude deciduous forest represents a response to a continental climate in which precipitation is adequate throughout the growth season. There is a strong annual temperature cycle with a cold winter season and a warm summer. Precipitation is markedly greater in the summer months (especially so in eastern Asia) and thus increases at the time of year when evapotranspiration is great and the soil-water demand is high. Only a small water deficit is incurred in the summer, whereas a large surplus normally develops in spring. The winter is exceptionally dry in eastern Asia but this factor is compensated for by cold.

Soils most closely identified with the midlatitude deciduous forests are Udalfs and Boralfs, suborders of the Alfisols. Some forest areas in the subhumid climate subtype are underlain by Udolls, the udic-regime suborder of the Mollisols. All these suborders are soils of high base

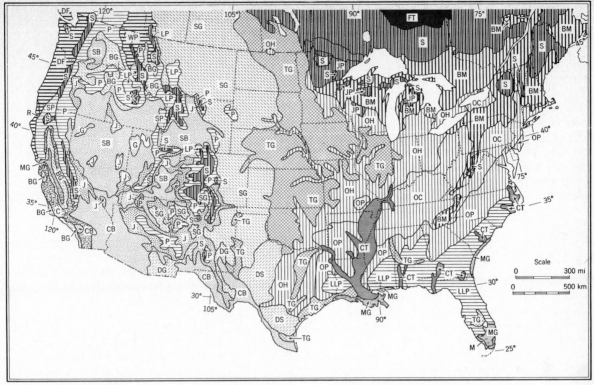

**Figure 16.13 Vegetation of the contiguous 48 United States and southern Canada.
(Modified and simplified from maps of H. L. Shantz and Raphael Zon, in *Atlas of
American Agriculture*, 1929, and Canada Department of Forestry, Bulletin 123, 1963.)**

status. Favorable parent materials of glacial origin, including till plains, lake plains, and loess covers underlie
wide areas where the midlatitude deciduous forest occurs.

Needleleaf forest (Fc, Fl, Fsp, Fbo, Fbl) is composed
largely of straight-trunked, conical trees with relatively
short branches and small, narrow, needlelike leaves.
These trees are conifers. Where evergreen, the needleleaf

forest provides continuous and deep shade to the ground
so that lower layers of vegetation are sparse or absent,
except for a thick carpet of mosses in many places (Plate
F.2). Species are few, and large tracts of forest consists
almost entirely of but one or two species. The world
vegetation map (Plate D.6) includes five units within the
needleleaf forest. These subdivide the needleleaf forest

by floristic composition.

Boreal forest (Fbo, Fbl) predominates in two great continental belts, one in North America and one in Eurasia, which span the landmasses from west to east in latitudes 45° to 75°. The boreal forest (Fbo) of North America, Europe, and western Siberia is composed of evergreen conifers, such as spruce (*Picea*), fir (*Abies*), and pine (*Pinus*), whereas that of north-central and eastern Siberia is boreal forest dominated by larch (Fbl). The larch (*Larix*) sheds its needles in winter and is thus a deciduous forest (Figure 16.14). Associated with the needleleaf trees is mountain ash. Aspen and balsam poplar (*Populus*), willow, and birch tend to take over rapidly in areas of needleleaf forest that have been burned over, or they can be found bordering streams and in open places.

Needleleaf evergreen forest (Fbo) extends into lower latitudes wherever mountain ranges and high plateaus exist. Thus in western North America, this formation class extends southward into the United States on the Sierra Nevada and Rocky Mountain ranges and over parts of the higher plateaus of the southwestern states. In Europe, needleleaf evergreen forests flourish on all the higher mountain ranges.

The *coastal needleleaf evergreen forest (Fc)* of British Columbia and California is of particular interest. Under a regime of heavy orographic precipitation and prevailing high humidities, those areas have perhaps the densest of all coniferous forests, with the world's largest trees. Forests of coastal redwood (*Sequoia sempervirens*), big tree (*Sequoiadendron giganteum*), and Douglas fir (*Pseudotsuga menziesii*) are particularly remarkable (Figure 16.15). Individuals of redwood and big tree attain heights of over 100 m (325 ft) and girths of over 20 m (65 ft), see Plate F.2.

Lake forest (Fl), another needleleaf evergreen forest type, occurs in the Great Lakes area of North America. It is dominated by white pine (*Pinus strobus*), red pine (*P. resinosa*), and eastern hemlock (*Tsuga canadensis*). Little

Figure 16.15 **This mixed stand of white pine, Douglas fir, and Englemann spruce is on the western slopes of the Cascade Range in Washington. (U.S. Forest Service.)**

Figure 16.14 **Woodland of larch *(Larix dahurica)*, Tompo River region, about latitude 64°N, Yakutsk, USSR. This scene shows the larch tree near the northern limit of its growth. (Komarov Botanical Institute, Leningrad.)**

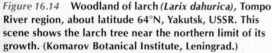

of this commercially valuable forest remains.

Another needleleaf forest type unique to North America is the *southern pine forest (Fsp)*, consisting of a number of different pine species. It is found in sandy soils of the southeastern coastal plain, where it appears to be a specialized vegetation type dependent on fast-draining sandy soils and frequent fires for its preservation.

In the northernmost range, the boreal needleleaf forest grades into **cold woodland.** This form of vegetation is limited to the very cold subarctic climate. Trees are low in height and spaced well apart; a shrub layer may be well developed. The ground cover of lichens and mosses is distinctive (see Plate F.2). Cold woodland is often referred to as **taiga.** It is transitional into the treeless tundra and arctic heath at its northern fringe.

Because much of the area of needleleaf evergreen forest in North America and Europe was subjected to glaciation in the Wisconsin stage of the Pleistocene epoch, lakes and poorly drained depressions abound. These bear the hygrophytic vegetation, described in Chapter 14 as forming a bog succession and leading to large, thick peat accumulations known in Canada as *muskeg.*

Boreal forest (Fbo) is closely identified with the boreal forest climate (11) throughout North America and Eurasia. Both northern and southern boundaries of the formation class coincide rather well with the climatic boundaries, defined in terms of annual water need. Much of the boreal forest region is underlain by Spodosols, acid soils of low base status. In interior areas of western Canada, Russia, and Siberia, the soils are Boralfs, the cold-regime suborder of the Alfisols. Also important are large areas of Histosols and Cryaquepts. The Cryaquepts are a cold-climate great soil group within the Aquepts, which is a suborder of the Inceptisols characterized by poor drainage and saturated soil. Lake forest (Fl) and coastal forest (Fcl) are closely associated with Spodosols. The southern pine forest (Fsp), which lies in the moist subtropical climate (6), is in a region of Udults and Aquults. The Aquults, a suborder of the Ultisols characterized by poor drainage, occupy much of the coastal fringe of the southeastern United States (see Figure 14.1).

Sclerophyll forest (Fsm, Fss, Fsa) consists of low trees with small, hard, leathery leaves. Typically the trees are low-branched and gnarled, with thick bark. The formation class includes much **woodland,** an open forest in which the canopy coverage is only 25 to 60 percent. Also included are extensive areas of **scrub,** a plant formation type consisting of shrubs having a canopy coverage of perhaps 50 percent. The trees and shrubs are evergreen, their thickened leaves being retained despite a severe annual drought. There is little stratification in the sclerophyll forest and scrub, although there may be a spring herb layer.

Sclerophyll forest is closely associated with the Mediterranean climate (7) and quite narrowly limited in geographic extent — primarily to west coasts between 30° and 40° or 45° latitude (Plate D.6). In the Mediterranean lands the sclerophyll forest forms a narrow peripheral coastal belt. Here the *Mediterranean evergreen mixed forest (Fsm)* consists of such trees as cork oak (*Quercus suber*), live oak (*Quercus ilex*), Aleppo pine (*Pinus halepensis*), stone pine (*Pinus pinea*), and olive (*Olea europaea*). What may have once been luxuriant forests of such trees were greatly disturbed by Man over the centuries and reduced to woodland or entirely destroyed. Today, large areas consist of dense scrub, locally called **maquis,** which includes many species, some of them very spiny.

The other northern hemisphere region of sclerophyll forest is that of the California coast ranges. Some of this is a woodland of live oak (*Quercus agrifolia*) and white oak (*Quercus lobata*) (Figure 16.16). Much of the vegetation is *sclerophyllous scrub (Fss)* or "dwarf forest," known as **chaparral** (Plate F.3). It varies in composition with elevation and exposure. Chaparral may contain wild lilac (*Ceanothus*), manzanita (*Arctostaphylos*), mountain mahogany (*Cercocarpus*), "poison oak" (*Rhus diversiloba*), and live oak.

The sclerophyll forest is represented in central Chile and in the Cape region of South Africa by a scrub vegetation that is of quite different flora from scrub of the northern hemisphere. Important areas of sclerophyll forest, woodland, and scrub are found in southeast, southcentral, and southwest Australia, including several species of eucalypts and acacias (Figure 16.17). This formation type is designated *Australian sclerophyll forest (Fsa)* on the world climate map.

The Mediterranean regime of the sclerophyll forest is one of great environmental stress because the severe drought season coincides with high air temperatures. Thus a large water deficit occurs in the summer. The wet, mild winter is, by contrast, highly favorable to rapid plant growth.

The soils of the sclerophyll forest, woodland, and

Figure 16.16 **Heavily grazed woodland of oak and open grassland, Monterey County, California. (U.S. Forest Service.)**

Figure 16.17 **Eucalyptus woodland, Darling Range, Western Australia, about 1930. This woodland has been largely destroyed since the photo was taken. (Agent General for Western Australia.)**

scrub are closely associated with the Xeralfs, a xeric-regime suborder of the Alfisols. These soils have high base status. Also important locally are Xerolls and Xerults, xeric-regime suborders of the Mollisols and Ultisols, respectively.

Savanna Biome

Savanna woodland (Sw) consists of trees spaced rather widely apart, permitting development of a dense lower layer, which usually consists of grasses. The woodland has an open, parklike appearance. Although plant formations of this general description can be found in a wide range of latitudes, most geographers associate savanna woodland with the wet-dry tropical climate (3) of Africa and South America. There the soil-water regime shows a seasonal decline in soil-water storage to an amount too small to sustain a closed-canopy forest. Savanna woodland typically forms a belt adjacent to equatorial rainforest in which soil-water storage remains high throughout the year.

In the tropical savanna woodland of Africa, the trees are of medium height, the crowns are flattened or umbrella-shaped, and the trunks have thick, rough bark (Plate F.3). Some species of trees are xerophytic forms with small leaves and thorns; others are broad-leaved deciduous species that shed their leaves in the dry season. In this respect, savanna woodland is akin to monsoon forest, into which it grades in some places.

Fire is a frequent occurrence in the savanna woodland during the dry season. The trees of the savanna are of species that are particularly resistant to fire. Many biogeographers hold the view that periodic burning of the savanna grasses is responsible for the maintenance of the grassland against the invasion of forest. Fire does not kill the un-derground parts of grass plants, but it limits tree growth to a few individuals of fire-resistant species. The browsing of animals, which kills many young trees, is also a factor in maintaining grassland at the expense of forest. Many rainforest tree species that might otherwise grow in the wet-dry climate regime are prevented by fires from invading.

In Africa, the savanna woodland grades into a belt of **thorntree–tall grass savanna (Stg),** a formation class transitional into the desert biome. The trees, largely of thorny species, are more widely scattered and the open grassland is more extensive than in the savanna woodland. One characteristic tree is the flat-topped acacia (*Acacia*), another is the grotesque baobab (*Adansonia digitata*), with its barrel-shaped water-storing trunk (Figure 16.18). Elephant grass (*Pennisetum purpureum*) is a common species; it may grow to a height of 5 m (16 ft) to form a thicket impenetrable to humans. Because of the dominance of grassland over woodland, this formation class may equally well be placed in the grassland biome in terms of structure. Because it is geographically and climatically intergradational with savanna woodland, inclusion in the savanna biome seems the more desirable alternative.

The thorntree–tall grass savanna is closely identified with the semiarid subtype of the dry tropical and subtropical climates (4s, 5s). In Africa and India, the formation class includes areas of the semidesert climate subtype (4sd, 5sd). In the semiarid climate, soil-water storage is very low in 10 out of the 12 months. Only during the brief rainy season is soil water adequate for the needs of plants. Onset of the rains is quickly followed by greening of the trees and grasses. For this reason, vegetation of the savanna biome is described as "raingreen," an adjective that applies also to the monsoon forest.

Soils of the savanna biome include Ustalfs, Ultisols, Oxisols, and Vertisols. The distribution of these orders varies according to parent matter of the soil and past climatic history. Ustalfs of the savanna biome in West Africa are believed to owe their high base status to a rain of fine dust particles carried from the Sahara Desert on the north. These dusts replenish primary minerals rich in base cations. The same process may explain the occurrence of Ustalfs in savanna regions of other parts of Africa, in South America, and in India.

The savanna biome in South America is exemplified in the *campo cerrado* of the interior Brazilian Highlands. Here the trees are largely broad-leaved and evergreen. They are deep-rooted and capable of tapping lower levels of soil water not available to the grasses during the dry season.

A different expression of the savanna biome is found in eastern Australia. Here the *Australian sclerophyllous tree savanna (Ssa)* forms a north-south belt in which trees are evergreen species of *Eucalyptus*. The climate is a semidesert subtype of the dry subtropical climate (5sd). According to the world soils map, Plate D.5, the soils here are Vertisols.

Of particular interest to American geographers is a small patch of rather specialized tropical savanna in southern Florida — the Everglades (Plate F.3). Underlain by a limestone formation, the extremely low, flat plain of

Figure 16.18 **Baobab tree** *(Adansonia digitata)*, **near Nairobi, Kenya, Africa. (Akeley Expedition, American Museum of Natural History.)**

the Everglades is flooded by a shallow layer of runoff from summer rains and becomes a swamp. In winter the area becomes extremely dry. Coarse saw grass *(Cladium effusum)* covers much of the plain. Scattered trees are represented by palms and, on higher ground, by pines.

Grassland Biome

Prairie (Gp) consists of tall grasses, comprising the dominant herbs and subdominant forbs (Plate F.3). Trees and shrubs are almost totally absent but may occur in the same region as forest or woodland patches in valley bottoms. The grasses are deeply rooted and form a continuous and dense sward. The grasses flower in spring and early summer; the forbs, in late summer. In Iowa, a representative region of tall-grass prairie, typical grasses are big bluestem *(Andropogon gerardi)* and little bluestem *(A. scoparius)*; a typical forb is black-eyed susan *(Rudbeckia nitida)*.

The largest areas of tall-grass prairie, those of North and South America, are closely associated with the subhumid subtype of the moist continental climate (10sh) and moist subtropical climate (6sh). There the soil-water budget shows a moderate soil-water deficiency, ranging from 0 to 15 cm. The tall-grass prairie also extends far into the area of the humid subtype of the same climates (10h, 6h). Details of the distribution of tall-grass prairie in North America are shown in Figure 16.13. Areas of deciduous forest are interfingered with areas of prairie over a broad transition belt. In Uruguay and eastern Argentina, the prairie plains region is known as the *Pampa*. A small but important European tall-grass prairie region is the *Puszta* of Hungary.

Tall-grass prairie is closely identified with Udolls, the udic-regime suborder of the Mollisols. Ustolls, the ustic-regime suborder of the Mollisols, underlie areas of prairie near its western limit in North America. These soils are exceptionally rich in base cations important to grasses.

Steppe (Gs), also called **short-grass prairie,** is a formation class consisting of short grasses occurring in sparsely distributed clumps or bunches (Plate F.3). Scattered shrubs and low trees may also be found in the steppe. Ground coverage is small and much bare soil is exposed. Many species of grasses and other herbs occur. A typical grass of the American steppe is buffalo grass *(Buchloe dactyloides)*; other typical plants are the sunflower *(Helianthus rigidus)* and loco weed *(Oxytropis lambertii)*. Steppe grades into semidesert in dry environments and into prairie where rainfall is higher.

The world vegetation map shows that steppe grassland is largely concentrated in vast midlatitude areas of North America and Eurasia. The only southern hemisphere occurrence shown on the map is the *veldt* of South Africa, formed on a highland surface in Orange Free State and Transvaal. We have excluded from the steppe formation class many occurrences of grasslands in low latitudes; these are, however, included in formation classes of transitional nature within the savanna and desert biomes.

The steppe formation class coincides quite closely with the semiarid subtype of the dry midlatitude climate (9s). The soil-water budget shows a substantial soil-water deficiency, and there is no water surplus. Soil-water storage is far below the soil storage capacity in all months. During a spring period of soil-water recharge, a substantial amount of water is made available to grasses and results in rapid growth into early summer. By midsummer the grasses are usually dormant, although occasional summer rainstorms cause periods of revived growth.

Soils of the steppe grasslands are largely within the order of Mollisols. Borolls are found in the colder northerly areas, Ustolls in the warmer southerly areas. Xerolls underlie the steppe grasslands west of the Rocky Mountains. Aridisols are represented in some of the more arid steppe lands, transitional into the desert biome.

Desert Biome

The desert biome includes several formation classes that are transitional from grassland and savanna biomes into vegetation of the more severe desert environment. We recognize two basic transitional formation classes: thorntree semidesert and semidesert. The first of these, **thorntree semidesert,** is found in low latitudes and is associated with tropical and subtropical climates. On the world vegetation map, Plate D.6, thorntree semidesert includes two lesser formation classes: *thorn forest and thorn woodland (Dtw)* and *thorntree–desert grass savanna (Dtg)*.

Thorntree semidesert consists of xerophytic trees and shrubs adapted to a climate with a very long, hot dry season and only a very brief but intense rainy season. These conditions are found in the semidesert and desert subtypes of the dry tropical and dry subtropical climates (4sd, 4d, 5sd, 5d). The thorny trees and shrubs are locally known as *thorn forest, thornbush,* and *thornwoods*. These are deciduous forms that shed their leaves in the dry season (Figure 16.19). The plants may be closely inter-

Figure 16.19 **Thornbush in the region of the Tona River, central Africa. (Akeley Expedition, American Museum of Natural History.)**

grown to form impenetrable thickets. Cactus plants are present in some localities. A lower layer of herbs may consist of annuals, which largely disappear in the dry season, or grasses that become dormant in the dry season. Examples of thorntree semidesert are the *caatinga* of northeastern Brazil and the *dornveldt* of South Africa. Aridisols are the dominant soil order of the thorntree semidesert, but Ustalfs are present in some areas.

Semidesert is a transitional formation class of very wide latitudinal distribution; it ranges from the tropical zone to the midlatitude zone and is identified with the semidesert and desert subtypes of all three dry climates. On the world vegetation map, Plate D.6, semidesert includes two secondary formation classes: *semidesert scrub and woodland (Dsd)* and *semidesert scrub (Dss)*.

Semidesert is a xerophytic shrub vegetation with a poorly developed herbaceous lower layer. An example is the sagebrush (*Artemisia tridentata*) vegetation of the middle and southern Rocky Mountain region and Colorado Plateau (Plate F.4). Semidesert shrub vegetation seems recently to have expanded widely into areas of the western United States that were formerly steppe grasslands as a result of overgrazing and trampling by livestock. Soils of the semidesert are largely Aridisols. Also present are many areas of Entisols, particularly the Psamments developed on dune sands and sandy alluvium.

Dry desert is a formation class of xerophytic plants widely dispersed and providing no important degree of ground cover. On the world vegetation map, Plate D.6, dry desert is represented by two formation classes: *desert (D)* and *desert alternating with porcupine grass semidesert (Dsp)*. The latter type is limited in occurrence to Australia.

The visible vegetation of dry desert consists of small hard-leaved or spiny shrubs, succulent plants (cacti), or hard grasses. Many species of small annuals may be present, but appear only after a rare but heavy desert downpour.

Desert floras differ greatly from one part of the world to another. In the Mojave-Sonoran deserts of the southwestern United States, plants are often large and in places give a near-woodland appearance (Plate F.4). Well known are the treelike saguaro cactus (*Carnegiea gigantea*), the prickly-pear cactus (*Opuntia imbricata*), the ocotillo (*Fouquiera splendens*), the creosote bush (*Larrea tridentata*) and the smoke tree (*Dalea spinosa*).

In the Sahara desert (most of it very much drier than the American desert), a typical plant is a hard grass (*Stipa*); another, found along the dry beds of watercourses, is tamarisk (*Tamarix*). The coastal desert of southwest Africa is known for the strange tumboa plant (*Weltwitschia mirabilis*), which has strap-shaped leaves radiating from a central tap root that penetrates deep into the ground (Figure 16.20).

Much of the world map area assigned to desert vegetation has no plants of visible dimensions because the surface consists of shifting dune sands or sterile salt flats. Soils of the dry desert belong to the order of Aridisols or to Entisols, suborder Psamments.

The soil-water budget of the dry desert vegetation usually shows a greater quantity of water need than of precipitation in every one of the twelve months. The annual total soil-water shortage is therefore very large. When rain falls in rare, torrential rainstorms, the soil water is used to maximum advantage by the plants and is rapidly depleted. During long periods when soil-water storage is near zero, the plants must maintain a nearly dormant state to survive.

Tundra Biome

Arctic tundra (T) and **alpine tundra (Ta)** are formation classes of the tundra climate (12). Tundra of the arctic regions flourishes under a regime of long summer days during which the ground ice melts only in a shallow surface layer. The frozen ground beneath (permafrost) remains impermeable and meltwater cannot readily escape. Consequently, in summer a marshy condition prevails for at least a short time over wide areas. Humus accumulates in well-developed layers because bacterial action is very slow. Size of plants is partly limited by the mechanical rupture of roots during freeze and thaw of the surface layer of soil, producing shallow-rooted plants. In winter, drying winds and mechanical abrasion by wind-driven snow tend to reduce any portions of a plant that project above the snow.

Plants of the arctic tundra are low and mostly herbaceous, although dwarf willow (*Salix herbacea*) occurs in places. Sedges, grasses, mosses, and lichens dominate the tundra in a low layer (Plate F.4). Typical species are ridge sedge (*Carex bigelowii*), arctic meadow grass (*Poa arctica*), cotton grasses (*Eriophorum*), and snow lichen (*Cetraria nivalis*). There are also many species of forbs that flower brightly in the summer. Tundra composition varies greatly because soils range from wet to well drained.

Figure 16.20 **The tumboa plant** *(Weltwitschia mirabilis)* **on a sandy plain in the Kalahari Desert of southwestern Africa. (Robert J. Rodin.)**

One form of tundra consists of sturdy hummocks of plants with low, water-covered ground between. Some areas of arctic scrub vegetation composed of willows and birches are also found in tundra.

In all latitudes, where altitude is sufficiently high, alpine tundra is developed above the limit of tree growth and below the vegetation-free zone of barren rock and perpetual snow. Alpine tundra resembles arctic tundra in many physical respects.

Soils of the arctic tundra include representatives from the Inceptisols, Entisols, and Histosols. Large areas of arctic soils are classified as Cryaquepts, a great soil group within the suborder of Aquepts, order Inceptisols. Cryaquepts are poorly drained mineral soils. Other soil areas are classified as Cryorthents, a great soil group within the suborder of Orthents, order Entisols. Histosols are found in filled lake basins.

Altitude Zones of Vegetation

In earlier chapters we described the effects of increasing elevation on climatic factors, particularly air temperature and precipitation, and on soils. Vegetation also responds to an increase in elevation, as we showed in the case of the transition of rainforest into montane forest in low latitudes. To round out this concept, we turn to the dry midlatitude climate of the southwestern United States because there altitude zonation of vegetation is particularly striking.

Figure 16.21 shows the vegetation zones of the Colorado Plateau region of northern Arizona and adjacent states. Zone names, elevations, dominant forest trees, and annual precipitation data are given in the figure. Ecologists have set up a series of life zones, whose names suggest the similarities of these zones to latitude zones encountered in poleward travel on a meridian. The Hudsonian zone, 2900 to 3500 m (9500 to 11,500 ft), bears a needleleaf forest essentially similar to needleleaf boreal forest of the subarctic zone. Here Spodosols lie beneath the forest. As the limit of forest, or tree line, is approached, the coniferous trees take on a stunted appearance and decrease in height to low shrublike forms (Plate F.4).

A zone of alpine tundra lies above tree line. The snow line is encountered at about 3500 to 4000 m (11,500 to 13,000 ft) in midlatitudes, which is of course much lower than at the equator.

Global Vegetation in Review

The leading theme of this chapter has been that the structure of the natural plant cover is a response to environments dictated by climate. Each of the life forms favored by green plants has a capability for suvival and a limit to survival. Forests exist where the environment is most favorable to net productivity of the biomass because of the abundance of heat and soil water during a long growth season. Where soil water is in short supply but heat remains adequate, forest gives way to the savanna structure in which shrubs and herbs have the upper hand over trees (Figure 16.22). The savanna woodland grades

Figure 16.21 **Altitude zoning of vegetation in the arid southwestern United States. Grand Canyon – San Francisco Mountain district of northern Arizona. (After G. A. Pearson, C. H. Merriam, and A. N. Strahler.)**

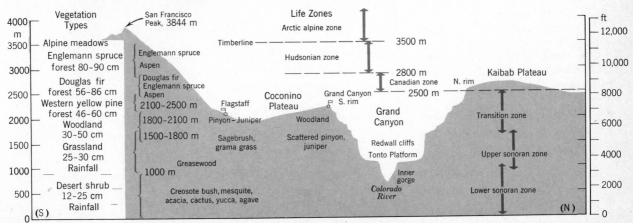

| S | Equatorial rainforest | Savanna woodland | Savanna grassland | | Tropical scrub | Tropical desert | N |

Profile from equator to tropic of cancer, Africa

| S | Tropical desert | Subtropical steppe | Sclerophyll forest | Midlatitude deciduous forest | Subarctic needleleaf forest | Subarctic woodland (taiga) | Arctic tundra | N |

Profile from tropic of cancer to Arctic circle, Africa-Eurasia

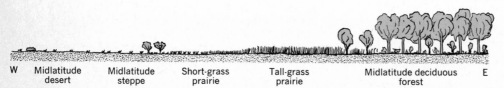

| W | Midlatitude desert | Midlatitude steppe | Short-grass prairie | Tall-grass prairie | Midlatitude deciduous forest | E |

Profile across United States, 40°N, Nevada to Ohio

Figure 16.22 **Three schematic profiles showing the succession of plant formation classes across climatic gradients.**

into grassland and this, in turn, grades into desert, where only those plants capable of living through long drought periods can survive. Traced into colder, higher latitudes, where heat supply becomes reduced below the optimum level, the forest gives way to arctic tundra. There plants contend with the effects of prolonged freezing of soil water and have only a short, cold annual period of growth.

We have found a significant correspondence of vegetation structure with soil type because soil-forming processes are also strongly dominated by climate. But organic processes also shape soil character, so the close associations of vegetation formation classes and soil types are partly a result of interaction.

References for Further Study

Dansereau, Pierre (1957), *Biogeography; an ecological perspective,* Ronald Press, New York.

Polunin, Nicholas (1960), *Introduction to plant geography,* McGraw-Hill Book Co., New York.

Küchler, A. W. (1964), *Potential natural vegetation of the conterminous United States* (map and manual), American Geographical Society, Special Publication No. 36.

Eyre, S. R. (1968), *Vegetation and soils; a world picture,* second edition, Aldine Publishing Co., Chicago.

Walter, Heinrich (1973), *Vegetation of the earth.* Translated from the second German edition by J. Wieser, Springer-Verlag, New York.

Review Questions

1. What two criteria are important in classification of the world's vegetation? Explain and describe these criteria.
2. What six categories of information are included in a structural description of vegetation?
3. What is meant by life form? Define and describe seven different life-forms, giving examples of each.
4. Explain how size and stratification, coverage, and periodicity are used in the structural description of vegetation.
5. Name and describe the various leaf forms and leaf textures.
6. Name the major biomes. Describe the characteristic vegetation of each biome and its general relationship to climate.
7. What is a formation class? On what basis are formation classes distinguished?

8. Describe the equatorial rainforest in terms of structure. What is the importance of lianas in this rainforest? Of epiphytes? Comment on the number of species of trees found in the equatorial rainforest. Name two respresentative trees. What climate types and soil orders are associated with the equatorial rainforest?
9. What coastal vegetation forms may be expected in regions of equatorial rainforest? Describe particularly the mangrove coasts.
10. How does tropical rainforest differ from equatorial rainforest, and in what respects is it similar? In which geographic regions is tropical rainforest found? Explain. Describe montane forest.
11. What are the unique characteristics of the monsoon forest as

compared with other low-latitude forests? What accounts for the deciduousness of the trees? Relate the monsoon forest to a particular climate.

12. How does broadleaf evergreen forest differ in structure and geographic location from equatorial and tropical rainforests? Name three areas in which broadleaf evergreen forest is found. With what climate types and soil orders is broadleaf evergreen forest associated?

13. Describe the midlatitude deciduous forest as to structure and seasonal development. Name common trees of the deciduous forests of eastern North America and Europe. In what ways does the midlatitude deciduous forest represent a response to the continental climate? With what soil orders is midlatitude deciduous forest associated?

14. Describe needleleaf forest in terms of vegetation structure. Define the major world regions of occurrence of boreal forest and name common trees of these belts. With what climate types and soil orders is boreal forest associated?

15. Describe the needleleaf forest of the Pacific coastal regions of North America. What are its distinctive features?

16. What two other needleleaf forest types occur in North America? Describe their species composition and their geographic location.

17. What is taiga? How is it related to the boreal forest?

18. Describe the sclerophyll forest in terms of vegetation structure. With what climate is this formation class closely associated? Name some representative sclerophyll forest plants of the Mediterranean region, California, and Australia.

19. What is savanna woodland? With what climate types and soil orders is it most commonly associated? What are the life-form characteristics of trees and shrubs of the tropical savanna woodland?

20. How does thorntree–tall grass savanna differ from savanna woodland? Describe the vegetation, soils, climate, and animal life associated with tropical savanna, with particular reference to Africa. Where can the savanna biome be found in North America?

21. What plants comprise prairie as a vegetation formation class? Name representative examples from the tall-grass prairie of Iowa. With what climate types and soil orders is prairie vegetation associated? What is the Puszta of Hungary? the Pampa of Argentina?

22. Compare steppe (short-grass prairie) with tall-grass prairie in terms of vegetation structure and typical species. With what climate types and soil orders is steppe associated?

23. Describe thorntree semidesert as a vegetation class. With what climates is the class associated? What is the caatinga of Brazil? The dornveld of South Africa?

24. What is semidesert vegetation? Describe semidesert in the southwestern United States.

25. Describe dry desert as a formation class. Name some typical desert plants of the southwestern United States, the Sahara, and southwest Africa. With what soil orders is dry desert associated?

26. Describe the tundra biome, giving names of typical plants. With what conditions of climate, drainage, and soil-forming process is tundra associated? With what soil orders is tundra associated?

27. Describe the altitude zonation of vegetation in the southwestern United States.

Landforms of Weathering and Mass Wasting

17

Landforms are the distinctive configurations of the land surface—mountains, hills, valleys, plains, and the like. Landforms are environmentally significant because they influence the place-to-place variation in ecological factors, such as water availability and exposure to radiant solar energy. Through varying qualities of slope, aspect, and relief, landforms directly influence hydrologic and soil-forming processes, which in turn act as direct controls on ecosystems. These influences have been stressed in earlier chapters.

Landforms are also of primary concern in geography because they exert far-reaching influences on the patterns of human activity. The direct environmental influences are obvious. A mountain chain is an effective barrier between groups of people who live in adjacent lowlands. A plain, on the other hand, may be densely populated, rich in agricultural resources, and unified culturally and politically by a network of good roads and railroads that permit people with common interests to communicate freely. One coastline, deeply indented with excellent natural harbors but bordered by a rocky rugged coastal belt, may have favored a seafaring community adapted to fishing, ocean commerce, and shipbuilding. Another coast with a straight shoreline and shallow bottom may not provide a single good natural harbor, but may be bordered by a low, fertile coastal plain. There human activity turned naturally to agriculture.

Geomorphology

Geomorphology is the study of landforms, including their history and processes of origin. Geomorphology deals largely with the action of **fluid agents** that erode, transport, and deposit mineral and organic matter. The four fluid agents are (1) running water in surface and underground flow systems; (2) waves, acting with currents in oceans and lakes; (3) glacial ice, moving sluggishly in great masses; and (4) wind, blowing over the ground.

Of the four agents, three are forms of water. Consequently, the science of hydrology is inseparably interwoven with geomorphology. One might not be far wrong in

saying simply that the hydrologist is preoccupied with "where water goes," the geomorphologist with "what water does." Hydrology concerns itself with the hydrologic cycle in an attempt to calculate the water balance and measure rates of flow of water in all parts of that cycle (Chapter 10). Geomorphology concerns itself with geologic work that the water in motion performs on the land.

Denudation is a useful term for the total action of all processes by which the exposed rocks of the continents are worn away and the resulting sediments are transported to the sea by the fluid agents. Denudation is an overall lowering of the land surface. Denudation tends toward reducing the continents to nearly featureless sea-level surfaces and, ultimately, through wave action, to submarine surfaces. If it had not been repeatedly counteracted by crustal uplifts throughout geologic time, denudation would have eliminated all terrestrial environments.

An important point that emerges as we look back through geologic time is that terrestrial life environments have been in constant change, even as plants and animals have undergone their evolutionary development. The varied denudation processes have produced, maintained, and changed a wide variety of landforms, which have been the habitats for evolving life forms. In turn, the life forms have become adapted to those habitats and have diversified to a degree that matches the diversity of the landforms themselves.

Geomorphic and hydrologic systems have long been subjected to radical modification by the works of Man. Agriculture has for centuries altered the surface properties of areas of subcontinental size. Agriculture has modified the action of running water and the water balance, to say nothing of having radically changed the character of the soil. Urbanization is an even more radical alteration, seriously upsetting hydrologic processes. Engineering and mining activities, such as strip-mining and the construction of highways, dams, and canals, not only upset hydrologic systems but can completely destroy or submerge entire assemblages of landforms. Of the four fluid agents of landform sculpture, only glaciers of ice have so far successfully resisted changes of activity imposed by Man.

Invariably, Man's attempts to control the action of running water, waves, and currents produce unpredicted and undesirable side effects, some of which are physical, others ecological. An important reason to study geomorphology is to predict the consequences of Man-made changes and to plan wisely for management of the environment.

Weathering and Mass Wasting

In Chapter 11, processes of physical and chemical weathering were explained to promote an understanding of the accumulation of regolith and the production of various kinds of sediments as parent materials of the soil. Weathering also plays a major role in denudation. Disintegration and decomposition of various kinds of hard bedrock greatly facilitate the work of the fluid agents of erosion. Besides this function, weathering leads to a number of distinctive landforms, which we shall describe in this chapter. Although we have dealt with weathering as a one-way series of changes leading to chemical stability of minerals in the surface environment, it is important to keep in mind that the physical weathering processes continue to agitate and move soil and regolith. Frost action, alternate wetting and drying, growth and decay of roots, and various other repetitive processes cause the soil and regolith to expand and contract in ceaseless daily and seasonal cycles.

The spontaneous downward movement of soil, regolith, and rock under the influence of gravity (but without the dynamic action of moving fluids) is included under the general term **mass wasting.** In Chapter 1 we emphasized the role of gravity as a pervasive environmental factor. All processes of the life layer take place in the earth's gravity field, and all particles of matter tend to respond to gravity. Movement to lower levels takes place when the internal strength of a mass of regolith, sediment, or rock declines to a critical point below which the force of gravity cannot be resisted. This failure of

strength under the ever-present force of gravity takes many forms and scales. We shall see that Man is a major agent in causing several forms of mass wasting.

The Wasting of Slopes

As used in geomorphology, the term *slope* designates some small element or area of the land surface that is inclined from the horizontal. Thus we speak of "mountain slopes," "hillslopes," or "valley-side slopes" with reference to the inclined ground surfaces extending from divides and summits down to valley bottoms.

Slopes guide the flow of surface water under the influence of gravity. Slopes fit together to form drainage systems in which surface-water flow converges into stream channels; these, in turn, conduct the water and rock waste to the oceans to complete the hydrologic cycle. Natural processes have so completely provided the earth's land surfaces with slopes that perfectly horizontal or vertical surfaces are extremely rare.

Figure 17.1 shows a typical hillslope forming one wall of the valley of a small stream. Soil and regolith mantle the bedrock except in a few places where the bedrock is particularly hard and projects in the form of **outcrops. Residual regolith** is derived from the rock beneath and moves very slowly down the slope toward the stream. Beneath the valley bottom are layers of transported regolith—alluvium transported and deposited by the stream. This sediment had its source in regolith prepared on hillslopes many kilometers or tens of kilometers upstream. All terrestrial accumulations of sediment, whether deposited by streams, waves and currents, wind, or glacial ice, can be designated **transported regolith** in contrast to residual regolith.

Geometry of Rock Breakup

As weathering processes attack bedrock, both the shapes of landforms that result and the kinds of sediment particles produced are strongly influenced by the texture of the

Figure 17.1 **Soil, regolith, and outcrops on a hillslope. Alluvium lies in the floor of an adjacent stream valley.**

Granular disintegration

Exfoliation

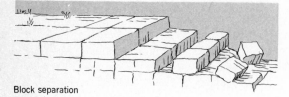

Block separation

Figure 17.3 This cliff of sandstone shows excellent jointing in two sets of vertical planes. (J. R. Stacy, U.S. Geological Survey.)

Shattering

Figure 17.2 **Rock breakup takes various forms, depending on rock type and presence of jointing.**

rock and the presence or absence of joints. In terms of geometry, we can identify at least four different modes of rock breakup.

Rocks composed of coarse mineral grains (intrusive igneous rocks of coarse texture and coarse clastic sedimentary rocks) commonly fall apart grain by grain, a form of breakup termed **granular disintegration** (Figure 17.2). The product is a gravel or coarse sand in which each grain consists of a single mineral particle separated from others along the original crystal or grain boundaries.

Exfoliation is the formation of curved rock shells that separate in succession from the original rock mass, leaving behind successively smaller spheroidal bodies. This type of breakup is also called **spalling.**

If a rock has well-developed sets of joints, produced by mountain-making pressures or by shrinkage during cooling from a magma, the common form of breakup is by **block separation.** Obviously, comparatively weak stresses can separate joint blocks, whereas strong stresses are required to make fresh fractures through solid rock. In sedimentary rocks the planes of stratification, or bedding planes, comprise a horizontal set of planes of weakness cut at right angles by sets of vertical joints. Figure 17.3 shows joint blocks being separated by weathering forces. Of course, it is quite possible that a single, solid joint block will later break up either by granular disintegration or by exfoliation.

Shattering is the disintegration of rock along new surfaces of breakage in otherwise massive, strong rock, to produce highly angular pieces with sharp corners and

edges (Figure 17.2). The surface of fracture may pass between individual mineral crystals or grains or may cut through them. The blocks seen in Plate G.1 are joint blocks, many of which have been shattered into smaller pieces.

Features Produced by Frost Action and Ground Ice

As we explained in Chapter 11, the growth of ice crystals in soil, regolith, and openings in bedrock exerts powerful expansive stresses, believed to be capable of rupturing even extremely hard rocks. Frost action produces a number of conspicuous effects and forms in all climates with cold winters. Features caused by ground-ice accumulation and frost action are particularly conspicuous in the tundra climate of arctic coastal fringes and islands, and above the timberline in high mountains. Frost shattering of hard rocks exposed above timberline leads to surface accumulations of large angular fragments, including boulders up to 3 m (10 ft) or more across. The German name **felsenmeer** (rock-sea) has been given to large expanses of such fragments (Plate G.1).

In fine-textured soils and sediments, dominantly composed of silt and clay, freezing of soil water takes place in horizontal layers, or lenses. As these ice layers thicken, the overlying soil layer is heaved upward. Prolonged soil heaving can produce minor irregularities and small mounds on the soil surface. If a rock fragment lies on the surface, perpendicular ice needles grow beneath the fragment, raising it above the surface (Plate G.1). The same process acting on a rock fragment below the soil surface will eventually bring the fragment to the surface.

A related process acting in coarse-textured regolith and sediment of the barren tundra causes the coarsest fragments—pebbles and cobbles—to move horizontally and to become sorted out from the finer particles. This

Figure 17.4 **Stone rings near Thule, Greenland. (A. E. Corte, Geology Department, Universidad Nacional del Sur, Bahia Blanca, Argentina.)**

type of sorting produces ringlike arrangements of coarse fragments. Linked with adjacent rings, the gross pattern becomes netlike to form a system of **stone polygons** (also called "stone rings," or "stone nets") (Figure 17.4).

In fine-textured alluvium, such as that formed on river floodplains and deltaic plains in the arctic environment, ice has accumulated in vertical wedge-forms in deep cracks in the sediment. These **ice wedges** are interconnected into a system of polygons, called **ice-wedge**

polygons (Figure 17.5). Subsurface ice of the tundra is a permanent feature of the arctic **permafrost,** or permanently frozen ground. Only a shallow surface layer, the **active layer,** experiences summer melting. (The distribution and environmental aspects of permafrost are described in Chapter 28.) Ice wedges are thought to originate as shrinkage cracks, formed during extreme winter cold. During the spring melt, water enters the cracks and becomes frozen. Repeated cracking and the addition of new ice causes the ice wedge to thicken until it becomes as wide as 3 m (10 ft) and as deep as 10 m (30 ft), see Figure 17.6. Both stone polygons and ice-wedge polygons belong to a class of features called **patterned ground.** These features also form in the alpine tundra zone of high mountains. Another remarkable ice-formed feature of the arctic tundra is a conspicuous conical mound, called a **pingo** (Figure 17.7). The pingo has a core of ice and grows in height as more ice accumulates, forcing up the overlying sediment. In extreme cases, pingos reach a height of 50 m (165 ft) and a basal diameter as great as 600 m (2000 ft).

Features Produced by Salt-Crystal Growth

The breakup of rock by the growth of salt crystals in dry climates was explained in Chapter 11. Cliffs of sandstone are specially susceptible to disintegration by this activity. As Figure 17.8 shows, slow seepage of water takes place at the cliff base, above a dense, impervious layer of shale. Continual surface evaporation leaves the dissolved salts behind in pore spaces in the sandstone. The pressure of growing crystals disrupts the sandstone into scales and flakes. Sand grains thus released are swept away by wind

Figure 17.5 **Ice-wedge polygons in fine-textured floodplain silts near Barrow, Alaska. The dark areas within polygons are lakes. In the middle distance is a meandering river channel. (R. K. Haugen, U.S. Army Cold Regions Research and Engineering Laboratory.)**

Figure 17.6 **This vertical riverbank exposure near Livengood, Alaska, in the subarctic climate region reveals a V-shaped ice wedge surrounded by layered silt of alluvial origin. (T. L. Péwé, U.S. Geological Survey.)**

Figure 17.7 A large, isolated pingo on the arctic coastal plain of Alaska, about 32 km (20 mi) south of Prudhoe Bay. Patterned ground and ice-covered lakes can be seen in the background. Notice the tracks made by vehicles used in petroleum exploration. (Steve McCutcheon, Alaska Pictorial Service.)

gusts or by sheets of rainwater flowing down the cliff face. Recession of the cliff base eventually produces a niche, or shallow cave. In the southwestern United States, many such niches were occupied by Indians, who built masonry walls to enclose the natural openings. These cliff dwellings were protected not only from the elements, but also from attack by hostile Indians, (Plate G.2).

Sheeting Structure and Exfoliation Domes

A curious but widespread process related to physical weathering results from **unloading,** the relief of confining pressure, as rock is brought nearer the earth's surface

Figure 17.8 Seepage of water from the cliff base localizes the development of niches through rock weathering. (Drawn by A. N. Strahler.)

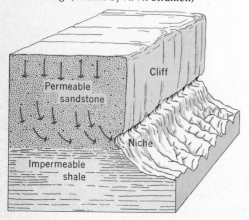

through the erosional removal of overlying rock. Rock formed at great depth beneath the earth's surface (particularly igneous and metamorphic rock) is in a slightly contracted state because of the tremendous load of the overlying rock. On being brought to the surface, the rock expands slightly and, in the process, thick shells of rock break free from the parent mass below. The new surfaces of fracture form a joint system described as **sheeting structure.** The structure shows best in massive rocks, such as granite, because in a closely jointed rock the expansion would be taken up among the blocks.

The rock sheets, or shells, produced by unloading generally parallel the ground surface and therefore tend to be inclined downward toward valley bottoms. On granite coasts, the shells are found to dip seaward at all points along the shore. Sheeting structure is well seen in quarries, where it greatly facilitates the removal of large stone blocks.

Where sheeting structure has formed over the top of a single large body of massive rock, an **exfoliation dome** is produced (Plate G.2). These domes are among the largest of the landforms due primarily to weathering. In the Yosemite Valley region of California, where domes are spectacularly displayed, the individual rock shells may be as thick as 6 to 15 m (20 to 50 ft).

Other varieties of large, smooth-sided rock domes lacking in shells are not true exfoliation domes; they are formed by granular disintegration of a single body of hard, coarse-grained intrusive igneous rock lacking in joints. Examples are the Sugar Loaf of Rio de Janeiro and Stone Mountain in Georgia (see Figure 26.27). These smooth domes rise prominently above surrounding areas of weaker rock.

Features Produced by Chemical Weathering

We investigated chemical weathering processes in Chapter 11 under the heading of chemical alteration. Recall that the dominant processes of chemical change affecting silicate minerals are oxidation, carbonic acid action, and hydrolysis. Feldspars and the mafic minerals—particularly olivine—are very susceptible to chemical decay. On the other hand, quartz is a highly stable mineral, almost immune to decay. One result of this fact is that, where felsic and mafic igneous rocks are exposed side by side, the felsic rock may show little alteration while the mafic rock may be in an advanced state of decomposition (Plate G.1).

As chemical alteration penetrates the bedrock, joint blocks are attacked from all sides. Decay progresses inward to make concentric shells of soft rock (Plate G.1). This form of change is called **speroidal weathering** and produces onionlike bodies from which thin layers can be peeled away from a spherical core.

Decomposition by hydrolysis and oxidation changes strong igneous rock into very weak regolith. This change allows erosion to operate with great effectiveness wherever the regolith is exposed. Weakness of the regolith also makes it susceptible to natural forms of mass wasting.

In warm, moist climates of the equatorial, tropical, and subtropical zones, hydrolysis and oxidation often result in

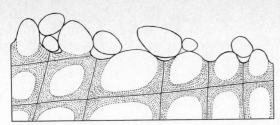

Figure 17.9 **Stages in the development of egg-shaped boulders from rectangular joint blocks. (After W. M. Davis.)**

the decay of igneous and metamorphic rocks to depths as much as 90 m (300 ft). Geologists who first studied this deep rock decay in the Southern Appalachians named the rotted layer **saprolite** (literally "rotten rock"). To the civil engineer, deeply weathered rock is of major importance in constructing highways, dams, or other heavy structures. Although the saprolite is soft and can be removed by power shovels with little blasting, there is serious danger of failure of foundations under heavy loads. This regolith may also have undesirable plastic properties because of a high content of certain clay minerals, such as montmorillonite, which have the property of swelling during absorption of water.

The hydrolysis of granite is accompanied by granular disintegration, the grain-by-grain breakup of the rock. This process creates many interesting boulder and pinnacle forms by the rounding of angular joint blocks (Figure 17.9 and Plate G.1). These forms are particularly conspicuous in arid regions. There is ample moisture in most deserts for hydrolysis to act, given sufficient time. The products of granular disintegration of felsic igneous rock form a coarse desert gravel, called **grus,** which consists largely of grains of quartz and partially decomposed feldspars.

Carbonate sedimentary rocks (limestone, marble) are particularly susceptible to the action of carbonic acid in rainwater and soil water. Mineral calcium carbonate (calcite) is dissolved, yielding calcium ions and bicarbonate ions. In regions of large water surplus, these ions are carried away in solution in the water of streams.

Carbonic acid reaction with limestone produces many interesting surface forms, mostly of small dimensions. Outcrops of limestone typically show cupping, rilling, grooving, and fluting in intricate designs (Plate G.3). In a few places the scale of deep grooves and high, wall-like rock fins reaches proportions that prevent the passage of humans and animals.

Carbonic acid action is a major agent of denudation in regions of moist climate underlain by limestone (Plate G.3). For example, in one study of a limestone valley in Pennsylvania, it was concluded that the ground surface is being lowered at an average rate of 0.3 m (1 ft) in 10,000 years through carbonic acid action alone. In moist climates, as we might expect, carbonate rocks underlie valleys and lowlands in contrast to other, less susceptible rock types that form adjacent ridges or uplands. Quite the reverse is true in dry climates, where limestone and dolomite are strongly resistant to weathering and form high ridges and plateaus. For example, the rim of the Grand Canyon and the surrounding high plateau are underlain by beds of dolomite (see Plate H.1). Sandstone strata formed of quartz grains cemented by calcium carbonate are also highly resistant to weathering in a dry climate.

The removal of limestone by carbonic acid action deep beneath the land surface leads to the formation of cavern systems. Because this activity takes place in the zone of ground water, we will describe limestone caverns and related features in Chapter 18.

Mass Wasting

Everywhere on the earth's surface, gravity pulls continually downward on all materials. Bedrock is usually so strong and well supported that it remains fixed in place but, where a mountain slope becomes too steep, bedrock masses break free, falling or sliding to new positions of rest. In cases involving huge masses of bedrock, the result can be catastrophic in loss to life and property in towns and villages in the path of the slide. Such slides are a major form of environmental hazard in mountainous regions. Because soil, regolith, and many forms of sediment are held together poorly, they are much more susceptible to being moved by the force of gravity than is bedrock. Abundant evidence shows that on most slopes at least a small amount of downhill movement is going on constantly. Although this motion is imperceptible, the regolith sometimes slides or flows rapidly.

Taken altogether, the various kinds of downhill movements occurring under the pull of gravity, collectively called mass wasting, constitute an important process in denudation of the continental surfaces.

Talus Cones

Steep rock walls of gorges and high mountains shed countless rock fragments under the attack of physical weathering processes, particularly frost action. These fragments accumulate in a distinctive landform, the **talus slope** (Plate G.4). A talus slope, or **scree slope,** as it is often called, has a nearly constant slope angle, usually in the range of 34° to 36° from the horizontal.

Most cliffs are notched by narrow ravines that funnel the fragments into individual tracks, so as to produce **talus cones** arranged side by side along the cliff. Where a large range of sizes of particles is supplied, the larger pieces, because of their greater momentum and ease of rolling, travel to the base of the cone, whereas the tiny grains lodge in the apex. This process tends to sort the fragments by size, progressively finer from base to apex (Figure 17.10).

Most fresh talus slopes are unstable, so the disturbance created by walking across the slope or by the dropping of a large rock fragment from the cliff above will easily set off a sliding of the surface layer of particles. The upper limiting angle to which coarse, hard, well-sorted rock fragments will stand is termed the **angle of repose.** Unstable slopes at the angle of repose are also found on the steep lee sides of sand dunes.

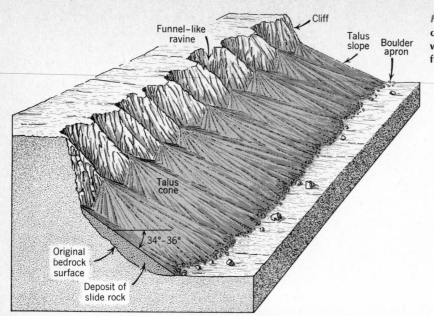

Figure 17.10 Idealized diagram of talus cones formed at the base of a cliff, which might be 60 to 150 m (200 to 500 ft) high. (Drawn by A. N. Strahler.)

Soil Creep

On almost any steep, soil-covered slope, evidence can be found of extremely slow downhill movement of soil and regolith, a process called **soil creep.** Figure 17.11 shows some of the evidence that the process is going on. Joint blocks of distinctive rock types are found moved far downslope from the outcrop. In some layered rocks, such as shales or slates, edges of the strata seem to bend in the downhill direction. This is not true plastic bending, but is the result of slight movement on many small joint cracks (Figure 17.12). Fence posts and telephone poles lean downslope and even shift measurably out of line. Retaining walls of road cuts lean and break outward under pressure of soil creep from above.

What causes soil creep? Heating and cooling of the soil, growth of frost needles, alternate drying and wetting of the soil, trampling and burrowing by animals, and shaking by earthquakes all produce some disturbance of the soil and regolith. Because gravity exerts a downhill pull on every such rearrangement, the particles are urged slowly downslope.

Creep also affects rock masses enclosed in the soil or lying on bare bedrock. Huge boulders that have gradually crept down a mountain side in large numbers may accumulate at the mountain base to produce a boulder field.

Earthflow

In regions of humid climate, a mass of water-saturated soil, regolith, or weak clay or shale layers may move down a steep slope during a period of a few hours in the form of an **earthflow.** Figure 17.13 is a sketch of an earthflow showing how the material slumps away from the top, leaving a steplike terrace bounded by curved, wall-like scarp. The saturated material flows sluggishly to form a bulging toe.

Shallow earthflows, affecting only the soil and regolith, are common on sod-covered and forested slopes that have been saturated by heavy rains. An earthflow may

Figure 17.12 Slow creep has caused this downhill bending of steeply dipping sandstone layers. (Ward's Natural Science Establishment, Inc., Rochester, N.Y.)

Figure 17.11 Slow, downhill creep of soil and weathered overburden. (After C. F. S. Sharpe.)

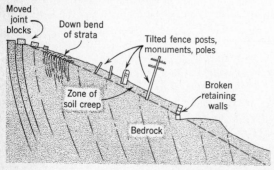

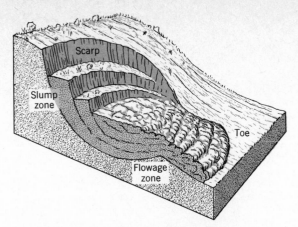

Figure 17.13 An earthflow with slump features well developed in the upper part. (Drawn by A. N. Strahler.)

affect a few square meters or it may cover an area of several hectares (Figure 17.14). If the bedrock of a mountainous region is rich in clay (shale or deeply weathered igneous rocks), earthflow sometimes involves millions of tons of bedrock, moving by plastic flowage like a great mass of thick mud.

Earthflows are a common cause of the blockage of highways and railroad lines, usually during periods of heavy rains. Usually the rate of flowage is slow, so the flows are not often a threat to life. Property damage to buildings, pavements, and utility lines is often large where construction has taken place on unstable soil slopes.

Examples of both large and small earthflows induced or aggravated by Man's activities are found in the Palos Verdes Hills of Los Angeles County, California. These movements occur in shales that tend to become plastic when water is added. The upper part of the earthflow undergoes a subsiding motion with backward rotation of the down-sinking mass, as illustrated in Figure 17.13. The interior and lower parts of the mass move by slow flowage, and a toe of extruded flowage material may be formed.

Largest of the earthflows in the Palos Verdes Hills area was the Portuguese Bend "landslide" that affected an area of about 160 hectares (400 acres). The total motion over a three-year period was about 20 m (70 ft). Damage to residential and other structures totaled some $10 million. The most interesting observation, from the point of view of assessing the impact of Man on the environment, is that the slide has been attributed by geologists to infiltration of

water from cesspools and from irrigation water applied to lawns and gardens. A discharge of over 115,000 liters (30,000 gallons) of water per day from some 150 homes is believed to have sufficiently weakened the shale beneath to start and sustain the flowage.

Earthflows Involving Quick Clays

One special form of earthflow has proved to be a major environmental hazard in parts of Norway and Sweden and along the St. Lawrence River and its tributaries in Quebec Province of Canada. In all these areas the flowage involves horizontally layered clays, sands, and silts of late Pleistocene age that form low, flat-topped terraces adjacent to rivers or lakes. Over a large area, which may be 600 to 900 m (2000 to 3000 ft) across, a layer of silt and sand 6 to 12 m (20 to 40 ft) thick begins to move toward the river, sliding on a layer of soft clay that has spontaneously turned into a near-liquid state. The moving mass also settles downward and breaks into steplike masses. Carrying along houses or farms, the layer ultimately reaches the river, into which it pours as a great, disordered mass of mud.

Figure 17.15 shows an earthflow of this type that occurred in 1898 in Quebec, along the Rivière Blanche. Beyond is the scar of a much older earthflow of the same type. The Rivière Blanche earthflow involved about 3 million cu m (3.5 million cu yd) of material and required three to four hours to move into the river through a narrow, bottleneck passage.

Disastrous earthflows have occurred a number of times since the occupation of Quebec by Europeans. A particularly spectacular example was the Nicolet earthflow of 1955, which carried a large piece of the town into the Nicolet River (Figure 17.16). Fortunately only three lives were lost, but the damage to buildings and a bridge ran into the millions of dollars.

Clays that spontaneously change from a solid condition to a near-liquid condition are called **quick clays.** The process is called **spontaneous liquefaction.** A sudden shock or disturbance will often cause a layer of quick clay to begin to liquefy; once begun, the process cannot be

Figure 17.14 Earthflows in a mountainous region. (After W. M. Davis.)

Figure 17.15 Block diagram of the 1898 earthflow near St. Thuribe, Quebec. (After C. F. S. Sharpe, 1938, *Landslides and Related Phenomena*, Columbia Univ. Press, New York.)

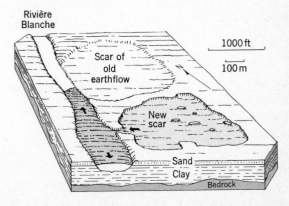

Figure 17.16   A portion of the city of Nicolet, Quebec, Canada, was carried into the channel of the Nicolet River by an earthflow that occurred in November 1955. The flow moved downvalley (right), passing beneath the bridge. (Raymond Drouin.)

stopped. A good example comes from the city of Anchorage, Alaska. Severe ground shaking by the Good Friday earthquake of 1964 set off the liquefaction of quick clays underlying an extensive flat area, or terrace, which was the site of a housing development. Clay flowage allowed the overlying sediment layer to subside and break into blocks (see Figure 25.27).

A study of quick clays in North America and Europe has led to an explanation of their strange behavior. The clays are marine glacial deposits, meaning that they are

Figure 17.17   Solifluction lobes cover this Alaskan mountain slope in the tundra climate region. (U.S. Geological Survey.)

sediments laid down in shallow saltwater estuaries during the Pleistocene Epoch. The thin, platelike particles of clay accumulated in a heterogeneous arrangement, often described as a "house of cards" structure. There is a very large proportion of water-filled void space between clay particles. It is thought that the original salt water saturating the clay acted as an electrolyte to bind the particles together, giving the clay layer strength. Some areas where quick clays occur have experienced a crustal uplift since they were laid down. The sediment deposit has been raised from immersion in salt water and the salt solution has been gradually replaced by fresh ground water. Now the clay is no longer bound by the electrolyte action and becomes sensitive. A mechanical shock causes the house-of-cards structure to collapse. Because a large volume of water is present (from 45 to 80 percent water content by volume), the mixture behaves as a liquid, with almost no strength remaining.

Because many towns, cities, and farms presently occupy land underlain by quick clays, an environmental hazard can exist locally. Such areas require identification and mapping so that urban planning can avoid them, limiting the land use to farming with a dispersed resident population.

Solifluction in the Arctic Tundra

A special variety of earth flowage characteristic of arctic permafrost regions is **solifluction** (from Latin words meaning "soil" and "to flow".) It is active in early summer, when thawing has penetrated the upper few decimeters. At this time soil is fully saturated with water that cannot escape downward because of the impermeable frozen mass. Flowing almost imperceptibly, this saturated soil forms terraces and lobes that give a mountain slope a stepped appearance (Figure 17.17).

Mudflow

One of the most spectacular forms of mass wasting and one that is potentially a serious environmental hazard is the **mudflow,** a mud stream that travels down canyons in mountainous regions (Figure 17.18). In deserts, where vegetation does not protect the mountain soils, local thunderstorms produce rain much faster than it can be absorbed by the soil. As the water runs down the slopes it forms a thin mud, which flows down to the canyon floors. Following stream courses, the mud continues to flow until it becomes so thickened that it must stop. Large boulders, buoyed up in the mud, are carried with the flow (Plate G.4). Roads, bridges, and houses in the canyon floor are engulfed and destroyed.

When a mudflow emerges from the canyon and spreads across a plain, property damage and loss of life can result. In desert regions, the plains lie close to the mountain ranges that supply irrigation water. These favored land surfaces are often densely populated and are sites of urban development, as well as intensive agriculture.

Mudflows also occur on the slopes of erupting volcanoes. Heavy rains can turn freshly fallen volcanic ash and dust into mud, which then flows down the volcano slopes. Herculaneum, a city at the base of Mount Vesuvius, was destroyed by such a mudflow during the eruption of A.D. 79, while the neighboring city of Pompeii was being buried under volcanic ash.

Mudflows show varying degrees of consistency, from a mixture about like the concrete that emerges from a mixing truck to thinner consistencies which are little different than the mixture in turbid stream floods. The watery type of mudflow is commonly called a **debris flood** in the western United States, particularly in southern California, where it occurs commonly and with disastrous effects.

In Los Angeles County, California, real estate

development has been carried out on very steep hillsides and mountainside by the process of bulldozing roads and homesites out of the deep regolith. The excavated regolith is pushed into adjacent embankments, where its instability poses a threat to slopes and stream channels below. When saturated by heavy winter rains, these embankments can give way, producing earthflows, mudflows, and debris floods that travel far down the canyon floors and spread out on lowland surfaces, burying streets and yards in bouldery mud. Many debris floods of this area are also produced by heavy rains falling on undisturbed mountain slopes denuded by the burning of vegetative cover in the preceding dry summer; some fires are set by Man, whether inadvertently or deliberately. Disturbance of slopes by construction practices is simply an added source of debris and serves to enhance what is already an important environmental hazard.

Landslides

The term **landslide** is used widely in a general sense to mean any downslope movement of a mass of regolith or bedrock under the influence of gravity. Many geologists and civil engineers refer to earthflows as landslides. We shall limit our use of the term "landslide" to mean the rapid sliding of large rock masses beginning their descent as unit blocks, without internal flowage. Softening of clay minerals to produce a plastic mass, as in the case of an earthflow, is not involved in the behavior of a typical landslide. A landslide mass is rigid, rather than plastic in its behavior, traveling like a massive sled over a plane of slippage. In most cases, however, the sliding block breaks up into many smaller blocks, and these in turn may disintegrate into a rubble. In its final stages, the mass of rubble often shows a gross flowage motion.

Two basic forms of landslides are (1) rockslides and (2) slump blocks. A **rockslide** consists of a bedrock mass slipping on a sloping rock plane, as shown in Figure 17.19A. The plane of slippage is usually a fault plane, joint plane, or bedding plane. A **slump block** is a bedrock mass that moves down a curved slip surface, as shown in the Figure 17.19B. As a slump block moves down, it also rotates on a horizontal axis, so the upper surface of the block becomes tilted toward the cliff that remains. A rockslide can travel a long distance down a mountainside, far from its original position, whereas a slump block remains close to its original position.

Figure 17.18 **Thin, streamlike mudflows commonly issue from canyon mouths in arid regions, spreading out over the alluvial fan slopes at the base of a mountain range. (Drawn by A. N. Strahler.)**

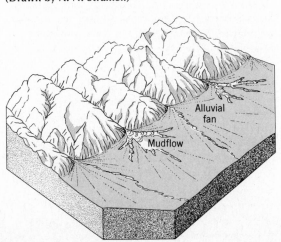

Figure 17.19 **Landslides may involve (A) slip on a nearly plane surface or (B) slump with rotation on a curved plane.**

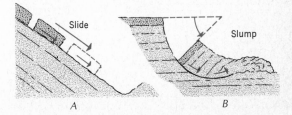

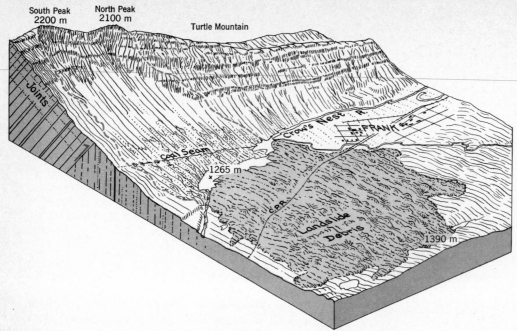

Figure 17.20   A classic example of an enormous, disastrous landslide is the Turtle Mountain slide, which took place at Frank, Alberta, in 1903. A huge mass of limestone slid from the face of Turtle Mountain between South and North peaks, descended to the valley, then continued up the low slope of the opposite valley side until it came to rest as a great sheet of bouldery rock debris. (Data from Canadian Geological Survey, Department of Mines. Drawn by A. N. Strahler.)

Rockslides

Wherever steep mountain slopes occur, there is a possibility of large, disastrous rockslides. In Switzerland, Norway, or the Canadian Rockies, for example, villages built on the floors of steep-sided valleys have been destroyed and their inhabitants killed by the sliding of millions of cubic yards of rock, set loose without any warning.

The Turtle Mountain landslide of 1903 in Alberta, Canada, is shown in Figure 17.20. It involved the sliding of an enormous mass of limestone, its volume estimated at 27 million cu m (35 million cu yd), through a descent of 900 m (3000 ft). The debris buried part of the town of Frank, with a loss of 70 persons.

Many of the world's great rockslides have occurred in mountains deeply carved by glacial action in the Pleistocene Epoch. The dominant landforms are U-shaped troughs, abundant in the European Alps, the Cordilleran Range of North America, the Andes Range of South America, and the Himalayas of southern Asia (Figure 17.21).

Glacial troughs in coastal locations are typically inundated by the ocean, becoming steep-walled fiords (see Figure 21.6). The environmental hazard of rockslides is extremely high for towns and cities located on the floors of glacial troughs and at the heads of fiords. A rockslide entering the deep water of a fiord generates an enormous wave of water that runs quickly to the head of the fiord, inundating and destroying habitations located on the low-lying trough floor.

In many places glacial troughs are occupied by deep trough lakes, which may be many kilometers long. A rockslide entering a trough lake sets off a great wave, traveling to both ends of the lake and riding some distance up the trough floor. In the 1963 Vaiont Reservoir disaster in Italy, a Man-made reservoir occupied a glacial trough. A huge wave produced by a large rockslide destroyed the high concrete dam, releasing an enormous volume of floodwater that quickly destroyed towns located at lower valley levels. Within a short time, 2600 lives were lost.

Closely related to the phenomenon of rockslide is **rockfall,** the free falling or rolling of single masses of rock from a steep cliff. Individual fragments may be as small as a boulder or as large as a city block, depending on the overall scale of the cliff and the manner in which the rock breaks up. Large blocks disintegrate upon falling, strewing the slope below with rubble and leaving a conspicuous scar on the upper cliff face.

Figure 17.21   A great rockslide at *S* has formed a dam, *D*, in the floor of a glacial trough and has produced a lake, *L*. (After W. M. Davis.)

A similar phenomenon, the *alpine debris avalanche,* occurs in high, alpine mountain chains. There glacial erosion has produced valleys with extremely steep gradients and has left large quantities of glacial rock rubble (moraine) and relict glaciers perched precariously in high positions. This kind of avalanche is a sudden rolling of a mixture of rock waste and glacial ice; it can produce a tongue of debris traveling downvalley at a speed which is little less than that of a freely falling body.

Slump Blocks

Slump blocks consist of great masses of bedrock sliding downward from a high cliff and, at the same time, rotating backward on a horizontal axis (Figure 17.22). Wherever massive sedimentary strata, usually sandstones or limestones, or lava beds, rest on weak clay or shale formations, a steep cliff tends to be formed by erosion. As the weak rock is eroded from the cliff base, the cap rock is undermined. When a point of failure is reached, a large block breaks off, slides down, and tilts back along a curving plane of slip. Slump blocks may be as much as 2 to 3 km (1 to 2 mi) long and 800 m (500 ft) thick. A single block appears as a ridge at the base of the cliff. A closed depression or lake basin may lie between the block and the cliff.

Slumping of large bedrock masses is rarely a threat to human life because of the slow rate of movement and the limited distance of travel. The environmental importance of slumping, in both bedrock and regolith, lies in the potential for disruption of such Man-made structures as buildings and highways. If a house happens to lie over the upper part of an active earthflow where slump scarps are forming, it can be sheared into sections. Because slumping is difficult to stabilize, abandonment of the property may be the only economically feasible solution. Large bedrock slump blocks located along the base of a cliff should be identified and avoided in the construction of highways, dams, and power plants. Even though the slump block may appear to be stabilized and immobile, excavation of material at the base of the block may lead to reactivation of the block. The recognition of ancient earthflows and slump blocks is an important aspect of environmental protection because serious future economic losses can often be avoided in the planning stages of engineering projects.

Man as a Geomorphic Agent

Man in a modern technological society is an important geomorphic agent, producing many new landforms. These result from the moving from one place to another of enormous masses of soil, regolith, and bedrock for two basic purposes: (1) to extract mineral resources; and (2) to reorganize terrain into suitable configurations for highway grades, airfields, building foundations, dams, canals, and various other large structures. Both activities involve removal of earth materials, which destroys entirely the preexisting ecosystems and habitats of plants and animals. The building up of new land on adjacent surfaces, using those same earth materials, is a process that also destroys

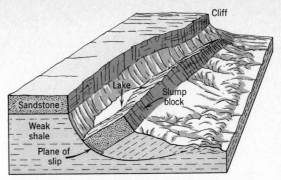

Figure 17.22 **Slump blocks rotate backward as they slide from a cliff. (Drawn by A. N. Strahler.)**

by burial the preexisting ecosystems and habitats.

Scarification is a general environmental-impact term for excavations and other land disturbances produced for purposes of extracting mineral resources. Scarification includes the nearby accumulation of mineral waste (spoil, tailings). Among the forms of scarification are open-pit ore mines, coal strip mines, quarries for structural materials, borrow pits along highway grades, sand and gravel pits, clay pits, phosphate pits, scars from hydraulic mining, and stream gravel deposits reworked by dredging.

Scarification is on the increase because demands for coal to meet energy requirements are on the rise and because of increased demands for industrial minerals used in manufacturing and construction. At the same time, as the richer and more readily available mineral deposits are consumed, industry turns to poorer grades of ores and to less easily accessible coal deposits; the result is that the rate of scarification is further increased.

An example of mass wasting associated with an accumulation of mine spoil is a disaster that occurred at Aberfan, Wales. A hill, built of coal-mine waste (culm) accumulated to a height of 180 m (600 ft) close to the town (Figure 17.23). The culm hill had been constructed on a steep hillslope and on a spring line as well, making a potentially unstable configuration. Following heavy rains, a large mass of the saturated culm began to move as an earthflow. The debris tongue overwhelmed part of the town, destroying a school and taking over 150 lives. Phenomena of this type are often called "mudslides" in the news media.

Perhaps the major form of land scarification both today and in years to come, is associated with the mining of coal. For deep-lying coal seams, vertical shafts are driven down from the surface to reach the coal, which is mined by extension of horizontal **drifts** and rooms into the face of the seam. In mountainous terrain, drifts can be driven directly into the seam where it is exposed on the mountainside. Subterranean mining of coal is often followed by damage to the overlying land surface through mine collapse leading to subsidence, which can be considered a form of mass wasting. One example comes from the city of Scranton, Pennsylvania. Here the collapse of abandoned anthracite mine workings has repeatedly caused settling and fracturing of the ground, damaging

Figure 17.23 **Debris flow at Aberfan, Wales. Sketched from a photograph. (From A. N. Strahler, 1972, *Planet Earth*, Harper and Row, New York.)**

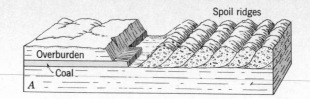

Figure 17.24 **Area strip mining (*A*) and contour strip mining (*B*). (Drawn by A. N. Strahler.)**

where it accumulates in valley bottoms (Chapter 19). On steep mountain slopes of the Appalachian area, mass wasting in the form of earthflows and mudflows affects the unstable spoil banks, while the presence of sulfur-bearing minerals in the coal seams gives rise to acid water drainage, severely impacting the streams that drain the area.

Figure 17.25 **Contour strip-mining in Wise County, Virginia. (Kenneth Murray, Kingsport, Tenn., from Nancy Palmer Photo Agency, Inc.)**

streets and houses. Near Hanna, Wyoming, the cave-in of abandoned shallow coal mines has produced many deep pits, while underground burning of the coal seam has added to progressive collapse.

Where the coal seams lie close to the surface or actually outcrop along hillsides, the **strip mining** method is used. Earth-moving equipment removes the covering strata (overburden) to bare the coal, which is lifted out by power shovels. There are two kinds of strip mining, each adapted to the given relationship between ground surface and coal seam. **Area strip mining** is used in regions of nearly flat land surface under which the coal seam lies horizontally (Figure 17.24*A*). After the first trench is made and the coal removed, a parallel trench is made, the overburden of which is piled as a spoil ridge into the first trench. Thus the entire seam is gradually uncovered, and a series of parallel spoil ridges remains. Phosphate beds are mined extensively by the area strip mining method in Florida, and the method is also used for mining clay layers. The **contour strip mining** method is used where a coal seam outcrops along a steep hillside (Figure 17.24*B*). The coal is uncovered as far back into the hillside as possible, and the overburden is dumped on the downhill side. The result is a bench bounded on one side by a steep rock cliff, or "high wall," and on the other side by a ridge of loose spoil with a steep outer slope leading down into the valley bottom. The benches form sinuous patterns following the plan of the outcrop (Figure 17.25).

Spoil ridges left by strip mining are highly susceptible to rapid erosion under torrential rainfall, yielding large amounts of sediments. This material is carried downvalley,

Weathering and Mass Wasting in Review

In this chapter we have compared natural processes of wasting of the continental surfaces with Man-induced changes of a similar nature. Although the processes of weathering and soil creep are for the most part slow-acting and produce effects that are visible only when accumulated over centuries, the natural mass-wasting processes also include catastrophic events. Indeed, some large landslides dwarf anything that Man has accomplished in equal time by earth-moving with machines and explosives. But Man works constantly with enormous quantities of energy, and the cumulative environmental damage and destruction continues to mount, almost unchecked. Only now are the conflicts of interest beginning to emerge and to be squarely faced. Only now are we designing and implementing environmental protection measures necessary to control the spread of scarification and its many harmful side effects.

References for Further Study

Sharpe, C. F. S. (1938), *Landslides and related phenomena,* Columbia University Press, New York.

Highway Research Board (1958), *Landslides and engineering practice,* Special Report 29, NAS—NRC Publication 544. Washington, D.C.

Leopold, L. B., M. G. Wolman, and J. P. Miller (1964), *Fluvial processes in geomorphology,* W. H. Freeman, San Francisco. See Chapters 4, 8.

Price, L. W. (1972), *The periglacial environment, permafrost, and Man,* Commission on College Geography, Resource Paper No. 14, Association of American Geographers, Washington, D.C.

Review Questions

1. What are landforms? What is geomorphology? Of what environmental and geographic importance are landforms?
2. Name the fluid agents active in geomorphic processes. How does hydrology become involved with geomorphology? Define denudation. What is the ultimate goal of denudation?
3. Of what importance are weathering processes in geomorphology? What is mass wasting? What role do slopes play in the total process of denudation?
4. By what four modes does the physical breakup of rock take place? What role does jointing of rock play in the disintegration process?
5. What effects are produced by the formation of ice in fine-textured soils and sediments? Describe the kinds of patterned ground formed by frost action in arctic and alpine environments. What is a felsenmeer?
6. What conspicuous features are produced by salt-crystal growth? Explain the process. Under what climatic conditions does this process operate most effectively?
7. What is meant by unloading? How does it lead to the development of sheeting structure? What prominent landforms are produced by this process?
8. How do felsic rocks and mafic rocks differ in their response to chemical weathering? What is spheroidal weathering? Describe saprolite and explain its environmental significance. What forms are produced by granular disintegration in a dry climate?
9. What surface features are the result of carbonic acid action on limestone? Under what climate conditions are carbonate rocks subject to rapid removal by carbonic acid action?
10. Describe a talus slope and the materials of which it is formed. Where do talus cones form? In what natural materials does the angle of repose determine surface slope?
11. Describe some common evidences of soil creep. What causes soil creep?
12. What is an earthflow? Describe the condition of the material of which an earthflow is composed. Describe the form of an earthflow. Give an example of an earthflow set off by human activity.
13. What is distinctive about earthflows involving quick clays? Where do such earthflows occur? Explain the behavior of quick clays.
14. Describe solifluction of the arctic tundra. How does the presence of permafrost promote solifluction?
15. How are mudflows produced? Describe the conditions of climate and relief associated with mudflows. How are mudflows and debris floods induced by human activity in urban areas?
16. Define landslide, making clear the distinction between landslide and such mass-wasting forms as earthflows and mudflows. What are the two basic forms of landslides?
17. What conditions of relief and geologic process favor the occurrence of large rockslides? Describe a rockfall and an alpine debris avalanche.
18. Describe the form and motion of a large slump block. What is the environmental importance of slump blocks?
19. Describe the role of Man as a geomorphic agent. How does human activity create new landforms? What is meant by scarification? Give an example of an earthflow on a deposit of mine wastes.
20. How is strip mining carried out? What two basic forms of strip mining are practiced? What harmful effects to the environment are directly related to strip mining?

Ground Water and Surface Water

18

In this chapter we return to the hydrologic cycle. Chapter 10 dealt with a phase of the hydrologic cycle in which soil water is recharged by precipitation and returned directly to the atmosphere by evapotranspiration. Recall that many soil-water budgets show a substantial water surplus, to be disposed as runoff. Our investigation now deals with that water surplus and the paths it follows as subsurface water and surface water.

There are two basic paths of escape for surplus water. First, surplus water may percolate through the soil, traveling downward under the force of gravity to become part of the underlying ground water body. Following subterranean flow paths, this water emerges to become surface water, or it may emerge directly in the shore zone of the ocean. Second, surplus water may flow over the ground surface as runoff to lower levels. As it travels, the dispersed flow becomes collected into streams, which eventually conduct the runoff to the ocean. In this chapter, we trace both the subsurface and surface pathways of flow of surplus water. In so doing, we will complete the hydrologic cycle.

Surplus water, as runoff, is a vital part of the environment of terrestrial life forms and of Man in particular. Surface water in the form of streams, rivers, ponds, and lakes constitutes one of the distinctive environments of plants and animals.

Our heavily industrialized society requires enormous supplies of fresh water for its sustained operation. Urban dwellers consume water in their homes at rates of 150 to 400 liters (40 to 100 gallons) per person per day. Huge quantities of water are used for cooling purposes in air-conditioning units and power plants.

In view of projections based on existing rates of increase in water demands, we shall be hard put in the future to develop the needed supplies of pure fresh water. Water pollution also tends to increase as populations grow and urbanization advances over broader areas. A disconcerting concept is that the available resource of pure fresh water is shrinking while demands are rising. Knowledge of hydrologic processes enables us to evaluate the total water resource, to plan for its management, and to protect it from pollution.

Ground Water

Ground water is that part of the subsurface water that fully saturates the pore spaces of the bedrock and regolith. The ground water occupies the **saturated zone** (Figure 18.1). Above it is the **unsaturated zone,** in which water does not fully saturate the pores. The upper surface of the saturated zone is the **water table.** Water is held in the unsaturated zone by capillary force in tiny films adhering to the mineral surfaces. The unsaturated zone may be absent or very shallow where the saturated zone is at or close to the surface in low, flat regions.

At the base of the unsaturated zone is the **capillary fringe,** a thin layer in which the water has been drawn upward from the ground water table through capillary force. The action is much like the rise of kerosene in a lamp wick or of water in a blotter. Water in the capillary fringe largely fills the soil pores and is thus continuous with the ground water body. Thickness of the capillary fringe depends on the soil texture because capillary rise is higher when the openings are smaller. In a silty material, the capillary fringe may be 1 m (3 ft) thick, but as little as 1 cm thick in coarse sand or fine gravel with large pore spaces.

The true position of the water table is shown by the

Figure 18.1 **Zones of subsurface water.**

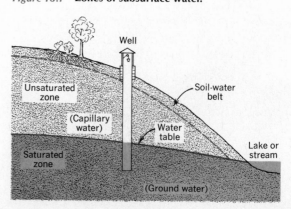

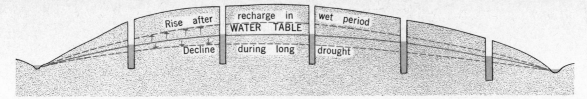

Rise after | recharge in | wet period
WATER TABLE
Decline | during long | drought

Figure 18.2 **The water table conforms roughly to surface topography.**

level of standing water in a well drilled or dug to some depth below the water table. Where wells are numerous in an area, the position of the water table can be mapped in detail by plotting the water heights and noting the trends in elevation from one well to the other. The water table is highest under the highest areas of surface — hilltops and divides — but descends toward the valleys, where it appears at the surface close to streams, lakes, or marshes (Figure 18.2). The reason for such a configuration of the water table is that water percolating down through the unsaturated zone tends to raise the water table, whereas seepage into streams, swamps, and lakes tends to draw off ground water and lower its level.

Ground Water Movement

Because ground water moves extremely slowly, a difference in water-table level, or **hydraulic head,** is built up and maintained between areas of high elevation and those of low elevation. In periods of water surplus accompanying abnormally high precipitation, this head is increased by a rise in the water-table elevation under divide areas; in periods of water deficit, occasioned by drought, the water table falls (Figure 18.2).

In humid climates having a strong seasonal cycle of precipitation, and in climates in which soil water is frozen for several months of the year, a seasonal cycle of rise and fall of the water table is produced. This cycle is illustrated by a graph of water-table fluctuations in an observation well on Cape Cod, Massachusetts (Figure 18.3). Percolating water reaches the water table in abundant quantities from late winter to late spring, a phenomenon

known as **ground water recharge.** This recharge period, which reflects the period of water surplus in the soil-water budget, causes a seasonal rise in the water table. Recharge declines to very small amounts from midsummer to early winter, and during this time the water table steadily declines as water moves to lower levels under the force of gravity. The effect of a major drought shows in the graph by a downward trend of the average elevation in 1965 and 1966.

The subsurface phase of the hydrologic cycle is completed when the ground water emerges in places where the water table intersects the ground surface. Such places are the channels of streams and the floors of marshes and lakes. By slow seepage and spring flow, the water emerges fast enough to balance the rate at which water enters the ground water table by percolation from above.

Figure 18.4 shows paths of flow of ground water. Flow takes paths curved concavely upward. Water entering the hillside midway between divide and stream flows rather directly. Close to the divide point on the water table, however, the flow lines go almost straight down to great depths, from which they recurve upward to points under the streams. Progress along these deep paths is incredibly slow; that near the surface is much faster. The most rapid flow is close to the place of discharge in the stream, where the arrows are shown to converge.

Ponds, Bogs, and Marshes

The various events of geologic history, such as crustal movements or erosion and deposition by wind and ice sheets, have created many natural depressions with floors below the elevation of the water table. Water stands in these depressions in the form of ponds and lakes. In humid

Figure 18.3 **Hydrograph of an observation well on Cape Cod, Massachusetts, showing the characteristic annual cycle of rise and fall of the water table. (Data from U.S. Geological Survey.)**

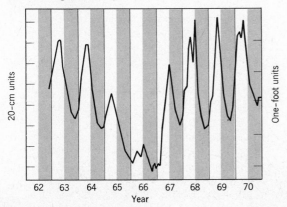

Figure 18.4 **Theoretical paths of ground-water movement under divides and valleys (Data from M. K. Hubbert. Drawn by A. N. Strahler.)**

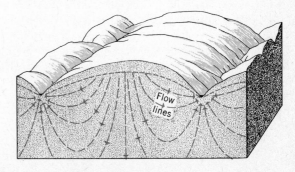

Ponds, Bogs, and Marshes **271**

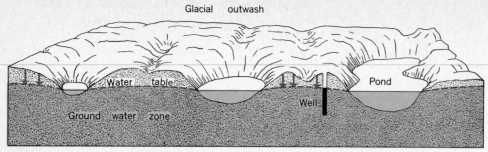

Glacial   outwash

Figure 18.5   Water table ponds in glacial deposits, largely sand, on Cape Cod, Massachusetts. (From A. N. Strahler, *A Geologist's View of Cape Cod,* Copyright © 1966 by Arthur N. Strahler. Reproduced by permission of Doubleday and Co.)

climates, water level coincides closely with the water table in the surrounding area. Seepage of ground water, as well as direct runoff of precipitation, maintains these free water surfaces permanently throughout the year. Examples of such freshwater ponds are found widely distributed in North America and Europe, where plains of glacial sand and gravel contain natural pits and hollows left by the melting of stagnant ice masses (see Chapter 21). Figure 18.5 is a block diagram showing small freshwater ponds on Cape Cod. The surface elevation of these ponds coincides closely with the level of the surrounding water table.

Many former freshwater water-table ponds have become partially or entirely filled by the organic matter from growth and decay of water-loving plants. The result is a bog with a surface close to the water table. (Bog succession was described in Chapter 15; see Figure 15.9.)

Freshwater marshes and swamps, in which water stands at or close to the ground surface over broad areas, represent the appearance of the water table at the surface. Such areas of poor surface drainage have a variety of origins. For example, the broad, shallow freshwater swamps of the Atlantic and Gulf coastal plain represent regions only recently emerged from the sea. Other marshes are created by the shifting of river channels on floodplains.

Aquifers and Aquicludes

Using our knowledge of rocks and sediments gained in Chapter 11, we can anticipate that the water-holding capacity and water-transmitting ability of materials of the saturated zone can vary greatly. Dense unaltered rocks — whether igneous, sedimentary, or metamorphic — are in many cases so tightly bonded or cemented that open pore space is almost nil. The capacity of such rocks to store water will be very small, and only when they are strongly jointed can ground water move freely through them. In contrast, clean, well-sorted sand and gravel — such as that found in beaches, dunes, or stream alluvium — can hold water in an amount equal to one-third of its bulk volume, in favorable cases. Ground water also moves freely through such coarse materials; the **permeability** is high. A permeable layer of sand or gravel, or of porous sandstone with little cementing material, is referred to as an **aquifer.** Other permeable materials that form good aquifers are

volcanic tephra and some kinds of frothy lava (scoria). Limestone that has been rendered cavernous by the action of carbonic acid may also form an excellent aquifer. Layers of clay and shale are usually good barriers to the movement of ground water. They have low permeability and are known as **aquicludes.**

Where rock layers lie nearly horizontal, or in gently inclined positions, the ground water flow paths may be quite different from the simple pattern illustrated in Figure 18.4. Suppose, for example, that the region is one of sedimentary strata with beds of sandstone alternating with beds of shale (Figure 18.6). These sandstone layers are good aquifers. In the particular case shown in Figure 18.6, a thin bed of shale has effectively blocked the downward percolation of water to the main water table below, creating a **perched water table.** Where the perched water table meets the valley side, a line of springs emerges. Most natural springs are mere trickles of water, unseen and unnoticed under a cover of dense vegetation. A few springs discharge enormous volumes of water where an unusually good aquifer is exposed (Figure 18.7).

Where strata are inclined, a favorable situation may exist for the development of **artesian flow,** in which the water can flow upward toward the surface through its own pressure. In Figure 18.8 we see a highly diagrammatic representation of artesian conditions; the vertical exaggeration is great merely to show the principle. The eroded edge of a sandstone aquifer is exposed to intake of water at a high position. Water entering here passes deep underground to a position below the valley floor. Here the water is under a strong hydraulic head from the weight of

Figure 18.6   A perched water table, producing a line of springs high on the valleyside slope.

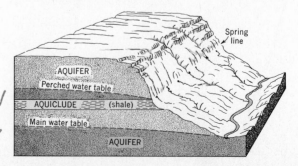

Figure 18.7 **Thousand Springs, Idaho, emerges from the north side of the Snake River Canyon, opposite the mouth of the Salmon River. The spring issues from a porous basalt layer. (U.S. Geological Survey.)**

the overlying water. This pressure is sufficient to force water up to the surface in an artesian well, penetrating the shale aquiclude. Flow as an artesian spring may occur naturally if there are fractures (faults) in the strata that permit water to leak upward through the shale layer.

Use of Ground Water as a Resource

Man's withdrawal of ground water has begun to make serious impacts on the environment in many places. The drilling of vast numbers of wells, from which water is forced out in great volumes by powerful pumps, has profoundly altered nature's balance of ground water recharge and discharge. Increased urban populations and industrial developments require larger water supplies, needs that cannot always be met from construction of new surface-water reservoirs.

In agricultural lands of the semiarid and desert climates, heavy dependence is placed on irrigation water from pumped wells, especially because many of the major river systems have already been fully developed for irrigation from surface supplies. Wells can be drilled within the limits of a given agricultural or industrial property and can provide immediate supplies of water without any need to construct expensive canals or aqueducts.

Formerly, the small well needed to supply domestic and livestock needs of a home or farmstead was actually dug by hand as a large cylindrical hole, lined with masonry where required. By contrast, the modern well put down to supply irrigation and industrial water is drilled by powerful machinery that may bore a hole 40 cm (16 in.) or more in diameter to depths of 300 m (1000 ft) or more. Drilled wells, often called **tube wells,** are sealed off by metal casings that exclude impure near-surface water and prevent clogging of the tube by caving of the walls. Near the lower end of the hole, where it enters the aquifer, the casing is perforated to admit the water. The yields of single wells range from as low as a few liters per day in a domestic well to many millions of liters per day for large, deep industrial or irrigation wells.

As water is pumped from a well, the level of water in the well drops. At the same time, the surrounding water table is lowered in the shape of a conical surface, termed the **cone of depression.** The difference in height between the cone base and the original water table is the **drawdown** (Figure 18.9). By producing a steeper gradient of the water table, the flow of ground water toward the well is also increased, so the well will yield more water. This increase holds only for a limited amount of drawdown, beyond which the yield fails to increase. The cone of depression may extend as far out as 16 km (10 mi) or more from a well where heavy pumping is continued. Where many wells are in operation, their intersecting cones produce a general lowering of the water table.

Depletion often greatly exceeds the rate at which the ground water of the area is recharged by percolation from rain or from the beds of streams. In an arid region, much of the ground water for irrigation is from wells driven into thick alluvial sands and gravels. Recharge of these deposits depends on the seasonal flows of water from

Figure 18.8 **Ground water in a sandstone aquifer is held under pressure beneath a capping aquiclude of shale. Water rises above the ground surface from an artesian well.**

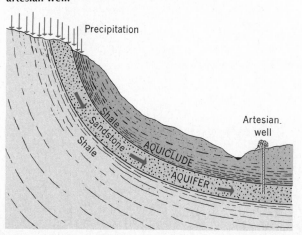

Figure 18.9 **Drawdown and cone of depression in a pumped well.**

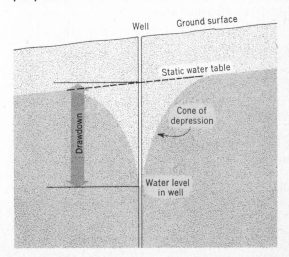

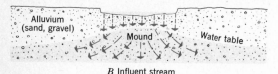
A Effluent stream

B Influent stream

Figure 18.10 **Effluent and influent streams. (From A. N. Strahler, 1971, *The Earth Sciences,* 2nd ed., Harper and Row, New York.)**

streams originating high in adjacent mountain ranges.

Streams in a dry climate, flowing on plains underlain by coarse alluvium, lose water by seepage through the channel floor. This water recharges the ground water body below and causes the water table to be raised in the form of a mound (Figure 18.10*B*). Streams of this type are called **influent streams.** In contrast, in moist climates where the water table is high and slopes toward the channels, streams receive flow by out-seepage of ground water; these are **effluent streams** (Figure 18.10*A*).

In dry climates, particularly, the extraction of ground water by pumping can greatly exceed the recharge by stream flow. Cones of depression deepen and widen; deeper wells and more powerful pumps are then required. Overdrafts of water accumulate, and the result is exhaustion of a natural resource not renewable except by long lapses of time.

In areas of moist climates with a large annual water surplus, natural recharge is by general percolation over the entire ground area surrounding the well. Here the prospects of achieving a balance of recharge and withdrawal are highly favorable through the control of pumping. An important recycling measure is the return of waste waters or stream waters to the ground water table by means of **recharge wells,** in which water flows down rather than up.

Pollution of Ground Water

Disposal of solid wastes poses a major environmental problem in heavily populated areas of North America because the advanced industrial economy produces an endless source of liquid waste (sewage) and solid waste

(garbage and trash). Traditionally, the solid wastes were trucked to the municipal dump and burned there in continually smoldering fires that emitted foul smoke and gases. The partially consumed residual waste was later buried under a cover of compacted earth.

In recent years, there has been a major improvement in solid-waste disposal methods. One method is high-temperature incineration. Another is the **sanitary landfill** method in which waste is not allowed to burn. Instead, the waste is continually covered by a protective layer of sand or clay available at the landfill site. The waste is thus buried in the unsaturated zone. Here it is subject to reaction with percolating rainwater that has infiltrated the ground surface. This water picks up a variety of ions from the waste body and carries these down as **leachate** to the water table. Once in the water table, the leachate follows the flow paths of the ground water.

As shown in Figure 18.11, a mounding of the water table develops beneath the disposal site. Loose soil of the disposal area facilitates infiltration of precipitation, while lack of vegetation reduces the evapotranspiration. Consequently, the recharge here is greater than elsewhere, and the mound is maintained. After leachate has moved vertically down by gravity percolation to the water table mound, it moves radially outward from the mound to surrounding lower points on the water table. As shown in Figure 18.12, a supply well with its cone of depression draws ground water from the surrounding area. Linkage between outward flow from the waste disposal site and inward flow to the well can bring leachate into the well, polluting the ground water supply. Pollution of supply wells by partially treated effluent infiltrating the ground at sewage disposal plants can occur in a basically similar manner.

An important step in guarding against this form of pollution is to place a monitor well (or several monitor wells) on a line between the disposal site and the well. Chemical tests for presence of leachate are made regularly, while the slope of the water table can also be determined. Movement of leachate toward the supply well may be blocked by placement of a recharge well (or wells), building a freshwater accumulation (actually an inverted cone) that will oppose the movement of the leachate.

It is not necessary for a ground water mound to be present for pollutants to travel to distant points. Where the water table has a pronounced slope, as it generally does everywhere except near the summit of a broad ground water divide, leachate or any pollutant introduced at a given point migrates as a pollution plume along the flow

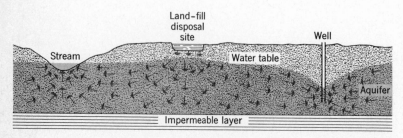

Figure 18.11 **Leachate from a waste disposal site moves toward a supply well (right) and a stream (left). (From A. N. Strahler, 1972, *Planet Earth,* Harper and Row, New York.)**

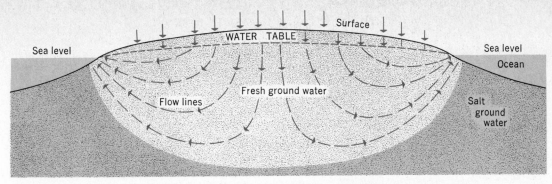

Figure 18.12 **Freshwater and saltwater relationships in an island or peninsula.**
(After G. Parker.)

paths of the ground water.

Another potential source of pollution of ground water supplies is from highways and streets, through spillage of chemicals and from deicing salt applied during the winter months. Spillage of large volumes of liquids from tank trucks and tank cars as a result of highway and railroad accidents poss a serious threat because a large slug of pollutant can be injected into the ground water recharge system. The commonest serious pollutants are automobile fuel and heating oil, but many toxic industrial chemicals are also transported in tank trucks and cars. Leakage of fuels from underground storage tanks, used in all gasoline stations, is a related source of possible pollution.

Saltwater Intrusion

A serious consequence of sustained, heavy ground water withdrawal in coastal zones is that wells near the shore eventually draw salt ground water and must be abandoned. To understand how this happens, we must examine the relationship between salt and fresh ground water.

Figure 18.12 shows an idealized diagram of the relationship in an island or a narrow peninsula. The body of fresh ground water takes the shape of a gigantic lens with convex faces, except that the upper surface has only a broad curvature whereas the lower surface, in contact with the salt ground water, bulges deeply downward. Because fresh water is less dense than salt, we can think of

this fresh water lens as floating on the salt water, pushing it down much as the hull of an ocean liner pushes aside the surrounding water. The ratio of densities of fresh water to salt water is 40 to 41. Hence, if the water table is, say, 10 m above sea level, the bottom of the freshwater lens will be located 400 m below sea level, or 40 times as deep as the water table is high with respect to sea level.

The fresh ground water extends seaward some distance beyond the shoreline. Although the salt ground water is stagnant, the fresh water travels in the curved paths shown by arrows in Figure 18.12. If water is pumped excessively from wells close to the coast, the contact of salt with fresh ground water shifts landwater, where it may intersect the wells and contaminate the fresh water. When such a condition exists, the only cure is to cease pumping and allow the fresh water to push the contact back toward its original seaward position or to create a freshwater recharge barrier between the wells and the coastline. Resumed pumping must then be regulated to a lower rate.

Limestone Caverns

Most persons are familiar with the names of famous caverns, such as Mammoth Cave or Carlsbad Caverns. Millions of Americans have visited these famous tourist attractions. **Caverns** are interconnected subterranean cavities in limestone, formed by carbonic acid carried in circulating ground water. Figure 18.13 suggests how

Figure 18.13 **Cavern development in the ground-water zone, followed by travertine deposition in the unsaturated zone. (From A. N. Strahler, 1971, *The Earth Sciences*, 2nd ed., Harper and Row, New York.)**

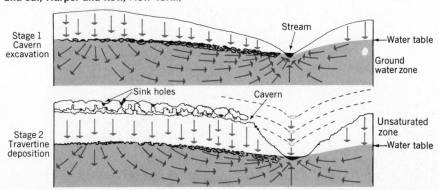

Figure 18.14 Outcrops of horizontal limestone strata show in the walls of this deep sinkhole on the Kaibab Plateau of northern Arizona. Because of the high elevation (2500 m, 8500 ft), the climate is cool and moist, favoring limestone solution. (A. N. Strahler.)

caverns may develop. In the upper diagram the action of carbonic acid is shown to be particularly concentrated in the saturated zone just below the water table. Products of solution are carried along in the ground water flow paths to emerge in streams. In a later stage, shown in the lower diagram, the stream has deepened its valley and the water table has been correspondingly lowered to a new position. The cavern system previously excavated is now in the unsaturated zone. Evaporation of percolating water on exposed rock surfaces in the caverns now begins the deposition of carbonate matter, known as **travertine.** Encrustations of travertine take many beautiful forms stalactites, stalagmites, columns, drip curtains, and terraces (Plate G.3).

The environmental importance of caverns is felt in several ways. Throughout Man's early development, caverns were an important habitation. We find the skeletal remains of humans, together with their implements and cave drawings, preserved through the centuries in caverns in many parts of the world. Today caverns are being used as storage facilities, living quarters, and factories.

Caverns have provided some valuable deposits of guano, the excrement of birds or bats, which is rich in nitrates. Guano deposits have been used in the manufac-

ture of fertilizers and explosives. Bat guano was taken from Mammoth Cave for making gunpowder during the War of 1812.

Karst Landscapes

Where limestone solution is very active, we find a landscape with many unique landforms. This is especially true along the Dalmatian coastal area of Yugoslavia, where the landscape is called **karst.** The term is used by geographers for the topography of any limestone area where sinkholes are numerous and small surface streams are nonexistent. A **sinkhole** is a surface depression in the limestone of a cavernous region. Some sinkholes are filled with soil washed from nearby hillsides (Plate G.3). Others are steep-sided, deep holes (Figure 18.14).

Development of a karst landscape is shown in Figure 18.16. In an early stage, funnellike sinkholes are numerous. Later, the caverns collapse, leaving open, flat-floored valleys. Some important western-hemisphere regions of karst topography are the Mammoth Cave region of Kentucky, the Yucatan Peninsula, and parts of Cuba and Puerto Rico.

Forms of Overland Flow

We have now traced the subsurface movements of surplus water beneath the lands. We turn next to trace the surface flow paths of surplus water.

Figure 18.16 shows what happens to rain falling on a hillslope that has its natural soil and plant cover intact. Some rain is caught and held above the ground on leaves and stems, a process called **interception.** Intercepted water may be returned directly to the atmosphere by

Figure 18.15 Features of a karst landscape. (*A*) Rainfall enters the cavern system through sinkholes in the limestone. (*B*) Extensive collapse of caverns reveals surface streams flowing on shale beds beneath the limestone. The flat-floored valleys can be cultivated. (Drawn by E. Raisz.)

A

B

(left) **Frost-shattered blocks of quartzite on the summit of the Snowy Range, Wyoming, elevation 12,000 ft (3700 m). The expanse of broken rock is a felsenmeer (rock-sea). (Arthur N. Strahler)**

(above) **Frost heaving above timberline. Needle ice, attached to the underside of a rock fragment, lifted the mass as the crystals lengthened. Below is the cavity in which the ice formed. (Mark A. Melton)**

(below) **Mafic igneous rock undergoing decay in place to produce a thick regolith. White dikes of felsic rock are little affected. Sangre de Cristo Mts., New Mexico. (Arthur N. Strahler)**

(above) **Granular disintegration of a boulder of coarse granite is forming hollows beneath a hardened surface layer covered by desert varnish. Sacaton Mountains, Arizona. (Mark A. Melton)**

(above) **Spheroidal weathering in mafic igneous rock, Spanish Peaks, Colorado. (Orlo E. Childs)**

(right) **Joint blocks of pink granite, rounded by granular disintegration in a semiarid climate. Granite Dells, near Prescott, Arizona. (Arthur N. Strahler)**

Weathering processes   Plate G.1

(above) **Cavernous weathering in cliffs of consolidated volcanic ash (tuff), Frijoles Canyon, New Mexico.** (Arthur N. Strahler)

(below) **A double rock arch formed by weathering of a massive sandstone layer (Entrada formation), Arches National Monument, Utah.** (Arthur N. Strahler)

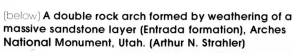

(above) **Montezuma Castle in the Verde Valley, Arizona, is an Indian cliff dwelling occupying niches weathered from limy strata of an ancient lake deposit.** (Donald L. Babenroth)

(above) Yosemite Valley, California. A large exfoliation dome with thick granite shells (sheeting structure). (Orlo E. Childs)

(left) **The White House Ruin, a former Indian habitation, occupies a great niche in sandstone in the lower wall of Canyon de Chelly, Arizona.** (Mark A. Melton)

Plate G.2   Weathering forms

(right) **Limestone showing cavernous weathering by solution. These boulders were removed by power shovel from an enclosing regolith of reddish clay, near Roanoke, Virginia. (Arthur N. Strahler)**

(below) **Cavities and pits produced by carbonic acid action on limestone, Everglades National Park, Florida. (Arthur N. Strahler)**

(left) **This large sinkhole in limestone, its floor choked with soil, has been used as a small cornfield, near Hilton, Virginia. (Arthur N. Strahler)**

(above) **Travertine deposits — stalactites and stalagmites — in Carlsbad Caverns, New Mexico. (Orlo E. Childs)**

(left) **A shallow sinkhole on the Kaibab Plateau, Arizona. Clay impedes the drainage of water, allowing a pond to form. (Arthur N. Strahler)**

(below) **Travertine deposits, Howe's Cavern, New York. Stalactites and drip curtains hang from joint fractures in the ceiling rock. (Orlo E. Childs)**

Limestone solution and caverns   **Plate G.3**

(left) **Talus cones of angular quartzite boulders built against the steep headwall of a glacial cirque. Snowy Range, Wyoming. (Arthur N. Strahler)**

(above) **A small earthflow made up of saturated glacial till, Vermont. The flow occurred during heavy summer rains. (Orlo E. Childs)**

(left) **Upper part of a coastal slide set off by an earthquake, San Pedro, California. Large blocks slumped down to different levels, displacing the roadway. (Ned L. Reglein)**

(one above) **This mudflow issued from a steep mountain canyon and buried a highway near Farrnington, Utah. The boulders were carried many miles, suspended in the soft mud. (Orlo E. Childs)**

(above) **This great landslide occurred in the Gros Ventre River canyon in 1925, blocking the river and creating a large lake. The mass descended about 2000 ft (600 m). (Orlo E. Childs)**

(right) **A large, prehistoric landslide underlies the hummocky ground in the middle distance. Pavilion, British Columbia. (Mark A. Melton)**

Plate G.4   Mass wasting

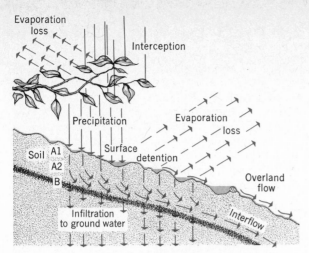

Figure 18.16 Precipitation, interception, and overland flow.

evaporation. Precipitation reaching the soil surface enters the soil by infiltration; but when the rate of precipitation exceeds the rate of infiltration, water begins to accumulate in minor surface depressions in the soil or behind small check dams of plant litter. The phenomenon is called **surface detention** (or *depression storage*). Some of this water is evaporated directly into the atmosphere; the remainder eventually infiltrates the soil.

Runoff that flows down the slopes of the land in broadly distributed sheets is referred to as **overland flow** (Figure 18.17). We distinguish overland flow from **stream flow (channel flow),** in which the water occupies a narrow channel confined by lateral banks. Overland flow can take several forms. It may be a continuous thin film, called **sheet flow,** where the soil or rock surface is smooth. Flow may take the form of a series of tiny rivulets connecting one water-filled hollow with another where the ground is rough or pitted. On a grass-covered slope, overland flow is subdivided into countless tiny threads of water passing around the stems. Even in a heavy and prolonged rain, you might not notice overland flow in progress on a sloping lawn. On heavily forested slopes, overland flow may pass entirely concealed beneath a thick mat of decaying leaves. At the base of a hillslope, overland flow is disposed of by passing into a stream channel or lake. In some cases, overland flow disappears into a permeable layer of colluvium.

Overland flow is measured in centimeters or inches of water per hour, just as for precipitation and infiltration. Therefore a simple formula expresses the rate at which overland flow will be produced by a given unit of ground surface as follows:

Rate of production of overland flow = rate of precipitation − rate of infiltration

For example, if the rate of infiltration became constant at a value of 1 cm/hr and the rate of rainfall was a steady 2

cm/hr (a heavy rain), the runoff would be produced at a rate of 1 cm/hr, assuming none to be returned to the atmosphere by evaporation.

Interflow

We learned in Chapters 13 and 14 that a mature soil consists of horizons and that the B horizon, in particular, is often the site of accumulation (by illuviation) of clay, humus colloids, and sesquioxides carried down from the A2 horizon (by eluviation). In Spodosols, this dense horizon of accumulation is the spodic horizon; in Alfisols, it is an argillic horizon. Ultisols, too, have a B horizon of clay accumulation. For these three soil orders, a forest cover is the natural vegetation. Forest litter and an organic epipedon (O horizon) protect the mineral soil from being sealed by impact of raindrops, so these soils can usually infiltrate rainfall at a high rate. When the infiltrating water reaches the B horizon, its downward passage is greatly impeded. The soil water then backs up into the more permeable A horizon and begins to move in the downslope direction, parallel with the ground surface (Figure 18.16). This lateral flow is called **interflow** (or *throughflow*). Eventually interflow reaches the base of the hillslope and seeps into a stream channel. Interflow is thus a pathway of runoff intermediate in position and speed of travel between overland flow and ground-water flow. Interflow is also important where a permeable soil rests on a dense, impermeable bedrock mass.

Drainage Systems

Overland flow, interflow, and ground-water flow eventually contribute to a stream, which is a much faster, more concentrated form of runoff (Figure 18.18). We

Figure 18.17 Overland flow running down a moderate slope following a heavy thunderstorm. The ditch in the foreground receives the runoff and conducts it away as channel flow. (Soil Conservation Service.)

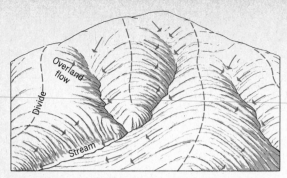

Figure 18.18 Overland flow from slopes in the headwater area of a stream system supplies water and rock debris to the smallest elements of the channel network.

define a **stream** as a long, narrow body of flowing water occupying a trenchlike depression, or channel, and moving to lower levels under the force of gravity.

The total system of downslope water flow from the point of arrival at the ground surface comprises the **drainage system.** It consists of a branched network of stream channels, as well as the sloping ground surfaces that contribute overland flow and interflow to those channels. The entire system is bounded by a drainage divide, outlining a more or less pear-shaped **drainage basin.** The ground slopes and channels are adjusted to dispose of, as efficiently as possible, the runoff and its contained load of mineral particles.

A typical stream network contributing to a single outlet is shown in Figure 18.19. Note that each fingertip tributary receives runoff from a small area of land surface surrounding the channel. This area may be regarded as the unit cell of the drainage system. The entire surface within the outer divide of the drainage basin constitutes the watershed for overland flow. A drainage system is a converging mechanism funneling and integrating the weaker and more diffuse forms of runoff into progressively deeper and more intense paths of activity.

Stream Channel Geometry

A **stream channel** is a narrow trough, shaped by the forces of flowing water to be highly effective in moving the quantities of water and sediment supplied to the stream. Channels may be so narrow that a person can jump across them, or as wide as 1.5 km (1 mi) for a great river — such as the Mississippi.

Hydraulic engineers who measure stream dimensions and flow rates have adopted a set of terms to describe channel geometry (Figure 18.20). Depth, d, in meters, is measured at any specified point in the stream as the vertical distance from surface to bed. Width, w, is the distance in meters across the stream from one water's edge to the other. The cross-sectional area, A, is the area in square meters of a vertical slice across the stream. The wetted perimeter, P, is the length of line of contact between the water and the channel, as measured from the cross section.

The rate of fall in altitude of the stream surface in the downstream direction is the **stream gradient;** it is often stated in meters per kilometer or feet per mile. A gradient of 5 m/km means that the stream surface undergoes a vertical drop of 5 m for each km of horizontal distance downstream. Slope can also be given in terms of percent grade, a common practice in engineering. A grade of 3 percent, or 0.03, means that the stream drops 3 m for every 100 m of horizontal distance.

Stream Flow

As a stream flows under the influence of gravity, the water encounters resistance — a form of friction — with the channel walls. As a result, water close to the bed and banks moves slowly; that in the deepest and most centrally located zone flows fastest. Figure 18.20 indicates by arrows the speed of flow at various points in the stream. The single line of maximum velocity is located in midstream, where the channel is straight and symmetrical.

Our statement about velocity needs to be qualified. Actually in all but the most sluggish streams, the water is

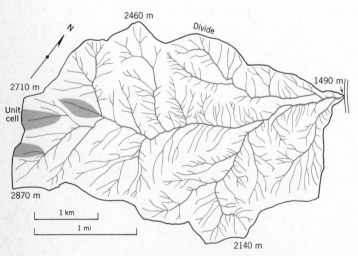

Figure 18.19 Channel network of the basin of Pole Canyon, Utah. (Data from U.S. Geological Survey and Mark A. Melton.)

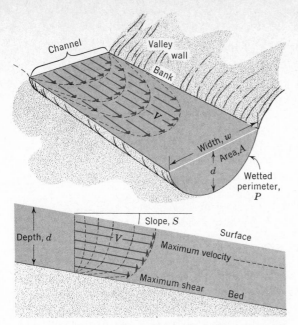

Figure 18.20 Stream flow within a channel is most rapid near the center and just below the water surface.

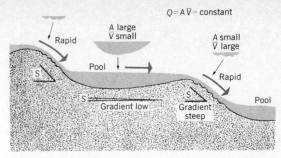

Figure 18.21 Schematic diagram of the relationships among cross-sectional area, mean velocity, and gradient. (From A. N. Strahler, 1971, *The Earth Sciences,* 2nd ed., Harper and Row, New York.)

affected by **turbulence,** a system of innumerable eddies that are continually forming and dissolving. A particular molecule of water, if we could keep track of it, would describe a highly irregular, corkscrew path as it is swept downstream. Motions include upward, downward, and sideward directions. Turbulence in streams is extremely important because of the upward elements of flow that lift and support fine particles of sediment. The murky, turbid appearance of streams in flood is ample evidence of turbulence, without which sediment would remain near the bed. Only if we measure the water velocity at a certain fixed point for a long period of time, say several minutes, will the average motion at that point be down-stream and in a line parallel with the surface and bed. Average values are shown by the arrows in Figure 18.20.

Because the velocity at a given point in a stream differs greatly according to whether it is being measured close to the banks and bed or out in the middle line, a single value, the **mean velocity,** is needed. Mean velocity is computed for the entire cross section to express the activity of the stream as a whole.

Stream Discharge

A most important measure of stream flow is **discharge,** Q, defined as the volume of water passing through a given cross section of the stream in a given unit of time. Discharge is stated in cubic meters per second (abbreviated to cms); in English units, cubic feet per second (cfs). Discharge may be obtained by taking the mean velocity, V, and multiplying it by the cross-sectional area, A. This relationship is stated by the important equation, $Q = AV$.

We realize that water will flow faster in a channel of steep gradient than one of gentle gradient because the component of gravity acting parallel with the bed is

stronger for the steeper gradient. As shown in Figure 18.21, velocity increases quickly where a stream passes from a wide pool of low gradient to a steep stretch of rapids. As V increases, A must decrease; otherwise their product, AV, would not be held constant. In the pool, where velocity is low, A is correspondingly increased.

Stream channels differ in the amount of frictional resistance that the bed and banks offer to the flow of water. Resistance is large in a broad but shallow channel and is much less in a deep, narrow channel. The optimum channel would be semicircular in cross section. Streams are, however, usually required to have broad, shallow channels because of the load of mineral particles that must be carried or because the banks are weak and will not hold a steep attitude.

Stream Gauging

Information on daily discharges of streams is vital to a nation, not only as a measure of the surface-water resource, but also for the design of flood-protection structures and for the prediction of stream floods as they progress down a river system. ("River" is a common noun applied to any large stream; we will use "river" in the popular sense without attempting a strict definition.)

In the United States, the measurement of stream flow, or **stream gauging,** is under the jurisdiction of the U.S. Geological Survey. In cooperation with states and municipalities, that organization maintains over 11,000 gauging stations on principal streams and their tributaries. Figures on discharge are published by the Geological Survey in a series of Water-Supply Papers.

A stream-gauging station requires a device for measuring the height of the water surface, or **stream stage.** Simplest to install is a **staff gauge,** which is simply a graduated vertical scale attached to a post or bridge pier. This must be read directly by an observer whenever the stage is to be recorded. More useful is an automatic-recording gauge, which is mounted in a stilling tower built beside the river bank (Figure 18.22). The tower is simply a hollow masonry shaft or sheet-metal tube filled by water that enters through a pipe at the base. By means of a float connected by a wire to a recording mechanism above, a continuous ink-line record of the stream stage is made on a graph paper attached to a slowly rotating drum.

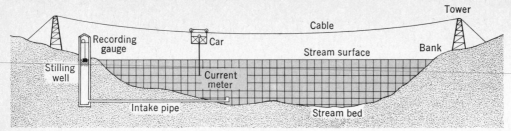

Figure 18.22 **Idealized diagram of stream-gauging installation.**

To measure stream discharge, one must determine both the area of cross section of the stream and the mean velocity. Velocity measurement requires a **current meter** (Figure 18.23). The meter is lowered into the stream at closely spaced intervals so the velocity can be read at a large number of points evenly distributed in a grid pattern through the stream cross section (Figure 18.22). A bridge often serves as a convenient means of crossing over the stream; otherwise, a cable car or small boat is used. Revolving cups on the current meter turn at a rate proportional to current velocity. As the velocities are being measured from point to point, a profile of the riverbed is

Figure 18.23 **In gauging large rivers, the current meter is lowered on a cable by a power winch. Earphones, connected by wires to the meter, receive a series of clicks whose frequency indicates water velocity. (U.S. Geological Survey.)**

also being made by sounding the depth. From these readings, a profile is drawn and the cross-sectional area is measured from the profile. Mean velocity is computed by summing all individual velocity readings and dividing by the number of readings. Discharge can then be computed using the formula $Q = AV$.

Stream Flow and Precipitation

It seems obvious that the discharge of a stream will increase in response to a period of heavy rainfall or snowmelt. The response is delayed, of course, but the length of delay depends on a number of factors. The most important factor is the size of the drainage basin feeding the stream above the place where the gauge is located. The relationship between stream discharge and precipitation is best studied by means of a simple graph, called a **hydrograph.**

Figure 18.24 is a hydrograph for a drainage basin about 800 sq km (300 sq mi) in area, located in Ohio in a region of moist continental climate. The graph gives data for a two-day summer storm. Rainfall is shown by a bar graph giving the precipitation (cm) in each two-hour period. Also plotted on the graph (smooth line) is the discharge (cms) of Sugar Creek, the trunk stream of the drainage basin. The average total rainfall over the watershed of Sugar Creek was about 15 cm (6 in.); of this amount about half passed down the stream within three days' time. Some rainfall was held in the soil as soil water, some evaporated, and some infiltrated to the water table to be held in long-term storage in the ground-water body.

Studying the rainfall and runoff graphs in Figure 18.24, we see that prior to the onset of the storm, Sugar Creek was carrying a small discharge. This flow, being supplied by the slow seepage of ground water into the channel, is termed **base flow.** After the heavy rainfall began, several hours elapsed before the stream gauge at the basin mouth began to show a rise in discharge. This interval, called the **lag time,** indicates that the branching system of channels was acting as a temporary reservoir. As the stage was rising, water was soaking into permeable bank materials, where it was being temporarily stored.

Lag time is measured as the difference between center of mass of precipitation (CMP) and center of mass of runoff (CMR), as labeled in Figure 18.24. The peak flow of Sugar Creek was reached almost 24 hours after the rain began; the lag time was about 18 hours. Note also that the rate of decline in discharge was much slower than the rate of rise because overland flow was followed by contributions

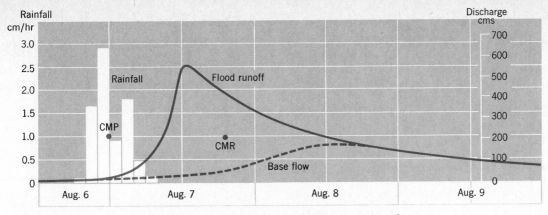

Figure 18.24 **Four days of flow of Sugar Creek, Ohio. (After William G. Hoyt and Walter B. Langbein FLOODS; copyright 1955 by Princeton University Press; Figs. 8 and 13, pp. 39 and 45. Reprinted by permission of Princeton University Press.)**

from interflow and, finally, from ground-water seepage.

In general, the larger a watershed, the longer is the lag time between peak rainfall and peak discharge and the more gradual is the rate of decline of discharge after the peak has passed. Notice that the flow of Sugar Creek showed a slow but distinct rise in the amount of discharge contributed by base flow.

Base Flow and Overland Flow

In regions of moist climates, where the water table is high and normally intersects the important stream channels, the hydrographs of larger streams will show clearly the effects of three sources of water: overland flow, interflow, and base flow. Figure 18.25 is a hydrograph of the Chattahoochee River, Georgia, a large stream draining a watershed of some 8700 sq km (3350 sq mi), much of it in the moist southern Appalachian Mountains. The sharp, abrupt fluctuations in discharge are produced by overland flow and interflow following rain periods of one to three days duration. These are each similar to the hydrograph of

Figure 18.24, except that they are shown in Figure 18.25 much compressed by the time scale.

After each rain period, the discharge falls off rapidly; but, if another storm occurs within a few days, the discharge rises to another peak. The enlarged inset graph shows details for the month of January. When a long period intervenes between storms, the discharge falls to a low value, the base flow, where it levels off.

Throughout the year the base flow, which represents ground water inflow into the stream, undergoes a marked annual cycle. During the period of recharge (winter and early sping), water-table levels are raised and the rate of inflow into streams is increased. For the Chattahoochee River, the rate of base flow during January, February, March, and April holds uniform at about 110 cms (4000 cfs). The base flow begins to decline in spring, when heavy evapotranspiration losses reduce soil water and therefore cut off the recharge of ground water. The decline continues throughout the summer, reaching a low of about 30 cms (1000 cfs) by the end of October.

Finally, examine the hydrograph of the Missouri River at Omaha, Nebraska, for a two-year period of record

Figure 18.25 **Flow peaks of the Chattahoochee River. (Data from U.S. Geological Survey in E. E. Foster, *Rainfall and Runoff.*)**

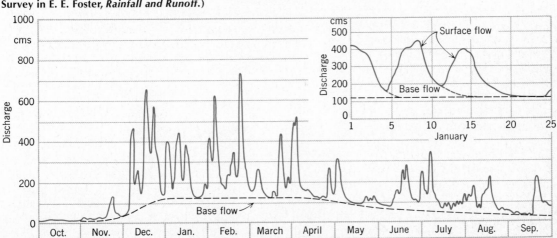

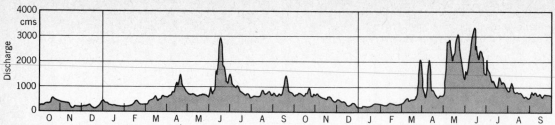

Figure 18.26 **Discharge of the Missouri River. (Data from U.S. Geological Survey in E. E. Foster, *Rainfall and Runoff*.)**

starting in October (Figure 18.26). This great river, draining 840,000 sq km (322,800 sq mi) of watershed, is a major tributary of the Mississippi River. Note that the discharge, ranging from 280 to 2800 cms (10,000 to 100,000 cfs), is many times greater than the discharges of smaller streams considered thus far. High rates of flow are chiefly from snowmelt, which occurs on the High Plains in spring and in the Rocky Mountain headwater areas in early summer. This event explains the sudden high discharges from April through June. During midwinter, when soil water is frozen and total precipitation is small over the watershed as a whole, the discharge rises little above the base flow. Ground water recharge occurring in the spring raises summer levels of base flow to about 570 cms (20,000 cfs), or two to three times the winter base flow.

Floods

Everyone has seen enough news-media photos of river floods to have a good idea of the appearance of floodwaters and the havoc wrought by their erosive power and by the silt and clay they leave behind. Even so, it is not easy to define the term **flood.** Perhaps it is enough to say that a condition of flood exists when the discharge of a river cannot be accommodated within the margins of its normal channel, so the water spreads over adjoining ground on which crops or forests are able to flourish.

Most larger streams of moist climates have a **floodplain,** a belt of low, flat ground bordering the channel on one or both sides inundated by stream waters about once a year. This flood usually occurs in the season when

Figure 18.27 **The city of Hartford was partly inundated by the Connecticut River flood of March 1936. The river channel is to the left, its banks marked by a line of trees. (Official Photograph, 8th Photo Section, A. C., U.S. Army.)**

abundant supplies of surface water combine with the effects of a high water table to supply more runoff than can stay within the channel. Such an annual inundation is considered a flood, even though its occurence is expected and does not prevent the cultivation of crops after the flood has subsided. The seasonal inundation does not interfere with the growth of dense forests, which are widely distributed over low, marshy floodplains in all moist regions of the world. Still higher discharges of water, the rare and disastrous floods that may occur as infrequently as three to five decades, inundate ground lying well above the floodplain (Figure 18.27).

For practical purposes, the National Weather Service, which provides a flood-warning service, designates a particular stage of gauge height at a given place as the **flood stage;** this implies that the critical level has been reached above which overbank flooding may be expected to set in. Immediately at or below flood stage, the river may be described as being in the **bank-full stage,** the flow being entirely within the limits of the heavily scoured channel.

Downstream Progress of a Flood Wave

The rise of a river stage to its maximum height, or **flood crest,** followed by a gradual lowering of stage, is termed the **flood wave.** The flood wave is simply a large-sized rise and fall of river discharge of the type already analyzed in earlier paragraphs, and it follows the same principles. Figure 18.28A shows the downstream progress of a flood on the Chattooga-Savannah river system. In the Chattooga River near Clayton, Georgia, the flood peak, or crest, was reached quickly — one day after the storm — and subsided quickly. On the Savannah River, 105 km (65 mi) downstream at Calhoun Falls, South Carolina, the peak flow occurred a day later; but the discharge was much larger because of the larger area of watershed involved. Downstream another 153 km (95 mi), near Clyo, Georgia, the Savannah River crested five days after the initial storm, with a discharge of over 1700 cms (60,000 cfs). This set of three hydrographs shows that (1) the lag time in occurrence of the crest increases downstream, (2) the entire period of rise and fall of flood wave becomes longer downstream, and (3) the discharge increases greatly downstream as watershed area increases.

Figure 18.28B is a somewhat different presentation of the same flood data in that the discharge is given in terms of a common unit of area (cms/sq km), thus eliminating the effect of increase in discharge downstream and showing us only the shape, or form, of the flood crest.

Flood Prediction

The National Weather Service operates a River and Flood Forecasting Service through 85 offices located at strategic points along major river systems of the United States. Each office issues river and flood forecasts to the communities within the associated district, which is laid out to cover one or more large watersheds. Flood warnings are publicized by every possible means. Close cooperation is maintained with various agencies to plan the evacuation

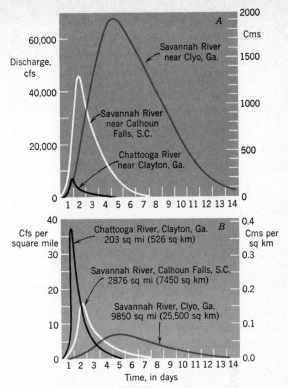

Figure 18.28 The downstream progress of a flood wave. (After William G. Hoyt and Walter B. Langbein FLOODS; copyright 1955 by Princeton University Press; Figs. 8 and 13, pp. 39 and 45. Reprinted by permission of Princeton University Press.)

of threatened areas and the removal or protection of vulnerable property.

Hydrologists use sophisticated statistical methods to arrive at estimates of the probability that a flood of a given stage of discharge will occur in any given year. Estimates are, of course, based on gauge records of past years. Figure 18.29 is a graph on which the greatest discharge of each year of record is plotted as a dot. Discharge in units of cms is plotted on the vertical axis in equally spaced units (arithmetic scale). Numbers on the bottom horizontal scale tell the probability that the given discharge will be equaled or exceeded in a given year. For example, a discharge associated with the value of 20 percent is interpreted to mean "a discharge of this magnitude (2000 cms) can be expected to be equaled or exceeded in 20 out of 100 years." The numbers on the scale at the top of the graph tell the return period (or recurrence interval), which is simply the inverse value of the probability percentage on the bottom scale. In the example just cited, probability 20 percent, the return period is five years, meaning that on the average the stated discharge (2000 cms) can be expected to be equaled or exceeded once every five years. The horizontal scale on the graph has been adjusted to obtain the best possible fit of the plotted points to a straight line. The fitted line can be expressed by a mathematical equation; it is one of several estimating equations that might be applied to the same data in efforts to secure the best predictions.

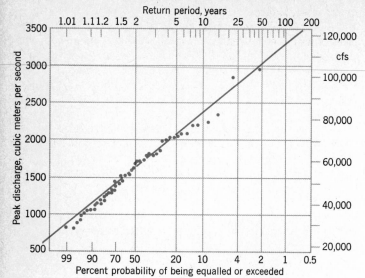

Figure 18.29 **Flood frequency data for the Clearwater River at Kamaiah, Idaho. Each dot is a measured maximum yearly discharge in a 53-year record. (Data from R. K. Linsley, M. A. Kohler, and J. L. Paulhus, 1975, *Hydrology for Engineers,* second edition, McGraw-Hill, New York, p. 347, Figure 11.2.)**

A much simpler analysis is used in Figure 18.30 to suggest the likelihood that a given maximum stage will occur in a given calendar month of the year. Each bar divides all observed maximum stages of the calendar month into quartiles, or groups of 25 percent of the occurrences. The dots at top and bottom of the bar give the absolute maximum and minimum stages observed during the entire period of record.

The graph for the Mississippi River at Vicksburg illustrates a great river responding largely to spring floods so as to yield a simple annual cycle. All floods have occurred in the first six months of the year. The Colorado River at Austin, Texas, illustrates a river draining largely semiarid plains. Summer floods are produced directly by torrential rains from invading moist tropical air masses. Floods of the late summer and fall are often attributable to tropical storms (hurricanes) moving inland from the Gulf of Mexico. The Sacramento River at Red Bluff, California, has a winter flood season when rains are heavy, but a sharp dip to low stages in late summer, which is the very dry period in the Mediterranean climate. The flood expectancy graph for the Connecticut River at Hartford shows two seasons of floods. The more reliable of the two is the early spring, when snowmelt is rapid over the mountainous New England terrain. The second season is in the fall, when rare but heavy rainstorms, some of hurricane origin, bring exceptional high stages.

The regulation of floods of large rivers is described in Chapter 19 in connection with landforms of floodplains.

Hydrologic Effects of Urbanization

Hydrology of a watershed is altered in two ways by urbanization. First, an increasing percentage of the surface is rendered impervious to infiltration by the construction of roofs, driveways, walks, pavements, and parking lots. It has been estimated that in residential areas, for a lot size of 1400 sq m (15,000 sq ft), the impervious area amounts to about 25 percent; for a lot size of 560 sq m (6000 sq ft), the impervious area amounts to about 80 percent.

An increase in proportion of impervious surface reduces infiltration and increases overland flow from the urbanized area. An important result is to increase the frequency and height of flood peaks during heavy storms. There is also a reduction of recharge to the ground water body beneath; this reduction, in turn, decreases the base-flow contribution to channels in the area. Thus the full range of stream discharges, from low stages in dry periods to flood stages, is made greater by urbanization.

A second change caused by urbanization is the introduction of storm sewers that allow storm runoff from paved areas to be taken directly to stream channels for discharge. Runoff travel time to channels is shortened at the same time that the proportion of runoff is increased by expansion in impervious surfaces. The two changes together conspire to reduce the lag time, as shown by the schematic hydrographs in Figure 18.31.

Many rapidly expanding suburban communities are now finding that certain low-lying residential areas, formerly free of flooding, are being subjected to inundation by overbank flooding of a nearby stream. The need for careful terrain study and land-use planning is obvious in such cases to protect the unwary home buyer from locating in a flood-prone neighborhood.

One partial solution to the problem created by storm sewering is to return storm runoff to the ground water body by means of infiltrating basins. This program has been adopted on Long Island, in New York, where infiltration rates are high in sandy glacial materials. Another method of disposal of storm runoff is by recharge wells. At Orlando, Florida, storm runoff enters wells that penetrate cavernous limestone. The capacity of the system to absorb runoff without clogging appears to be adequate.

Figure 18.30 **(below and right) The highest stage that occurred in each month is given in terms of percentages on these graphs of four rivers. (National Weather Service.)**

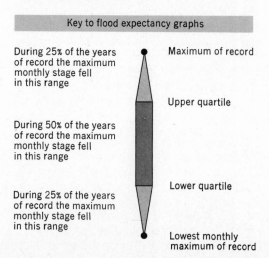

Key to flood expectancy graphs

During 25% of the years of record the maximum monthly stage fell in this range — Maximum of record

— Upper quartile

During 50% of the years of record the maximum monthly stage fell in this range

— Lower quartile

During 25% of the years of record the maximum monthly stage fell in this range — Lowest monthly maximum of record

In Fresno, California, a large number of gravel-packed wells of 76-cm (30-in.) diameter receive runoff from streets. The system has proved successful in disposing of storm drainage.

Water Pollution

In this chapter we have examined the flow of water itself, largely from the hydrologist's point of view. Streams, lakes, bogs, and marshes are specialized habitats of plants and animals; their ecosystems are particularly sensitive to changes induced by Man in the water balance and in water chemistry. Not only does our industrial society make radical physical changes in water flow by construction of engineering works (dams, irrigation systems, canals, dredged channels), but we also pollute and contaminate our surface waters with a large variety of wastes. Some of these wastes are in the form of ions in solution. The subject of water quality is appropriate to discuss in this chapter. Introduction of mineral sediment into surface waters as a result of land disturbances is a suitable topic

for the next chapter.

Chemical pollution by direct disposal into streams and lakes of wastes generated in industrial plants is a phenomenon well known to the general public and can be seen firsthand in almost any industrial community in the United States. Direct outfall of sewage, whether raw (untreated) or partially treated, is another form of direct pollution of streams and lakes.

In urban and suburban areas, pollutant matter entering streams and lakes includes deicing salt, lawn conditioners (lime and fertilizers), and sewage effluent. In agricultural regions, important sources of pollutants are fertilizers and the body wastes of livestock. Major sources of water pollution are associated with the mining and processing of mineral deposits. The possibility of contamination by radioactive substances released from nuclear processing plants also exists.

Among the common chemical pollutants of both surface water and ground water are sulfate, nitrate, phosphate, chloride, sodium, and calcium ions. Sulfate ions enter runoff by fallout from polluted urban air and as

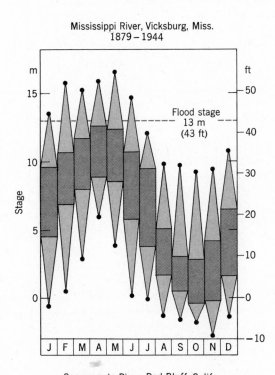

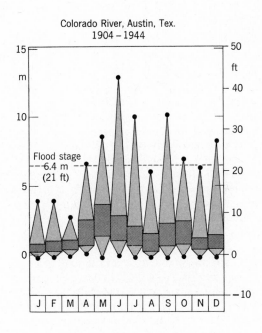

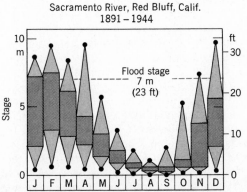

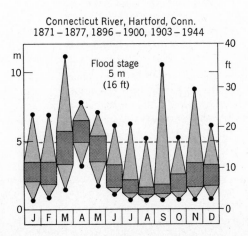

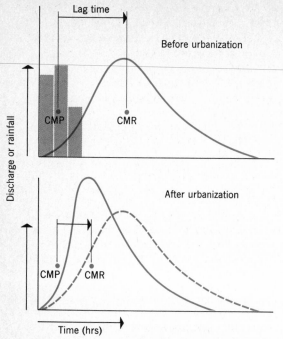

Figure 18.31 **Schematic hydrographs showing the effect of urbanization on lag time and peak discharge. Points CMP and CMR are centers of mass of rainfall and runoff, respectively, as in Figure 18.24. (After L. B. Leopold, 1968, U.S. Geological Survey, Circular 554.)**

sewage effluent. Important sources of nitrate ions are fertilizers and sewage effluent. Excessive concentrations of nitrate in water supplies are highly toxic to humans but, at the same time, the removal of nitrate is difficult and expensive. Phosphate ions are contributed in part by fertilizers and by detergents in sewage effluent. Phosphate and nitrate are plant nutrients and can lead to excessive growth of algae in streams and lakes. This process is called **eutrophication.** Chloride and sodium ions are contributed both by fallout from polluted air and by deicing salts used on highways. Instances are recorded in which water supply wells close to highways have become polluted from deicing salts.

A particular form of chemical pollution of surface water goes under the name of **acid mine drainage.** It is an important form of enviromental degradation in parts of Appalachia where abandoned coal mines and strip mine workings are concentrated. Ground water emerging from abandoned mines and seeping from strip mine spoil banks is charged with sulfuric acid and various salts of metals, particularly of iron. The sulfuric acid is formed by the reaction of water with iron sulfides, particularly the mineral pyrite (Fe_2S), which is a common constituent of coal seams. Acid of this origin is stream waters can have adverse effects on animal life. In sufficient concentrations, it is lethal to certain species of fish and has at times caused massive fish kills. The acid waters also cause corrosion to boats and piers. Government sources have estimated that nearly 10,000 km (6000 mi) of streams in this country, together with almost 100 sq km (40 sq mi) of reservoirs and lakes are seriously affected by acid mine drainage. One particularly undesirable by-product of acid mine drainage is precipitation of iron to form slimy red and yellow deposits in stream channels.

Sulfuric acid may also be produced from the drainage of mines from which sulfide ores are being extracted and from tailings produced in plants where the ores are processed. Such chemical pollution also includes salts of various toxic metals, among them zinc, lead, arsenic, copper, and aluminum. Yet another source of chemical pollution of streams is by radium derived from the tailings of uranium ore processing plants.

Toxic metals, among them mercury, along with pesticides and a host of other industrial chemicals, are introduced into streams and lakes in quantities that are locally damaging or lethal to plant and animal communities. In addition, sewage introduces live bacteria and viruses that are classed as biological pollutants; these pose a threat to the health of Man and animals.

Thermal pollution is a term applied generally to the discharge of heat into the environment from the combusion of fuels and from nuclear energy conversion into electric power. Thermal pollution of water takes the form of heavy discharges of heated water locally into streams, estuaries, and lakes. The thermal environmental impact may thus be quite drastic in a small area.

Fresh Water as a Natural Resource

Fresh water is a basic natural resource essential to Man's varied and intense agricultural and industrial activities. Runoff held in reservoirs behind dams provides water supplies for great urban centers, such as New York City and Los Angeles; diverted from large rivers, it provides irrigation water for highly productive lowlands in arid lands, such as the Imperial Valley of California and the Nile Valley of Egypt. To these uses of runoff are added hydroelectric power, where the gradient of a river is steep, or routes of inland navigation, where the gradient is gentle.

Unlike ground water, which represents a large water storage body, fresh surface water in the liquid state is stored only in small quantities. (An exception is the Great Lakes system.) Referring back to Figure 10.4, note that the quantity of available ground water is about 20 times as large as that stored in freshwater lakes, while the water held in streams is only about 1/100 that in lakes. Because of small natural storage capacities, surface water can be drawn only at a rate comparable with its annual renewal through precipitation. Dams are built to develop useful storage capacity for surplus runoff that would otherwise escape to the sea; but, once the reservoir has been filled, water use must be scaled to match the natural supply rate averaged over the year. The development of surface-water supplies brings on many environmental changes, both physical and biological, and these must be taken into account in planning for future water developments.

Studies of the total U.S. water resource have led to the conclusion that for the nation as a whole, there is an adequate freshwater supply for at least several decades.

For individual regions of the country, however, we can anticipate severe shortages that can be met only by the transfer of water from regions of surplus. The enormous concentration of persons and industry into a small area (e.g., in Los Angeles County and the New York City region), require the development of surface-water facilities in distant uplands, along with elaborate water-transfer systems using expensive and vulnerable aqueducts. It is only by continual and costly expansion of these facilties that shortages can be avoided. Compounding distribution problems is water pollution. If the quality of our freshwater supplies continues to be degraded, we must resort more and more to expensive water treatment procedures necessary to keep the water usable.

In recent years, much emphasis has been placed on alternatives to massive water transfers to meet rising demands. The basic alternative is a reduction in water use, particularly reductions in wasteful use in urban systems and in the huge demands for irrigation in arid lands. It is obvious that problems of water-resource management involve a broad spectrum of economic, political, and cultural factors.

References for Further Study

Foster, E. E. (1948), *Rainfall and runoff,* Macmillan Co., New York.

Hoyt, W. G., and W. B. Langbein (1955), *Floods,* Princeton University Press, N.J.

U.S. Dept. of Agriculture (1955), *Water; yearbook of agriculture 1955,* Government Printing Office, Washington, D.C.

Todd, D. K. (1959), *Ground water hydrology,* John Wiley and Sons, New York.

Chorley, R. J., ed. (1971), *Introduction to fluvial processes,* Methuen and Co., London. See Chapters 2, 5, 7.

Kazmann, R. G. (1972), *Modern hydrology,* second edition, Harper and Row, New York.

Review Questions

1. What is the environmental importance of runoff? How can a knowledge of hydrology be useful in water-resource management?

2. Define ground water. Distinguish between the saturated zone and the unsaturated zone. What is the water table? Explain the presence of a capillary fringe above the water table. How can the depth and configuration of the water table be plotted?

3. How is a hydraulic head produced within the ground water body? How does ground water recharge occur? How does it vary with season? Describe the flow paths of ground water. Where does ground water emerge?

4. Explain how freshwater ponds, bogs, and marshes are related to the water table.

5. Explain how storage capacity and permeability can vary according to the kinds of subsurface materials present. Describe a good aquifer. What materials form aquicludes? What is a perched water table? Explain the principle of artesian flow.

6. Describe the various kinds of wells used to withdraw ground water. How is a cone of depression produced by pumping? What is drawdown?

7. Distinguish between influent and effluent streams. What problems are associated with ground water use in dry climates? How are recharge wells used?

8. What are the major sources of ground water pollution? Describe the movement of pollutants in the ground water zone.

9. Describe and explain the relation of fresh ground water to salt ground water beneath an island. What physical principle governs the relation of water-table elevation to depth of fresh water? How is saltwater contamination of wells caused?

10. How are limestone caverns excavated? What is the relationship between caverns and the water table? What is travertine? Describe a karst landscape. What is a sinkhole?

11. Describe the various ways precipitation is held or disposed of at the surface. What losses can occur in this zone? Distinguish between interception and surface detention.

12. Distinguish between overland flow and stream flow. What are the common forms of overland flow? What is interflow, and how is it related to soil horizons?

13. Describe a drainage system and its parts. What forms the unit cell of a drainage system? To what functions must the form of a drainage system be adjusted?

14. Describe a stream channel. How can the size of a stream be stated? What is stream gradient? In what units is it stated?

15. Describe the distribution of velocity in a stream. What is mean velocity? What is turbulence? Why is turbulence important in the activity of a stream? What is discharge? How is it related to velocity and cross-sectional area?

16. What are the practical applications of stream-gauging programs? Describe the methods and equipment used in stream gauging. How is mean velocity determined?

17. What is a hydrograph? Describe the hydrograph of a stream in a small drainage basin in response to a local rainstorm. What is the lag time? Why does it occur? What is base flow? Where does it come from? How does base flow change throughout a yearly cycle? Give an example.

18. How is a flood defined? What is the bank-full stage?

19. Describe the progress of a flood crest down a large river. How does discharge change with increasing distance downstream? How does the form of the flood wave change?

20. What government agency supplies river and flood forecasts? How are its activities organized and what is their function?

21. How can the probability of floods of various magnitudes be predicted? What data are used? Explain the meaning of probability of recurrence. What is meant by recurrence interval?

22. Describe the annual cycle of change in stage of the Mississippi River. At what season do flood stages occur? Explain. Compare this annual cycle with that of a river in the Mediterranean climate of the Pacific coast.

23. Describe and explain the hydrologic effects of urbanization. How are flood peaks affected? How is lag time affected? What can be done to alleviate the undesirable effects?

24. Describe the various classes of pollutant substances found in surface water and ground water. What is the origin of these substances? What is eutrophication? Acid mine drainage? Thermal pollution?

25. Make a general statement about the basic problems of freshwater resources and their management.

Landforms Made by Running Water

19

In this chapter we will deal with the activity of overland flow and stream flow as agents of landform sculpture. But, before we embark on our study of the work of the fluid agents, it is important to look at landforms in general in broad perspective. The configuration of continental surfaces reflects the balance of power, so to speak, between internal earth forces, acting through volcanic and crustal-breaking (tectonic) processes, and external forces, acting through the agents of denudation. Seen in this perspective, landforms in general fall into two basic categories.

Landforms produced directly by volcanic and tectonic activity are **initial landforms** (Figure 19.1). These activities and the landforms they produce are explained in Chapter 24. Initial landforms include volcanoes and lava flows, fault blocks and rift valleys, and alpine ranges elevated in zones of recent crustal deformation. The energy for lifting molten rock and rigid crustal masses to produce the initial landforms has an internal heat source. This heat is believed to be produced by natural radioactivity in rock of the earth's outer layers.

Figure 19.1 **Initial and sequential landforms (Drawn by A. N. Strahler.)**

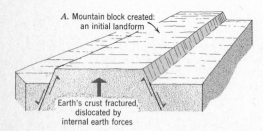

A. Mountain block created: an initial landform

Earth's crust fractured, dislocated by internal earth forces

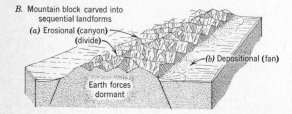

B. Mountain block carved into sequential landforms
(a) Erosional (canyon)
(divide)
(b) Depositional (fan)
Earth forces dormant

Landforms shaped by processes and agents of denudation belong to the class of **sequential landforms,** meaning that they follow in sequence after the initial landforms are created and a crustal mass—a landmass—has been raised to an elevated position. As shown in Figure 19.1, a single uplifted crustal block (an initial landform) is set on by agents of denudation and carved up into a large number of sequential landforms.

Any landscape is really nothing more than the existing stage in a great contest. As thick crustal masses collide or pull apart, the internal earth forces spasmodically elevate parts of the crust to create initial landforms. The external agents persistently keep wearing these masses down and carving them into vast numbers of smaller sequential landforms.

All stages of this struggle can be seen in various parts of the world. High alpine mountains and volcanic chains exist where the internal earth forces have recently dominated. Rolling plains of the continental interiors reflect the ultimate victory of agents of denudation. All intermediate stages can be found. Because the internal earth forces act repeatedly, new landmasses keep coming into existence as old ones are subdued.

Fluvial Processes and Landforms

Landforms shaped by running water are conveniently described as **fluvial landforms** to distinguish them from landforms made by the other fluid agents—glacial ice, wind, and waves. Fluvial landforms are shaped by the **fluvial processes** of overland flow and stream flow. Weathering and the slower forms of mass wasting, such as soil creep, operate hand in hand with overland flow and cannot be separated from the fluvial processes.

Fluvial landforms and fluvial processes dominate the continental land surfaces the world over. Throughout geologic history, glacial ice has been present only in comparatively small global areas located in the polar zones and in high mountains. Landforms made by wind action occupy only trivially small parts of the continental surfaces, and landforms made by waves and currents are

Figure 19.2 Erosional and depositional landforms. (Drawn by A. N. Strahler.)

restricted to a very narrow contact zone between oceans and continents. In terms of area, the fluvial landforms are dominant in the environment of terrestrial life and are the major source areas of Man's food resources through the practice of agriculture. Almost all lands in crop cultivation and almost all grazing lands have been shaped by fluvial processes.

Fluvial processes perform the geologic activities of erosion, transportation, and deposition. Consequently, there are two major groups of fluvial landforms: erosional landforms and depositional landforms (Figure 19.1). Valleys are formed where rock is eroded away by fluvial agents. Between the valleys are ridges, hills, or mountain summits representing unconsumed parts of the crustal block. All such sequential landforms shaped by the progressive removal of the bedrock mass are **erosional landforms.**

Fragments of soil, regolith, and bedrock that are removed from the parent mass are transported by the fluid agents and deposited elsewhere as sediment to make an entirely different set of surface features, the **depositional landforms.** Figure 19.2 illustrates the two groups of landforms. The ravine, canyon, peak, spur, and col are erosional landforms; the fan, built of coarse sediment below the mouth of the ravine, is a depositional landform. The floodplain, built of alluvium transported by a stream, is also a depositional landform.

Normal and Accelerated Slope Erosion

Fluvial action starts on the uplands of drainage basins. Overland flow, by exerting a dragging force over the soil surface, picks up particles of mineral matter ranging in size from fine colloidal clay to coarse sand or gravel, depending on the speed of the flow and the degree to which the particles are bound by plant rootlets or held down by a mat of leaves. Added to this solid matter is dissolved mineral matter in the form of ions produced by acid reactions or direct solution. This slow removal of soil is part of the natural geologic process of landmass denudation; it is both inevitable and universal. Under stable natural conditions, the erosion rate in a moist climate is slow enough that a soil with distinct horizons is formed and maintained, enabling plant communities to maintain themselves in a stable equilibrium. Soil scientists refer to this state of activity as the **geologic norm.**

By contrast, the rate of soil erosion may be enormously speeded up by Man's activities or by rare natural events to result in a state of **accelerated erosion,** in which the soil is removed much faster than it can be formed. This condition comes about most commonly from a forced change in the plant cover and in the physical state of the ground surface and uppermost soil horizons. The destruction of vegetation by the clearing of land for cultivation or by forest fires sets the stage for a series of drastic changes. The interception of rain by foliage is ended; protection afforded by a ground cover of fallen leaves and stems is removed. Consequently, the rain falls directly on the mineral soil.

The direct force of a falling raindrop (Figure 19.3) causes a geyserlike splashing in which soil particles are lifted and then dropped into new positions, a process termed **splash erosion.** It is estimated that a torrential rainstorm has the ability to disturb as much as 225 metric tons of soil per hectare (100 tons per acre). On a sloping ground surface, splash erosion shifts the soil slowly downhill. A more important effect is to cause the soil surface to become much less able to infiltrate water because the natural soil openings become sealed by particles shifted by raindrop splash. Reduced infiltration permits a much greater proportion of overland flow to occur from a given amount of rain. Increased overland

Figure 19.3 A large raindrop (above) lands on a wet soil surface, producing a miniature crater (below). Grains of clay and silt are thrown into the air, and the soil surface is disturbed. (Official U.S. Navy photograph.)

flow intensifies the rate of soil removal.

Another effect of the destruction of vegetation is to reduce greatly the resistance of the soil surface to the force of erosion under overland flow. On a slope covered by grass sod, even a deep layer of overland flow causes little soil erosion because the energy of the moving water is dissipated in friction with the grass stems, which are tough and elastic. On a heavily forested slope, countless check dams made by leaves, twigs, roots, and fallen tree trunks take up the force of overland flow. Without such vegetative cover the eroding force is applied directly to the bare soil surface, easily dislodging the grains and sweeping them downslope.

Man-Induced Changes in Infiltration Capacity

Use of the land surface by Man for purposes of crop cultivation and the grazing of domesticated animals can have a profound effect on the infiltration of precipitation through the soil surface. We can imagine the soil surface to be a fine sieve that receives rainfall of a given intensity and is capable of transmitting water downward at a given rate (see Figure 10.5). Rainfall intensity is usually stated for short time periods, for example, cm per 10 min. or cm/hr. Infiltration is also stated as depth per unit of time. When rainfall intensity exceeds infiltration rate, the excess water leaves the surface as overland flow. Overland flow may also be stated in intensity units, as water depth per unit of time. The following equation then applies:

$$P - I = R_0$$

where   P is precipitation rate (intensity),
          I is infiltration rate,
and R_0 is overland flow,

all stated as cm/hr or in./hr. Evaporation is assumed to be zero during the rainfall period. Obviously infiltration rate will equal precipitation rate until the limit of the infiltration rate, or **infiltration capacity,** is reached.

Now, it is an important fact about soils that their infiltration capacity is usually great at the start of a rain which has been preceded by a dry spell, but drops rapidly as the rain continues to fall and to soak into the soil. After two or three hours the infiltration capacity becomes almost constant. This change with time is shown in the graphs of Figure 19.4. In the first quarter-hour or so, infiltration capacity drops sharply; then the curve flattens rapidly. One reason for the high starting value and its rapid drop is that the larger soil openings that allow water to flow downward rapidly become clogged by particles brought from above, or tend to close up as the colloidal clays take up water and swell. From this effect we can easily reason that a sandy soil with little or no clay will not undergo a great drop in infiltration capacity, but will continue to let the water through indefinitely at a generous rate. In contrast, a clay-rich soil is quickly sealed to the point that it allows only a very slow rate of infiltration. This principle is illustrated by Figure 19.4A, which shows the infiltration curves of two soils, one sandy, one rich in

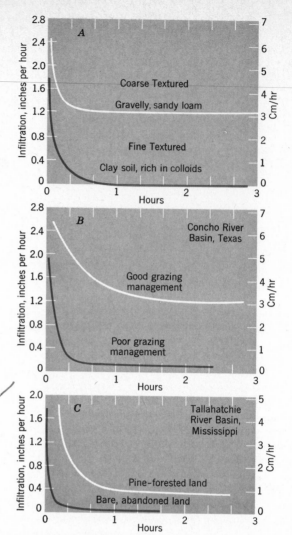

Figure 19.4   **Infiltration rates vary greatly according to soil texture and land use. (Data from Sherman and Musgrave; Foster.)**

clay. A sandy soil may be able to infiltrate water of a heavy, long-continued rain without any overland flow occurring, whereas the clay soil must divert much of the rain into overland flow.

Certain forms of Man-made disturbance of soils tend to decrease the infiltration capacity and increase the amount of surface runoff. Trampling by livestock will tamp the porous soil into a dense, hard layer (Figure 19.4B). Forest cutting followed by crop cultivation tends to leave the soil exposed so that raindrop impact quickly seals the soil pores (Figure 19.4C). Man-made fires, by destroying the protective vegetation and surface litter, also expose the soil to raindrop impact and cause reduced infiltration capacity. It is little wonder, then, that Man has, through unwise farming and grazing practices, radically changed the original proportions of infiltration to runoff. As a result of reduced infiltration, severe erosion damage has occurred in many areas.

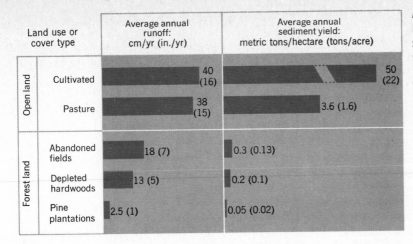

Land use or cover type		Average annual runoff: cm/yr (in./yr)	Average annual sediment yield: metric tons/hectare (tons/acre)
Open land	Cultivated	40 (16)	50 (22)
	Pasture	38 (15)	3.6 (1.6)
Forest land	Abandoned fields	18 (7)	0.3 (0.13)
	Depleted hardwoods	13 (5)	0.2 (0.1)
	Pine plantations	2.5 (1)	0.05 (0.02)

Figure 19.5 **The bar graph shows that both runoff and sediment yield are much greater for open land than for land covered by shrubs and forest. Cultivated land has an enormous sediment yield, as compared with any of the other types. (Data of S. J. Ursic, 1965, Department of Agriculture.)**

Land Use and Sediment Yield

We can get a good appreciation of the contrast between normal and accelerated erosion rates by comparing the quantity of sediment derived from cultivated surfaces with that derived from naturally forested or reforested surfaces within a given region in which climate, soil, and topography are fairly uniform. **Sediment yield** is a technical term for the quantity of sediment removed by overland flow from a unit area of ground surface in a given unit of time. Yearly sediment yield is stated in metric tons per hectare, or tons per acre.

Figure 19.5 gives data on annual average sediment yield and runoff by overland flow from several types of upland surfaces in northern Mississippi. Notice that both surface runoff and sediment yield decrease greatly with increased effectiveness of the protective vegetative cover. Sediment yield from cultivated land undergoing accelerated erosion is over ten times greater than that from pasture and about one thousand times greater than from pine plantation land. The reforested land has a sediment-yield rate representing the geologic norm of soil erosion for this region; it is about the same as for mature pine and hardwood forests that have not experienced cultivation.

The distinction between normal and accelerated slope erosion applies to regions in which the water balance shows an annual surplus. Under a midlatitude semiarid climate with summer drought, the natural plant cover consists of short-grass prairie. Although sparse and providing rather poor ground cover of plant litter, the grass cover is strong enough that the geologic norm of erosion can be sustained. In these semiarid environments, however, the natural equilibrium is highly sensitive to upset. Depletion of plant cover by fires or the grazing of herds of domesticated animals can easily set off rapid erosion. These sensitive, marginal environments require cautious use because they lack the potential to recover rapidly from accelerated erosion once it has set in.

Erosion at a very high rate by overland flow is actually a natural geologic process in certain favorable localities in semiarid and arid lands; it takes the form of **badlands.** One well-known area of badlands is the Big Badlands of South Dakota, along the White River. Badlands are underlain by clay formations, easily eroded by overland flow. Erosion rates are too fast to permit plants to take hold, and no soil can develop. A maze of small stream channels is developed; ground slopes are very steep (Figure 19.6). Badlands such as these are self-sustaining and have been in existence at one place or another on continents throughout much of geologic time.

Forms of Accelerated Soil Erosion

Humid regions of substantial water surplus, for which the natural plant cover is forest or prairie grasslands,

Figure 19.6 **Badlands, such as these in the Petrified Forest National Monument, Arizona, are like miniature mountains formed on bare clay formations. (B. Mears, Jr.)**

Figure 19.7 **Shoestring rills on a barren slope. (Soil Conservation Service.)**

experience accelerated soil erosion when Man expends enough energy to remove the plant cover and keep the land barren by annual cultivation. With fossil fuels to power machines of plant and soil destruction, Man has easily overwhelmed the restorative forces of nature over vast expanses of continental surfaces. We now consider the consequences of these activities.

When a plot of ground is first cleared of forest and plowed for cultivation, little erosion will occur until the action of rain splash has broken down the soil aggregates and sealed the larger openings. Overland flow then begins to remove the soil in rather uniform thin layers, a process termed **sheet erosion.** Because of seasonal cultivation, the effects of sheet erosion are often little noticed until the upper horizons of the soil are removed or greatly thinned. Reaching the base of the slope, where the surface slope is sharply reduced to meet the valley bottom, soil particles come to rest and accumulate in a thickening layer of colluvium. This sediment deposit, too, has a sheetlike distribution and may be little noticed, except where it can be seen that fence posts or tree trunks are being slowly buried.

Material that continues to be carried by overland flow to reach a stream in the valley axis is then carried farther downvalley and may accumulate as alluvium in layers on the valley floor. Deposition of alluvium results in the burial of fertile floodplain soils (Inceptisols, Entisols) under infertile, sandy layers. This form of valley sedimentation chokes the channels of small streams, causing the water to flood broadly over the valley bottoms.

Where slopes are steep, runoff from torrential rains produces a more intense activity, **rill erosion,** in which innumerable, closely spaced channels are scored into the soil and regolith (Figure 19.7). If these rills are not destroyed by soil tillage, they may soon begin to integrate into still larger channels, called **gullies.** Gullies are steep-walled, canyonlike trenches whose upper ends grow progressively upslope (Figure 19.8 and Plate H.2). Ultimately, a rugged, barren topography, like the badland forms of the dry climates, results from accelerated soil erosion that is allowed to proceed unchecked.

The natural soil, with its well-developed horizons, is a nonrenewable natural resource. The rate of soil formation is extremely slow in comparison with the rate of its destruction once accelerated erosion has begun and is allowed to go unchecked. Soil erosion as a potentially disastrous form of environmental degradation was brought to public attention decades ago in the United States.

Curative measures developed by the Soil Conservation Service have proved effective in stopping accelerated soil erosion and in permitting the return to slow erosion rates approaching the geologic norm. These measures include the construction of terraces to eliminate steep slopes, permanent restoration of overly steep slope belts to dense vegetative cover, and the healing of gullies by placing check dams in the gully floors.

Geologic Work of Streams

The geologic work of streams consists of three closely interrelated activities: erosion, transportation, and deposition. **Stream erosion** is the progressive removal of mineral material from the floor and sides of the channel, whether bedrock or regolith. **Transportation** consists of movement of the eroded particles by dragging along the bed, by suspension in the body of the stream, or in solution. **Deposition** is the accumulation of transported particles on the streambed and floodplain, or on the floor of a standing body of water into which the stream empties. Deposition is synonymous with sedimentation. Obviously, erosion cannot occur without some transportation taking place, and the transported particles must eventually come to rest. Erosion, transportation, and deposition are simply three phases of a single activity.

Stream Erosion

Streams erode in various ways, depending on the nature of the channel materials and the tools with which the current

Figure 19.8 **This great gully, eroded into thick saprolite, was typical of some parts of the Piedmont region of South Carolina and Georgia before remedial measures were applied. (Soil Conservation Service.)**

is armed. The force of the flowing water alone, exerting impact and a dragging action on the bed and banks, can erode poorly consolidated alluvial materials, such as gravel, sand, silt, and clay. This erosion process, called **hydraulic action,** is capable of excavating enormous quantities of unconsolidated materials in a short time (Figure 19.9). The undermining of the banks causes large masses of alluvium to slump into the river, where the particles are quickly separated and become part of the stream's load. This process of **bank caving** is an important source of sediment during high river stages.

Where rock particles carried by the swift current strike against bedrock channel walls, chips of rock are detached. The rolling of cobbles and boulders over the streambed will further crush and grind smaller grains to produce an assortment of grain sizes. These processes of mechanical wear constitute **abrasion;** it is the principal means of erosion in bedrock too strong to be affected by simple hydraulic action.

Finally, the chemical processes of rock weathering —acid reactions and solution—are effective in the removal of rock from the stream channel and may be referred to as **corrosion.** The effects of corrosion are conspicuous in limestone, which develops cupped and fluted surfaces.

One interesting form produced by stream abrasion is the **pothole,** a cylindrical hole carved into the hard bedrock of a swiftly moving stream (Figure 19.10 and Plate H.1). Often a spherical or discus-shaped stone is found in the pothole and is apparently the tool, or grinder, with which the pothole was deepened. A spiraling flow of water in the pothole rotates the grinder at the base of the hole, boring gradually into the rock. Many other features of abrasion, such as plunge pools, chutes, and troughs lend variety to the rock channel of a mountain stream.

Figure 19.10 **These potholes have been carved in granite in the channel of the James River, Henrico County, Virginia. (U.S. Geological Survey.)**

Stream Transportation

The solid matter carried by a stream is the **stream load;** it is carried in three forms. Dissolved matter is transported invisibly in the form of chemical ions. All streams carry some dissolved salts resulting from mineral alteration. Clay and silt particles are carried in **turbulent suspension;** that is, they are held up in the water by the upward elements of flow in turbulent eddies in the stream. This fraction of the transported matter is the **suspended load.** Sand, gravel, and cobbles move as **bed load** close to the channel floor by rolling or sliding and an occasional low leap (Plate H.2).

The load carried by a stream varies enormously in the total quantity present and the size of the fragments, depending on the discharge and stage of the river. In flood, when velocities of 6 m (20 ft) per second or more are produced in large rivers, the water is turbid with suspended load. Large boulders may be moving over the streambed where the river gradient is steep.

Channel Changes in Flood

We think of a river in flood as changing largely through increase in height of water surface, which causes channel overflow and inundation of the adjoining floodplain. Because of the turbidity of the water, we cannot see important changes taking place on the streambed; but these can be determined by sounding the river depth while stream-gauging measurements are being taken (Figure 19.11). These changes apply to a channel formed in thick alluvial materials, not in bedrock. At first the bed may be built up by large amounts of bed load supplied to the stream during the first phases of heavy runoff. This upbuilding is soon reversed, however, and the bed is actively deepened by scour as stream stage rises. Thus, in the period of highest stage, the riverbed is at its lowest elevation. When the discharge then starts to decline, the level of the stream surface drops and the bed is built back up by the deposition of alluvium. In the example shown in Figure 19.11, a layer of alluvium about 3 m (10 ft) thick was reworked, that is, moved about in the complete cycle of rising and falling stages.

Figure 19.9 **A river in flood eroded this huge trench at Cavendish, Vermont, in November 1927. An area 1.6 km (1 mi) wide and 5 km (3 mi) long, once occupied by eight farms, was cut away by the floodwater. The sand and gravel offered little resistance. (Wide World Photos.)**

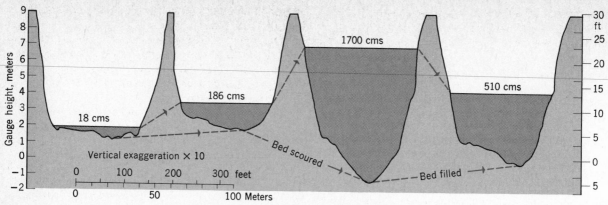

Figure 19.11 Changes in channel form of the San Juan River near Bluff, Utah, during the progress of a flood. (After Leopold and Maddock, U.S. Geological Survey.)

Alternate deepening by scour and shallowing by deposition of load are responses to changes in the ability of the stream to transport its load. The maximum load of debris that can be carried by a stream is a measure of **stream capacity.** Load is stated as weight of material moved through a given stream cross section in a given unit of time; common units are metric tons per day. Total load includes both the bed load and the suspended load.

Where a stream is flowing in a channel of hard bedrock, it may not be able to pick up enough alluvial material to supply its full capacity for bed load. Bedrock channels are typical of streams occupying deep gorges and having steep gradients. When a flood occurs in a bedrock channel, the channel cannot be quickly deepened in response. In an alluvial river, where thick layers of silt, sand, and gravel underlie the channel, the rising river easily picks up and sets in motion all the material that it is capable of moving. In other words, the increasing capacity of the stream for load is easily satisfied.

Capacity for bed load increases sharply with stream velocity because the swifter the current, the more intense the turbulence and the stronger the dragging force against the bed. Capacity to move bed load goes up approximately as the third to fourth power of the velocity. Thus, when stream velocity is doubled in flood, capacity for bed load is increased from eight to sixteen times. It is small wonder, then, that most of the conspicuous changes in a stream channel, including sidewise shifting of the course, occur in flood stage. Few important changes occur in low stages.

Suspended load also increases greatly as discharge increases and stage rises. Increased turbulence is the cause. Figure 19.12 gives an example of the relationship between suspended load and discharge, measured at a single gauging station. The scales are logarithmic and show that the suspended load in flood stages may be over 10,000 times as great as in the lowest stages. The line fitted to the plotted points shows that a 10-fold increase in discharge brings a 100-fold increase in suspended load. Suspended load may be derived from erosion by overland flow on watershed surfaces and from bank caving.

When the flood crest has passed and the discharge begins to decrease, the capacity to transport load also

declines. Therefore some of the particles that are in motion must come to rest on the bed in the form of sand and gravel bars. First the largest cobbles will cease to roll, then the pebbles and gravel, then the sand. Particles of fine sand and silt, carried in suspension, can no longer be sustained and settle to the bed. Clay particles of colloidal size continue to travel far downstream with the flood

Figure 19.12 Increase of suspended sediment load with increase in discharge. Both scales are logarithmic. The dots show how individual observations are scattered about the line of best fit. Data are for the Powder River, Arvada, Wyoming. (After L. B. Leopold and T. Maddock, 1953, U.S. Geological Survey, Professional Paper 252.)

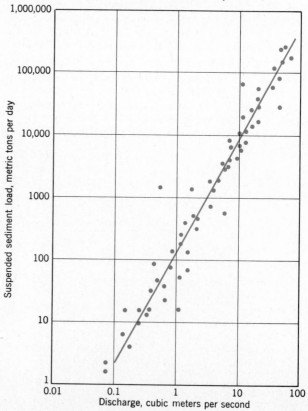

wave. In this way the falling stream adjusts to its decreasing capacity. When restored to low stage, the water may become clear.

Suspended Sediment Load of Large Rivers

The great rivers of the world show an enormous range in the quantity of suspended sediment load and in sediment yield. Data for seven selected major rivers (Table 19.1) reveal some interesting relationships among load, climate, and land-surface properties.

The Yellow River (Hwang Ho) of China heads the world list in annual suspended sediment load, while its sediment yield is one of the highest known for a large river basin. The explanation lies in a high soil-erosion rate on intensively cultivated upland surfaces of wind-deposited silt (loess) in Shensi and Shansi provinces. Much of the drainage area is in a semiarid climate having dry winters; vegetation is sparse, and the runoff from heavy summer rains entrains a large amount of sediment.

The Ganges River derives its heavy sediment load from steep mountain slopes of the Himalayas and from intensively cultivated lowlands, all subjected to torrential rains of the tropical wet monsoon season. The Colorado River represents an exotic stream, deriving its runoff largely from snowmelt and precipitation on high mountain watersheds of the Rockies; but most of its suspended sediment is from tributaries in the semiarid plateau lands through which it passes.

The sediment load of the Mississippi River comes largely from subhumid and semiarid grassland watersheds of its great western tributary, the Missouri River. The Amazon River, a colossus in discharge and basin area, has a very low sediment yield because, like the Congo River, much of its basin lies within a wet equatorial climate where the land surface bears a highly protective rainforest. The Yenisei River of Siberia has a remarkably low sediment load and sediment yield for its vast drainage area, most of which is in the needleleaf forest, or taiga, of the subarctic and high midlatitude zones.

Although it is difficult to assess the importance of Man-induced soil disturbance on the sediment load of major rivers, most investigators are generally agreed that cultivation has greatly increased the sediment load of rivers of eastern and southeastern Asia, Europe, and North America. The increase due to Man's activities is thought to be greater, by a factor of 2½, than the geologic norm for the entire world land area. For the more strongly affected river basins, the factor may be ten or more times larger than the geologic norm.

The sediment load carried by a large river is of considerable importance in planning for the construction of large water-storage dams and canal systems for irrigation. Sediment will be trapped in the reservoir behind a dam, eventually filling the entire basin and ending the useful life of the reservoir as a storage body. At the same time, depriving the river of its sediment in the lower course below the dam may cause serious upsets in river activity. Resulting deep scour of the bed and lowering of river level may render irrigation systems inoperable. In designing for canal systems, the forms of artificial channels must be adjusted to the size and quantity of sediment carried by the water; otherwise obstruction by deposition or abnormal scour may follow.

Concept of Equilibrium and the Graded Stream

A stream system, fully developed within its drainage basin of contributing valleyside slopes, has undergone a long period of adjustment of its geometry so it can discharge, through the trunk exit, not only the surplus water produced by the basin but also the solid load with which the channels are supplied. A purely hydraulic system could operate without a gradient because accumulated surplus water can generate its own surface slope and is capable of flowage on a horizontal surface. The transport of bed load requires a gradient, and it is in response to this requirement that a stream channel system has adjusted its gradient and has achieved an average steady state of operation, year in and year out and from decade to decade. In this condition, the stream is referred to as a **graded stream;** it is considered to have achieved an equilibrium in its state of operation.

As an idealized concept we consider that in an equilibrium condition, the supply of load furnished a

---

Table 19.1 / Suspended Sediment Loads and Sediment Yields of Selected Large Rivers

River	Drainage Area (sq km × 10³)	Average Discharge (cu m/sec)	Average Annual Sediment Load (metric tons × 10³)	Average Annual Sediment Yields (metric tons/sq km)
Yellow (Hwang Ho), China	715	1.6	1,900,000	2,600
Ganges, India	960	12	1,500,000	1,400
Colorado, U.S.A.	640	0.17	140,000	380
Mississippi, U.S.A.	3,200	19	310,000	97
Amazon, Brazil	6,100	190	360,000	60
Congo, Congo	4,000	42.	65,000	16
Yenisei, USSR	2,500	18	11,000	4

Data source: J. N. Holeman (1968), "The sediment yield of major rivers of the world," *Water Resources Research,* vol. 4, no. 4, pp. 737–47.
Note: Data rounded to two digits.

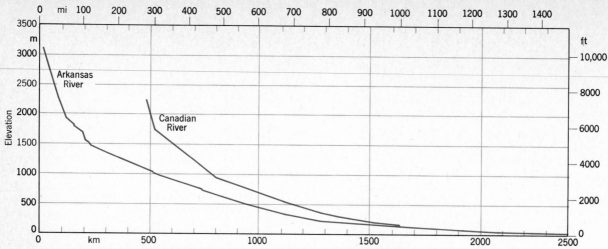

Figure 19.13 Longitudinal profiles of the Arkansas and Canadian rivers. The middle and lower parts of the profiles are for the most part smoothly graded, whereas the poorly graded upper parts reflect rock inequalities and glacial modifications within the Rocky Mountains. (From Gannett, U.S. Geological Survey.)

stream from its watershed exactly matches the stream's capacity for transport. In nature, the balance between load and a stream's capacity exists only as an average condition over periods of many years. As already explained, streams scour their channels in flood and deposit load when in the falling stage. Thus in terms of conditions of the moment, a stream is rarely in equilibrium; but over long periods of time, the graded stream maintains its level by restoring those channel deposits temporarily removed by the excessive energy of flood flows.

The channel gradient of a graded stream diminishes in steepness in the downstream direction. The **longitudinal profile** of a stream is shown in Figure 19.13 by a graph in which altitude (vertical scale) is plotted against downstream distance (horizontal scale). One important cause of the downstream decrease in gradient is that, as discharge increases and channel cross section becomes larger, the stream becomes more efficient in its operation. Frictional resistance is disproportionately less for a larger channel, so the same work can be accomplished on a lower gradient. In some cases, also, the average size of particles of the bed load diminishes in the downstream direction; the finer load can be carried on a lower gradient.

Evolution of a Graded Stream

It will be helpful in developing the concept of a graded-stream system to investigate the changes that will take place along a stretch of stream that is initially poorly adjusted to the transport of its load. An ungraded channel is illustrated in Figure 19.14. The starting profile is imagined to be brought about by crustal uplift in a series of fault steps, bringing to view a surface that was formerly beneath the ocean and exposing it to fluvial processes for the first time. Runoff collects in shallow depressions, which fill and overflow from higher to lower levels. In this way a through-going channel originates and begins to conduct runoff to the sea.

Figure 19.15 illustrates the gradation process in a series of block diagrams. Over **waterfalls** and **rapids,** which are simply steep-gradient portions of the channel, flow velocity is greatly increased and abrasion of bedrock is most intense (Block A). As a result, the falls are cut back and the rapids trenched, while the ponded reaches are filled by sediment and lowered in level as the outlets are cut down. In time the lakes disappear and the falls are transformed into rapids. Erosion of rapids reduces the

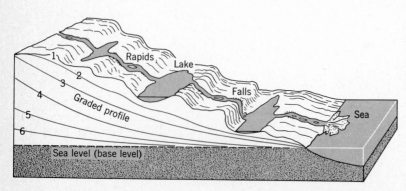

Figure 19.14 Schematic diagram of gradation of a stream. Originally, the channel consists of a succession of lakes, falls, and rapids. (Drawn by A. N. Strahler. From A. N. Strahler, 1972, *Planet Earth,* Harper and Row, New York.)

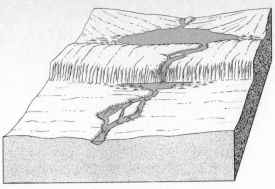

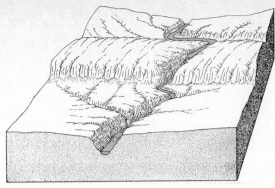

(A) Stream established on a land surface dominated by landforms of recent crustal activity.

(B) Gradation in progress; lakes and marshes drained; deepening gorge; tributary valleys extending.

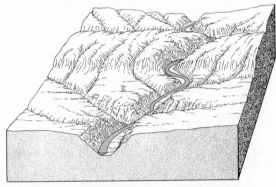

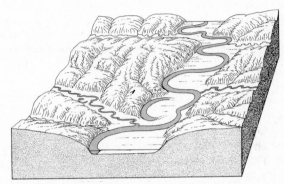

(C) Graded profile attained; beginning of floodplain development; valley widening in progress.

(D) Floodplain widened to accommodate meanders; floodplains extended up tributary valleys.

Figure 19.15   **Evolution of a stream and its valley. (Drawn by E. Raisz.)**

gradient to one more closely approximating the average gradient of the entire stream (Block *B*). At the same time, branches of the stream system are being extended into the landmass, carving out a drainage basin and transforming the original landscape into a fluvial landform system.

In the early stages of gradation and extension, the capacity of the stream exceeds the load supplied to it, so little or no alluvium accumulates in the channels. Abrasion continues to deepen the channels, with the result that they come to occupy steep-walled **gorges** or **canyons** (Plate H.1). Weathering and mass wasting of these rock walls contributes an increasing supply of rock debris to the channels. Debris shed from land surfaces contributing overland flow to the newly developed branches is also on the increase.

We can anticipate a gradual decrease in stream capacity for bed load, resulting from the gradual reduction in the channel gradient. This decrease will be converging on the increasing load with which it is being supplied. There will come a point in time at which the supply of load exactly matches the capacity of the stream to transport it. At this point the stream has achieved the graded condition and possesses a **graded profile,** descending smoothly and uniformly in the downstream direction (Figure 19.14). After attaining this state of balance, the stream continues to cut sidewise into its banks on the

outsides of bends. This **lateral cutting** does not appreciably alter the gradient and therefore does not materially affect the equilibrium.

The first indication that a steam has attained a graded condition is the beginning of floodplain development. On the outside of a bend, the channel shifts laterally into a curve of larger radius and thus undercuts the valley wall. On the inside of the bend, alluvium accumulates in the form of a sandbar. Widening of the bar deposit produces a crescentic piece of low ground, which is the first stage in floodplain development. This stage is illustrated in Block C of Figure 19.15. As lateral cutting continues, the floodplain strips are widened and the channel develops sweeping bends, called **alluvial meanders** (Block *D*). The floodplain is then widened into a continuous belt of flat land between steep valley walls.

Floodplain development reduces the frequency with which channel scour attacks and undermines the adjacent valley wall. Weathering, mass wasting, and overland flow then act to reduce the steepness of the valleyside slopes (Figure 19.16). As a result, in a moist climate, the gorgelike aspect of the valley gradually disappears and eventually gives way to an open valley with soil-covered slopes protected by a dense plant cover.

With the passage of long spans of time, the graded profile is slowly lowered throughout its entire length, as

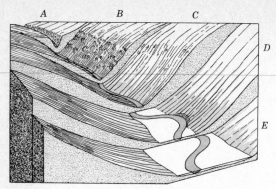

Figure 19.16 **After a stream has become graded, its valley walls become more gentle in slope and the bedrock is covered by soil and weathered rock. (Drawn by W. M. Davis.)**

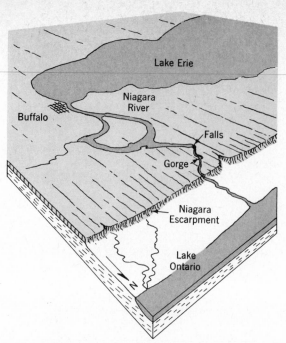

Figure 19.17 **A bird's-eye view of the Niagara River, with its falls and gorge carved in strata of the Niagara Escarpment. View is toward the southwest from a point over Lake Ontario. Redrawn from a sketch by G. K. Gilbert, 1896. (From A. N. Strahler, 1971,** *The Earth Sciences,* **2nd ed., Harper and Row, New York.)**

shown by numbered profiles in Figure 19.14. At its mouth, the stream surface merges smoothly with sea level. The bottom of the stream channel lies below sea level for some distance upstream, a condition that may persist many tens of kilometers inland in the case of a large river. Sea level, projected inland beneath the continent as an imaginary surface, is the **base level** of stream activity. It is the theoretical lower limit to which the profile of the stream surface can be reduced. Because the overall rate of denudation diminishes gradually with time, later stages of reduction of the graded profile and the contributing slopes of the drainage basin become immeasurably prolonged. Although base level represents the theoretical limit to which the stream profile can be reduced, the achievement of that limit would require an infinitely long span of time. In reality, crustal movements and changes of sea level repeatedly disturb the relationship between stream profile and base level. We shall return to this subject at the end of this chapter.

Environmental Significance of Gorges and Waterfalls

Deep gorges and canyons of major rivers exert environmental controls in a variety of ways. There is little or no room for roads or railroads between the stream and the valley sides, so roadbeds must be cut or blasted at great expense and hazard from the sheer rock walls. Maintenance is expensive because of flooding or undercutting by the stream and the sliding and falling of rock, which can damage or bury the roadbed. But a gorge may afford the only feasible passage through a mountain range. A deep canyon is also a barrier to movement across it and may require the construction of expensive bridges.

Although small waterfalls are common features of alpine mountains carved by glacial erosion (Chapter 21), large waterfalls on major rivers are comparatively rare the world over. New river channels resulting from flow diversions caused by ice sheets of the Pleistocene Epoch provide one class of falls and rapids of large discharge. Certainly the preeminent example is Niagara Falls (Plate H.1). Overflow of Lake Erie into Lake Ontario happened to

be situated over a gently inclined layer of limestone, beneath which lies easily eroded shale (Figure 19.17). As Figure 19.18 shows, the fall is maintained by continual undermining of the limestone by erosion in the plunge-pool at the base of the fall. The height of the falls is now 52 m (170 ft), and its discharge is about 17,000 cms (200,000 cfs). The drop of Niagara Falls is utilized for the production of hydroelectric power by the Niagara Power Project, in which water is withdrawn upstream from the

Figure 19.18 **Niagara Falls is formed where the river passes over the eroded edge of a massive limestone layer. Continual undermining of weak shales at the base keeps the fall steep. (Drawn by E. Raisz.)**

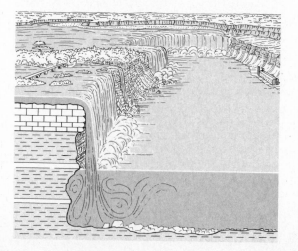

falls and carried in tunnels to generating plants located 6 km (4 mi) downstream from the falls. Capability of this project is 2400 megawatts of power, making it the largest single producer in the western hemisphere.

Most large rivers of steep gradient do not possess falls, and it is necessary to build dams to create artificially the vertical drop necessary for turbine operation. An example is the Hoover Dam, behind which lies Lake Mead, occupying the canyon of the Colorado River. With a dam height of 220 m (726 ft), the generating plant of Hoover Dam is capable of producing 1345 megawatts of power, about half as much as the Niagara Project.

We should not lose sight of the aesthetic and recreational values of gorges, rapids, and waterfalls of major rivers. The Grand Canyon of the Colorado River, probably more than any single product of fluvial processes, highlights the scenic value of a great river gorge. Against the advantages in obtaining hydroelectric power and fresh water for urban supplies and irrigation by construction of large dams, we must weigh the permanent loss of some large segments of our finest natural scenery, along with the destruction of ecosystems adapted to the river environment. It is small wonder, then, that new dam projects are meeting with stiff opposition from concerned citizen groups who fear that the harmful environmental impacts of such structures far outweigh their future benefits.

Aggradation and Alluvial Terraces

A graded stream, delicately adjusted to its supply of water and rock waste from upstream sources, is highly sensitive to changes in those inputs. Changes in climate and in surface characteristics of the watershed bring changes in discharge and load at downstream points, and these changes in turn require channel readjustments.

Consider first the effect of an increase in bed load beyond the capacity of the stream. At the point on a channel where the excess load is introduced, the coarse sediment accumulates on the streambed in the form of bars of sand, gravel, and pebbles. These deposits raise the elevation of the streambed. The process is called **aggradation;** it is a form of sediment deposition. As more bed materials accumulate, the stream channel gradient is increased; the increased flow velocity enables bed materials to be dragged downstream and spread over the channel floor at progressively more distant downstream reaches.

Aggradation typically changes the channel cross section from one of narrow and deep form to a wide, shallow cross section. Because bars are continually being formed, the flow is divided into multiple threads, and these rejoin and subdivide repeatedly to give a typical **braided channel** (Figure 19.19). The coarse channel deposits spread across the former floodplain, burying fine-textured alluvium with coarse material. We have already seen that alluvium accumulates as a result of accelerated soil erosion.

How is aggradation induced in a stream system by natural processes? One natural cause of aggradation that has been of major importance in stream systems of North America and Eurasia is glaciation during the Pleistocene

Figure 19.19 **The braided stream in the foreground is aggrading the valley floor. A shrunken glacier in the distance provides the stream discharge and bed load. Peters Creek, Chugach Mountains, Alaska. (Steve McCutcheon, Alaska Pictorial Service.)**

Epoch (Chapter 21). The advance of a valley glacier results in the input of a large quantity of coarse rock debris at the head of a drainage system. Valley aggradation was widespread in a broad zone marginal to the great ice sheets of the Pleistocene Epoch; the accumulated alluvium filled most valleys to depths of many tens of meters. Figure 19.20 shows a valley filled in this manner by an aggrading stream. The case could represent any one of a large number of valleys in New England or the Middle West.

Suppose, next, that the source of bed load is cut off or greatly diminished. In the case illustrated in Figure 19.20, the ice sheets have disappeared from the headwater areas and, with them, the supplies of coarse rock debris. Reforestation of the landscape has restored a protective cover to valley walls and hillslopes of the region, holding back coarse mineral particles from entrainment in overland flow. Now the streams have copious water discharges but little bed load. In other words, they are operating below capacity. The result is channel scour and deepening. The channel form becomes deeper and narrower. Gradually, the stream profile level is lowered, a process called **degradation.** Because the stream is very close to being in the graded condition at all times, its

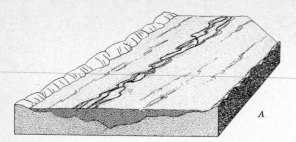

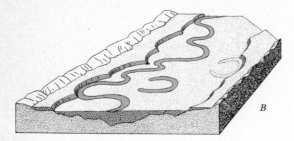

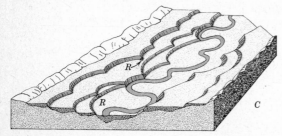

Figure 19.20 **Alluvial terraces form when a graded stream slowly cuts away the alluvial fill in its valley. (Drawn by A. N. Strahler.)**

laid out on the flat ground of a terrace, and roads and railroads are easily run along the terrace surfaces parallel with the river.

Aggradation Induced by Man

Channel aggradation is a common form of environmental degradation brought on by Man's activities. Accelerated soil erosion following cultivation, lumbering, and forest fires is the most widespread source of sediment for valley aggradation. Aggradation of channels has also been a serious form of environmental degradation in coal-mining regions, along with water pollution (acid mine drainage), mentioned in Chapter 18. Throughout the Appalachian coalfields, channel aggradation is widespread because of the huge supplies of coarse sediment from mine wastes (Figure 19.21). Strip mining has enormously increased the aggradation of valley bottoms because of the vast expanses of broken rock available to entrainment by runoff.

Urbanization and highway construction are also major sources of excessive sediment, causing channel aggradation. Large earth-moving projects are involved in creating highway grades and in preparing sites for industrial plants and housing developments. Although these surfaces are eventually stabilized, they are vulnerable to erosion for periods ranging from months to years. The regrading involved in these projects often diverts overland flow into different flow paths, further upsetting the activity of streams in the area.

Mining, urbanization, and highway construction not only cause drastic increases in bed load, which cause channel aggradation close to the source, but also increase the suspended load of the same streams. Suspended load travels downstream and is eventually deposited in lakes,

dominant activity is lateral cutting by growth of meander bends, as shown in Block *B* of Figure 19.20. The valley alluvium is gradually excavated and carried downstream, but it cannot all be removed because the channel encounters hard bedrock in many places. Consequently, as shown in Block *C*, steplike surfaces remain on both sides of the valley. The treads of these steps are **alluvial terraces.** As each terrace is cut, the width of the next one above it is reduced.

All older terraces would be destroyed were it not that bedrock of the valley wall here and there projects through the alluvium and protects higher terraces. In Block *C* of Figure 19.20, the valley alluvium has been largely removed; but some **rock-defended terraces** remain on the valley sides, protected from stream attack by the rock that outcrops at the points labeled *R*. Notice that the scarps separating terraces are curved in broad arcs concave toward the valley. The curvature is easily explained as the result of cutting of the scarps by curved meander bends.

Alluvial terraces have always attracted occupation by Man because of their advantages over both the valley-bottom floodplains, which are subject to annual flooding, and the hillslopes beyond, which may be too steep and rocky to cultivate. Besides, the soils of terraces are easily tilled and made prime agricultural land. Towns are easily

Figure 19.21 **This valley bottom in Kentucky is choked with coarse debris from a nearby strip mine area. The natural channel has been completely buried under the rising alluvium. (W. M. Spaulding, Jr., Fisheries and Wildlife, U.S. Dept. of the Interior.)**

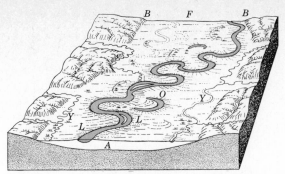

Figure 19.22   Floodplain landforms of an alluvial river.
L=natural levee; O= oxbow lake; Y = yazoo stream; A =
alluvium; B = bluffs; F = floodplain. (Drawn by E. Raisz.)

reservoirs, and estuaries far from the source areas. This
sediment is particularly damaging to the bottom environments of aquatic life. The accumulation of fine sediment
reduces the capacity of reservoirs, limiting the useful life
of a large reservoir to perhaps a century or so. Excess
sediment also results in rapid filling of tidal estuaries,
requiring increased dredging of channels.

Landforms of Floodplains

An **alluvial river** is one that flows on a thick accumulation
of alluvial deposits constructed by the river itself in earlier
stages of its activity. A characteristic of an alluvial river is
that it experiences overbank floods with a frequency
ranging between annual and biennial occurrence during
the season of large water surplus over the watershed.
Overbank flooding of an alluvial river normally inundates
part or all of a floodplain that is bounded on either side by
rising slopes, called **bluffs.**

Typical landforms of an alluvial river and its floodplain
are illustrated in Figure 19.22. Dominating the floodplain
is the meandering river channel itself, plus abandoned
reaches of former channel (Figure 19.23). Meanders
develop narrow necks, which are cut through, shortening
the river course and leaving a meander loop abandoned.
This event is called a **cutoff.** It is followed quickly by
deposition of silt and sand across the ends of the abandoned channel, producing an **oxbow lake.** The oxbow
lake is gradually filled in with fine sediment brought in
during high floods and with organic matter produced by
aquatic plants. Eventually the oxbows are converted into
swamps, but their identity is retained indefinitely.

The growth of alluvial meanders leaves distinctive
marks on the floodplain. As shown in Figure 19.24, the
channel in a meander bend is deep close to the outside or
undercut bank, which yields by caving and allows the
bend to grow in radius. Here the channel has its greatest
depth, forming a pool. As flow passes from one bend to
the next, the threads of swiftest current cross the channel
diagonally in a zone known as the "crossing." This
element of the channel, which is shallow and has many
shifting bars, constitutes a riffle. Thus pools and riffles
occur in alternation, corresponding with each meander

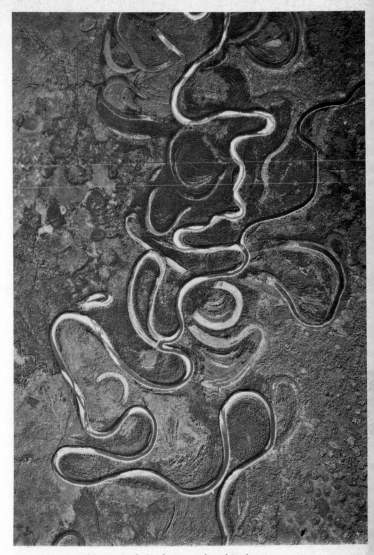

Figure 19.23   This vertical air photograph, taken from an
altitude of about 6 km (20,000 ft), shows meanders,
cutoffs, oxbow lakes and swamps, and floodplain of the
Hay River in Alberta. (Department of Energy, Mines and
Resources, Canada.)

bend. The meander bend not only grows laterally, but also
shifts slowly downvalley in a migratory movement known
as *downvalley sweep.* Figure 19.25 shows both meander
growth and sweep on the Mississippi River over a
century's time. The combined effect of meander growth
and sweep gives to the point-bar deposits nested arcuate
patterns consisting of bars (embankments of bed material)
and swales (troughs between bars). **Bar-and-swale** topography thus produced is clearly visible in Figure 19.23.

During periods of overbank flooding, when the entire
floodplain is inundated, water spreads from the main
channel over adjacent floodplain deposits. As the current
rapidly slackens, sand and silt are deposited in a zone
adjacent to the channel. The result is an accumulation
known as a **natural levee.** Because deposition is heavier
closest to the channel and decreases away from the
channel, the levee surface slopes away from the channel

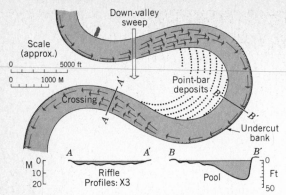

Figure 19.24 **An idealized map and cross-profiles of a meander bend of a large alluvial river, such as the lower Mississippi River. Small arrows show position of the swiftest current. (From A. N. Strahler, 1971, *The Earth Sciences,* 2nd ed., Harper and Row, New York.)**

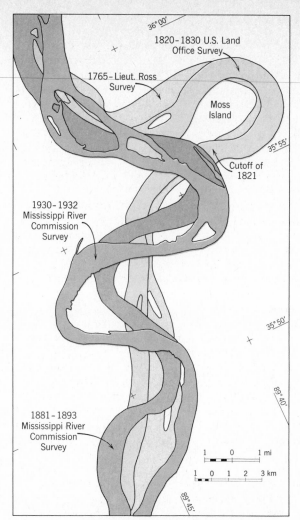

Figure 19.25 **A series of four surveys of the Mississippi River shows considerable changes in the position of the channel and the form of the meander bends. Note that one meander cutoff has occurred (1821) and new bends are being formed. (After U.S. Army Corps of Engineers.)**

(Figure 19.22 and Plate H.3). In times of overbank flooding, the higher ground of the natural levees is often revealed by a line of trees on either side of the channel. Between the levees and the bluffs is lower ground, the **backswamp** (Figure 19.22 and Plate H.3).

After a flood wave has passed, the water from the backswamps drains back into the river by means of smaller streams that flow downvalley, parallel with the main channel (see Figure 19.22). These parallel streams are prevented from joining the large stream because of the natural levees. At some downvalley point where the main channel is undercutting the floodplain bluffs, the parallel tributary is forced into a junction with the main channel.

A famous example is the Tallahatchie-Yazoo River system in Mississippi. For a distance of about 280 km (175 mi), this stream flows parallel with the Mississippi River, to make its junction at Vicksburg, where the big river is cutting into the eastern bluffs of the floodplain (Plate H.3). Many years ago, a leading geographer selected the Tallahatchie-Yazoo as the type example of a floodplain stream with its junction deferred far downvalley by the presence of a natural levee; he named it a **yazoo stream**. This quaint term is still widely used.

Overbank flooding results not only in the deposition of a thin layer of silt on the floodplain, but also brings an infusion of dissolved mineral substances that enter the soil. As a result of the resupply of nutrient base cations, floodplain soils retain their remarkable fertility in regions of soil-water surplus from which these bases are normally leached away. Important areas of Inceptisols in the Mississippi River floodplain and the lowland of the Ganges River maintain a high base status by this process.

Alluvial Rivers and Man

Alluvial rivers attracted Man's habitations long before the dawn of recorded history. Early civilizations arose in the period 4000 to 2000 B.C. in alluvial valleys of the Nile River in Egypt, the Tigris and Euphrates Rivers in Mesopotamia, the Indus River in what is now Pakistan,

and the Yellow River in China. Fertile and easily cultivated alluvial soils, situated close to rivers of reliable flow, from which irrigation water was easily lifted or diverted, led to intense utilization of these fluvial zones. With respect to human culture, the alluvial river and its floodplain are sometimes referred to as the *riverine environment.* Today about half the world's population lives in southern and southeastern Asia; most of these persons are small farmers who cultivate alluvial soils of seven great river floodplains.

To the agricultural advantages of alluvial zones is added the value of the rivers themselves as arteries of transportation. Navigability of large alluvial rivers led to the growth of towns and cities, many situated at the outsides of meander bends where deep water lies close to the bank, or at points on the floodplain bluffs where the river is close by (e.g., Memphis, Tennessee, and Vicksburg, Mississippi).

Flood-Abatement Measures

In the face of repeated disastrous floods, vast sums of money have been spent on a variety of measures to reduce flood hazards on the floodplains of alluvial rivers. The economic, social, and political aspects of flood abatement are beyond the scope of physical geography; we shall review only the engineering principles applied to the problem. Two basic forms of regulation are to (1) detain and delay runoff by various means on the ground surfaces and in smaller tributaries of the watershed and (2) modify the lower reaches of the river where floodplain inundation is expected.

The first form of regulation aims at treatment of watershed slopes, usually by reforestation or planting of other vegetative cover to increase the amount of infiltration and reduce the rate of overland flow. This type of treatment, together with construction of many flood-storage dams in the valley bottoms, can greatly reduce the flood crests and allow the discharge to pass into the main stream over a longer period of time.

Under the second type of flood control, designed to protect the floodplain areas directly, two quite different theories can be practiced. First, the building of **artifical levees,** parallel with the river channel on both sides, can function to contain the overbank flow and prevent inundation of the adjacent floodplain (Figure 19.26). Artificial levees, also known as *dikes,* are broad embankments built of earth; they must be high enough to contain the greatest floods, or they will be breached rapidly by great gaps at the points where water spills over. Under the control of the Mississippi River Commission, which was created in 1879, a vast system of levees was built along the Mississippi River in the expectation of containing all floods. Levees have been continuously improved and now total more than 4000 km (2500 mi) in length and in places are as high as 10 m (30 ft).

Because of natural and artificial levees, the river channel is slowly built up to a higher level than the floodplain; this makes these "bottomlands," as they are called, subject to repeated inundations and consequent heavy loss of life and property. Floodplains of many of the world's great rivers present serious problems of this type. In China in 1887, the Hwang Ho inundated an area of 130,000 sq km (50,000 sq mi), causing the direct death of a million people and the indirect death of a still greater number through ensuing famine.

The second theory, practiced in more recent years on the Mississippi River by the U.S. Army Corps of Engineers, is to shorten the river course by cutting channels directly across the great meander loops to provide a more direct river flow. Shortening has the effect of increasing the river slope, which in turn increases the mean velocity. Greater velocity enables a given flood discharge to be moved through a channel of smaller cross-sectional area; the flood stage is correspondingly reduced. Channel improvement, begun in the 1930s, initially had a measurable effect in reducing flood crests along the lower Mississippi, and the levees were not in such great danger of being overtopped. New meander bends grew rapidly and proved difficult to control, however.

Figure 19.26 **This air view, photographed in April 1952, shows a break in the artifical levee adjacent to the Missouri River in western Iowa. Water is spilling from the high river level at right to the lower floodplain level at left. (U.S. Department of Agriculture.)**

Certain parts of the floodplain are also set aside as temporary basins into which the river is to be diverted according to plan to reduce the flood crest. In the delta region, floodways are designed to conduct flow from river channels to the ocean by alternate routes.

The flood abatement measures we have described constitute a structural approach to the problem of coping with flood hazards. An alternative lies in a nonstructural approach that recognizes floods as natural phenomena and accepts the fact that even the most elaborate and costly engineering structures may not prevent flood damage. Nonstructural measures include zoning the floodplain for uses that can accommodate occasional floods (e.g., agricultural use and recreational use) without exorbitant financial loss and danger to human life. Under wise floodplain zoning, new urban housing and industry are kept out of flood-prone areas.

Unfortunately, the history of settlement of our nation has been such that flood-prone areas have attracted towns and cities and the industrial plants and highways that go with them. Cheap river transport was, of course, a potent factor in encouraging the early growth of these urban concentrations close to riverbanks on low floodplain sites. A river not only supplied large amounts of water needed for domestic use and industry, but also served as a convenient flow system by means of which sewage and industrial wastes could be disposed of cheaply.

Continued urban and industrial development on floodplains of major rivers began to be questioned in the 1930s. In 1937 the *Engineering News-Record* posed this

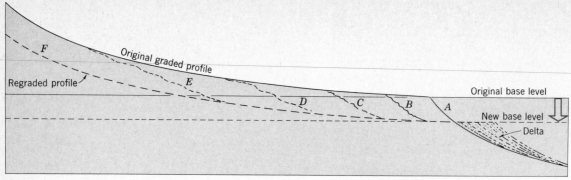

Figure 19.27 A drop in base level brings on rejuvenation and regrading of the stream profile, starting at point *A* and progressing upstream. (From A. N. Strahler, 1971, *The Earth Sciences,* 2nd ed., Harper and Row, New York.)

question: "Is it sound economics to let such property be damaged year after year, to rescue and take care of the occupants, to spend millions for their local protection, when a slight shift of location would assure safety?" A new, broader outlook on the relationship of Man to floods was presented in 1945 by a geographer, Professor Gilbert White, in a treatise titled *Human Adjustment to Floods.* Since that time, geographers have played a leading role in analyzing problems of flood hazards and in suggesting broad policies for Man's adaptation to floods.

Stream Rejuvenation and Entrenched Meanders

A graded stream experiences a major disruption in activity when the crust beneath it is raised with respect to sea level. During and following crustal uplift, the stream undergoes channel degradation in an attempt to reestablish grade at a lower level. This process, termed **rejuvenation,** begins as a series of rapids near the mouth, where the flow passes from the former mouth down to the lowered sea level. The rapids quickly shift upstream, and soon the entire stream valley is being trenched to form a new valley (Figure 19.27).

If rejuvenation affects an alluvial river with a floodplain, the effect is to give a steep-walled inner gorge. On either side lies the former floodplain, now a flat terrace high above river level (Figure 19.28). This feature is called a **rock terrace** to distinguish it from an alluvial terrace.

Rejuvenation may cause the meanders to become impressed into the bedrock and give the inner gorge a meandering pattern. These sinuous bends are termed

entrenched meanders to distinguish them from the floodplain meanders of an alluvial river (Figure 19.29).

Although entrenched meanders are not free to shift about as floodplain meanders, they can enlarge slowly so as to produce cutoffs. Cutoff of an entrenched meander leaves a high, round hill largely surrounded by the deep abandoned river channel and separated from the valley wall by the shortened river course (see rear part of Figure 19.30). As you might guess, such hills formed ideal natural fortifications. Many European fortresses of the Middle Ages were built on such cutoff meander spurs. A good example is Verdun, near the Meuse River.

Under unusual circumstances, where the bedrock includes a strong, massive sandstone formation, meander cutoff leaves a **natural bridge** formed by the narrow meander neck (Figure 19.30). One well-known example is Rainbow Bridge at Navajo Mountain, in southeastern Utah. Other fine examples can be seen in Natural Bridges National Monument at White Canyon in San Juan County, Utah.

Fluvial Processes in Review

Running water as a fluid agent of denudation dominates the sculpturing of the continents. We have examined fluvial systems in which overland flow becomes organized into channel systems, converging the flow into larger and larger streams. As streams erode, transport, and deposit rock material, they shape a variety of secondary landforms. Gorges and canyons dominate the earlier stages in valley evolution. These transient forms finally give way to broad floodplains on which meandering rivers move on

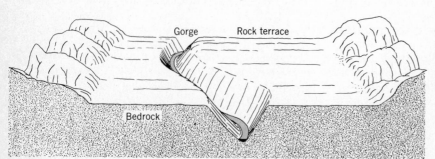

Figure 19.28 Following rejuvenation, a winding gorge has been carved into a former floodplain, which has become a high rock terrace. (Drawn by A. N. Strahler. From A. N. Strahler, 1971, *The Earth Sciences,* 2nd ed., Harper and Row, New York.)

Figure 19.29 The Goose-Necks of the San Juan River in Utah are entrenched river meanders in horizontal sedimentary strata. (Spence Air Photos.)

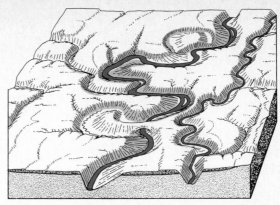

Figure 19.30 Rejuvenation of a meandering stream has produced entrenched meanders. One meander neck has been cut through, forming a natural bridge. (Drawn by E. Raisz.)

low gradients. More subtle landforms, such as the faint natural levees, now dominate the fluvial landscape. Overbank flooding and widespread inundation bring major environmental impact to these flat alluvial lands, where they are intensively occupied by Man. Efforts to control flooding by engineering works often fail, despite enormous inputs of money. Doubts are now raised as to the wisdom of continuing to protect lands naturally subject to river inundations.

The impact of Man on the action of running water is severe in agricultural and urban areas. Soil erosion is easily induced, but controlled only with great difficulty. Channels are damaged by aggradation from upstream sources of sediment. Reservoirs impounded by great dams are being steadily filled with sediment, and we have good reason to question the desirability of building more of these expensive works.

In this chapter, we have made a number of brief references to the role of climate in varying the intensity of fluvial processes. The control of climate over soil-forming processes and ecosystems has been stressed in earlier chapters. In the next chapter, we turn to a systematic analysis of the way denudation is influenced by climate.

References for Further Study

Powell, J. W. (1875), *Exploration of the Colorado River of the West and its tributaries,* Government Printing Office, Washington, D.C.

Lobeck, A. K. (1939), *Geomorphology,* McGraw-Hill Book Co., New York. See Chapters 5–7.

Leopold, L. B., M. G. Wolman, and J. P. Miller (1964), *Fluvial processes in geomorphology,* W. H. Freeman and Co., San Francisco.

Morisawa, M. (1968), *Streams; their dynamics and morphology,* McGraw-Hill Book Co., New York.

Smith, G. H., ed. (1971, *Conservation of natural resources,* fourth edition, John Wiley and Sons, New York. See Chapters 5, 13.

Chorley, R. J., ed. (1971), *Introduction to fluvial processes,* Methuen and Co., London. See Chapters 5, 7.

Review questions

1. Distinguish between initial and sequential landforms. What are fluvial landforms? What activities are included under fluvial processes? Distinguish between erosional and depositional landforms.
2. Compare the geologic norm of erosion with accelerated soil erosion. What causes accelerated soil erosion to set in? Describe the effects of splash erosion.
3. How does infiltration capacity change with time after the onset of heavy rain? Explain. What physical factors affect infiltration capacity? How does land use cause changes in infiltration capacity, and how do these changes affect susceptibility of the surface to erosion?
4. Define sediment yield. How do factors of vegetation cover and land use influence sediment yield? Give an example. What are badlands? Where and why do they occur?

5. Describe the various forms produced by accelerated erosion. In what ways is sedimentation, associated with accelerated erosion, detrimental to the quality of the land surface? How can accelerated soil erosion be arrested?
6. Describe the processes of stream erosion. Of what importance is bank caving? What forms are produced by abrasion of a bedrock channel?
7. In what three forms is stream load carried? Distinguish between suspended load and bed load. What mechanism causes particles to be suspended in the stream?
8. Describe the changes that a channel undergoes during passage of a flood wave. What is meant by stream capacity? How does capacity for bed load vary with changing velocity? How does the suspended load vary with changing discharge, as measured at a gauging station?

9. Using several major rivers as examples, show how climate, vegetation cover, and parent matter of the soil influence sediment yield and average annual sediment load. To what extent has Man been a factor in increasing sediment loads of rivers?

10. Explain the concept of stream equilibrium. Describe the activity of a graded stream. How are load and capacity related in a graded stream? What is the normal form of the longitudinal profile of a graded stream?

11. Describe the evolutionary stages of a stream from an ungraded initial condition to the graded state. What landforms are developed during this evolution? How does the relationship of capacity to load change with time? What landforms are developed by a graded stream?

12. What is meant by base level? Explain why the profile of a graded stream can never be reduced to base level throughout its length.

13. What environmental significance have river gorges and waterfalls? Describe and explain Niagara Falls. What conflicts of interest arise because of dam building and the utilization of river gorges?

14. What causes a stream to aggrade its channel? Describe a channel experiencing aggradation. For what reason might aggradation be followed by degradation? How are alluvial terraces formed? What environmental factors favor the use of terraces by Man?

15. In what ways is aggradation caused by Man's activities? How does urbanization promote aggradation and downstream sediment deposition?

16. Describe the landforms found on the floodplain of an alluvial river. Describe the change in cross-profile of a river from one meander bend to the next. What is downvalley sweep? What is bar-and-swale topography? How are natural levees formed? What is a yazoo stream?

17. For what reasons has Man been attracted to the riverine environment throughout history?

18. Describe the various forms of flood-abatement measures applied to control or reduce floods on large alluvial rivers. To what extent are such measures successful? Where and why do they fail? Discuss the nonstructural approach to flood hazards. What nonstructural adaptations are available?

19. What causes stream rejuvenation? What landforms are produced following rejuvenation? Distinguish between entrenched meanders and alluvial meanders.

Denudation and Climate

20

In the last three chapters we examined the processes of weathering, mass wasting, overland flow, and stream flow. These processes act in concert and cannot be considered in isolation. Fluvial landforms reflect the simultaneous action of all the contributing processes. In this chapter we turn to a wider outlook on fluvial denudation, broadening our focus to cover an entire region of subcontinental dimensions and emphasizing the landscape changes that take place over long spans of time.

The concept of a **cycle of denudation** has been strongly rooted in geographic thought, ever since its introduction in the late 1800s by a leading American geographer, William Morris Davis. The denudation cycle, as envisioned by Davis, has its initial stage at a time when the crust of a continent is rapidly elevated by internal earth forces. Previously, this portion of crust may have been the floor of a shallow inland sea or of a shallow continental shelf. Once raised high above sea level as a landmass, the fluvial agents begin their attack. Streams come into existence and evolve through the stages we examined in the last chapter, eventually becoming graded in their trunk sections. In a stage of youth, branching tributaries carve up the elevated landmass into a myriad of erosional landforms. When the landscape assumes a rugged, mountainous aspect, it has entered the stage of maturity.

As time passes, land surfaces are lowered in altitude and the slopes decline in steepness. As relief diminishes, thick regolith mantles the hillslopes and divides and the landscape assumes a subdued aspect — a stage of late maturity. Eventually relief diminishes to the point that only a low, undulating land surface remains. In this stage of old age, the land surface has become a **peneplain.** In coining this word, Professor Davis had in mind two words: "penultimate" and "plain." Realizing that erosion processes must act with increasing slowness in the latter stages of the denudation cycle, Davis postulated that the cycle would be terminated through disruption by a geologic event in which the land surface is abruptly lifted or submerged. If the peneplain is uplifted, a new cycle will be initiated.

We no longer use Davis's stage names — youth,

maturity, old age — because they imply a biologic model of growth stages in individuals of an animal species. We do, however, incorporate many of Davis's concepts into a useful physical model of landmass denudation:

Concept of the Available Landmass

That part of the continental crust lying above sea level, which is the base level for erosion by streams, constitutes the **available landmass.** It is a mass of bedrock subject to denudation and available for ultimate removal. As denudation takes place, however, the crust gradually and spontaneously rises, so the total available landmass is actually of a much greater thickness than appears at any one instant. To understand what happens, we must investigate a principle of geology.

Figure 20.1 is a schematic diagram showing the earth's outer shell of hard, brittle rock — the lithosphere — resting on a highly heated plastic rock layer — the asthenosphere. When we speak of the earth's crust, it is with reference to the uppermost layer of the lithosphere.

Figure 20.1 **Isostatic compensation for landmass erosion and sediment deposition. (From A. N. Strahler, 1971, *The Earth Sciences,* 2nd ed., Harper and Row, New York.)**

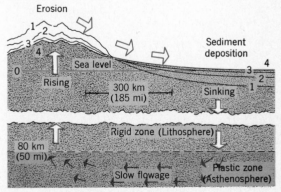

(Lithosphere and asthenosphere will be explained more fully in Chapter 24.) An important geologic principle, established for many decades, is that the lithosphere literally floats on the plastic asthenosphere, much as a layer of sea ice or an iceberg floats on denser seawater. In the case of an iceberg, if you were to remove the part showing above water the iceberg would rise and bring more ice above water level. If you placed an added load of ice on top of the iceberg, it would sink lower in the water and come to a new position of rest. This flotation principle, applied to the lithosphere, is called **isostasy.** The word is derived from Greek words: *isos,* "equal," and *stasis,* "standing."

As Figure 20.1 shows, when bedrock is eroded from a continent (left), the lithosphere rises beneath it; but the amount of rise is somewhat less than the thickness of material removed. When the eroded sediment accumulates in layers (right), the added load causes the lithosphere to sink lower; but the amount of sinking is somewhat less than the thickness of sediment accumulating. A rising or sinking of the crust in response to denudation or sediment accumulation is referred to as **isostatic compensation.** In the plastic zone below, a lateral movement of mass by very slow flowage accompanies the rising and sinking of the lithosphere. This activity occurs at a depth averaging about 80 km (50 mi), where rock is close to its melting point and has little strength to resist flowage.

Because of slow crustal uplift due to isostasy, the available landmass is actually several times greater than the mass lying above base level at any one instant. As a satisfactory working figure, we can say that the removal of 5000 m of rock would be accompanied by a compensating crustal uplift of 4000 m, so the net lowering of the land surface would amount to only 1000 m. In setting up a model denudation cycle, the net lowering figure must be used.

Rates of Tectonic Uplift

A geologic phenomenon quite different from crustal rise due to isostasy is a rapid upward movement associated with local belts of mountain-making. This form of crustal activity is described as tectonic activity, meaning processes that dislocate the crust by bending and fracturing (faulting). We shall explain tectonic processes in detail in Chapter 24.

Our model of denudation begins with an event of tectonic uplift. It is reasonable to assume that the rate of tectonic uplift is far more rapid than the rate of fluvial denudation. Based on direct geologic evidence, rates of recent tectonic uplift in various parts of the world are on the order of 4 to 12 m per 1000 years (15 to 40 ft/1000 yr). Using a rate of 6 m/1000 yr (20 ft/1000 yr), a mountain summit might rise from sea level to an elevation of 6 km (20,000 ft) in about one million years, assuming that none of the mass were lost by erosional reduction during the uplift. Even with erosion in progress during uplift, a full-sized mountain mass could be created in a time span vastly shorter than the time required for its reduction to a low plain by denudation.

Rates of Denudation

Geologists and hydrologists have made a number of attempts to estimate rates of continental denudation. One method is to take available data on the load of streams, combining both suspended load and solid load. Another form of data consists of the measured accumulations of sediment in reservoirs. As we showed in Chapter 19, sediment yield varies enormously from place to place, depending on climate, vegetation cover, and parent material of the soil. Relief is another major factor; some areas of high relief and steep slopes have enormously high rates of sediment yield compared with undulating plains under a cover of forest. We shall investigate these regional differences later in this chapter, in connection with the influences of climate on geomorphic processes.

For our model denudation cycle, which presupposes a moist climate with a substantial water surplus, we shall use an initial denudation rate comparable to that observed today in high mountains: 1 to 1.5 m/1000 yr (3 to 5 ft/1000 yr). Notice that this rate is only a small fraction of the assumed rate of tectonic uplift.

We reason that denudation rates are strongly influenced by average summit elevation of the landmass. The higher a mountain block rises, the steeper will be the gradients, on the average, of streams carving into that mass. Both intensity of stream abrasion and the ability of a stream to transport load (stream capacity) increase strongly as channel gradients increase. Valley wall slopes are also steeper, on the average, where stream gradients are steeper, so mass wasting and erosion by overland flow will also be more intense. Thus, as gradients diminish with time, the rate of denudation will diminish steadily from the high initial value we have assumed.

We anticipate that, by the time the landmass has been lowered to an average summit elevation of about 300 m (1000 ft), the denudation rate will have decreased to a value comparable to that found today for the eastern United States or the Amazon River basin: about 4 cm/1000 yr (0.2 ft/1000 yr). This rate is corrected for greatly increased erosion in the United States caused by Man's activities.

A Model Denudation System

Using the assumptions made thus far about rates of tectonic uplift, isostatic compensation, and denudation, we can devise a model of the denudation process and from it perhaps obtain some idea of the order of magnitude of time required to reduce the landmass to a peneplain. First, we image that a large crustal mass is raised as a single block from a previous position close to sea level. An arbitrary width of about 100 km (60 mi) is assigned to the uplifted block because this dimension is about the order of magnitude of width of a number of present-day mountain ranges, including the European Alps, Carpathians, Pyrenees, Caucasus, Alaska Range, Sierra Nevada, Northern Cascades, Rocky Mountains, and Appalachians. Length of the uplifted block is not important for landform analysis; a segment some tens of kilometers will do. As Figure 20.2 shows, the uplifted block (right) **Falls of the Yellowstone River, Yellowstone National Park, Wyoming.** The gorge is carved into volcanic rocks, colored through mineral alteration by rising hydrothermal solutions. (Orlo E. Childs)

(below) **The American Falls of Niagara Falls, New York.** A great mass of broken limestone lies at the base of the falls. (Orlo E. Childs)

(left) **Potholes worn in granite, Sand Creek, Wyoming.** (Arthur N. Strahler)

(below) **Grand Canyon of the Colorado River at Toroweap, Arizona.** The view is downstream, toward the west. Basaltic lavas have poured down the canyon walls at the right; a cinder cone lies at the left. The narrow gorge is about 3000 ft (900 m) deep at this point. (Donald L. Babenroth)

Gorges and falls   **Plate H.1**

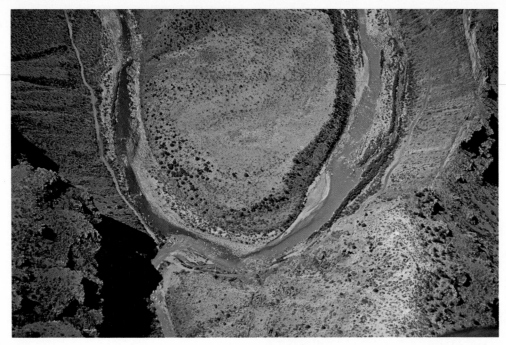

(left) **In this vertical air view of the gorge of the Salt River, Arizona, the turbid storm runoff of a small tributary contrasts sharply with the clear water of the main stream. (Mark A. Melton)**

(left) **Green irrigated areas of floodplain (right) and low alluvial terrace (left) border a small meandering river, fed from mountain snowmelt, near Greybull, Montana. (Arthur N. Strahler)**

(above) **Rounded cobbles of volcanic rock comprise the bed load of this ephemeral desert stream in Arizona. (Arthur N. Strahler)**

(right) **Deep branching gullies have carved up an overgrazed pasture near Shawnee, Oklahoma. Contour terracing and check dams have halted the headward growth of the gullies. (Mark A. Melton)**

Plate H.2 Stream channels

(above) **Part of an oxbow lake from an abandoned section of the Mississippi River near Mound Bayou, Louisiana. Soil color differences in the cultivated area reveal old bars and swales. (Arthur N. Strahler)**

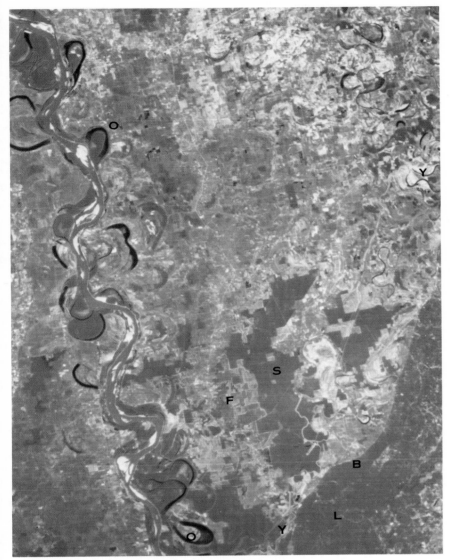

(above) Landsat image of the Yazoo Basin, north of Vicksburg, Mississippi, taken in late August, 1973. Blue represents water surfaces, such as the Mississippi River (left) and patches of floodwater (F) remaining from the great flood of March-April, 1973. Oxbow lakes (O) appear as dark blue crescents. Forested land forms solid red patches on backswamp areas (S) of the floodplain. The eastern bluffs (B) of the floodplain show as a sharp line, with forested upland underlain by loess (L) at the right. The Yazoo River (Y) follows the eastern edge of the floodplain. Sand bars in the Mississippi River channel show as pure white. Other white areas are cultivated fields with a large proportion of bare soil. Highways and railroad embankments form thin white lines. (NASA)

No. AS9-26A-3741   V—Vicksburg
O—oxbow lake   Y—Yazoo River
S—backswamp   B—bluffs
N—natural levee

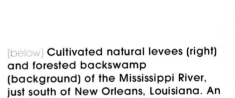

(below) **Cultivated natural levees (right) and forested backswamp (background) of the Mississippi River, just south of New Orleans, Louisiana. An artificial levee runs close to the river channel. (Orlo E. Childs)**

The Mississippi River   **Plate H.3**

(right) **This barren desert landscape near Death Valley, California, is dominated by knife-edged divides and V-shaped ravines carved by running water. (Mark A. Melton)**

(below) **Bedrock is exposed in places on this pediment, carved upon ancient schist. Fragments of schist form a desert pavement. Cabeza Prieta Range, Arizona. (Mark A. Melton)**

(left) **A flash flood fills a desert stream channel with raging, turbid waters. A distant thunderstorm produced the runoff. Coconino Plateau, Arizona. (Arthur N. Strahler)**

(left) **Badlands are sustained on weak clay formations where no protective plant cover can gain hold. Mancos formation, Henry Mountains region, Utah. (Alan H. Strahler)**

(above) **In this air view across the Harquehala Mountains, Arizona, broad rock pediments slope gently away from the mountains toward a central zone of alluvium. (Mark A. Melton)**

(left) **Great alluvial fans extend out upon the floor of Death Valley, California. (Mark A. Melton)**

Plate H.4   Desert fluvial processes

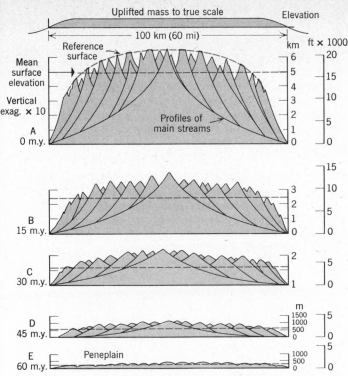

Figure 20.2 Schematic diagram of landmass denudation. In this model, the average surface elevation is reduced by one-half every 15 million years.

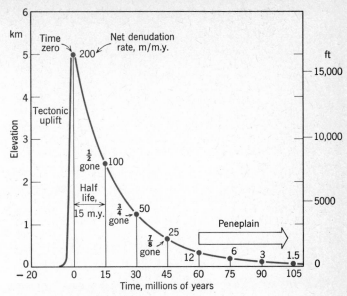

Figure 20.3 Graph of decrease in average surface elevation with time, as shown in Figure 20.2.

has a very broadly domed summit and steep sides; it is bordered by low areas, at or below sea level, that serve as receiving sites for sediment carried out from the mountain block by streams.

The initial surface or reference, a plain close to sea level, is raised rapidly to an elevation of 6 km (20,000 ft); see Figure 20.2. Although this initial surface is carved up by streams during uplift, its position in space is shown by a dashed line in cross section (A); it is labeled "reference surface." The rapid tectonic uplift is shown on a graph of elevation versus time, Figure 20.3, by a steeply rising curve. We have allowed 5 million years (5 m.y.) for uplift, but most of this occurs within a 2-m.y. period. Uplift tapers off in rate and ceases at a point labeled "time-zero" on the horizontal scale of the graph.

Denudation in progress during uplift increased sharply in intensity as average elevation increased so that, at time-zero, the mass has already been carved into a maze of steep-walled gorges, organized into a fluvial system of steep-gradient streams. The profile of the rugged mountain mass and the main stream system are suggested in greatly exaggerated scale in Stage A of Figure 20.2. At time-zero the average elevation of the eroded surface lies at 5 km (16,400 ft). Thus about 1 km (3300 ft) of rock has been removed during uplift.

Starting at time-zero, an average denudation rate of 1 m/1000 yr is assumed for the entire surface. An additional assumption is that isostatic compensation occurs continuously so that the net lowering rate is only one-fifth of the

denudation rate, or 0.2 m (20 cm)/1000 yr. In million-year units, used on the graph, this net rate is 200 m/m.y. (about 650 ft/m.y.). With decreasing elevation, the rate of denudation itself diminishes in a constant ratio such that one-half the available landmass is removed in each 15-m.y. period. We may call this time unit the **half-life** of the available mass. After a lapse of 15 m.y. (one half-life), the average elevation will have been reduced to 2.5 km (8200 ft), and the net rate of lowering will have fallen to 100 m/m.y. At the end of 30 m.y. (two half-lives), average elevation and denudation rate are again reduced by one-half: to one-quarter of the initial values. Now three-quarters of the mass is gone and the average altitude is down to 1.25 km (4100 ft).

The program of decrease in elevation of landmass with time, described here, is known as **exponential decay**. This is a common program in natural and physical science. For example, it applies to the rate at which a radioactive element decays spontaneously by radioactivity. Exponential decay is exactly the inverse of a program of exponential increase seen in an annual percentage growth rate, such as a savings account at compound interest, a gross national product (GNP), or a percentage annual population increase. The exponential decay rate, as applied to the denudation of a landmass, simply reflects the principle that the energy available to perform the work of denudation is proportionate to the height of a mass above sea level. Therefore, as average elevation declines, the intensity of change is proportionately diminished.

In our model denudation program, an average elevation of about 300 m (1000 ft) is reached after about 60 m.y. By now, the rate of net lowering has fallen to 12 m/m.y. or 1.2 cm/1000 yr. The land surface can now be described as a peneplain. No exact definition of a peneplain exists in terms of average elevation or relief. We can only say that hillslopes are very gentle, divides are very broadly rounded, and the region has the aspect of an undulating plain. Peneplain reduction continues to follow the exponential decay schedule, which we have shown carried out to seven half-lives, or 105 m.y. Most likely, crustal stability would not endure nearly this long and the simple denudation curve would be terminated abruptly by another tectonic event.

Denudation and Spasmodic Uplift

Our simple model of denudation needs an important modification to bring it closer to reality, as judged from the evidence of geologic history. We have assumed that isostatic compensation occurs uniformly and constantly as denudation proceeds, but this is probably not the case. It is more reasonable to suppose that isostatic compensation occurs spasmodically. Because the lithosphere is a massive, strong layer, it resists being lifted until a certain minimum thickness of rock has been removed. Then, beyond the critical point of strength, any further removal triggers a rapid uplift, restoring the equilibrium.

We must now build a program of spasmodic isostatic compensation into our denudation model. Figure 20.4 shows this program. As each critical thickness of mass is removed, an isostatic uplift occurs. Uplift is shown as instantaneous and is followed by denudation on the exponential decay schedule. The result is a sawtooth curve. It has been plotted on the assumption that full isostatic compensation occurs abruptly after 300 m (1000 ft) of rock has been removed. The uplift is, however, only 80 percent of the amount removed, or 240 m (800 ft).

Our revised model shows that isostatic uplifts are triggered at short time intervals during the early stages of denudation, when gradients are steep and relief is great. Because the abrupt changes in elevation are only a small relative fraction of the total elevation, there will be no observable changes in the profiles of the slopes and stream channels. In contrast, the effects of spasmodic uplift are very conspicuous in the part of the graph representing the peneplain. Each sudden isostatic uplift brings a great relative increase in the available landmass. Depending on the average elevation at which this uplift occurs, the available landmass may be suddenly doubled or tripled in thickness.

Figure 20.5 **Stages in the denudation cycle caused by uplift of a peneplain and a rejuvenation of the stream system. Another uplift of the peneplain, shown in Block E, will bring a return to conditions shown in Block A and the cycle will be repeated. (Drawn by A. N. Strahler.)**

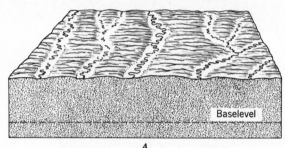

A

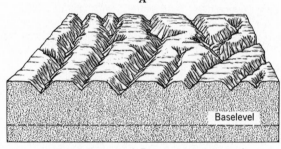

B

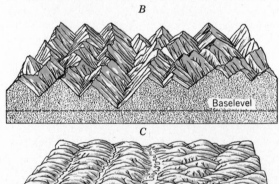

C

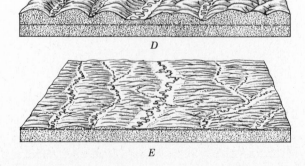

D

Figure 20.4 **Repeated sudden uplifts because of isostatic compensation modify the curve shown in Figure 20.3.**

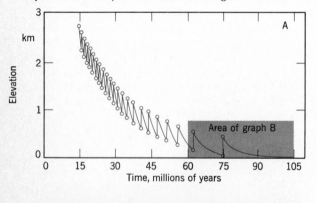

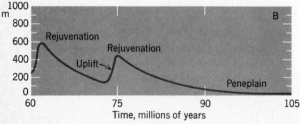

E

Figure 20.6 **The gently rolling upland surface in the distance is part of the St. John peneplain; its elevation is about 600 m (2000 ft). The peneplain is deeply trenched by Canyon de San Cristobal (foregound), carved by the Rio Usabon, Puerto Rico. (U.S. Geological Survey.)**

To make our denudation model more realistic, the sawtooth graph must be softened by inserting the effects of denudation during the period of rapid isostatic uplift. In any case, the isostatic uplift will not be instantaneous, and it will taper off in rate toward the end of the uplift period. A portion of the revised denudation curve is shown in Graph *B* of Figure 20.4.

Landmass Rejuvenation

A series of block diagrams, Figure 20.5, shows the evolution of the landscape following uplift. Block *A* shows the elevated peneplain still intact because the small area shown lies in an inland location, far from the coastline where changes begin. What follows is **landmass rejuvenation.** (We explained stream rejuvenation in Chapter 19.) The trenching of previously graded streams begins near their mouths and progresses up the valleys, finally reaching the area of our block diagram. A system of

narrow, steep-walled gorges spreads throughout the landmass (Block *B*). The peneplain is now dismembered into isolated upland patches (Figure 20.6). The landscape is in the stage that Professor Davis regarded as "youthful." Steep valley-wall slopes continue to increase in height and number; they gradually replace the last peneplain remnants. In the stage shown in Block *C,* an "early-mature stage," the landscape is extremely rugged (Figure 20.7). Ridges are sharp-crested; little or no flat land can be found anywhere. Relief has reached its maximum in this stage. By this time, the principal streams have again attained the condition of grade, so further lowering of the valley floors proceeds very slowly. Floodplains begin to appear. As time passes, relief diminishes over the area, and divides are reduced in elevation and become broadly rounded (Block *D*). Regolith accumulates in greater thickness on divides and hillslopes. Erosion rates become slower, and sediment supplied to the streams becomes smaller in quantity and finer in texture. Floodplains widen and the major streams

Figure 20.7 **A great plutonic mass of granite, exposed in the Sawtooth Mountains of south central Idaho. View is northward. (From *Geology Illustrated,* by John S. Shelton, W. H. Freeman and Company, © 1966.)**

take on the characteristics of alluvial rivers. The subdued landscape shown in Block *D* would, in Professor Davis's language, be in the "late mature stage." Then, with the passage of many tens of millions of years, a peneplain is again formed with respect to the stable base level (Block *E*). This event marks the "old stage" in the denudation cycle. At this point in time, another isostatic uplift is triggered off and the peneplain is uplifted, bringing conditions back to those shown in Block *A*.

In Chapter 26 we shall return to this model of rejuvenation by spasmodic isostatic uplift. It is typical of large interior regions of the continents, far removed from those belts of tectonic activity in which mountains are created by volcanic activity and severe crustal deformation.

The landform evolution we have described is placed in the setting of a moist climate having a substantial annual water surplus and little or no soil-water shortage at any time of year. Streams receive important water contributions from interflow and ground-water seepage, so the larger streams usually are sustained in flow throughout the entire year. These conditions are typical of moist subtropical and continental climates of midlatitudes, as well as the wet equatorial and trade-wind littoral climates. Associated with these climates is the production of thick regolith by chemical weathering beneath a forest cover. The well-protected hillslopes yield little coarse sediment, even during periods of heavy rainfall. Thus the streams carry most of their load as suspended fine sediment and as dissolved matter (ions). Although some of the silt accumulates on floodplains during overbank floods, most of the suspended sediment is carried to the ocean, where it becomes marine sediment. Thus the products of denudation are largely exported from the region. Let us now compare these environmental conditions with those of dry climates.

Fluvial Processes in an Arid Climate

The general appearance of a desert landscape is strikingly different from that of a humid region, reflecting differences in both vegetation and landforms. Torrential rainfalls occur in dry as well as in moist climates, and most landforms of desert regions are formed by running water. A particular locality in a dry desert may experience heavy rain only once in several years; but, when it does fall, stream channels carry water and perform important work as agents of erosion, transportation, and deposition. Although running water is a rather rare phenomenon in dry deserts, it works with more spectacular effectiveness on the few occasions when it does act. This is explained by the meagerness of vegetation in dry deserts. The few small plants that survive offer little or no protection to soil or bedrock. Without a thick plant cover to protect the ground and hold back the swift downslope flow of water, large quantities of sediment are swept into the streams. A dry channel is transformed in a few minutes into a raging flood of muddy water, heavily charged with rock fragments (Plate H.4). The bedload left behind when the flood has passed consists of coarse, bouldery debris, heaped into massive channel bars (Figure 20.8).

Recall from Chapter 18 that, in a dry climate, streams

Figure 20.8 **A boulder bar, built during a brief desert flood. Gonzales Pass Canyon, Arizona. (Mark A. Melton.)**

flowing across thick accumulations of permeable alluvium are of the influent class (see Figure 18.10). Stream water infiltrates the bed and contributes to a ground-water mound beneath. Because of loss of discharge through influent seepage and direct surface evaporation, streams of arid lands usually show braided channels and are actively engaged in aggradation. These streams are often short and disappear on alluvial deposits or end on shallow, dry lake floors.

Alluvial Fans

A very common landform built by braided, aggrading streams is the **alluvial fan,** a low cone of alluvial sands and gravels resembling in outline an open Japanese fan (Figure 20.9). The apex, or central point of the fan, lies at the mouth of a canyon or ravine. The fan is built out on an adjacent plain. Alluvial fans are of many sizes; some desert fans are many kilometers across (Plate H.4).

Fan-building streams carry heavy bed loads of coarse rock waste from a mountain or upland region. The braided channel shifts constantly, but its position is firmly fixed at the canyon mouth. The lower part of the channel, below the apex, sweeps back and forth. This activity accounts for the semicircular fan form and the downward slope in all radial directions from the apex.

Ground Water and Alluvial Fans

Large, complex alluvial fans also include mudflows (see Figure 17.18). As shown in Figure 20.10, mud layers are interbedded with sand and gravel layers. Water infiltrates the fan at its head, making its way to lower levels along sand layers that serve as aquifers. The mudflow layers act as barriers, or aquicludes. Ground water trapped beneath an aquiclude is under hydraulic pressure from water accumulated higher in the fan apex. When a well is drilled into the lower slopes of the fan, water rises spontaneously as artesian flow.

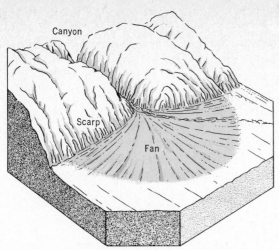

Figure 20.9 **A simple alluvial fan. (Drawn by A. N. Strahler.)**

Alluvial fans are the dominant class of ground-water reservoirs in the southwestern United States. Sustained heavy pumping of these reserves for irrigation has lowered the water table severely in many fan areas. The natural rate of recharge is extremely slow in comparison. However, efforts are made to increase this recharge by means of water-spreading structures and infiltrating basins on the fan surfaces.

A serious side effect of excessive ground water withdrawal is that of subsidence of the ground surface. Important examples of subsidence are found in alluvial valleys filled to great depth with alluvial fan and lake-deposited sediments. For example, in one locality in the San Joaquin Valley of California, the water table has been drawn down over 30 m (100 ft). The resulting ground subsidence has amounted to about 3 m (10 ft) over a 35-year period. Subsidence produces cracks in pavements and buildings and may dislocate wells, pipelines, and water drains.

The Landscape of Mountainous Deserts

Where tectonic activity has been active in an area of continental desert, the assemblage of fluvial landforms is

particularly diverse. The Basin and Range Province of the western United States is such an area; it includes large parts of Nevada and Utah, southeastern California, southern Arizona and New Mexico, and adjacent parts of Mexico. Between uplifted and tilted crustal blocks are down-dropped tectonic basins.

Under internal earth forces, the brittle crust becomes dislocated along faults — breakage surfaces on which motion occurs. (In Chapters 24 and 25, faulting is explained and fitted into the global pattern of crustal movements.) Block *A* of Figure 20.11 shows two lifted fault blocks with a down-dropped block between. Although denudation acts on the uplifted blocks as they are being raised, we have shown them as very little modified at the time tectonic activity has ceased. Our model landscape thus has a very great local relief; that is, there is a very great elevation difference between block summits and the nearby basin floors. This elevation difference might be on the order of 2 to 3 km (6,500 to 10,000 ft). At first, the faces of the fault block are extremely steep. They are scored with deep ravines. Many cones of talus have been built along the base of each block.

Block *B* shows a later stage of denudation in which streams have carved up the mountain blocks to produce a rugged terrain of deep, steep-walled canyons and high, sharp divides. Plate H.4 shows this type of landscape as it appears today along the western side of Death Valley, California. Large alluvial fans have been built side by side; the fan deposits form a continuous apron extending far out into the basin (Figure 20.12).

In a central position in the basins lie **playas,** the flat floors of lakes that are dry much of the time or contain only shallow lakes. Fine silts and clays lie under the playa surfaces, along with layers of soluble salts. (These salts were described in Chapter 11 as evaporites.) Soils of fans and playas belong to the order of Aridisols (Chapter 14). Salt lakes of playas represent the terminal sites of stream flow. In many cases, large alluvial fans, built across the entire basin, serve as barriers between individual depressions in the basin floors. Because evaporation greatly exceeds precipitation, these lakes do not rise above the limits of the closed depressions in which they lie.

As the mountain masses are lowered in height and reduced in extent, the fan slopes encroach on the mountain bases. Close to the receding mountain front,

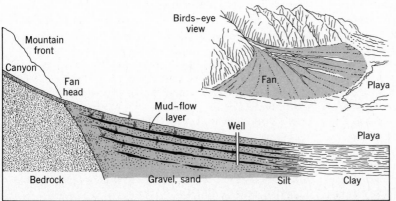

Figure 20.10 **An idealized cross section of an alluvial fan showing mudflow layers (aquicludes) interbedded with sand layers (aquifers). (From A. N. Strahler, 1972, *Planet Earth*, Harper and Row, New York.)**

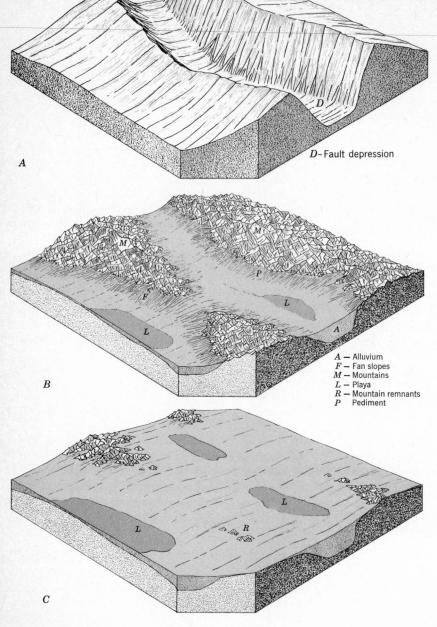

Figure 20.11 **Stages in the denudation of mountain blocks in a desert environment. (Drawn by A. N. Strahler.)**

D- Fault depression

A — Alluvium
F — Fan slopes
M — Mountains
L — Playa
R — Mountain remnants
P Pediment

erosion processes develop a sloping rock floor called a **pediment** (Figure 20.13 and Plate H.4). As the remaining mountains continue to shrink in extent, the pediments widen to form broad rock platforms thinly veneered with alluvium. In the deserts of the southwestern United States and northern Mexico, the graded slope extending from mountain base to playa is called a **bajada.** The term includes both alluvial fan and pediment surfaces.

Block C of Figure 20.11 shows an advanced stage of denudation in the desert environment. The once lofty mountains are reduced to small remnants, but these retain their steep slopes and sharp ridge crests. A small mountain mass, isolated from the main body of the range and completely surrounded by pediment or fan surfaces is an **inselberg** (German term meaning "island mountain").

An advanced stage of denudation, comparable to that shown in Block C of Figure 20.11, is shown in the side-looking airborne radar (SLAR) image of Figure 4.19. A remnant mountain mass with several inselbergs around it is seen at the left. Irrigated Aridisols on the alluvial fan slopes occupy the central part of the frame.

Throughout the denudation stages we have described, fluvial processes are limited to local transport of rock particles from a mountain range to the nearest basin. The basin floor, which receives all the sediment, must be gradually rising in elevation as the mountains are diminishing in elevation. Because there is no outflow to the sea, the concept of a base level of denudation has no meaning and a peneplain does not represent the penultimate stage of denudation, as it does in the humid

Figure 20.12   **This air view of Death Valley, California, shows a desert landscape comparable to that shown in Figure 20.11*B*. (Spence Air Photos.)**

environment. Each arid basin becomes a closed system so far as mass transport is involved. Only the hydrologic system is open, with water entering as precipitation and leaving as evaporation. The desert land surface produced in an advanced stage of fluvial activity is an undulating plain, called a **pediplain.** As an idealized landscape, it consists of areas of pediment merging with surrounding alluvial fan and playa surfaces.

Hillslope Evolution in Humid and Arid Climates

In describing the evolution of landscapes in both humid and arid environments, we have indicated two quite different modes of change in the forms of land slopes. Collectively, all land surfaces contributing runoff and sediment to stream channels are referred to as **hillslopes,**

Figure 20.13   **Bedrock is widely exposed over this eroded pediment at the foot of the Dragoon Mountains, near Benson, Arizona. (Douglas Johnson.)**

whether these be land surfaces of high mountains or low hills, or the walls of canyons. An individual element of a hillslope begins at a point on a drainage divide, or watershed, and follows the path of overland flow to the nearest stream channel. We visualize this hillslope element as a narrow strip to which a unit width is assigned — one meter or one foot, for example. A **hillslope profile** is a plot of elevation versus horizontal distance, following the midline of the hillslope element.

Hillslope profiles show a great variety of shapes. The meaning of each shape and the way it changes with time are subject to interpretation. Alternative possibilities are strongly debated among geomorphologists. One difficulty in drawing firm conclusions about the evolution of hillslopes is that they change so very slowly in terms of the human life span. Systematic, evolutionary changes must be inferred rather than observed.

We present two contrasting models of hillslope evolution, one for moist climates in low and middle latitudes, the other for arid climates in tropical and midlatitude zones. These models agree rather well with slopes observed in regions underlain by bedrock that is uniform in resistance to weathering and erosion. For a simple model, we can assume that the bedrock is a plutonic mass, such as granite or diorite, and that it is fractured into small blocks of uniform dimensions by three sets of uniformly spaced joints. Our ideal rock mass is therefore homogeneous in its physical and chemical composition.

Figure 20.14 is a schematic diagram of hillslope evolution as denudation progresses in a humid climate (left) and in an arid climate (right). The hillslope profile ends in a stream channel, which is assumed to be undergoing a continuous lowering of elevation. For the humid climate, the profile is shown with a rounded summit. Here the effects of rain-splash erosion and soil creep are dominant and tend to produce an upwardly convex profile. The remainder of the profile is straight to the point of junction with the channel. On this straight, steep slope, denudation proceeds by a combination of processes, including overland flow and soil creep. Material is removed both as sediment and as dissolved solids (ions). Rapid channel lowering ensures the removal of all material brought down the slope and maintains the steepness of the profile. For some time, the hillslope may be uniformly lowered, as shown by profiles a and b, which are parallel lines. As channel lowering slows with time, the rounded summit is expanded in area, while the lower portion of the profile declines in angle (profiles c and d). Relief is diminishing with time. Now an up-concavity begins to form at the base of the slope (profile e). In some cases a deposit of colluvium forms here. The slope profile is now S-shaped in outline and can be described as a sigmoid profile. The sigmoid form persists to the peneplain stage (profile f).

In the arid environment (Figure 20.14, right) the initial profile form is assumed to be steep and straight from the base to a sharp-edged summit. Throughout all stages of denudation, the angle of this steep profile element remains constant; but the slope base retreats toward the divide, as shown in successive profiles. The slope is mantled with a thin layer of coarse, bouldery regolith and

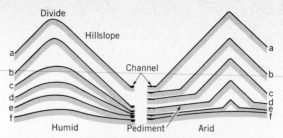

Figure 20.14 **Hillslope evolution with decreasing relief in humid and arid climates. (From A. N. Strahler, 1971, *The Earth Sciences*, 2nd ed., Harper and Row, New York.)**

bedrock outcrops are numerous. From the channel, a pediment begins to form at an early stage (profile b). The pediment widens as the base of the steep element retreats. The sharp break between pediment and mountain front is a characteristic feature of the desert of the southwestern United States and northern Mexico; it is visible in the background of Figure 20.13. Profile e of Figure 20.14 represents an inselberg surrounded by a broad pediment. The steep slopes disappear entirely when the inselberg is removed, as shown in profile f; now only a pediment remains.

In summarizing the two idealized programs of slope evolution, we can make this generalization: Hillslopes evolving in humid climates follow a program of **reclining retreat,** in which the entire profile between divide and channel diminishes in angle with time. Slopes evolving in an arid climate show **parallel retreat,** in which the initial slope angle is retained but the base of the steep element retreats from the stream channel and is replaced by a pediment.

Keep in mind that the concepts we have presented are generalizations and that differences of opinion exist as to many details of slope evolution. Few large areas of the continents are underlain by homogeneous rock, as our model required. Highly varied bedrock structures are usually present locally, and these features exert strong controls over slope form, as we will explain in Chapter 26. Rejuvenation can be expected to produce complex slope profiles in humid climates, as a new generation of steeper slopes replaces an older generation of gentler slopes. There is also the possibility that a climate change has occurred so that, for example, slope forms of an earlier arid environment are undergoing modification in a humid environment.

Our slope models do not apply closely to landscapes found in the tropical wet-dry climates or to landscapes of the arctic tundra climate. This observation leads to the conclusion that several models may be needed to encompass the full range of denudational landscapes found over the globe.

Climate and Landscape Regions

For decades, some geomorphologists have harbored a vision of a world system of several basic landscape types, each type uniquely dictated by climate and each clearly different from the others in the assemblage of landforms present. World maps have been drawn to show **climogenetic regions** (or *morphogenetic regions*). These maps appear to be closely correlated with climate systems and are little more than world climate maps with new sets of labels. If one were to travel through all parts of a single climogenetic region, the landscape would actually prove greatly different from place to place and from continent to continent.

Obviously, climate is not the sole factor determining landform characteristics. The climogenetic concept omits from consideration the factor of geologic control acting through differences in bedrock composition and structure and through the geologic history of the region in terms of volcanic and tectonic activity. In short, although climate is a potent control over landform development, as we have already established, it is but one of two basic controls of landform development; the other is geology. We shall develop the factor of geologic control in Chapters 25 and 26.

Climate-Process Systems

The role of climate in shaping landforms can be expressed through recognition of **climate-process systems.** Each system comprises a unique combination of levels of intensity of several basic denudation processes. The major process categories are weathering, mass wasting, fluvial processes, and wind action. The action of waves and currents in shaping coastlines is usually omitted from consideration because it affects only a thin line bordering the continents. Glacial processes are, of course, under direct climate control, but the landforms made beneath glacial ice by erosion and deposition are best considered separately. (Chapter 21 covers this subject.) Processes of weathering and mass wasting acting in the cold environment close to bodies of glacial ice comprise a **periglacial** climate-process system.

Two of the important climate-process systems have been examined in detail earlier in this chapter: those of moist climates and those of desert climates. Other climate-process systems usually recognized are identified with the wet-dry tropical climate, the semiarid (steppe) subtype of the dry climates, and the tundra climate.

Within each climate-process system, certain landforms can be expected to occur in abundance. For example, the alluvial fan and the pediment are conspicuous landforms in deserts, but only where mountainous relief is present. Large fans would not be found in many desert areas of the world — for example, the Kalahari Desert in South Africa or the Great Australian Desert; nor would rugged mountain masses in which slope retreat follows the model we described for arid regions be found. On the other hand, a feature universal to deserts is the broad, shallow stream channel of braided form with bars of coarse bed material. Desert channels look much the same in Algeria or Australia as in Arizona.

As another example, consider solifluction terraces of the arctic tundra. These features do not occur in other climates because the permafrost that governs their development simply does not exist elsewhere. On the other hand, solifluction terraces appear only where slopes are quite steep and regolith is thick, so they would not be present in all arctic landscapes. Climate-process systems can only suggest what landforms are most likely to be found in abundance, geologic controls permitting.

Climate-process systems are fitted into place in the final two chapters of this book, in which we recognize and describe global environmental regions based on climate. Geomorphic processes and landforms typical of each environmental region will take their place along with the other two basic elements of physical geography that comprise the total physical environment: soils and vegetation. But before that synthesis can be attempted, we must investigate the role of geologic processes in shaping landforms. When the geologic processes and structures are well understood, we are in a much stronger position to recognize those geomorphic effects that are genuinely related to climate acting independently of geologic events and the geologic structure of the underlying rock mass.

References for Further Study

Davis, W. M. (1909, 1954), *Geographical essays,* Ginn and Co., Boston, reprinted by Dover Publications, New York. See pp. 249–78, 296–322, 381–412.

Cotton, C. A. (1941), *Landscape,* Cambridge Univ. Press, England. See Chapters 7, 14, 17–20.

Cotton, C. A. (1942), *Climatic accidents in landscape-making,* Whitcombe and Tombs, Christchurch, New Zealand. See Section I.

Thornbury, W. D. (1969), *Principles of geomorphology,* second edition, John Wiley and Sons, New York. See Chapters 5, 6, 8, 11.

Carson, M. A., and M. J. Kirkby (1972), *Hillslope process and form,* Cambridge Univ. Press.

Derbyshire, E., ed. (1973), *Climatic geomorphology,* Harper and Row, New York.

Review questions

1. Explain the original concept of a cycle of denudation, as developed by W. M. Davis. From what words was the term "peneplain" coined? In what way is this an apt derivation?
2. What is the available landmass? With respect to what surface is it measured? Explain how isostatic compensation affects the total available landmass.
3. Compare estimated rates of tectonic uplift with rates of fluvial denudation. What factors affect the denudation rate?
4. Describe a simple denudation model based on rapid uplift of a mountain block. Explain the principle of exponential decay and give examples. Apply the exponential decay schedule to the denudation model. How long a time span is required for a peneplain to be formed?
5. Explain how spasmodic isostatic uplift complicates a simple denudation model. Why does isostatic uplift probably not go on continuously during denudation?
6. Describe the stages in denudation following uplift of a peneplain. How does landmass rejuvenation take place?

How does relief change with time? Under what climatic conditions does this sequence of changes take place?

7. Describe the special characteristics of fluvial denudation in an arid climate. How are streams of dry climates affected by influent seepage?

8. Describe an alluvial fan. Explain the outline of a fan. How does internal structure of a large alluvial fan influence the movement and storage of ground water? What undesirable environmental changes can be caused by excessive withdrawal of ground water from beneath alluvial fans?

9. Describe the evolution of a landscape through fluvial denudation in a mountainous desert region. What is a pediment and where is it formed? What is an inselberg? What is a pediplain? Compare a pediplain with a peneplain.

10. Describe and compare the evolution of hillslopes in humid and arid climates. How is reclining slope retreat different from parallel retreat?

11. What concept is behind the designation of climogenetic regions? Why is the concept difficult to apply? What factor besides climate determines the shapes of landforms?

12. What are climate-process systems? What major categories of process are used in defining climate-process systems?

Landforms Made by Glaciers

21

Glacial ice has played a dominant role in shaping landforms of large areas in midlatitude and subarctic zones. Glacial ice also exists today in two great accumulations of continental dimensions and in many smaller masses in high mountains. In this sense, glacial ice is an environmental agent of the present, as well as of the past, and is itself a landform. Glacial ice of Greenland and Antarctica strongly influences the radiation and heat balances of the globe. Moreover, these enormous ice accumulations represent water in storage in the solid state; they constitute a major component of the global water balance. Changes in ice storage can have profound effects on the position of sea level with respect to the continents.

Coastal environments of today have evolved with a rising sea level following the melting of ice sheets of the last ice advance in the Pleistocene Epoch, or Ice Age. When we examine the evidence of former extent of those great ice sheets, we need to keep in mind that the evolution of modern Man as an animal species occurred during a series of climatic changes that placed many forms of environmental stress on all terrestrial plants and animals in the midlatitude zone. There were also important climatic effects extending into the tropical zone and even into the equatorial zone.

Glaciers

Most of us know ice only as a brittle, crystalline solid because we are accustomed to seeing it only in small quantities. Where a great thickness of ice exists — 100 m (300 ft) or more — the ice at the bottom behaves as a plastic material. The ice will then slowly flow in such a way as to spread out the mass over a larger area, or to cause it to move downhill, as the case may be. This behavior characterizes a **glacier,** defined as any large natural accumulation of land ice affected by present or past motion.

Conditions necessary for the accumulation of glacial ice are simply that snowfall of the winter shall, on the average, exceed the amount of ablation of snow that occurs in summer. The term **ablation** includes both

evaporation and the melting of snow and ice. Thus each year a layer of snow is added to what has already accumulated. As the snow compacts by surface melting and refreezing, it turns into a granular ice; then it is

Figure 21.1 **The Eklutna Glacier, Chugach Mountains, Alaska, seen from the air. A deeply crevassed ablation zone with a conspicuous medial moraine (foreground) contrasts with the smooth-surfaced firn zone (background). (Steve McCutcheon, Alaska Pictorial Service.)**

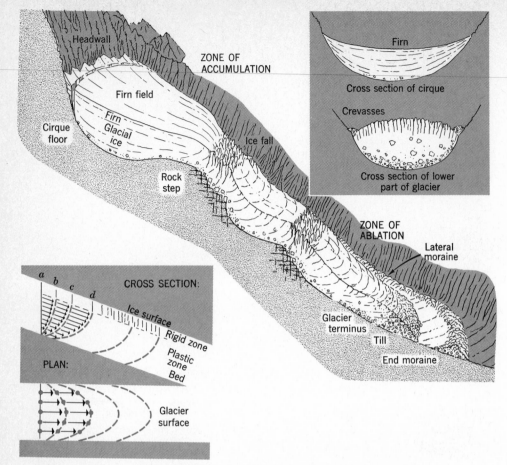

Figure 21.2 **Structure and flowage of a simple alpine glacier. (Drawn by A. N. Strahler.)**

compressed by overlying layers into hard crystalline ice. When the ice becomes so thick that the lower layers become plastic, outward or downhill flow commences and an active glacier has come into being.

At sufficiently high altitudes, even in the tropical and equatorial zones, glaciers form because air temperature is low enough and mountains receive heavy orographic precipitation. Glaciers that form in high mountains are characteristically long and narrow because they occupy former stream valleys. These mountain glaciers bring the plastic ice from small collecting grounds high on the range down to lower altitudes and, consequently, to warmer temperatures. Here the ice disappears by ablation. These **alpine glaciers** are a distinctive type (Figure 21.1 and Plate I.1).

In arctic and polar zones, prevailing temperatures are low enough that ice can accumulate over broad areas, wherever uplands exist to intercept heavy snowfall. As a result, the uplands become buried under enormous plates of ice whose thickness may reach several thousand meters. An ice plate limited to high summit areas of mountains or plateaus is referred to as an **icecap.** During glacial periods, an icecap may expand greatly, spreading over surrounding lowlands and enveloping all landforms it encounters. This extensive type of ice layer is called a **continental glacier** or an **ice sheet.**

Alpine Glaciers

Figure 21.2 illustrates a number of features of alpine glaciers. The center illustration is of a simple glacier occupying a sloping valley between steep rock walls. Snow is collected at the upper end in a bowl-shaped depression, the **cirque.** The upper end lies in the **zone of accumulation.** Layers of snow in the process of compaction and recrystallization constitute the **firn** (or *névé*). The smooth **firn field** is slighly concave-up in profile. Flowage in the glacial ice beneath carries the excess ice out of the cirque, downvalley. An abrupt steepening of the grade forms a **rock step,** over which the rate of ice flow is accelerated and produces deep **crevasses** (gaping fractures), which form an **ice fall.** The lower part of the glacier lies in the **zone of ablation.** Here the rate of ice wastage is rapid and old ice is exposed at the glacier surface, which is extremely rough and deeply crevassed (Plate I.2). The **glacier terminus** is heavily charged with rock debris. The lower portion of the glacier has an upwardly convex cross-profile, shown in the upper-right drawing of Figure 21.2.

The uppermost layer of a glacier, about 60 m (200 ft) in thickness, is brittle and fractures readily into crevasses, whereas the ice beneath behaves as a plastic substance and moves by slow flowage (Figure 21.2, lower left). A line

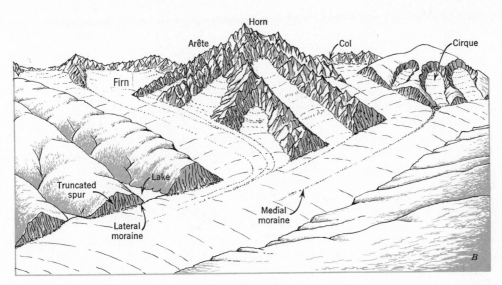

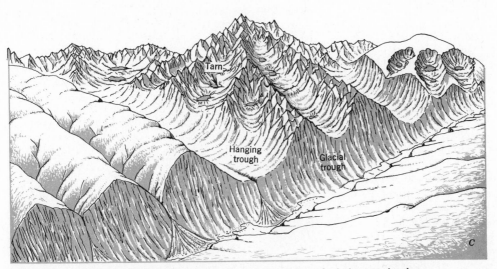

Figure 21.3 **Landforms produced by alpine glaciers. (*A*) Before glaciation sets in, the region has smoothly rounded divides and narrow, V-shaped stream valleys. (*B*) After glaciation has been in progress for thousands of years, new erosional forms are developed. (*C*) With the disappearance of the ice, a system of glacial troughs is exposed. (Drawn by A. N. Strahler.)**

of stakes placed across the glacier surface would be gradually deformed into a parabolic curve by glacier flow, indicating that rate of movement is fastest in the center and diminishes toward the sides. Rate of continual flowage of glaciers is very slow, indeed, amounting to a few centimeters per day for large ice sheets and the more sluggish alpine glaciers, but ranging up to several meters per day for an active alpine glacier.

A simple glacier readily establishes a dynamic equilibrium in which the rate of accumulation at the upper end balances the rate of ablation at the lower end. Equilibrium is easily upset by changes in the average annual rates of nourishment or ablation.

Besides showing the continual slow internal flowage we have described, some glaciers experience episodes of very rapid movement, described as **surges.** When surging occurs, the glacier may develop a sinuous pattern of movement, not unlike the open bends of an alluvial river (see Plate I.1). A surging glacier (sometimes called a "galloping glacier" in the news media) may travel downvalley several kilometers in a few months.

Glacial Erosion

Glacial ice is usually heavily charged with rock fragments, ranging from pulverized rock flour to huge angular boulders of fresh rock. Some of this material is derived from the rock floor on which the ice moves. In alpine glaciers, rock debris is also derived from material that slides or falls from valley walls. Glaciers are capable of great erosive work. **Glacial abrasion** is erosion caused by ice-held rock fragments that scrape and grind against the bedrock. By **plucking,** the moving ice lifts out blocks of bedrock that have been loosened by the freezing of water in joint fractures. Rock debris obtained by erosion must eventually be left stranded at the lower end of a glacier when the ice disappears by ablation. Both erosion and deposition result in distinctive landforms.

Landforms Made by Alpine Glaciers

Landforms made by alpine glaciers are shown in a series of diagrams in Figure 21.3. Previously unglaciated mountains are attacked and modified by glaciers, after which the glaciers disappear and the remaining landforms are exposed to view. Block A shows a region sculptured entirely by weathering, mass wasting, and streams. The mountains have a smooth, full-bodied appearance, with rather rounded divides. Soil and regolith are thick. Imagine now that a climatic change results in the accumulation of snow in the heads of most of the valleys high on the mountain sides.

An early stage of glaciation is shown at the right side of Block B, where snow is collecting and cirques are being carved by the outward motion of the ice and by intensive frost shattering of the rock near the masses of compacted snow. Glaciers fill the valleys and are integrated into a system of tributaries that feed a trunk glacier just as in a stream system. Tributary glaciers join the main glacier with smooth, accordant junctions; but, as we shall see later, the rock floors of their channels are strongly discordant in level.

Vigorous freezing and thawing of meltwater from snows lodged in crevices high on the walls of a cirque shatters the bare rock into angular fragments; these fall or creep down upon the snowfield and are incorporated into the glacier. Frost shattering also affects the rock walls against which the ice rests. The cirques thus grow steadily larger. Their rough, steep walls soon replace the smooth, rounded slopes of the original mountain mass (Figure 21.3, Block B, center).

Where two cirque walls intersect from opposite sides, a jagged, knifelike ridge, called an **arête,** results. Where three or more cirques grow together, a sharp-pointed peak is formed by the intersection of the arêtes. The name **horn** is applied to such peaks (Plate I.2). One of the best known is the striking Matterhorn of the Swiss Alps. Where opposed cirques have intersected deeply, a pass or notch, called a **col,** is formed.

Glacier flow constantly deepens and widens its channel so that after the ice has finally disappeared there remains a deep, steep-walled **glacial trough,** characterized by a relatively straight or direct course and by the U-shape of its cross-profile (Plate I.3). Tributary glaciers also carve U-shaped troughs. But they are smaller in cross section, with floors lying high above the floor level of the main trough; these are called **hanging troughs.** Streams, which later occupy the abandoned trough systems, form scenic waterfalls and cascades where they pass down from the lip of a hanging trough to the floor of the main trough. These streams quickly cut a small V-shaped notch in the trough bottom.

Valley spurs that formerly extended down to the main stream before glaciation occurred have been beveled off by ice abrasion and are termed **truncated spurs** (Block B of Figure 21.3). The bedrock is not always evenly excavated

Figure 21.4 **Moraines of a former valley glacier appear as curved embankments marking successive positions of the ice margins. (After W. M. Davis.)**

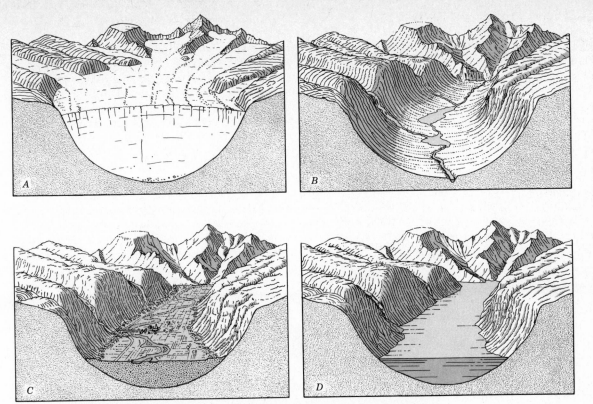

Figure 21.5 **Development of a glacial trough. (Drawn by E. Raisz.) (*A*) During maximum glaciation, the U-shaped trough is filled by ice to the level of the small tributaries. (*B*) After glaciation, the trough floor may be occupied by a stream and lakes. (*C*) If the main stream is heavily loaded, it may fill the trough floor with alluvium. (*D*) Should the glacial trough have been deepened below sea level, it will be occupied by an arm of the sea, or fiord.**

under a glacier, so that floors of troughs and cirques may contain **rock basins** and rock steps. Cirques and upper parts of troughs thus are occupied by small lakes, called **tarns** (Plate I.3). The major troughs usually contain large, elongate **trough lakes,** sometimes referred to as **finger lakes.** Landslides are numerous because glaciation leaves oversteepened trough walls. In glaciated countries such as Switzerland and Norway, slides are a major type of natural disaster because towns and cities lie in the trough floors, where they are readily destroyed by debris avalanches, rockfalls, and landslides. (Chapter 17).

Debris may be carried by an alpine glacier within the ice or it may be dragged along between the ice and the valley wall as a **lateral moraine** (Figures 21.1 and 21.2). Where two ice streams join, this marginal debris is dragged along to form a **medial moraine,** riding on the ice in midstream (Figures 21.1 and 21.3). At the terminus of a glacier, debris accumulates in a heap known as a **terminal moraine,** or **end moraine.** This heap is usually in the form of a curved embankment lying across the valley floor and bending upvalley along each wall of the trough to merge with the lateral moraines (Figures 21.2 and 21.4). As the end of the glacier recedes, scattered debris is left behind. Successive halts in ice retreat produce successive moraines, termed **recessional moraines.**

Glacial Troughs and Fiords

Many large glacial troughs now are nearly flat-floored because aggrading streams that issued from the receding ice front were heavily laden with rock fragments. The deposit of alluvium extending downvalley from a melting glacier is the **valley train.** Figure 21.5 shows a comparison between a trough with little or no fill (*B*) and another with an alluvial-filled bottom (*C*).

When the floor of a trough open to the sea lies below sea level, the seawater will enter as the ice front recedes, producing a narrow estuary known as a **fiord** (Figure 21.5*D*). Fiords may originate either by submergence of the coast or by glacial erosion to a depth below sea level. Most fiords are explained in the second way. When floating, from three-fourths to nine-tenths of a body of glacial ice lies below water level. Therefore a glacier several hundred meters thick can erode to considerable depth below sea level before the buoyancy of the water reduces its erosive power where it enters the open water.

Fiords are observed to be opening up today along the Alaskan coast, where some glaciers are melting back rapidly and the fiord waters are being extended along the troughs. Fiords are found largely along mountainous coasts in lat. 50° to 70°N and S (Figure 21.6).

Figure 21.6 **This Norwegian fiord has the steep rock walls of a deep glacial trough. (Mittet and Co.)**

ference between the two ice sheets is their position with reference to the poles. The antarctic ice rests almost squarely on the south pole, whereas the Greenland Ice Sheet is considerably offset from the north pole, with its center at about lat. 75°N. This position illustrates a fundamental principle: A large area of high land is essential to the accumulation of a great ice sheet. No land exists near the north pole; ice accumulates there only as sea ice.

The surface of the Greenland Ice Sheet has the form of a very broad, smooth dome. From a high point at an altitude of about 3000 m (10,000 ft), there is a gradual slope outward in all directions. The rock floor of the ice sheet lies near sea level under the central region, but is higher near the edges. Accumulating snows add layers of ice to the surface, while at great depth, the plastic ice flows slowly outward toward the edges. At the outer edge

Environmental Aspects of Alpine Glaciation

In past centuries the high, rugged terrain produced by alpine glaciation was sparsely populated and formed impassable barriers between settlements or between nations. This difficulty of access protected the alpine terrain from the impact of Man and has left a legacy of wilderness areas of striking scenic beauty. Mountainsides steepened by glaciation make ideal ski slopes. Ski resorts have mushroomed by the dozens in glaciated mountains, and this form of recreation is now a major industry. In the summer, these same areas are invaded by tens of thousands of campers and hikers. Rock climbing draws hundreds more to scale the near-vertical rock walls of glacial troughs, cirques, and horns.

The floors of glacial troughs have served for centuries as major lines of access deep into the heart of glaciated alpine ranges. For example, in the Italian Alps, several major flat-floored glacial troughs extend from the northern plain of Italy deep into the Alps. The principal Alpine passes between Italy on the south and Switzerland and Austria on the north lie at the heads of these troughs. An example is the Brenner Pass, located at the head of the Adige trough.

Ice Sheets of the Present

Two enormous accumulations of glacial ice are the Greenland and Antarctic ice sheets. These are huge plates of ice, over 3000 m (10,000 ft) thick in the central areas, resting on landmasses of subcontinental size. The Greenland Ice Sheet has an area of 1,740,000 sq km (670,000 sq mi) and occupies about seven-eighths of the entire island of Greenland (Figure 21.7). Only a narrow, mountainous coastal strip of land is exposed.

The Antarctic Ice Sheet covers 13 million sq km (5 million sq mi) (Figure 21.8). A significant point of dif-

Figure 21.7 **Generalized map of Greenland. (After R. F. Flint, *Glacial and Pleistocene geology.*)**

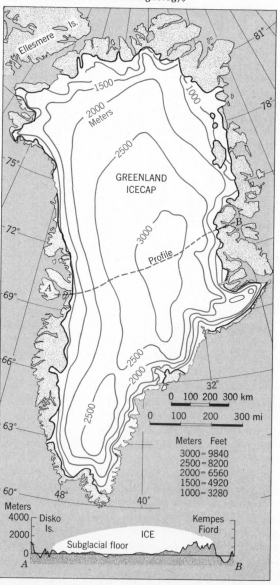

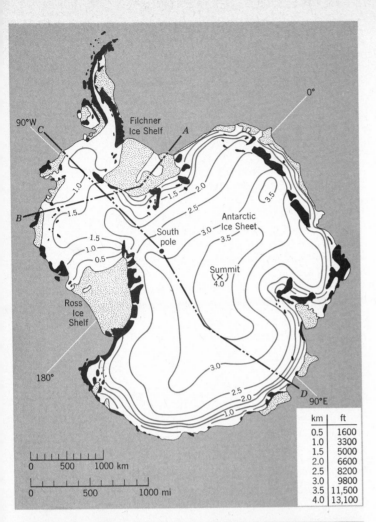

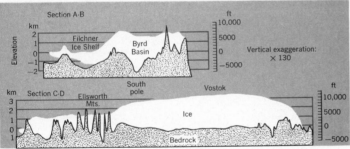

Figure 21.8 **The Antarctic Ice Sheet and its ice shelves. (Based on data from American Geophysical Union.)**

measured. Here the rock floor lies 2000 m (6500 ft) below sea level.

An important glacial feature of Antarctica is the presence of great plates of floating glacial ice, called **ice shelves** (Figure 21.8). The largest of these is the Ross Ice Shelf, with an area of about 520,000 sq km (200,000 sq mi) and a surface height averaging about 70 m (225 ft) above the sea. Ice shelves are fed by the ice sheet, but they also accumulate new ice through the compaction of snow.

Sea Ice

The phenomenon of floating sea ice began to assume major scientific importance during and following World War II, when submarine operations began to be extended in the Arctic Ocean and floating ice masses were occupied as scientific observing stations. Discovery of large petroleum deposits beneath the Arctic Slope of Alaska and Canada has recently added a new interest in sea ice as a serious environmental hazard in the development of those oil resources.

The oceanographer distinguishes **sea ice,** formed by direct freezing of ocean water, from icebergs and ice islands, which are bodies of land ice broken free from tide-level glaciers and continental ice shelves. Aside from differences in origin, a major difference between sea ice and floating masses of land ice is in thickness. Sea ice, which begins to form when the surface water is cooled to temperatures of about −2°C (28½°F), is limited in thickness to about 5 m (15 ft).

Pack ice is the name given to ice that completely covers the sea surface (Figure 21.9). Under the forces of winds and currents, pack ice breaks up into individual patches, termed **ice floes.** The narrow strips of open water between such floes are **leads.** Where ice floes are forcibly

Figure 21.9 **The U.S. Coast Guard icebreaker Northwind forces a passage through McClure Strait, Banks Island, Canadian Northwest Territory, in mid-August. To the left is an open lead. Cutting across from left to right is a rugged zone of pressure ridges. (Official U.S. Coast Guard photo.)**

of the sheet, the ice thins down to a few hundred meters. Continual loss through ablation keeps the position of the ice margin relatively steady where it is bordered by a coastal belt of land. Elsewhere the ice extends in long tongues, called **outlet glaciers,** to reach the sea at the heads of fiords. From the floating glacier edge, huge masses of ice break off and drift out to open sea with tidal currents to become icebergs.

Ice thickness in Antarctica is even greater than that of Greenland (Figure 21.8). For example, on Marie Byrd Land, an ice thickness of 4000 m (13,000 ft) was

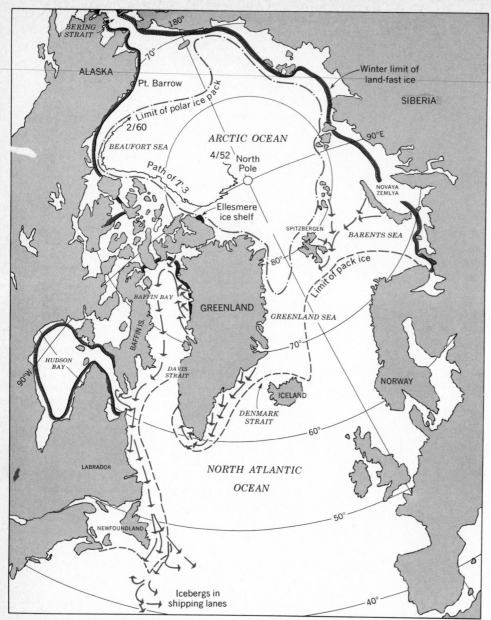

Figure 21.10 **Sea ice of the Arctic Ocean. Common tracks of icebergs are shown by arrows. A tilted homolographic projection is used for this map. (Based on data from the National Research Council.)**

brought together by winds, the ice margins buckle and turn upward into pressure ridges resembling walls or irregular hummocks. These rugged features made travel on foot across the sea ice a terrible ordeal for early explorers of the polar sea. The surface zone of sea ice is composed of fresh water because the salt has been excluded in the process of freezing. The Arctic Ocean, which is surrounded by landmasses, is largely covered by pack ice throughout the year, although open leads are numerous in the summer (Figure 21.10). The relatively warm North Atlantic drift maintains an ice-free zone off the northern coast of Norway.

The situation is quite different in the antarctic region, where a vast open ocean bounds the sea ice zone on the equatorward margin (Figure 21.11). Because the ice floes can drift freely north into warmer waters, the antarctic ice pack does not spread far beyond lat. 60°S in the cold season. In March, close to the end of the warm season, the ice margin shrinks to a narrow zone bordering the Antarctic continent.

Icebergs and Ice Islands

Icebergs, formed by the breaking off, or **calving,** of blocks from a valley glacier or tongue of an icecap, may be as thick as several hundred meters. Being only slightly less dense than seawater, the iceberg floats very low in the water, about five-sixths of its bulk lying below water level

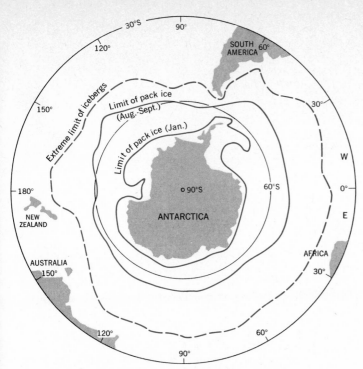

Figure 21.11 **Sea ice of the antarctic region. (Data from the National Academy of Sciences and American Geographical Society.)**

(Figure 21.12). The ice is fresh, of course, because it is formed of compacted and recrystallized snow.

In the northern hemisphere, icebergs are derived largely from glacier tongues of the Greenland Ice Sheet (Figure 21.10). They drift slowly south with the Labrador and Greenland currents and may find their way into the North Atlantic in the vicinity of the Grand Banks of Newfoundland. Icebergs of the antarctic are distinctly different. Whereas those of the North Atlantic are irregular in shape and present rather peaked outlines above water, the antarctic icebergs are commonly tabular in form, with flat tops and steep, clifflike sides (Figure 21.13). This is because tabular bergs are parts of ice shelves. In dimensions, a large tabular berg of the antarctic may be many kilometers broad and over 600 m (2000 ft) thick, with an ice wall rising 60 to 90 m (200 to 300 ft) above sea level.

Somewhat related in origin to the tabular bergs of the antarctic are **ice islands** of the Arctic Ocean. These huge plates of floating ice may be 30 km (20 mi) across and have an area of 800 sq km (300 sq mi). The bordering ice cliff, 6 to 10 m (20 to 30 ft) above the surrounding pack ice, indicates an ice thickness of 60 m (200 ft) or more.

The few ice islands known are derived from a shelf of land-fast glacial ice attached to Ellesmere Island, about lat. 83°N (Figure 21.10). The ice islands move slowly with the water drift of the Arctic Ocean, and a charting of their tracks reveals much about circulation in that ocean. (See path of T-3 in Figure 21.10). Ice islands serve as permanent and sturdy platforms for scientific research stations, from which observations of oceanography, meteorology, and geophysics can be carried out over long periods.

Ice Sheets of the Pleistocene Epoch

Much of North America and Europe and parts of northern Asia and southern South America were covered by enormous ice sheets in a unit of time named the **Pleistocene Epoch.** This ice age ended 10,000 to 15,000 years ago with the rapid wasting away of the ice sheets. Landforms made by the last ice advance and recession are very fresh today and show little modification by erosion.

Figures 21.14 and 21.15 show the extent to which North America and Europe were covered at the maximum known spread of the last advance of the ice. Canada was engulfed by the vast Laurentide Ice Sheet. It spread south into the United States, covering most of the land lying north of the Missouri and Ohio rivers, as well as northern Pennsylvania and all of New York and New England. Alpine glaciers of the Cordilleran ranges coalesced into a single ice sheet that spread to the Pacific shores and met the Laurentide sheet on the east.

In Europe, the Scandinavian Ice Sheet centered on the Baltic Sea, covering the Scandinavian countries. It spread south into central Germany and far eastward to cover much of Russia. In north-central Siberia, large icecaps formed over the northern Ural Mountains and highland areas farther east. At one time ice from these centers grew into a large ice sheet covering much of central Siberia. The British Isles were almost covered by a small ice sheet that had several centers on highland areas and spread outward to coalesce with the Scandinavian Ice Sheet. The Alps at the same time were heavily inundated by enlarged alpine glaciers, fused into massive icecaps.

South America, too, had a Pleistocene ice sheet. This grew from icecaps on the southern Andes range south of about lat. 40°S westward to the Pacific shore, as well as eastward to cover a broad belt of Patagonia. It covered all

Figure 21.12 **This great iceberg in the east Arctic Ocean dwarfs the U.S. Coast Guard icebreaker Eastwind. (Official U.S. Coast Guard photo.)**

Figure 21.13 **A tabular iceberg near the Bay of Whales, Little America, Antarctica. (Official U.S. Coast Guard photo.)**

of Tierra del Fuego, the southern tip of the continent. The South Island of New Zealand, which today has a high spine of alpine mountains with small relict glaciers, developed a massive icecap in late Pleistocene time. All high mountain areas of the world underwent greatly intensified alpine glaciation at the time of maximum ice-sheet advance. Today only small alpine glaciers remain and, in less favorable locations, the glaciers are entirely gone.

Since the middle of the nineteenth century, when the great naturalist Louis Agassiz first defended the theory of continental glaciation, careful study of the deposits left by the ice has led to the knowledge that at least four major ice-sheet advances and retreats occurred. These events were spaced over a total period of about one million years. Deposits of the last advance, known as the **Wisconsin Stage,** form fresh and conspicuous landforms. For the most part, then, our discussion of glacial landforms will concern these most recent deposits.

Figure 21.14 **Pleistocene ice sheets of North America at their maximum spread reached as far south as the present Ohio and Missouri rivers. (After R. F. Flint.)**

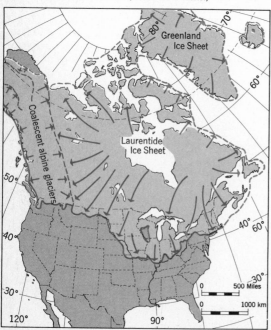

Glacial Stages

The four principal glacial stages of North America are matched by corresponding stages in Europe. Between each glacial stage was an interglacial stage, a period of mild climate during which the ice sheets disappeared. The American glacial stages in order of increasing age are shown in Table 21.1.

Figure 21.16 shows the limit of southerly spread of ice of each glacial stage in the north-central United States. Where older limits were overridden by younger limits, the boundaries are conjectural. Notice that an area in Wisconsin (the Driftless Area) was surrounded by ice but never inundated.

The absolute age and duration in years of the Pleistocene Epoch and its stages are extremely difficult to fix, even though the relative order of events is well established. The last ice disappeared 10,000 to 15,000 years ago from the north-central United States. The Wisconsin Stage may have endured 60,000 years. The earliest onset of glacial advance (Nebraskan) may have occurred 300,000 to 600,000 years ago. The total duration of the Pleistocene Epoch was 1 to 2 million years.

Figure 21.15 **The Scandinavian Ice Sheet dominated northern Europe during the Pleistocene glaciations. The solid line shows limits of ice in the last glacial stage; the dotted line on land shows maximum extent at any time. (After R. F. Flint.)**

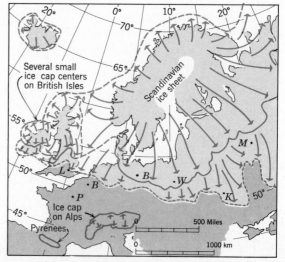

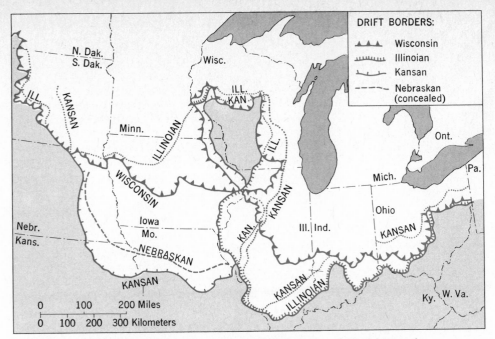

Figure 21.16 **Drift borders in the north-central United States. In each glacial stage, the ice sheet reached a different line of maximum advance. (After R. F. Flint,** *Glacial and Pleistocene geology.)*

Cause of Continental Glaciation

There is excellent geologic evidence of occasional periods of glaciation in the early part of the earth's geologic history. The underlying cause of climate change necessary to bring on glaciations has been subject to a wide range of speculation. Obviously, a prolonged period of colder climate with ample snowfall initiates the growth of ice sheets.

One possible initiating cause of glaciation is a decrease in the intensity of solar radiation intercepted by the earth. If this rate of energy input diminished somewhat at the beginning of the Pleistocene Epoch, the average temperature of the earth's atmosphere would have dropped to a lower level. Providing that this change did not also diminish the quantity of snowfall, the reduced ablation of snowfields would have increased the quantity of snow turned into glacial ice. Evidence from deep-sea sediment cores has established quite soundly that glaciations were directly correlated with periods of colder-than-average sea-surface temperatures.

A decrease in intensity of solar radiation intercepted by the earth — a reduction in value of the solar constant, that is — might represent reduced energy output from the sun's surface. Firm evidence is, as yet, lacking to show that reduced solar energy output affected the earth's climate during Pleistocene glaciations. Another solar energy hypothesis states that known cyclic variations in the geometry of the earth's orbit and in the tilt of the earth's axis have generated periods of reduced insolation sufficiently intense to initiate glaciations.

Three cycles are involved in the **astronomical hypothesis.** First, the time at which perihelion occurs shifts gradually through our calendar year in a cycle of 21,000 years' duration. Whereas the earth is now closest to the sun near the first day of January, in about 10,500 years the earth will be farthest from the sun in January. This event will tend to give slightly decreased average winter temperatures in the northern hemisphere. A second cycle is that of a change in tilt of the earth's axis, which is now 23½° from the perpendicular to the plane of the ecliptic (Chapter 2). In a 41,000-year cycle, tilt angle varies through a total range of about 2°. When the tilt angle is greatest, winter temperatures will tend to be colder and summer temperatures warmer in middle and high latitudes of both hemispheres. A third cycle, 92,000 years in duration, marks a change in the degree of eccentricity (oblateness) of the earth's orbit and thus varies the distance between sun and earth. When the effects of all three cycles are combined, there emerges a history of cyclic change in intensity of solar radiation (insolation), showing strong peaks and valleys for northern latitudes in the range from 45° to 65°N.

In 1976 a group of scientists studying global climatic

Table 21.1 / Glacial and Interglacial Stages of North America and Europe

North American Stages *(North-central U.S.)*	*European Stages* *(Alps)*
Wisconsin glacial	Würm glacial
Sangamon interglacial	Riss-Würn interglacial
Illinoian glacial	Riss glacial
Yarmouth interglacial	Mindel-Riss interglacial
Kansan glacial	Mindel glacial
Aftonian interglacial	Günz-Mindel interglacial
Nebraskan glacial	Günz glacial

variation announced that the record of changing seawater temperatures interpreted from submarine sediment cores is closely correlated with the astronomical cycle of insolation changes.* They expressed the opinion that doubt no longer remains that the combined astronomical cycles have been the initiating and regulating causes of glacial and interglacial stages, at least as far back in time as 500,000 years. An interesting result of this study is that the long-term temperature trend over the next 20,000 years is toward extensive cooling of climate in the northern hemisphere and possibly another glaciation.

A geologic factor, tending to work in harmony with any one of several causes of glaciation, is the known increase in elevation of large parts of the continents just prior to the Pleistocene Epoch. This height increase was a result of widespread mountain making, as well as broad up-arching of some parts of the continental interiors. We know that mountains intercept large quantities of precipitation by the orographic mechanism. The combined effects of reduced insolation and increased altitude of continents could have brought on increased snowfall over favorable highland areas of the continents, such as the Laurentian Upland of eastern Canada and the Scandinavian peninsula.

An important hypothesis of past decades attributed glaciation to a reduction of the carbon dioxide content of the atmosphere. The role of carbon dioxide in absorbing longwave radiation and warming the atmosphere was explained in Chapter 4. It has been estimated that if the carbon dioxide content of the atmosphere were reduced by half, the earth's average surface temperature would drop about 4C° (7F°). A reduction in carbon dioxide content coinciding by chance with an increase in height of continental masses might thus bring on the growth of ice sheets.

Other hypotheses have invoked quite different mechanisms. It has been postulated that increased quantities of volcanic dust in the atmosphere might bring on glaciation because more solar energy would be absorbed in the stratosphere, permitting less to enter the lower atmosphere. Along with the reduced air temperature, there would be an increase in the number of tiny dust particles to serve as nuclei for the condensation of moisture, favoring increased precipitation.

Another geologic event favoring glaciation is a shifting in the positions of the continents with respect to the poles, bringing various parts of the landmasses into favorable geographic positions for the growth of ice sheets. Still another hypothesis requires that changes in oceanic currents, specifically the diversion or blocking of warm currents such as the Gulf Stream, would have brought colder climates to subarctic regions.

Erosion by Ice Sheets

Like alpine glaciers, ice sheets proved to be highly effective eroding agents. The slowly moving ice scraped and ground away much solid bedrock. Left behind were smoothly rounded rock masses bearing countless minute

abrasion marks. Scratches, called **striations,** trend in the general direction of ice movement (Figure 21.17); but variations in ice direction from time to time resulted in intersecting lines. Certain kinds of rock were susceptible to deep grooving (Plate I.4).

Where a strong, sharp-pointed piece of rock was held by the ice and dragged over the bedrock surface, the moving point produced a series of curved cracks fitted together along the line of ice movement. These **chatter marks,** and closely related **crescentic gouges,** whose curvature is the opposite, are good indicators of the direction of ice movement (Figure 21.17). Some very hard rocks acquired highly polished surfaces from the rubbing of fine clay particles against the rock. The evidences of ice erosion described here are common throughout eastern United States and Canada. They may be seen on almost any exposed hard rock surface.

Bearing abrasion marks is a type of conspicuous knob of solid bedrock that has been shaped by the moving ice (Figure 21.17). One side, that from which the ice was approaching, is characteristically smoothly rounded and shows a striated and grooved surface; this is the stoss side. The other, or lee side, where the ice plucked out angular joint blocks, is irregular, blocky, and steeper than the stoss side. The quaint name **roches moutonnées,** which probably originated in the European Alps, has been applied to such glaciated rock knobs.

Vastly more important than the minor abrasion forms are enormous excavations that the ice sheets made in some localities of weak bedrock. In many places, the ice current was accentuated by the presence of a valley paralleling the direction of ice flow. Under such conditions the ice sheet behaved much as a valley glacier, scooping out a U-shaped trough (Plate I.4). Fine examples are the Finger Lakes of western New York State. Here a set of former stream valleys lay parallel to southward spread

Figure 21.17 **Features of ice abrasion. (Above) A roche moutonnée with smooth stoss side and blocky lee side. (Below) Chatter marks and crescentic gouges; glacial grooves paralleling the ice flow (arrow). (Drawn by A. N. Strahler.)**

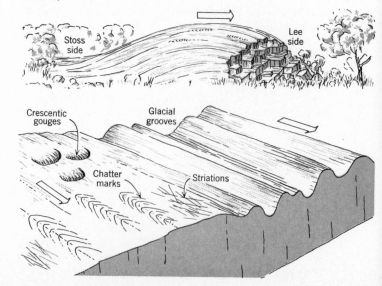

*J. D. Hays, J. Imbrie, and N. J. Shackleton (1976), *Science,* vol. 194, pp. 1121–32.

of the ice, which scooped out a series of deep troughs. Blocked at the north ends by glacial debris, the basins now hold elongated lakes. Many hundreds of lake basins were created in a similar manner all over the glaciated portion of North America and Europe. Countless small lakes of Minnesota, Canada, and Finland occupy rock basins scooped out by ice action. Irregular debris deposits left by the ice are also important in causing lake basins.

Deposits Left by Ice Sheets

The term **glacial drift** includes all varieties of rock debris deposited in close association with glaciers. Drift is of two major types: (1) **Stratified drift** consists of layers of sorted and stratified clays, silts, sands, or gravels deposited by meltwater streams or in bodies of water adjacent to the ice. (2) **Till** is a heterogeneous mixture of rock fragments ranging in size from clay to boulders, deposited directly from the ice without water transport (see Chapter 11 and Figure 11.10).

Over those parts of the United States formerly covered by Pleistocene ice sheets, glacial drift averages from 6 m (20 ft) thick over mountainous terrain, such as New England, to 15 m (50 ft) and more thick over the lowlands of the north-central United States. Over Iowa, drift is from 45 to 60 m (150 to 200 ft) thick; over Illinois, it averages more than 30 m (100 ft) thick. In some places, where deep stream valleys existed prior to glacial advance, as in Ohio, drift is a few hundred meters deep.

To understand the form and composition of deposits left by ice sheets, one must consider the conditions prevailing at the time of existence of the ice. Block A of Figure 21.18 shows a region partly covered by an ice sheet with a stationary front edge. This condition occurred when the rate of ice ablation balanced the amount of ice brought forward by the spreading of the ice sheet. Although the Pleistocene ice fronts advanced and receded in many minor and major fluctuations, there were long periods when the front was essentially stable and thick deposits of drift accumulated.

The transportational work of an ice sheet is like that of a huge conveyor belt. Anything carried on the belt is dumped off at the end and, if not constantly removed, will pile up in increasing quantity. Rock fragments brought within the ice are deposited at the edge as the ice evaporates or melts. There is no possibility of return transportation.

Glacial till that accumulated at the immediate ice edge formed an irregular, rubbly heap, the terminal moraine. After the ice disappeared (Figure 21.18, Block B), the moraine became a belt of knobby hills interspersed with basin like hollows, some of which now hold small lakes. The name **knob and kettle** is often applied to moraine belts. Terminal moraines form great curving patterns; the convex curvature is southward and indicates that the ice advanced as a series of great **ice lobes,** each with a curved front (Figure 21.19). Where two lobes come together, the moraines curve back and fuse together into a single moraine pointed northward. This deposit is an **interlobate moraine.** In its general recession accompanying disappearance, the ice front paused for some time along a

number of lines, causing morainal belts similar to the terminal moraine belt to be formed. These belts are recessional moraines (Figures 21.18 and 21.19). They run roughly parallel with the terminal moraine but are often thin and discontinuous.

Figure 21.18, Block B, shows a smooth, sloping plain lying in front of the ice margin. This is the **outwash plain,** formed of stratified drift left by braided streams issuing from the ice. The plain was built of layer upon layer of sands and gravels. Outwash materials were described in Chapter 11 as a form of **glaciofluvial sediment.**

Large streams issued from tunnels in the ice, particularly when the ice — for many kilometers back from the front — became stagnant, without forward movement. Tunnels then developed throughout the ice mass, serving to carry off the meltwater. Now that the ice has disappeared, the former ice tunnel is represented by a long, sinuous ridge known as an **esker.** The esker is the deposit of sand and gravel formerly laid on the floor of the ice tunnel (Plate I.4). After the ice has melted away, only the streambed deposit remains, forming a ridge (Figure 21.20). Eskers are often many kilometers long; a few are more than 150 km (100 mi) long.

Another common glacial form is the **drumlin,** a smoothly rounded, oval hill resembling the bowl of an inverted teaspoon. It consists of glacial till (Figure 21.21). Drumlins invariably lie in a zone behind the terminal moraine. They commonly occur in groups or swarms, which may number in the hundreds. The long axis of each drumlin parallels the direction of ice movement, and the drumlins thus point toward the terminal moraines and serve as indicators of direction of ice movement. Drumlins were formed under moving ice by a plastering action in which layer upon layer of bouldery clay was spread on the drumlin.

Between moraines, the surface overridden by the ice is overspread by a cover of glacial till known as **ground moraine.** This cover is often inconspicuous because it forms no prominent landscape feature. The ground moraine may be thick and may obscure or entirely bury the hills and valleys that existed before glaciation. Where thick and smoothly spread, the ground moraine forms a level **till plain.** Plains of this origin are widespread throughout the Middle West.

Between the ice front and rising ground, valleys that may have opened out northward were blocked by ice. Under such conditions, **marginal glacial lakes** formed along the ice front (Figure 21.18A). These lakes overflowed along the lowest available channel between the ice and the rising ground slope, or over some low pass along a divide. Streams of meltwater from the ice built sand deposits, called **glacial deltas,** into these marginal lakes. When the ice disappeared, the lake drained away, exposing layers of fine clay and silt accumulated on the bottom. These fine-grained strata, which had settled out from suspension in turbid lake waters, are called **glacio-lacustrine sediments** and are a variety of stratified drift (see Chapter 11 and Figure 11.12). Glacial lake plains are extremely flat, with meandering streams and extensive areas of marshland.

Deltas, built with a flat top at what was formerly the

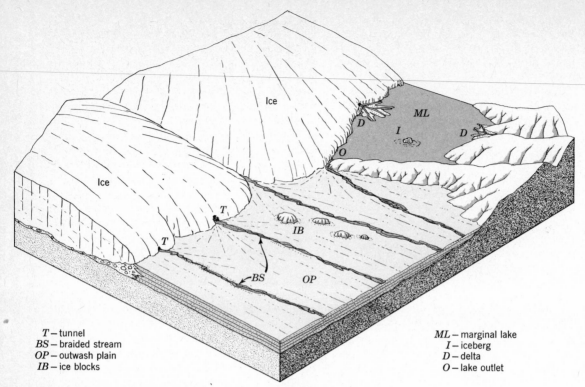

T – tunnel	*ML* – marginal lake
BS – braided stream	*I* – iceberg
OP – outwash plain	*D* – delta
IB – ice blocks	*O* – lake outlet

(A) With the ice front stabilized and the ice in a wasting, stagnant condition, depositional features are built by meltwater.

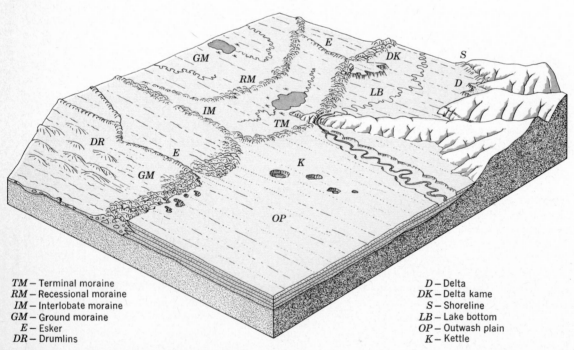

TM – Terminal moraine	*D* – Delta
RM – Recessional moraine	*DK* – Delta kame
IM – Interlobate moraine	*S* – Shoreline
GM – Ground moraine	*LB* – Lake bottom
E – Esker	*OP* – Outwash plain
DR – Drumlins	*K* – Kettle

(B) After the ice has wasted completely away, many new landforms made under the ice are exposed.

Figure 21.18 **Marginal landforms of continental glaciers. (Drawn by A. N. Strahler.)**

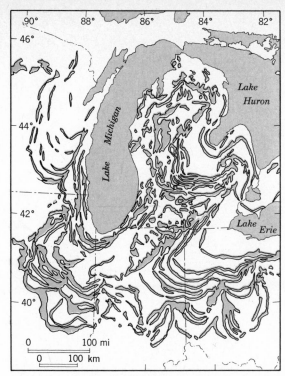

Figure 21.19 Moraine belts of the north-central United States have a festooned pattern left by ice lobes. (After R. F. Flint and others, *Glacial Map of North America*.)

lake level, are now curiously isolated, flat-topped landforms known as **delta kames** (Figure 21.18). Delta and stream channel deposits built between a stagnant ice mass and the wall of a valley are now **kame terraces,** whose steep scarps are ice-contact slopes (Figure 21.22). Kame terraces are difficult to distinguish from the uppermost member of a series of alluvial terraces, but most kames have undrained depressions or pits produced by the melting of enclosed ice blocks. Built of very well washed

Figure 21.20 Seen from the air, this esker in the Canadian shield area appears as a narrow embankment crossing the terrain of glacially eroded lake basins. (Canadian Department of Mines, Geological Survey.)

Figure 21.21 **This small drumlin, located south of Sodus, New York, shows a tapered form, which indicates that the ice moved from upper right to lower left (north to south). (Ward's Natural Science Establishment, Inc., Rochester, N.Y.)**

and sorted sands and gravels, kames commonly show the dipping stratification planes characteristic of deltas (Plate I.4).

Environmental and Resource Aspects of Glacial Deposits

Because much of Europe and North America was glaciated by the Pleistocene ice sheets, landforms associated with the ice are of major environmental importance and the deposits constitute a natural resource as well. Agricultural influences of glaciation are both favorable and unfavorable, depending on preglacial topography and whether the ice eroded or deposited heavily.

In hilly or moutainous regions, such as New England, the glacial till is thinly distributed and extremely stony. Spodosols developed on glacial deposits of the northern United States and Canada are acid and low in fertility. Extensive bogs are floored by Histosols unsuited to agriculture unless transformed by water-drainage systems. Early settlers found cultivation of till difficult because of countless boulders and cobbles in the soil. Till accumulations on steep mountain slopes are subject to mass movements in the form of earthflows. Clays in the till become weakened after absorbing water from melting snows and spring rains. Where slopes have been oversteepened by excavation for highways, movement of till is a common phenomenon.

The steep slopes, the irregularity of knob and kettle topography, and the abundance of boulders along moraine belts conspired to prevent crop cultivation but invited use as pasture. These same features, however, make moraine belts extremely desirable as suburban residential areas. Pleasing landscapes of hills, depressions, and small lakes made ideal locations for large estates.

Flat till plains, outwash plains, and lake plains, on the other hand, comprise some of the most productive agricultural land in the world. Alfisols and Mollisols, both of high base status, formed on these till plains and on lake plains bordering the Great Lakes. We must not lose sight of the fact that in the prairie areas, wind-deposited silt (loess) forms a blanket over clay-rich till and sandy outwash. This loess formed an ideal parent material for the

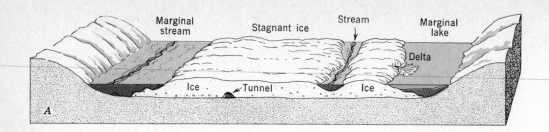

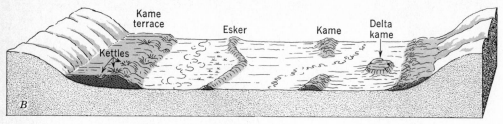

Figure 21.22 **Kames may originate as stream or lake deposits laid between a stagnant ice mass and the valley sides. (Drawn by A. N. Strahler.)**

development of the Mollisols.

Stratified drift deposits are of great economic value. The sands and gravels of outwash plains, delta kames, and eskers provide the aggregate necessary for concrete and the base courses beneath highway pavements. The purest sands may be used for molds, which are needed for metal castings.

Stratified drift, where thick, forms an excellent aquifer and is a major source of ground-water supplies. Deep accumulations of stratified sands in preglacial valleys are capable of yielding ground water in adequate quantities for municipal and industrial uses. Water development of this type is widespread in Ohio, Pennsylvania, and New York.

Origin of the Great Lakes

The Great Lakes of the United States and Canada are truly remarkable environmental features. By an accident of nature — the Pleistocene glaciation — these huge freshwater bodies happen to lie in a midlatitude continental area that has proved enormously productive both agriculturally and industrially. Taken together, the surface area of the Great Lakes is about 246,000 sq km (95,000 sq mi), much greater than any single freshwater lake in the world. Other large freshwater lakes of the world lie in less favorable environments from the human standpoint: Great Bear and Great Slave lakes in the far north of Canada; Lakes Victoria and Tanganyika in equatorial Africa; and Lake Baikal in Siberia.

In preglacial time, lowlands existed where the Great Lakes now stand. These lowlands were occupied by major streams. Repeated ice advances of four glaciations extensively and deeply eroded weak rocks of the former lowlands and carried the debris south to form blocking lines of moraines.

Stages in Great Lakes evolution are shown in the series

of six maps in Figure 21.23. These maps are simplified and show only a few representative stages. Map *A* shows the earliest lakes beginning to form as the ice front receded. Lakes Chicago and Maumee were marginal glacial lakes, ponded between the ice front and higher ground to the south. Both lakes overflowed southward by streams draining into the Mississippi River system.

Map *B* shows continued retreat and the diversion of one lake into another by a marginal stream following the ice front. Map *C* catches the action at a point when drainage was established eastward along the ice front to enter the Hudson River system by way of the Mohawk valley. In Map *D,* final ice recession was under way and part of ancestral Lake Superior had opened up, draining south into the Mississippi system while the other lakes drained east to the Hudson system.

Map *E* shows the very last of the ice disappearing in Ontario and opening an outlet along what is now the Ottawa River. This outlet led directly into the St. Lawrence valley, which was then an estuary of salt water. Map *F* shows a stage of maximum extent of the Great Lakes. Crustal tilting caused the Ottawa River outlet to be abandoned, and lowering of lake levels caused abandonment of the drainage of Lake Michigan into the Mississippi system.

One result of this complex history of changing lake levels and areal extents is that today there are broad marginal zones of lake plains along the shores of Lakes Michigan, Huron, Erie, and Ontario. These plains are intensively developed as agricultural lands and have absorbed the urban expansion of such major lake cities as Chicago, Toledo, Detroit, Cleveland, Toronto, Buffalo, and Erie. Many serious environmental problems beset the Great Lakes. One reason is that the water supplies for a large number of industrial cities are drawn from the same lakes into which are fed, in return, enormous quantities of pollutant wastes.

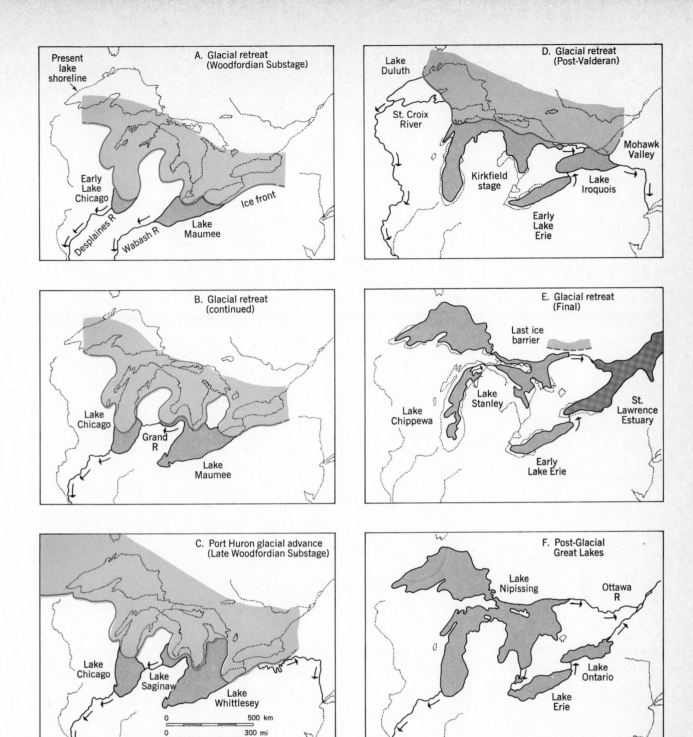

Figure 21.23 **Six stages in evolution of the Great Lakes. (After J. L. Hough, 1958,** *Geology of the Great Lakes,* **Univ. of Illinois Press, pp. 284–96. © 1958 by the Board of Trustees of the University of Illinois.)**

Holocene Environments

The elapsed time span of about 10,000 years since the Wisconsin glaciation ended is called the **Holocene Epoch;** it began with the rapid warming of ocean surface temperatures. Continental climate zones shifted rapidly poleward. Soil-forming processes began to act upon new parent matter of glacial deposits in midlatitudes. Plants became reestablished in glaciated areas in a succession of climate stages. The first of these was known as the **Boreal stage.** Boreal refers to the present subarctic region where needleleaf forests dominate the vegetation. The history of climate and vegetation throughout the Holocene time has been interpreted through a study of spores and pollens found in layered order from bottom to top in postglacial bogs (This study is called palynology.) Plants can be identified and ages of samples can be determined. A dominant tree was spruce. Interpretation of pollens indicates that the Boreal stage in midlatitudes had a vegetation similar to that now found in the region of boreal forest climate.

There followed a general warming of climate until the **Atlantic climatic stage** was reached about 8000 years ago (−8000 years). Lasting for about 3000 years, the Atlantic stage had average air temperatures somewhat higher than those of today — perhaps on the order of 2.5C° (4.5F°) higher. We call such a period a **climatic optimum** with reference to the midlatitude zone of North America and Europe. There followed a period of temperatures below average, the **Subboreal climatic stage,** in which alpine glaciers showed a period of readvance. In this stage, which spanned the age range −5000 to −2000 years, sea level — drawn far down from the level during glaciation — had returned to a position close to that of the present, and coastal submergence of the continents was largely completed (Chapter 22).

The past 2000 years, from the time of Christ to the present, shows climatic cycles on a finer scale than those we have described as Holocene climatic stages. This refinement in detail of climatic fluctuations is a consequence of the availability of historical records and of more detailed evidence generally. A secondary climatic optimum occurred in the period A.D. 1000 to 1200 (−1000 to −800 years). This warm episode was followed by the **Little Ice Age** (A.D. 1450–1850; −550 to −150 years). During this time valley glaciers made new advances to lower levels. In the process, the ice overrode nearby forests and thus left a mark of its maximum extent.

Man and Glaciations

In Chapter 5 we examined conflicting predictions as to Man's role in causing global climate change. Cycles of glaciation and interglaciation demonstrate the power of natural forces to make drastic swings from cold to warm climates. Lesser climatic cycles of the Holocene Epoch also occurred through natural causes. Only following the Industrial Revolution do we recognize possible linkages between global air-temperature change and Man's combustion of hydrocarbon fuels on a massive scale. There is general agreement that increased carbon dioxide tends to cause a rise in average temperatures and that an increase in input of industrial dusts reaching the upper atmosphere tends to lower average temperatures near the earth's surface. But we do not know to what extent observed changes in global temperatures are parts of a natural cycle and to what extent they are influenced by the impacts of an industrial society. Whether Man is hastening or delaying the onset of another glaciation will continue to be debated.

References for Further Study

Davis, W. M. (1954), *Geographical essays,* Dover Publications, New York. See pp. 617–34.

Hough, J. L. (1958), *Geology of the Great Lakes,* Univ. of Illinois Press, Urbana.

Wright, H. E., and D. G. Frey, eds. (1965), *The Quaternary of the United States,* Princeton University Press, N.J.

Paterson, W. S. B. (1969), *The physics of glaciers,* Pergamon Press, Oxford, England.

Flint, R. F. (1971), *Glacial and Quarternary geology,* John Wiley and Sons, New York.

Strahler, A. N. (1971), *The earth sciences,* second edition, Harper and Row, New York. See Chapters 40, 41.

Review Questions

1. What is a glacier? What conditions are necessary for the formation of glaciers? Define ablation. What is firn? What is the distinction between alpine and continental glaciers?

2. Describe a simple alpine glacier and explain how it operates. What is glacier equilibrium? How fast does a glacier flow?

3. How do glaciers erode their channels? Compare alpine glaciers with streams in terms of the activities of erosion and transportation.

4. How do alpine glaciers modify mountain topography? Describe and explain the following features: cirque, arête, horn, col, glacial trough, hanging trough, rock basin, rock step, tarn, and finger lake.

5. What is the form of a glacial trough? How does this form differ from that of a normally eroded stream valley? Explain how a fiord is formed. Where are fiords found?

6. Describe the various kinds of deposits made by a valley glacier. What are the location and form of the following types of moraines: medial, lateral, terminal, and recessional? What is a valley train?

7. Discuss the environmental aspects of alpine landforms. In what way is alpine scenery a valuable resource today?

8. Where do great ice sheets exist today? How thick is the ice in these ice sheets? Explain how an ice sheet is fed and how the ice moves. What are ice shelves? Outlet glaciers?

9. How is sea ice different from the ice of icebergs and ice islands? Why is the thickness of sea ice limited? Describe pack ice, ice floes, leads, and pressure ridges.

10. Explain how the distribution of sea ice differs between the Arctic Ocean and the Antarctic Ocean.

11. How are icebergs formed? What proportion is submerged? Compare the form and source of bergs found in the North Atlantic with those of the antarctic. What is an ice island?

12. Describe the extent of ice sheets of the Pleistocene Epoch in North America and Eurasia. Describe the southern glacial limit in the United States.

13. Name in order the glacial and interglacial stages of North America. How long ago did the last glacial ice disappear from the United States?

14. Review the principal hypotheses that attempt to explain the occurrence of glaciations.

15. What erosional features on rock surfaces give evidence of the former presence of an ice sheet? How can the direction of ice movement be inferred?

16. How can the direction of ice movement be inferred from the shapes of glaciated rock knobs (roches moutonnées)? How were the Finger Lakes of western New York State formed?

17. What is glacial drift? What is the distinction between stratified drift and till?
18. Describe and explain the various depositional forms associated with the margin of an ice sheet.
19. What kinds of moraines are left by ice sheets? What is a glacial outwash plain? Describe the surface topography of an outwash plain.
20. How is an esker formed? How long are eskers? Of what material are they composed?
21. What is a drumlin? Of what material is it composed? How is it formed? Where are most drumlins found in relation to the terminal moraine?
22. What is a till plain? How is it formed?
23. Explain how marginal glacial lakes are formed. What types of deposits are formed in them? What is a delta kame? A kame terrace?
24. Discuss the environmental and resource aspects of glacial landforms.
25. Briefly describe the changes in lake areas and outlets during the evolution of the Great Lakes.
26. Name the major climatic events in the Holocene Epoch. How did alpine glaciers respond to these changes?

Landforms Made by Waves and Currents

22

The continental shoreline, where the salt water of the oceans contacts fresh water and the solid mineral base of the continents, is a complex environmental zone of great importance to Man. Humans have occupied the shore zone for a number of reasons. First, there are food resources of shellfish, finfish, and waterfowl to be had in the shallow waters and estuaries. Second, the shoreline is a base from which ships embark to seek marine food resources farther from land and to transport people and goods between the continents; in time of war the shoreline is a critical barrier to be defended from invading forces arriving by sea. Third, the coastal zone is a recreational facility, with its sea breezes, bathing beaches, and opportunities for surfing, skin diving, sport fishing, and boating.

Along with its opportunities, the shore zone imposes restraints and hazards on Man and Man-made structures. Some coasts are rocky and cliffed; they provide little or no shelter in the form of harbors. Along other coasts the enormous energy of storm wave can cut back the shore, undermining buildings and roads. High water levels in time of storm can cause inundation of low-lying areas and bring the force of breaking waves to bear on ground many meters above the normal levels reached by seawater. For centuries, Man has been at war with the sea, building fortresslike walls to keep out the sea and even forcing the sea to give up coastal land so that more crops for food and forage could be produced.

Great segments of our coastlines face environmental degradation and destruction as urbanization of the coastal zone demands more land and expanded port facilities. The pressures of a growing population to seek its holiday pleasures along the shore threaten to destroy the very benefits the shore offers. Management of the environment of the continental shorelines requires a knowledge of the natural forms and processes of this sensitive zone; the purpose of this chapter is to provide that basic knowledge.

Throughout this chapter we will use the term **shoreline** to mean the shifting line of contact between water and land. The broader term **coastline,** or simply **coast,** refers to a zone in which coastal processes operate or have a strong influence. The coastline includes the shallow water zone in which waves perform their work, as well as beaches and cliffs shaped by waves and coastal dunes.

Waves in Deep Water

Waves of the oceans and lakes are generated by winds. Energy is transferred from the atmosphere to the water surface by a rather complex mechanism involving both friction of the moving air on the water surface and direct wind pressure.

Wind-generated ocean waves belong to a type known as **oscillatory waves** because the wave form travels through the water and causes an oscillatory water motion. A simple terminology applied to waves is illustrated in

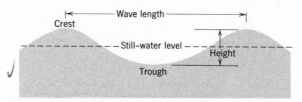

Figure 22.1 **Terminology of water waves.**

Figure 22.2 **Orbital motion in deep-water waves of relatively low height.**

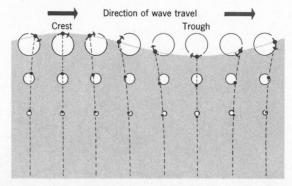

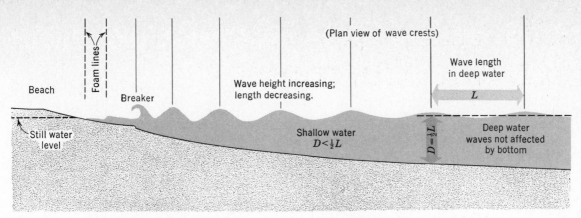

Figure 22.3   As waves enter shallow water, the form changes until breaking occurs.

Figure 22.1. Wave height is the vertical distance between trough and crest. Wave length is the horizontal distance from trough to trough, or crest to crest.

In the oscillatory wave, a tiny particle, such as a drop of water or a small floating object, completes one vertical circle, or wave orbit, with the passage of each wave length (Figure 22.2). Particles move forward on the wave crest, backward in the wave trough. At the sea surface, the orbit is of the same diameter as the wave height, but dies out rapidly with depth. The water particles return to the same starting point at the completion of each orbit. In ideal waves of this type, there is no net motion of water in the direction of the wind.

Shoaling Waves and Breakers

Most shore zones have a fairly smooth, sloping bottom extending offshore into deeper water. As a train of waves enters progressively shallower water depths, there comes a point at which the orbital motion of the waves encounters interference with the bottom. As a general rule this critical depth is about equal to one-half the wave length (Figure 22.3).

As the waves continue to travel shoreward, the wave length decreases while the wave height increases, as shown in Figure 22.3. Consequently, the wave is steepened and becomes unstable. Rather suddenly the crest of the wave moves forward and the wave is transformed into a **breaker,** which then collapses (Figure 22.4). The turbulent water mass then rides up the beach as the **swash,** or **uprush.** This powerful surge causes a landward movement of sand and gravel on the beach. When the force of the swash has been spent against the slope of the

beach, the return flow, or **backwash,** pours down the beach; but much of the water disappears by infiltration into the porous beach sand. Sand and gravel are swept seaward by the backwash.

Marine Erosion

The forward thrust of water produced by storm breakers is a powerful eroding agent along those coasts where the land is high close to the shore (Figure 22.5). The swash of storm waves carves a steep wall, or **marine cliff,** into a bedrock mass. Wave erosion is extremely slow in hard bedrock, so environmental changes by natural processes is comparatively unimportant along rocky coasts. On the other hand, wave erosion is an important environmental concern along coasts made up of soft sedimentary strata, regolith, alluvium, glacial deposits, or sand dunes.

In a few places along the coasts of North America and northern Europe, marine cliffs are being rapidly eroded back in weak glacial deposits. For example, the Atlantic Ocean (eastern) shore of Cape Cod has a 24-km, (15-mi) stretch of marine cliff ranging in height from 18 to 52 m (60 to 170 ft) carved into unconsolidated glacial sands and gravels. This feature is best described as a **marine scarp** because the loose sand holds a slope angle of about 30° to 35° from the horizontal and is by no means a sheer wall (Figure 22.6). Here the rate of shoreline retreat has averaged nearly 1 m (3 ft) per year for the past century, and a single winter storm will cut the scarp back a distance of several feet. Shoreline retreat is called **retrogradation.**

Some details of a marine cliff, or **sea cliff,** formed in hard bedrock are shown in Figure 22.7. A deep basal indentation, the **wave-cut notch,** marks the line of most

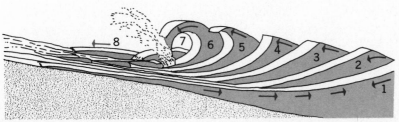

Figure 22.4   A breaking wave. (Drawn by W. M. Davis.)

Figure 22.5 Tremendous forward thrust is evident in these storm waves breaking against a seawall at Hastings, England. (Photographer not known.)

Figure 22.7 Landforms of sea cliffs. A = arch; S = stack; C = cave; N = notch; P = abrasion platform. (Drawn by E. Raisz.)

intense wave erosion. The waves find points of weakness in the bedrock, penetrating deeply to form **crevices** and **sea caves.** More resistant rock masses project seaward and are cut through to make picturesque **arches** (Plate J.1). After an arch collapses, the remaining rock column forms a **stack,** which is ultimately leveled (Figure 22.7).

As the sea cliff retreats landward, continued wave abrasion forms an **abrasion platform.** This sloping rock floor continues to be eroded and widened by abrasion beneath the breakers. If a beach is present, it is little more than a thin layer of gravel and cobblestones.

Retreat of a coastline may take place so rapidly that a stream emptying into the sea is unable to lower its channel fast enough to maintain a sea-level junction. The result is a **hanging valley,** in which the stream channel ends abruptly near the top of the cliff and the valley cross-profile is abruptly truncated (Figure 22.8). Where bedrock is weak, the marine cliff may be unstable. Yielding occurs spontaneously, and mass movements take the form of earthflows and slump blocks (Figure 22.9).

Sea cliffs are features of spectacular beauty as well as habitats for many forms of life, including sea mammals and shore birds. Only in recent years has the need to preserve these cliffed coasts in their natural state been appreciated fully. Although the strength of these rocky features resists Man-made alterations, the cliff line is vulnerable to heavy use for summer homes, motels, and restaurants. Intensive use not only destroys the pristine scenery, but also adds pollution.

Beaches

Sediment enters the shore zone of breaking waves from a number of possible sources. Sediment may be derived directly from a marine cliff or scarp that is being actively eroded. Sediment entering the ocean or a lake from the mouth of a stream is another major source of supply. Storm waves may scour the offshore zone, dragging sand and gravel landward to reach the breaker zone.

Whatever the origin of the sediment, it is shaped by swash and backwash into a wedge-shaped sediment deposit, familiar to everyone as a **beach.** Sediment composing beaches ranges from fine sand to cobblestones several centimeters in diameter. Within a given stretch of

Figure 22.6 This wave-eroded scarp of glacial sand at Highland Light, Cape Cod, recedes at a rate of about 1 m (3 ft) per year. (A. N. Strahler.)

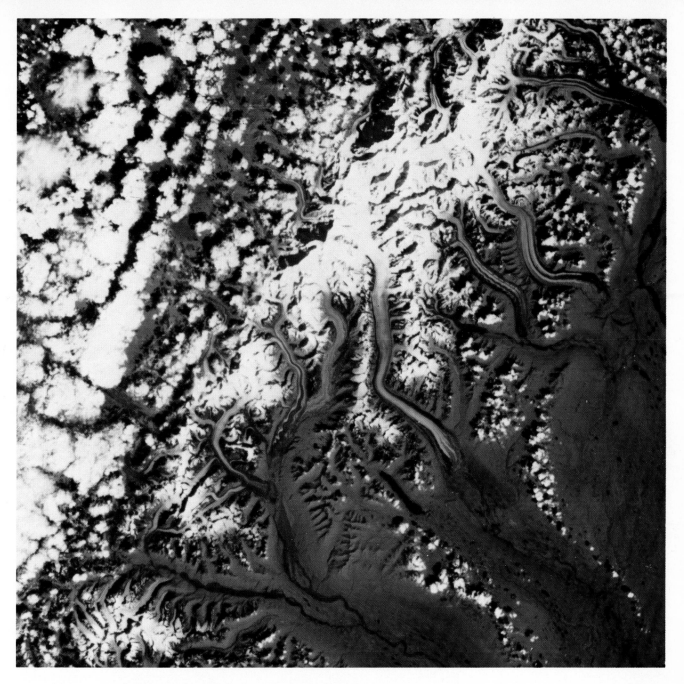

Photographed by ERTS-1 orbiting satellite from a height of about 570 mi (920 km) glaciers of the Alaska Range in south central Alaska appear as blue curving bands, emerging from high collecting grounds on a snow-covered mountain axis trending from northeast to southwest across the center of the photo. The white patches at the left and right are clouds. Darker lines running down the length of a glacier are medial moraines, formed of rock debris on the ice surface. Where these moraines have been distorted into a sinuous pattern, the glacier has experienced a rapid downvalley surge at rates up to 4 ft (1.2 m) per hour. Those glaciers with smooth moraines, paralleling the banks, are experiencing very slow, uniform flow throughout their entire length. The group of high mountain peaks in the upper, central part of the area includes Mt. McKinley, highest point in North America. This is a false-color image, created by combining data of three narrow spectral bands, each assigned a primary color for printing. The area shown in about 65 mi (105 km) across; north is toward the top. (NASA — EROS Data Center, No. 81033210205A2)

(below) **Cirques and horns of the lofty Hindu Kush range, Afghanistan. Glaciers occupying the cirques hang far above the empty main glacial troughs. (Kenneth Boche)**

(above) **Looking down from a crest in the Hindu Kush range in Afghanistan we see a large, debris-covered alpine glacier (left) joined by small tributary glacier (center) with a deeply crevassed ice fall. (Shelly Sack)**

(left) **This wasting ice surface is in the lower, ablation zone of the Capps Glacier, Wrangell Mountains, Alaska. The man is standing in the channel of a meltwater stream. (Lynne Dozier)**

(right) The Mer de Glace, in the French Alps, seen from an aerial tramway. The glacier is greatly shrunken in both depth and width as compared with its full dimensions two centuries ago, during the Little Ice Age. A massive lateral moraine runs along the left side of the glacier. (Paul W. Tappan)

Plate I.2   Alpine glaciers

(below) Chamonix, France, occupies the bottom of a great glacial trough of the French Alps. Alluvial deposits have formed a flat valley floor. (Paul W. Tappan)

(right) **This abraded surface of granite has been striated and polished by an alpine glacier. (Mark A. Melton)**

(above) **Delta Lake occupies the floor of a cirque in the Grand Teton Range, Wyoming. In the background is a mass of morainal debris burying small relict glacial ice bodies. (Mark A. Melton)**

(left) **A U-shaped glacial trough in the Beartooth Plateau, near Red Lodge, Montana. Talus cones have been built out from the steep trough walls. (Alan H. Strahler)**

Alpine glacial landforms   **Plate I.3**

(above) **A Pleistocene continental ice sheet carved this finger-lake trough in granite on Mt. Desert Island, Maine. Beyond is the smoothly rounded ice-abraded profile of Mt. Sargent. (Arthur N. Strahler)**

(right) **This esker in southern Michigan is being excavated as a source of sand and gravel. The coarse boulders are screened out. (Orlo E. Childs)**

(above) **These striations on a glacially polished surface of marble, near West Rutland, Vermont, prove the former presence of an ice sheet over northern New England. (Orlo E. Childs)**

(below) **Seen close up, these layers of sand and gravel in a glacial delta kame are nicely sorted by grain size. Curvature of the upper layers shows flowage accompanying deposition. (Orlo E. Childs)**

(above) **Cattle are grazing on a glacial kame, adjacent to a steep-sided kettle, near Guilford, in southern New York State. (Arthur N. Strahler)**

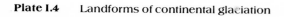

Plate I.4   Landforms of continental glaciation

(above) **Slabs of pink granite dip seaward on the Atlantic coast of Mt. Desert Island, Maine. Sea caves in the middle distance; a cobblestone pocket beach in the far distance. (Arthur N. Strahler)**

(above) **Rock arches have been carved by waves from this sea cliff of horizontal sandstone strata on the Lake Superior shore. (Orlo E. Childs)**

(right) **Each property owner has built his own protective sea wall to suit his style along this stretch of Lake Michigan shoreline of northern Indiana. (Ned L. Reglein)**

(below) **A white sandspit, growing in a direction toward the observer, leaves only a narrow inlet (foreground) for tidal currents. Sandy shoals lie in the bay at the right. Martha's Vineyard, Massachusetts. (Donald W. Lovejoy)**

(above) **The flat land surface at the left, above the rocky cliffs, is a marine terrace, a former abrasion platform now elevated above the sea. South of Lucia, California. (Orlo E. Childs)**

(left) **The ria coastline of Tasmania stretches beyond the city of Hobart. Submergence, partly a result of postglacial rise of sea level, has drowned many stream valleys. (Paul W. Tappan)**

Coastal landforms   **Plate J.1**

(right) The barrier island coast of North Carolina, seen from Apollo 9 spacecraft. The white barrier beach of sand — the Outer Banks — projects sharply seaward in two points: Cape Hatteras (H) and Cape Lookout (L). Pamlico Sound (P) lies between the barrier and the mainland shore, which is deeply embayed as a result of postglacial submergence. Extensive areas of salt marsh (S) make up the low coastal zone. Through two inlets (I) between Hatteras and Lookout sediment is being carried seaward in plumes of lighter color. The Gulf Stream boundary (G) is sharply defined. The area shown is about 100 mi (160 km) across. (NASA — No. AS9-20-3128)

(below) A close look at the barrier beach, Cape Hatteras. A dune ridge, well protected by grass cover, lies to the left of the beach. The view is north; Atlantic Ocean at the right. (Mark A. Melton)

(left) Near Cape Hatteras this stretch of barrier beach has been swept over by storm swash, topping the dune barrier and carrying sand in sheets toward the bay at the left (Mark A. Melton)

Plate J.2   Barrier island coast

(right) The Nile Delta appears in true color as a dark blue-green triangle bounded by pale desert areas. Two major distributaries (N) of the Nile River end in prominent cusps: the Damietta Mouth (D) and the Rosetta Mouth (R). Alexandria (A) is just within view in the foreground. Sand carried by littoral drift from the river mouths has accumulated in a broad barrier beach (B), separated from the delta plain by an open lagoon (L). The Suez Canal (C) connects the Mediterranean Sea (left) with the Gulf of Suez (right). Beyond lies the Sinai Peninsula. (NASA — No. S-65-34776)

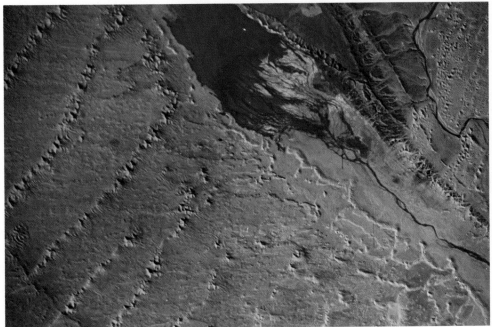

(left) Star dunes of reddish-yellow sand cover much of the area in this Gemini VII photo of the Sahara Desert in western Algeria. The upper bluish zone is a shallow lake. Water and sediment are brought to the lake basin by a major stream, Wadi Saura, entering from the lower right. Bedrock is exposed in narrow hogback ridges representing a deeply eroded fold structure in sedimentary rocks. The area shown is about 30 mi (50 km) across. (NASA — No. S65-6380)

Nile Delta and Saharan dunes   **Plate J.3**

(left) **A great sand sea of transverse dunes near Yuma, Arizona. Slip faces tell us that the prevailing wind blows from right to left. (Arthur N. Strahler)**

(below) **This vertical road cut in thick loess, east of Vicksburg, Mississippi, has sustained little erosion over many years' time. (Orlo E. Childs)**

(right) **Closeup of the loess in the same cliff as in the photograph at right. The material is a soft, friable silt, deposited by wind during the late Pleistocene. (Orlo E. Childs)**

(above) **Deflation of this desert floor winnowed away the fines, leaving a pavement of pebbles, closely fitted together to protect the soft material beneath. Big Bend National Park, Texas. (Mark A. Melton)**

(above) **Coastal dunes advance landward along the Pacific shore near Morro Bay, California. Morro Rock, marine stack of plutonic rock, rises offshore in the distance. (Orlo E. Childs)**

(right) **This coastal blowout dune is advancing over a forest, the slip face gradually burying the tree trunks. Lake Michigan coast, near Michigan City, Indiana. (Barry Voight)**

Plate J.4 Dunes and loess

Figure 22.8 Hanging valleys appear as notches in a marine cliff. The large stream at the right has been able to maintain an accordant junction with the sea level and provides a small harbor. (Drawn by W. M. Davis.)

Figure 22.10 A multiple-crested shingle beach, Smith Cove, Guysboro, Nova Scotia. The beach forms a crescent between rocky headlands. (Maurice L. Schwartz.)

beach, the sediment is usually quite well sorted into a particular size grade. Thus there are beaches of fine sand, of coarse sand, of gravel, or of cobbles. As a rule, beaches composed of fine sand are broad and have a very gentle seaward slope, whereas beaches of coarse sand or gravel are quite steep (Figure 22.10). Beaches formed of cobble-stones are very steep and show a high crest, or ridge form. Particles of silt and clay do not form beaches, but instead are easily carried away from the shoreline in suspension in currents, to settle out in deep water.

The Beach Profile

A broad sand beach with its bordering shallow-water zone of breaking waves, on the one side, and a zone of wind action with dune development forming a border on the landward side represents a succession of unique life habitats in which each assemblage of plant and animal forms is adapted to a different environment.

Figure 22.11 is an idealized profile across a typical broad beach developed by exposure to waves of the open ocean in a midlatitude location experiencing a strong contrast between wave action of summer and winter seasons. The profile shows summer conditions, in which waves are of low height and comparatively low levels of energy.

During the summer, accumulation of sand takes place,

Figure 22.9 Coastal landslides in weak sedimentary strata are the result of oversteepening of a marine cliff by wave attack. (Drawn by W. M. Davis.)

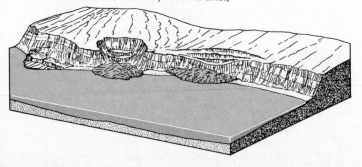

building a **summer berm,** which is a benchlike structure. A higher **winter berm** lies behind the summer berm. The winter berm can be cut back deeply by a storm, but is rebuilt following the storm. The **foreshore** is the sloping beach face in the zone of swash and backwash. Beneath the breaker zone is a low underwater bar, called an **off-shore bar.** It lies in the **offshore** region of the beach, in the zone of shoaling waves and below the level of low tide. Actually, there are many variations in beach-profile forms from place to place and season to season, depending on wave form, wave energy, and the composition of the beach.

Beaches can experience a widening, or out-building process called **progradation,** as shown in Figure 22.12A. The sand that is added to the beach may come from deeper water, or it may be brought along the shore from another part of the coast. A common indication of progradation is the presence of numerous parallel **beach ridges,** each of which represents a former berm crest. Beaches can also experience a narrowing, or back-cutting retrogradation, shown in Figure 22.12B. In this case the sand may be moved to deeper water in the offshore zone, or it may be transported along the shoreline to another part of the coast.

Littoral Drift

In an idealized situation in which waves approach a straight shoreline, their crests parallel with that line, a given wave breaks at the same instant at all points and the swash rides up the beach at right angles to the shoreline. The backwash returns along the same line. Consequently, particles move up and down the beach slope along a fixed line.

Along most shorelines most of the time, however, waves approach the coast at an oblique angle, as shown in Figure 22.13. As these waves travel toward a shoreline

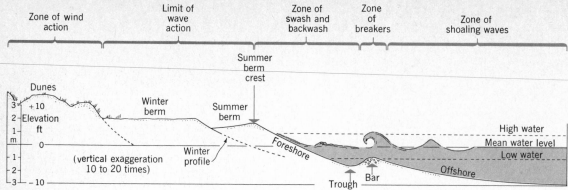

Figure 22.11   Typical forms and zones of a sand beach in the midlatitude zone. (From A. N. Strahler, 1971, *The Earth Sciences*, 2nd ed., Harper and Row, New York.)

over a gently shoaling ocean bottom, they undergo a gradual decrease in velocity of forward travel. As a result, the wave crests become curved in plan and tend to become more nearly parallel with the shoreline. This wave-bending phenomenon is called **wave refraction.**

Despite some refraction, however, the wave crest arrives in the breaker zone with an oblique approach to the shoreline. The swash of the breaker then rides obliquely up the foreshore, as shown in Figure 22.14. As a result the sand, pebbles, and cobbles are moved obliquely up the slope. After the swash has spent its energy, the backwash flows down the slope of the beach, being controlled by the pull of gravity, which moves it in the most direct downhill direction. Now the particles are dragged directly seaward and come to rest at positions to one side of the starting points. Because wave fronts approach consistently from the same direction on a particular day, this movement is repeated many times. Individual rock particles thus travel a considerable distance along the shore. Multiplied many thousands of times to include the numberless particles of the beach, this form of mass transport, called **beach drift,** is a major process in shoreline development.

Rarely is beach drift not taking place, in one direction or the other, along a marine shoreline. Usually, a given stretch of shoreline is subjected to a dominant direction of wave approach throughout a given season of the year, or throughout the entire year. Consequently, beach drift can be assigned a single direction of net transport as the seasonal or yearly average.

A process related to beach drifting is **longshore drift.** When waves approach a shoreline under the influence of strong winds, the water level is slightly raised near shore by a slow shoreward drift of water. An excess of water,

Figure 22.13   **Wave refraction along a straight shoreline. A longshore current will run from right to left, carrying sand in that direction, parallel with the shoreline.**

Figure 22.12   **Progradation and retrogradation.**

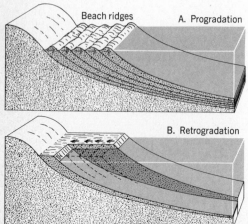

Figure 22.14   **Beach drift of sand, caused by an oblique approach of the swash.**

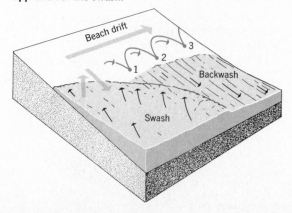

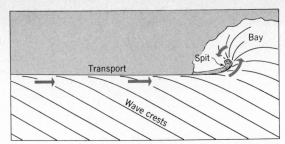

Figure 22.15 Along a straight coast, littoral drift carries sand in one direction to reach the mouth of a bay, where it is formed into a sandspit. (From A. N. Strahler, 1971, *The Earth Sciences,* 2nd ed., Harper and Row, New York.)

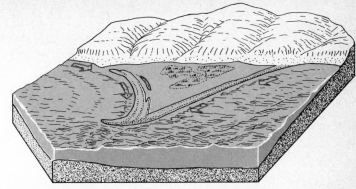

Figure 22.17 This cuspate bar, which has enclosed a triangular lagoon, receives drifted beach materials from both sides. (Drawn by E. Raisz.)

which must escape, is thus pushed shoreward. A **longshore current** is set up parallel to shore in a direction away from the wind (Figure 22.15). When wave and wind conditions are favorable, this current is capable of moving sand along the bottom in the breaker zone in a direction parallel to the shore.

Both beach drifting and longshore drifting move particles in the same direction for a given set of onshore winds and oblique wave fronts and therefore supplement each other's influence in sediment transportation. The combined transport by beach and longshore drift is termed **littoral drift.** Let us now apply the principle to the evolution of beach deposits.

Littoral drift along a straight section of shore is illustrated in Figure 22.15. Where an embayment occurs, drift continues along the line of the straight shore, with the result that an embankment of sediment is constructed along that line. This narrow beach deposit, extending out into open waters, is called a **sandspit,** or simply a **spit** (Plate J.1). Because of wave refraction in shallow water of the bay, sediment is carried around the spit end by littoral drift. Thus the spit develops a landward curvature. A sandspit may grow in length and become extended across the entire mouth of a bay, forming a **baymouth bar** (see Figure 22.30D). The word "bar" refers to any long, narrow sand embankment formed by wave action. Littoral drift

from an island may form a **tombolo,** which is a sand bar connecting the island with the mainland (Figure 22.16).

Where littoral drift converges from opposite directions upon a given point on a shoreline, sediment accu-

Figure 22.18 Sketch map of Cape Canaveral (renamed Cape Kennedy), Florida, as it was in 1910, before Man-induced modification. Ridges near the shore *(right)* are beach ridges; those farther inland are dune ridges built on older beach ridges. (After Douglas Johnson, 1919.)

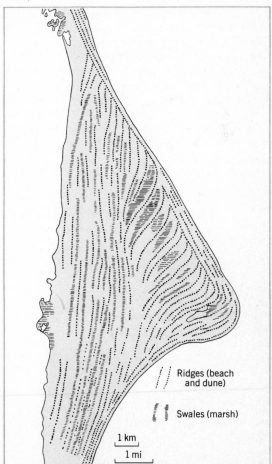

Ridges (beach and dune)

Swales (marsh)

1 km
1 mi

Figure 22.16 Two tombolos have connected this cliffed island with the mainland. (Drawn by W. M. Davis.)

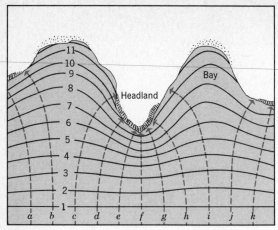

Figure 22.19 Wave refraction on an embayed coast.

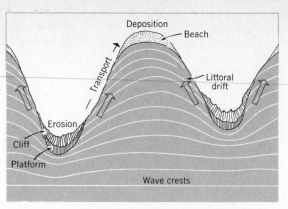

Figure 22.21 Littoral drift moves sediment from promontories to bay heads. (From A. N. Strahler, 1971, *The Earth Sciences*, 2nd ed., Harper and Row, New York.)

mulates in the form of a **cuspate bar** (Figure 22.17). When progradation continues over a long period of time, a much larger feature, called a **cuspate foreland,** is constructed (Figure 22.18). It consists of many beach ridges separated from one another by narrow belts of low, marshy ground, called **swales.** Wind action may transform the sandy beach ridges into dunes, but the original plan of the ridges continues to be prominent.

Wave Refraction on an Embayed Coast

On a coastline with promontories and bays, such as that shown in Figure 22.19, wave refraction concentrates wave energy upon the promontories. Successive positions of a wave crest are shown by lines numbered 1, 2, 3, and so on. In deep water, the wave fronts are parallel. As the shore is neared, the retarding influence of shallow water is felt first in the areas in front of the promontories. Shallowing of water reduces the speed of wave travel at those places; but in the deeper water in front of the bays, the retarding action has not yet occurred. Consequently, the wave front is refracted in rough conformity with the shoreline. The waves break first on the promontory and on

the bay head last, as shown in Figure 22.19.

Particularly important in understanding the development of embayed shorelines is the distribution of wave energy along the shore. On Figure 22.19, dashed lines (lettered *a, b, c, d,* etc.) divide the wave at position 1 into equal parts, which may be taken to include equal amounts of energy. Along the headlands, the energy is concentrated into a short piece of shoreline; along the bays, it is spread over a much greater length of shoreline. Consequently, the breaking waves act as powerful erosional agents on the promontories, but are relatively weak and ineffective at the bay heads. The important principle is that headlands and promontories are rapidly eroded back. The process tends to produce a simple, broadly curving shoreline as a stable form.

Concentrated wave attack on promontories produces a high marine cliff and a broad abrasion platform (Figure 22.20). Waves breaking obliquely on the bay shores cause littoral drift of the detritus toward the bay heads, as shown in Figure 22.21. This sediment accumulation becomes a crescent-shaped **pocket beach** (Figure 22.22).

Figure 22.20 Marine cliffs bordered by a broad abrasion platform. A pocket beach lies at lower left. Pacific coast, south of Cape Flattery, Washington. (Photographer not known.)

Figure 22.22   **Stacks, arches, and sea caves reflect the intense erosion of a promontory of the chalk cliffs of Normandy. A pocket beach has formed in the shelter of a small bay. This photo was taken in the early 1900s. (Photographer not known.)**

Littoral Drift and Shore Protection

Along stretches of shoreline affected by retrogradation, the beach may be seriously depleted or even entirely destroyed. When this occurs, cutting back of the coast can be rapid in weak materials, destroying valuable shore property. Protective engineering structures designed for direct resistance to frontal wave attack, such as seawalls, are prone to failure and are extremely expensive. In some circumstances, a successful alternative strategy is to install structures that will cause progradation, building a broad protective beach. The principle here is that the excess energy of storm waves will be dissipated in reworking the beach deposits. Cutting back of the beach in a single storm will be restored by beach-building between storms.

Progradation requires that sediment moving as littoral drift be trapped by the placement of baffles across the path of transport. To achieve this result, groins are installed at close intervals along the beach. A **groin** is simply a wall or embankment built at right angles to the shoreline; it may be constructed of huge rock masses, reinforced concrete, or wooden pilings. Figure 22.23 shows the shoreline changes induced by groins. Sand accumulates on the up-drift side of the groin, developing a curved shoreline. On the down-drift side of the groin, the beach will be depleted because of the cutting off of the normal supply of drift sand. The result may be harmful retrogradation and cutting back of the coast. For this reason groins must be closely spaced so the trapping effect of one groin will extend to the next (Figure 22.24). Ideally, when the groins have trapped the maximum quantity of sediment, beach drift will be restored to its original rate for the shoreline as a whole.

In some instances the source of beach sand is from the mouth of a river. Construction of dams far upstream on the river may drastically reduce the sediment load and therefore also cut off the source of sand for littoral drift. Retrogradation may then occur on a long stretch of shoreline. The Mediterranean shoreline of the Nile Delta has suffered retrogradation because of reduction in sediment supply following dam construction far up the Nile.

Tidal Currents

Most marine coastlines are influenced by the **ocean tide,** a rhythmic rise and fall of sea level under the influence of changing attractive forces of moon and sun on the rotating earth. Where tides are great, the effects of changing water level and the currents set in motion are of major importance in shaping coastal landforms.

The tidal rise and fall of water level is graphically represented by the **tide curve.** We can make half-hourly observations of the position of water level against a tide

Figure 22.23   **Construction of a groin causes marked changes in configuration of the sand beach. (From A. N. Strahler, 1972, *Planet Earth,* Harper and Row, New York.)**

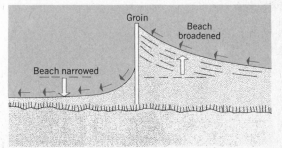

Figure 22.24 A system of groins for trapping beach sand, Willoughby Spit, Virginia. Littoral drift is from lower left to upper right. (Department of the Army, Corps of Engineers.)

staff (measuring stick) attached to a pier or sea wall. We then plot the changes of water level and draw the tide curve. Figure 22.25 is a tide curve for Boston Harbor covering a day's time. The water reached a maximum height, or **high water,** of about 4 m (12 ft), then fell to a minimum height, or **low water,** of about 1 m (3 ft). This occurred about 6¼ hours after high water. A second high water occurred about 12½ hours after the previous high

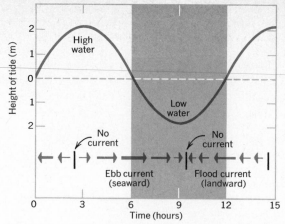

Figure 22.26 The ebb current flows seaward as the tide level falls; the flood current flows landward as the tide level rises.

water, completing a single tidal cycle. In this example the **range of tide,** or difference between heights of successive high and low waters, is about 3 m (9 ft).

The rising tide sets in motion currents of water, known as **tidal currents,** in bays and estuaries. The relationships between tidal currents and the tide curve are shown in Figure 22.26. When the tide begins to fall, an **ebb current** sets in. This flow ceases about the time the tide is at its lowest point. As the tide begins to rise, a landward current, the **flood current,** begins to flow.

Tidal currents may be active in the lower reaches of large rivers that have experienced partial drowning because of recent sea-level rise. The tidal section of the river is a form of estuary, in which fresh water and salt water are mixed. River discharge augments the ebb current but opposes the flood current, so the ebb current attains a higher speed than the flood current. Thus a floating object would travel farther seaward than landward with each tidal cycle and would eventually reach the open ocean. This process provides a natural flushing action that carries Man-made wastes, such as sewage, to the ocean.

Tidal Current Deposits

Ebb and flood currents generated by tides perform several important functions along a shoreline. For one, the currents that flow in and out of bays through narrow inlets are very swift and can scour an inlet strongly. Scour keeps the inlet open despite the tendency of shore drifting processes to close it with sand.

Also, tidal currents carry much fine silt and clay in suspension, derived from streams that enter the bays or from bottom muds agitated by storm wave action against the outer shoreline. This fine sediment settles to the floors of the bays and estuaries, where it accumulates in layers and gradually fills the bays. Much organic matter is present in this sediment.

In time, tidal sediments fill the bays and produce **mud**

Figure 22.25 Height of water at Boston Harbor measured every half hour.

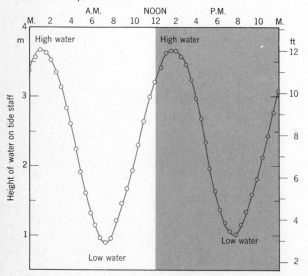

flats, which are barren expanses of silt and clay exposed at low tide but covered at high tide. Next, a growth of salt-tolerant plants takes hold on the mud flat. The plant stems entrap more sediment and the flat is built up to approximately the level of high tide, becoming a **salt marsh** (Figure 22.27). A thick layer of peat is eventually formed at the surface. Tidal currents maintain their flow through the salt marsh by means of a highly complex network of sinuous tidal streams (Figure 22.28).

Salt marsh is important to geographers because it is land that can be drained and made agriculturally productive. The salt marsh is first cut off from the sea by the construction of an embankment of earth (a dike), in which gates are installed to allow the freshwater drainage of the land to exit during flood flow. Gradually, the salt water is excluded and soil water of the diked land becomes fresh. Such diked lands are intensively developed in Holland (where they are called *polders*) and southeast England (*fenlands*).

Over many decades, the surface of reclaimed salt marsh subsides because of the compaction of underlying peat layers and may come to lie well below mean sea level. The threat of flooding by salt water, when storm waves breach the dikes, hangs constantly over the inhabitants of such low areas. The reclamation of salt marsh by dike constructon and drainage was practiced by New World settlers in New England and Nova Scotia. In the industrial era, large expanses of salt marsh have been destroyed by landfill, a practice only recently inhibited by new legislation.

Common Kinds of Coastlines

There are many different kinds of coastlines, each kind unique because of the distinctive landmass against which the ocean water has come to rest. One group of coastlines derives its qualities from **submergence,** the partial drown-

Figure 22.28　**This broad tidal marsh along the east coast of Florida is laced with serpentine tidal channels. (Laurence Lowry.)**

Figure 22.27　**Salt marsh at South Wellfleet, Massachusetts. Late winter stubble and dead leaves of salt-marsh grass cover the peat layer. A small tidal channel lies drained empty at low tide. (A. N. Strahler.)**

ing of a coast by a rise of sea level or a sinking of the crust. Another group derives its qualities from **emergence,** the exposure of submarine landforms by a falling of sea level or a rising of the crust. Another group of coastlines results when new land is built out into the ocean by volcanoes and lava flows, by the growth of river deltas, or by the growth of coral reefs.

A few important types of coastlines are illustrated in Figure 22.29. The **ria coast** (*A*) is a deeply embayed coast resulting from submergence of a landmass dissected by streams. This coast has many offshore islands. A **fiord coast** (*B*) is deeply indented by steep-walled fiords, which are submerged glacial troughs (Chapter 21). The **barrier-island coast** (*C*) is associated with a recently emerged coastal plain. The offshore slope is very gentle, and a barrier island of sand is usually thrown up by wave action at some distance offshore. Large rivers build elaborate deltas, producing **delta coasts** (*D*). The **volcano coast** (*E*) is formed by the eruption of volcanoes and lava flows, partly constructed below water level. Reef-building corals create new land and make a **coral-reef coast** (*F*). Down-faulting of the coastal margin of a continent can allow the shoreline to come to rest against a fault scarp, producing a **fault coast** (*G*).

A. Ria coast

B. Fiord coast

C. Barrier-island coast

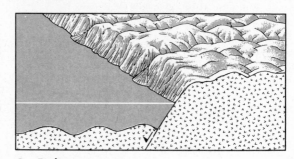

D. Delta coast

E. Volcano coast (left); F. Coral-reef coast (right)

G. Fault coast

Figure 22.29 **Seven common kinds of coastlines are illustrated here. These examples have been selected to illustrate a wide range in coastal features. (Drawn by A. N. Strahler.)**

Development of a Ria Coast

The ria coast is formed when a rise of sea level or a crustal sinking (or both) brings the shoreline to rest against the sides of valleys previously carved by streams. This event is illustrated in Frame *A* of Figure 22.30. Soon wave attack forms cliffs on the exposed seaward sides of the islands and headlands (Frame *B*). Sediment produced by wave action then begins to accumulate in the form of beaches along the cliffed headlands and at the heads of bays. This sediment is carried by littoral drift and is built into sandspits across the bay mouths (baymouth bars) and as connecting links (tombolos) between islands and mainland (Frame *C*). Finally, all outlying islands are planed off by wave action and a nearly straight shoreline develops in which the sea cliffs are fully connected by baymouth bars (Frame *D*). Now the bays are sealed off from the open ocean, although narrow tidal inlets may persist, kept open by tidal currents. Frame *E* shows a much later stage in which the coastline has receded beyond the inner limits of the original bays.

The influence of ria coastlines on human activity has

been strong throughout the ages. The deep embayments of the ria shoreline make splendid natural harbors. Much of the ria coastline of Scandinavia, France, and the British Isles is provided with such harbor facilities. Consequently, these peoples have a strong tradition of fishing, shipbuilding, ocean commerce, and marine activity generally. Mountainous relief of ria and fiord coasts made agriculture difficult or impossible, forcing the people to turn to the sea for a livelihood. New England and the Maritime Provinces of Canada have a ria coastline with abundant good harbors. The influence of this environment was to foster the same development of fishing, whaling, ocean commerce, and shipbuilding seen in the British Isles and Scandinavian countries.

Barrier-Island Coasts

In contrast to ria and fiord coasts, with their bold relief and deeply embayed outlines, we find low-lying coasts from which the land slopes gently beneath the sea. The coastal plain of the Atlantic and Gulf coasts of the United States presents a particularly fine example of such a gently slop-

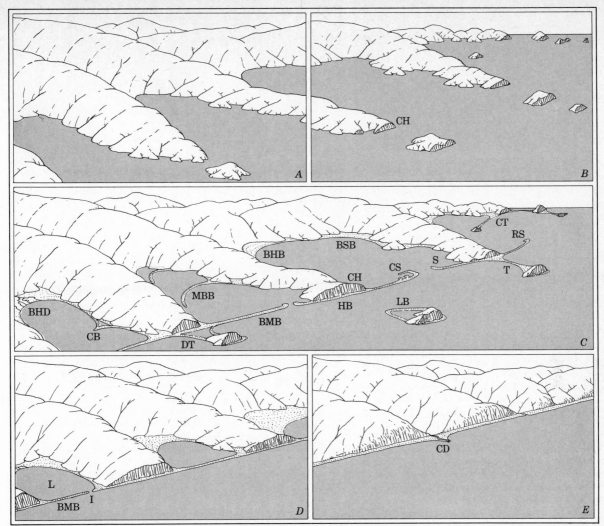

Figure 22.30 **Stages in the evolution of a ria coastline. (Drawn by A. N. Strahler.)**
T = tombolo; S = spit; RS = recurved spit; CS = complex spit; CT = complex tombolo;
LB = looped bar; CH = cliffed headland; DT = double tombolo; HB = headland beach;
BMB = baymouth bar; CB = cuspate bar; BHB = bayhead beach; BSB = bayside beach;
BHD = bayhead delta; L = lagoon; I = inlet; CD = cuspate delta.

ing surface. A coastal plain is a belt of relatively young
sedimentary strata, formerly accumulated beneath the sea
as deposits on the continental shelf (see Figure 24.9).
Emergence as a result of repeated crustal uplifts has
characterized this coastal plain during the latter part of the
Cenozoic Era and into recent time.

Along much of the Atlantic and Gulf coast, a **barrier
island** exists; it is a low ridge of sand built by waves and
further increased in height by the growth of sand dunes
(Figure 22.31). Behind the barrier island lies a **lagoon,**
which is a broad expanse of shallow water, often several
kilometers wide and in places filled largely with tidal
deposits.

A characteristic feature of most barrier islands is the
presence of gaps, known as **tidal inlets.** Through these
gaps, strong currents flow alternately seaward and landward as the tide rises and falls. In heavy storms, the barrier

may be breached by new inlets (Figure 22.32). Tidal
currents will subsequently tend to keep a new inlet open,
but it may be closed by shore drifting of sand.

Shallow water results in generally poor natural harbors
along barrier-island shorelines. The lagoon itself may
serve as a harbor if channels and dock areas are dredged
to sufficient depths. Ships enter and leave through one of
the passes in the barrier island, but artificial seawalls and
jetties are required to confine the current and keep sufficient channel depth. In some cases, major port cities
are located where a large river empties into the lagoon.
The lower courses of large rivers provide tidal channels
that may be dredged to accommodate large vessels and
thus make seaports of cities many kilometers inland.

One of the finest examples of a barrier island and
lagoon is along the Gulf Coast of Texas (Figure 22.33).
Here the island is unbroken for as much as 160 km (100

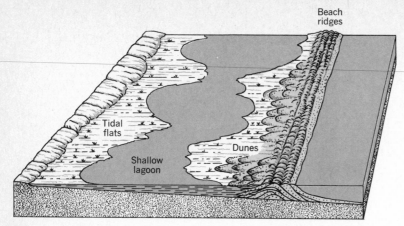

Beach ridges

Tidal flats

Shallow lagoon

Dunes

Figure 22.31 **A barrier island is separated from the mainland by a wide lagoon. Sediments fill the lagoon, while dune ridges advance over the tidal flats. (Drawn by A. N. Strahler.)**

mi) at a stretch, and passes are few. The lagoon is 8 to 16 km (5 to 10 mi) wide. Galveston is built on the barrier island adjacent to an inlet connecting Galveston Bay with the sea. Most other Texas ports, however, are located on the mainland shore. Corpus Christi, Rockport, Texas City, Port Lavaca, and other ports are located along the shores of river embayments.

Barrier islands of the coast of North Carolina are characterized by prominent cuspate headlands — Cape

Hatteras and Cape Lookout — and by a large lagoon — Pamlico Sound (Plate J.2). The inner shoreline is deeply embayed, indicating recent submergence of the coastal plain. The barrier beach has been moving landward through the process of **overwash,** in which storm swash travels across the entire barrier and sweeps sand into the lagoon (Plate J.2). This barrier beach is a subject of intense environmental controversy. Those who seek to develop the beach as a recreation facility and to protect private

Figure 22.32 **East Moriches Inlet was cut through Fire Island, a barrier island off the Long Island shoreline, during a severe storm in March 1931. This aerial photograph, taken a few days after the breach occurred, shows the underwater tidal delta being built out into the lagoon (right) by currents. The entire area shown is about 1.6 km (1 mi) long. North is to the right; the open Atlantic Ocean is on the left. (U.S. Army Air Forces Photograph.)**

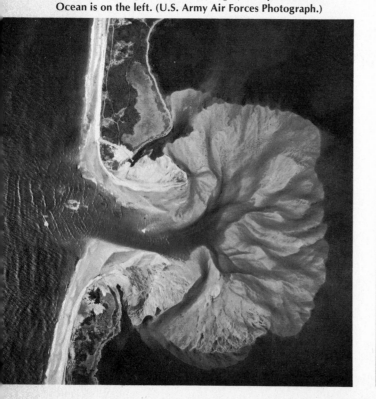

Figure 22.33 **The Gulf Coast of Texas is dominated by its offshore barrier island.**

350   Landforms Made by Waves and Currents

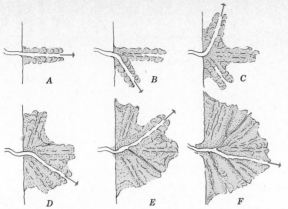

property from destruction are opposed by conservationists and National Park Service officials, who advocate that natural processes should be allowed to proceed unchecked.

Delta Coasts

The deposit of clay, silt, and sand made by a stream where it flows into a body of standing water is known as a **delta** (Figure 22.34). Deposition is caused by rapid reduction in velocity of the current as it pushes out into the standing water. Typically, the river channel divides and subdivides into lesser channels called **distributaries.** The coarse particles settle out first, forming foreset beds. Silts and clays continue out farthest and come to rest as bottomset beds (Figure 22.35). Contact of fresh with salt water causes the finest clays to clot (flocculate) into larger aggregates, which settle to the seafloor.

Deltas show a variety of shapes, both because of the configuration of the coastline and because of wave action. The Nile delta, whose resemblance to the Greek letter "delta" suggested the name for this type of landform, has distributaries that branch out in a radial arrangement

(Figure 22.36*A*). Because of its broadly curving shoreline, causing it to resemble in outline an alluvial fan, this type may be described as an **arcuate delta** (Plate J.2). The Mississippi delta presents a very different sort of picture (Figure 22.36*B*). It is a **bird-foot delta,** with long, projecting fingers growing far out into the water at the ends of each distributary.

Where wave attack is vigorous, the sediment brought out by the stream is spread along the shore in both directions from the river mouth, giving a pointed delta with curved sides. Because of its resemblance to a sharp tooth, this type is called a **cuspate delta** (Figure 22.36*C*). Where a river empties into a long, narrow estuary, the delta is confined to the shape of the estuary (Figure 22.36*D*). This type can be called an **estuarine delta.**

Deltas of large rivers have been of environmental importance from earliest historical times because their extensive flat fertile lands support dense agricultural populations. Important coastal cities, linking ocean and river traffic, are often situated on or near deltas. Examples are Alexandria on the Nile, Calcutta on the Ganges-Brahmaputra, Amsterdam and Rotterdam on the Rhine, Shanghai on the Yangtze, Marseilles on the Rhone, and New Orleans on the Mississippi.

Delta growth is often rapid, ranging from about 3 m (10 ft) per year for the Nile to 60 m (200 ft) per year for the Po and Mississippi rivers. Some cities and towns that were at river mouths several hundred years ago are today several kilometers inland. An important engineering problem is to keep an open channel for ocean-going vessels that have to enter the delta distributaries to reach port. The mouths of the Mississippi River delta distributaries, known as passes, have been extended by the construction of jetties. The narrowed stream is forced to move faster, scouring a deep channel.

Coral-Reef Coasts

Coral-reef coasts are unique in that the addition of new land is made by organisms: corals and algae. Growing together, these organisms secrete rocklike deposits of

Figure 22.35 **Structure of a simple delta shown in a vertical section. (After G. K. Gilbert.)**

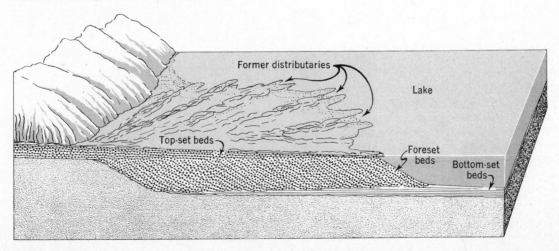

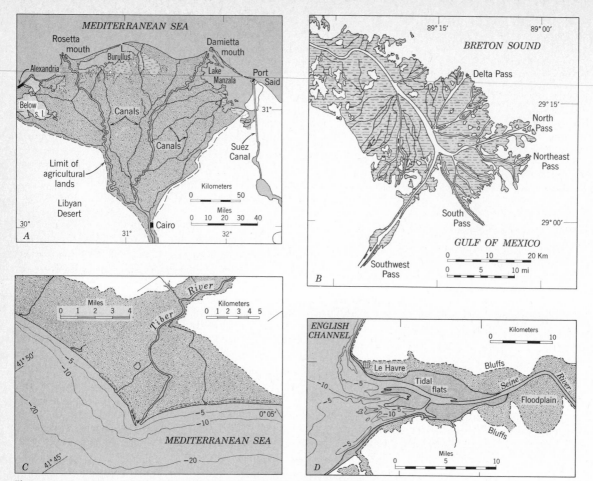

Figure 22.36 Deltas. (*A*) The Nile delta has an arcuate shoreline and is triangular in plan. (*B*) The Mississippi delta is of the branching, bird-foot type with long passes. (*C*) The Tiber delta on the Italian coast is pointed, or cuspate, because of strong wave and current action. (*D*) The Seine delta is filling in a narrow estuary.

mineral carbonate, called **coral reefs.** As coral colonies die, new ones are built on them, accumulating as limestone. Coral fragments are torn free by wave attack, and the pulverized fragments accumulate as sand beaches.

Coral-reef coasts occur in warm, tropical and equatorial waters between the limits lat. 30°N and 25°S. Water temperatures above 20°C (68°F) are necessary for dense reef-coral growth. Reef corals live near the water surface. Water must be free of suspended sediment and well aerated for vigorous coral growth. For this reason corals thrive in positions exposed to wave attack from the open sea. Because muddy water prevents coral growth, reefs are missing opposite the mouths of muddy streams. Coral reefs are remarkably flat on top (Figure 22.37). They are exposed at low tide and covered at high tide.

Three general types of coral reefs may be recognized: (1) fringing reefs, (2) barrier reefs, and (3) atolls. **Fringing reefs** are built as platforms attached to shore (Figure 22.37). They are widest in front of headlands where wave attack is strongest, and the corals receive clean water with abundant food supply.

Barrier reefs lie out from shore and are separated from the mainland by a lagoon (Figure 22.38). Narrow gaps

Figure 22.37 A fringing reef on the south coast of Java forms a broad bench between the surf zone (left) and a white coral-sand beach. Inland is rainforest. (Luchtvaart-Afdeeling, Bandung.)

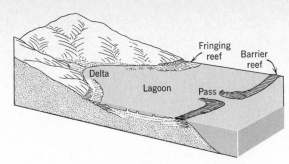

Figure 22.38 A barrier reef is separated from the mainland by a shallow lagoon. (Drawn by W. M. Davis.)

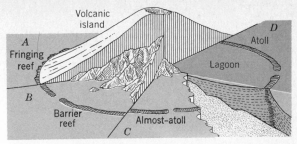

Figure 22.40 The subsidence theory of barrier-reef and atoll development is shown in four stages, beginning with a fringing reef attached to a volcanic island and ending with a circular reef. (Drawn by W. M. Davis.)

occur at intervals in barrier reefs. Through these openings, excess water from breaking waves is returned from the lagoon to the open sea.

Atolls are more or less circular coral reefs enclosing a lagoon, but without any land inside (Figure 22.39). On large atolls, parts of the reef have been built up by wave action and wind to form low island chains, connected by the reef. Most atolls have been built on a foundation of volcanic rock. These foundations are thought to have been basaltic volcanoes, built on the deep ocean floor. According to Charles Darwin's **subsidence theory,** the extinct volcanoes slowly subsided after being planed off by wave erosion. At the same time the fringing coral reef continued to build upward (Figure 22.40).

Figure 22.39 Photographed from an altitude of about 240 km (150 mi) by astronauts aboard the Gemini V spacecraft, Rongelap Atoll in the Marshall Islands, Pacific Ocean, appears as a closed loop among cloud patches. (NASA photograph.)

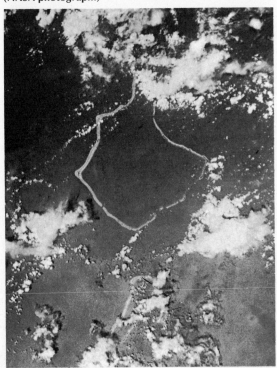

The environmental aspects of atoll islands are unique in some respects. For example, there is no rock other than coral limestone, composed of calcium carbonate. This means that trees requiring other minerals, such as silica, cannot be cultivated without the aid of fertilizers or some outside source of rock from a larger island composed of volcanic or other igneous rock.

The palm tree is native to atoll islands because it thrives on brackish water, and the seed, or palm nut, is distributed widely by floating from one island to another. Native inhabitants cultivated the cocoanut palm to provide food, clothing, fibers, and building materials. Fresh water is scarce on small atoll islands. Rainfall must be caught in open vessels or catchment basins and carefully conserved. Fish and other marine animals are an important part of the human diet on atoll islands. Calm waters of the lagoon make a good place for fishing and for beaching canoes.

Coral islands of the western Pacific stand in continual danger of devastation by tropical cyclones (typhoons). Breaking waves wash over the low-lying ground, sweeping away palm trees and houses and drowning the inhabitants. There is no high ground for refuge.

Raised Shorelines and Marine Terraces

The active life of a shoreline is sometimes cut short by a rapid rise to the coast. When this event occurs, a **raised shoreline** is formed. The marine cliff and abrasion platform are abruptly raised above the level of wave action. The former abrasion platform has now become a **marine terrace** (Figure 22.41). Fluvial denudation begins to destroy the terrace, and it may also undergo partial burial under alluvial fan deposits.

Marine terraces along mountainous coasts are important to Man because they offer strips of flat ground extending for tens of kilometers parallel with the shoreline. Highways and railroads follow these terraces, and they are excellent sites for coastal towns and cities. Agriculture makes use of the flat terrace surfaces where the soil is good.

Raised shorelines are common along the continental and island coasts of the Pacific Ocean because here tectonic processes are active along coastal mountain ranges and island chains. Repeated uplifts result in a series of

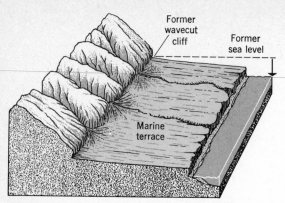

Figure 22.41 A raised shoreline becomes a cliff parallel with the newer, lower shoreline. The former abrasion platform is now a marine terrace. (Drawn by A. N. Strahler.)

raised shorelines in a steplike arrangement. Fine examples of these multiple marine terraces are seen on the western slope of San Clemente Island, off the California coast (Figure 22.42).

Coastal Landforms in Review

Variety is the keyword in describing the landforms produced by waves, currents, and organisms along the world's shorelines. We have dealt with only a few representative examples of the kinds of coastlines to be found on our planet. Each kind of coastal zone offers a different habitat for life forms, and each offers a different situation to confront Man as development and exploitation of coastal resources continues.

In past centuries the coastal zone was used principally as a source of food or as a place from which to embark on the ocean in search of food or to trade with peoples in

other lands. Today the situation is changing. The coastal zone is now in great demand as a recreation zone, and its economic value lies mostly in the worth of the waterfront land as real estate. Marinas spring up at every available sheltered spot, and recreational boating and sport fishing dwarf the shrinking commercial-fishing activity. Luxury condominiums rise on filled land where the tidal flat and salt marsh once supported a complex marine ecosystem. Industry, too, presses for its share of the coastal zone for use as sites of nuclear power plants and oil refineries.

Wise decisions on coastal-zone management often depend partly on accurate knowledge of the operation of coastal processes, such as wave erosion and the littoral drift of sediment. Man-made shoreline changes often set off unwanted retrogradation or progradation elsewhere along the coast, or result in the filling of channels and harbors. These changes also profoundly affect the shallow-water ecosystems.

Transportation and deposition of sand by wind is an important process in shaping coastal landforms. We have made references in this chapter to coastal sand dunes derived from beach sand. In the next chapter we shall investigate the transport of sand by wind and the shaping of dune forms. In so doing we complete the linkage between wind action and wave action in controlling coastal environments.

References for Further Study

Johnson, D. W. (1919, 1965), *Shore processes and shoreline development,* John Wiley and Sons, New York; reprinted by Hafner Publishing Co., New York.

Bascomb, W. (1964). *Waves and beaches,* Doubleday and Co., New York.

King, C. A. M. (1972) *Beaches and coasts,* second edition, St. Martin's Press, New York.

Shepard, F. P. (1975), *Submarine geology,* third edition, Harper and Row, New York.

Strahler, A. N. (1966), *A geologist's view of Cape Cod,* Natural History Press (Doubleday and Co.), Garden City, N.Y.

Figure 22.42 Marine terraces on the western slope of San Clemente Island, off the southern California coast. More than 20 terraces have been identified in this series; the highest has an elevation of about 400 m (1300 ft). (John S. Shelton.)

Review Questions

1. Distinguish between a shoreline and a coastline, or coast. Which includes the greater variety of landforms?
2. Describe the motion of water particles in oscillatory waves. What causes waves of oceans and lakes?
3. Describe the breaking of a wave. What causes a shoaling wave to steepen? Distinguish between swash and backwash. What work is performed by swash and backwash?
4. In what materials is marine erosion most effective? What is a marine scarp, and how does it differ from a marine cliff? Name and describe the erosional features of a marine cliff. What is a hanging valley?
5. Of what grade sizes of material are beaches formed? How does grade size affect beach slope? Describe a typical beach profile. What is a berm and how is it formed? What features are produced by beach progradation?
6. Explain how wave refraction on a straight shoreline can produce littoral drift. Distinguish between beach drift and longshore drift. What landform is produced by littoral drift?
7. How are cuspate bars and cuspate forelands produced by littoral drift? Describe the surface features of a cuspate foreland.
8. Describe wave refraction along an embayed coast. Where is wave energy concentrated? What landforms are developed by wave erosion on promontories or headlands? What landform is produced by sediment accumulation at the head of a bay? What is a shingle beach? What is a tombolo?
9. What engineering measures are taken to prevent beach destruction by retrogradation? How do groins alter the shoreline? What effects may dam-building on large rivers have on a shoreline? Explain.
10. Describe a typical tide curve. How much time elapses between one high water and the next low water? How are tidal currents generated? Distinguish between the ebb current and the flood current. Which has the higher speed and why?
11. Describe the process of accumulaton of tidal sediments. How is salt marsh formed? What environmental problems arise from reclamation of salt marsh?
12. Distinguish between coastal submergence and coastal emergence. How does each of these events influence the form of a shoreline? Name and describe seven common types of coastlines.
13. Describe the evolution of a ria coast and the shoreline landforms that are developed. What are the environmental influences of a ria coast?
14. What is a barrier-island coast? How is a barrier island formed? What is a lagoon? What are tidal inlets? How are they maintained? Describe the barrier-island coast of Texas, using it as an example of the environmental influences of this type of coast. How does overwash affect a barrier island?
15. How is a simple delta formed? How do shoreline processes modify the outline of a delta? Give examples. Compare the outline of the Nile delta with that of the Mississippi delta. Explain why they differ.
16. How are coral reefs formed? In what latitudes do they thrive? Describe fringing reefs and barrier reefs. What is an atoll? How is it formed? Describe an atoll island as an environment for human habitation. What natural hazards affect atoll islands?
17. What is a raised shoreline? How are marine terraces formed? How are marine terraces utilized by Man?
18. Discuss current attitudes and conflicts of interest in the continued exploitation of coastlines. In what ways is the coastline a nonrenewable natural resource?

Landforms Made by Wind

23

Wind blowing over the solid surface of the lands is another active agent of landform development. Ordinarily, wind is not strong enough to dislodge mineral matter from the surfaces of hard rock, or of moist, clay-rich soils, or of soils bound by a dense plant cover. Instead, the action of wind in eroding and transporting sediment is limited to land surfaces where small mineral and organic particles are in the loose state. Such areas are typically deserts and semi-arid lands (steppes). An exception is the coastal environment, where beaches provide abundant supplies of loose sand, even where the climate is moist and the land surface inland from the coast is well protected by a plant cover.

Landforms shaped and sustained by wind erosion and deposition represent distinctive life environments, often highly specialized with respect to the communities of animals and plants they support. In climates with barely sufficient soil water, there is a contest between wind action and the growth of plants that tends to stabilize landforms and protect them from wind action. We shall find precarious balances in the ecosystems of certain marginal climatic zones. These balances are not only altered by natural changes in climate, but are easily upset by Man's activities, often with serious consequences. To

understand these environmental changes, we need to acquire a working knowledge of the physical processes of wind action on the land surfaces.

Erosion by Wind

Wind performs two kinds of erosional work. Loose particles lying on the ground surface may be lifted into the air or rolled along the ground. This process is **deflation.** Where the wind drives sand and dust particles against an exposed rock or soil surface, causing it to be worn away by the impact of the particles, the process is **wind abrasion.** Abrasion requires cutting tools carried by the wind; deflation is accomplished by air currents alone.

Deflation acts wherever the ground surface is thoroughly dried out and is littered with small, loose particles of soil or regolith. Dry river courses, beaches, and areas of recently formed glacial deposits are highly susceptible to deflation. In dry climates, almost the entire ground surface is subject to deflation because the soil or rock is largely bare. Wind is selective in its deflational action. The finest particles, those of clay and silt sizes, are lifted most easily and raised high into the air. Sand grains

Figure 23.1 **This blowout hollow on the plains of Nebraska contains a remnant column of the original material, thus providing a natural meter stick for the depth of material removed by deflation. (N. H. Darton, U.S. Geological Survey.)**

Figure 23.2 **Front of an approaching dust storm, Coconino Plateau, Arizona. (D. L. Babenroth.)**

are moved only by moderately strong winds and travel close to the ground. Gravel fragments and rounded pebbles can be rolled over flat ground by strong winds, but they do not travel far. They become easily lodged in hollows or between other large grains. Consequently, where a mixture of sizes of particles is present on the ground, the finer sizes are removed and the coarser particles remain behind.

A landform produced by deflation is a shallow depression called a **blowout,** or **deflation hollow.** This depression may be from a few meters to a kilometer or more in diameter, but it is usually only a few meters deep. Blowouts form in plains regions in dry climates. Any small depression in the surface of the plain, particularly where the grass cover is broken through, may develop into a blowout. Rains fill the depression, creating a shallow pond or lake. As the water evaporates, the mud bottom dries out and cracks, forming small scales or pellets of dried mud that are lifted out by the wind. In grazing lands, animals trample the margins of the depression into a mass of mud, breaking down the protective grass-root structure and facilitating removal when dry. In this way the depression is enlarged (Figure 23.1). Blowouts are also found on rock surfaces where the rock is being disintegrated by weathering.

In the great desert of the southwestern United States, the floors of intermontane basins are vulnerable to defla-

tion. The flat floors of shallow playas have in some places been reduced by deflation as much as several meters over areas of many square kilometers.

Where deflation has been active on a ground surface littered with loose fragments of a wide range of sizes, the pebbles that remain behind accumulate until they cover the entire surface. By rolling or jostling about as the fine particles are blown away, the pebbles become closely fitted together, forming a **desert pavement** (Plate J.4). The armored surface is well protected against further deflation, but it is easily disturbed by the wheels of trucks or motorcycles.

Dust Storms and Sandstorms

Strong, turbulent winds, blowing over desert surfaces, lift great quantities of fine dust into the air, forming a dense, high cloud called a **dust storm.** In semiarid grasslands a dust storm is generated where ground surfaces have been stripped of protective vegetative cover by cultivation or grazing. A dust storm approaches as a dark cloud extending from the ground surface to heights of several kilometers (Figure 23.2). Within the dust cloud, there is deep gloom or even total darkness. Visibility is cut to a few meters, and a fine choking dust penetrates everywhere.

It has been estimated that as much as 1000 metric tons of dust may be suspended in a cubic kilometer of air (4000 tons/cu mi). On this basis, a large dust storm can carry more than 100 million metric tons of dust — enough to make a hill 30 km (100 ft) high and 3 km (2 mi) across the base. Dust travels long distances in the air. Dust from a single desert dust storm is often traceable as far as 4000 km (2500 mi).

The true desert **sandstorm** is a low cloud of moving sand that rises usually only a few centimeters and at most 2 m (6 ft) above the ground. It consists of sand particles driven by a strong wind. Those who have experienced sandstorms report that the head and shoulders of a person standing upright may be entirely above the limits of the sand cloud. The reason the sand does not rise higher is that the individual particles are engaged in a leaping motion, termed **saltation** (Figure 23.3). Grains describe a curved trajectory and strike the ground with considerable force, but at a low angle. The impact with other grains causes the sand grains to rebound into the air. At the same time, the surface layer of sand grains creeps downwind as the result of the constant impact of the bouncing sand grains.

Figure 23.3 **Sand particles travel in a series of long leaps. (After R. A. Bagnold.)**

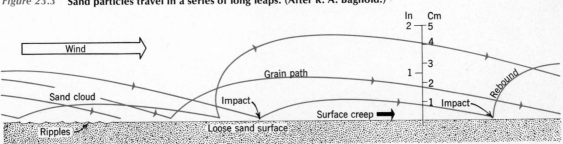

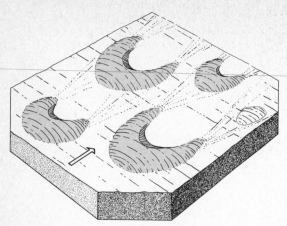

Figure 23.4 Barchans, or crescentic dunes. Arrow indicates prevailing wind direction. (Drawn by A. N. Strahler.)

The erosional effect of sand in saltation is thus concentrated on surfaces exposed less than a meter (3 ft) above the flat ground surface. Fence posts and telephone poles on wind-swept sandy plains are quickly cut through at the base unless a protective metal sheathing or heap of large stones is placed around the base.

Sand Dunes

A **dune** is an accumulation of sand shaped by the wind and capable of movement over the underlying ground surface. Dunes that are bare of vegetation and constantly changing form under wind currents are **live dunes,** or active dunes. They may be inactive, as **fixed dunes,** covered by vegetation that has taken root and prevented further shifting of the sand.

Most dune sands are composed of the mineral quartz. Quartz grains become highly spherical in shape through abrasion (see Figure 11.7). Rarely, dunes are composed of tephra (volcanic sand), of shell fragments that accumulated in beaches, of grains of gypsum, or of heavy minerals such as magnetite.

One common type of live sand dune is an isolated heap of free sand called a **crescent dune,** or **barchan.** As the name suggests, this dune has a crescentic outline; the points of the crescent are directed downwind (Figure 23.4). On the windward side of the crest, the sand slope is gentle and smoothly rounded. On the lee side of the dune, within the crescent, is a steep curving dune slope, the **slip face.** This face maintains an angle of about 35° from the horizontal (Figure 23.5). Sand grains slide down the steep face after being blown free of the sharp crest. When a strong wind is blowing, the flying sand makes a perceptible cloud at the crest.

Crescent dunes rest on a flat, pebble-covered ground surface. The sand heap may originate as a drift in the lee of some obstacle, such as a small hill, rock, or clump of brush. Once a sufficient mass of sand has formed, it begins to move downwind, taking the form of a crescent dune. For this reason the dunes are usually arranged in chains extending downwind from the sand source.

Where sand is so abundant that it completely covers the ground, dunes take the form of wavelike ridges separated by troughlike furrows. The dunes are called **transverse dunes** because, like ocean waves, their crests trend at right angles to the direction of wind (Figure 23.6). The entire area may be called a **sand sea** because it resembles a storm-tossed sea suddenly frozen to immobility. The sand ridges have sharp crests and are asymmetrical, the gentle slope being on the windward, the steep slip face on the lee side. Deep depressions may lie between the dune ridges. Sand seas require enormous quantities of sand, often derived from the weathering of a sandstone formation underlying the ground surface or from adjacent alluvial plains. Still other transverse dune belts form adjacent to beaches that supply abundant sand and have strong onshore winds (see Figure 23.8).

Another group of dunes belongs to a family in which the curve of the dune crest is bowed convexly downwind, the opposite of the curvature of crests in the crescent dune. These dunes are parabolic in outline and are classed as **parabolic dunes.** A common representative of this class is the **coastal blowout dune,** formed adjacent to beaches. Here large supplies of sand are available and are blown landward by prevailing winds (Figure 23.7A). A saucer-shaped depression is formed by deflation; the sand is heaped in a curving ridge resembling a horseshoe in plan.

Figure 23.5 Barchan dunes at Biggs, Oregon. (G. K. Gilbert, U.S. Geological Survey.)

Figure 23.6 This air photograph of a sand-dune field between Yuma, Arizona, and Calexico, California, shows a sand sea of transverse dunes in the background and a field of crescentic barchan dunes in the foreground. (Spence Air Photos.)

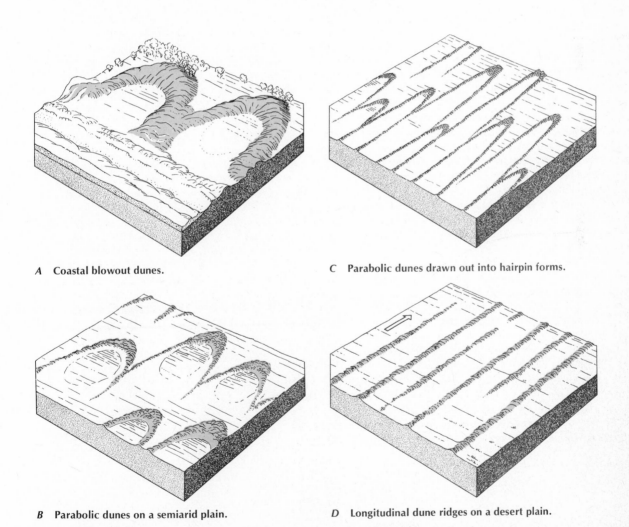

A Coastal blowout dunes.

C Parabolic dunes drawn out into hairpin forms.

B Parabolic dunes on a semiarid plain.

D Longitudinal dune ridges on a desert plain.

Figure 23.7 Four types of dunes. Types *A, B,* and *C,* are of the parabolic class. The prevailing wind direction, shown by an arrow in Block *D,* is the same for all four examples. (Drawn by A. N. Strahler.)

Figure 23.8 **The arrows on this photograph point to elongate blowout dunes of hairpin form, which once advanced from the beach and have since become stabilized by vegetation. Active transverse dunes are overriding the blowout dunes in a fresh wave. San Luis Obispo Bay, California. (Spence Air Photos.)**

On the landward side is a steep slip face that advances over the lower ground and buries forests, killing the trees (Plate J.4). Coastal blowout dunes are well displayed along the southern and eastern shore of Lake Michigan. Dunes of the southern shore have been protected for public use as the Indiana Dunes State Park.

On semiarid plains, where vegetation is sparse and winds strong, groups of **parabolic blowout dunes** develop to the lee of shallow deflation hollows (Figure 23.7*B*). Sand is caught by low bushes and accumulates in a broad, low ridge. These dunes have no steep slip faces and may remain relatively immobile. In some cases the dune ridge migrates downwind, drawing the parabola into a long, narrow form with parallel sides resembling a hairpin in outline (Figure 23.7*C*). These **hairpin dunes,** stabilized by vegetation, are seen in Figure 23.8.

Another class of dunes is described as longitudinal because the dune ridges run parallel with the wind direction. On desert plains and plateaus, where sand supply is meager but winds are strong from one direction, **longitudinal dunes** are formed (Figure 23.7*D*). These are usually only a few meters high, but may be several kilometers long. In some areas, the longitudinal dune is produced by extreme development of the hairpin dune, such that the parallel side ridges become the dominant form.

Longitudinal dune ridges, oriented parallel with dominant winds, occupy vast areas of central Australia; they are referred to as **sand-ridge deserts** (Figure 23.9). Ridges average 10 to 15 m (30 to 50 ft) in height and run in continuous length as much as 80 km (50 mi).

Coastal Dunes and Man

Landward of sand beaches we usually find a narrow belt of dunes in the form of irregularly shaped hills and depressions; these are the **foredunes.** They normally bear a cover of beachgrass and a few other species of plants capable of survival in the severe environment. Dunes

developed under a partial cover of plants are called **phytogenic dunes.**

On coastal foredunes, the cover of beachgrass and other small plants, sparse as it seems to be, acts as a baffle to trap sand moving landward from the adjacent beach

Figure 23.9 **This oblique air photograph shows longitudinal sand-dune ridges reaching as far as the eye can see. Simpson Desert, southeast of Alice Springs, Australia. (George Silk, *Life Magazine,* © Time, Inc.)**

Figure 23.10 **Foredunes protected by beachgrass, Provincelands of Cape Cod. (A. N. Strahler.)**

(Figure 23.10). As a result, the foredune ridge is built up as a barrier rising several meters above high tide level. For example, dune summits of the Landes coast of France reach heights of 90 m (300 ft) and span a belt 10 km (6 mi) wide.

The swash of storm waves cuts away the upper part of the beach, and the dune barrier is attacked. Between storms the beach is rebuilt and, in due time, the dune ridge is also restored if a plant cover is maintained. In this way the foredunes form a protective zone for tidal lands lying on the landward side of a barrier island.

If the plant cover of the dune ridge is depleted by vehicular and foot traffic or by the bulldozing of sites for approach roads and buildings, a blowout will rapidly develop. The new cavity may extend as a trench across the dune ridge. With the onset of a storm with high water levels, swash is funneled through the gap and spreads out on the tidal marsh or tidal lagoon behind the ridge. This activity, called overwash, was mentioned in Chapter 22. Sand swept through the gap is spread over the tidal deposits.

Breaching of dune barriers along the North Sea Coast has caused severe losses in lives and property. These barriers are part of the system to exclude seawater from reclaimed polders and fenlands. Protection of the dunes of the Netherlands coast assumes vital importance in view of the loss of life and property that a series of storm breaches can bring.

Another form of environmental damage related to dunes is the rapid downwind movement of sand when the dune status is changed from one of fixed, plant-controlled forms to that of active dunes of free sand. When plant cover is depleted, wind rapidly reshapes the dunes to produce crests and slip faces. Blowout dunes are developed, and the free sand slopes advance on forests, roads, buildings, and agricultural lands. In the Landes region of coastal dunes on the southwestern coast of France, landward dune advance has overwhelmed houses and churches and even caused entire towns to be abandoned.

A striking case of Man's interference with a dune environment is that of the Provincelands of Cape Cod, located at the northern tip of that peninsula, making up the fist of the armlike outline of the Cape (Figure 23.11). The Provincelands has been constructed of beach sand built into a succession of beach ridges, and these have been modified in form and increased in height by dune

Figure 23.11 **Sketch map of the Provincelands of Cape Cod, Massachusetts, as mapped in 1887. Dunes show as stippled pattern. (Drawn by A. N. Strahler.)**

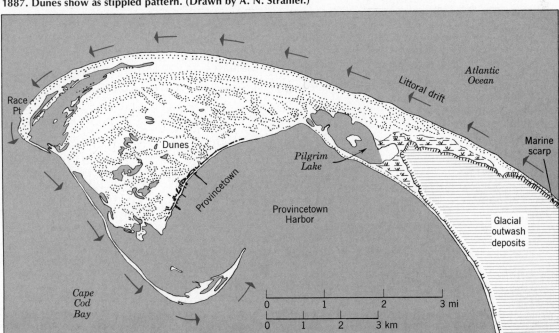

building. When the first settlers arrived in Provincetown, a city now occupying the south shore of the Provincelands, the dunes were naturally stabilized by grasses and other small plants covering the dune summits and by pitch pine and other forest trees on lower surfaces and in low swales between dune ridges, although dunes were probably active then, as now, on ridges close to the northern shore. Inhabitants grazed their livestock on the dune summits and rapidly cut the forests for fuel, with the result that the dunes were activated and began to move southward.

By 1725 Provincetown was being overwhelmed by drifting sand. Some buildings were partially buried and some had to be abandoned. Sand was carted away from the streets in large volumes. By the early 1800s the major dune ridges were moving southward at a rate estimated to be 27 m (90 ft) per year. Fortunately, beachgrass plantings, begun in 1825, and the rigorous enforcement of laws forbidding grazing and tree cutting resulted finally in the stabilization of all but the northernmost dunes. The area is today a part of the Cape Cod National Seashore. Park authorities have made extensive new plantings of dune grass and have minimized vehicular traffic in an attempt to bring further control to sand movement.

Loess

Wind-deposited silt, called loess, was described in Chapter 11 as an important class of parent matter of soils. Although loess is a variety of sediment, it lacks horizontal stratification. Instead, because of slight shrinkage during compaction, loess has a tendency to break away along vertical planes. This structure, called **cleavage,** is similar to vertical jointing in sedimentary rocks. Because of cleavage, deep natural exposures of loess in the upper walls of stream valleys tend to form vertical cliffs. Cliff retreat occurs by the collapse of loess columns. A Man-made perpendicular highway cut made in loess is remarkably stable and may resist erosion for decades (Plate J.4).

The thickest deposits of loess are in northern China, where a layer over 30 m (100 ft) thick is common and a maximum of 90 m (300 ft) has been measured. Loess covers many hundreds of square kilometers and appears to have been derived as dust from the interior of Asia. Loess deposits are also of major importance in the United States, central and eastern Europe, central Asia, and Argentina.

In the United States, thick loess deposits lie in the Missouri-Mississippi valley (Figure 23.12). Much of the prairie plains region of Indiana, Illinois, Iowa, Missouri, Nebraska, and Kansas is underlain by a loess layer ranging in thickness from 1 to 30 m (3 to 100 ft). Extensive deposits also occur along the lands bordering the lower Mississippi River floodplain on its east side, throughout Tennessee and Mississippi. Other important loess deposits are the Palouse region of southeast Washington, western Idaho, and northeast Oregon.

The American and European loess deposits are directly related to the continental glaciers of the Pleistocene

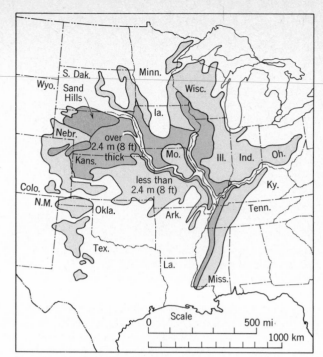

Figure 23.12 **Map of loess distribution in the central United States. (Data from Map of Pleistocene Eolian Deposits of the United States, Geological Society of America, 1952.)**

Epoch. At the time when the ice covered much of North America and Europe, a generally dry winter climate prevailed in the land bordering the ice sheets. Strong winds blew southward and eastward over the bare ground, picking up silt from the floodplains of braided streams that discharged the meltwater from the ice. This dust settled on the ground between streams, gradually building up to produce a smooth, level ground surface. The loess is particularly thick along the eastern sides of the valleys because of prevailing westerly winds, and it is well exposed along the bluffs of most streams flowing through the region today.

Loess is of major importance in world agricultural resources. Loess forms the parent matter of the Mollisols, black and brown soils especially suited to the cultivation of grains because of their high base status. The highly productive plains of southern Russia, the Argentine Pampa, and the rich wheat-producing region of north China are underlain by loess. In the United States, corn is extensively cultivated on the loess plains in such states as Iowa and Illinois, where rainfall is sufficient; wheat is grown farther west on loess plains of Kansas and Nebraska and in the Palouse region of eastern Washington.

Because loess forms vertical walls and is easily excavated, it has been widely used for cave dwellings both in China and in central Europe. In China, old trails and roads in the loess have become deeply sunken into the ground as a result of the pulverization of the loess of the roadbed and its removal by wind and water (Figure 23.13).

Man as an Agent in Inducing Deflation

Cultivation of vast areas of short-grass prairie under a climate of substantial seasonal water shortage is a practice that invites deflation of soil surfaces. Much of the Great Plains region of the United States is such a marginal region. In past centuries these plains have experienced many dust storms generated by turbulent winds. Strong cold fronts frequently sweep over this area, lifting dust high into the troposphere at times when soil water is low.

Man's activities in the very dry, hot deserts have contributed measurably to the raising of dust clouds. In the desert of northwest India and Pakistan (the Thar Desert bordering the Indus River), the continued trampling of fine-textured soils by the hooves of grazing animals and by human feet produces a blanket of dusty hot air that hangs over the region for long periods and extends to a height of 9 km (30,000 ft).

In other deserts, such as those of North Africa and the southwestern United States, ground surfaces in the natural state contribute comparatively little dust because of the presence of desert pavements and sheets of coarse sand from which fine particles have already been winnowed. This protective layer is easily destroyed by wheeled vehicles, exposing finer-textured materials and allowing deflation to raise dust clouds. The disturbance of large expanses of North African desert by tank battles during World War II caused great dust clouds; dust from this source was identified as far away as the Caribbean region. Dust clouds, visible on Landsat images, have been noted in recent years in the Mojave Desert of southern California. They result from various types of Man-induced ground disturbances, including the grinding action of wheels of off-road vehicles, such as trail motorcycles and four-wheel-drive recreation vehicles.

Landforms and Man

The group of seven chapters we have now completed has reviewed geomorphic processes that operate on the surface of the continents. The great variety and complexity of landforms we have described is not difficult to comprehend when each agent of denudation is examined separately. Landforms produced by glaciers, waves, and wind are localized in distinctive environments — alpine and arctic regions, coastlines, and dry regions, respectively. The landforms of fluvial denudation are the most widespread and complex features of the continental landscape. Fluvial denudation integrates weathering, mass wasting, and fluvial processes into highly complex systems of denudation responding to climate controls.

Man's influence on landforms is most strongly felt on surfaces of fluvial denudation because of the severity of surface changes caused by agriculture and urbanization. Landforms shaped by wind and by waves and currents are also highly sensitive to changes induced by human activity. Only glaciers maintain their integrity and are thus far undisturbed by human activity. Perhaps even this last realm of Nature's superiority will eventually fall prey to Man-induced climate changes.

Figure 23.13   **Road sunken deeply into loess, Shensi, China. (Frederick G. Clapp, *The Geographical Review*, American Geographical Society.)**

In the next three chapters we change our perspective somewhat, emphasizing first the large crustal features of the earth and their geologic differences as continents and ocean basins. Through a revolutionary concept of earth processes — plate tectonics — the earth's major crustal features find a coherent and unified explanation. Using plate tectonics as a new frame of reference, we can take a second look at landforms. Two major processes are powered by internal energy sources: volcanism and tectonic activity. These processes create many kinds of bold landforms of mountainous proportions. Rock structure within these initial mountain masses exerts a strong control over the sequential landforms of fluvial denudation. Structural control adds a new dimension of complexity to the denudational landforms. In the next three chapters we hope to achieve a global perspective on landforms comparable to that achieved in earlier chapters by recognizing global patterns of climate, soils, and vegetation.

References for Further Study

Gautier, E. F. (1935), *Sahara: The great desert,* translated by D. F. Mayhen, Columbia University Press, New York.

Lobeck, A. K. (1939), *Geomorphology,* McGraw-Hill Book Co., New York. See Chapter 11.

Bagnold, R. A. (1941), *The physics of blown sand and desert dunes,* Methuen and Co., London.

Cotton, C. A. (1942), *Climatic accidents in landscape-making,* Whitcombe and Tombs, Christchurch, New Zealand. See Chapters 1 and 9.

Thornbury, W. D. (1969), *Principles of geomorphology,* second edition, John Wiley and Sons, New York. See Chapter 12.

Flint, R. F. (1971), *Glacial and Quaternary geology,* John Wiley and Sons, New York, See Eolian features, pp. 243–266.

Review Questions

1. Describe the processes of deflation and abrasion by wind. What conditions favor deflation? What landforms are produced ?

2. What is a desert pavement? How does it form? What forms does wind abrasion produce?

3. How do dust storms originate? How much material might be carried in a single dust storm? How far does the dust travel?

4. How do sand grains travel in a sandstorm? How high do the grains rise? At what level is abrasion concentrated?

5. What are sand dunes? What is the distinction between live dunes and fixed dunes?

6. Describe a crescentic, or barchan, dune. In which direction does it move?

7. What are transverse dunes? What is a sand sea? What is the source of the sand? What places would be most favorable for the development of dune areas?

8. Describe a coastal blowout dune, a parabolic blowout dune, and a hairpin dune. How does each develop?

9. Describe longitudinal sand dunes. What do they indicate as to the direction of prevailing winds?

10. How do Man's activities result in environmental damage to coastal foredunes and their bordering tidal marshes? How can the migration of live coastal dunes be controlled?

11. What is loess? How is it formed? What structure has loess?

12. Describe the distribution of loess throughout the world. What origin has the loess of northern China?

13. What states of the Mississippi-Missouri river region have loess deposits? How is loess important as a natural resource?

14. In what ways do Man's activities induce deflation on a large scale in arid lands?

The Lithosphere and Plate Tectonics

24

Environmental regions of the globe depend for their distribution on configurations of the earth's crust dictated by geologic processes. These processes are powered by energy sources deep within the earth. The internal earth forces have shaped the continents and ocean basins without conforming in the least to the orderly latitude zones of climate. We can think of latitude zones of temperature, winds, and precipitation as concentric color bands painted on a circular dinner plate as the potter's wheel spins. Suppose that the dinner plate falls to the floor and is shattered and that then the fragments are put back into place and cemented together. The fracture patterns cut across the circular color zones in a discordant and random pattern.

This same unique combination of disorder on order characterizes the earth's solid surface. We find volcanoes erupting today in the cold desert of Antarctica as well as near the equator in Africa. An alpine mountain range has been pushed up in the cold subarctic zone of Alaska, where it trends east-west; but another lies astride the equator in South America and runs north-south. Both ranges lie in belts of intense crustal activity, where numerous strong earthquakes are generated. Although the geologic processes that create high mountains are quite insensitive to latitude and climate, climates do respond to mountain ranges through the orographic effect. In this way, the chance configurations of the earth's relief features bring diversity to the global climate.

In this chapter we shall survey the major geologic features of our planet, starting with its deep interior as a layered structure. We then examine the outermost layer, or crust, comparing crustal forms of the continents with those of the ocean basins. Only within the past decade have geologists provided a unified theory to explain the differences between continents and ocean basins and to interpret the major forms of crustal unrest in a meaningful way. Fortunately, we are reviewing geologic processes just at the moment in history when a new revolution in geology has accomplished a scientific upheaval. The revolutionary findings can now furnish us with a complete scenario of earth history on a grand scale of both time and spatial dimensions.

The Earth's Interior

Figure 24.1 is a cutaway diagram of the earth to show its major parts. The earth is an almost spherical body approximately 6400 km (4000 mi) in radius. The center is occupied by the **core,** a spherical zone about 3500 km (2200 mi) in radius. Because of the sudden change in behavior of earthquake waves upon reaching this zone, it has been concluded that the outer core has the properties of a liquid, in abrupt contrast to the solid state of the rock shell that surrounds it. The innermost part of the core, with a radius of 1255 km (780 mi), is probably in the solid state.

Astronomical calculations show that the earth has an average density of about 5½ gm/cc, whereas the surface rocks average 3 gm/cc or less (Chapter 11). This observation must mean that density increases greatly toward the earth's center, where it may be about 10–15 gm/cc. Iron,

Figure 24.1   **Concentric zones of the earth's interior.**

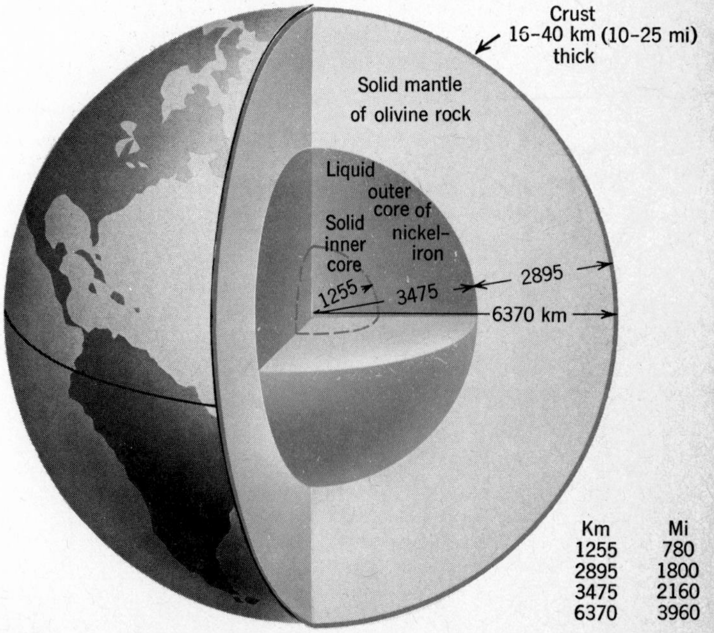

Km	Mi
1255	780
2895	1800
3475	2160
6370	3960

with a small proportion of nickel, is considered to be the substance comprising the entire core. This conclusion is supported by the fact that many meteorites, representing disrupted fragments of our solar system, are of iron-nickel composition. Temperatures in the earth's core may lie between 2200° and 2750°C (4000° and 5000°F); pressures are as high as three to four million times the pressure of the atmosphere at sea level.

Outside the core lies the **mantle,** a layer about 2895 km (1800 mi) thick, composed of mineral matter in a solid state. Judging from the behavior of earthquake waves, the mantle is probably composed largely of the mineral olivine (magnesium iron silicate). A comparatively rare surface rock of this composition is dunite (see Figure 11.2). The upper zone of the mantle is probably of the composition of peridotite, another ultramafic rock described in Chapter 11.

The Earth's Crust

Outermost and thinnest of the earth zones is the **crust,** a layer 5 to 40 km (3 to 25 mi) thick, formed largely of igneous rocks. The base of the crust, where it contacts the mantle, is sharply defined. This contact is established from the way in which earthquake waves change velocity abruptly at that level (Figure 24.2A). The surface of separation between crust and mantle is called the **Moho,** a simplification of the name of the seismologist who discovered it.

From a study of earthquake waves it is concluded that the crust beneath the continents consists of two rock layers: a lower, continuous rock layer of mafic composition; and an upper layer of granitic rock. There is no

sharply defined surface of separation between these layers. The crust of the ocean basins consists of basalt; the felsic continental layer is absent. Figure 24.2B shows schematically a small part of the crust near the margin of a continent.

The crust is much thicker beneath the continents than beneath the ocean floors, as Figure 24.2B shows. Whereas 40 km (25 mi) is a good average figure for crustal thickness beneath the continents, 5 km (3 mi) is an average figure for thickness of the basaltic crust beneath the deep ocean floors. The reasons for differences in both thickness and rock composition between continental and oceanic crust are explained later in terms of processes that have created the crust.

Lithosphere and Asthenosphere

A new concept must now be added to what has been said thus far about the earth's structure. Evidence derived from the study of earthquake waves shows that within the upper mantle is a **soft layer** in which the mantle rock is at a temperature close to the melting point. A good analogy would be a bar of cast iron heated at one end until it is white hot, but not quite hot enough to melt. The cold part of the bar is brittle and will snap with a sharp fracture if it is struck. The hot part of the bar is quite soft and can be shaped under pressure of a vise or by hammer blows. Beneath the continental crust, the soft layer of the mantle sets in at an average depth of about 80 km (50 mi), which is well below the base of the continental crust (see Figure 24.2). Beneath the oceanic crust, the soft layer sets in at a depth of about 40 km (25 mi).

In modern geologic language, the soft layer of the upper mantle is the **asthenosphere;** the rigid layer above it is the **lithosphere.** Geologists have thus restricted the meaning of lithosphere to an outer rock layer; but this does not conflict with our use of the word to mean the solid earth realm. The asthenosphere, or soft layer, extends to a depth of about 300 km (185 mi). The upper and lower limits are gradational. The weakest portion of the asthenosphere is at a depth of roughly 200 km (125 mi).

The important concept we derive from these facts is that the rigid lithosphere has the capability of moving bodily over the asthenosphere. The asthenosphere yields by slow plastic flowage distributed through a thickness of many tens of kilometers.

If the lithosphere formed a single continuous shell over the entire earth, that shell would—in theory at least—be capable of rotating bodily over the deeper mantle and core. Instead, the lithosphere is broken into large units called **lithospheric plates.** A single plate is of continental dimensions and is capable of moving independently of the plates that surround it. Plate motions and boundaries are a topic we shall elaborate on later in this chapter.

Distribution of Continents and Ocean Basins

The first-order relief features of the earth are the continents and ocean basins. Using a globe, we can compute that about 29 percent of the globe is land, 71 percent is oceans.

Figure 24.2 The earth's crust is much thicker under continents than beneath the ocean basins.

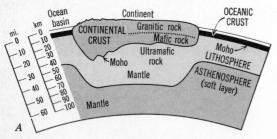

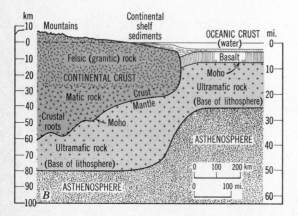

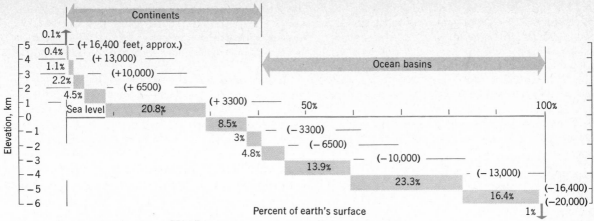

Figure 24.3 **Distribution of the earth's solid surface in successively lower elevation zones.**

If the seas were to drain away, however, it would become obvious that broad areas lying close to the continental shores are actually covered by shallow water, less than 180 m (600 ft) deep. From these relatively shallow continental shelves the ocean floor drops rapidly to depths of thousands of meters. In a way, then, the ocean basins are brimful of water. The oceans have even spread over the margins of ground that would otherwise be assigned to the continents. If the ocean level were to drop by 180 m (600 ft), the surface area of continents would increase to 35 percent; the ocean basins would decrease to 65 percent. We can use these figures as representative of the true relative proportions.

Figure 24.3 shows graphically the percentage distribution of the earth's surface area with respect to elevation both above and below sea level. Note that most of the land surface of the continents is less than 1 km (3300 ft) above sea level. There is a rapid drop off from about −1 to −3 km

(−3000 to −10,000 ft) until the ocean floor is reached. A predominant part of the ocean floor lies between 3 and 6 km (10,000 and 20,000 ft) below sea level. Disregarding the earth's curvature, the continents can be visualized as platformlike masses, the oceans as broad, flat-floored basins.

Scale of the Earth's Relief Features

Before turning to a description of the major subdivisions of the continents and ocean basins, we need to grasp the true scale of the earth's relief features in comparison with the earth as a sphere (Figure 24.4). Most relief globes and relief maps are greatly exaggerated in vertical scale.

For a true-scale profile around the earth, we might draw a chalk-line circle 6.4 m (21 ft) in diameter, representing the earth's circumference on a scale of 1:2,000,000. A chalk line 1 cm (⅜ in.) wide would include

Figure 24.4 **These profiles show the earth's great relief features in true scale, with sea-level curvature fitted to a globe 6.4 m (21 ft) in diameter.**

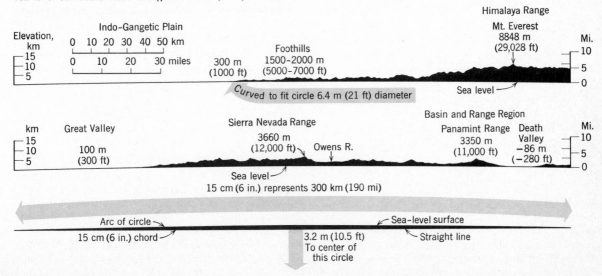

Table 24.1 / The Geologic Time Scale

Era	Period	Epoch	Duration m.y.	Age m.y.	Orogenies
CENOZOIC		Holocene	(10,000 yr)		
		Pleistocene	2		
				2	
		Pliocene	3		Cascadian
				5	
		Miocene	18		
				23	
		Oligocene	15		
				38	
		Eocene	16		
				54	
		Paleocene	11		
				65	
MESOZOIC	Cretaceous		71		Laramian
				136	
	Jurassic		54		Nevadian
				190	
	Triassic		35		
				225	
PALEOZOIC	Permian		55		Appalachian (Hercynian)
				280	
	Carboniferous	Pennsylvanian	45		
				325	
		Mississippian	20		
				345	Acadian
	Devonian		50		
				395	(Caledonian)
	Silurian		35		
				430	Taconian
	Ordovician		70		
				500	
	Cambrian		70		
				570	

			Duration b.y.	Age b.y.	
PRECAMBRIAN TIME	Upper Precambrian		0.3–0.4		
				0.9–1.0 -------- Grenville	
	Middle Precambrian		0.6–0.8		
				1.6–1.7 -------- Hudsonian	
			0.7–0.9		
				2.4–2.5 -------- Kenoran	
	Lower Precambrian		0.9–1.0		

-------------------------- Oldest dated rocks --------------------------------- 3.5±0.1 ---------------------------

----------------------------- Earth accretion completed ------------------------------------- 4.6–4.7 -------------------------

----------------------------- Age of universe -- 17–18 -------------------------

Data source: D. Eicher (1976), *Geologic time,* 2nd ed., Prentice-Hall, Englewood Cliffs, N.J., end paper.

within its limits not only the highest point on the earth, Mount Everest (8840 m; 29,000 ft), but also the deepest known ocean trenches, somewhat deeper than 11,000 m (36,000 ft).

Figure 24.4 shows profiles correctly curved and scaled to fit a globe whose diameter is 6.4 m (21 ft). The surface profile is drawn to natural scale, without vertical exaggeration. Although the most imposing landscape features of Asia and North America are shown, they are only trivial irregularities on the great global circle.

The Geologic Time Scale

To place crustal rocks and structures in their positions in time, we need to refer to some major units in the scale of geologic time. Table 24.1 is a greatly abbreviated list of the

major time divisions and their ages. All time older than 600 million years is **Precambrian time.** Three **eras** of time follow: **Paleozoic, Mesozoic,** and **Cenozoic.** These eras saw the evolution of life forms in the oceans and on the lands.

Geologic eras are subdivided into **periods.** Name, duration, and age of each period is given in Table 24.1. Individual periods will be mentioned on later pages in connection with the movements of lithospheric plates and the breakup of the ancient continents. Throughout the entire course of geologic time, brief but intense episodes of crustal deformation occurred in which strata were crumpled and broken by tectonic activity. These events of geologic history are called **orogenies.** The names of several important orogenies are shown in Table 24.1. Precambrian orogenies named in the table are those of the

Canadian Shield region of North America. Names of orogenies throughout the three younger eras also apply to North America; equivalent European names are given in parentheses.

The Cenozoic Era is particularly important in terms of continental landscapes because nearly all landforms seen today were produced in the 65 million years (m.y.) since that era began. (Recall that our model of fluvial denudation called for the completion of a peneplain within a 60-m.y. time span.) Because the Cenozoic Era is comparatively short in duration—scarcely more than the average duration of a single period within older eras—it is subdivided directly into **epochs.** Details of the Pleistocene and Holocene epochs were given in Chapter 21.

Ages given in Table 24.1 have been determined through chemical analysis of radioactive elements enclosed in rocks of all ages. Although the method of age determination is subject to a small range of error, ages are well established and are accepted by geologists. Events of Precambrian time are subject to greater uncertainties because these ancient rocks have been repeatedly deformed and intruded by magmas.

Second-Order Relief Features of the Continents

When the continents and ocean basins are considered as first-order relief features, we can then recognize subdivisions within each that are relief features of a second order of magnitude. We shall deal first with the relief features of the continents, familiar to most persons from direct experiences of travel and from the medium of photography.

Broadly viewed, the continental masses consist of two basic subdivisions: (1) active belts of mountain-making and (2) inactive regions of old rocks. The growth of mountain ranges occurs through one of two very different geologic processes. First is **volcanism,** the formation of massive accumulations of volcanic rock by extrusion of magma (Chapter 11). Many lofty mountain ranges consist of chains of volcanoes built of lava and tephra. Second of the mountain-building processes is **tectonic activity,** the breaking and bending of the earth's crust under internal earth forces. Crustal masses that are raised by tectonic activity form mountains and plateaus; masses that are lowered form crustal depressions. In many instances, volcanism and tectonic activity have combined to produce a mountain range. In Chapter 19, landforms created directly by volcanic and tectonic activity were classified as initial landforms.

Alpine Chains

Active mountain-making belts are narrow zones; most lie along continental margins. These belts are sometimes referred to as **alpine chains** because they are characterized by high, rugged mountains, such as the Alps of central Europe. These mountain belts were formed in the Cenozoic Era, and activity has continued to the present day in many places. The alpine chains are characterized by broadly curved patterns on the world map (Figure 24.5). Each curved section of an alpine chain is referred to as a **mountain arc;** the arcs are linked in sequence to form the two principal mountain belts. One is the **circum-Pacific belt;** it rings the Pacific Ocean basin. In North and South America, this belt is largely on the continents and includes the Andes and Cordilleran ranges. In the western part of the Pacific basin, the mountain arcs lie well offshore from the continents and take the form of **island arcs,** running through the Aleutians, the Kuriles, Japan, the Philippines, and many lesser islands. Between the large islands, these arcs are represented by volcanoes rising above the sea as small, isolated islands. The island arcs, which are of volcanic origin, properly belong to the ocean basins; we shall return to them later.

The second chain of major mountain arcs forms the **Eurasian-Melanesian belt,** starting in the west at the Atlas

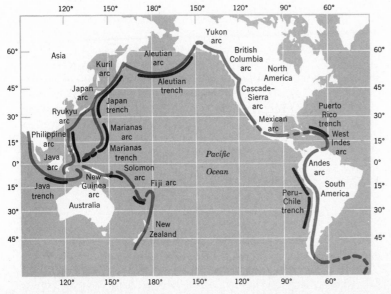

Figure 24.5 **The circum-Pacific ring of mountain arcs and island arcs. Deep trenches lie offshore in many parts of the ring.**

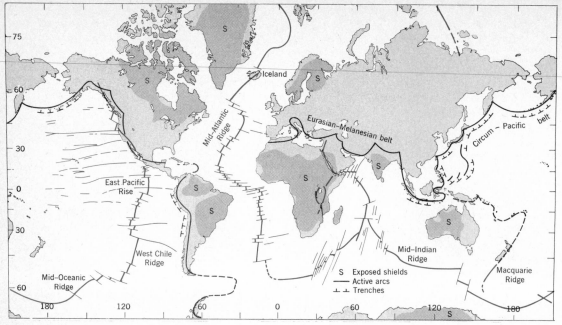

Figure 24.6 Generalized world map of major crustal features of the continents and ocean basins. The Mercator projection is used to show true compass directions of all linear features.

Mountains of North Africa and running through the Near East and Iran to join the Himalayas. The belt then continues through Southeast Asia into Indonesia, where it joins the circum-Pacific belt in a T-junction. Figure 24.6 shows the location of this second belt. We shall return later to the location of these active belts of mountain-making, explaining them in terms of lithospheric plate motions.

A world map of structural regions, Figure 24.7, recognizes the alpine chains as belonging to an **Alpine system.** Because the Alpine system also includes some adjacent inactive regions produced by orogenies of the Mesozoic Era, it appears on the world map in broad belts rather than as the narrow, linear arcs of late Cenozoic activity suggested in Figure 24.6.

Continental Shields and Mountain Roots

Belts of recent and active mountain-making account for only a small portion of the continental crust. The remainder consists of inactive regions of much older rock. Within these inactive regions, we recognize two structural types of crust: shields and mountain roots. **Continental shields** are low-lying continental surfaces beneath which lie igneous and metamorphic rocks in a complex arrangement (Figures 24.6 and 24.7). The rocks are very old, mostly of Precambrian age, and have had a very involved geologic history. For the most part, the shields are regions of low hills and low plateaus, although there are some exceptions where large crustal blocks have been uplifted. Many thousands of meters of rock have been eroded from the shields during their exposure throughout a half-billion years. In these areas, peneplains have been

formed and rejuvenated repeatedly, as we explained in Chapter 20.

Large areas of the continental shields are under a cover of younger sedimentary layers, ranging in age from the Paleozoic through the Cenozoic eras. These strata accumulated at times when the shields subsided and were inundated by shallow seas. Marine sediments were laid down on the ancient shield rocks in thicknesses ranging from hundreds to thousands of meters. These shield areas were then arched and again became land surfaces. Fluvial denudation has since removed large sections of the sedimentary cover, but it remains intact over vast areas. We refer to such areas as **covered shields** to distinguish them from the **exposed shields,** in which the Precambrian rocks lie bare. Figure 24.7 shows the covered and exposed shields in relation to the mountain belts. An example of an exposed shield is the Canadian Shield of North America. Exposed shields are also extensive in Scandinavia, South America, Africa, peninsular India, and Australia.

Crustal movements of the shields in later geologic time have been of a type known as **epeirogenic movements,** that is, rising or sinking of the crust over broad areas without appreciably breaking or bending the rocks. Epeirogenic movements reflect crustal stability generally, in contrast to tectonic activity affecting mountain arcs. One cause of epeirogenic uplift is isostatic compensation,

---

Figure 24.7 (right) **World map of structural regions.** (Adapted from R. E. Murphy, 1968, *Annals, Association of American Geographers,* Map Supplement No. 9. Based on Goode Base Map.)

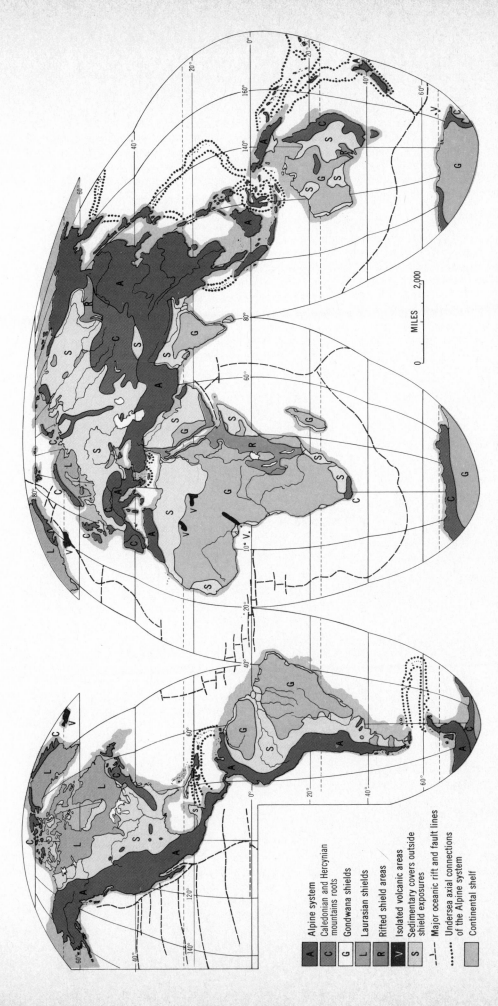

A **Alpine system.** World-girdling system of mountain chains formed since the Jurassic period. Faulted areas, plateaus, basins, and coastal plains enclosed by such ranges are included.

C **Caledonian (or Hercynian or Appalachian) mountain roots.** Remains of mountain chains and ranges formed during the Paleozoic and Mesozoic eras, prior to the Cretaceous period and experiencing no orogeny since then, although epeirogenic movements may (and often have) occurred. Faulted areas, plateaus, basins, and coastal plains enclosed by these remnants are included with them.

G **Gondwana shields.** Areas of stable, massive blocks of continental crust. They lie south of the great east-west portion of the Alpine system, where Precambrian rocks form either the entire surface rock or form an encircling enclosure with no gap of more than 320 km (200 mi) between outcroppings or covering extrusives and within which crystalline rocks form more than 50 percent of the surface rock. These shields have not been subject to orogeny since the Cambrian period.

L **Laurasian shields.** Areas of stable, massive blocks of continental crust lying north of the great east-west portion of the Alpine system. (Remainder of definition same as for G.)

R **Rifted shield areas.** Block-faulted areas of shields forming grabens, together with associated horsts and volcanic features. Rifting is a result of the extension and thinning of continental lithospheric plates.

V **Isolated volcanic areas.** Areas of volcanoes—active or extinct—with associated volcanic features, lying outside the Alpine or older mountain systems and outside the rifted shield areas. Volcanism is an expression of hot spots above mantle plumes.

S **Sedimentary covers.** Areas of sedimentary layers that have not been subjected to orogeny and lie outside the crystalline rock enclosures of the shields or outside the enclosing mountains and hills of the Alpine or older orogenic systems. These areas of sedimentary rock form continuous covers over underlying structures.

A	Alpine system
C	Caledonian and Hercynian mountains roots
G	Gondwana shields
L	Laurasian shields
R	Rifted shield areas
V	Isolated volcanic areas
S	Sedimentary covers outside shield exposures
	Major oceanic rift and fault lines
	Undersea axial connections of the Alpine system
	Continental shelf

0 MILES 2,000

explained in Chapter 20.

Remains of older mountain belts lie within the shields in many places. These **mountain roots** are formed mostly of Paleozoic and Mesozoic sedimentary rocks that have been intensely deformed and locally changed into metamorphic rocks.

One important system of mountain roots was formed in the Caledonian orogeny, occurring about the end of the Silurian Period (Table 24.1). Often referred to as Caledonides, these roots are found in a belt extending from northern Ireland, through Scotland, into Scandinavia (Figure 24.7). A second important root system was formed in the Appalachian orogeny that closed the Paleozoic Era. In North America this system is represented by the Appalachian Mountains. In Europe, a roughly equivalent orogeny, called the Hercynian orogeny, formed a mountain belt (the Hercynides) extending across the southern British Isles and northern Europe. Our world map of structural regions, Figure 24.7, recognizes the Paleozoic mountain roots of both Caledonian and Appalachian (Hercynian) orogenies as a distinct region. Thousands of meters of overlying rocks have been removed from these old tectonic belts, so only basal structures remain. Roots appear as chains of long, narrow ridges, rarely rising over a thousand meters above sea level. Landforms of mountain roots are described in Chapter 26.

A somewhat younger generation of mountain belts was produced by orogenies of the Mesozoic Era. Named in Table 24.1 are the Nevadian and Laramian orogenies that affected strata of Mesozoic age in the western Cordilleran ranges, including the Rocky Mountains. Although these ranges have experienced a great amount of erosion throughout the Cenozoic Era, they remain lofty mountains because of a series of later uplifts. They cannot be classified as ancient mountain roots and are best described as being intermediate in evolution between the ancient mountain roots of Paleozoic age and the alpine belts of late Cenozoic age. On the world map of structural regions (Figure 24.7), the Mesozoic mountain belts are included within the Alpine system.

Second-Order Relief Features of the Ocean Basins

Crustal rock beneath the ocean floors consists almost entirely of basalt, covered over large areas by a comparatively thin accumulation of sediments. Age determinations of the basalt and its sediment cover show that the oceanic crust is quite young, geologically speaking. Much of that crust was formed during the Cenozoic era and is less than 60 million years old. Over large areas, the rock age is Mesozoic and falls into the Cretaceous Period (−65 to −136 m.y.). Small areas belong to the Jurassic Period (−136 to −190 m.y.). These data mean that the ocean basins began to come into existence early in the Mesozoic Era. When we consider that the bulk of the continental crust is of Precambrian age—mostly over one billion years old—the young age of the oceanic crust is all the more remarkable. We will need to fit this fact into the general theory of global tectonics.

In overall plan, the ocean basins are characterized by a central ridge structure that divides the basin about in half. This feature is shown by schematic diagram in Figure 24.8. The **mid-oceanic ridge** consists of submarine hills rising gradually to a mountainous central zone (Plate K). Precisely in the center of the ridge, at its highest point, is an **axial rift,** which is a trenchlike feature. The form of this rift suggests that the crust is being pulled apart along the line of the rift axis. On either side of the mid-oceanic ridge are broad, deep plains and hill belts belonging to the **ocean basin floor.** Their average depth is about 5 km (17,000 ft). The flat surfaces are called **abyssal plains;** they are extremely smooth because they have been built up of fine sediment.

Nearing the continental margins, the ocean floor begins to slope gradually upward, forming the **continental rise.** The floor then steepens greatly in the **continental slope.** At the top of this slope we arrive at the brink of the **continental shelf,** a gently sloping platform some 120 to 160 km (75 to 100 mi) wide along the eastern margin of North America. Water depth is about 180 m (600 ft) at the outer edge of the shelf.

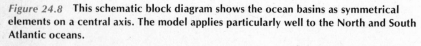

Figure 24.8 **This schematic block diagram shows the ocean basins as symmetrical elements on a central axis. The model applies particularly well to the North and South Atlantic oceans.**

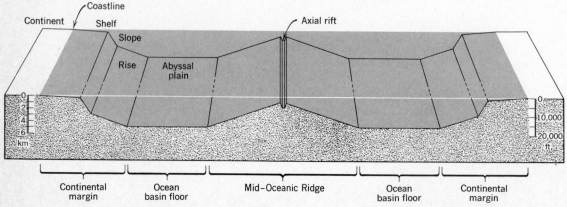

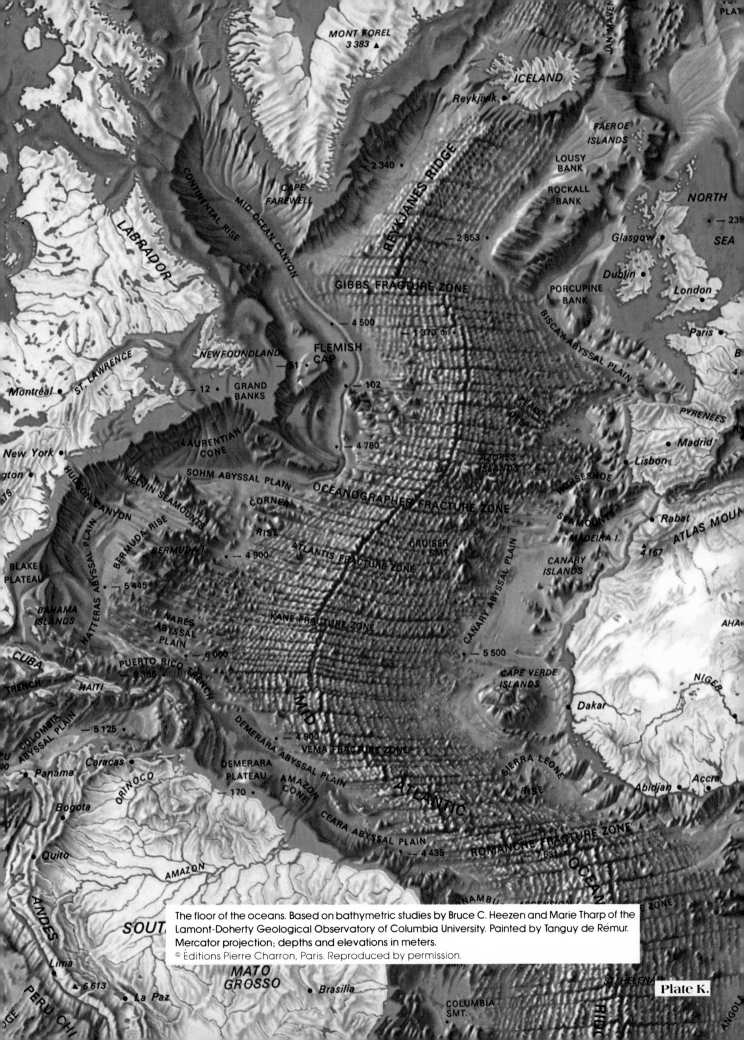

The floor of the oceans. Based on bathymetric studies by Bruce C. Heezen and Marie Tharp of the Lamont-Doherty Geological Observatory of Columbia University. Painted by Tanguy de Rémur. Mercator projection; depths and elevations in meters.

© Éditions Pierre Charron, Paris. Reproduced by permission.

Plate K.

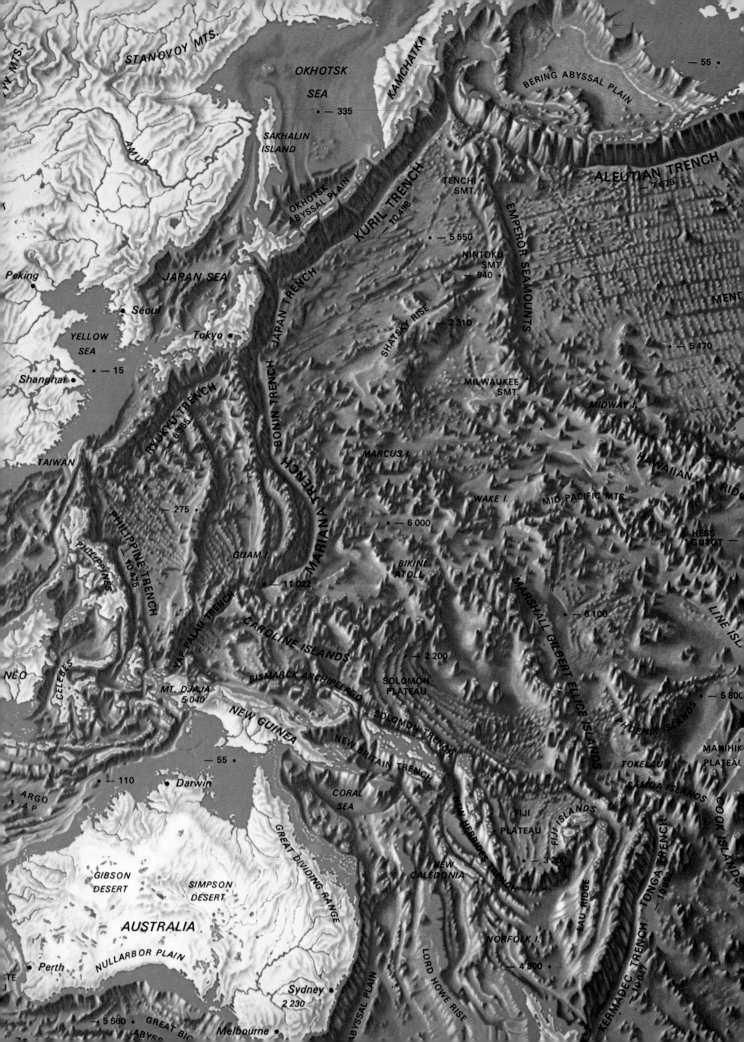

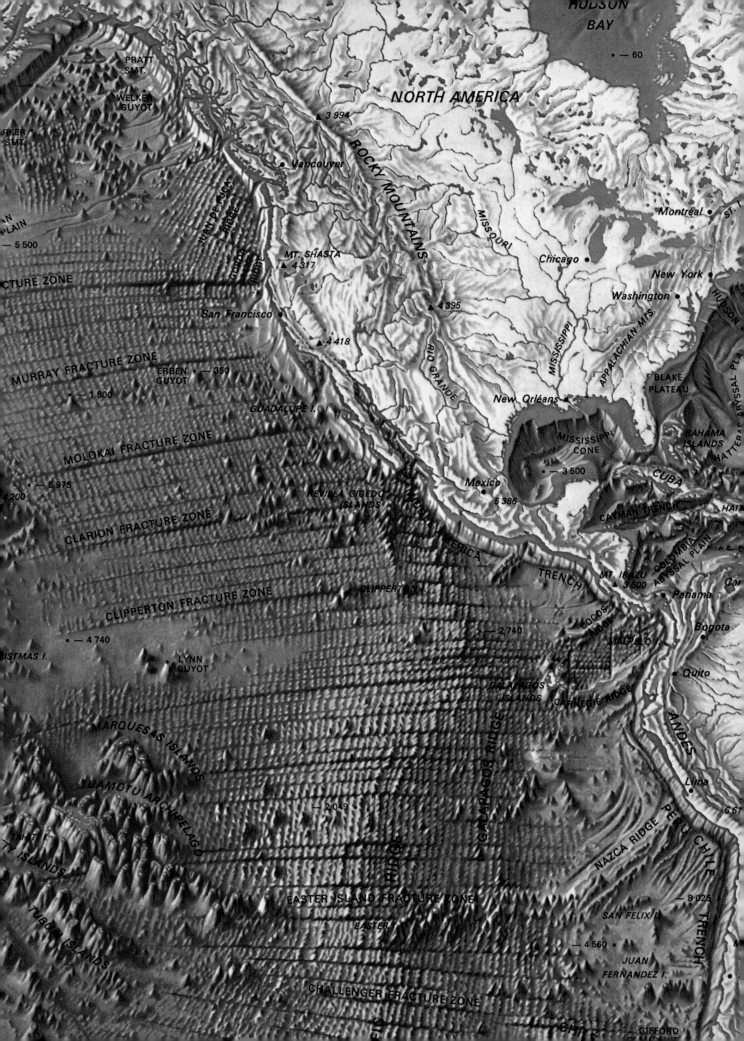

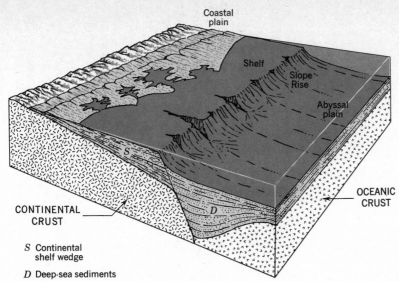

Coastal
plain

Shelf

Slope
Rise

Abyssal
plain

OCEANIC
CRUST

CONTINENTAL
CRUST

S Continental
shelf wedge

D Deep-sea sediments

Figure 24.9 **This block diagram shows an inner wedge of sediments beneath the continental shelf and an outer wedge of deep-sea sediments beneath the continental rise and abyssal plain. (Drawn by A. N. Strahler.)**

The continental shelf is underlain by a great thickness of sedimentary strata derived from the continent. These strata form a large wedge-shaped deposit, thinning landward and feathering out over the continental shield. Figure 24.9 is a block diagram showing the wedge of continental shelf sediments. These sediments are brought from the land by rivers and spread over the shallow seafloor by currents. A great deal of attention is now being paid to the continental-shelf wedge as a potential source of rich petroleum accumulations, reached only from offshore drilling platforms. Below the continental rise and its adjacent abyssal plain is another thick sediment deposit; it is formed of deep-sea sediments carried down the continental slope by swift muddy currents, called **turbidity currents.**

Sea-Floor Spreading

The mid-oceanic ridge and its axial rift can be traced through the ocean basins for a total distance of about 64,000 km (40,000 mi). Figure 24.6 shows the extent of the ridge. From the South Atlantic, the ridge turns east and enters the Indian Ocean. There, one branch penetrates Africa while the other continues east between Australia and Antarctica, then swings across the South Pacific. Nearing South America, it turns north and penetrates North America at the head of the Gulf of California.

The axial rift of the mid-oceanic ridge is the origin of numerous earthquakes, indicating continued tectonic activity. Unmistakable evidence has been accumulated to show that the crust is spreading apart along the rift. The rate of separation is from 2.5 to 5 cm (1 to 2 in.) per year. As this **sea-floor spreading** occurs, basaltic lava rises from beneath the rift, solidifying and forming new oceanic crust. This continuing process explains why the ocean basin crust is so young in geologic age. The axial rift is broken across at right angles by numerous fractures, which are a type of fault. Segments of the rift are offset along these fractures, as shown on Figure 24.6.

Oceanic Trenches

The circum-Pacific belt is characterized by a chain of deep **oceanic trenches** lying offshore from the mountain arcs and running close to the island arcs. The positions of these trenches are shown in Figures 24.5 and 24.6. Floors of the narrow trenches reach depths of 7 km (23,000 ft) and even more (Plate K). Many lines of evidence show that the oceanic crust is sharply downbent to form these trenches. Numerous earthquakes, many of them very intense, occur in the trenches and adjacent mountain arcs, attesting to continuing tectonic activity.

Plate Tectonics

Crustal spreading along the axial rift of the mid-oceanic ridge and crustal downbending beneath the oceanic trenches involves entire lithospheric plates. The total system of plate motions is referred to as **plate tectonics.** Moving over the weak asthenosphere, individual lithospheric plates glide slowly over the surface of the globe, much as pack ice of the Arctic Ocean drifts slowly under the dragging force of currents and winds.

Figure 24.10 is a schematic cross section of the lithosphere and asthenosphere showing the fundamentals of plate tectonics. Curvature of the earth's surface is removed in this drawing for the sake of simplicity. Sea-floor spreading at the axial rift of the mid-oceanic ridge represents a line along which two lithospheric plates are moving apart. In contrast, a deep oceanic trench represents a line along which two lithospheric plates are coming together. In this case, one plate is forced to plunge down into the mantle, a process called **subduction.** Figure 24.11 is a three-dimensional representation of lithospheric plates and their relative motions.

Besides coming into contact along lines of sea-floor spreading and subduction, lithospheric plates can slide past one another along simple faults, called **transform faults** (Figure 24.11). Motion on a transform fault is

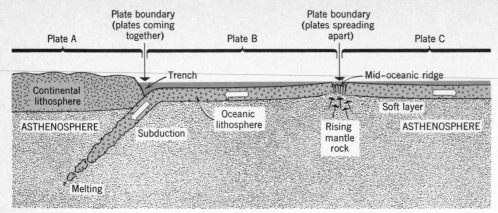

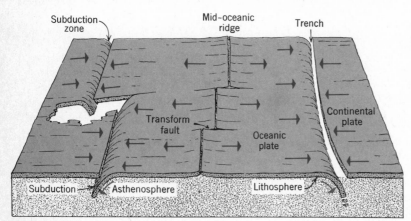

Figure 24.10 **The lithosphere, a rigid upper plate, glides over the soft asthenosphere. While one edge of a lithospheric plate is being formed during seafloor spreading on an axial rift, the opposite edge is disappearing into the mantle by subduction.**

Figure 24.11 **This schematic diagram shows major features of plate tectonics. Earth curvature has been removed. (From A. N. Strahler, 1971, *The Earth Sciences,* 2nd ed., Harper and Row, New York.)**

horizontal and always in the same direction as the fault itself. Consequently, on a transform fault the plates do not separate, nor does one plate pass beneath the other. On the sea floor, transform faults appear as long, straight walls or narrow trenches. Many of the faults offsetting the mid-oceanic ridge are transform faults (Figure 24.6).

The Origin of Volcanic Arcs

Applying concepts of plate tectonics, we can now explain chains of volcanic mountain ranges adjacent to deep oceanic trenches. Consider first the case of a volcanic chain, such as the Andes Mountains of South America, lying on the edge of a continent and a bordering trench a short distance offshore. As shown in Figure 24.12, sub-duction of the plate forming the ocean basin is the cause of the trench, while melting of the downbent plate produces magma pockets beneath the edge of the adjacent plate. Being less dense than the surrounding rock, the magma bodies rise slowly through the continental crust and emerge as extrusive igneous rocks to form the volcanic mountain range. Typically, the extrusive rock is of felsic composition and produces lava consisting of rhyolite and andesite rock (see Figure 11.2).

Notice in Figure 24.12 that the lithosphere is much thicker under the continent than under the ocean basin. Geologists recognize two varieties of lithosphere: **oceanic lithosphere,** which is thin, and **continental lithosphere,** which is thick. Because it has a comparatively thick upper

Figure 24.12 **Subduction of oceanic lithosphere beneath the margin of a continental lithospheric plate, producing a volcanic mountain chain. Landforms are greatly exaggerated in vertical scale. This case illustrates the origin of the Andes Mountains of South America. (From A. N. Strahler, 1977, *Principles of Physical Geology,* Harper and Row, New York.)**

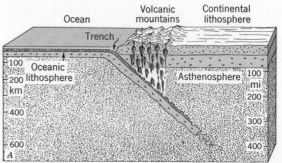

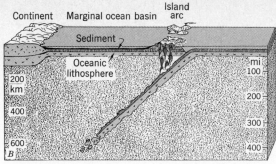

Figure 24.13 Subduction of oceanic lithosphere of one plate beneath oceanic lithosphere of another plate, leading to the formation of a volcanic island arc. (From A. N. Strahler, 1977, *Principles of Physical Geology,* Harper and Row, New York.)

crust of felsic rock, continental lithosphere is comparatively buoyant. Thus, when a plate consisting of oceanic lithosphere is forced to move against a plate of continental lithosphere, the former must bend down beneath the latter. An interesting conclusion from this relationship is that oceanic lithosphere will be disposed of by melting as it disappears into the upper mantle, whereas the continental lithosphere will remain intact and will gain in bulk by additions of rising igneous rock.

Figure 24.13 shows the case of an island arc, typical of arcs of the western Pacific Ocean. These island arcs are

Figure 24.14 This map of the western Pacific shows trenches (solid black), island arcs (dashed lines), active volcanoes (black dots), and deep-focus earthquake centers (color dots). (After H. H. Hess.)

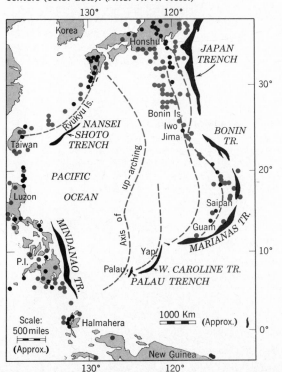

shown on a map, Figure 24.14. In this case, oceanic lithosphere lies on both sides of the subduction zone. Volcanic islands rise by the upbuilding of volcanoes from the ocean floor, adjacent to the trench. Between the island arc and the continent lies a broad **marginal ocean basin.** Sediment, derived by erosion of the volcanic islands, accumulates on the floor of the marginal ocean basin.

Continental Collision and Alpine Chains

Alpine mountain chains typically consist of intensely deformed sedimentary strata of marine origin. Some details of alpine structure are shown in Figure 24.15. The strata are compressed into wavelike structures. These **folds** become overturned and take on a recumbent attitude. Accompanying folding is a form of faulting in which slices of rock move over underlying rock along fault surfaces of low inclination; these are **overthrust faults.** Individual slices, called **thrust sheets,** are carried many tens of kilometers over the overlying rock.

Plate tectonics explains alpine structure through the process of **continental collision,** illustrated in Figure 24.16. In stage *A,* subduction is in progress. The lithospheric

Figure 24.15 In alpine structure, the tightly folded strata are broken by overthrust faults; one slice is heaped on the next. (From A. N. Strahler, 1971, *The Earth Sciences,* 2nd ed., Harper and Row, New York.)

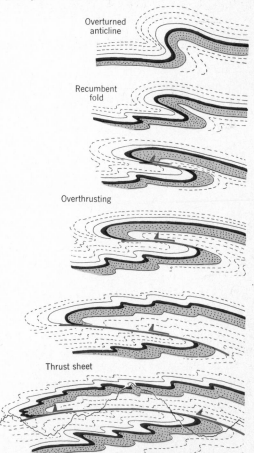

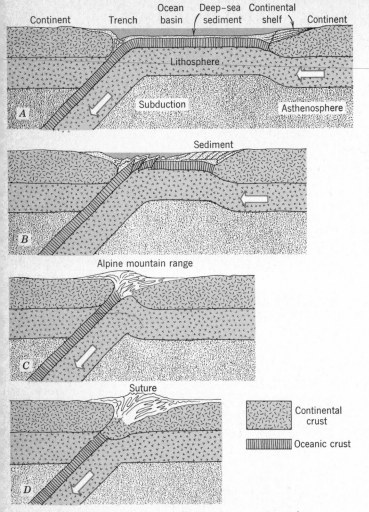

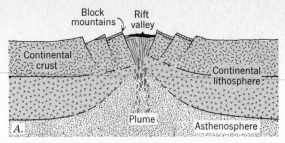

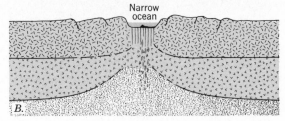

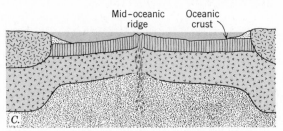

Figure 24.16 **Stages in collision of two continental lithospheric plates. (After J. F. Dewey and J. M. Bird, 1970.)**

Figure 24.17 **Stages in continental rupturing and the formation of a new ocean basin. (From A. N. Strahler, 1976, *Principles of Earth Science*, Harper and Row, New York.)**

plate on the right is moving toward the left, bringing the two continents closer together, while the ocean basin is being reduced in width. Sediment is accumulating both on the deep ocean floor and at the continental margin, under the continental shelf. In stage *B*, narrowing of the ocean basin continues. Sediment is being scraped off the descending plate margin and is becoming crumpled and faulted in the trench. In stage *C*, the two continents have collided, squeezing the sediment mass strongly and throwing it into complicated folds with thrust faults. Now the ocean basin has disappeared entirely and a high alpine mountain range has come into existence. In stage *D*, the oceanic crust has been eliminated entirely from the crust. The alpine zone remaining is called a **suture;** the process by which it is produced is called **continental suturing.**

Examples of alpine mountain ranges that have been formed by continental suturing in the Cenozoic Era are the European Alps and the Himalaya Range of southern Asia. Suturing also explains ancient mountain roots, such as the Caledonides and Appalachians. These belts represent deeply eroded suture zones resulting from continental

collisions that occurred in the Paleozoic Era. Each orogeny recognized by geologists turns out to be a continental collision. With each collision, new felsic rock was added to the margin of a continent. Thus the continents grew in area through geologic time.

Continental Rupturing

If lithospheric plates collided repeatedly through the geologic eras, they must also have broken apart and separated many times. **Continental rupture** is illustrated in Figure 24.17. Diagram *A* shows the start of rupture. From deep beneath the continental plate, a column of heated mantle rock begins to rise. This column is called a **mantle plume.** The plume lifts the plate above it, causing the plate to fracture. The fracture system spreads rapidly across the entire plate, dividing it into two parts, which begin to

Figure 24.18 (right) **World map of lithospheric plates.**

Plate boundaries	
═══ Spreading	──── Transform fault
▲▲▲▲▲ Subduction	----- Uncertain or inactive

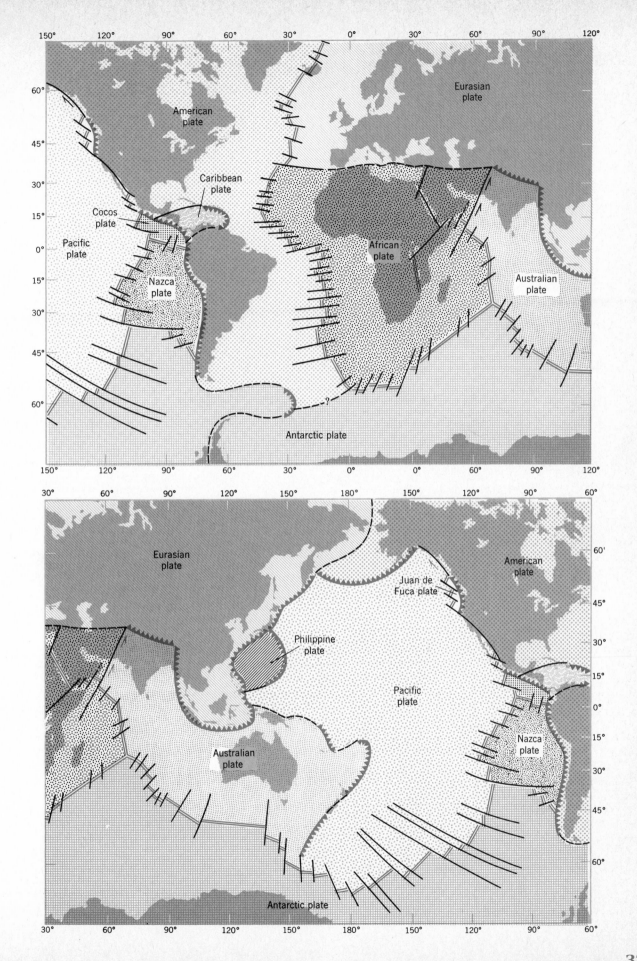

separate. An axial **rift valley** is now formed. Basalt magma rises to pour out on the floor of the rift valley. As spreading continues, the rift widens and deepens, admitting ocean water and becoming a narrow ocean, as shown in Diagram B. New oceanic lithosphere is being formed in the widening ocean basin. As shown in Diagram C, a broad ocean basin has been produced, with a characteristic mid-oceanic ridge.

The Red Sea, a narrow straight-sided trough lying between Africa and the Arabian Peninsula, is interpreted by geologists as a young ocean being formed by continental rupturing. We shall investigate the landforms of rupturing in Chapter 25.

The Global System of Lithospheric Plates

The culmination of the new revolution in geology has been the recognition of a global system of lithospheric plates. Figure 24.18 shows the major plates of the globe. The American plate includes the North American and South American continental crust and all of the oceanic crust as far east as the Mid-Atlantic Ridge. The Pacific plate is the only unit consisting entirely of oceanic lithosphere. It occupies all of the Pacific region west and north of the mid-oceanic ridge. It undergoes subduction beneath the American plate along the Alaskan-British Columbia coastal zone. The Antarctic plate occupies the globe south of the mid-oceanic ridge system. The African plate consists of the African continental lithosphere and a zone of surrounding oceanic lithosphere limited by the mid-oceanic ridge. A single Eurasian plate, which consists largely of continental lithosphere, is bounded on the east and south by subduction zones of the great alpine mountain chains and island arcs. The Australian plate consists of continental lithosphere of India and Australia, as well as oceanic lithosphere of the Indian Ocean and part of the southwestern Pacific. Other smaller plates are shown and named on the map.

Although the driving force for plate motions is not known, a leading hypothesis attributes the motion to convection currents deep in the mantle (Figure 24.19). In some respects this model convection current system in the mantle resembles the atmospheric circulation at low

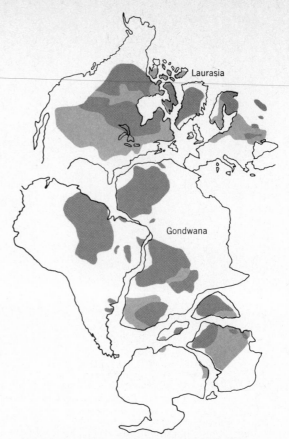

Figure 24.20 **Reassembled continents, prior to start of continental drift. Dark pattern: areas of oldest shield rocks (older than −1.7 b.y.). Light pattern: rocks in the age range −0.8 to −1.7 b.y. (Redrawn from P. M. Hurley and J. R. Rand, 1969, *Science,* vol. 164, p. 1237, Figure 8. Copyright © 1969 by the American Association for the Advancement of Science.)**

latitudes within the Hadley cell (Chapter 6). Unequal heating of mantle rock leads to differences in rock density, and this in turn may set in motion very slow rising and sinking motions. Between the zone of rising mantle rock (mid-oceanic ridge) and sinking rock (beneath the subduction zones), horizontal mantle motion exerts a dragging force on the lithospheric plate. Many variations of the convection hypothesis have been set forth, and various other energy sources have been proposed as well. While the reality of plate motions seems well established, we are far from understanding the impelling forces.

Continental Drift

As early as the late 1800s, the proposal had been made that the continents of North and South America, Europe, Africa, Australia, and Antarctica and the subcontinent of India (south of the Himalayas), along with Madagascar, had originated as a single supercontinent. Throughout early decades of the 1900s, the principal advocate of this hypothesis was a German meteorologist, Alfred Wegener, who named the supercontinent *Pangaea* (see Figure 24.21).

Using several lines of sound geologic evidence,

Figure 24.19 **Simplified model of a convection system in the mantle causing motion of the overlying lithospheric plates.**

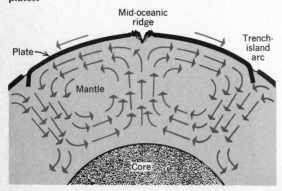

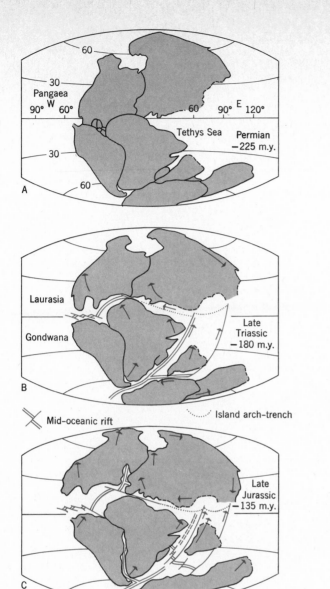

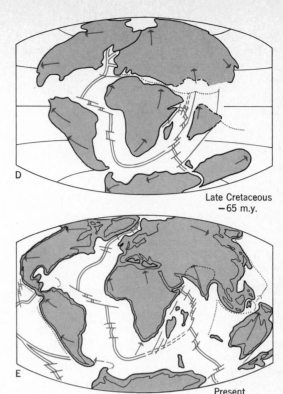

Figure 24.21 The breakup of Pangaea is shown in five stages. Inferred motion of lithospheric plates is indicated by arrows. (Redrawn and simplified from maps by R. S. Dietz and J. C. Holden, 1970, *Journal of Geophysical Research,* vol. 75, pp. 4943–51, Figures 2 to 6, copyrighted by the American Geophysical Union.)

Wegener argued the existence of Pangaea and postulated that it began to break apart late in the Mesozoic Era. Wegener's hypothesis, known as **continental drift,** required that the granitic (felsic) crust of the continents moved through basaltic (mafic) crust, much as a tabular iceberg might drift through the ocean. The hypothesis was widely rejected in Wegener's time on valid grounds: It could be shown that Wegener's drift mechanism was physically impossible. Nearly 30 years were to elapse before sea-floor spreading could be demonstrated beyond doubt and an acceptable mechanism of continental separation could be worked out. When lithospheric plate motions became fully accepted in the 1960s, the scenario of drifting apart of the continents from a united Pangaea was revived.

Geologists made new reconstructions of Pangaea, based on new data. Figure 24.20 shows how the present continental shields have been fitted together so that the oldest portions, called **continental nuclei,** are brought into

two groupings. A northern group comprises the landmass of *Laurasia;* the southern group comprises *Gondwana* (or *Gondwanaland*). These two subcontinents were probably contiguous across what is now Africa and South America. Today's coastlines are shown only for reference.

Continental rupturing of Pangaea began about in mid-Triassic time with the opening of the North Atlantic basin. Figure 24.21 shows one reconstruction of the breakup. South America became separated from Africa in Cretaceous time. In the meantime, Antarctica pulled away from Africa, while India traveled toward Southeast Asia and eventually collided with Asia, producing the Himalayas. Australia and New Zealand had drifted far to the east, and the Indian Ocean was then fully opened up.

The Rock Cycle in a New Light

Plate tectonics has brought new understanding of the rock transformation cycle described in Chapter 11. As Figure

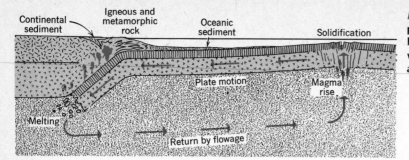

Continental sediment — Igneous and metamorphic rock — Oceanic sediment — Solidification — Plate motion — Magma rise — Melting — Return by flowage

Figure 24.22 **The moving lithospheric plate works like a conveyor belt, bringing sediment to a subduction zone where it is transformed into new igneous and metamorphic rock.**

24.22 shows, a moving lithospheric plate bearing oceanic crust can be visualized as a great conveyor belt. The oceanic sediment deposited on the oceanic crust is carried toward a subduction zone, where it is scraped off, along with continental sediment. During continental collision, sediment is altered into metamorphic rock. Some sediment is transformed into igneous rock deep within the subduction zone and is added to the continental crust. The descending plate is melted and its substance returned to the mantle. By very slow drift, the same igneous material is eventually returned to the axial rift zone, where it rises to solidify as new lithosphere.

References for Further Study

Heezen, B. C., M. Tharp, and M. Ewing (1959), *The floors of the oceans,* Geological Society of America, Special Paper 65.

Wilson, J. T., ed. (1970), *Continents adrift,* W. H. Freeman and Co., San Francisco.

Hallam, A. (1973), *A revolution in the earth sciences,* Clarendon Press and Oxford Univ. Press, London.

Sullivan, W. (1974), *Continents in motion; the new earth debate,* McGraw-Hill Book Co., New York.

Eicher, D. L. (1976), *Geologic time,* 2nd ed., Prentice-Hall, Englewood Cliffs, N.J.

Strahler, A. N. (1977), *Principles of physical geology,* Harper and Row, New York, See Chapters 2, 5, 7–9.

Review Questions

1. Describe the earth's internal structure, including the core and mantle. Give approximate dimensions of each zone. Of what material is the core composed? In what physical state is the core? What is the mineral composition of the mantle?

2. What layers comprise the continental crust? Of what rock is the oceanic crust composed? Give approximate thicknesses of both continental and oceanic crust.

3. Describe the soft layer, or asthenosphere, of the upper mantle. Why is strength low in this layer? What properties has the lithosphere? How thick is the lithosphere? What are lithospheric plates? How can they move over the asthenosphere?

4. What percentage of the global area is occupied by continents? By ocean basins? In what depth range is the floor of the ocean basin concentrated? Drawn to true scale on a perfect circle 6.4 m (21 ft) in diameter, how high would Mount Everest appear?

5. What time units are used in the scale of geologic time? What is Precambrian time? Name the eras that followed the Precambrian. What is an orogeny? Into what units is the Cenozoic Era subdivided?

6. What are the two basic subdivisions of continents in terms of geologic history and process? What processes are involved in the growth of mountains? Define tectonic activity.

7. Describe the alpine chains. What is their geologic age? Name the two principal mountain arcs. What is the difference between a mountain arc and an island arc?

8. Describe the continental shields. What is the age of rocks in shields? What kinds of rock make up the shields? What are covered shields? How do epeirogenic movements affect shields?

9. Describe ancient mountain roots of the continents. Give the geologic ages and locations of the Caledonides and the Appalachians (Hercynides).

10. Give a general description of the second-order relief features of the ocean basins, using the North Atlantic Ocean basin as a model. What is the range in geologic age of crust of the ocean basins? What feature marks the center line of the mid-oceanic ridge?

11. Describe the submarine features of the continental margins. Describe the geologic structure beneath the continental shelves. What kind of deposit lies beneath the continental rise? How is it formed?

12. Where is sea-floor spreading occurring? At what rate? What submarine landforms are produced by sea-floor spreading?

13. Describe the oceanic trenches. How deep are they? What is their typical outline form on a map? How are they related to island arcs and mountain arcs?

14. What is plate tectonics? Describe the major processes and structures involved in plate tectonics. What is subduction and where does it occur? How are transform faults produced?

15. Describe the formation of volcanic mountain chains in terms of plate tectonics. Where does the magma originate? Distinguish between oceanic lithosphere and continental lithosphere. Which is thicker? What is a marginal ocean basin?

16. Describe alpine structure. How does continental collision produce alpine structure? What is a suture? Name two ancient mountain ranges that represent sutures.

17. How does continental rupturing occur? What landforms are the first to appear at the start of rupturing? Give an example. What is a mantle plume?

18. Describe the global system of lithospheric plates. Name the major plates. Which plate consists entirely of oceanic lithosphere? What driving mechanism has been proposed for plate motions?

19. Describe the original concept of continental drift, as envisioned for Pangaea by Alfred Wegener. On what grounds was it rejected? What were Laurasia and Gondwana? Describe the breakup of Pangaea.

20. How can plate tectonics be fitted into the concept of a cycle of rock transformation? How is recycling of crustal rock carried out by plate motions?

Volcanic and Tectonic Landforms

25

Initial landforms, produced by internal earth forces, are one subject of this chapter. Plate tectonics provides the basic understanding of the initial landforms, which are produced by volcanism and tectonic activity. Even while the initial landforms are coming into existence, they are subject to modification by fluvial denudation. Thus we need to analyze not only the stark outlines of volcanoes and fault blocks, but also the modified forms converted by fluvial processes into sequential landforms. In this chapter and the next, the principles of fluvial denudation, covered in Chapters 19 and 20, will be indispensable to understanding how the erosional landforms are controlled by various rock structures.

Forms of Igneous Rock Bodies

In Chapter 11, we recognized that magma may solidify below the earth's surface to form intrusive igneous rock bodies, or it may reach the surface and emerge as lava, which solidifies as extrusive igneous rock. Intrusive igneous rock typically accumulates in enormous subterranean bodies, called **batholiths.** Figure 25.1 shows the relationship of a batholith to the overlying rock. While forcing its way upward, the magma makes room for its bulk by dissolving and incorporating the older rock above it. Batholiths are several kilometers in depth and may extend beneath an area of several thousand square kilometers.

Figure 25.1 shows two other common forms of intrusive rock bodies. One is a **sill,** a platelike layer formed when magma forced its way between two horizontal rock layers, lifting the overlying rock to make room. A second is the **dike,** a near-vertical wall-like body formed by the spreading apart of a vertical rock fracture. The dike rock is fine textured because of rapid cooling. Dikes are commonly the conduits through which magma reaches the surface. Magma entering small fractures in the overlying rock solidifies in a branching network of thin veins.

Volcanoes are conical or dome-shaped structures built by the emission of lava and its contained gases from a restricted vent in the earth's surface (Figure 25.1). The magma rises in a narrow, pipelike conduit from a magma reservoir far below. Reaching the surface, igneous material may pour out in tonguelike lava flows or may be ejected

as tephra under pressure of confined gases. Form and dimensions of a volcano are quite varied, depending on the type of lava and the presence or absence of tephra. The nature of volcanic eruption, whether explosive or quiet, depends on the type of magma.

An important point is that the felsic lavas (rhyolite and andesite) have a high degree of viscosity (property of tackiness, resisting flowage) and hold large amounts of gas under pressure. As a result, these lavas produce explosive eruptions. In contrast, mafic lava (basalt) is highly fluid (low viscosity) and holds little gas, with the result that the eruptions are quiet and the lava can travel long distances to spread out in thin layers.

Composite Volcanoes

Tall, steep-sided volcanic cones are produced by felsic lavas. These cones usually steepen toward the summit, where a depression, the **crater,** is located. Tephra in these volcanic eruptions takes the form of volcanic ash, which falls on the area surrounding the crater and contributes to the structure of the cone (Figure 25.2). The interlayering of ash layers and lava streams produces a **composite volcano.** Included are volcanic bombs. These solidified masses of lava range up to the size of large boulders and fall close to the crater (Figure 25.3). Very fine volcanic dust rises high

Figure 25.1 **Forms of igneous rock bodies. (Drawn by A. N. Strahler.)**

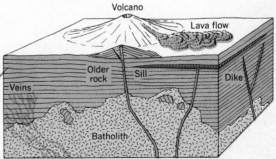

381

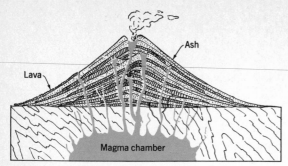

Figure 25.2   Idealized cross section of a composite volcanic cone with feeders from magma chamber beneath. (From A. N. Strahler, 1972, *Planet Earth,* Harper and Row, New York.)

into the troposphere and stratosphere, where it remains suspended for years.

Another important form of emission from the explosive types of volcanoes is a cloud of incandescent gases and fine ash. Known as a **nuée ardente** (French for "glowing cloud"), this intensely hot cloud travels rapidly down the flank of the volcanic cone, searing everything in its path.

On the island of Martinique, in 1902, a glowing cloud issued without warning from Mount Pelée; it swept down on St. Pierre, destroying the city and killing all but two of its 30,000 inhabitants.

Most lofty conical volcanoes, well known for their scenic beauty, are of the composite type. Examples are Mount Hood in the Cascade Range (Plate L.1). Fujiyama in Japan, Mount Mayon in the Philippines, and Mount Shishaldin in the Aleutian Islands (Figure 25.4).

Many of the world's active composite volcanoes lie above subduction zones. In Chapter 24 we explained the rise of andesitic magmas beneath volcanic arcs of continental borders and island arcs (Figures 24.12 and 24.13). One good example is the volcanic arc of Sumatra and Java, lying over the subduction zone between the Australian plate and the Eurasian plate; another is the Aleutian volcanic arc, located above the subduction zone between the Pacific plate and the North American plate.

Calderas

One of the most catastrophic of natural phenomena is a volcanic explosion so violent that it destroys the entire central portion of the volcano. Only a great central de-

Figure 25.3   Sakurajima, a Japanese volcano, erupted violently in 1914. These pictures show various scenes from the eruption. (T. Nakasa.)

(A)   This distant view shows the great cauliflower cloud of volcanic gases and condensed steam.

(C)   Reaching the sea, the hot lava makes clouds of steam.

(B)   A blocky lava flow is advancing slowly over a ground surface littered with volcanic bombs and ash.

(D)   Volcanic ash has buried this village. Torrential rains have carved rills in the ash.

Figure 25.4 **Mount Shishaldin, an active composite volcano on Unimak Island in the Aleutian Islands, rises to an elevation just over 2800 m (9300 ft). A plume of condensed steam marks the summit crater. (U.S. Navy Department, from The National Archives.)**

pression, named a **caldera,** remains. Although some of the upper part of the volcano is blown outward in fragments, most of it subsides into the ground beneath the volcano. Vast quantities of ash and dust are emitted and fill the atmosphere for many hundreds of square kilometers.

Krakatoa, a volcanic island in Indonesia, exploded in 1883, leaving a huge caldera. It is estimated that 75 cu km (18 cu mi) of rock disappeared during the explosion. Great seismic sea waves generated by the explosion killed many

thousands of persons living on low coastal areas of Sumatra and Java. Another historic explosion was that of Katmai, on the Alaskan Peninsula, in 1912. A caldera more than 3 km (2 mi) wide and 1000 m (3300 ft) deep was produced. The explosion was heard at Juneau, 1200 km (750 mi) distant, while at Kodiak, 160 km (100 mi) away, the ash formed a layer 25 cm (10 in.) deep.

A classic example of a caldera produced in prehistoric times is Crater Lake, Oregon (Figure 25.5). Mount Maza-

Figure 25.5 **Crater Lake, Oregon, is an outstanding illustration of a caldera, now holding a lake. (Drawn by E. Raisz.)**

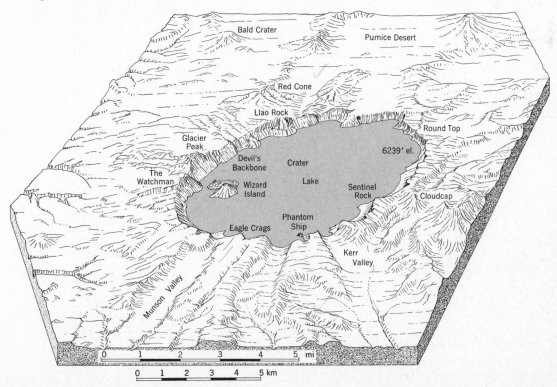

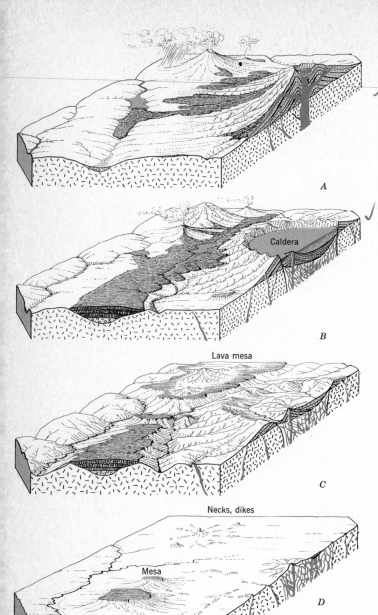

landforms. Lava flows issuing from the volcanoes have spread down into a stream valley, following the downward grade of the valley and forming a lake behind the lava dam.

In Block B, some changes have taken place, the most conspicuous of which is the destruction of the largest volcano to produce a caldera. A lake occupies the caldera, and a small cone has been built inside. One of the other volcanoes, formed earlier, has become extinct. It has been dissected by streams, losing the original conical form. Smaller, neighboring volcanoes are still active, and the contrast in form is marked.

The drainage pattern of streams on a volcanic cone is of necessity a **radial pattern,** shown in Figure 25.7. It is often possible to recognize volcanoes from a drainage map alone because of the perfection of the radial pattern. Where a well-formed crater exists, small streams flow from the crater rim toward the bottom of the crater. Here the water is absorbed in the permeable layers of ash within the cone or is conducted outward by means of a single gap in the crater rim. This inward drainage, described as centripetal, often adds to the certainty of interpreting a volcano from a drainage pattern.

In Block C of Figure 25.6, all volcanoes are extinct and have been deeply eroded. The caldera lake has been drained and the rim worn to a low, circular ridge. The lava flows that formerly flowed down stream valleys have been able to resist erosion far better than the rock of the surrounding area and have come to stand as platforms (mesas) high above the general level of the region.

An example of a dissected volcano is Mount Shasta in the Cascade Range (Figure 25.8). A smaller subsidiary cone of more recent date, named Shastina, is attached to the side.

Block D of Figure 25.6 shows an advanced stage of erosion of volcanoes. There remains now only a small

Figure 25.6 **Stages in the erosional development of volcanoes and lava flows. (Drawn by E. Raisz.)**

ma, the former volcano, is estimated to have risen 1200 m (4000 ft) higher than the present rim. Valleys previously cut by streams and glaciers into the flanks of Mount Mazama were beheaded by the explosive subsidence of the central portion and now form distinctive notches in the rim. The event occurred about 6600 years ago.

Erosional Landforms of Composite Volcanoes

Figure 25.6 shows successive stages in the erosion of volcanoes, lava flows, and a caldera. In Block A, active volcanoes are in the process of building. These are initial

Figure 25.7 **Radial drainage patterns of volcanoes in the East Indies. The letter C shows the location of a crater.**

Figure 25.8 Mount Shasta in the Cascade Range is a dissected volcano. The bulge on the left-hand slope is a more recent subsidiary cone, Shastina. (Infrared photograph by Eliot Blackwelder.)

sharp peak, or **volcanic neck,** representing the solidified lava in the pipe, or neck, of the volcano. Radiating from this are wall-like dikes, formed of lava that previously filled fractures around the base of the volcano (Figure 25.9). Perhaps the finest illustration of a volcanic neck with radial dikes is Ship Rock, New Mexico (Plate L.1). Because the central neck and radial dikes extend to great depths in the rock below the base of the volcano, they may persist as landforms long after the cone and its associated flows have been removed.

Flood Basalts and Shield Volcanoes

As explained in Chapter 24, a mantle plume is a column of heated rock rising within the asthenosphere. Above a plume, crustal basalt can be heated to the point of melting and produce a magma pocket. The site of magma is called a **hot spot.** Magma of basaltic composition makes its way through the overlying lithosphere to emerge at the surface as lava.

Where a mantle plume lies beneath a continental lithospheric plate, the hot spot may generate enormous volumes of basaltic lava that accumulate layer upon layer. The basalt may ultimately attain a thickness of thousands of meters and may cover thousands of square kilometers. These accumulations are called **flood basalts.** An important example is found in the Columbia Plateau region of southeastern Washington, northeastern Oregon, and westernmost Idaho (Figure 25.10); basalts of Cenozoic age cover an area of about 130,000 sq km (50,000 sq mi), about the same area as the state of New York.

Hot spots also form above mantle plumes in the oceanic lithosphere. The emerging basalt builds a class of volcanoes known as **shield volcanoes.** These are constructed on the deep ocean floor, far from plate boundaries, and may be built high enough to rise above sea level as volcanic islands. As a lithospheric plate drifts slowly over a mantle plume beneath, a succession of shield volcanoes is formed. Thus a chain of volcanic islands comes into existence. Several chains of volcanic islands exist in the Pacific Ocean basin. Best known of

the island chains is the Hawaiian group.

Basaltic volcanoes also occur along the mid-oceanic ridge, where sea-floor spreading is in progress. Perhaps the outstanding example is Iceland, in the North Atlantic Ocean. Iceland is constructed entirely of basalt flows superimposed on other basaltic rocks in the form of dikes and sills that entered the spreading rift at deeper levels. Mount Hekla, an active volcano on Iceland, is a shield volcano somewhat similar to those of Hawaii. Farther south along the Mid-Atlantic Ridge are other islands consisting of basaltic volcanoes: the Azores, Ascension Island, and Tristan da Cunha.

Shield volcanoes of the Hawaiian Islands are characterized by gently rising, smooth slopes that flatten near the top, producing a broad-topped volcano (Figure 25.11 and Plate L.1). Domes on the island of Hawaii rise to summit elevations 4000 m (13,000 ft) above sea level.

Figure 25.9 **A vertical dike of volcanic rock exposed by erosion. West Spanish Peak, Colorado. (G. W. Stose, U.S. Geological Survey.)**

Figure 25.10 **Flood basalts of the Columbia Plateau region. The basalt layers have been eroded to produce steep cliffs, rimming broad, flat-topped mesas. Dry Falls, Grand Coulee, central Washington. (John S. Shelton.)**

Including the basal portion lying below sea level, they are more than twice that high. In width they range from 16 to 80 km (10 to 50 mi) at sea level and up to 160 km (100 mi) wide at the submerged base.

Explosive behavior and the emission of tephra are not as important in shield volcano eruptions as they are for composite cones built of felsic magmas. The basalt lava in the Hawaiian domes is highly fluid and travels far down the low slopes, which do not usually exceed angles of 4 or 5°.

Lava domes have a wide, steep-sided central depression that may be 3 km (2 mi) or more wide and several hundred meters deep (Figure 25.12 and Plate L.1). These large depressions are a type of caldera produced by subsidence accompanying the removal of molten lava from beneath. Molten basalt is actually seen in the floors of

Figure 25.11 **This air view of Mauna Loa, Hawaii, shows a chain of pit craters leading up to the great central depression at the summit. (U.S. Army Air Force.)**

Figure 25.12 Halemaumau, a pit crater on Mauna Loa, Hawaii. A fire fountain of molten basalt is erupting from the crater floor. (National Park Service, U.S. Department of Interior.)

Figure 25.14 A fresh cinder cone and a recent basaltic lava flow (left), Dixie State Park, near St. George, Utah. (Frank Jensen.)

Figure 25.13 Lava domes in various stages of erosion make up the Hawaiian Islands. (Drawn by A. N. Strahler.) (A) Initial dome with central depression and fresh flows issuing from radial fissure lines. (B) Early stage of erosion, with deeply eroded valley heads. (C) Advanced erosion stage, with steep slopes and mountainous relief.

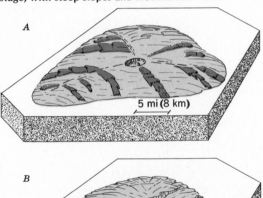

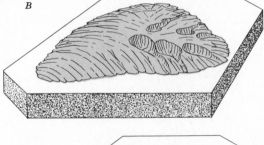

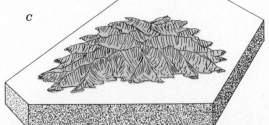

deep pit craters that occur on the floor of the sink or elsewhere over the surface of the lava dome. Most lava flows issue from fissures on the sides of the volcano.

Shield volcanoes show erosion features quite different from those of the composite volcanoes. Figure 25.13 shows the stages of erosion of Hawaiian shield volcanoes, beginning in Diagram A with the active volcano and its central depression. Radial streams cut deep canyons into the flanks of the extinct shield volcano, and these canyons are opened out into deep, steep-walled amphitheaters. Eventually, as Diagram C shows, the original surface of the shield volcano has been entirely obliterated, and a rugged mountain mass made up of sharp-crested divides and deep canyons remains (Plate L.1).

Cinder Cones

At points where frothy basalt magma is ejected under high pressure from a narrow vent, tephra accumulates in a steep-sided conical hill, called a **cinder cone** (Figures 25.14 and 25.15). Groups of cinder cones are common in

Figure 25.15 A cinder cone with its lava flows has dammed a valley, making a lake. Farther downvalley, another lava mass has made a second dam. (After W. M. Davis.)

areas where basaltic lava flows are emerging from fissures. The cones rarely grow to heights over a few hundred meters. The loose material absorbs heavy rain without permitting surface runoff. Erosion is delayed until weathering produces a soil that fills the interstices.

Lava flows sometimes issue from the same vent as a cinder cone; they may burst apart the side of the cone. Cinder cones may erupt in almost any conceivable topographic location — on ridges, on slopes, and in valleys (Figure 25.15). Cinder cones usually occur in groups, often many dozens in an area of a few tens of square kilometers.

Environmental Aspects of Volcanic Activity and Landforms

Eruptions of volcanoes and lava flows are environmental hazards of the severest sort, often taking a heavy toll of plant and animal life and the works of Man. What natural phenomenon can compare with the Mount Pelée disaster, in which thousands of lives were snuffed out in seconds? Perhaps only an earthquake or storm surge of a tropical cyclone is equally disastrous. The wholesale loss of life and the destruction of towns and cities are frequent in the history of peoples who live near active volcanoes. Loss occurs principally from sweeping clouds of incandescent gases that descend the volcano slopes like great avalanches; from lava flows whose relentless advance engulfs whole cities; from the descent of showers of ash, cinders, and bombs; from violent earthquakes associated with the volcanic activity; and from mudflows of volcanic ash saturated by heavy rain. For habitations along lowlying coasts, there is the additional peril of great seismic sea waves, generated elsewhere by the explosive destruction of volcanoes.

The surfaces of volcanoes and lava flows remain barren and sterile for long periods after their formation. Certain types of lava surfaces are extremely rough and difficult to traverse; the Spaniards who encountered such terrain in the southwestern United States named it "malpais" (bad ground). Most volcanic rocks in time produce highly fertile soils that are intensively cultivated.

Volcanoes are a valuable natural resource in terms of recreation and tourism. Few landscapes can rival in beauty the mountainous landscape of volcanic origin. National parks have been made of Mount Rainier, Mount Lassen, and Crater Lake in the Cascade Range. Hawaii Volcanoes National Park recognizes the natural beauty of Mauna Loa and Kilauea; their displays of molten lava are a living textbook of igneous processes.

Landforms of Tectonic Activity

Our introduction to tectonic activity in Chapter 24 brought out the essential distinction between two basically different expressions of tectonic activity. Along converging lithospheric plate boundaries, tectonic activity is basically that of compression, illustrated schematically in Figure 25.16. In subduction zones, sedimentary layers of the ocean floor are subject to compression within a trench as the descending plate forces them against the fixed plate.

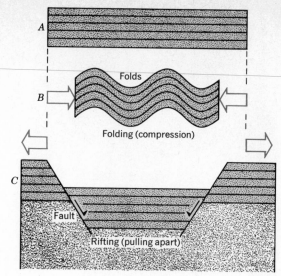

Figure 25.16   **Folding occurs under compression; rifting occurs where the crust is pulled apart.**

In continental collision, compression is of the severest kind, producing alpine structure with thrust faulting (see Figure 24.15).

In zones of rifting of continental plates, explained in Chapter 24, the brittle continental crust is pulled apart and yields by faulting. In the simple model shown in Figure 25.16, rifting is expressed in a pair of opposite-facing faults. The **fault block** between them moves down to form a depressed area. In the remainder of this chapter, we shall examine the landforms of faulting and the phenomenon of earthquakes generated during the faulting process.

Faults and Fault Landforms

A **fault** is a break in the brittle surficial rocks of the earth's crust as a result of sudden yielding under unequal stresses. Faulting is accompanied by a slippage or displacement along the plane of breakage, or **fault plane.** Faults are often of great horizontal extent, so the surface trace, or **fault line,** can sometimes be followed along the ground for many kilometers. Little is known of what happens to faults at depth, but in all probability most extend down for at least several thousands of meters.

Faulting occurs in sudden slippage movements that generate earthquakes. A particular fault movement may result in a slippage of as little as a centimeter or as much as 15 m (50 ft). Successive movements may occur many years apart, even many hundreds of years apart, but accumulate into total displacements of hundreds or thousands of meters. In some places clearly recognizable sedimentary rock layers are offset on opposite sides of a fault, and the amount of displacement can be accurately measured.

According to the angle of inclination and relative direction of the displacement, four basic types of faults are recognized. A **normal fault** has a steep or nearly vertical fault plane (Figure 25.17A). Movement is predominantly in a vertical direction, so one side is raised, or upthrown, relative to the other side, which is downthrown. A normal

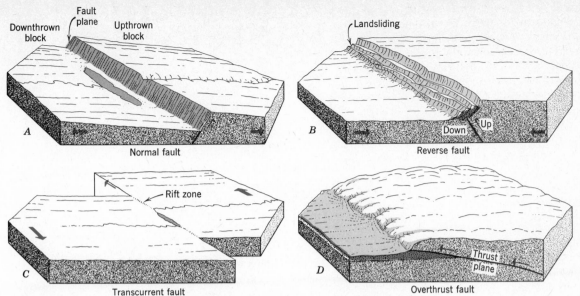

Figure 25.17 Four types of faults and their expression as landforms. (Drawn by A. N. Strahler.)

fault results in a steep **fault scarp,** whose height is an approximate measure of the vertical element of displacement (Figure 25.18). Fault scarps range in height from a few meters to several hundred meters (Plate L.2). Their length is measurable in kilometers; often they attain lengths of 100 km (60 mi) or more. Normal faulting is an expression of the pulling apart (extension) of the earth's outer crust. Sliding on an inclined surface of the type indicated in Figure 25.17A must result in a horizontal separation of points situated on opposite sides of the fault.

In a **reverse fault,** the inclination of the fault plane is such that one side rides up over the other and a crustal shortening occurs (Figure 25.17B). Reverse faults produce fault scarps similar to those of normal faults, but the possibility of landsliding is greater because an overhanging scarp tends to be formed. The San Fernando, California, earthquake of 1971 was generated by slippage on a reverse fault, but this is a rare type of earthquake-producing motion.

A **transcurrent fault** is unique in that the movement is predominantly in a horizontal direction (Figure 25.17C). On an ideal flat plain, no fault scarp would be produced. Instead, only a thin line would be traceable across the surface (Figure 25.19). Streams sometimes turn and follow the fault line for a short distance (Figure 25.20). Sometimes a narrow trench, or rift, marks the fault line.

A low-angle overthrust fault (Figure 25.17D) also involves predominantly horizontal movement, but the fault plane is in a nearly horizontal position. One slice of rock rides over the adjacent ground surface. A thrust slice may be up to 50 km (30 mi) wide.

Faults are rarely isolated features. Normal faults usually occur in multiple arrangements, commonly as a parallel series of faults. This gives rise to a belted terrain pattern. A narrow block dropped down between two normal faults is a **graben** (Figure 25.21 and Plate L.2). A narrow block elevated between two normal faults is

a **horst.** Grabens make conspicuous trenches, with straight, parallel walls. Horsts make blocklike plateaus or mountains, often with a flat top but steep, straight sides. An example of a very large graben, produced by continental rifting, is the Red Sea graben separating the Arabian Peninsula from Africa (Plate L.2). Two smaller grabens form the Gulf of Suez and the Gulf of Aqaba.

Figure 25.18 **This fresh fault scarp in alluvial materials was formed during the Hebgen Lake earthquake of August 17, 1959, in Gallatin County, Montana. Displacement was about 6 m (19 ft) at the maximum point. The vehicle stands on the upthrown side of the fault. (J. G. Stacy, U.S. Geological Survey.)**

Figure 25.19   In this vertical air view, the nearly straight trace of the San Andreas fault, a transcurrent fault, contrasts sharply with the sinuous lines of stream channels. Carrizo Plain, Kern County, California. (Aero Service Division, Western Geophysical Company of America.)

Erosional Development of a Fault Scarp

Landforms shaped from normal faults throughout the erosion period that follows their formation are illustrated in Figure 25.22. At the rear part of Block *A* is the original fault scarp, produced directly by crustal movement. Although eroded by running water, the scarp base is straight. A few stream-cut canyons notch the scarp; some

Figure 25.20   Schematic map of streams offset by long-continued movement along a transcurrent fault. (From A. N. Strahler, 1976, *Principles of Earth Science,* Harper and Row, New York.)

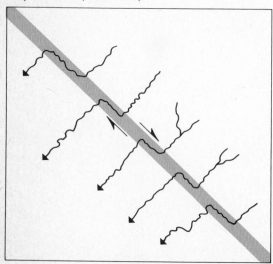

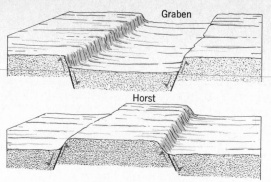

Figure 25.21   **Graben and horst. (Drawn by A. N. Strahler.)**

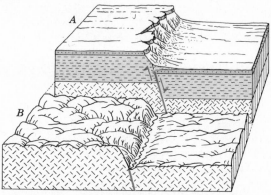

Figure 25.22   **(*A*) Fault scarp. (*B*) Fault-line scarp. (Drawn by A. N. Strahler. From A. N. Strahler, 1971, *The Earth Sciences,* 2nd ed., Harper and Row, New York.)**

Figure 25.23   **Fault-line scarp, MacDonald Lake, near Great Slave Lake, Northwest Territories, Canada. (Canadian Forces Photograph No. 5120-105R.)**

Figure 25.24 **Fault-block mountains may be tilted (left) or lifted (right). (Drawn by W. M. Davis.)**

talus cones and alluvial fans have been built along the scarp base.

Block *B* shows a scarp resulting entirely through erosion; it is a **fault-line scarp.** The scarp appears because a weak shale formation has been stripped from both sides of the fault, revealing a basement of resistant igneous rocks upon which the shale was once deposited. Fault-line scarps are common features of the continental shields (Figure 25.23). These scarps have persisted hundreds of millions of years since the faults ceased to be active because the fault planes penetrated deeply into the crust.

Fault-Block Mountains

As continental rifting begins to occur on a vast scale within a continental lithospheric plate, enormous crustal blocks are raised or depressed by faulting. This rifting process has dominated the landscape of the Basin and Range Province of the southwestern United States and northern Mexico. In Chapter 20, we used large fault blocks as the geologic base upon which to illustrate the progress of fluvial denudation in an arid climate. In the Basin and Range Province, mountain blocks are of two basic types (Figure 25.24). A tilted block has one steep side (the fault scarp) and one gentle side. A lifted block is a form of horst; it has a flattened summit and two steep boundary fault scarps.

Figure 25.25 shows some erosional features of a large, tilted fault block. The freshly uplifted block has a steep mountain face that is rapidly dissected into deep canyons. The upper mountain face reclines in angle as it is removed, while rock debris accumulates in the form of alluvial fans adjacent to the fault block. Vestiges of the fault plane are preserved as **triangular facets,** the snubbed ends of ridges between canyon mouths.

Environmental and Resource Aspects of Faults and Block Mountains

Faults are of environmental and economic significance in several ways. Fault planes are usually zones along which the rock has been pulverized, or at least considerably fractured. This breakage has the effect of permitting ore-forming solutions to rise along fault planes. Many important ore deposits lie in fault planes or in rocks that faults have broken across.

Another related phenomenon is the easy rise of ground water along fault planes. Springs, both cold and hot, are commonly situated along fault lines. They occur along the base of a fault-block mountain. Examples are Arrowhead Springs along the base of the San Bernardino Range and Palm Springs along the foot of the San Jacinto Mountains, both in southern California.

Petroleum, too, finds its way along fault planes where the rocks have been rendered permeable by crushing, or it becomes trapped in porous beds that have been faulted against impervious shale beds. Some of the most intensive searches for oil center about areas of faulted sedimentary strata because of the great production that has come from pools of this type.

Fault scarps and fault-line scarps can form imposing topographic barriers across which it is difficult to build roads and railroads. The great Hurricane Ledge of southern Utah is a feature of this type, in places a steep wall 760 m (2500 ft) high.

Grabens may be so large as to form broad lowlands. An illustration is the Rhine graben of Western Germany. Here a belt of rich agricultural land, 32 km (20 mi) wide and 240 km (150 mi) long, lies between the Vosges and Black Forest ranges, both of which are block mountains faulted up in contrast to the downdropped Rhine graben block.

Earthquake — an Environmental Hazard

The news media bring detailed accounts of disastrous earthquakes. Nearly everyone has seen pictures of their destructive effects. Californians know about severe earthquakes from first hand experience; but many other areas in North America have also experienced earthquakes, and a few of these have been severe. An **earth-**

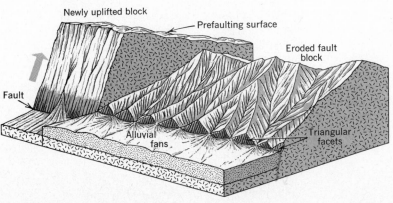

Newly uplifted block

Prefaulting surface

Eroded fault block

Fault

Alluvial fans

Triangular facets

Figure 25.25 **Erosion of a tilted fault block produces a rugged mountain range. A line of triangular facets at the mountain base marks the position of the original fault plane. (Drawn by A. N. Strahler.)**

Figure 25.26 **Evidences of lateral earth movement accompanying the San Francisco Earthquake of 1906, Marin County, California. (Photographs by G. K. Gilbert, U.S. Geological Survey.)** *Above:* **A fence offset 2.4 m (8 ft) along the main fault near Woodville.** *Below:* **A road offset 6 m (20 ft) along the main fault near Point Reyes Station. The shear zone is 18 m (60 ft) wide.**

quake is a motion of the ground surface, ranging from a faint tremor to a wild motion capable of shaking buildings apart and causing gaping fissures to open up in the ground.

The earthquake is a form of energy of wave motion transmitted through the surface layer of the earth in widening circles from a point of sudden energy release — the **earthquake focus.** Like ripples produced when a pebble is thrown into a quiet pond, these **seismic waves** travel outward in all directions, gradually losing energy.

As already noted, earthquakes are produced by sudden movements along faults; commonly these are normal faults or transcurrent faults. The famed San Andreas fault passes through the San Francisco Bay Area. The devastating earthquake of 1906 resulted from slippage along this fault, which is of the transcurrent type. The fault is 1000 km (600 mi) long and extends into southern California, passing about 60 km (40 mi) inland of the Los Angeles metropolitan area. In places the fault is expressed as a rift valley (Figure 25.19). Associated with the San Andreas

fault are several other important transcurrent faults, all capable of generating severe earthquakes.

We shall not go into the details of the mechanics of faults and how they produce earthquakes. It must be enough to say that rock on both sides of the fault is slowly bent over many years as tectonic forces are applied in the movement of lithospheric plates. Energy accumulates in the bent rock, just as it does in a bent crossbow. When a critical point is reached, the strain is relieved by slippage on the fault and a large quantity of energy is instantaneously released in the form of seismic waves. Slow bending of the rock takes place over many decades. Its release then causes offsetting of features that formerly crossed the fault in straight lines, for example, a roadway or fence (Figure 25.26). Faults of this type can also show a slow, steady displacement known as *fault creep,* which tends to reduce the accumulation of stored energy.

The **Richter scale** of earthquake magnitudes was devised in 1935 by the distinguished seismologist, Charles F. Richter, to indicate the quantity of energy released by a single earthquake. Scale numbers range from 0 to 9, but there is no upper limit except for nature's own limit of energy release. A value of 8.6 was the largest observed between 1900 and 1950.

One of the great earthquakes of recent times was the Good Friday earthquake of March 27, 1964, with the surface point of origin located about 120 km (75 mi) from Anchorage, Alaska. Its magnitude was 8.4 to 8.6 on the Richter scale, approaching the maximum known. Of particular interest in connection with earthquakes as environmental hazards are their secondary effects. At Anchorage, most of the damage was from secondary effects because buildings of wooden-frame construction often experience little damage by shaking alone when built on solid rock areas. Damage was largely from earth movements in the weak clays underlying the city (Figure 25.27). These quick clays developed liquid properties upon being shaken and allowed great segments of ground to subside and pull apart in a succession of steps, tilting and rending houses. Other secondary effects were from the rise of water level and landward movement of large waves, destroying ships and low-lying structures.

Seismic Sea Waves

Another important secondary effect of a major earthquake is the **seismic sea wave,** or **tsunami,** as it is known to the Japanese. A train of these waves is often generated in the ocean at a point near the earthquake source by a sudden movement of the sea floor. The waves travel over the ocean in ever-widening circles, but they are not perceptible at sea in deep water. When a wave arrives at a distant coastline, however, the effect is to cause a rise of water level. Wind-driven waves, superimposed on the heightened water level, allow the surf to attack places inland that are normally above the reach of waves. For example, the particularly destructive seismic sea wave of 1933 in the Pacific Ocean caused waves to attack ground as high as 9 m (30 ft) above normal tide level, causing widespread destruction and many deaths by drowning in low-lying coastal areas. It is thought that coastal flooding which oc-

Figure 25.27 **Slumping and flowage of unconsolidated sediments, resulting in property destruction after the Good Friday earthquake of March 27, 1964, at Anchorage, Alaska. (U.S. Army Corps of Engineers.)**

curred in Japan in 1703, with an estimated life loss of 100,000 persons, may have been caused by seismic sea waves.

Earthquakes and Urban Planning — the San Fernando Earthquake

About 6:00 A.M., on February 9, 1971, a violent earthquake shook the San Fernando Valley area, lying northwest of the city of Los Angeles. The quake lasted only about 60 seconds, but during this single minute 64 persons lost their lives. The toll in human lives and the severe structural effects of the San Fernando earthquake shocked the entire Los Angeles community into renewed awareness of the need for urban planning to minimize or forestall the damaging effects of a major earthquake.

Although the earthquake was not in the really severe category according to the Richter scale (it measured 6.6, which is moderate in severity), local areas experienced a ground motion of acceleration as high as or higher than any previously measured in an earthquake. Fortunately, the ground shaking was of brief duration; had it persisted for a longer time, structural damage would have been much more severe than it was. Particularly disconcerting was the collapse of the Olive View Hospital in Sylmar, a new structure supposedly conforming with earthquake-resistant standards. The Veterans Hospital in Sylmar also suffered severe damage; several buildings collapsed and 42 persons were killed (Figure 25.28).

A crack produced in the Van Norman Dam prompted authorities to drain that reservoir to prevent dam collapse and disastrous flooding of a densely built-up area. The Sylmar Converter Station, one of the key elements in the electrical power transmission system of the Los Angeles area, was severely damaged. Collapse of a freeway overpass blocked the highway beneath, and freeway pavements were cracked and dislocated (Figure 25.29).

Fortunately, the time of the quake was early morning, when most persons were at home and few were traveling the major arteries.

The fault movement that set off the San Fernando earthquake was not on the great San Andreas fault; rather, it was from an epicenter some 25 km (15 mi) from that fault, along a system of relatively minor faults. In this area the San Andreas fault is capable of producing an earthquake of far greater intensity than the 1971 San Fernando earthquake. Although the year of such an earthquake is not yet predictable to the nearest decade, the progress of urbanization will have greatly expanded the structures and population subject to devastation.

Soon after the San Fernando disaster, the National Academy of Sciences and the National Academy of Engineering set up a joint panel of experts to study the earthquake effects and draw up recommendations. The panel concluded: "It is clear that existing building codes do not provide adequate damage control features. Such codes should be revised." The panel further recommended that public buildings, such as hospitals, schools, and buildings housing police and fire departments and other emergency services should be so constructed as to withstand the most severe shaking to be anticipated. Fortunately, most school buildings that were built after the severe Long Beach earthquake of the 1930s showed no structural damage. Many older school buildings, however, were rendered unfit for use.

Damage from the San Fernando earthquake of 1971 has been estimated at $500 million, but experts think that an earthquake as severe as that of 1906 at San Francisco would cause damage on the order of $20 billion if it occurred now in a large metropolitan area. Perhaps, then, one beneficial effect of the San Fernando earthquake will be the appropriation of adequate funds for research on many important aspects of earthquakes and their effects in an urban environment.

Figure 25.28 Severe structural damage and the collapse of buildings of the Veterans Administration Hospital, in Sylmar, Los Angeles County, caused by the San Fernando earthquake of February 1971. (Wide World Photos.)

Figure 25.29 Collapsed pavement and overpass on the Golden State Freeway at the northern end of the San Fernando Valley, California, resulting from the earthquake of February 1971. (Wide World Photos.)

References for Further Study

Williams, H. (1941), *Crater Lake, the story of its origin,* University of California Press, Berkeley.

Cotton, C. A. (1944), *Volcanoes as landscape formers,* Whitcombe and Tombs, Christchurch, New Zealand.

Macdonald, G. A., and A. T. Abbott (1970), *Volcanoes in the sea: the geology of Hawaii,* Univ. of Hawaii Press, Honolulu.

Iacopi, R. (1971), *Earthquake country,* Lane Books, Menlo Park, Calif.

Staff of U.S. Geological Survey and National Oceanographic and Atmospheric Administration (1971), *The San Fernando earthquake of February 9, 1971,* U.S. Geological Survey, Professional Paper 733, Government Printing Office, Washington, D.C.

Bullard, F. M. (1976), *Volcanoes of the earth,* University of Texas Press, Austin.

Panel on Earthquake Prediction (1976), *Predicting earthquakes: a scientific and technical evaluation with implications for society,* National Academy of Sciences, Washington, D.C.

Review Questions

1. Describe the various forms of intrusive igneous rock bodies. What is a batholith? How does a sill differ from a dike?

2. What is a volcano? What factors influence the shape and dimensions of a volcano? How does the composition of the lava influence the form and explosiveness of a volcano?

3. Describe a composite volcano, both as to outward form and internal structure. Give examples of three large composite volcanoes. What is a nuée ardente? In what way are composite volcanoes related to subduction zones? How is a caldera formed?

4. Describe the stages in erosion of a composite volcano. What drainage pattern is produced by streams dissecting a volcano? What erosion features remain after the cone is entirely removed?

5. How are flood basalts related to mantle plumes and hot spots? Give an example of a region of thick flood basalts. How are shield volcanoes related to hot spots and moving lithospheric plates? Why are basaltic volcanoes associated with the mid-oceanic ridge?

6. Describe the shield volcanoes of the Hawaiian Islands. How large are these shield volcanoes? Explain the central depression of a shield volcano. Describe the changes in form of extinct shield volcanoes as fluvial erosion progresses.

7. What are cinder cones? Describe their form and composition.

8. Discuss the environmental aspects of volcanic activity. In what ways are volcanic eruptions severe environmental hazards? What value have volcanoes as natural resources? Give examples of volcanic areas that are national parks.

9. What is a fault? A fault plane? A fault line? Describe normal, reverse, transcurrent, and low-angle overthrust faults. What is the relative motion of the fault blocks in each type? What landforms are associated with each type? Give an example of a large graben.

10. Explain the difference between a fault scarp and a fault-line scarp. In what regions are fault-line scarps abundant? Describe the erosional features formed from a tilted fault block. What is the meaning of triangular facets?

11. Discuss the environmental and resource aspects of faults, fault lines, and fault-block mountains.

12. What causes an earthquake? How do seismic waves travel? What is the Richter scale? Describe the destructive effects of the Good Friday earthquake in Alaska. Where did the most severe damage to buildings occur? Explain.

13. What are seismic sea waves, or tsunamis? In what way are they a hazard to humans?

14. Use the San Fernando earthquake of 1971 as an example of the impacts of a major earthquake on a large urban area. What kinds of structural damage occurred? What lessons can be learned from the San Fernando earthquake?

Landforms and Rock Structure

26

In this chapter we turn to erosional landforms developed on the stable crust of the continental lithosphere, far removed from active plate boundaries where subduction and rifting are taking place. We shall examine geologic relics of past continental collisions and of deep-seated tectonic and intrusive activity that once took place over subduction zones. These continent-building activities started early in Precambrian time, at least 3 billion years ago, and persisted through the Paleozoic and Mesozoic Eras as one orogeny after another left its mark. Since the close of the Cretaceous Period, about −65 m.y., the stable continental shields have experienced fluvial denudation, punctuated by occasional epeirogenic downwarping that brought marine submergence to large interior regions, and by upwarping that raised the submerged areas to become land again. True, some minor faulting has locally affected the shields in the Cenozoic Era, but this activity is scarcely perceptible in the landscape. There has also been sporadic and isolated volcanic activity in hot spots within the shields, as the continental lithosphere has drifted over mantle plumes. These isolated volcanic features have been treated in the previous chapter; they include flood basalts and some isolated groups of volcanoes, such as the San Francisco Peaks of northern Arizona.

Structural Types Within Continental Plates

Within the stable continental lithospheric plates we find several distinct structural types of landmasses, each of which exerts a strong control over the erosional landforms that develop on it. Three basic groups of structures can be recognized: (1) undisturbed strata; (2) deformed strata affected by doming, folding, overthrusting, and metamorphism; and (3) plutons (batholiths) uncovered by prolonged fluvial denudation. Deeply eroded faults and volcanoes were described in Chapter 25; they should be added to the list of erosional landforms controlled by rock structure.

Rock Structure as a Landform Control

As denudation takes place, landscape features develop in close conformity with patterns of bedrock composition and structure. Figure 26.1 shows four types of sedimentary rock, together with a mass of much older igneous rock on which the sediments were deposited. The diagram shows the usual landform habit of each rock type, whether to form valleys or mountains. The cross section shows conventional rock symbols used by geologists. These rock strata have been strongly tilted and deeply eroded.

Shale is a weak rock and is reduced by fluvial action to form the lowest valley floors of the region. Limestone, easily dissolved by carbonic acid action, also forms valleys in humid climates. In arid climates, on the other hand, limestone is highly resistant and usually forms high, prominent landforms. Sandstone is usually a resistant rock; it forms ridges and uplands. As a group, the felsic igneous rocks are resistant to denudation. Occurring as plutons, they typically form uplands rising above adjacent sedimentary strata.

The metamorphic rocks are, as a group, more resistant to denudation than their sedimentary parent types. As shown in Figure 26.24, however, there are conspicuous landform differences among the types of metamorphic rocks.

Dip and Strike

Natural layers and planes of weakness are characteristic of the structure of each kind of rock. A system of geometry is

Figure 26.1 **Many landscape features originate through the slow erosional removal of weaker rock, leaving the more resistant rock standing as ridges or mountains. (Drawn by A. N. Strahler.)**

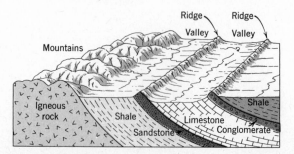

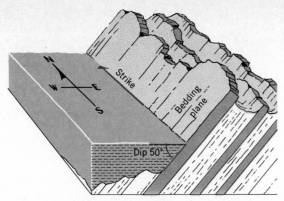

Figure 26.2 **Strike and dip. (Drawn by A. N. Strahler.)**

needed to measure and describe the attitude of these natural planes and to indicate them on maps. Examples of such planes are the bedding layers of sedimentary strata, the sides of a dike, and the joint sets in granite. Rarely are these planes truly horizontal or vertical.

The acute angle formed between a natural rock plane and an imaginary horizontal plane is the **dip.** The amount of dip is stated in degrees ranging from 0° for a horizontal plane to 90° for a vertical plane. Figure 26.2 shows the dip angle for an outcropping layer of sandstone, against which a horizontal water surface is at rest. The compass direction of the line of intersection between the inclined rock plane and an imaginary horizontal plane is the **strike.** In Figure 26.2, the strike is north.

Undisturbed Sedimentary Strata

Undisturbed sedimentary strata of more or less horizontal attitude cover large parts of the continental shields. Although there is no set limit to the age of strata in these

covers, almost all are of Paleozoic, Mesozoic, and Cenozoic ages. Rarely are Precambrian strata found undisturbed by tectonic activity. The undisturbed sedimentary strata fall into two major classes: coastal plains and interior shield covers.

Coastal Plains

Coastal plains are undisturbed accumulations of marine strata lapping over stable continental margins. Refer to Figure 24.9 for the geologic interpretation of coastal plain strata. Where continental and oceanic lithosphere are in contact within a single lithospheric plate, sediment accumulates in a continental shelf wedge. The margin of the continental plate subsides because of isostatic compensation (Chapter 20) as the sediment load accumulates. Water depth over the continental shelf remains shallow, but the sediment wedge can reach thicknesses of thousands of meters over a span of 50 to 100 million years. Stable continental margins with thick continental shelf deposits occur on continental margins enclosing the North and South Atlantic Ocean basins. (The world map of lithospheric plates, Figure 24.18, shows that the continental margins are well inside the lithospheric plates.) As these continental margins experienced episodes of epeirogenic downsinking, they became submerged inland extensions of the continental shelves. During submergence, marine clay, marl (lime mud), and sand were deposited in thin, wedge-shaped layers. Epeirogenic uplift of the continental margin has, in many places, brought a broad zone of these marine strata above sea level as coastal plains.

Block *A* of Figure 26.3 shows a recently emerged coastal plain. It is a smooth, almost featureless surface, sloping gently toward the coastline. Streams that drain the older landmass extend their courses across the plain to

Figure 26.3 **Development of a broad coastal plain. (*A*) Early stage; plain recently emerged. (*B*) Advanced stage; cuestas and lowlands developed. *S* = subsequent stream; *C* = consequent stream. (Drawn by A. N. Strahler.)**

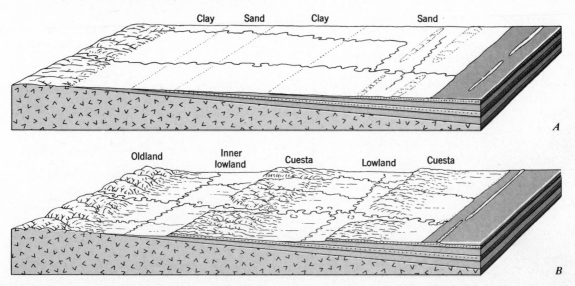

reach the sea. These are known as **consequent streams** because they follow the initial slope of the new land surface. (The radial streams that originate on the slopes of a new volcano also qualify as consequent streams by this same definition.) The diagram shows that parallel lines of uplifted beach ridges formed a partial barrier to the consequent streams, causing one of them to turn parallel with the shoreline and join the other.

In an advanced stage of coastal-plain erosion, a new series of streams and topographic features has developed (Block *B* of Figure 26.3). Where more easily eroded strata (usually clay, marl, or shale) are exposed, denudation is rapid, making **lowlands.** Between them rise broad belts of hills called **cuestas.** Cuestas are commonly underlain by sand, sandstone, or limestone. The lowland lying between the area of older rock (*oldland*) and the first cuesta is called the **inner lowland.**

Streams that develop along the trend of the lowlands, parallel with the shoreline, are of a class known as **subsequent streams.** They take their positions along zones of weak rock and therefore follow closely the patterns of rock structure. Subsequent streams occur in many regions, and we shall mention them again in the discussion of domes and folds. The drainage lines on a fully dissected coastal plain combine to form a **trellis drainage pattern.** In this pattern, the subsequent streams trend at about right angles to the consequent streams.

Coastal Plain of North America

The coastal plain of eastern North America is a major geographic region, ranging in width from 160 to 500 km (100 to 300 mi) and extending for 3000 km (2000 mi) along the Atlantic and Gulf coasts. The coastal plain starts at Long Island, New York, as a partly submerged cuesta, and widens rapidly southward so as to include much of New Jersey, Delaware, Maryland, and Virginia (Figure 26.4). Throughout this portion, the coastal plain has but one cuesta. The inner lowland is a continuous broad valley developed on weak clay strata.

In Alabama and Mississippi, the coastal plain is fully dissected. Cuestas and lowlands run in belts roughly parallel with the coast (Figure 26.5). The cuestas are underlain by sandy formations and support pine forests. Limestone forms lowlands such as the Black Belt in Alabama, named for its dark, fertile soils.

Environmental and Resource Aspects of Coastal Plains

Lowlands of the broad coastal plain of the eastern United States show intensive agricultural development because of the natural fertility of soils derived from carbonate strata. Sandy cuestas, bearing soils poor in plant nutrients, provide valuable pine forests in the southern United States. Transportation tends to follow the lowlands and to connect the larger cities located there. For example, important highways and railroads connect New York City with Trenton, Philadelphia, Baltimore, Washington, and Richmond, all of which are situated in the inner lowland (Figure 26.4). The inner limit of coastal plain strata is known as the Fall Line, because here the major rivers have developed falls and rapids on the resistant oldland rocks. Cities grew at the Fall Line, where the rivers become navigable tidal estuaries. Thus geology has played a part in the location of the nuclei of Megalopolis.

The seaward dip of sedimentary strata in a coastal plain provides a structure favorable to the development of artesian water supplies (Figure 26.6). Water penetrates

Figure 26.4   **The coastal plain of the Atlantic seaboard states shows little cuesta development except in New Jersey. The inner limit of the coastal plain is marked by a series of fall-line cities. (After A. K. Lobeck.)**

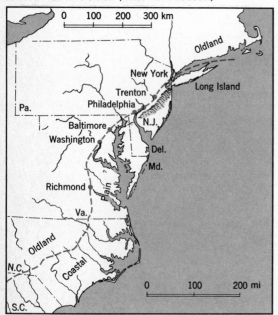

Figure 26.5   **The Alabama-Mississippi coastal plain is belted by a series of sandy cuestas and shale or marl lowlands. (After A. K. Lobeck.)**

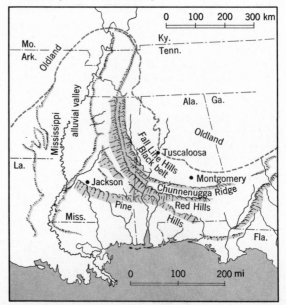

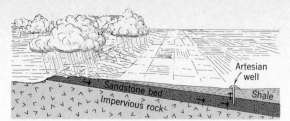

Figure 26.6 An artesian well derives its flow from a gently dipping sandy aquifer. (Drawn by E. Raisz.)

Figure 26.8 This old photograph shows a butte of horizontal sandstone beds capped by a gypsum layer, near Cambria, Wyoming. (N. H. Darton, U.S. Geological Survey.)

deeply into sandy strata of a cuesta, overlain by aquicludes of shale or clay. When a well is drilled into the sand formation considerably seaward of its surface exposure, water under hydraulic pressure rises toward the surface. Artesian water in large quantities is available in many parts of the Atlantic and Gulf coastal plains, although it is no longer sufficient to supply the demands of densely populated and industrialized communities.

Of the economic mineral resources of broad coastal plains, the most important are petroleum and natural gas. The Gulf Coastal Plain is a major oil-producing region. Here the strata beneath the coastal plain and offshore continental shelf are extremely thick. Lignite, a low grade of coal, occurs beneath a large area of Alabama, Mississippi, Louisiana, and Texas (see Figure 26.10).

Other mineral deposits of economic importance in coastal plains include sulfur, occurring in the coastal plain of Louisiana and Texas; phosphate beds, found in Florida; and clays, used in the manufacture of pottery, tile, and brick in New Jersey and the Carolinas.

Interior Shield Covers

Extensive areas of the continental shields are covered by thick sequences of nearly horizontal sedimentary strata. At various times in the 600 million years following the end of Precambrian time, these strata were deposited in shallow inland seas. Following crustal uplift, with little disturbance other than minor warping or faulting, these areas became continental surfaces undergoing denudation.

In arid climates, where vegetation is sparse and the action of overland flow is especially effective, sharply defined landforms develop on horizontal sedimentary strata (Figure 26.7). The dominant landform is a sheer rock

Figure 26.7 In arid climates, a distinctive set of landforms develops in flat-lying sedimentary formations. (Drawn by A. N. Strahler.)

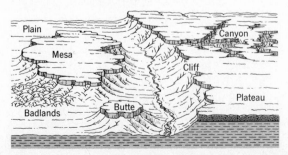

wall, or **cliff.** At the base of the cliff is an inclined slope. This slope flattens out to make a bench, terminated at the outer edge by the cliff of the next lower set of forms. These forms are wonderfully displayed in the walls of the great canyons of the Colorado Plateau region (see Plates H.1 and L.3).

In these arid regions, erosion strips away successive rock layers, leaving behind **plateaus** capped by hard rock layers. Cliffs retreat as near-perpendicular sufaces because the weak shale formations exposed at the cliff base are rapidly washed away by storm runoff. When undermined, the rock in the upper cliff face repeatedly breaks away along vertical joint fractures.

Cliff retreat produces **mesas,** table-topped plateaus bordered on all sides by cliffs (Figure 26.7). Mesas represent the remnants of a formerly extensive layer of resistant rock. As a mesa is reduced in area by retreat of the rimming cliffs, it maintains its flat top. Before its complete consumption, the final landform is a small steep-sided hill known as a **butte** (Figure 26.8 and Plate L.3).

Areas underlain by weak shales, clays, or marls erode rapidly and may be entirely barren of plant cover. The weak matérial is dissected by a network of small streams to produce badlands, a landscape of miniature valleys and divides. (Badlands were explained in Chapter 19; see Figure 19.6.)

Regions of horizontal strata have branching stream networks formed into a **dendritic drainage pattern** (Figure 26.9 and Plate L.3). The smaller streams in this pattern take a variety of directions. This treelike branching pattern is also found in areas of exposed plutons (see Figure 26.28).

Shield covers may include areas of flood basalts, in which erosional landforms resemble those of horizontal strata. One such occurrence is the Columbia Plateau region of eastern Washington and Oregon, described in Chapter 25. Another is the Deccan Plateau of western India. In both areas, basalt lavas cover thousands of square kilometers and are several thousand meters thick.

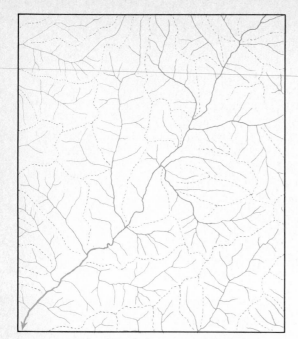

Figure 26.9 **A dendritic drainage pattern formed on horizontal strata. Divides are shown in dotted line.**

Interbedded with the lavas are lake and stream deposits of sands, gravels, and clays. Canyons, mesas, and buttes carved from lava flows are similar to those formed of sedimentary strata.

Resources of Shield Covers

Horizontal strata of shield covers contain many minerals and rocks of economic value. Building stone, such as the Bedford limestone in Indiana or the Berea sandstone in Ohio, is a valuable resource. Limestone is quarried for use in the manufacture of portland cement or as flux in iron smelting. Some important deposits of lead, zinc, and iron ores occur in sedimentary rocks. For example, the lead and zinc mines of the Tristate district (Missouri, Kansas, and Oklahoma) are in horizontal limestones. Uranium ores are important in strata of the Colorado Plateau.

Perhaps the greatest mineral resources occurring in sedimentary strata are coal and petroleum. We described these fossil fuels in Chapter 11. Recall from Chapter 17 that the strip mining of coal is an extreme form of environmental degradation, having various undesirable side effects besides the massive scarification of the landscape. As the map of North American coal fields shows (Figure 26.10), large reserves of coal lie close to the surface in areas of horizontal strata in the Great Plains region west of the Mississippi, in the northern High Plains still farther west, in sedimentary basins of the Rocky Mountains, and in the Colorado Plateau near the common corner of the states of Utah, Arizona, Colorado, and New Mexico. The great demands for this coal in decades to come make it likely that large areas will be strip mined and that power-generating plants of large capacity will be built close to the supplies of coal. Environmentalists are greatly concerned with the land scarification and air pollution that such developments can bring (Figure 26.11).

Figure 26.10 **Coal fields of the United States and southern Canada. The lignite area of the southern states lies within the coastal plain. Anthracite fields are in folded strata.**

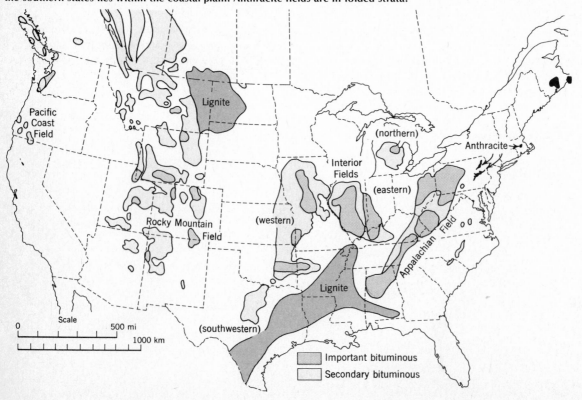

Figure 26.11 **This great coal seam, ranging from 18 to 27 m (60 to 90 ft) thick, is being strip mined at Wyodak, Wyoming. The upper power shovel is removing the overburden; the shovel at the base of the seam is removing the coal. (Bureau of Mines, U.S. Department of the Interior.)**

Deformed Strata

A second structural class of landmasses found within the continental lithospheric plates consists of deformed strata. The class includes two structural types in which deformation is relatively simple: **domes** and **open folds.** A third structural type includes belts of intensely deformed strata showing tight folding and thrust faulting. The strata have been largely transformed into metamorphic rock.

Deformed strata can be related to the tectonic process of suturing described in Chapter 24. Belts of open folds usually lie parallel with belts of highly deformed strata, as shown in Figure 26.12. (The Appalachian Mountains illustrate this relationship particularly well.) The diagram shows a belt of intense folding and overthrusting in which sediments have been converted into metamorphic rocks — schist, slate, marble, and gneiss. This belt represents the core of a continental suture, produced in an early continental collision. In the case of the Appalachian Mountains, final suturing occurred at the close of the Paleozoic Era. Melting of a large mass of squeezed sediments deep in the suture produced a large pluton of felsic (granitic) rock; this is shown exposed by denudation at the extreme right of the block diagram. Farther into the continental interior, at the left, the belt of overthrust and overturned folds gives way to a belt of open folds, which are simple, wavelike undulations of the sediment cover. A fold belt of this type forms an important subdivision of the

Figure 26.12 **A schematic block diagram of basic structural types of deformed strata. (Drawn by A. N. Strahler.)**

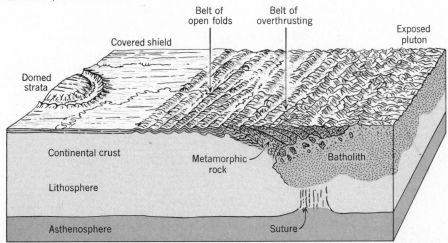

Appalachian Mountains. Bordering the fold belt, at the far left, is an area in which the shield cover lies horizontal, unaffected by the tectonic activity. In the Appalachian region, a highland zone of deeply dissected horizontal strata called the Appalachian Plateau borders the fold belt on the west.

Within the interior region of the shield cover, we have shown in Figure 26.12 a broad, low dome structure, not related to the suture zone or to any tectonic feature of lithospheric plate boundaries. Some domes are nearly circular in outline, others are elliptical. Ancient shield rock lies beneath these domes, and they seem to represent a local uplift of a thick layer of crust. Most show no signs of igneous activity at the time of doming, and it appears likely that deep-seated mechanical forces were responsible for uplift. The relationship of broad sedimentary domes to plate tectonics remains to be established.

Sedimentary Domes

Erosion features of a sedimentary dome are illustrated in the two block diagrams of Figure 26.13. The diameter of the dome is on the order of 50 to 100 km (30 to 60 mi). Strata are first removed from the summit region of the dome, exposing older strata beneath. Eroded edges of

Figure 26.14 **A hogback on the flank of the Virgin anticline in southwestern Utah. (Frank Jensen.)**

Figure 26.13 **Erosion of strata from the summit of a sedimentary dome structure. (A) Strata partially removed, forming an encircling hogback ridge. (B) Strata removed from center of dome, revealing a core of much older igneous or metamorphic rock. (Drawn by A. N. Strahler.)**

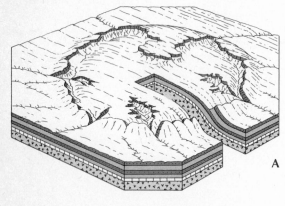

A

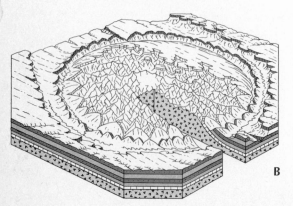

B

steeply dipping strata form sharp-crested sawtooth ridges called **hogbacks** (Figure 26.14 and Plate L.4). When the last of the strata has been removed, the ancient shield rock is exposed in the central core of the dome. Here a mountainous terrain develops on plutons lacking in layered structure.

Figure 26.15 is an idealized block diagram to show the relationship of landforms to dip of strata. In the foreground are hogbacks formed on steeply dipping sandstone layers. Weak shale forms narrow valleys between the hogbacks and between exposed shield rock and a hogback. Subsequent streams occupy the valleys of weak rock. Toward the central part of the block diagram, the angle of dip of the strata is shown to lessen, so the hogback form gives way to broader ridges of cuesta form. The cuesta has a steep, clifflike face on one side and a gentle, descending surface on the opposite side. Cuestas on coastal plains

Figure 26.15 **Hogbacks gradually merge into cuestas and the cuestas into plateaus and mesas, where the dip of the strata becomes less from one place to another. S = subsequent stream; H = hogback ridge; Cu = cuesta; M = mesa; Cl = cliff; P = plateau. (Drawn by W. M. Davis.)**

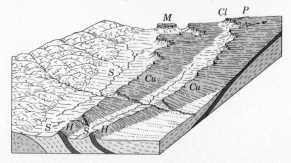

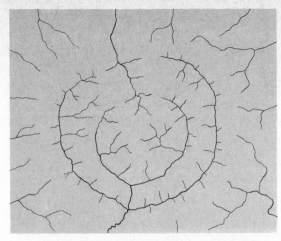

Figure 26.16 **The drainage pattern on an eroded dome combines annular and radial elements.**

may show this asymmetrical form if the strata underlying the cuesta are well indurated. Cuestas are found on flanks of sedimentary domes where the dip is low. Toward the rear of the diagram, the strata are shown to flatten in attitude and become nearly horizontal. Here, erosion produces plateaus, cliffs, and mesas.

The stream network on a deeply eroded dome shows dominant subsequent streams forming a circular system, called an **annular drainage pattern** (Figure 26.16). The shorter tributaries make a radial arrangement. The total pattern resembles a trellis pattern bent into a circular form.

The Black Hills Dome

A classic example of a large sedimentary dome is the Black Hills dome of western South Dakota and eastern Wyoming (Figure 26.17). Valleys that encircle this dome

Figure 26.17 **The Black Hills dome. (Drawn by A. N. Strahler.)**

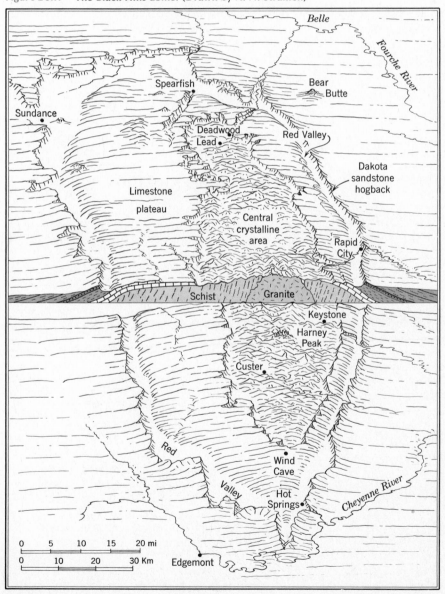

are ideal locations for railroads and highways, so it is natural that towns and cities should have grown in these valleys. One valley in particular, the Red Valley, is continuously developed around the entire dome and has been called the Race Track because of its shape. It is underlain by a weak shale, which is easily washed away. The cities of Rapid City, Spearfish, and Sturgis are located in the Red Valley. On the outer side of the Red Valley is a high, sharp hogback of Dakota sandstone, known simply as Hogback Ridge. It rises some 150 m (500 ft) above the level of the Red Valley. Farther out toward the margins of the dome the strata are less steeply inclined and form a series of cuestas. Artesian water is obtained from wells drilled in the surrounding plain.

The eastern central part of the Black Hills consists of a mountainous core of intrusive and metamorphic rocks. These mountains are richly forested, whereas the intervening valleys are beautiful open parks. Thus the region is attractive as a summer resort area. Harney Peak, elevation 2207 m (7242 ft), is the highest peak of the core. In the northern part of the central core, in the vicinity of Lead and Deadwood, are valuable ore deposits. At Lead is the fabulous Homestake Mine, formerly one of the world's richest gold-producing mines.

The western central part of the Black Hills consists of a limestone plateau deeply carved by streams. The original dome had a flattened summit. The limestone plateau represents one of the last remaining sedimentary rock layers to be stripped from the core of the dome.

Landforms of Open Folds

Parallel, wavelike folds pass through a series of erosion stages, illustrated in Figure 26.18. A downfold of strata is known as a **syncline;** an upfold is known as an **anticline** (Plate L.4). In an early stage of the development of folds, anticlines are identical with the mountains, or ridges; synclines are identical with the valleys.

Block *A* of Figure 26.18 shows erosion of anticlines occurring during the last stages of folding. Synclines are being filled with alluvial fan materials swept down from the adjacent anticlines. After folding has ceased the upper layers of soft, unconsolidated rock are removed until, as shown in Block *B*, a hard, well-cemented sandstone layer is exposed, reflecting the full amplitude of the folding. Coinciding with synclines are **synclinal valleys;** coinciding with anticlines are **anticlinal mountains.**

Streams that drain the flanks of the anticlines quickly cut deep ravines, exposing the underlying layers. Breaching spreads rapidly to the crest of the anticline, where a long, narrow valley is opened out along the summit (Block *B*). This valley is occupied by a subsequent stream excavating a belt of weak rock; it is an **anticlinal valley** because it lies on the center line of the anticline. As this valley grows in length, depth, and breadth, it replaces the original anticlinal mountain. A reversal of topography thus occurs. The synclinal valley, which originally contained the major stream, is now shrunken between the growing anticlinal valleys on either side. The anticlinal valleys are now rapidly deepened because the core of weak rock is exposed to attack. Eventually the syncline

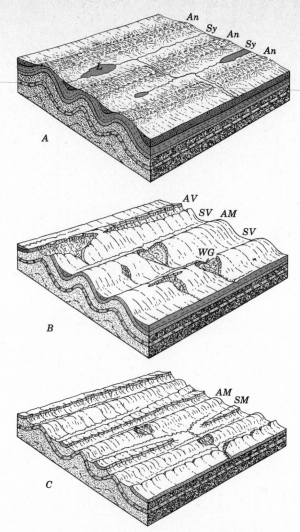

Figure 26.18 **Stages in the erosional development of folded strata. (A) While folding is still in progress, erosion cuts down the anticlines; alluvium fills the synclines, keeping relief low.** *An* = **anticline;** *Sy* = **syncline;** *L* = **lake. (B) Long after folding has ceased, erosion exposes a highly resistant layer of sandstone or quartzite.** *AV* = **anticlinal valley;** *SV* = **synclinal valley;** *WG* = **watergap. (C) Continued erosion partly removes the resistant formation but reveals another below it.** *AM* = **anticlinal mountain;** *SM* = **synclinal mountain. (Drawn by A. N. Strahler.)**

becomes a mountain ridge, called a **synclinal mountain** (Block C). At this stage, the original topography has been completely reversed. The drainage pattern is a trellis type, similar in most respects to the trellis pattern of a mature coastal plain, but different in that the major subsequent streams are more closely spaced and their tributaries are shorter (Figure 26.19).

Here and there in a belt of folds, a major stream crosses several folds at nearly right angles, passing through the sharply defined ridges by narrow **watergaps.** These streams may have existed prior to the folding and were able to maintain themselves as the folds were (right) **Mt. Hood, a recently formed composite volcano of the Cascade Range, Oregon, with a summit elevation just over 11,000 ft (3350 m). Glaciers have carved the sides of the cone. (Orlo E. Childs)**

(below) **A fresh cinder cone near Bend, Oregon. The rough surface at the base of the cone is a recent basaltic lava flow. (Orlo E. Childs)**

(above) **Ship Rock, New Mexico, is a volcanic neck enclosed by a weak shale formation. The peak rises 1700 ft (520 m) above the surrounding plain. (Mark A. Melton)**

(left) **Basaltic shield volcanoes of the Island of Hawaii. Closest is snow-capped, extinct Mauna Kea. Distant and left is Mauna Loa, an active shield volcano. Both have summit elevations over 13,000 ft (4000 m). (Arthur N. Strahler)**

(below) **Waimaea Canyon, over 2500 ft (760 m) deep, has been eroded into the flank of an extinct shield volcano on Kauai, Hawaii. Basaltic lava flows are exposed. (Arthur N. Strahler)**

(above) **Halemaumau fire pit, a steep-walled crater within the central depression (caldera) of Kilauea volcano, Hawaii. (Arthur N. Strahler)**

Volcanic landforms   **Plate L.1**

(above) **This sheer rock wall 1000 ft (300 m) high is a fault scarp. Uplift along the fault raised a plateau of massive limestone, Big Bend National Park, Texas.** (Mark A. Melton)

(right) **Photographed by astronauts aboard Gemini XII space vehicle, the Red Sea stretches southeastward toward the horizon. The Red Sea is about 125 mi (200 km) wide; its straight, parallel shores reveal its origin as a widening belt of ocean floor between two separating lithospheric plates. The Arabian Peninsula lies to the left; Egypt to the right. The Gulf of Suez (bottom, center) and Gulf of Aqaba (left) are narrower rift depressions — grabens — bounded by normal faults. The triangular Sinai Peninsula lies between them.** (NASA — No. S66-63481)

(left) **Small-scale normal faulting in red shale beds, Kaibab Plateau, Arizona.** (Arthur N. Strahler)

(above) **A great block mountain forms the east wall of Death Valley, California. At the base of the mountain is an active fault scarp, partly buried by alluvial fans.** (Mark A. Melton)

(right) **A small graben. The valley represents a block faulted down between two normal faults. Coconino Plateau, Arizona.** (Donald L. Babenroth)

Plate L.2   Landforms of faulting

(right) **Intricately carved strata of Eocene age, Bryce Canyon National Park, Utah. These limestone beds were laid down in a former freshwater lake, then uplifted to form a high plateau. (Orlo E. Childs)**

(above) **A butte of sandstone (Mesa Verde formation) not far from Ship Rock, northwestern New Mexico. Weak shales (Mancos formation) form the base of the butte. (Arthur N. Strahler)**

(one above) **Sandstones and shales carved into deep canyons in a semiarid climate. View from Dead Horse Point, Canyon Lands National Park, Utah. (Alan H. Strahler)**

(above) **Horizontal flows of basalt lava, piled one upon the other, form these cliffs in the south wall of the Columbia River gorge, Washington. Note the columnar jointing. (Orlo E. Childs)**

(right) **In this Gemini VII space photo, canyons carved into horizontal limestone strata show a dendritic drainage pattern. The area is part of the Hadramawt Plateau, near the southern coast of the Arabian peninsula. At the upper left of Wadi Hadramawt is a broad dry valley floored with sandy alluvium. Width of the area is about 100 mi (160 km). (NASA—No. S65-64010)**

Horizontal strata   **Plate L.3**

(below) **A broad anticlinal fold in Death Valley, California. The soft, easily eroded strata show sharp upturning around the sides and end of the anticline. (Mark A. Melton)**

(below) **Flatirons of massive Jurassic sandstone have been carved from upturned strata near Marsh Pass, Arizona. (Alan H. Strahler)**

(above) **Hogbacks of steeply dipping sandstone, Paria, Utah. Weak shale underlies the valleys between hogbacks. (Donald L. Babenroth)**

(above) **A near-vertical bed of massive sandstone gives the Garden of the Gods a distinctive appearance. This locality is near Colorado Springs, Colorado. (Arthur N. Strahler)**

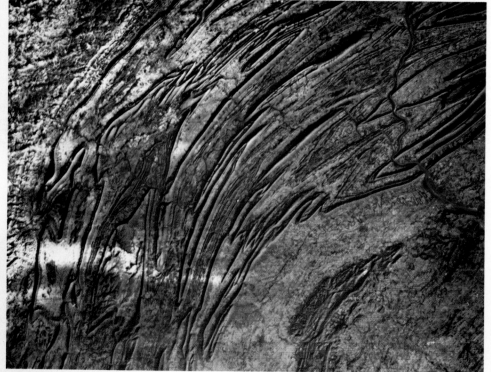

(left) **Zigzag ridges, formed on deeply eroded plunging folds of Paleozoic strata, characterize the Newer Appalachians of southcentral Pennsylvania. In this image, obtained by ERTS-1 orbiting satellite, the Susquehanna River cuts across the narrow ridges in a series of watergaps north of Harrisburg. The area shown is about 100 mi (160 km) across. (NASA — EROS Data Center, No. 81171152455N)**

Plate L.4   Domes and folds

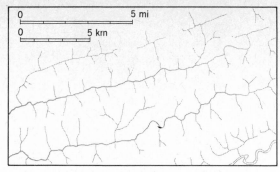

Figure 26.19 A trellis drainage pattern on folds.

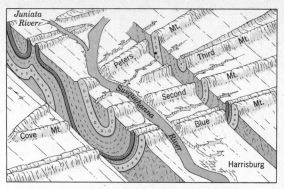

Figure 26.21 A great synclinal fold, involving three resistant quartzite formations and thick intervening shales, has been eroded to form bold ridges through which the Susquehanna River has cut a series of watergaps. (Based on a drawing by A. K. Lobeck.)

formed. Streams that maintain their courses across a rising barrier, whether an anticline or an uplifted fault block, are called **antecedent streams.**

As the dissection of the folded strata progresses, there is a continuous change in the form and position of the various types of ridges and valleys. In time the synclinal ridges are completely removed by erosion. Meanwhile, new ridges are appearing in the centers of the anticlinal valleys. These form as a result of the uncovering of still older resistant strata that were folded along with the rest. The new ridges, which may be thought of as second-generation anticlinal mountains, grow in height as the weak rock is stripped from both sides. Two such anticlinal mountains are shown in Block C of Figure 26.18.

Zigzag Ridges and Plunging Folds

The folds illustrated in Figure 26.18 are continuous and level crested; they produce ridges that are approximately parallel and continue for great distances. In some fold regions, however, the fold crests rise or descend from place to place. Folds of this form are called **plunging folds.** When eroded, they give rise to a zigzag line of ridges (Figure 26.20).

The topographic form of a plunging syncline differs from that of an anticline which plunges in the same direction, when both are dissected. Figure 26.20 illustrates landforms of plunging folds. The plunging syncline is represented by a ridge with a slightly concave summit but steeply descending cliffs on the end and sides. Along

the direction of plunge of the fold center line, or axis, this ridge develops an increasing concavity, then separates into two diverging ridges. The plunging anticline has a smoothly rounded summit that descends gradually to the level of the valley in the direction of plunge.

Zigzag ridges formed on plunging folds are remarkably well developed in central Pennsylvania. They are clearly shown on the remote sensing image reproduced in Plate L.4. (For most viewers, the ridges will appear correctly as raised features if the page is inverted, placing north toward the bottom.) The series of watergaps of the Susquehanna River, shown in geologic detail in Figure 26.21, can be identified in the eastern part of the frame.

Environmental and Resource Aspects of Fold Regions

Some of the environmental and resource aspects of dissected fold regions are illustrated by the Appalachians of south-central and eastern Pennsylvania (Figure 26.21). The ridges, of resistant sandstones and conglomerates, rise boldly to heights of 600 m (2000 ft) above broad lowlands underlain by weak shales and limestones. Major highways run in the valleys, crossing from one valley to another through the watergaps of streams that have cut through the ridges. Important cities are situated near the watergaps of major streams. An example is Harrisburg, located where the Susquehanna River issues from a series of watergaps. Where no watergaps are conveniently located, roads must climb in long, steep grades over the ridge crests. The ridges are heavily forested; the valleys are rich agricultural belts.

In various fold regions of the world, an important resource is anthracite, or hard coal. This coal occurs in strata that have been tightly folded. Pressure has converted the coal from bituminous into anthracite. Because of extensive erosion, all coal has been removed except that which lies in the central parts of synclines. Figure 26.22 shows coal-bearing synclines in Pennsylvania. The coal seams dip steeply; workings penetrate deeply to reach the coal that lies in the bottoms of the synclines. Seams near the surface are worked by strip-mining. Although workable anthracite seams are now exhausted

Figure 26.20 Plunging folds give zigzag ridges when deeply eroded. (Drawn by E. Raisz.)

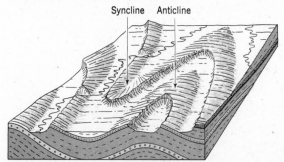

Syncline   Anticline

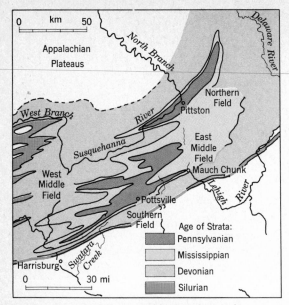

Figure 26.22 Anthracite coal basins of central Pennsylvania correspond with areas of Pennsylvanian strata, downfolded into long synclinal troughs.

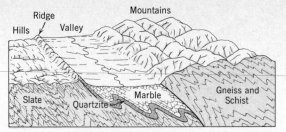

Figure 26.24 Metamorphic rocks tend to form elongate, parallel belts of valleys and mountains. (Drawn by A. N. Strahler.)

Metamorphic Belts

Where strata have been tightly folded and altered into metamorphic rocks in sutures, denudation develops a landscape with a strong grain of ridges and valleys paralleling the strike of the folds and thrust faults. These features lack the sharpness and parallelism of ridges and valleys in belts of open folds, but each belt clearly reflects the greater or lesser resistance to denudation offered by each type of metamorphic rock.

Figure 26.4 shows the control of metamorphic rocks over landforms in the humid climate of eastern North America, where metamorphic rocks of the Older Appalachians form a narrow zone continuous from Georgia to the Maritime Provinces. Marble tends to form open valleys; slate and schist make belts of medium to strong relief; quartzite usually stands out boldly and may produce conspicuous narrow hogback ridges. Most metamorphic belts have been broken by overthrust faults that run parallel with the belt and often separate one rock type from another. Subsequent stream valleys occupying these fault lines accentuate the grain of the topography.

Much of New England, particularly the Taconic and Green Mountains, illustrates these principles well. The larger valleys trend north and south and are underlain by marble. These are flanked by ridges of gneiss, schist, slate, or quartzite. The highlands of the Hudson and of northern New Jersey continue this belted pattern southward, where it joins the Blue Ridge. Near Harpers Ferry, Maryland, quartzite ridges rise prominently above broad valley belts of schist.

Exposed Plutons

The third structural group of landmasses consists of exposed plutons originating as batholiths of igneous rock that solidified under a thick cover of overlying rocks (Figure 26.25). Most batholiths are composed of felsic igneous rock, which is commonly granite or a closely related rock type. Batholiths are produced along active lithospheric plate boundaries, but details are obscure and the origin of the igneous material is subject to highly varied interpretations. One hypothesis considers granite to be derived from thick masses of metamorphic rocks — schists and slates, for example. Melting and recrystallization in tectonic zones converted the parent materials into felsic igneous rock.

in the Pennsylvania field, environmental damage persists through continued subsidence and cracking of the ground surface above abandoned workings.

Many broad, gentle anticlinal folds, as well as sedimentary domes, form important traps for the accumulation of petroleum. The principle is shown in Figure 26.23. Oil migrates in permeable sandstone beds to reach the anticlinal crest or dome summit, where it is trapped by an impervious cap rock of shale. Many of the oil and gas pools of western Pennsylvania, where petroleum production first succeeded, are on low anticlines. Many elongate domes of elliptical outline, resembling anticlines plunging down at both ends, form important oil-bearing structures in the large sedimentary basins of the middle Rocky Mountain region. Good examples are the Rock Springs Dome and Teapot Dome in Wyoming. Low domes are also major oil producers in the Los Angeles Basin.

Figure 26.23 Idealized cross section of an oil pool in an anticline or dome in sedimentary strata. Well *A* will draw gas; well *B* will draw oil; and well *C* will draw water. The cap rock is shale; the reservoir rock is sandstone. (From A. N. Strahler, 1972, *Planet Earth,* Harper and Row, New York.)

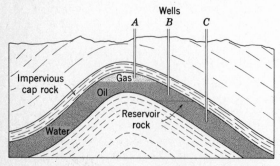

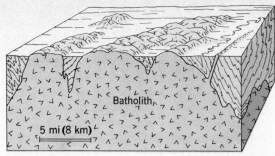

Figure 26.25 Deep-seated igneous rocks appear at the surface only after long-continued erosion has removed thousands of meters of overlying rocks. (Based on a drawing in C. R. Longwell, A. Knopf, and R. F. Flint, 1941, *Outlines of Physical Geology,* 2nd ed., John Wiley and Sons, New York.)

Where the rock mass of the batholith is quite uniform and free of strong faults, it is eroded into a maze of canyons and drainage divides that follow no predominant trend (see Figure 20.7). The drainage pattern is dendritic and resembles that developed on horizontal sedimentary strata (Figure 26.26).

Exposed batholiths occur widely over the continental lithosphere. Good examples are found throughout the Cordilleran ranges of western North America. During the Cretaceous Period, large batholiths were formed from Lower (Baja) California to Alaska (Figure 26.27). One of these rock masses most familiar to Americans is the Sierra

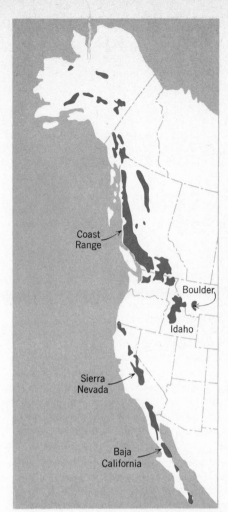

Figure 26.27 Map of Cretaceous batholiths of western North America. The word "Batholith" has been omitted from each label. (From C. O. Dunbar and K. M. Waage, *Historical Geology,* 3rd ed., p. 376, Figure 16.7. Copyright © 1969 by John Wiley and Sons. Reprinted by permission of John Wiley & Sons, Inc.)

Figure 26.26 This dendritic drainage pattern is developed on the dissected Idaho batholith.

Batholith; it comprises the bulk of the Sierra Nevada fault block in California. Another example is the Idaho Batholith, exposed over an area of about 40,000 sq km (16,000 sq mi). Largest of the group is the Coast Range Batholith of British Columbia. An example of a small body of massive granite, intruded into older metamorphic rocks of the Piedmont upland of Georgia, is Stone Mountain (Figure 26.28).

Landforms and Plate Tectonics in Review

In two chapters we have attempted to relate landforms of the continents to newly established principles and definitions of plate tectonics. We began with forms of current and recent volcanic and tectonic activity, concentrated close to active lithospheric plate boundaries. Volcanic mountain arcs built largely of felsic magma are found

Figure 26.28   **Stone Mountain, Georgia, is a striking erosion remnant about 2.4 km (1.5 mi) long and rising 193 m (650 ft) above the surrounding Piedmont peneplain surface. The rock is granite, almost entirely free of joints, and has been rounded into a smooth dome by weathering processes. (U.S. Army Air Service.)**

along continental plate margins, beneath which oceanic crust is disappearing by subduction. Isolated volcanic islands built of basaltic lavas are largely situated along the plate-spreading axis of the mid-oceanic ridge system; but some basaltic volcanoes occur in chains, built from the ocean floor as the oceanic lithosphere moves over a mantle plume. Hot spots beneath continental lithosphere have produced a few large but isolated areas of thick flood basalts. Suturing occurring very recently during continental collision has produced the great alpine chains, formed of intensely folded and thrust-faulted strata. Totally different are the landforms of rifted continental lithosphere where a pulling apart of the plate has begun and may presage the separation of a continental plate into two separate masses. Rifting produces a bold landscape of block-faulted mountains, or a long, narrow graben system.

Turning to landform types within the inactive areas of continental lithosphere, we examined first the coastal plains of stable continental margins where thick continental shelf wedges have been forming for millions of years. More extensive are the undisturbed sedimentary covers found over ancient shield rocks. Here the varying resistance of alternating weak and resistant strata dominates landform expression. Locally, the cover is elevated in dome structures that generate circular hogback ridges. Exposed shields exhibit a subdued landscape of highly varied aspect. In places, exposed plutons produce a highly uniform landscape; elsewhere, bands and belts of meta-

morphic rocks produce linear relief features. Mountain roots, not yet subdued to the level of a low shield surface, rise as linear ranges traceable for thousands of kilometers. Where these roots consist of open folds of sedimentary strata, the landscape is one of narrow-crested ridges and intervening valleys, often forming zigzag patterns. Where the roots consist of metamorphic rock, ridge patterns show a less distinct banded pattern. These metamorphic zones represent the deep roots of continental sutures produced in earlier geologic eras.

Our aim in these chapters has been to emphasize the role of geology in determining the nature of landforms. Earlier, in our chapter on denudation and climate (Chapter 20), we referred to attempts by some geographers to recognize morphoclimatic regions in which a particular climate is assumed to dominate the nature of the landforms. If our analysis of landforms and rock structure has been successful, it can lead only to the conclusion that geology is in equal partnership with climate to form a dual control system over the form, size, and distribution of landform types.

With this chapter we conclude the systematic, structured study of the separate ingredients of physical geography: climate, soil, vegetation, and landform. In the remaining two chapters we attempt a synthesis of this information through the recognition of several important global environmental regions. In the discussion to follow, climate as a landform control will reemerge strongly through the concept of climate-process systems.

References for Further Study

Dutton, C. E. (1882), *Tertiary history of the Grand Canyon district,* U.S. Geological Survey, Monograph 2.

Davis, W. M. (1909, 1954), *Geographical essays,* Ginn and Co., Boston; reprinted by Dover Publications, New York. See pp. 413–84, 725–72.

Johnson, Douglas (1931), *Stream sculpture on the Atlantic slope,* Columbia Univ. Press, New York.

Lobeck, A. K. (1939), *Geomorphology,* McGraw-Hill Book Co., New York. See Chapters 13–17.

Thornbury, W. D. (1965), *Regional geomorphology of the United States,* John Wiley and Sons, New York.

Hunt, C. B. (1974), *Natural regions of the United States and Canada,* W. H. Freeman and Co., San Francisco.

Review Questions

1. Explain how rock structure controls the development of landforms produced by fluvial denudation. What kinds of rock tend to form valleys? What kinds tend to form uplands and mountains? Define dip and strike.

2. Describe the structure and composition of strata beneath a coastal plain. How are coastal plains related to continental shelf deposits of stable continental plate margins?

3. Describe the sequence of landforms developed during denudation of a coastal plain. What is the difference between a consequent stream and a subsequent stream? What kind of drainage pattern develops on a coastal plain?

4. Describe the coastal plain of eastern North America. How has the form of the coastal plain influenced the placement of major cities of the Atlantic seaboard? What is the Fall Line? What natural resources are found in the North American coastal plain?

5. Describe landforms produced by the erosion of horizontal strata in an arid climate. Describe the process of cliff retreat. What is a mesa? A butte? How are badlands formed? What kind of drainage pattern is formed on horizontal strata? What natural resources are found in sedimentary shield covers?

6. Where are belts of open folds produced with respect to a suture belt? Use the Appalachian Mountains to illustrate these geologic relationships.

7. Describe the landforms produced during fluvial erosion of a sedimentary dome. How does a hogback differ from a cuesta? What kind of drainage pattern forms on a dissected dome? Using the Black Hills as an example, discuss the environmental and resource aspects of a large sedimentary dome.

8. What is the difference between a syncline and an anticline? Describe the landforms produced during denudation of a belt of open folds. How are watergaps formed? What ridge forms develop on plunging folds?

9. Discuss the environmental and resource aspects of fold belts, using the Appalachians as an example. How do anticlines and domes serve as traps for petroleum and natural gas?

10. What kinds of landforms develop on eroded belts of metamorphic rocks? Which types of rock underlie the valleys? Illustrate with an example from the Appalachians.

11. Describe a batholith. Of what kind of rock is it formed? Name an important batholith in the Cordilleran ranges of North America. Describe a typical landscape formed by erosion of a large batholith. What type of drainage pattern is present?

12. Summarize the most important relationships between plate tectonics and structural classes of landmasses. Are major classes and types of landforms accounted for by plate tectonics acting through geologic time?

Global Environmental Regions: I. The Low-Latitude Environments

27

Physical geography contributes useful knowledge to Mankind through the recognition of a global system of **environmental regions.** Each region presents its unique set of physical environmental qualities and supports a distinctive ecosystem. Physical and biological factors interact to produce the total set of environmental qualities that make up each region.

Distinctive environmental regions existed for eons of geologic time before *Homo sapiens* came on the scene. Until late in the Pleistocene Epoch, the sparse populations of genus *Homo* functioned environmentally in a manner little different from other higher mammals. Throughout Holocene time, Man's modification of natural ecosystems increased in intensity and total impact. Through the use of fire, the grazing of animals, and land tillage, vegetation of large areas became radically changed, while geomorphic processes were significantly altered in intensity. In this respect, some ecosystems fared quite differently from others. For example, the Mediterranean lands, with a climate precariously balanced between dry and moist regimes, suffered widespread destruction of the forest and woodland cover, along with devastating soil erosion and valley sedimentation. In central and western Europe, the deciduous forest biome was largely destroyed through agricultural activities and little remains today to show us what that landscape might have looked like without Man's intervention. On the other hand, the equatorial rainforest biome remained, up to very recent decades, in nearly pristine condition over large areas. Until machinery, powered with petroleum energy, could be used to mount a massive attack on the great trees of the rainforest, Amazonia held out as a vast natural ecologic fortress. Deserts also resisted Man's occupancy on a vast scale, except on floodplains close to exotic rivers.

Thus each environmental region has shown its own set of opportunities for and restraints on expanded utilization by Man. Especially crucial are the opportunities that Man has to modify further the natural terrestrial ecosystems into food-producing ecosystems. We have already made the point that this transformation has been remarkably suc-

cessful in some parts of the world, but highly unsuccessful in others. Can the production of food be further extended into regions now clothed in equatorial rainforests or savanna grasslands? Can more deserts be made to bloom by massive applications of irrigation water? Partial answers to these questions lie in the environmental factors relating to physical geography. Other parts of the answers lie in the application of technology and the availability of energy supplies. Ethical issues and political policies are also factors to be weighed in seeking these answers.

Environmental Regions

An environmental region is best identified in location and extent in terms of climate types and subtypes, following the system of climate classification based on the soil-water balance (Chapter 10). An environmental region should also be closely identified with a biome, or with one or more vegetation formation classes. We have already established the role of climate as the dominant driving factor in shaping the characteristics of vegetation. Climate has an important but often secondary role in governing soil characteristics and the intensity of various solar-powered geomorphic processes. We are thus attempting a synthesis of physical and biotic factors and forms within a climatic framework. The product of this synthesis serves both as a review of physical geography and as a point of departure to evaluate questions of planning for expanded exploitation and management of natural resources.

Although environmental regions are climatically defined, they are most effectively treated by combining and rearranging the 13 climate types and their subtypes to maximize correspondence of the environmental regions with the major biomes and vegetation formation classes. Table 27.1 lists 12 environmental regions, designated for convenience by capital letters A through L. Names of environmental regions are, in some instances, modified from the parent climate names. The table also lists climate

Table 27.1 / Environmental regions

Name of region	Climates and climate subtypes	Biome	Formation class	Soil order or suborder
LOW LATITUDE				
A Rainforest	(1) Wet equatorial (2) Monsoon and trade-wind littoral	Forest	Equatorial and tropical rainforest (Fe, Ft) Montane forest (Fm)	Oxisols Ultisols
B Savanna	(3) Wet-dry tropical (4s) Semiarid subtype of dry tropical	Savanna	Monsoon forest (Fmo) Savanna woodland (Sw) Thorntree–tall grass savanna (Stg)	Oxisols Ultisols Ustalfs Vertisols
C Tropical desert	(4sd, 4d, 4dw) Semidesert, desert, and western littoral subtypes of dry tropical (5sd, 5d) Semidesert and desert subtypes of dry subtropical (7sd) Semidesert subtype of Mediterranean	Desert	Thorn forest and thorn woodland (Dtw) Thorntree–desert grass savanna (Dtg) Semidesert scrub and woodland (Dsd, Dss) Desert (D, Dsp)	Aridisols Psamments
MIDLATITUDE				
D Moist subtropical forest	(6h, 6p) Humid and perhumid subtypes of moist subtropical	Forest	Broadleaf evergreen forest (Fbe) Southern pine forest (Fsp) Midlatitude deciduous forest (Fd)	Ultisols
E Mediterranean	(7s, 7sh, 7h) Semiarid, subhumid, and humid subtypes of Mediterranean	Forest	Mediterranean mixed evergreen forest (Fsm) Sclerophyllous scrub (Fss) Australian sclerophyll forest (Fsa)	Xeralfs Xerolls Xerults
F Marine west-coast forest	(8h, 8p) Humid and perhumid subtypes of marine west-coast	Forest	Coastal forest (Fc) Midlatitude deciduous forest (Fd) Broadleaf evergreen forest (Fbe)	Spodosols Alfisols Entisols Inceptisols
G Moist continental forest	(10h, 10p) Humid and perhumid subtypes of moist continental	Forest	Midlatitude deciduous forest (Fd) Mixed boreal and deciduous forest (Fbd) Lake forest (Fl)	Udalfs Spodosols
H Prairie and steppe grassland	(10sh, 10h) Subhumid and humid subtypes of moist continental (9s, 5s) Steppe subtype of dry midlatitude and dry subtropical (6h) Humid subtype of moist subtropical (South America)	Grassland	Tall-grass prairie (Gp) Short-grass prairie (steppe) (Gs)	Mollisols Aridisols
I Midlatitude desert	(9sd, 9d) Semidesert and desert subtypes of dry midlatitude	Desert	Semidesert scrub and woodland (Dsd) Desert (D)	Aridisols
HIGH LATITUDE				
J Boreal forest	(11) Boreal forest	Forest	Boreal forest (Fbo, Fbl) Arctic woodland (Taiga)	Spodosols Histosols Boralfs
K Tundra	(12) Tundra	Tundra	Arctic tundra (T)	Cryaquents Cryorthents Histosols
L Ice sheet	(13) Ice sheet			

types and subtypes, biomes and vegetation formation classes, and soil orders and suborders associated with each region.

In this chapter and the next, each environmental region is introduced with a brief summary of climate, vegetation, and soils. You may wish to review the pertinent background information given in detail in earlier chapters. Applying the concept of climate-process systems in geomorphology, mentioned in Chapter 20, we shall comment on the relative intensity of various geomorphic processes within each environmental region. Other topics of interest are plant resources and agriculture and the special problems associated with further development of agricultural and water resources.

Rainforest Environment (A)

Climate, Vegetation, and Soils

The rainforest environment is represented by all world areas of the wet equatorial climate (1) and the monsoon and trade-wind littoral climate (2). The soil-water regimes of these climates show very large annual total rainfall, very large annual water surplus, and high soil-water storage throughout all or most months of the year. Equally important is the extreme uniformity of monthly mean air temperature, which is typically between 26° and 29°C (79° and 84°F) for stations at low elevation in the equatorial zone. The equatorial thermal regime is illustrated by graphs of minimum and maximum daily temperatures for two months in the city of Panama, lat. 9°N (Figure 27.1). The selected months — July and February — represent the extremes of the yearly temperature cycle, but the means differ by only 0.16 C° (0.3 F°). The daily temperature range, in contrast, is usually from 8 to 11 C° (15 to 30 F°). Thus the daily range greatly exceeds the annual range.

Because of the warm and extremely uniform temperature and high levels of soil-water storage, rainforest is the dominant formation class of natural vegetation. Net primary production of the equatorial and tropical rainforest is extremely high, exceeding that of any other terrestrial biome and comparing favorably with the most productive of the aquatic ecosystems (Chapter 12).

A major difference between the low-latitude rainforest and forests of higher latitudes is the great diversity of species that it possesses. The rainforests may have as many as 3000 different tree species in an area of a few square kilometers; midlatitude forests would be considered very diverse indeed with only one-tenth that number. The fauna of the rainforest is also very rich. A 16-sq-km (6-sq-mi) area in the Canal Zone, for example, contains about 20,000 species of insects, whereas there are only a few hundred in all of France. This large number of species occurs because the rainforest environment is so uniform and free of physical stress.

Animal life of the rainforest is most abundant in the upper layers of the vegetation. Above the canopy, birds and bats are important carnivores, feeding largely on insects above and within the topmost canopy. Below this level live a wide variety of birds, mammals, reptiles, and invertebrates; they feed on the leaves, fruit, and nectar abundantly available in the main part of the canopy. Ranging between the canopy and the ground are the climbing (scansorial) mammals, which forage in both layers. At the surface are the larger ground mammals; these include herbivores, which graze the low leaves and fallen fruits, and carnivores, which prey on the abundant vertebrates and invertebrates found at the surface.

Streams flow copiously throughout most of the year, and river channels are lined along the banks with dense forest vegetation (Figure 27.2). Natives of the rainforest travel the rivers in dugout canoes, and they now enjoy self-propulsion by attaching outboard motors. Larger shallow-draft river craft have turned the major waterways into the main arteries of trade; towns and cities are situated on the river banks. Aircraft have added a new dimension in mobility, crisscrossing the almost trackless green sea of forest to find landings in clearings or on broad reaches of the larger rivers.

In the rainforest environment, the large water surplus and prevailing warm soil temperatures promote decay and decomposition of rock to great depths, so a thick regolith is present over large areas. Mineral alteration has reached an advanced stage, in which clays of high cation-exchange capacity have almost entirely disappeared from the soil and regolith. Silica has also been depleted. A heavy concentration of sesquioxides of aluminum and iron remains. Oxisols are thus the dominant soil order of the rainforest environment, along with Ultisols of a somewhat less advanced state of oxide accumulation.

Geomorphic Processes in the Rainforest

Under the rainforest environment, chemical weathering is dominant in intensity over physical weathering. Note that this is a frost-free climate, so mechanical weathering by frost action is totally absent. A thick cover of vegetation, soil, and regolith protects bedrock from direct insolation and from extremes of daily temperature change.

Of the total mineral load carried by streams, by far the greater portion is in the form of dissolved solids. Little coarse sediment moves as bed load because the cover of vegetation holds back all but the finest soil particles from transport by overland flow. Streams of steep gradient have bouldery channels, but the boulders experience little downstream movement. Mafic rock, particularly basaltic lava, undergoes rapid removal under attack by soil acids and produces landforms quite similar to those formed by carbonic acid on massive limestones in moist climates of higher latitudes. The effects of solution removal of basaltic lava are displayed in spectacular grooves, fins, and spires on the walls of deep alcoves in parts of the Hawaiian

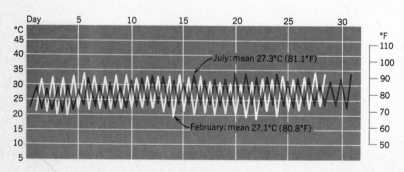

Figure 27.1 **July and February temperatures at Panama City, lat. 9° N. The sawtooth graph shows daily maximum and minimum readings for each day of the month. (Data from Mark Jefferson.)**

Figure 27.2   This air view over the rainforest of the Amazon basin shows the Rio Negro, a tributary of the Amazon River. The view is southward from a point about lat. 0°23'S, long. 64°05'W. (Colonel Richmond, American Geographical Society.)

Islands (Figure 27.3).

Mass wasting, involving rapid sliding and flowage of moist soil and regolith, is especially conspicuous as a denudation process where the terrain is mountainous and slopes are steep. Slides and avalanches of soil are common and can strip away the forest and soil to expose regolith and bedrock. These conditions are characteristic of those parts of the rainforest that happen to lie on the recent alpine and volcanic chains where subduction is active on plate boundaries. Rugged, mountainous terrain is thus associated with rainforest in Central America, the West Indies, the Andes range of South America, and the East Indies. On stable shield areas, slopes are low or moderate over vast areas; examples are the Amazon-Orinoco lowland of South America and the Congo Basin of Africa. Gradients of major rivers crossing these lowlands are extremely low. For example, at Iquitos, Peru, the elevation of the Amazon River is only 116 m (280 ft) above sea level; but the airline distance from Iquitos to the mouth of the Amazon is almost 3000 km (1850 mi).

Plant Products and Food Resources of the Rainforest

Several forest products are of economic value. Rainforest lumber, such as mahogany, ebony, or balsawood, is an important export. Quinine, cocaine, and other drugs come from the bark and leaves of tropical plants; cocoa comes from the seed kernel of the cacao plant. Natural rubber is made from the sap of the rubber tree. The tree comes from South America, where it was first exploited. Rubber trees are also widely distributed throughout the rainforest of Africa. Today the principal production is from plantations in Indonesia, Malaya, Thailand, Vietnam, and Sri Lanka (Ceylon).

An important class of food plants native to the wet low-latitude environment are starchy staples; some are root structures, others are fruits. Manioc, also known as cassava, is one of these staples. The plant has a tuberous root — something like a sweet potato — that reaches lengths over 0.3 m (1 ft) and may weigh several kilograms (Figure 27.4). When properly prepared to remove a poisonous cyanide compound, the manioc roots yield a starchy food with very little (1% or less) protein. Manioc was used as a food in the Amazon basin of Brazil; it was taken to Africa in the sixteenth century by the Portuguese. In the present century, its use in Africa has increased widely. Manioc is now also important in Indonesia. The food unfortunately contributes to malnutrition because it contains too little protein. Manioc is cultivated in small plantations placed in forest clearings. In Brazil, plans are being made to cultivate manioc on a large scale as a source of starch from which to produce alcohol, to be

used as a partial substitute for gasoline.

Yams are another starchy staple of the wet equatorial regions. Like the manioc, the yam is a large underground tuber. Yams are a major source of food in West Africa. The plant was introduced into the Caribbean region during the era of slavery and is an important food in the region today. The yam has a higher protein content than the manioc.

Taro is another starchy staple of wet low-latitude climates. The taro plant has large leaves, which are edible, but the food value lies mostly in an enlarged underground portion of the plant, called a "corm." Visitors to Hawaii know the taro plant through its transformation into poi, a fermented paste. Few mainland tourists eat poi a second time, but it has long been a favored food of the native Hawaiians. Taro corms consist of about 30 percent starch, 3 percent sugar, and only about 1 percent protein. Taro was imported into Africa from Southeast Asia and eventually reached the Caribbean region.

An important starchy staple in the form of a fruit is the breadfruit. A single breadfruit can attain a weight of 5 kg (10 lb) and is rich in carbohydrate. A native of the Pacific Islands, the breadfruit has long been a staple of the diet of the Polynesians. In 1789 Captain Bligh attempted to bring a cargo of small breadfruit trees in his ship, H.M.S. *Bounty,* to the West Indies to be cultivated as a food supply for African slaves.

The banana and plantain are starchy staples familiar to everyone. The banana plant lacks woody tissue and is, in fact, a perennial herb; the fruit is classed as a berry by botanists. The banana was first cultivated for food in Southeast Asia, then spread to Africa. It was imported into the Americas in the sixteenth century and quickly became well established. The plantain is a coarse variety of the banana that is starchy, has little sugar, and requires cooking.

Perhaps the plant most important to Man in the low latitudes is the coconut palm. Besides being a staple food, it provides a multitude of useful products in the form of fiber and structural materials. The coconut palm flourishes on islands and coastal fringes of the wet equatorial and tropical climates (see Plate F.1). Copra, the dried meat of the coconut, is a valuable source of vegetable oil. Copra and coconut oil are major products of Indonesia, the Philippines, and New Guinea. Palm oil and palm kernels of other palm species are an important product of the equatorial zone of West Africa and the Congo River basin.

Exploitation of the Rainforest Ecosystem

In the past, Man has farmed the low-latitude rainforest by the **slash-and-burn** method—cutting down all the vegetation in a small area, then burning it (Figure 27.5). Most of the nutrients in a rainforest ecosystem are tied up in the biomass rather than in the soil. Burning the slash on the site releases the trapped nutrients, returning them to the soil. The supply of nutrients derived from the original biomass is small, however, and the harvesting of crops rapidly depletes the nutrients. After a few seasons of cultivation, the soil loses much of its productivity. A new field is then cleared in another area, and the old field is

Figure 27.4 **Tuberous manioc roots, brought in dugout canoes from villages in the Brazilian rainforest, are being loaded on a riverboat for transport to market. At one time manioc, in the form of tapioca, was popular in the United States as a dessert pudding. (Charles Perry Weimer.)**

abandoned. Plants of the rainforest are able to reinvade the abandoned area because fruiting species and animal seed carriers are close by, and so the rainforest soon returns to its original state. In this way the primitive slash-and-burn agriculture is compatible with the maintenance of the rainforest ecosystem.

On the other hand, modern intensive agriculture uses large areas of land and is not compatible with the rainforest ecosystem. When such lands are abandoned, seed sources are so far away that the forest species cannot take hold. Instead, secondary species dominate, often accompanied by species from other vegetation types. The dominance of these secondary species is permanent, at least on the human time scale. Thus the rainforest ecosystem is, in this sense, a nonrenewable genetic resource of many, many species of plants and animals. Once displaced by large-scale cultivation, those species can never return to reoccupy the area. Ecologists warn that the disappearance of thousands of species of organisms from the rainforest environment would mean the loss of millions of years of evolution, together with the destruction of the most complex ecosystem on the globe.

Future Prospects for the Rainforest Environment

Now we can take up a question raised in the opening paragraphs of this chapter: Can the rainforest environment be exploited as a new source of food for Mankind? Some agricultural specialists are highly optimistic; but others, specializing in ecology, are pessimistic, or at best dubious.

What we can do here is to point out restraints on such expansion when it relates to raising such staples as rice, soybeans, corn, or sugar cane, which are field crops harvested seasonally in other environmental regions.

First, the low content of nutrients in the Oxisols and Ultisols will require massive and repeated applications of fertilizers. These applied nutrients are not held in storage in substantial quantities in the soil and are quickly exported from the land in runoff because of the large water surplus. Second, there is no dry season in which crops can reach maturity and be harvested under dry conditions, as is the case in the wet-dry tropical climates and the sub-humid and semiarid continental climates of midlatitudes. Third, the Oxisols are capable, in some areas, of becoming lithified — literally turning to rock — when denuded of plant cover and exposed to the atmosphere. This has actually happened on agricultural test farms in the rainforest of Brazil.

It is easy to say that agricultural technology can overcome these and other difficulties. Even if this conclusion is valid, no one questions the fact that the cost will be enormous in terms of energy input through all parts of the agricultural system. Fertilizers, pesticides, machinery, and fuel are only part of that energy input. Large amounts of capital are needed to set up efficient management and marketing systems. In view of the pending destruction of the rainforest ecosystem as a consequence of agricultural expansion, we can justifiably raise the question as to whether benefits will outweigh losses. Would not the human effort and energy expenditure be more effectively applied in those environments where many natural factors favor the increased cultivation of protein-rich food staples?

Figure 27.5 **Slash and burn agriculture in the Amazon rainforest. The circular plot at left, ringed by second-growth vegetation, is ready for abandonment. Burning of felled trees in the new plot at right will release a meager supply of nutrients. (Loren McIntyre/Woodfin Camp.)**

Savanna Environment (B)

Climate, Vegetation, and Soils

The savanna environment is represented by all world areas of the wet-dry tropical climate (3) and adjacent belts of the semiarid subtype of the dry tropical climate (4s). Savanna areas of major importance are found in South America, Africa, India, Indochina, and northern Australia. The savanna region typically occupies a position between the rainforest, on the equatorward side, and a great belt of tropical desert on the poleward side. The outstanding feature of the soil-water regime of the savanna environment is the alternation of a very wet season, in the high-sun period, with a very dry season, in the low-sun period. The temperature regime of the savanna environment is characterized by a very hot period immediately preceding the onset of the rainy season and by a comparatively cool season at time of low sun. Alternation of a wet-season water surplus with a dry-season soil-water deficit is the unique mark of the savanna environment in its moister equatorward portion, where monsoon forest (Fmo) and savanna woodland (Sw) are the typical plant formation classes. The semiarid portion, across which is a rapid transition to semidesert, has too short a wet season to generate a water surplus. Here the vegetation consists of thorntree–tall grass savanna (Stg).

A sweeping generalization about the natural vegetation of the savanna environment is that it is **raingreen vegetation.** The broad-leaved savanna trees shed their leaves in the cool, dry season, while the grasses turn to dry straw. This long period of dormancy terminates almost abruptly with the first torrential rains of the wet season. The grasses quickly become a green carpet and the trees put forth new leaves.

Soils of the savanna region belong to several soil orders. Large upland areas are underlain by very old Oxisols and Ultisols. Vertisols, soils of high base status, are important where parent materials favor their development (e.g., in western India). Ustalfs, the ustic-regime suborder of the Alfisols, are found over large areas of savanna in West Africa, eastern Africa, India, Indochina, northeastern Brazil, and northern Australia. In contrast to the Oxisols and Ultisols, the Ustalfs are soils of high base status, basically capable of high agricultural yields. Also important are Aquepts (suborder of Inceptisols), which are found over agriculturally productive river floodplains and deltaic plains, for example, plains of the Ganges-Brahmaputra, Irrawaddy, and Mekong river systems in Southeast Asia.

Geomorphic Processes and Landforms of the Savanna Environment

Geomorphic processes of the savanna environment are closely tied to pedogenic processes — not only processes acting under present climates but also processes that may have acted during long periods in the late Cenozoic Era, in which a moister climate may have prevailed. Large areas of savanna landscape lie on stable shield areas of the former subcontinent of Gondwana (Chapter 24). Bedrock

of these shield areas is extremely complex in rock composition and structure. Exposures of plutonic igneous masses (batholiths) are numerous, while metamorphic belts of great complexity are also abundant. Some parts of the shields bear sedimentary covers; elsewhere are expanses of ancient flood basalts. Thus the landforms of the savanna regions are to a large degree structurally controlled, making difficult the recognition of distinctive landforms governed solely by processes unique to the wet-dry tropical climate.

The nature and intensity of weathering processes acting under the savanna soil-water regime are poorly understood. Chemical weathering during the wet season is considered by some investigators to be a very important process affecting igneous and metamorphic rocks rich in aluminosilicate minerals. The intensity of erosion by running water is also considered to be extremely high because of the rather poor protection afforded the soil surface by savanna vegetation, particularly where it has been burned over prior to the rains and heavily grazed and trampled by large herds of wild herbivores or cattle. Sediment loads of streams — both suspended load and bed load — are very high during periods of peak flow. In contrast, stream discharges fall to low values or to zero during the long dry season, when broad braided channels of sand and silt are exposed to view.

Two landscape features of the savanna region deserve notice: (1) prominent, isolated rock knobs and (2) bench-like upland surfaces. The rock knobs are of a special type, called **bornhardts;** they are found throughout the savanna region of East Africa and South Africa. Typically, a bornhardt is a steep-sided knob of granite or similar plutonic rock having a rounded summit and often showing exfoliation shells (Chapter 17). Surrounding the bornhardt is a more or less level land surface, or plain, underlain by regolith. Some geomorphologists consider this type of plain to be a pediplain formed by processes similar to pediment-producing processes acting in deserts (Chapter 20). Figure 27.6 is a schematic cross section showing one interpretation of the origin of the rock knob. Deep chemical weathering has taken place beneath the deep regolith surrounding the knob, where the bedrock is closely jointed or consists of more susceptible rock. Related to bornhardts are **tors,** which consist of rounded joint blocks. Heaps of rounded boulders rising above a plain are a common landform of the savanna landscape of the Indian

Figure 27.6 **Schematic diagram of a bornhardt with its surrounding plain bearing a laterite crust.**

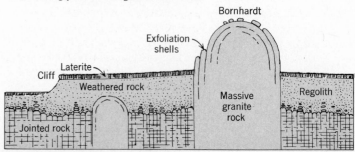

peninsula. To imply that bornhardts and tors are unique to the savanna environment, as some authors have done, would be inaccurate. Similar landforms can be found in a wide range of climates, including the desert and the moist midlatitude climates. For example, the rock tors of the Dartmoor region of southwestern England bear a close resemblance to the tors of Africa.

The second landscape feature, and one that is undoubtedly associated with the tropical environment, is a stepping or benching of the land surface caused by the presence of a layer or layers of rocklike material derived from the soil. Recall from Chapter 13 that plinthite is a soil horizon consisting of dense sesquioxides of iron and aluminum, which are residual products of weathering. Plinthite in Oxisols hardens irreversibly into a rocklike layer, becoming **laterite,** when the surface is exposed to repeated wetting and drying. An alternate term for laterite is **ferricrete,** which recognizes the presence of iron oxide as the cementing material.

Laterite represents a plinthite horizon that has become exposed to the surface environment through erosional removal of overlying soil horizons of an Oxisol. The same process is carried out by Man in Southeast Asia, where plinthite is quarried and cut into building blocks that harden into enduring laterite blocks (Figure 27.7). Figure 27.6 shows a laterite layer forming a cap layer or crust over the plain surrounding the bornhardt. Where streams have carved valleys into the laterite layer, it forms a small cliff. Laterite crusts are not necessarily related to bornhardts or other landforms and occur widely over the uplands of savanna regions. Two, three, or more laterite crusts may be formed in succession at lower levels, as denudation proceeds. The benched landscape resembles in some ways the landscape of plateaus, cliffs, and mesas developed in flat-lying sedimentary strata of covered shields (Chapter 26). Elsewhere in tropical environments, crusts consist of weathered materials of soil or regolith cemented by silica; this material is called **silcrete** to distinguish it from ferricrete.

A general hypothesis for the development of crusts of ferricrete and silcrete is that they originated as soil horizons during long periods of moister climate representing a former geographic expansion of the wet monsoon climate into what is today the region of wet-dry tropical climate. Exposure of the soil horizon and its conversion into a laterite crust accompanied a change to the existing climate, as the belt of wet equatorial climate contracted and withdrew into lower latitudes.

Animal Life of the African Savanna

Closely geared to vegetation and climate is the natural animal life of the savanna grasslands and woodlands. These are the regions of the carnivorous game animals and the multitude of grazing animals on which they feed (Figure 27.8). The savannas of Africa are the natural home of such herbivores as the wildebeest, gazelle, deer, antelope, buffalo, rhinoceros, zebra, giraffe, and elephant. The lion, leopard, hyena, wild dog, and jackal feed on these herbivores. Some of the herbivores depend on fleetness of foot to escape their predators. Others, such as the

Figure 27.7 **This Cambodian temple was built of blocks of plinthite cut from the soil while in a soft condition. The material has hardened into laterite upon exposure to the atmosphere. (Georg Gerster/Rapho-Photo Researchers.)**

rhinoceros, buffalo, and elephant, defend themselves by their size, strength, or armor-thick hide. The giraffe is a peculiar adaptation to savanna woodlands; its long neck permits browsing on the higher foliage of scattered trees.

The dry season brings a severe struggle for existence to animals of the African savanna. As streams and hollows dry up, the few muddy waterholes must supply all drinking water. Danger of attack by carnivores is greatly increased.

The savanna ecosystem in Africa faces the prospect of widespread destruction through the impact of Man. Parks set aside to preserve the ecosystem confine the grazing animals to a narrow range and prevent their seasonal migrations in search of food. In some instances the growth in animal populations has been phenomenal because they have been protected from hunting. In confined preserves, they rapidly consume all available vegetation in futile attempts to survive. Rapidly growing human populations are bringing increasing pressure on management agencies to allow encroachment on game preserves to expand cattle grazing and agriculture.

Food Crops of the Savanna Environment

Of the staple food crops grown widely in the areas of wet-dry tropical climate, rice is perhaps the most important and the most closely linked to the wet-dry cycle. About one-third of the human race subsists on rice, and most of this rice-eating mass of humanity is crowded into the arable lands of Southeast Asia. Most of the rice cultivation is within the monsoon variety of the wet-dry tropical climate and in the moist subtropical climate that lies at

Figure 27.8   **Giraffes and zebras roam free on the highland savanna of Kenya in East Africa. (Marc & Evelyne Bernheim, Woodfin Camp.)**

higher latitudes in China and Japan. Rice requires flooding of the ground at the time the seedling plants are cultivated, and this activity has been traditionally timed to coincide with the peak of the rainy monsoon season (Figure 27.9). The crop matures and is harvested in the dry season. Sugar cane is another important crop that grows rapidly during the rainy season and is harvested in the dry season.

Sorghum, which goes by other names, such as kaffir corn and guinea corn, is an important food crop of the savanna environment in Africa and India. It is a grain capable of survival under conditions of a short wet season and a long, hot dry season.

Wheat is extensively cultivated in northern India and Pakistan, within the semiarid part of the dry tropical climate. This agricultural region includes the Punjab of northern Pakistan and northwestern India and the Indian state of Uttar Pradesh, which lies mostly within the broad

lowland of the Ganges River (Figure 27.10). Much of the region is irrigated to supplement the monsoon rains, and some areas produce two crops per year. Peanuts are another major food crop of the savanna environment in India, and they have been introduced into a corresponding climate zone of West Africa.

The Green Revolution

Prospects for expoliting the savanna environment for increased food production rest today on concerted application of techniques embodied in the term **green revolution.** A major ingredient in that revolution has been the development of new genetic strains of rice and wheat capable of greatly increased production per unit of area of land now under cultivation. To achieve the potential of these new plant strains, substantial applications of fertilizers are required. Changes in agricultural practices and

Figure 27.9   **In this scene in the northern Philippines, rice seedlings are being transplanted into a paddy field flooded by monsoon rains. By the time this rice crop has been harvested, a single hectare acre may have consumed over 2000 person-hours of hand labor. (Department of Tourism, Philippines Board of Travel and Tourist Industry.)**

increased use of machines are also part of the total picture. While striking increases of rice and wheat yields have been achieved in Southeast Asia by the green revolution, the recent sharp increases in the price of fertilizers and fuel have dealt a staggering blow to the new agriculture.

Because of the green revolution, wheat production in the Punjab of India was doubled in the five years between 1966 and 1971. It was hoped that this phenomenal increase would sustain the growing needs of India's population, which increases by some 17 million persons per year. Wheat production leveled off however, and in 1974 was forced into sharp decline by a number of negative factors. These included partial failure of the rains and the enormous increase in cost of fertilizers and fuels. Tube wells, drilled into the sand and gravel to reach the ground water, require electricity for pumps, and this energy source was also severely cut.

One hazard of the green revolution lies in the requirement that the genetic strains bred for high yields must be used to the exclusion of a variety of native strains. Should the high-yield strain prove vulnerable to an epidemic plant disease, the entire crop of a whole nation could be wiped out in one season. Current research in crop breeding is now stressing the development of more individualized varieties that respond better to local conditions. These varieties are less subject to epidemic diseases and can still produce improved yields with less fertilizer input.

A second hazard lies in the need to increase the size of

Figure 27.10 **Here in the Punjab province of north India, wheat sheaves are being loaded on a wagon for transport to the threshing site. The tractor, a European import, symbolizes the technology and input of fossil fuels required to gain full benefits of the green revolution. (Jehangir Gazdar; copyright © Woodfin Camp.)**

fields by merging many small plots into large ones to allow mechanized agriculture to work most efficiently. In so doing, a variety of food crops is no longer grown and dependence for survival comes to rest on the single crop. Under traditional practices, the Asiatic farmer planted several food crops to ensure that if some failed, others would yield enough food to prevent starvation. The small farmer is therefore reluctant to move to more efficient, single-crop agriculture.

It seems obvious now that future increases in yields will be achieved only by modifying the techniques of the green revolution to become less dependent on industrial midlatitude technology. New crop strains and methods of cultivation and fertilization will have to be more compatible with ancient agricultural systems and local cultural patterns.

Soil scientists have pointed out that large areas of Vertisols remain to be placed under cultivation. The Vertisols are rich in nutrient bases, but will require machine cultivation to overcome poor tillage. Most of the arable land of the savanna environment in Southeast Asia has already been intensively developed by existing standards. To hope for major expansion of agriculture into poor Oxisols and to semidesert zones marginal to the tropical deserts is, at best, unrealistic.

Tropical Desert Environment (C)

Climate, Vegetation, and Soils

The tropical desert as a global environmental region is sustained by subsiding air masses of the continental high-pressure cells. These cells dominate much of the earth's land area in a latitutde belt between 15° and 30°N and S, roughly centered on the tropics of cancer and capricorn. The vast desert belt extending across North Africa (Sahara Desert), Arabia, and Iran to Pakistan (Thar Desert) best exemplifies the tropical desert. Other important continental deserts of this belt are the Sonoran Desert of the southwestern United States and northern Mexico, the interior Kalahari Desert of South Africa, and the Australian Desert.

Also included in this environmental region are the western littoral deserts of tropical and equatorial latitudes. The Atacama Desert of Chile and the Namib Desert of coastal southwest Africa (Figure 27.11) are perhaps the most celebrated of these cool, foggy deserts; but they also exist in Lower (Baja) California, the Atlantic coast of North Africa, and the west coast of Australia.

The global extent of the tropical desert environment is represented by all world areas of semidesert and desert subtypes of the dry tropical and subtropical climates (4sd, 4d, 5sd, 5d), the western littoral subtype of the dry tropical climate (4wd), and the semidesert subtype of the Mediterranean climate (7sd). The soil-water balance of these dry climates is characterized by a very large annual soil-water deficiency, absence of any water surplus, and very low amounts of soil-water storage throughout the entire year. Annual potential evapotranspiration (water need) is very large, particularly in the dry tropical climate. High-sun temperatures are extremely high (Figure 27.12). There

Figure 27.11   In this air view of the barren coast of Peru, a fog bank can be seen in the distance, lying over the cold Peru current. (Ministerio de Fomento, Peru.)

Figure 27.12   Monthly air temperature data for **Bou-Bernous, Algeria, at lat. 27½°N in the heart of the Sahara Desert of North Africa.**

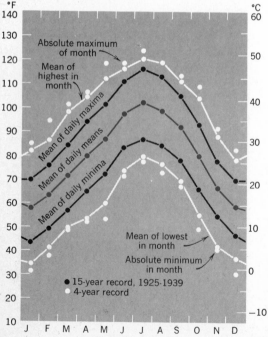

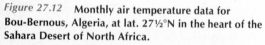

is a strong annual temperature cycle in the continental interiors, so the subtropical desert experiences a pronounced cool season at time of low sun. The western littoral deserts are distinctive because of the comparative coolness of the air temperatures throughout the year and the very small annual temperature range. Figure 27.13 shows the monthly mean temperatures for a desert station on the Atlantic coast of Africa compared with a station in the heart of the Sahara Desert. Both are at the same latitude. Both have about the same minimum temperature but the summer maximum is much higher in the interior desert.

Figure 27.13   These two temperature graphs for stations in the Republic of Mauretania, North Africa, show the low annual range of the west-coast desert (Port Etienne) in contrast with the large annual range within the interior desert (Tessalit).

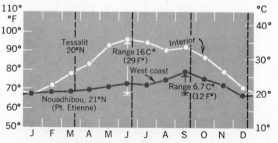

420   Global Environmental Regions: I

Rainfall variability is extremely great in the tropical desert. By variability, we mean the degree to which the measured rainfalls of individual years depart from the mean value over a long period of record. Figure 27.14 illustrates the principle and compares the variability of the tropical desert with that of the other low-latitude climates. Rainfall of the wettest month of the year is shown. The lowermost graph shows precipitation received at Abbassia, Egypt (near Cairo), through a 36-year period. Five of the years had more than twice the average, 2 years had four or more times the average, and 10 years had only a trace or none at all. This record is one of extreme variability. The middle graph shows a 46-year record for Bombay, India, in the wet-dry tropical climate. A moderately large variability is shown. Only one year had more than twice the average, and there was some rain in every year. The uppermost graph shows the corresponding record for Padang, Sumatra, a wet equatorial station. The variability is obviously small compared with the other records.

Rainfall variability for global land areas has been estimated in terms of percentage departure from the normal values (Figure 27.15). This map is highly generalized and is little better than a schematic diagram. The tropical deserts show the highest variability. Very low values are associated with the moist climates, both in the equatorial and midlatitude zones. The lowest values

shown are for coastal belts, for example, the marine west-coast climate of western Europe.

Adaptations of both plants and animals to the desert environment (xeric environment) were explained in Chapter 15. Vegetation of the tropical desert is closely correlated with climate subtypes through differences in the soil-water storage. In the semidesert climate belt, bordering the tropical savanna environment, plant formation classes include thorn forest and thorn woodland (Dtw) and thorntree–desert grass savanna (Dtg). In the semidesert climate belt transitional to moister climates on the poleward side, the vegetation is semidesert scrub and woodland (Dsd, Dss). For vast areas of true desert climate, the vegetation is classed as desert (D). Aridisols are the soil order associated with desert and semidesert climates. Soil differences represented by soil suborders of the Aridisols are determined by past climatic history and by factors of soil texture, drainage, and degree of salinity.

Geomorphic Processes of the Tropical Desert

Geomorphic processes associated with aridity of climate were developed in depth in Chapter 20. Keep in mind that geologic history plays a major role in determining the overall appearance and relief of the desert landscape. The Basin and Range structure of the mountainous desert of

Figure 27.14 These graphs for three locations illustrate the concept of variability of rainfall by showing the actual amount of rain received each year during the month that, on the average, is the rainiest month at each location. (Data from H. H. Clayton, Smithsonian Institution.)

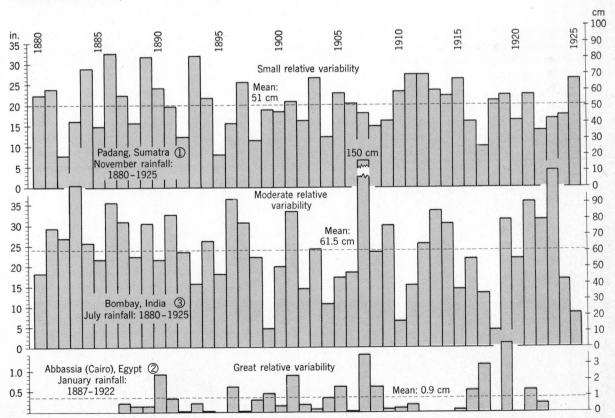

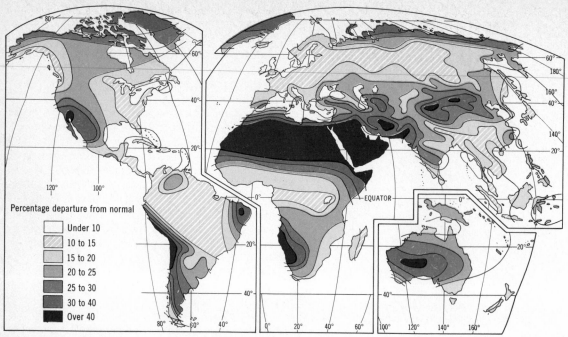

Figure 27.15 Precipitation variability map of the world. (After William Van Royen, 1954, *Atlas of the World's Agricultural Resources*, Prentice-Hall, Englewood Cliffs, N.J. Based on Goode Base Map.)

Percentage departure from normal

☐	Under 10
	10 to 15
	15 to 20
	20 to 25
	25 to 30
	30 to 40
■	Over 40

the American Southwest, associated with continental rifting and block faulting, is almost unique as a geologic type. The largest of the world's tropical deserts — those of Africa and Australia — occupy stable continental shields that are the fragments of the ancient sub-continent of Gondwana. A large variety of rocks and structures is exposed in these shields, while some large areas have covers of relatively undisturbed sedimentary strata. Other deserts are found in the mountainous terrain of alpine zones, where plate boundaries have experienced collision in recent geologic time. Deserts of Iran, Afghanistan, and Pakistan fall into this category. The coastal desert of Peru and Chile owes its position and narrow width to an active subduction zone along which the Pacific plate is descending beneath the American plate.

A common denominator of desert landscape evolution exists through a unique set of weathering processes, fluvial processes, and the work of wind. Mechanical weathering by salt-crystal growth is a dominant and pervasive process. Rock disruption by thermal expansion and contraction during the diurnal cycle of heating and cooling is also a universal process, but its effectiveness is questionable. Although chemical weathering is considered relatively less important in deserts than it is in moister climates, evidence is clear that hydrolysis and oxidation affect the silicate minerals wherever they are exposed throughout the desert environment. Surfaces of pebbles, boulders, and rock outcrops become darkened by thin films of oxides of iron and manganese. This black, iridescent coating is called **desert varnish** (Figure 27.16). In the absence of below-freezing temperatures, frost action can be considered ineffective in the warmer

desert regions.

Evaporation of soil water and ground water, brought toward the surface of soil and bedrock by capillary film movement, results in a variety of rocklike crusts. The surface layer of porous sandstone becomes indurated into a hard crust by carbonate deposition. Where the crust is broken away, crumbling of the softer rock inside through salt-crystal growth leads to the formation of deep pits and hollows, and these may be enlarged into shallow caves (Chapter 17). Within Aridisols developed on alluvial materials, a petrocalcic horizon is a common feature (Chapter 13). Where the overlying soil horizons have been removed by erosion, the petrocalcic horizon is exposed as a hard rocklike crust. Geographers refer to this layer as **calcrete** (also *caliche*) to distinguish it from ferricrete and silicrete crusts found in the savanna environment. As calcrete crusts are eroded, they give rise to stepped landforms resembling the cliffs, mesas, and buttes of eroded sedimentary strata. Crusts formed of gypsum are also common; they represent exposed gypsic horizons formed in Aridisols of extremely dry deserts.

Poorly drained, shallow basins (playas) accumulate highly soluble salts. These form sterile, white salt flats. On rare occasions these flats are covered by a shallow layer of water, brought by flooding streams heading in adjacent highlands. A number of desert salts are of economic value and have been profitably extracted. An example well known to most persons is borax (sodium borate), widely used as a water-softening agent. In shallow coastal estuaries in the desert climate, sea salt is commercially harvested by allowing it to evaporate in shallow basins. One well-known salt source of this kind is the Rann of

Figure 27.16 Petroglyphs, crude drawings made by Indians, were scratched into the natural coating of desert varnish on these boulders in the Mojave Desert. (U.S. Department of the Interior, Bureau of Land Management.)

Kutch, a coastal lowland in the tropical desert of westernmost India, close to Pakistan. Here the evaporation of shallow water of the Arabian Sea has long provided a major source of salt, the only product of value in the region.

Fluvial processes, as we learned in Chapter 19, are of major importance in shaping desert landforms, even though production of runoff in large volumes is sporadic. Large volumes of alluvium accumulate wherever topographic basins exist.

Landforms of the Sahara Desert illustrate some of the features to be expected in tropical deserts occupying stable shield areas (see Plate J.3). Structural features are brought out into sharp relief. These may be mesas and buttes, hogbacks, volcanic necks, or granite domes, depending on the composition of the rock and its structure. Pediments are developed locally at the bases of these residual relief features. Broad, shallow stream channels conduct alluvium and dissolved salts to shallow basins of accumulation. Here crusts of gypsum and other salts are formed through evaporation.

Wind is a major agent of landscape development in the Sahara Desert. Enormous quantities of reddish dune sand have been derived from the weathering of older sandstone formations. The sand is formed into a great expanse of free sand dunes, called an **erg.** Elsewhere there are vast flat-sufaced sheets of sand, armored against further deflation by a layer of pebbles. A surface of this kind is called a **reg.**

Some of the Saharan dunes are complex forms, not represented in the western hemisphere. One of these is the sword dune, a huge tapering sand ridge whose crestline rises and falls in alternate peaks and saddles. The side slopes are indented by crescentic slip faces. Sword dunes may be tens of kilometers long. Another Saharan type is

the **star dune** (heaped dune), a large hill of sand whose base resembles a many-pointed star when seen from above (Figure 27.17). Radial ridges of sand rise toward the dune center, culminating in sharp peaks as high as 100 m (300 ft) or more above the base. Star dunes remain fixed in position and have served for centuries as reliable landmarks for desert travelers.

Permanent human habitations are few and far between in the Sahara. They are localized in low places where ground water lies close to the surface and can be reached by dug wells. With fresh water assured, a productive oasis can thrive, with its groves of date palms, some citrus trees, and small plots of grains and vegetables (Figure 27.18). Often the oasis is surrounded by sand dunes, and these must be prevented from invading the oasis through the construction of barriers of brush on the dune crests. Modern tube wells, some of them tapping deep artesian sources of ground water, have been installed in recent decades at favorable locations.

Desert Irrigation

Man's interaction with the tropical desert environment is as old as civilization itself. Two of the earliest sites of civilization — Egypt and Mesopotamia — lie in the tropical deserts. The key to Man's successful occupation of the deserts lies in the availability of large supplies of water from nondesert sources. This is a concept so familiar to all that it scarcely needs to be stated. For Egypt and Mesopotamia, the water sources of ancient times were **exotic rivers** deriving their flow from regions having a water surplus

Figure 27.17 Seen from an altitude of 10 km (6 mi), these sand dunes of the Libyan desert appear as irregular patches that rise to star-shaped central peaks 90 to 180 m (300 to 600 ft) higher than the intervening flat ground. The width of the photograph represents about 11 km (7 mi). (Aero Service Division, Western Geophysical Company of America.)

Figure 27.18 **Looking straight down on the desert city of El Oued, Algeria, we see small groves of date palms and citrus, planted on the floors of hollows in dune sand. (Georg Gerster/Rapho-Photo Researchers, Inc., New York.)**

and flowing across the desert region because of geologic events and controls having nothing to do with climate.

A look at a global population map shows that population density is less than one person per square km (two persons per sq mi) over nearly all the area of the tropical deserts. Only where exotic streams cross the desert does the population density rise sharply. Valleys of the Nile, the Tigris and Euphrates, and the Indus are striking examples in the Old World. But in the coastal desert of Peru, we also find a substantial population long dependent on exotic streams fed from the Andes range and crossing the coastal desert to reach the Pacific Ocean.

Can Man increase the production of food by expanding agriculture into the tropical deserts? Making the desert bloom in a romantic concept fostered on the American scene for generations by bureaucrats, politicians, and land developers. Were these promoters of vast irrigation schemes working in the long-term public interest? Only in recent years have the undesirable environmental impacts of desert irrigation come to the forefront. But we could have read the modern scenario in the history of the rise and fall of the Mesopotamian civilization.

Irrigation systems in arid lands divert the discharge of a large river, such as the Nile, Indus, Jordan, or Colorado, into a distributary system that allows the water to infiltrate the soil of areas under crop cultivation. Ultimately, such irrigation projects suffer from two undesirable side effects: salinization and waterlogging of the soil.

The irrigated area is subject to very heavy soil-water losses through evapotranspiration. Salts contained in the irrigation water remain in the soil and increase in concentration. This process is called **salinization.** (Areas of salinization show as white surfaces on the remote sensing imagery of Figure 5, Plate A.) Ultimately, when salinity of the soil reaches the limit of tolerance of the plants, the land must be abandoned. Prevention or cure of salinization may be possible by flushing the soil salts downward to lower levels by the use of more water. This remedy requires greater water use than for crop growth alone.

Infiltration of large volumes of water causes a rise in the water table and may, in time, bring the zone of saturation close to the surface. This phenomenon is called **waterlogging.** Crops cannot grow in perpetually saturated soils. Furthermore, when the water table rises to the point that upward movement under capillary action can bring water to the surface, evaporation is increased and salinization is intensified.

One of the largest of the modern irrigation projects affected adversely by salinization and waterlogging lies within the basin of the lower Indus River in Pakistan. Here the annual rate of rise of the water table has averaged about 0.3 m (1 ft), while the annual increase in land area adversely affected is on the order of 20,000 hectares (50,000 acres).

The question of expanding irrigation agriculture in the southwestern United States was studied by the Committee on Arid Lands of the American Association for the Advancement of Science. In its 1972 report, this body recommended that additional large-scale importation of irrigation water to the American Southwest from distant sources should be made only where there are compelling reasons to do so. One such reason is to augment rapidly failing ground water supplies in districts already under irrigation. A second is to arrest the progress of salinization in areas already under irrigation. In short, the committee agreed that additional water should be imported only to prevent social and economic disruption in established irrigated areas. The lesson is that the search for new regions in which to expand agriculture should be directed to other, more favorable environments where the scales are not so heavily weighted by enormous evaporative water losses.

The Low-Latitude Environments in Review

No generalization is meaningful for the three low-latitude environmental regions considered together because they include great extremes of the climatic spectrum. Anyone who makes a sweeping statement about the "tropics" is very poorly informed. How can the extreme aridity of the tropical deserts be equated with the extreme wetness of the equatorial zone in reference to the thermal environment, the soil-water balance, or the food resource? If diversity is the outstanding characteristic of the low-latitude environments, shall we look to the midlatitudes for some sweeping simplicity of character? The midlatitude zone has for centuries been called the "temperate zone." How temperate is the midlatitude zone? Keep this question in mind as we turn next to the environmental regions of the midlatitude zone.

References for Further Study

Cressey, George B. (1957), Water in the desert, *Annals, Association of American Geographers,* vol. 47, pp. 105–124.

Paton, T. R., and M. A. J. Williams, (1972), The concept of laterite, *Annals, Association of American Geographers,* vol. 62, pp. 42–56.

Janzen, David H. (1973), Tropical agroecosystems, *Science,* vol. 182, pp. 1212–1219.

Sanchez, P. A., and S. W. Buol (1975), Soils of the tropics and the world food crisis, *Science,* vol. 188, pp. 598–603.

Strahler, A. N., and A. H. Strahler, (1977), *Geography and Man's environment,* John Wiley and Sons, New York. See Units 43, 52–54.

Review Questions

1. Explain the concept of global environmental regions. On what basis is an environmental region defined?
2. Review the outstanding features of climate, soil-water regime, and vegetation of the rainforest environment. What soil orders are dominant in the rainforest? How does climate affect the abundance of plant and animal species? Describe the animal life of the rainforest.
3. Which geomorphic processes are most intense in the rainforest environment? What unusual forms are produced by solution removal of mineral matter?
4. Describe the important forest products and food plants of the rainforest.
5. In what ways has the rainforest been exploited by Man? What is slash-and-burn agriculture? What future lies ahead for the rainforest environment? What difficulties may deter development of intensive crop agriculture?
6. Review the outstanding features of climate, soil-water regime, and vegetation of the savanna environment. What is raingreen vegetation? Which soil orders are most widespread in the savanna regions?
7. Which geomorphic processes are most active in the savanna environment? What is a bornhardt? How is it formed? Describe and explain the formation of laterite crusts in the savanna environment. How does a laterite (ferricrete) crust originate?
8. Describe the animal life of the African savanna. What management problems are involved in the preservation of wildlife of the African savanna?
9. What are the important food crops of the savanna environment? Explain the principles and techniques of the green revolution. What problems have been encountered in implementing the green revolution? What future modifications of the program are indicated?
10. Review the outstanding characteristics of climate, soil-water regime, and vegetation of the tropical desert environment. Compare the climate of the western littorals with that of interior deserts. Explain the concept of precipitation variability and apply it to the low-latitude climates. Which climate shows the greatest variability? Which shows the least?
11. Which geomorphic processes are most active in the desert environment? What is desert varnish? What is calcrete? What landforms are associated with calcrete crusts? What role do salts play in the desert environment?
12. Describe the distinctive and unusual landforms and other geomorphic features of the Sahara Desert. What is an erg? A reg? How does ground water play a role in human occupation of the Sahara Desert?
13. Along what exotic rivers is irrigation practiced in the tropical deserts? What problems are encountered after prolonged use of irrigation in a desert climate? What policies have been recommended in connection with plans for increased large-scale water importation into desert regions?

Global Environmental Regions: II. The Midlatitude and High-Latitude Environments

28

The midlatitude environmental regions are identified with climates of Group II; these occupy the polar front zone in which both tropical and polar air masses play an important climatic role. Strong climatic seasons are a characteristic of the midlatitude climates — seasons in which temperatures as well as precipitation show strong annual cycles.

Those who have not studied physical geography refer to the midlatitude region of the globe as the "temperate zone." Nothing could be more misleading. "Temperate" means a regime marked by moderation — no extremes or excesses. But intemperance is the rule of most of the midlatitude zone. Not only do seasons swing from one extreme to the other in heat and cold or in dryness and wetness, but the day-to-day swings in weather as cyclones pass by provide a pattern of intemperance often more dramatic than that of the seasons. Intemperance in the midlatitude zone is partly due to the occurrence of vast continents in that zone in the northern hemisphere. Continentality is synonymous with intemperance of climate. Only along the west coasts do we fine narrow strips of land where a truly temperate climate prevails.

Moist Subtropical Forest Environment (D)

Climate, Vegetation, and Soils

The moist subtropical forest environment is defined as a region by those global areas of humid and perhumid subtypes of the moist subtropical climate (6h, 6p) in which the forest biome prevails. The major areas are in the northern hemisphere: southeastern United States, southern China, northern Taiwan, and southern portions of Japan (Kyushu and southern Honshu). A narrow coastal strip of southeastern Australia qualifies for inclusion, but in some respects is not typical. Representative natural vegetation is broadleaf evergreen forest (temperate rainforest, laurel forest) (Fbe). The typical soils are Udults, the udic-regime suborder of the Ultisols.

The moist subtropical climate, by its very name, borders on the tropical zone and is transitional between the low-latitude and midlatitude environments. The climate has a mild winter season; but toward its low-latitude borders, the winter season weakens greatly in intensity. Here it is not easy to decide where the change to the trade-wind littoral climate sets in. This transitional character of climate is also reflected in soils and vegetation. In North America, Udults of the moist subtropical climate are transitional between Spodosols of the cold climates and Oxisols of the warm, wet climates. In North America and Southeast Asia, broadleaf evergreen forest gives way in the colder northern part to midlatitude deciduous forest. While being aware of the transitional nature of the environment, we can try to make some broad generalizations about the more typical parts of the moist subtropical region.

The moist subtropical climate is found on the eastern sides of landmasses at lat. 25° to 30°N and S (Figure 9.8). Figure 6.13 shows the oceanic high-pressure cells pumping air into the eastern sides of the continents at lat. 25° to 35°. This effect is strongest in the summer, when the highs are strong and persistent. The maritime tropical air mass involved in this motion has a high moisture content and produces ample summer rainfall. In Southeast Asia, under monsoon influence, rainfall peaks sharply in the summer. Tropical cyclones arrive on these shores after their long westward paths through the trades. As these storms turn inland, they bring very heavy rainfall and flooding conditions.

The soil-water regime of the moist subtropical environment shows a large annual water surplus, while there is only a very small soil-water shortage or none at all. The high levels of soil-water storage through the long warm summer favor intensive plant growth and support the forest biome.

Geomorphic Processes and Landforms

Because the moist subtropical forest environment is transitional from the wet low-latitude environment, many of the statements made in Chapter 27 about geomorphic processes in the latter region apply here as well. Intensity of chemical weathering is high and produces deep regolith (saprolite) on rocks rich in silicate minerals. Frost action as a form of mechanical weathering should be included as a process in the colder, northern limits of the region. Mass wasting in the form of flowage and sliding of moist soil and regolith is conspicuous. Dominance of suspended load over coarse bed load is typical of the fluvial process.

Denudation and slope evolution in the moist subtropical environment conforms closely to the idealized model presented in Chapter 20 for the uplift, rejuvenation, and dissection of a peneplain in a humid climate. The larger graded rivers occupy forested, marshy floodplains and show sinuous meanders, with development of natural levees and bar-and-swale landforms. Base flow is an important contributor to the total stream flow, so most larger streams are perennial.

Geologic controls over landform development are highly varied from place to place in this environmental region. In the southeastern United States — a stable shield region with a continental shelf margin — a low, flat coastal plain contrasts with subdued hill and mountain terrain developed on Appalachian mountain roots. In contrast, the mountainous islands of Japan show very steep slopes, typical of a mountain arc adjacent to deep trenches over an active subduction zone (see Figure 28.2).

Agricultural Resources

No simple statement can cover Man's agricultural adaptation to the moist subtropical forest environment in both North America and Southeast Asia. The differences in these two widely separated areas are partly historical and cultural; but in part they reflect the stronger monsoon effect in Asia, which causes a stronger concentration of precipitation in the summer. The human population is vastly denser in Southeast Asia than in the New World and subsists largely on rice, the dominant staple food crop. Two and even three rice crops are harvested annually in southern China. The rice crop is often followed by the planting of wheat, winter legumes, peas, or green fertilizer crops. Except in the Mississippi delta region, little rice is produced in the southern United States. Both American and Asiatic regions produce sugar cane, peanuts, tobacco, and cotton, though not in equal intensity. One striking difference is that tea is widely cultivated in Southeast Asia, but not at all in the southern United States (Figure 28.1). Corn is an important crop in the southern United States but is not important in southern China.

The potential of the moist subtropical climate to produce more food rests in more intensive land use instead of expansion of the area now under cultivation. The best land is already in use and, in Southeast Asia, elaborate terrace systems have been in use for centuries to allow farming of steep hillslopes (Figure 28.2). New genetic strains of rice and corn offer promise of increased yields when the necessary fertilizers are used. Whereas Japan and Taiwan apply very high levels of fertilizers and achieve high rice yields, the People's Republic of China is only now in the process of sharply increasing its use of fertilizers from comparatively low levels of the recent past.

Figure 28.1 **Tea leaves are carefully picked from the new growth of a cultivated evergreen shrub** *(Camellia)* **on this tea plantation near Kyoto, Honshu Island, Japan. (Rene Burri, © Magnum Photos.)**

China is also developing independently some new high-yielding strains of rice and wheat.

In the southern United States, cattle raising is another direction of increased food production; it makes use of soils too sandy for field crops. With soil water frequently replenished through the long, warm summer, pasture and range land can be continuously productive. Tree farming is also an important use of sandy soils. Pines are well adapted to rapid growth on sandy soils and thrive where nutrient bases are in short supply (Figure 28.3).

The large water surplus of the moist subtropical climate has important implications in terms of economic development. The large flows of rivers can furnish abundant freshwater resources for urbanization and industry without competition from irrigation demands. Evaporative losses from reservoirs are much less important than in arid lands. The maintenance of copious stream flows tends to reduce the dangers of severe water pollution and its adverse effects on ecosystems of streams and estuaries.

Mediterranean Environment (E)

Climate, Vegetation, and Soils

The Mediterranean environment is defined by those global areas having the semiarid, subhumid, and humid subtypes of the Mediterranean climate (7s, 7sh, 7h). This dry-summer, wet-winter climate is named for its occurrence in lands bordering the Mediterranean Sea. A second important occurrence is in central and southern California, extending from the Pacific shores inland into the Great Valley and other valleys within the coast ranges. In the southern hemisphere, the Mediterranean climate occurs in coastal Chile, the Cape Town area of South Africa, and coastal zones of western and southern Australia.

The Mediterranean climate spans a wide range from moist to dry. Whereas the soil-water shortage developed in summer is always substantial, the water surplus may be zero in the semiarid subtype or as great as 30 cm in the humid subtype. This spread in water surplus results in a corresponding variety in sclerophyll plant formation classes, ranging from Mediterranean mixed evergreen forest (Fsm) to woodland, dwarf forest, and scrub (Fss). In Australia, the formation class is Australian sclerophyll (Eucalyptus) forest (Fsa). The thermal regime is also quite different in the narrow western littorals — where summers are cool and the annual temperature range is small — than in interior valleys — where summers are hot and the annual range is substantial. Soils of the Mediterranean climate are assigned to the xeric-regime suborders. Most are Xeralfs, Xerolls, and Xerults. Of these, the Xeralfs and Xerolls are of high base status and are fertile soils when soil-water is adequate. Xerults, on the other hand, are soils of low base status.

Geomorphic Processes

In the Mediterranean environment, mechanical weathering by salt-crystal growth is probably important throughout the dry summer. Chemical weathering is particularly active in the rainy winter; but because this is the time of cool temperatures, mineral alteration is probably less rapid than in moist-summer climates. Chemical weathering has caused widespread occurrences of thick regolith over bodies of igneous and metamorphic rock rich in silicate minerals. Mass wasting in the form of earthflows and landslides occurs in such materials and in clay-rich bedrock where slopes are steep. These forms of mass wasting are a serious hazard in California, where urbanization has spread over steep slopes.

Because of the sparseness of plant cover in woodland and scrub plant formations, fluvial processes of the Mediterranean environment are closely akin to those of

semiarid and semidesert environments. When torrential rains occur in winter, large quantities of debris are swept downslope by overland flow and carried long distances by streams in flood. Mudflows and debris floods are produced at such times. Alluvial fans are a prominent landform of the Mediterranean environment in California. Fan surfaces and canyon floors are high-hazard areas, where the destruction of houses and roadways by stream floods, debris floods, and mudflows are an ever-present threat. The hazards of flooding are greatly increased when

Figure 28.3 **This plantation of slash pine grows on sandy soil of the Georgia coastal plain. A cup is attached to the base of each tree to catch sap exuding from a cut into the sapwood. Turpentine is distilled from the sap. (Grant Heilman.)**

fire has consumed the plant cover of a mountain watershed. The chaparral of southern California hillsides is extremely flammable during the long dry summer.

Accelerated soil erosion, induced by human activity over the past 2000 years or longer, has been a dominant geomorphic process throughout the Mediterranean lands. Many hillsides have been denuded of residual soils and present a rocky aspect today (Figure 28.4). Correspondingly, sedimentation has occurred in valley floors and in places aggradation has reached depths of several meters. Deforestation occurred widely in the days of ancient Rome, as firewood was consumed for the heating of baths and smelting of ores. Through later centuries, herds of grazing goats and sheep reduced the shrub cover, while seasonal burning of the dry grass and shrub vegetation was practiced by herders to induce the growth of fresh green foliage.

Agriculture

Lands bordering the Mediterranean Sea produce cereals — wheat, oats, and barley — where arable soils are extensive enough to be cultivated. However, we usually think of that region as an important source of citrus fruits, grapes, and olives for European markets. Cork from the bark of the cork oak is also a product of economic value (Figure 28.5). In central and southern California, citrus, grapes, avocados, nuts (almond, walnut), and deciduous fruits are grown extensively (Figure 28.6). Irrigated alluvial soils are also highly productive of such vegetable crops as carrots, lettuce, cauliflower, broccoli, artichokes, and strawberries, as well as sugar beets and forage crops (e.g., alfalfa). Cattle ranching and sheep grazing are of major importance on grassy hillslopes, which are unsuited to field crops and orchards, and on irrigated lowland pastures.

Because the Mediterranean environment is limited in global extent to comparatively small land areas, it offers little prospect for important additions to the world's food supply. Irrigation is essential for high productivity, and we have already stressed the hazards associated with heavy irrigation of lowland soils: salinization and waterlogging. Urbanization and industrial development also face the major problem of obtaining water supplies through importation over long distances by aqueduct. Nevertheless, the mild, sunny climate of southern California has proved a powerful population magnet and water importation has been developed on a mammoth scale.

Marine West-Coast Forest Environment (F)

Climate, Vegetation, and Soils

The marine west-coast forest environment is represented by all global areas having the humid and perhumid subtypes of the marine west-coast climate (8h, 8p). This climate is found along the western coast of North America from Oregon to British Columbia and in the British Isles, western Europe, Victoria, Tasmania, New Zealand, and southern Chile. The climate is quite mild in winter temperatures for the comparatively high latitude in which it

Figure 28.4   Only a goat could find nourishment on this rock-strewn Lebanese hillside. (United Nations.)

lies (35° to 60°). Ample precipitation occurs in all months. The soil-water balance usually shows a moderate to large annual water surplus, and there is only a small soil-water shortage or no shortage in summer.

Precipitation is strongly affected by topography. Whereas lowlands of Europe receive as little as 75 cm (30 in.) annually, coastal mountains, such as those of the northern Pacific coast, receive totals over 200 cm (80 in.). These perhumid mountainous areas support dense needle-leaf forests (coastal forest, Fcl); the typical soils are Entisols and Inceptisols related to the Spodosols. Under the lower precipitation regime of the British Isles and western Europe, midlatitude deciduous forest (Fd) is the

Figure 28.5   This bark, stripped from the cork oak (Quercus suber), will be ground up and cemented into corkboard and other structural products. Thick bark of choice quality is used for wine corks. Algeria, North Africa. (A. N. Strahler.)

Figure 28.6 **Lemon orchards in Ventura County, California, show two geometric patterns. Narrow terraces on steep hillsides (left) follow curving contours. The grid pattern (right) characterizes older orchards on gently sloping lowland surfaces. (U.S. Department of Agriculture.)**

prevailing formation class; Udalfs, a suborder of the Alfisols, are the dominant soil class. In the southern hemisphere, broadleaf evergreen forest (temperate rainforest), Fbe, is a dominant formation class in Tasmania and New Zealand, while mixed boreal and deciduous forest (Fbd) is found in the mountainous coastal belt of southern Chile.

Geomorphic Processes

Weathering processes of the marine west-coast forest environment include limited frost action as a mechanical process acting in winter in the colder and higher areas. Although lower in total intensity than in warmer climates, chemical weathering is a dominant process. Because of the recency of Pleistocene glaciation, however, bedrock subjected to ice abrasion shows little decomposition, while the parent materials of the soil have had insufficient time to undergo extensive alteration. Many landforms of glacial erosion and deposition are as yet little changed from their original configurations. Glacial troughs and fiords are striking landforms of the rugged coasts of British Columbia, Scotland, and Norway. Extensive lowlands of northwestern Europe consist of till plains, moraines, and outwash plains left by Pleistocene ice sheets. Where forest cover is largely undisturbed, sediment yield is low and denudation rates of mountain watersheds are probably very low. Locally, however, where clear-cutting of forests is practiced, severe soil erosion and high sediment yields occur.

Agricultural and Water Resources

The marine west-coast environment of western Europe and the British Isles has been intensively developed for centuries for such diverse uses as crop farming, dairying, orchards, and forests. It is an environment in most respects similar in agricultural character and productivity to the moist continental forest environment with which it merges on the east.

In North America the mountainous terrain of the coastal belt offers only limited valley floors for agriculture, which is generally diversified farming. Forests are the primary plant resource, and here they constitute perhaps the greatest structural and pulpwood timber resource on earth. Douglas fir, western cedar, and western hemlock are the principal lumber trees of the Pacific Northwest (Figure 28.7; see also Figure 16.15.)

Figure 28.7 **Old-growth forest of Douglas fir and western hemlock in the Olympic National Forest, Washington. Removal of all trees in isolated patches, a practice known as block cutting, completes the initial cycle of timber harvesting. (J. Prater, U.S. National Forest Service.)**

The same mountainous terrain that limits agriculture in the Pacific Northwest is a producer of enormous water surpluses that run to the sea in rivers. With the inclusion of the ranges of the northern Rocky Mountains with the Pacific coastal ranges, the potential for long-distance transfer of this excess water to dry regions of the western United States has not passed unnoticed.

Moist Continental Forest Environment (G)

Climate, Vegetation, and Soils

The moist continental forest environment is defined by global areas of the humid and perhumid subtype of the moist continental climate (10h, 10p). Also included are some areas of the subhumid subtype (10sh), where the forest biome prevails. Limited to the northern hemisphere landmasses, this environmental region occupies a large part of the northeastern United States and southeastern Canada in the latitude range 35° to 48°N. In Europe, this environmental region extends from western Europe across central Europe and Russia at least as far as the Urals. In eastern Asia, a large area covers much of northern China, Manchuria, Korea, and northern Japan.

The climate shows a high degree of continentality, with cold winters, warm summers, and a large annual temperature range. Precipitation is substantial in all months, but typically shows a strong summer maximum. Soil water is frozen throughout one to three winter months, at which time water need is effectively zero. Annual water surplus is moderate to large, and soil-water shortage is small or may be zero in summer. Forest is of two formation classes. Midlatitude deciduous forest (Fd) is typical of the southerly portions having less severe winters and longer summers; mixed boreal and deciduous forest (Fbd) is typical of the northerly portions having colder winters and cooler summers. In North America, lake forest (Fl) is an important formation class in the Great Lakes region. Spodosols are the dominant soil order in areas of mixed boreal and deciduous forest in North America, whereas Alfisols are associated with most other areas in North America and throughout Europe and east Asia. Suborders of the Alfisols are Boralfs and Udalfs, the former being typical of the colder northerly portions.

Geomorphic Processes

Weathering, mass-wasting, and fluvial processes of the moist continental forest environment are generally similar to those of the moist subtropical environment to the south, except that frost action is more intense and chemical weathering is less intense. Under natural conditions of undisturbed forest cover, fluvial erosion would be of low intensity; but little or no native forest remains over large parts of this environmental region, and Man-induced soil erosion is locally intense. Pleistocene continental glaciation has dominated the landforms over much of the region in North America and Europe; these include till plains and moraines, lake basins and lake plains, outwash plains, and sheets of loess over uplands. Many stream valleys in unglaciated areas lying south of the ice limits in

Figure 28.8 The Mosel River in its winding, entrenched meandering gorge through the Rhineland-Pfalz province of western Germany. The village of Uerzig is in the foreground. On the steep, undercut valley wall at the right are vineyards and forests. Agriculture flourishes on the gentle slipoff slope on the left side of the river channel. (Photograph by Fritz Henle, courtesy of The German National Tourist Office, New York.)

North America and Europe underwent aggradation during glacial stages, followed by trenching and alluvial terrace formation in postglacial time.

Agricultural Resources

Throughout Europe, large areas of the moist continental forest environment have been under field crops, pastures, and vineyards for centuries; at the same time forests have been carefully cultivated over large areas (Figure 28.8). In this environment, the potentials for both crop agriculture and forest culture (silviculture) have reached a near-optimum adjustment in terms of the soils and terrain.

Because of the availability of soil water throughout a warm summer growing season, the moist continental environment has an enormous potential for food production. The cooler, more northerly sections in North America and Europe support dairy farming on a large scale. A combination of acid Spodosols and unfavorable glacial terrain in the form of bogs and lakes, rocky hills, and stony soils has deterred crop farming in many parts. Farther south, plains formed on former lake floors and undulating uplands, bearing a loess cover, are ideally suited to crop farming, while Udalfs of high base status provide fertile soils. Cereals grown extensively in North America and Europe include corn, wheat, rye (especially in Europe), oats, and barley. Corn is also an important crop in Hungary and Rumania. Beet sugar is an important product of this environmental region in Europe, but not in North America. On the other hand, soybeans are intensively cultivated in the midwestern United States and in northern China and Manchuria, but very little in Europe. Rice is a dominant crop in both South Korea and Japan, much farther poleward than elsewhere in Asia. The rice seedlings can be planted in paddies flooded during the brief but copious rains of midsummer, then harvested in the dry autumn. Among geographers, this northern rice area is included in the region called "Monsoon Asia."

Prairie and Steppe Grassland Environment (H)

Climate, Vegetation, and Soils

The prairie and steppe environmental region includes all midlatitude global regions of the grassland biome. These are most clearly correlated with areas of the subhumid subtype of the moist continental climate (10sh) and the steppe subtype of the dry midlatitude and dry subtropical climates (9s, 5s). In North America, a small portion of the humid subtype of the moist continental climate (10h) is included. In South America, grassland of the Pampa occupies an area classified as humid subtype of the moist subtropical climate (6h).

In North America, areas of prairie and steppe grasslands extend continuously from the Mississippi River valley to the Rocky Mountains, while a more westerly region occupies the Columbia River basin. Tall-grass prairie (Gp) forms the eastern portion of this vast area; steppe, or short-grass prairie, forms the western portion. In the Ukraine of Russia is a large expanse of steppe grassland that projects eastward in a long narrow belt reaching into Siberia as far east as long. 90°E. Other areas of steppe occur in Mongolia and northern China. In South America the Pampa of eastern Argentina, Uruguay, and southern Brazil is a large expanse of tall-grass prairie.

In the northern hemisphere, climates associated with grasslands show extreme continentality, with cold winters and warm to hot summers and a very large annual temperature range. A substantial soil-water shortage (greater than 15 cm) is always found in the semiarid climate subtype, and there is no water surplus. In the subhumid subtype of the moist continental climate (10sh), the shortage ranges from 0 to 15 cm and there is little or no surplus. Tall-grass prairie also extends eastward into the humid subtype of the moist continental climate (10h), where a substantial water surplus occurs. The Pampa of South America, lying in the humid subtype of the moist subtropical climate (6h), shows a water surplus of from 20 to 30 cm; it is clear that this grassland area occupies a much moister environment than any of the northern hemisphere grasslands except the tall-grass prairie lying east of the Mississippi River. The subtropical climate of the Pampa region has exceptionally mild winters and lacks continentality because no large landmass lies on the poleward side.

The dominant soils of the grassland biome are suborders of the Mollisols. Borolls occupy the colder northerly portions in North America and Eurasia; Udolls and Ustolls occupy the warmer portions in North America. Xerolls are associated with steppe grasslands of the Columbia River basin. Udolls underlie much of the tall-grass Pampa of South America. Aridisols are associated with fringe areas of steppe bordering on semidesert.

Geomorphic Processes

Because the grasslands extend from moist to dry climates, no single statement about geomorphic processes and landforms can apply to the entire environmental region. We shall limit our comments to the semiarid climate subtype, where short-grass prairie occurs. Because of the large annual soil-water shortage and lack of water surplus, two processes are particularly important: disruption of rock by salt crystallization and accumulation of deposits of calcium carbonate. Carbonate matter accumulates in the soil as a calcic horizon, and this may harden into a petrocalcic horizon, called "caliche" in the American Southwest. Where the upper soil horizons are stripped away, crusts of calcrete resist erosion and tend to produce platforms, cliffs, and mesas on a small scale.

Locally, the grass cover and soil are removed entirely by the trenching of streams, so areas of badlands are formed and maintained. Sediment yield can rise to extremely high values from these rapidly eroding areas. Stream channels are of the broad, shallow form with braided appearance. Streamflow is highly seasonal and many channels are dry much of the year. Wind action is very important locally. Deflation raises dust storms, even on surfaces bearing a grass cover. Locally, blowout depressions are formed and maintained. Enlargement of these depressions is carried out by a combination of water erosion on the exposed side slopes of the depression and deflational removal of dried particles of mud and sand. Trampling and wallowing by grazing animals contributes to enlargement of the depressions.

Fixed sand dunes occur over large areas of the American steppes. These dunes were active in late Pleistocene time during climatic episodes in which the plant cover was depleted. An outstanding example is the Sand Hills region of Nebraska, where fixed sand dunes occupy an area of about one-quarter of the state (Figure 28.9). This area is now given over to cattle ranching. Water-table ponds in the deeper depression supply fresh water. The grass cover is easily broken, with rapid development of small blowouts.

Agricultural Resources

Agricultural productivity of the tall-grass prairie lands in the United States is now legendary under the name of the *corn belt*. Corn production is concentrated most heavily in the prairie plains of Illinois, Iowa, and eastern Nebraska (Figure 28.10). Wheat is also a major crop near the western limits of the tall-grass prairie in Kansas and Oklahoma (Figure 28.11). A remarkably favorable combination of fertile Mollisols formed on a loess cover and provided with adequate soil-moisture during the growing season largely accounts for the phenomenal productivity. The Pampa region of South America is perhaps the only true counterpart of the American prairie in terms of agricultural output. Corn and wheat are important crops of the Pampa;

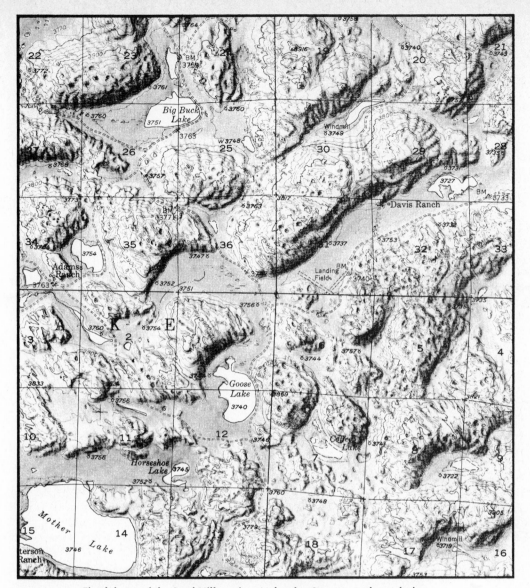

Figure 28.9 **Fixed dunes of the Sand Hills region, Nebraska. Steeper southeast-facing slopes suggest former slip faces and a prevailing northwesterly wind. Lakes represent level of water table. (Ashby, Nebraska, Quadrangle, U.S. Geological Survey.)**

but the raising of cattle on large ranches is still a dominant activity, not seen in the intensively farmed American tall-grass prairie.

Wheat is perhaps the most important single crop produced in unirrigated areas of short-grass prairie bordering the subhumid zone. One such important wheat-producing region lies in southern Alberta and Saskatchewan and in the northern border region of Montana. Here the crop is spring wheat, planted in the spring of the year. Using the soil water that has been recharged in early spring and the precipitation that falls in late spring and early summer, the crop is able to reach maturity for midsummer harvesting.

In Russia, the rich wheat region of the Ukraine is continued in a narrow zone far eastward across the steppes of Kazakhstan. In northern China, wheat is grown within a steppe region bordering the moist continental climate.

Wheat production of the midlatitude steppes is very much at the mercy of variations in seasonal rainfall. Good years and poor years follow cyclic variations. Soil water, not soil fertility, is the key to wheat production over these vast steppe lands that lie beyond the practical limits of upland irrigation systems.

Semiarid steppes form the great sheep and cattle ranges of the world. The steppes of Central Asia have for centuries supported a nomadic population, their sheep and goats finding subsistence on the scanty grassland (Figure 28.12). On the vast expanses of the High Plains, the American bison lived in great numbers until almost

Figure 28.10 Rich farmlands of the corn belt in Iowa are laid out on a grid pattern. There are about four farms per square mile in this area. (Aero Service Division, Western Geophysical Company of America.)

exterminated by hunters. The short-grass veldt of South Africa supported much game at one time.

Steppe grasses do not form a complete sod cover; loose, bare soil is exposed between grass clumps. For this reason, overgrazing during a series of dry years can easily reduce the hold of grasses enough to permit destructive deflation, followed in wet years by soil erosion and gullying.

On the Great Plains of Kansas, Oklahoma, and Texas, deflation and soil drifting reached disastrous proportions during a series of drought years in the middle 1930s, following a great expansion of wheat cultivation. These former grasslands are underlain by friable Mollisols. During the drought, a sequence of exceptionally intense dust storms occurred. Within their formidable black clouds, visibility declined to nighttime darkness, even at noonday. The area affected became known as the Dust Bowl. Many

centimeters of soil were removed from fields and transported out of the region as suspended dust, while the coarser silt and sand particles accumulated in drifts along fence lines and around buildings (Figure 28.13). The combination of environmental degradation and repeated crop failures caused widespread abandonment of farms and a general exodus of farm families.

Among geographers who have studied the Dust Bowl phenomenon, there is a difference of opinion as to how great a role soil cultivation and livestock grazing played in inducing deflation. The drought was a natural event over which Man had no control, but it seems reasonable that the natural grassland would have sustained far less soil loss and drifting if it had not been destroyed by the plow.

Although Man cannot prevent cyclic occurrences of drought over the Great Plains, measures can be taken to minimize the deflation and soil drifting that occur in

Figure 28.11 **Combines harvesting winter wheat on the High Plains of western Kansas. (Grant Heilman.)**

Figure 28.12 This scene in the steppes of Central Asia was taken over a half-century ago, when nomads lived in harmony with nature. The Mongol village in the foreground is easily moved to follow favorable conditions of water and pasture. (American Museum of Natural History.)

periods of dry soil conditions. Improved farming practices include the use of listed furrows (deeply carved furrows) that act as traps to soil movement. Stubble mulching will reduce deflation when land is lying fallow. Tree belts may have significant effect in reducing the intensity of wind stress at ground level.

Midlatitude Desert Environment (I)

Brief mention, only, is appropriate for the midlatitude desert environment because most of the basic properties of climate, vegetation, soils, and geomorphic processes of the tropical desert environment, described in Chapter 27, apply with little modification. The midlatitude desert occupies large areas only in Central Asia: Turkestan, the Tarim Basin, and the Gobi Desert of Mongolia. In North America a narrow belt of semidesert runs west-east across the Great Basin in southeastern California, southern Nevada, and western Utah.

The important distinguishing feature of the climate of the Asiatic midlatitude desert and semidesert is the extreme cold of the winters (see Figure 9.24). Because of the lower temperatures, water need is much less than in the hot low-latitude deserts. Further agricultural development of the midlatitude desert environment by irrigation faces much the same set of problems as does the low-latitude desert environment.

The Midlatitude Environments in Review

The midlatitude environments run the gamut from very wet to very dry climates and from mild marine coastal climates to strongly seasonal continental climates. Soils and natural vegetation cover an equally great range of

types befitting the spectrum of climate. No useful generalization is possible for such a diverse group of environments.

Production of food resources in the midlatitude regions spans as wide a range of intensities as the climates. Here we have the richest food-producing regions of the world, fully developed by the Western nations through

Figure 28.13 A typical scene in the Dust Bowl during the late 1930s. Drifts of sand and silt have accumulated around abandoned farm buildings and a fence in Dallam County, Texas. (Soil Conservation Service, U.S. Department of Agriculture.)

massive inputs of fertilizers, pesticides, and fuels and guided by the most advanced technology. But there are also unproductive deserts at the same latitudes. These midlatitude environments, taken as a whole, are neither temperate in climate nor uniform in plant resources. Perhaps this heterogeneous quality of the midlatitudes is just what we should expect of a global zone where polar and tropical air masses wage war incessantly over vast land areas that cut across the latitude zones and their prevailing westerly air flow.

The High-Latitude Environments

The high-latitude environments are identified with the climates of Group III, which are controlled by polar and arctic air masses. As a group, these climates have low air temperatures, a severe winter season, and only small amounts of precipitation. A condition of frozen soil water for several consecutive months of winter is also typical, so there is usually a long annual period with near-zero water use and near-zero water need.

Boreal Forest Environment (J)

The boreal forest environment is identified with global areas having the boreal forest climate (11); these are found only in the northern hemisphere. The boreal forest climate occupies two great zones, one extending across North America from Alaska to Labrador, the other across Eurasia from the Scandinavian Peninsula to the Bering Sea. The salient feature of this climate is the extreme continentality of the thermal regime. Long, bitterly cold winters dominate the climate and summers are brief. Soil water is frozen from five to seven consecutive months of the year, when water need is effectively zero. Precipitation is small, but accumulates as snow throughout the winter and is rapidly released as surface water in the spring thaw. Boreal forest, the dominant plant formation class, consists of needleleaf evergreen trees, except in Siberia, where deciduous forests of larch occupy semiarid areas of the boreal climate. Dominant soil orders are Spodosols and Histosols.

Most of the boreal forest region is dominated by landforms shaped beneath great Pleistocene ice sheets, which had their centers over the Hudson Bay–Labrador region, the northern Cordilleran Range , the Baltic region, and highland centers in Siberia. Much of the boreal forest region lies on the continental shield areas of varied rock type and complex geologic structure. Severe ice erosion exposed hard bedrock over vast areas, creating numerous irregularly shaped rock basins (Figure 21.20). Thin, bouldery till mantles the bedrock in places. Here and there are isolated glaciofluvial deposits—eskers and kames, and sometimes drumlins. Many of the shallower lake basins have been filled by bog materials and converted into muskeg. These materials are classed as Histosols. Discontinuous permafrost lies beneath the surface of most of the colder northern part of the boreal forest region, while sporadic permafrost is found throughout the less cold southern part (see Figure 28.17).

Crop farming in the continental subarctic environment is largely limited to lands surrounding the Baltic Sea, in bordering Finland and Sweden. Cereals grown in this area include barley, oats, rye, and wheat. Along with dairying, these crops primarily provide food for subsistence. The principal nonmineral economic product throughout the subarctic lands of eastern Canada is pulpwood from the needleleaf forests. Logs are carried down the principal rivers to pulp mills and lumber mills (Figure 28.14). Forests of pine and fir in Finland and European Russia are the primary plant resource. The wood products are exported in the form of paper, pulp, cellulose, and construction lumber.

Tundra Environment (K)

Climate and Soils

The tundra environmental region is defined by global areas of tundra climate (12) and is identified with the tundra biome. The tundra climate occupies northern continental fringes of North America and Eurasia from the arctic circle to about the 75th parallel (Figure 28.15). The distinctive feature of the tundra climate is coldness throughout all months of the year. Although there is a brief milder season corresponding to summer of lower latitudes, mean

Figure 28.14 **This sawmill at Hudson, Ontario, processes logs floated across the surface of a sprawling lake in the Canadian Shield. (Ronny Jaques Studio, Toronto, from Black Star.)**

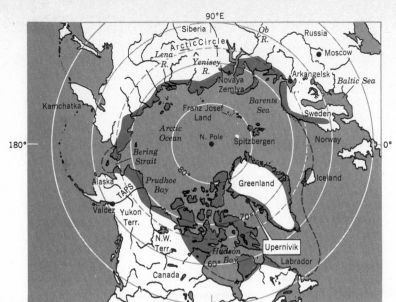

Figure 28.15   **World map of tundra climate.**

monthly temperatures are usually below 7° C (45° F). Winters are very long and extremely cold. Soil water is solidly frozen for eight or more consecutive months. Precipitation is small in all months; but because of poor soil drainage, saturated conditions follow thawing of the surface layer.

Soils of the tundra are poorly developed with respect to horizons and show little or no mineral alteration of parent materials. Much of the tundra region is underlain by Cryaquepts and Cryorthents, great groups respectively of the suborder of Aquents and Orthents. Bog areas are classified as Histosols. Many areas of broken rock have no identifiable soil.

Tundra Ecosystem

Vegetation of the treeless tundra consists of grasses, sedges, and lichens, along with shrubs of willow. Traced southward the vegetation changes into birch-lichen woodland, then into needleleaf forest. In some places a distinct tree line separates the forest and tundra. Coinciding approximately with the 10°C (50°F) isotherm of the warmest month, this line has been used by geographers as a boundary between boreal forest and tundra.

Vegetation is scarce on dry, exposed slopes and summits. The rocky pavement of these areas gives them the name **fell-field,** the Danish term meaning "rock desert" (Figure 28.16).

The number of species in the tundra ecosystem is small, but the abundance of individuals is high. Among the animals, vast herds of caribou in North America or reindeer (their Eurasian relatives) roam the tundra, lightly grazing the lichens and plants and moving constantly. A smaller number of musk-oxen are also primary consumers of the tundra vegetation. Wolves and wolverines, arctic foxes, and polar bears are predators. Among the smaller mammals, snowshoe rabbits and lemmings are important herbivores. Invertebrates are scarce in the tundra, except for a small number of insect species. Blackflies, deerflies, mosquitoes, and no-see-ums (tiny biting midges) are all abundant and can make July on the tundra most uncomfortable for Man and beast. Reptiles and amphibians are also rare. The boggy tundra, however, offers an ideal summer environment for many migratory birds, such as waterfowl, sandpipers, and plovers.

The food chain of the tundra ecosystem is simple and direct. The important producer is "reindeer moss," a lichen (see Figure 16.4). In addition to the caribou and reindeer, lemmings, ptarmigan (arctic grouse), and snowshoe rabbits are important lichen grazers. The important predators are the fox, wolf, lynx, and Man, although all these animals may feed directly on plants as well. During the summer, the abundant insects help support the migratory waterfowl populations. The directness of the tundra food chain makes it particularly vulnerable to fluctuations in the population of a few species.

Figure 28.16   **Vegetation of the arctic fell-field. The plants grow in a small patch of soil surrounded by a ring of cobbles. (American Museum of Natural History.)**

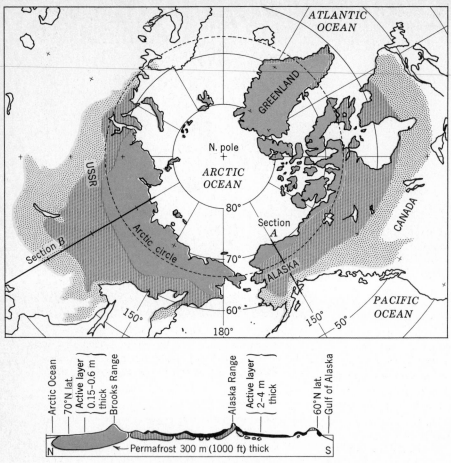

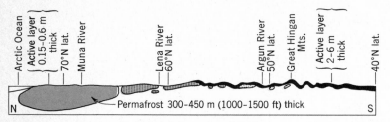

Section A: Alaska, on long. 150° W

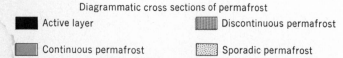

Section B: Asia, on long. 120° E
(Modified from I. V. Poiré)

Diagrammatic cross sections of permafrost

■ Active layer

▨ Discontinuous permafrost

▦ Continuous permafrost

⬚ Sporadic permafrost

Figure 28.17   **Distribution of permafrost in the northern hemisphere and representative cross sections in Alaska and Asia. (From Robert F. Black, "Permafrost," Chapter 14 of P. D. Trask's *Applied Sedimentation,* copyright 1950 by John Wiley & Sons. Reprinted by permission of John Wiley & Sons, Inc.)**

Arctic Permafrost

Perennially frozen ground, or **permafrost,** prevails over the tundra region and a wide bordering area of boreal forest climate. The active layer of seasonal thaw is from 0.6 to 4 m (2 to 14 ft) thick, depending on latitude and the nature of the ground. The distribution of permafrost in the northern hemisphere is shown in Figure 28.17. Three zones are recognized. Continuous permafrost, which extends without gaps or interruptions under all surface features, coincides largely with the tundra climate, but also includes a large part of the boreal forest climate in Siberia. Discontinuous permafrost, which occurs in patches separated by frost-free zones under lakes and rivers, occupies much of the boreal forest climate zone of North America and Eurasia. Sporadic occurrence of permafrost in small patches extends into the southern limits of the boreal forest climate.

Depth of permafrost reaches 300 to 450 m (1000 to 1500 ft) in the continuous zone near lat. 70°N. Much of this permanent frost is an inheritance from more severe conditions of the last ice age, but some permafrost bodies may be growing under existing climate conditions. Various features of the arctic permafrost, including ice wedges, patterned ground, and pingos, were described in Chapter 17 (see Figures 17.6 through 17.9).

Environmental degradation of permafrost regions arises from Man-made surface changes. The undesirable consequences are usually related to the destruction or removal of an insulating surface cover, which may consist of a moss or peat layer in combination with living plants of the tundra or arctic forest. When this layer is scraped off, the summer thaw is extended to a greater depth, with the result that ice wedges and other ice bodies melt in the summer and waste downward. This activity is called **thermal erosion.** Meltwater mixes with silt to form mud, which is then eroded and transported by water streams, with the destructive effects evident in Figure 28.18.

The consequences of disturbance of permafrost terrain became evident in World War II, when military bases, airfields, and highways were constructed hurriedly without regard for maintenance of the natural protective surface insulation. In extreme cases, scraped areas turned into mud-filled depressions and even into small lakes that expanded in area with successive seasons of thaw, engulfing nearby buildings. Engineering practices now call for placing buildings on piles with an insulating air space below or for the deposition of a thick insulating pad of coarse gravel over the surface prior to construction. Stream- and hot-water lines are placed above ground to prevent thaw of the permafrost layer.

Another serious engineering problem of arctic regions is the behavior of streams in winter. As the surfaces of streams and springs freeze over, the water beneath bursts out from place to place, freezing into huge accumulations of ice. Highways are thus made impassable.

The lessons of superimposing Man's technological ways on a highly sensitive natural environment were learned the hard way—by encountering unpleasant and costly effects that were not anticipated. The threat of environmental destruction will persist as oil continues to flow through the Trans-Alaska Pipeline. This facility carries hot oil from the northern shores of Alaska, across a permafrost landscape, to the port of Valdez on the south coast. Effects of this pipeline on permafrost and other elements of the environment were heavily debated and were the subject of intensive investigation. In addition to the prospects of thaw of the permafrost layer by heat of the pipeline, the possibility exists of damage to the ecosystem from spills caused by pipe breakage. The possible effect of the pipeline as a barrier to animal migration remains to be seen.

Figure 28.18 **After one season of thaw, this winter truck trail through the Alaskan arctic forest had suffered severe thermal and water erosion. (R. K. Haugen, U.S. Army Cold Regions Research and Engineering Laboratory.)**

Ice-Sheet Environment (L)

The ice-sheet environment is represented by areas of the ice sheet climate (13) over the Greenland and Antarctic ice sheets and sea ice of the Arctic Ocean. The physical features of ice sheets and floating sea ice were discussed in Chapter 21. Because of low monthly mean temperatures throughout the year over the ice sheets, this environment is devoid of vegetation and soils. The few species of animals found on the ice margins are associated with a marine habitat. In terms of habitation by humans, the ice-sheet environment is extremely hostile because of extreme cold, high winds, and a total lack of food resources. Enormous expenditures of energy are required to import these necessities of life and to provide shelter. These efforts are justified because of the need for scientific research, but in the foreseeable future there is little prospect that this icy environment will provide useful supplies of energy or minerals.

References for Further Study

Baird, Patrick D. (1964), *The polar world,* John Wiley and Sons, New York.
Mackay, R. Ross (1966), "Tundra and taiga," in *Future environments of North America,* F. F. Darling and J. P. Milton, eds., Natural History Press, Garden City, N.Y., pp. 156–171.
Borchert, John R. (1971), "The Dust Bowl in the 1970s," *Annals, Association of American Geographers,* vol. 61, pp. 1–22.
Hare, F. K., and J. C. Ritchie (1972), The boreal bioclimates, *Geographical Review,* vol. 62, pp. 333–365.
Mather, E. Cotton (1972), "The American Great Plains," *Annals, Association of American Geographers,* vol. 62, pp. 237–257.

Review Questions

1. Summarize the important characteristics of climate, soil-water regime, and vegetation of the moist subtropical forest environment. What soil orders and suborders are found in this region? Compare geomorphic processes of this environment with those of the rainforest.

2. What are the principal agricultural products of the moist subtropical region? Compare the southern United States with Southeast Asia in terms of dominant crops and agricultural practices.

3. Summarize the distinctive features of climate, soil-water regime, vegetation, and soils of the Mediterranean environment.

4. Which geomorphic processes are most intense in the Mediterranean environment? Describe the environmental problems caused by mass wasting and floods in southern California. What factors contributed to severe and prolonged soil erosion in the Mediterranean lands? What agricultural products are important in the Mediterranean environment?

5. Summarize the important characteristics of climate, soil-water regime, vegetation, and soils of the marine west-coast forest environment. Which geomorphic processes are most effective in this environment? What kinds of agriculture are practiced? In what ways are forest resources important in this environmental region?

6. Summarize the important characteristics of climate, soil-water regime, vegetation, and soils of the moist continental forest environment. What role did Pleistocene continental glaciation play in shaping the landforms of this region? How is agriculture affected by these conditions of landforms and soils?

7. Summarize the important characteristics of climate, soil-water regime, vegetation, and soils of the prairie and steppe environment. What is unusual about the climatic affiliation of the Pampa grassland region?

8. Which geomorphic processes are dominant in the steppe environment? How are blowout depressions formed and maintained? Comment on the fragile character of areas of fixed sand dunes.

9. Describe the principal agricultural resources of the prairie and steppe environment. Which are the dominant crops? Describe the events leading to the conditions of the Dust Bowl of the 1930s. To what extent was human activity a factor in this phenomenon?

10. Compare the midlatitude desert environment with that of the tropical desert environment. What are the most obvious differences in the two kinds of climates?

11. Summarize the important characteristics of climate, soil-water regime, vegetation, and soils of the boreal forest environment. What role did continental glaciation play in shaping the landscape of the boreal forest region? What plant resources are of primary importance in this region?

12. What is unique about the climate of the tundra environment? What forms of vegetation and soils are associated with the arctic tundra? Describe the tundra ecosystem, including the animal life.

13. Describe the three permafrost zones. How is each zone related to climate type and vegetation? What environmental problems have been encountered as a result of human occupation of permafrost areas?

14. Describe the ice-sheet environment in terms of climatic factors inhospitable to plant and animal life and to occupation by humans.

Appendix I   Climate Definitions and Boundaries

The following table summarizes the definitions and boundaries of climates and climate subtypes based on the soil-water balance, as described in Chapter 10 and shown on the world climate map, Plate D.2. All definitions and boundaries are provisional. (See Chapter 10 for definitions of symbols used.)

GROUP I / LOW-LATITUDE CLIMATES

1. Wet equatorial climate
   Ep ≥ 10 cm in every month, and
   S ≥ 20 cm in 10 or more months.

2. Monsoon and trade-wind littoral climate
   Ep ≥ 4 cm in every month, or
   Ep > 130 cm annual total, or both, and
   S ≥ 20 cm in 6, 7, 8, or 9 consecutive months, or, if
   S > 20 cm in 10 or more months, then Ep ≤ 10
       cm in 5 or more consecutive months.

3. Wet-dry tropical climate
   D ≥ 20 cm, and
   R ≥ 10 cm, and
   Ep ≥ 130 cm annual total, or Ep ≥ 4 cm in every
       month, or both, and
   S ≥ 20 cm in 5 months or fewer, or minimum
       monthly S < 3 cm.

4. Dry tropical climate
   D ≥ 15 cm, and
   R = 0, and
   Ep ≥ 130 cm annual total, or Ep ≥ 4 cm in every
       month, or both.

   Subtypes of dry climates (4, 5, 7, and 9)
   s   Semiarid subtype (Steppe subtype)
       At least 2 months with S ≥ 6 cm.
   sd Semidesert subtype (Steppe-desert transition)
       Fewer than 2 months with S > 6 cm, and at least
       1 month with S > 2 cm.
   d   Desert subtype
       No month with S > 2 cm.

GROUP II / MIDLATITUDE CLIMATES

5. Dry subtropical climate
   D ≥ 15 cm, and
   R = 0, and Ep < 130 cm annual total, and
   Ep ≥ 0.8 cm in every month, and
   Ep < 4 cm in 1 month.
   (Subtypes 5s, 5sd, 5d, defined under 4.)

6. Moist subtropical climate
   D < 15 cm when R = 0, and
   Ep < 4 cm in at least 1 month, and
   Ep ≥ 0.8 cm in every month.

Subtypes of moist climates (6, 8, 10, 11, 12)
   sh Subhumid subtype
       15 cm > D > 0 when R = 0, or
       D > R when R > 0.
   h Humid subtype
       R ≥ 0.1 cm, and
       R > D, and R < 60 cm.
   p Perhumid subtype R > 60 cm.

7. Mediterranean climate
   D ≥ 15 cm, and
   R ≥ 0, and
   Ep ≥ 0.8 cm in every month, and storage index
       > 75%, or P/Ea × 100 < 40%.

   Subtypes 7s, 7sd, defined under 4, and 7h Humid
       subtype R > 15 cm.

8. Marine west-coast climate
   D < 15 cm, and
   Ep < 80 cm annual total, and
   Ep ≥ 0.8 cm in every month.
   (Subtypes 8sh, 8h, 8p, defined under 6.)

9. Dry midlatitude climate
   D ≥ 15 cm, and
   R = 0, and
   Ep ≤ 0.7 cm in at least 1 month, and
   Ep > 52.5 cm annual total.
   (Subtypes 9s, 9sd, 9d, defined under 4.)

10. Moist continental climate
    D < 15 cm when R = 0, and
    Ep ≤ 0.7 cm in at least 1 month, and
    Ep > 52.5 cm annual total.
    (Subtypes 10sh, 10h, 10p, defined under 6.)

GROUP III / HIGH-LATITUDE CLIMATES

11. Boreal forest climate
    52.5 cm > Ep > 35 cm annual total, and
    Ep = 0 in fewer than 8 consecutive months.

    Subtypes
       11s Dry subtype D > 15 cm.
       (Moist subtypes 11sh, 11h, defined under 6.)

12. Tundra climate
    Ep < 35 cm annual total, and
    Ep = 0 in 8 or more consecutive months.

    Subtypes
       11s Dry subtype D > 15 cm.
       (Moist subtypes 12sh, 12h, defined under 6.)

13. Ice-sheet climate
    Ep = 0 in all months.

Appendix II Great Groups of the Soil Taxonomy

Suborders of the Comprehensive Soil Classification System are divided into **great groups,** of which about 185 are currently known to occur in the United States. The great groups are identified by examining all the soil horizons and their nature collectively, as well as the moisture and temperature regimes. Because moisture and temperature regimes influence soils, they can be considered properties of the whole soil rather than of specific horizons. At the great-group level, therefore, we try to consider the whole soil, the assemblage of horizons, and the most significant properties of the whole soil, selected on the basis of the numbers and importance of accessory properties.

Great groups are differentiated according to the following criteria: (1) close similarities in kind, arrangement, and degree of expression of horizons; (2) close similarities in soil moisture and temperature regimes; and (3) similarities in base status.

Names of great groups are given in Table A.II.1. Names of great groups make use of the formative elements given in Table A.II.2. Properties of each great group are indicated by formative elements. For example, the *Calciborolls* are Mollisols of the suborder Borolls, formed under a cold regime and possessing a calcic horizon.

A detailed description of each great group will be found in the following treatise: *Soil Taxonomy, A Basic System of Soil Classification for Making and Interpreting Soil Surveys,* Soil Survey Staff, Agriculture Handbook No. 436, U.S. Department of Agriculture, Government Printing Office, Washington, D.C., 1975.

Table A.II.1 / Names of Great Groups

Order	Suborder	Great group	Order	Suborder	Great group
Alfisols	Aqualfs	Albaqualfs.			Natrudalfs.
		Duraqualfs.			Paleudalfs.
		Fragiaqualfs.			Rhodudalfs.
		Glossaqualfs.			Tropudalfs.
		Natraqualfs.		Ustalfs	Durustalfs.
		Ochraqualfs.			Haplustalfs.
		Plinthaqualfs.			Natrustalfs.
		Tropaqualfs.			Paleustalfs.
		Umbraqualfs.			Plinthustalfs.
	Boralfs	Cryoboralfs.			Rhodustalfs.
		Eutroboralfs.		Xeralfs	Durixeralfs.
		Fragiboralfs.			Haploxeralfs.
		Glossoboralfs.			Natrixeralfs.
		Natriboralfs.			Palexeralfs.
		Paleboralfs.			Plinthoxeralfs.
	Udalfs	Agrudalfs.			Rhodoxeralfs.
		Ferrudalfs.	Aridisols	Argids	Durargids.
		Fragiudalfs.			Haplargids.
		Fraglossudalfs.			Nadurargids.
		Glossudalfs.			Natrargids.
		Hapludalfs.			Paleargids.

Order	Suborder	Great group	Order	Suborder	Great group
	Orthids	Calciorthids.		Aquepts	Andaquepts.
		Camborthids.			Cryaquepts.
		Durorthids.			Fragiaquepts.
		Gypsiorthids.			Halaquepts.
		Paleorthids.			Haplaquepts.
		Salorthids.			Humaquepts.
					Placaquepts.
Entisols	Aquents	Cryaquents.			Plinthaquepts.
		Fluvaquents.			Sulfaquepts.
		Haplaquents.			Tropaquepts.
		Hydraquents.		Ochrepts	Cryochrepts.
		Psammaquents.			Durochrepts.
		Sulfaquents.			Dystrochrepts.
		Tropaquents.			Eutrochrepts.
	Arents	Arents.			Fragiochrepts.
	Fluvents	Cryofluvents.			Ustochrepts.
		Torrifluvents.			Xerochrepts.
		Tropofluvents.		Plaggepts	Plaggepts.
		Udifluvents.		Tropepts	Dystropepts.
		Ustifluvents.			Eutropepts.
		Xerofluvents.			Humitropepts.
	Orthents	Cryorthents.			Sombritropepts.
		Torriorthents.			Ustropepts.
		Troporthents.		Umbrepts	Cryumbrepts.
		Udorthents.			Fragiumbrepts.
		Ustorthents.			Haplumbrepts.
		Xerorthents.			Xerumbrepts.
	Psamments	Cryopsamments.			
		Quartzipsamments.	Mollisols	Albolls	Argialbolls.
		Torripsamments.			Natralbolls.
		Tropopsamments.		Aquolls	Argiaquolls.
		Udipsamments.			Calciaquolls.
		Ustipsamments.			Cryaquolls.
		Xeropsamments.			Duraquolls.
Histosols	Fibrists	Borofibrists.			Haplaquolls.
		Cryofibrists.			Natraquolls.
		Luvifibrists.		Borolls	Argiborolls.
		Medifibrists.			Calciborolls.
		Sphagnofibrists.			Cryoborolls.
		Tropofibrists.			Haploborolls.
	Folists	Borofolists.			Natriborolls.
		Cryofolists.			Paleborolls.
		Tropofolists.			Vermiborolls.
	Hemists	Borohemists.		Rendolls	Rendolls.
		Cryohemists.		Udolls	Argiudolls.
		Luvihemists.			Hapludolls.
		Medihemists.			Paleudolls.
		Sulfihemists.			Vermudolls.
		Sulfohemists.		Ustolls	Argiustolls.
		Tropohemists.			Calciustolls.
	Saprists	Borosaprists.			Durustolls.
		Cryosaprists.			Haplustolls.
		Medisaprists.			Natrustolls.
		Troposaprists.			Paleustolls.
Inceptisols	Andepts	Cryandepts.			Vermustolls.
		Durandepts.		Xerolls	Argixerolls.
		Dystrandepts.			Calcixerolls.
		Eutrandepts.			Durixerolls.
		Hydrandepts.			Haploxerolls.
		Placandepts.			Natrixerolls.
		Vitrandepts.			Palexerolls.

Order	Suborder	Great group	Order	Suborder	Great group
Oxisols	Aquox	Gibbsiaquox.			Haplorthods.
		Ochraquox.			Placorthods.
		Plinthaquox.			Troporthods.
		Umbraquox.	Ultisols	Aquults	Albaquults.
	Humox	Acrohumox.			Fragiaquults.
		Gibbsihumox.			Ochraquults.
		Haplohumox.			Paleaquults.
		Sombrihumox.			Plinthaquults.
	Orthox	Acrorthox.			Tropaquults.
		Eutrorthox.			Umbraquults.
		Gibbsiorthox.		Humults	Haplohumults.
		Haplorthox.			Palehumults.
		Sombriorthox.			Plinthohumults.
		Umbriorthox.			Sombrihumults.
	Torrox	Torrox.			Tropohumults.
	Ustox	Acrustox.		Udults	Fragiudults.
		Eutrustox.			Hapludults.
		Sombriustox.			Paleudults.
		Haplustox.			Plinthudults.
Spodosols	Aquods	Cryaquods.			Rhodudults.
		Duraquods.			Tropudults.
		Fragiaquods.		Ustults	Haplustults.
		Haplaquods.			Paleustults.
		Placaquods.			Plinthustults.
		Sideraquods.			Rhodustults.
		Tropaquods.		Xerults	Haploxerults.
	Ferrods	Ferrods.			Palexerults.
	Humods	Cryohumods.	Vertisols	Torrerts	Torrerts.
		Fragihumods.		Uderts	Chromuderts.
		Haplohumods.			Pelluderts.
		Placohumods.		Usterts	Chromusterts.
		Tropohumods.			Pellusterts.
	Orthods	Cryorthods.		Xererts	Chromoxererts.
		Fragiorthods.			Pelloxererts.

Data source: Soil Conservation Service, U.S. Department of Agriculture.

Formative Element	Derivation	Mnemonicon	Connotation
Acr	Modified from Gr. *akros*, at the end	Acrolith	Extreme weathering.
Agr	L. *ager*, field	Agriculture	An agric horizon.
Alb	L. *albus*, white	Albino	An albic horizon.
And	Modified from ando	Ando	Andolike.
Arg	Modified from argillic horizon; L. *argilla*, white clay.	Argillite	An argillic horizon.
Bor	Gr. *boreas*, northern	Boreal	Cool.
Calc	L. *calcis*, lime	Calcium	A calcic horizon.
Camb	L.L. *cambiare*, to exchange	Change	A cambic horizon.
Chrom	Gr. *chroma*, color	Chroma	High chroma.
Cry	Gr. *kryos*, icy cold	Crystal	Cold.
Dur	L. *durus*, hard	Durable	A duripan.
Dystr, dys	Modified from Gr. *dys*, ill; dystrophic, infertile.	Dystrophic	Low base saturation.
Eutr, eu	Modified from Gr. *eu*, good; eutrophic, fertile.	Eutrophic	High base saturation.
Ferr	L. *ferrum*, iron	Ferric	Presence of iron.
Fluv	L. *fluvus*, river	Fluvial	Floodplain.
Frag	Modified from L. *fragilis*, brittle	Fragile	Presence of fragipan.
Fragloss	Compound of fra(g) and gloss		See the formative elements frag and gloss.
Gibbs	Modified from gibbsite	Gibbsite	Presence of gibbsite in sheets or nodules.
Gyps	L. *gypsum*, gypsum	Gypsum	Presence of a gypsic horizon.
Gloss	Gr. *glossa*, tongue	Glossary	Tongued.
Hal	Gr. *hals*, salt	Halophyte	Salty.
Hapl	Gr. *haplous*, simple	Haploid	Minimum horizon.
Hum	L. *humus*, earth	Humus	Presence of humus.
Hydr	Gr. *hydor*, water	Hydrophobia	Presence of water.
Luv	Gr. *louo*, to wash	Ablution	Illuvial.
Med	L. *media*, middle	Medium	Of midlatitude climates.
Nadur	Compound of na(tr) and dur		See the formative elements natr and dur.
Natr	Modified from *natrium*, sodium		Presence of natric horizon.
Ochr	Gr. base of *ochros*, pale	Ocher	Presence of ochric epipedon.
Pale	Gr. *paleos*, old	Paleosol	Excessive development.
Pell	Gr. *pellos*, dusky		Low chroma.
Plac	Gr. base of *plax*, flat stone		Presence of a thin pan.
Plagg	Modified from Ger. *plaggen*, sod		Presence of plaggen epipedon.
Plinth	Gr. *plinthos*, brick		Presence of plinthite.
Psamm	Gr. *psammos*, sand	Psammite	Sand texture.
Quartz	Ger. *quarz*, quartz	Quartz	High quartz content.
Rhod	Gr. base of *rhodon*, rose	Rhododendron	Dark red color.
Sal	L. base of *sal*, salt	Saline	Presence of salic horizon.
Sider	Gr. *sideros*, iron	Siderite	Presence of free iron oxides.
Sombr	F. *sombre*, dark	Somber	A dark horizon.
Sphagn	Gr. *sphagnos*, bog	Sphagnum	Presence of sphagnum.
Sulf	L. *sulfur*, sulfur	Sulfur	Presence of sulfides or their oxidation products.
Torr	L. *torridus*, hot and dry	Torrid	Torric moisture regime.
Trop	Modified from Gr. *tropikos*, of the solstice	Tropical	Humid and continually warm.
Ud	L. *udus*, humid	Udometer	Udic moisture regime.
Umbr	L. base of *umbra*, shade	Umbrella	Presence of umbric epipedon.
Ust	L. base of *ustus*, burnt	Combustion	Ustic moisture regime.
Verm	L. base of *vermes*, worm	Vermiform	Wormy, or mixed by animals.
Vitr	L. *vitrum*, glass	Vitreous	Presence of glass.
Xer	Gr. *xeros*, dry	Xerophyte	A xeric moisture regime.

Data source: Soil Conservation Service, U.S. Department of Agriculture.

Appendix III Map Reading

Topographic Maps

Several methods are used to show accurately the configuration of the land surface on topographic maps: plastic shading, altitude tints, hachures, and contours (see Figure A.III.1). The first three methods give a strong visual effect of three dimensions so that most persons can easily grasp the essential character of the landscape features. These methods of showing relief are, however, inadequate because they do not tell the reader the elevation above sea level of all points on the map or how steep the slopes are. Topographic contours give this information and make the most useful type of topographic map.

Plastic Shading

Maps using **plastic shading** to show relief look very much like photographs taken looking down on a plastic relief model of the land surface (Figure A.III.1). The effect of relief is produced by gray or brown tones applied according to the oblique illumination method. Light rays are imagined as coming from a point in the northwestern sky somewhere intermediate between the horizon and zenith. Thus all slopes facing southwest receive the heaviest shades and are darkest where the slopes are steepest (Figure A.III.2). Maps with plastic shading look much like photographs (Figure A.III.3).

Figure A.III.1 **Plastic shading combined with contours greatly enhances the visual effect of relief. (Portion of U.S. Geological Survey, Kitzmiller, Md. W. Va., topographic quadrangle, scale 1:24,000.)**

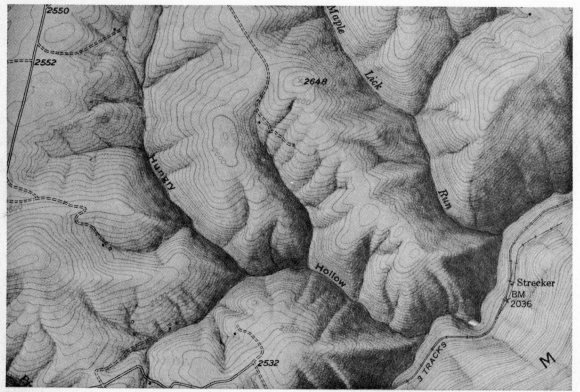

Altitude Tints

In its simplest form, **altitude tinting** consists of assigning a certain color, or a certain depth of tone of a color, to all areas on the map lying within a specified range of elevation. Maps in atlases and wall maps commonly show low areas in green, intermediate ranges of elevations in successive shades of buff and light brown, and high mountain elevations in darker shades of brown, red, or violet. Each shade of color is assigned a precise elevation range. The method is effective for small-scale maps that are viewed at some distance.

Hachures

Hachures are very fine, short lines arranged side by side into roughly parallel rows. Each hachure line lies along the direction of the steepest slope and represents the direction that would be taken by water flowing down the surface.

A precise system of hachures, adapted to representation of detailed topographic features on accurate, large-scale maps, was invented by J. G. Lehmann and was widely used on military maps of European countries throughout the nineteenth century. In the Lehmann system, steepness of slope is indicated by thickness of the hachure line (Figure A.III.4).

Because hachures do not tell the elevation of surface

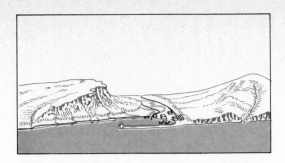

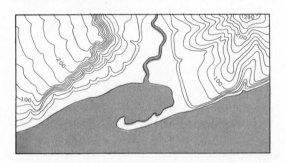
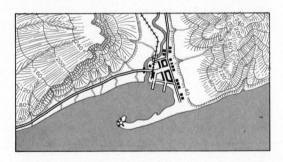
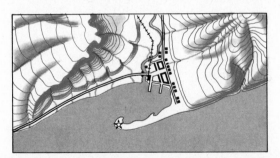

Figure A.III.2 **Various ways in which relief can be shown are, from top to bottom: (1) perspective diagram or terrain sketch, (2) hachures, (3) contours, (4) hachures and contours combined, and (5) plastic shading and contours combined. (Drawn by E. Raisz.)**

Figure A.III.3 **A vertical air photograph (above) is a kind of topographic map, showing all relief details but lacking elevation information. The contour topographic map (below) covers the same area and shows a small side canyon of the Grand Canyon. North is toward the bottom of these maps to give the proper effect of relief in the photograph. (U.S. Forest Service and U.S. Geological Survey.)**

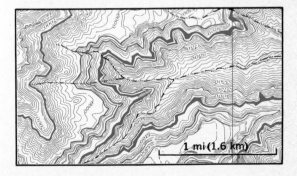

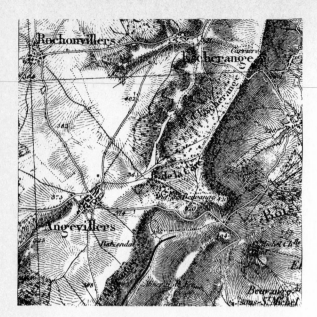

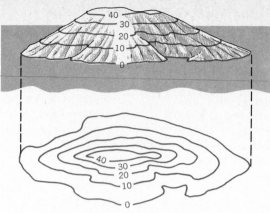

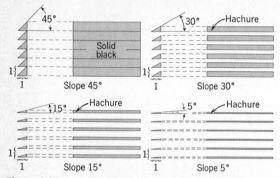

Figure A.III.4 A portion of the Metz sheet is shown above to correct scale. This map is one of the French 1:80,000 topographic series using black hachures and spot heights. The Lehmann system of hachuring, shown below, varies the thickness of hachure line according to ground slope.

Figure A.III.5 **Contours on a small island.**

contour line because it is a line connecting all points having zero elevation. Suppose that sea level could be made to rise exactly 10 meters (or that the island could be made to sink exactly 10 meters); the water would come to rest along the line labeled "10." This new water line would be the 10-meter contour because it connects all points on the island that are exactly 10 meters higher than the original shoreline. By successive rises in water level, each exactly 10 meters more than the last, the positions of the remaining contours would be fixed.

Contour Interval and Slope

Contour interval is the vertical distance separating successive contours. The interval remains constant over the entire map, except in special cases where two or more intervals are used on the map sheet.

Because the vertical contour interval is fixed, horizontal spacing of contours on a map varies with changes in land slope. The rule is that close crowding of contour lines represents a steep slope; wide spacing represents a gentle slope. Figure A.III.6 shows a small island, one side of which is a steep, clifflike slope. From the summit point *B* to the cliff base at *A*, the contours are crossed within a

points, it is necessary to print numerous elevation figures on hilltops, road intersections, towns, and other strategic locations. These numbers are known as **spot heights**. Without them a hachure map would be of little practical value.

Contour Lines

A **contour line** is an imaginary line on the ground, every point of which is at the same elevation above sea level. Contour lines on a map are simply the graphic representations of ground contours, drawn for each of a series of specified elevations, such as 10, 20, 30, 40, or 50 meters or feet above sea level or any other chosen base, known as a **datum plane**. The resulting line pattern not only gives a visual impression of topography to the experienced map reader, but also supplies accurate information about elevations and slopes.

To clarify the contour principle, imagine a small island, shown in Figure A.III.5. The shoreline is a natural

Figure A.III.6 **On the steep side of this island, the contours appear more closely spaced.**

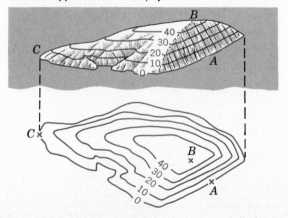

short horizontal distance and therefore appear closely spaced on the map. To travel from *B* to the shore at *C* requires the same total vertical descent; but, because the slope is gentle, the horizontal distance traveled is much greater and the contours between *B* and *C* are widely spaced on the map.

Selection of a contour interval depends both on relief of the land and on scale of the map. Topographic maps showing regions of strong relief require a large interval, such as 10, 25, or 50 m (50, 100, or 200 ft); regions of moderate relief require such intervals as 2 or 5 m (10 or 25 ft).

Because much of the earth's land surface is sculptured by streams flowing in valleys, special note should be made of how contours change direction when crossing a stream valley. Figure A.III.7 is a small contour sketch map illustrating some stream valleys. Notice that each contour is bent into a V whose apex lies on the stream and points in an upstream direction.

Determining Elevations by Means of Contours

Figure A.III.7 can be used to illustrate the determination of elevations. Point *B* is easy to determine because it lies exactly on the 1300-m contour. Point *C* requires interpolation. Because it lies midway between the 1100- and 1200-m lines, its elevation is estimated as the midvalue of the vertical interval, or 1150 m. Point *D* lies about one-fifth of the distance from the 1000-m to the 1100-m contour. One-fifth of the contour interval is 20 m. Thus we estimate that point *D* has an approximate elevation of 1020 m. If the ground is not highly irregular, the error of estimate will probably be small. Determination of the summit elevation, point *A*, involves more uncertainty. It is certain that the summit point is more than 1700 m and less than 1800 m because the 1700-m contour is the highest one shown. Because a fairly large area is included within the 1,700-m contour, we may suppose that the summit rises appreciably higher than 1700 m. A guess would

place the actual elevation at about 1750 m.

On many topographic maps the elevations of hilltops, road intersections, bridges, streams, and lakes are printed on the map to the nearest meter or foot. These spot heights do away with the need for estimating elevations at key points.

Government agencies, such as the U.S. Geological Survey, determine the precise elevation and position of convenient reference points. These points are known as **bench marks.** On the map they are designated by the letters B.M., together with the elevation stated to the nearest meter or foot.

Depression Contours

A special type of contour is used where the land surface has basinlike hollows, or **closed depressions,** which would make small lakes if they could be filled with water. This contour line is the **hachured contour,** or **depression contour.** Figure A.III.8 is a sketch of a depression in a gently undulating plain. Below the sketch is the corresponding contour map. Hachured contours have the same elevations and contour intervals as regular contours on the same map.

Map Scale

Distance between points shown on a map depends on the scale of the map: the ratio between map distances and the actual ground distances that the map represents (Chapter 1). The fractional scale (representative fraction, or R.F.) is converted to conventional units of length. For example, a map scale of 1:100,000 can be read as "one centimeter represents one kilometer." An ordinary centimeter scale can thus be used to measure map distances.

Most maps of small areas carry a **graphic scale** printed on the map margin. This device is a length of line divided off into numbered segments (Figure A.III.9). The units are in conventional terms of measurement, such as meters and

Figure A.III.7 **Stream valleys produce V-shaped indentations of the contours. Elevations are given in meters.**

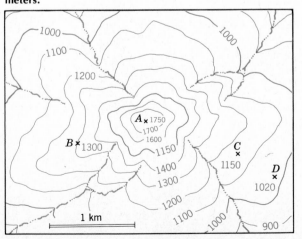

Figure A.III.8 **Contours that close in a circular pattern show either closed depressions or hills.**

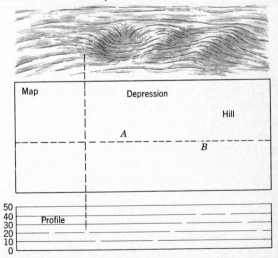

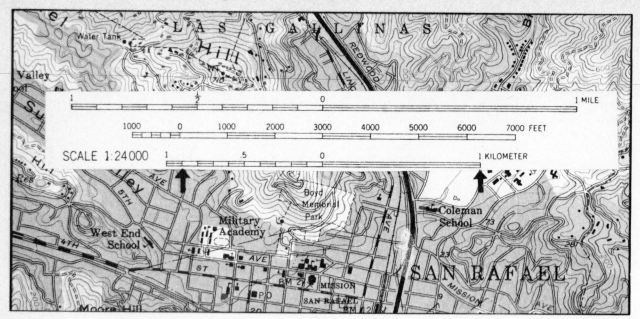

Figure A.III.9 **The distance between two points on a map is read directly from a graphic scale.**

kilometers or feet, yards, and miles. To use the graphic scale, hold the edge of a piece of paper along the line to be measured on the map and mark the distance on the edge of the paper. Then place the paper along the graphic scale and read the length of the line directly.

Topographic Profiles

To visualize the relief of a land surface, **topographic profiles** can be drawn. These are lines that show the rise and fall of the surface along a selected line crossing the map. Figure A.III.10 shows how to construct a profile. Draw a line, *XY*, across the map at the desired location. Place a piece of paper, ruled with horizontal lines, so that its top edge lies along the line *XY*.

Each horizontal line represents a contour level and is so numbered along the left-hand side. Starting at the left, drop a perpendicular from the point *a* where the map contour intersects the profile line, *XY*, down to the corresponding horizontal level. Mark the point *a'* on the horizontal line. Next, repeat the procedure for the 400-m contour at point *b,* and so on, until all points have been plotted. Then draw a smooth line through all the points, completing the profile. Where contours are widely spaced, some judgment is required in drawing the profile.

Figure A.III.10 shows two profiles, both of which are drawn along the same line *XY.* The difference is one degree of exaggeration of the vertical scale. In this illustration, horizontal map scale is 1 cm to 1000 m, or 1:100,000, whereas the vertical scale of the upper profile is 1 cm to 100 m, or 1:10,000. The vertical scale is thus ten times as large as the horizontal map scale, and the profile is said to have a **vertical exaggeration** of ten times. In the lower profile the horizontal scale remains the same, of course, but the vertical scale is 1 cm to 200 m, or 1:20,000. The

vertical exaggeration is therefore five times. Some degree of vertical exaggeration is usually needed to bring out the details of the topography.

Large-Scale and Small-Scale Maps

The relative size of two different scales is determined according to which fraction (R.F.) is the larger quantity. For example, a scale of 1:10,000 is twice as large as a scale of

Figure A.III.10 **Construction of a topographic profile.**

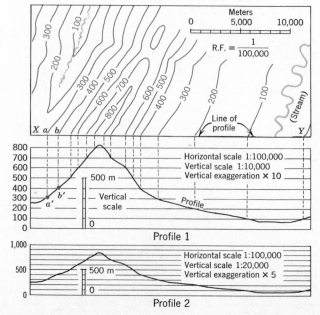

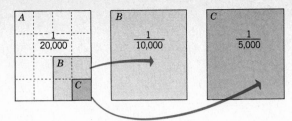

Figure A.III.11 **Map area decreases as scale increases.**

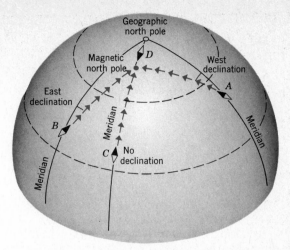

Figure A.III.12 **Whether declination is east or west depends on the observer's global position with respect to the magnetic and geographic north poles. (From A. N. Strahler, *The Earth Sciences*, Harper and Row, N.Y.)**

1:20,000. Maps with scales ranging from 1:600,000 down to 1:100,000,000 or smaller are known as small-scale maps; those of scale 1:600,000 to 1:75,000 are medium-scale maps; and those of scale greater than 1:75,000 are large-scale maps.

For representing details of the earth's surface, large-scale maps are needed and the area of land surface shown by an individual map sheet must necessarily be small. A topographic sheet measuring 40 cm by 50 cm, on a scale of 1:100,000 (1 cm = 1 km), would of course include an area 40 km by 50 km, or 2000 sq km. Of the common sets of topographic maps published by national governments for general distribution, most fall within the scale range of 1:20,000 to 1:250,000.

Relationship Between Scales and Areas

Assuming that two maps, each on a different scale, have the same dimensions, what is the relationship between the ground areas shown? Figure A.III.11 shows three maps, each having the same dimensions but representing scales of 1:20,000, 1:10,000, and 1:5,000, respectively, from left to right. Although map *B* is on twice the scale of map *A*, it shows a ground area only one-fourth as great. Map *C* is on four times as large a scale as map *A*, yet it covers a ground area only one-sixteenth as much. Thus the ground area that is represented by a map of given dimensions varies inversely with the square of the change in scale. For example, if the scale is reduced to one-third its original value, the area that can be shown on a map of fixed dimensions increases to nine times the original value.

Map Orientation and Declination of the Compass

By convention, large-scale topographic maps are oriented so that north is in a direction toward the top of the map and south is toward the bottom of the map.

The geographic north pole, to which all meridians converge, forms a reference point for **true north, or geographic north.** There is, however, another point on the earth, the **magnetic north pole,** to which magnetic compasses point (Figure A.III.12). The magnetic north pole is located in the Northwest Territories of Canada on Prince of Wales Island, at about lat. 73°N, long. 100°W. Most large-scale maps have printed on the margin two arrows stemming from a common point. One arrow designates true north, the other **magnetic north.** The angular distance

between the two directions is known as the **magnetic declination.**

Magnetic declination varies greatly in different parts of the world, depending principally on one's position relative to the geographic and magnetic poles. Lines on a map drawn through all places having the same compass declination are known as **isogonic lines** (Figure A.III.13). The line of zero declination (agonic line) runs through eastern North America. Anywhere along this line, the compass points to true geographic north and no adjustment is required.

Magnetic declination changes appreciably with the passage of years. The amount of annual change in declination is usually stated on the margin of the map.

Bearings and Azimuths

When a map is used in the field, it is often necessary to state the direction followed by a road or stream or to describe the direction that can be taken to locate a particular object with respect to some known reference point. For this purpose, the observer determines the horizontal angle between the line to the objective and a north-south line. The common unit of angular measurement is the degree, 360 of which comprise a complete circle. Other systems of angular measurement, such as the mil (of which there are 6400 in a complete circle), are sometimes preferable for special applications.

Two systems are used to state direction with respect to north. **Compass quadrant bearings** are angles measured eastward or westward of either north or south, whichever happens to be the closer. Examples are shown in Figure A.III.14A. The direction from a given point to some object on the map is written as "N49°E" or "S70°W." All bearings range between 0 and 90°. Compass bearings may be magnetic bearings, related to magnetic north, or true bearings, related to geographic north. Unless otherwise stated, a bearing should be assumed to be a true bearing.

Azimuths are used by military services and in air and

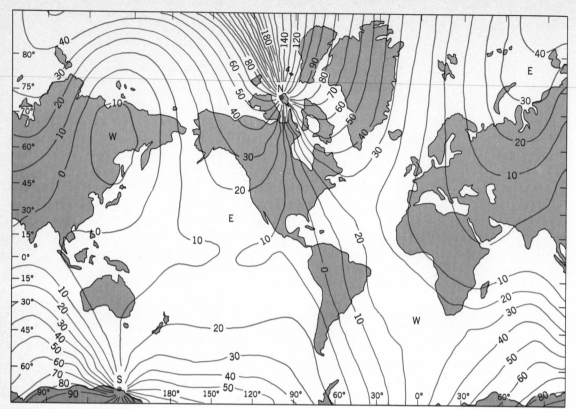

Figure A.III.13 **On this world isogonic map, declination is given in intervals of 10 degrees. (Data of U.S. Naval Oceanographic Office.)**

marine navigation generally. As shown in Figure A.III.14B, all azimuths are read in a clockwise direction from north and range between 0° and 360°. Azimuths are measured from either magnetic north or true north, referred to as **magnetic azimuth** and **true azimuth,** respectively.

Figure A.III.14 **Directions are expressed as bearings or azimuths.**

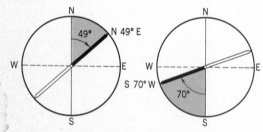

A. Compass quadrant bearings

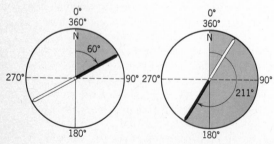

B. Azimuths

Topographic Quadrangles

Any system used to locate points on the earth's surface with reference to a fixed set of intersecting lines is a **coordinate system.** The system of geographic coordinates using parallels and meridians is described in Chapter 1. Two other coordinate systems are the military grid and the township grid of the U.S. Land Office Survey.

Most published sets of large-scale topographic maps use the geographic grid to determine the position and size of individual map sheets in a series. A single map sheet, or **quadrangle,** is bounded on the right- and left-hand margins by meridians, and on the top and bottom by parallels, which are a specified number of minutes or degrees apart.

Seven standard scales comprise the U.S. National Topographic Map Series. English units of length continue to be used, but a change to metric units has begun.

Series	R.F.	Unit Equivalents
7.5-minute	1:24,000	1 inch to 2000 feet
7.5-minute	1:31,680	1 inch to ½ mile
15-minute	1:62,500	1 inch to about 1 mile
Alaska	1:63,360	1 inch to 1 mile
30-minute	1:125,000	1 inch to about 2 miles
1:250,000	1:250,000	1 inch to about 4 miles
1:1,000,000	1:1,000,000	1 inch to about 16 miles (1 cm to 10 km)

Figures A.III.15 and A.III.16 compare the coverages and sizes of standard quadrangles on several of these series.

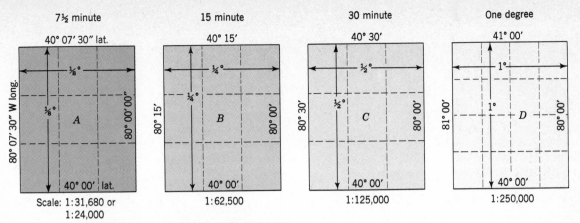

Figure A.III.15 **Large-scale maps of the U.S. Geological Survey are bounded by parallels and meridians to form quadrangles. The four quadrangles shown here represent the scales and areas commonly used in the United States, exclusive of Alaska.**

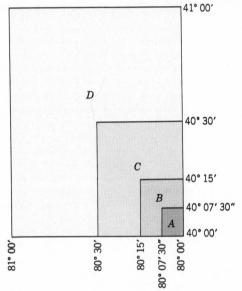

Figure A.III.16 **If the four quadrangles of Figure A.III.15 were reduced to the same scale, their areas would compare as shown here.**

Topographic Map Symbols

Large-scale topographic maps of the U.S. Geological Survey use a standard set of symbols to show many kinds of features that cannot be represented to true scale. Reproduced on the front end paper of this book is the set of symbols used by the U.S. Geological Survey on its recent series of large-scale topographic maps, together with a representative example of a map on the scale of 1:24,000.

In general, it is conventional to show relief features in brown, hydrographic (water) features in blue, vegetation in green, and cultural (Man-made) features in black or red.

The U.S. Land Office Survey

Topographic maps of the central and western United States show the civil divisions of land according to the U.S. Land Office Survey.

In 1785 Congress authorized a survey of the territory lying north and west of the Ohio River. To avoid the irregular and unsystematic type of land subdivision that has grown up in seaboard states during colonial times, Congress specified that the new lands should be divided into six-mile squares, now called **congressional townships,** and that the grid of townships should be based on a carefully surveyed east-west base line, designated the "geographer's line." Meridians and parallels laid off at six-mile intervals from the base line were to form the boundaries of the townships. This general plan, believed to have been proposed by Thomas Jefferson, was subsequently carried out to cover the balance of the central and western states.

The **principal meridians** and **base lines,** from which rows of townships were laid off, are shown in Figure A.III.17. Principal meridians run north or south, or both, from selected points whose latitude and longitude were originally calculated by astronomical means. Thirty-two principal meridians have been surveyed. Westward from the Ohio-Pennsylvania boundary, these are numbered from 1 through 6, beyond which they are designated by names.

Through the initial point selected for starting the principal meridian, an east-west base line was run, corresponding to a parallel of latitude through that point. North and south from the base line, horizontal tiers of townships were laid off and numbered accordingly. Vertical rows of townships, called **ranges,** were laid off to the right and left of the principal meridians and were numbered accordingly (Figure A.III.18).

The area governed by one principal meridian and its base line is restricted to a particular section of country, usually about as large as one or two states. Where two systems of townships meet, they do not correspond because each system was built up independently of the others.

Because the range lines, on eastern and western boundaries of townships, are meridians converging

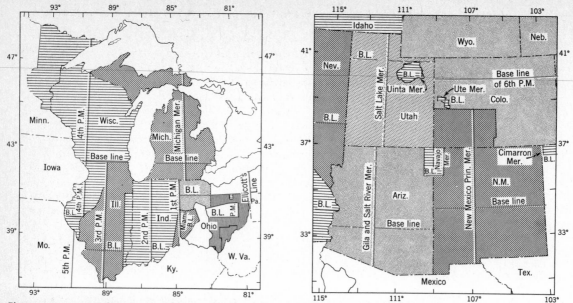

Figure A.III.17 Baselines and principal meridians of these two portions of the United States are representative of the system used by the U.S. Land Office. (After U.S. Department of Interior, General Land Office Map of the United States, 1937.)

slightly as they are extended northward, the width of townships is progressively diminished in a northward direction. To avoid a considerable reduction in township widths in the more northerly tiers, new base lines, known as **standard parallels,** have been surveyed for every four tiers of townships. They are designated 1st, 2nd, 3rd standard parallel north, and so on (Figure A.III.18). The ranges will be found to offset at the standard parallels; consequently, roads that follow range lines make an offset or jog when crossing standard parallels.

Subdivisions of the township are square-mile **sections,** of which there are 36 to the township. These are usually numbered as illustrated in Figure A.III.19. Each section

may be subdivided into halves, quarters, and half-quarters, or even smaller units. These divisions, together with the number of acres contained in each, are illustrated in Figure A.III.20.

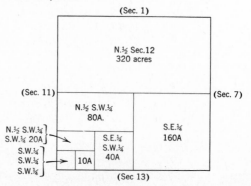

Figure A.III.19 A township is divided into 36 sections, each one a square mile.

Figure A.III.18 Designation of townships and ranges.

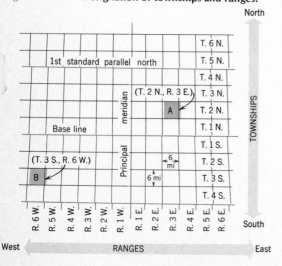

Figure A.III.20 A section may be subdivided into many units

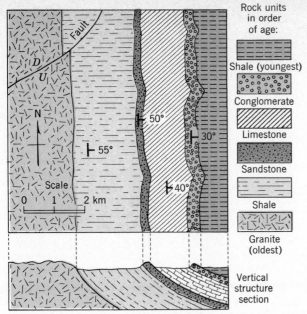

Rock units in order of age:

Shale (youngest)

Conglomerate

Limestone

Sandstone

Shale

Granite (oldest)

Vertical structure section

Figure A.III.21   **A geologic map shows the surface distribution of rocks and structures. The structure section shows rocks at depth.**

Geologic maps and structure sections

The **areal geologic map** shows, by means of colors or patterns, the surface distribution of each rock unit, with special emphasis on the lines of contact of rocks unlike in variety or age. Faults are shown as lines. Strikes and dips of strata are added by special symbols.

Figure A.III.21 is a simple geologic map of the same area shown by a perspective diagram in Figure 26.1. When map reproduction is limited to black and white, patterns are applied to differentiate the rock units. Shorthand letter combinations may be added as a code to set apart formations of different ages. Small T-shaped symbols, seen on the map, tell strike and dip. The long bar gives direction of strike; the short bar that abuts it at right angles shows direction of dip. Amount of dip in degrees is given by a figure beside the symbol. A small fault, cutting across the northwest corner of the map, is shown by a solid line. The letters *D* and *U* indicate which side moved downward, which upward.

To show the internal geologic structure of an area, a **structure section** is used. It is an imaginary vertical slice through the rocks. A structure section is illustrated in the lower part of Figure A.III.21. The upper line of the section is a topographic profile.

Glossary

This Glossary contains definitions of all terms printed in boldface type in the text. Italicized terms will be found as individual entries elsewhere in the Glossary.

ablation Wastage of glacial ice by both *melting* and *evaporation*.

abrasion Erosion of *bedrock* of a *stream channel* by impact of particles carried in a *stream* and by rolling of larger *rock* fragments over the stream bed. Abrasion is also an activity of glacial ice, waves, and wind.

abrasion platform Sloping, nearly flat *bedrock* surface extending out from foot of *marine cliff* under shallow water of breaker zone.

absolute temperature Temperature measured by the *Kelvin scale* on which absolute zero is equivalent to $-273°C$.

absorption of radiation Transfer of energy of *electromagnetic radiation* into heat energy within a *gas* or *liquid* through which the radiation is passing.

abyssal plain Large expanse of very smooth, flat ocean floor found at depths of 4600 to 5500 m (15,000 to 18,000 ft).

accelerated erosion *Soil erosion* occurring at a rate much faster than *soil horizons* can be formed from the parent *regolith*.

acid (acidity) A condition of the *soil* caused by the presence of readily exchangeable *acid-generating cations*, comprising from 5% to 60% of the total *cation exchange capacity* (CEC).

acid-generating cations *Cations*, mostly cations of aluminum and hydrogen, whose presence in large numbers in the *soil solution* produces an acid condition of the soil.

acid mine drainage Sulfuric acid effluent from *coal* mines, mine tailings, or spoil ridges made by *strip mining*.

active layer Shallow surface layer subject to seasonal thawing in *permafrost* regions.

active pools A type of *pool* in the *material cycle* in which the materials are in forms and places easily accessible to life processes. (See also *storage pools*.)

active systems *Remote sensor* systems that use a Man-made beam of wave energy as a source and measure the intensity of that energy reflected back to the source.

actual evapotranspiration (water use) Actual rate of *evapotranspiration* at a given time and place.

adiabatic lapse rate (See *dry adiabatic lapse rate, wet adiabatic lapse rate.*)

adiabatic process Change of temperature within a *gas* because of compression or expansion, without gain or loss of heat from the outside.

advection fog *Fog* produced by *condensation* within a moist basal air layer moving over a cold land or water surface.

advection process The horizontal mixing of cold and warm air in midlatitude *cyclones* and *anticyclones*, resulting in poleward meridional transport of heat to reach high latitudes.

aggradation Raising of *stream channel* altitude by continued deposition of *bed load*.

agric horizon An illuvial *soil horizon* formed under cultivation and containing significant amounts of illuvial *silt, clay,* and *humus*.

agricultural ecosystems *Ecosystems* modified and managed by Man for agricultural purposes.

A horizon *Mineral* horizon of the *soil*, overlying the *B horizon* and often characterized in the lower part by loss of *clay minerals* and oxides of iron and aluminum.

air mass Extensive body of air within which upward gradients of temperature and moisture are fairly uniform over a large area.

albedo Percentage of *electromagnetic radiation* reflected from a surface.

albic horizon Pale, often sandy *soil horizon* from which *clay* and free iron oxides have been removed.

Aleutian low Persistent center of low *atmospheric pressure* located in the area of the Aleutian Islands and strongly intensified in winter.

Alfisols A *soil order* within the *Soil Taxonomy* consisting of *soils* of humid and subhumid climates, with high *base status* and an *argillic horizon*.

alkaline Condition of the *soil solution* present when the large majority of *cations* held by *soil colloids* are *base cations*.

allophane Amorphous *mineral* substance consisting of varying proportions of *sesquioxide of aluminum* and *silica*, with varying amounts of water.

alluvial fan Gently sloping, conical accumulation of coarse *alluvium* deposited

by a *braided stream* undergoing *aggradation* below the point of emergence of the channel from a narrow *canyon* or *gorge*.

alluvial meanders Sinuous bends of a *graded stream* flowing in the alluvial deposit of a *floodplain*.

alluvial river River (*stream*) of low gradient flowing upon thick deposits of *alluvium* and experiencing approximately annual overbank flooding of the adjacent *floodplain*.

alluvial terrace *Terrace* carved in *alluvium* by a *stream* during *degradation*.

alluvium Any stream-laid *sediment* deposit found in a *stream channel* and in low parts of a stream valley subject to flooding.

alpine chains High mountain ranges that are narrow *tectonic* belts severely deformed by folding and thrusting in comparatively recent geologic time.

alpine glacier Long narrow mountain *glacier* on a steep downgrade, occupying the floor of a troughlike valley.

Alpine system Global system of *alpine chains*.

alpine tundra A plant *formation class* within the tundra *biome,* found at high altitudes above the limit of *tree* growth.

altitude tinting Application of varied color tints to altitude zones on a map.

altocumulus *Cloud* of middle height range formed into individual small masses within a single layer and fitted closely together in a geometrical pattern.

altostratus *Cloud* of middle height range with blanketlike form.

aluminosilicates *Silicate minerals* containing aluminum as an essential element.

amphibole group Complex *aluminosilicate minerals* rich in calcium, magnesium, and iron, dark in color, high in *density,* and classed as *mafic minerals*.

andesite *Extrusive igneous rock* of *diorite* composition; occurs as *lava*.

anemometer *Weather* instrument used to indicate *wind* speed.

aneroid barometer *Barometer* using a mechanism consisting of a partially evacuated air chamber and a flexible diaphragm.

angle of repose Natural angle (*dip*) of a *slope* consisting of loose, coarse, well-sorted *rock* fragments; for example, the *slip face* of a sand dune, a *talus slope,* or the sides of a cinder cone.

angstrom *Wavelength* unit equal to 0.000,000,01 cm (10^{-8} cm).

anion Negatively charged *ion.*

annual temperature range Difference between *mean monthly temperature* of warmest and coldest months of the year.

annular drainage pattern A *stream* network dominated by concentric (ringlike) major *subsequent streams.*

antarctic air mass Cold *air mass* with *source region* over Antarctica.

antarctic circle *Parallel of latitude* at 66½°S.

antarctic circumpolar current *Ocean current* system flowing eastward in the zone of *prevailing westerly winds* of the southern hemisphere.

antarctic zone Latitude zone in the *latitude* range 60° to 75° S (more or less), centered about on the *antarctic circle,* and lying between the *subantarctic zone* and the *polar zone.*

antecedent stream A *stream* that has maintained its course across a rising *rock* barrier, such as an anticlinal *fold* or a *fault block.*

anticlinal mountain Long, narrow ridge or mountain formed on an *anticline.*

anticlinal valley Valley eroded in weak *strata* on the central line or axis of an eroded *anticline.*

anticline Upfold of *strata* or other layered *rock* in an archlike structure; a class of *folds.* (See also *syncline.*)

anticyclone Center of high *atmospheric pressure.*

antipode That point which lies diametrically opposite to a given point on the surface of a globe or the earth.

anvil top Flattened top of a *cumulonimbus cloud,* produced as ice particles are carried downwind at high levels.

aphelion Point on the earth's elliptical *orbit* at which the earth is farthest from the sun.

aquatic ecosystems *Ecosystems* consisting of life forms of the marine environments and the freshwater environments of the lands.

aquiclude *Rock* mass or layer that impedes or prevents the movement of *ground water;* it has low *permeability.*

aquifer *Rock* mass or layer that readily transmits and holds *ground water;* possesses both high porosity and high *permeability.*

arch (marine) Arch of *bedrock* remaining when a *rock* promontory of a *marine cliff* has been cut through from two sides by wave action.

arctic air mass Cold *air mass* developed over a *source region* in the Arctic Ocean and fringing lands.

arctic circle *Parallel of latitude* at 66½° N.

arctic front Frontal zone of interaction between *arctic air masses* and *polar air masses.*

arctic front zone Belt in the *subarctic zone* and *arctic zone* in which the fluctuating *arctic front* is found.

arctic tundra Plant *formation class* within the *tundra biome,* consisting of low, mostly *herbaceous* plants, but with some very small stunted *trees,* associated with the *tundra climate.*

arctic zone Latitude zone in the latitude range 60° to 75° N (more or less), centered about on the *arctic circle,* and lying between the *subarctic zone* and the *polar zone.*

arcuate delta Variety of *delta* with *shoreline* curved convexly outward from the land.

areal geologic map Map that depicts the surface exposures and boundaries of various *rock* units and other geologic features, such as *faults.*

area strip mining Form of *strip mining* practiced in regions where a horizontal coal seam lies beneath a land surface that is approximately horizontal.

arête Sharp, knifelike divide or crest formed between two *cirques* by alpine glaciation.

argillic horizon Illuvial *soil horizon,* usually the *B horizon,* in which *lattice-layer clay minerals* have accumulated by *illuviation.*

Aridisols A *soil order* within the *Soil Taxonomy* consisting of *soils* of dry climates, with or without *argillic horizons,* and with accumulations of *carbonates* or soluble salts.

artesian flow Spontaneous rise of water in a well or fracture zone above the level of the surrounding *water table.*

artesian well Drilled well in which water rises under hydraulic pressure above the level of the surrounding *water table* and may reach the surface.

artificial levee Earth embankment built parallel with an *alluvial river* channel, usually on the crest of the *natural levee,* to contain the river in *flood stage.*

aspect Compass orientation of a *slope* as an inclined element of the ground surface.

asthenosphere Soft layer of the *mantle,* beneath the rigid *lithosphere.*

astronomical hypothesis Explanation for onset of a *glaciation* making use of cyclic variations in the form of the earth's *orbit* and the angle of inclination of the earth's axis.

astronomical seasons Spring, summer, autumn, and winter; subdivisions of the *tropical year* defined by the successive occurrences of *vernal equinox, summer solstice, autumnal equinox,* and *winter solstice.*

Atlantic climatic stage Warm climate stage of the *Holocene Epoch,* spanning a period about from −8000 to −5000 years.

atmosphere Envelope of gases surrounding the earth, held by *gravity.*

atmospheric pressure Pressure exerted by the *atmosphere* because of the force of *gravity* acting upon the overlying column of air.

atoll Circular or closed-loop *coral reef* enclosing an open *lagoon* with no island inside.

Australian sclerophyllous tree savanna Variety of the *savanna biome,* dominated by species of eucalyptus trees, widespread in eastern Australia.

autogenic succession Form of *ecological succession* that is self-producing, i.e., the result of the action of the plants and animals themselves. (See *old-field succession.*)

autumnal equinox *Equinox* occurring on September 22 or 23, when the *sun's declination* is 0°.

available landmass *Landmass* above *base level* that can be consumed by fluvial *denudation.*

axial rift Narrow, trenchlike depression situated along the center line of the mid-oceanic ridge and identified with active *sea-floor spreading.*

azimuth Direction referred to a circular scale of degrees read clockwise and ranging from 0° to 360°.

Azores high Persistent cell of high *atmospheric pressure* located over the *subtropical zone* in the North Atlantic Ocean.

backswamp Area of low, swampy ground on the *floodplain* of an *alluvial river* between the *natural levee* and the *bluffs.*

backwash Return flow of *swash* water under influence of *gravity.*

badlands Rugged land surface of steep *slopes,* resembling miniature mountains, developed on weak *clay* formations or clay-rich *regolith* by fluvial erosion too rapid to permit plant growth and *soil formation.*

bajada Graded *slope* that is an *alluvial fan* or a *pediment,* extending from mountain base to *playa.*

bank caving Incorporation of masses of *alluvium* or other bank materials into a *stream channel* because of undermining, usually in high flow stages.

bank-full stage *Stream stage* corresponding with *flood stage,* at which time *discharge* is contained entirely within the limits of the *stream channel.*

bar-and-swale Assemblage of *floodplain landforms* consisting of alternate low ridges (bars) and low troughs (swales), formed as *point-bar deposits* on the outsides of *alluvial meanders*.

barchan dune *Sand dune* of crescentic base outline with sharp crest and steep lee *slip face*, with crescent points (horns) pointing downwind.

barometer Instrument for measurement of *atmospheric pressure*.

barrier island Long narrow island, built largely of *beach sand* and *dune sand*, parallel with the mainland and separated from it by a *lagoon*.

barrier-island coast *Coast* with broad zone of shallow water offshore (a *lagoon*) shut off from the ocean by a *barrier island*.

barrier reef *Coral reef* separated from mainland *shoreline* by a *lagoon*.

basalt *Extrusive igneous rock* of *gabbro* composition; occurs as *lava*.

basal till Glacial *till* formed beneath moving glacial ice; a densely compacted form of till, often with a high *clay* content.

base cations (bases) Certain *cations* in the *soil solution* that are also plant *nutrients*; the most important are cations of calcium, magnesium, potassium, and sodium.

base flow That portion of the *discharge* of a *stream* contributed by *ground water* seepage and *interflow*.

base level Lower limiting surface or level that can ultimately be attained by a *stream* under conditions of stability of the earth's *crust* and sea level; an imaginary surface equivalent to sea level projected inland.

base line Reference *parallel* used in the U.S. Land Office Survey.

base status of soils Quality of a *soil* as measured by the *percentage base saturation (PBS)*; soils with PBS greater than 35% are soils of *high base status*; those with PBS less than 35% are soils of *low base status*.

batholith Large, deep *pluton* (body of *intrusive igneous rock*), usually with an area of surface exposure greater than 100 sq km (40 sq mi).

bauxite Mixture of several *clay minerals,* consisting largely of aluminum oxide and water with impurities; a principal ore of aluminum.

baymouth bar A low ridge of *sand* built above water level across the mouth of a bay by *littoral drift* and wave action; a variety of *beach*.

beach Thick, wedge-shaped accumulation of *sand, gravel,* or *cobbles* in the zone of breaking waves.

beach drift Transport of *sand* on a *beach* parallel with a *shoreline* by a succession of landward and seaward water movements at times when *swash* approaches obliquely.

beach ridges Low coastal *sand* ridges, representing *berms* produced in succession during *progradation* of a beach.

bedding planes Planes of separation between individual *strata* in a sequence of *sedimentary rocks*.

bed load That portion of the *stream load* moving close to the *stream* bed by rolling, sliding, and low leaps.

bedrock Solid *rock* in place with respect to the surrounding and underlying rock and relatively unchanged by *weathering* processes.

bench mark Surface point inscribed on a permanent monument to provide a fixed reference of elevation above the *datum plane*.

berm Low embankment or ridge on a *sand beach* constructed by *swash* of breaking waves. (See *summer berm, winter berm*.)

B horizon *Mineral soil horizon* located beneath the *A horizon*, and usually characterized by a gain of mineral matter (such as *clay minerals* and oxides of aluminum and iron) and organic matter (*humus*).

bioclimatic frontier Geographic boundary corresponding with a critical limiting level of climatic stress, beyond which a species cannot survive.

biogenic sediments *Organically-derived sediments*.

biogeochemical cycle (See *material cycle*.)

biogeography Study of the distribution patterns of plants and animals on the earth's surface and the processes that produce those patterns.

biomass Dry weight of living organic matter in an *ecosystem* within a designated surface area; units are grams of organic matter per square meter.

biome Largest recognizable subdivision of the *terrestrial ecosystems,* including the total assemblage of plant and animal life interacting within the *life layer*.

biosphere All living organisms of the earth and the environments with which they interact.

biotic communities Local associations of plants and animals that are interdependent and often found together.

bird-foot delta A *delta* with long projecting distributary fingers extending out into open water.

black body Ideal object or surface that is a perfect radiator of energy and will absorb all radiation falling upon it.

block separation Separation of individual *joint blocks* during the process of *physical weathering*.

blowout Shallow depression produced by continued *deflation;* also called a *deflation hollow*.

bluffs Steeply rising *slopes* marking the outer limits of a *floodplain*.

bog succession *Ecological succession* typical of a shallow glacial lake, in which the lake basin is eventually filled with *freshwater peat*.

bora Strong, cold wind occurring during the winter along the Adriatic coastal region when a strong *pressure gradient* exists.

boreal forest Variety of *needleleaf forest* found in the *boreal forest climate* regions of North America and Eurasia.

boreal forest climate Cold *climate* of the *subarctic zone* in the northern hemisphere with long, extremely severe winters and several consecutive months of zero *potential evapotranspiration (water need)*.

Boreal stage First (earliest) climate stage of the *Holocene Epoch*.

bornhardt Prominent knob of massive *granite* or similar plutonic *rock* with rounded summit and often showing *exfoliation* shells.

braided channel (See *braided stream*.)

braided stream *Stream* with shallow channel in coarse *alluvium* carrying multiple threads of fast flow, subdividing and rejoining repeatedly and continually shifting in position.

breaker Sudden collapse of a steepened water wave as it approaches the *shoreline*.

broadleaf Leaf form that is wide in relation to length, thin, and comparatively large.

broadleaf evergreen forest *Formation class* in the *forest biome* consisting of *broadleaf evergreen trees* and found in the *moist subtropical climate* and in parts of the *marine west-coast climate*. (Also known as *temperate rainforest*.)

bryoids Group of low, small plants that lie close to the ground or are attached to *tree* trunks; most are mosses and liverworts.

butte Prominent, steep-sided hill or peak, often representing the final remnant of a resistant *rock* layer in a region of *horizontal strata*.

caatinga Region of *thorntree semidesert* in northeastern Brazil.

calcic horizon *Soil horizon* of accumulation of calcium carbonate or magnesium carbonate.

calcification Accumulation of calcium carbonate in a *soil,* usually occurring in the *B horizon* or in the C horizon below the *soil solum*.

calcite *Mineral* having the composition calcium carbonate.

calcrete Rocklike layer rich in calcium carbonate, formed below the *soil solum* in the C horizon. (See *petrocalcic horizon, caliche.*)

caldera Large, steep-sided circular depression resulting from the explosion and subsidence of a large *composite volcano.*

caliche Name applied in the southwestern United States to the *petrocalcic horizon* of the *soil,* associated with *Aridisols.* (See also *calcrete.*)

calving Breaking off of blocks of glacial ice from a *glacier terminus* that extends into tidewater; the process by which *icebergs* are formed.

cambic horizon Altered *soil horizon* that has lost *sesquioxides* or *bases,* including *carbonates,* through *leaching.*

campo cerrado Region of *savanna biome* in the interior Brazilian highlands of South America.

Canadian high Region of above-average *atmospheric pressure* centered over central North America during the winter.

canyon (See *gorge.*)

capillary fringe Thin layer immediately above the *water table* in which water has been drawn upward from the *ground water* zone beneath by capillary force. (See also *capillary water.*)

capillary water Water clinging to a solid surface by means of the force of capillary film tension.

carbohydrate Class of organic compounds consisting of the elements carbon, hydrogen, and oxygen.

carbonates (carbonate minerals, carbonate rocks) *Minerals* that are carbonate compounds of calcium or magnesium or both, i.e., calcium carbonate or magnesium carbonate. (See *calcite.*)

carbonation Reaction of carbonic acid in rainwater, soil water, and *ground water* with *minerals;* most strongly affects *carbonates* (carbonate minerals and rocks); an activity of *chemical weathering.*

carbon cycle *Material cycle* in which carbon moves through the *biosphere;* includes both *gaseous cycles* and *sedimentary cycles.*

carbonic acid Weak acid formed by the solution of atmospheric carbon-dioxide gas in *surface water* and *ground water.*

cation Positively charged *ion.*

cation exchange Replacement of certain *cations* by other cations on the surfaces of *clay mineral* particles of colloidal dimensions, following a replacement order.

cation exchange capacity (CEC) Capacity of a given quantity of *soil* to hold and to exchange *cations,* stated in a unit known as the milliequivalent.

caverns (See *limestone caverns.*)

Celsius scale Temperature scale in which the freezing point of water is 0°, the boiling point 100°.

Cenozoic Era Last (youngest) of the eras of geologic time.

central eye Cloud-free central vortex of a *tropical cyclone.*

chalcedony *Mineral,* composition *silica* (silicon dioxide), lacking in visible crystalline structure.

channel flow (See *stream flow.*)

chaparral *Scherophyllous scrub* and dwarf *forest* plant *formation class* found throughout the coastal mountain ranges and hills of central and southern California.

chatter marks Curved fractures produced by ice pressure on the surface of *bedrock* subjected to *glacial abrasion* of moving glacial ice; usually associated with *glacial striations.*

chemical pollutants *Gases* introduced into the *atmosphere* from industrial activities, fuel combustion, and other human activities; not included are the normal gaseous constituents of pure dry air.

chemical precipitate *Sediment* consisting of *mineral* matter chemically precipitated from a water solution in which the matter has been transported in the dissolved state as *ions.*

chemical weathering Chemical change in rock-forming *minerals* through exposure to atmospheric conditions in the presence of water; mainly involving *oxidation, hydrolysis,* carbonic acid reaction, or direct solution.

chert *Sedimentary rock* composed largely of *chalcedony* and various impurities, in form of nodules and layers, often with *limestone* layers.

chinook winds One of a group of *local winds,* similar to the *foehn,* occurring at certain times to the lee of the Rocky Mountains; a very dry wind with a high capacity to evaporate *snow.*

cinder cone Conical hill built of coarse *tephra* ejected from a narrow volcanic vent; a type of *volcano.*

circle of illumination Great circle that divides the globe at all times into a sunlit hemisphere and a shadowed hemisphere.

circum-Pacific belt Chains of *andesite volcanoes* making up mountain belts and *island arcs* surrounding the Pacific Ocean basin.

cirrocumulus High *cloud* formed of ice particles and shaped into small patches arranged in geometrical patterns.

cirrostratus High *cloud* formed of ice particles appearing as a whitish veil and producing a halo of the sun or moon.

cirrus High *cloud* formed of ice and shaped into delicate white filaments, streaks, or narrow bands.

clastic sediment *Sediment* consisting of particles broken away physically from a parent *rock* source.

clay *Sediment* particles smaller than 0.004 mm in diameter.

clay minerals Class of *minerals* produced by alteration of *silicate minerals,* having plastic properties when moist.

cleavage of loess Natural vertical partings typical of thick deposits of *loess.*

cliff Sheer, near-vertical *rock* wall formed from flat-lying resistant layered rocks, usually *sandstone, limestone,* or *lava flows.* Cliff may refer to any near-vertical rock wall. (See *marine cliff.*)

climate Generalized statement of the prevailing *weather* conditions at a given place, based upon statistics of a long period of record and including mean values, deviations from those means, and the probabilities associated with those deviations.

climate-process systems Concept of a number of geomorphic systems each of which contains a unique combination of levels of intensity of several basic *denudation* processes.

climate types Varieties of *climate* recognized under a system of climate classification.

climatic optimum Past period of *climate* warmer than the present climate; associated with the *Atlantic climatic stage.*

climatology Science of *climate.*

climax Stable *biotic community* of plants and animals reached at the end point of *ecological succession;* end point of a *sere.*

climogenetic regions Concept of a classification of the global land surfaces into several basic landscape types, each having a set of *landforms* uniquely dictated by *climate.* (Same as *morphogenetic regions.*)

climograph A graph on which two or more climatic variables, such as *monthly mean temperature* and *monthly mean precipitation,* are plotted for each month of the year.

closed depression Area of ground that is upwardly concave with all of its enclosing surface sloping inward.

cloud families Groups of *cloud* varieties defined in terms of height range or degree of vertical development.

cloud reflection Reflection of incoming *shortwave radiation* from the upper surfaces of *clouds* to space.

clouds Dense concentrations of suspended water or ice particles in diameter range 20 to 50 *microns.* (See *cumiform cloud, stratiform clouds.*)

cloud seeding Fall of ice crystals from the anvil top of a *cumulonimbus cloud,* serving as *nuclei* of *condensation* at lower levels. (Seeding may also be carried out artificially.)

coal *Rock* consisting of hydrocarbon compounds, formed of compacted, lithified, and altered accumulations of plant remains (*peat*).

coastal blowout dune High *sand dune* of the *parabolic dune* class formed adjacent to a *beach,* usually with a deep *deflation hollow* enclosed within the dune ridge.

coastal plain Coastal belt, emerged from beneath the sea as a former *continental shelf,* underlain by *strata* with gentle *dip* seaward.

coastline (coast) Zone in which coastal processes operate or have a strong influence.

col Natural pass or low notch in an *arête* between opposed *cirques.*

colatitude Angular value in degrees of arc equal to 90° minus the *latitude.*

cold-blooded animal Animal whose body temperature passively follows the temperature of the environment.

cold front Moving weather *front* along which a cold *air mass* is forcing itself beneath a warm air mass, causing the latter to be lifted.

cold woodland Plant *formation class* consisting of *woodland* with low, widely spaced trees and a ground cover of *lichens* and mosses, found along the northern fringes of the region of *boreal forest climate;* also called *taiga.*

colloids (mineral) *Mineral* particles of extremely small size, capable of remaining indefinitely in *suspension* in water; typically in the form of thin plates or scales.

colluvium Deposit of *mineral* particles accumulating from *overland flow* at the base of a *slope* and originating from higher slopes where *sheet erosion* is in progress. (See also *alluvium.*)

compass quadrant bearings System of direction measurement using four divisions of the compass referred to the four cardinal points.

composite volcano *Volcano* constructed of alternate layers of *lava* and *tephra (volcanic ash).*

compound leaf Leaf form in which a single leaf consists of three or more separated leaflets, each joining a single main stem.

Comprehensive Soil Classification System (CSCS) Scheme of *soil* classification developed through the 1950s and 1960s by soil scientists of the U.S. Department of Agriculture and other scientists; formerly known as the Seventh Approximation. (See also *Soil Taxonomy.*)

condensation Process of change of matter in the *gaseous state (water vapor)* to the *liquid state* (liquid water) or *solid state* (ice).

cone of depression Conical configuration of the lowered *water table* around a well from which water is being rapidly withdrawn.

conformal projection Map projection that preserves the true shape or outline of any small surface feature of the earth.

congressional townships Units of land area, 6 miles on a side, used in the U.S. Land Office survey.

conic projections A group of *map projections* in which the *geographic grid* is transformed to lie on the surface of a developed cone.

consequent stream *Stream* that takes its course down the slope of an *initial landform,* such as a newly emerged *coastal plain* or a *volcano.*

consumers Animals in the *food chain* that live on organic matter formed by *primary producers* or by other consumers.

continental air mass *Air mass* developed over a continental *source region.*

continental collision Event in *plate tectonics* in which *subduction* brings two segments of the *continental crust* into contact, leading to *suturing* and formation of a *suture.*

continental drift Hypothesis, introduced by Alfred Wegener and others early in the 1900s, of the breakup of a parent continent, *Pangaea,* starting near the close of the Mesozoic Era, and resulting in the present arrangement of *continental shields* and intervening *ocean basin floors.*

continental glacier (See *ice sheet.*)

continentality Tendency of large land areas in midlatitudes and high latitudes to impose a large *annual temperature range* on the air temperature cycle.

continental lithosphere *Lithosphere* bearing *continental crust.*

continental nuclei Oldest masses or patches of the *continental shields,* generally older than about 2½ billion years.

continental rise Gently sloping sea floor lying at the foot of the *continental slope* and leading gradually into the *abyssal plain.*

continental rupture *Crustal spreading* affecting the *continental lithosphere,* so as to cause a *rift valley* to appear and to widen, eventually creating a new belt of *oceanic lithosphere.*

continental shelf Shallow, gently sloping belt of sea floor adjacent to the continental *shoreline* and terminating at its outer edge in the *continental slope.*

continental shields Ancient crustal *rock* masses of the continents, largely *igneous rock* and *metamorphic rock,* and mostly of *Precambrian* age.

continental slope Steeply descending belt of sea floor between the *continental shelf* and the *continental rise.*

continental suturing Process of formation of a *suture* during *continental collision.*

contour interval Vertical distance between successive *contour lines* on a *topographic map.*

contour line Line drawn on a topographic map to pass through all surface points having the same altitude.

contour strip mining Form of *strip mining* practiced in hilly or mountainous regions where coal seams form natural *outcrops* along the *contour* of the hillslopes.

convection (atmospheric) Air motion consisting of strong updrafts taking place within a *convection cell.*

convection cell Individual column of strong updrafts produced by atmospheric *convection.*

coordinate system In cartography, any system of fixed intersecting lines used to designate the position of points. Example: geographic coordinates of the *geographic grid.*

coral reef Rocklike accumulation of *carbonate mineral* secreted by corals and algae in shallow water along a marine *shoreline.*

coral-reef coast *Coast* built out by accumulations of *limestone* in *coral reefs.*

corange lines Lines on a map (isopleths) drawn through all points having the same *annual temperature range.*

core of earth Spherical central mass of the earth composed largely of iron and consisting of an outer liquid zone and an interior solid zone.

Coriolis effect Effect of the earth's rotation tending to turn the direction of motion of any object or fluid toward the right in the northern hemisphere and to the left in the southern hemisphere.

corrosion Erosion of *bedrock* of a *stream channel* (or other *rock* surfaces) by chemical reactions between solutions in stream water and *mineral* surfaces.

counterradiation *Longwave radiation* of *atmosphere* directed downward to the earth's surface.

covered shields Areas of *continental shields* in which the ancient *rocks* are covered beneath a veneer of sedimentary *strata.*

crater Central summit depression associated with the principal *vent* of a *volcano.*

crescent dune (See *barchan dune.*)

crescentic gouge Form of curved fracture on a *bedrock* surface affected by *glacial abrasion;* it is convex toward the downstream direction of ice motion.

crevasse Gaping crack in the brittle surface ice of a *glacier.*

crevices (marine) Narrow, cleftlike

cavities eroded into a *marine cliff* by the action of breaking storm waves.

crust of earth Outermost solid shell or layer of the earth, composed largely of *silicate minerals.*

cuesta *Erosional landform* developed on resistant *strata* having low to moderate *dip* and taking the form of an asymmetrical low ridge or hill belt with one side a steep scarp and the other a gentle slope; usually associated with a *coastal plain.*

cultural energy Energy in forms exclusive of solar energy of *photosynthesis* that is expended upon the production of raw food or feed crops in *agricultural ecosystems.*

cumuliform cloud *Cloud* of globular shape, often with extended vertical development.

cumulonimbus cloud Large, dense *cumuliform cloud* yielding *precipitation.*

cumulus *Cloud* type consisting of low-lying, white cloud masses of globular shape well separated from one another.

current meter A device to measure the velocity of flow of water at a given point in a *stream.*

cuspate bar Low coastal *sand* ridge projecting seaward in a toothlike form; a form of *beach.*

cuspate delta *Delta* with a *shoreline* sharply pointed toward the open water.

cuspate foreland Accumulation of *beach ridges* projecting seaward in a toothlike or arcuate form, formed by *progradation.*

cutan Thin film or coating on a *soil ped* or an individual coarse *mineral* grain; commonly such cutans are clay skins (argillans).

cuticle Outermost protective cell layer of a leaf.

cutoff Cutting-through of a *meander neck,* so as to bypass the *stream* flow in an *alluvial meander* and cause it to be abandoned.

cut-off high Upper-air *anticyclone* formed by occlusion of an upper-air wave, or *Rossby wave.*

cut-off low Upper-air *cyclone* formed by occlusion of the trough of an upper-air wave, or *Rossby wave.*

cycle of denudation Concept of an orderly sequence of evolutionary stages of fluvial *denudation* in which *relief* of the *available landmass* declines with time to reach a late stage when the landscape becomes a peneplain.

cycle of rock transformation (See *rock transformation cycle.*)

cyclone Center of low *atmospheric pressure.* (See *tropical cyclone, wave cyclone.*)

cyclone family Succession of *wave cyclones* tracking eastward along the *polar front* while developing from open

stage to occluded stage.

cyclonic storm Intense weather disturbance within a moving *cyclone* generating strong *winds,* cloudiness, and *precipitation.*

cylindric projections A group of *map projections* in which the *geographic grid* is transformed to lie on the surface of a developed cylinder.

datum plane Horizontal plane of reference or zero altitude base, usually sea level, applied to a topographic map.

daylight saving time Time system under which time is advanced by one hour with respect to the *standard time* of the prevailing *standard meridian.*

debris flood Streamlike flow of muddy water heavily charged with *sediment* of a wide range of size grades, including boulders, generated by sporadic torrential rains upon steep mountain watersheds.

decalcification Removal of calcium carbonate from a *soil horizon* or *soil solum* as carbonic acid reacts with *carbonate mineral* matter.

deciduous plant *Tree* or *shrub* that sheds its leaves seasonally, i.e., a *tropophyte.*

decomposers Organisms that feed on dead organisms from all levels of the *food chain;* most are microorganisms and bacteria that feed on decaying organic matter.

deep environment Environment of high pressure and high temperature to which *rock* is subjected deep within the earth's crust.

deflation Lifting and transport in *turbulent suspension* by *wind* of loose particles of *soil* or *regolith* from dry ground surfaces.

deflation hollow (See *blowout.*)

delta *Sediment* deposit built by a *stream* entering a body of standing water and formed of the *stream load.*

delta coast *Coast* bordered by a *delta.*

delta kame Flat-topped hill of *stratified drift* representing a glacial *delta* constructed adjacent to an *ice sheet* in a *marginal glacial lake.*

dendritic drainage pattern Drainage pattern of treelike branched form, in which the smaller *streams* take a wide variety of directions and show no parallelism or dominant trend.

denitrification Biochemical process in which nitrogen in forms usable to plants is converted into molecular nitrogen in the gaseous form and returned to the atmosphere, a process that is part of the *nitrogen cycle.*

density Quantity of mass per unit of volume, stated in gm/cc.

denudation Total action of all processes whereby the exposed *rocks* of the con-

tinents are worn down and the resulting *sediments* are transported to the sea by the *fluid agents;* includes also *weathering* and *mass wasting.*

deposition (See *stream deposition.*)

depositional landforms *Landforms* that are *sequential landforms* created by the deposition of *sediment* from a fluid medium.

depression contour (See *hachured contour.*)

depression storage (See *surface detention.*)

desalinization Natural process of removal of soluble salts from a *soil.*

desert biome *Biome* of the dry climates consisting of thinly dispersed plants that may be *shrubs,* grasses, or perennial *herbs,* but lacking in *trees.*

desert climate subtype Subtype of the *dry climate* in which no month has *soil-water storage* exceeding 2 cm.

desert pavement Surface layer of closely fitted pebbles or coarse *sand* from which finer particles have been removed by *deflation.*

desert varnish Dark iridescent coating found on *rock* surfaces in a desert climate.

dew-point temperature Temperature of *saturated air.*

diagnostic horizons Certain *soil horizons,* rigorously defined, that are used as diagnostic criteria in classifying *soils* within the *Soil Taxonomy.*

diffuse reflection Form of *scattering* in which solar rays are deflected or reflected by minute dust particles or *cloud* particles.

dike Thin layer of *intrusive igneous rock,* often near-vertical or with steep *dip,* occupying a widened fracture in the *rock,* and typically cutting across older rock planes.

diorite *Intrusive igneous rock* consisting dominantly of *plagioclase feldspar* and *pyroxene,* with some *amphibole* and biotite; a *felsic igneous rock,* occurs as a *pluton.*

dip Acute angle between an inclined natural *rock* plane or surface and an imaginary horizontal plane of reference; always measured perpendicular to the *strike.* Also a verb, meaning to incline toward.

discharge Volume of flow moving through a given cross section of a *stream* in a given unit of time; commonly given in cubic meters (feet) per second.

distributaries Branching *stream channels* that cross a *delta* to discharge into open water.

doldrums Belt of calms and variable *winds* occurring at times along the *equatorial trough.*

dome (See *sedimentary dome.*)

dornveldt   Local name for a region of *thorntree semidesert* in South Africa.

down-scatter   Scattered *shortwave radiation* directed earthward within the *atmosphere*.

drainage basin   Total land surface occupied by a *drainage system*, bounded by a drainage divide or watershed.

drainage system   A branched network of *stream channels* and adjacent land *slopes*, bounded by a drainage divide and converging to a single channel at the outlet.

drainage winds   Winds, usually cold, that flow from higher to lower regions under the direct influence of *gravity*; synonymous with *katabatic winds*.

drawdown   Difference in height between the base of a *cone of depression* and the original *water table* surface.

drifts (in mines)   Horizontal tunnels or passageways in mines.

drumlin   Hill of glacial *till*, oval or elliptical in basal outline and with smoothly rounded summit, formed by plastering of till beneath moving, debris-ladened glacial ice.

dry adiabatic lapse rate   Rate at which rising air is cooled by expansion when no *condensation* is occurring; 1.0 C°/100 m (5.5 F°/1000 ft).

dry climate   Climate in which the total annual *soil-water shortage* is 15 cm (5.9 in.) or greater, and there is no *water surplus*.

dry desert   Plant *formation class* in the *desert biome* consisting of widely dispersed xerophytic plants that may be small, hardleaved or spiny *shrubs*, succulent plants (cacti), or hard grasses.

dry midlatitude climate   *Dry climate* of the *midlatitude zone* with a strong annual cycle of *potential evapotranspiration (water need)* and cold winters.

dry subtropical climate   *Dry climate* of the *subtropical zone*, transitional between the *dry tropical climate* and the *dry midlatitude climate*.

dry tropical climate   *Dry climate* of the *tropical zone* with large total annual *potential evapotranspiration (water need)*.

dune   (See *sand dune*.)

dune sand   Sand of which *sand dunes* are composed, usually consisting of well-rounded *quartz* grains.

duripan   Dense, hard subsurface *soil horizon* cemented by *silica* that does not soften during prolonged soaking.

dust storm   Heavy concentration of dust in a turbulent *air mass*, often associated with a *cold front*.

earthflow   Moderately rapid downhill flowage of masses of water-saturated *soil*, *regolith*, or weak *shale*, typically forming a steplike terrace at the top and a bulging toe at the base.

earthquake   A trembling or shaking of the ground produced by the passage of *seismic waves*.

earthquake focus   (See *focus*.)

easterly wave   Weak, slowly moving trough of low pressure within the belt of *tropical easterlies*; causes a weather disturbance with *rain* showers.

ebb current   Outward (seaward) flow of a *tidal current*.

ecological succession   Time-succession (sequence) of distinctive plant and animal communities occurring within a given area of newly formed land or land cleared of plant cover by burning, clear cutting, or other agents.

ecology   Science of interactions between life forms and their environment; the science of *ecosystems*.

ecosystem   Group of organisms and the environment with which the organisms interact.

edaphic factors   Factors influencing a *terrestrial ecosystem* that are related to the *soil*.

effluent stream   *Stream* that receives an increment of flow by out-seepage of *ground water*.

electromagnetic radiation   Wavelike form of energy radiated by any substance possessing heat; travels through space at the speed of light.

elfin forest   High-altitude variety of *forest* found in the *tropical zone* and *equatorial zone*, consisting of dwarfed *trees* festooned with mosses.

eluviation   *Pedogenic process* consisting of the downward transport of fine particles, particularly the colloids (both mineral and organic), carrying them out of an upper *soil horizon*.

emergence   Exposure of submarine *landforms* by a lowering of the sea level or a rise of the *crust*, or both.

end moraine   (See *terminal moraine*.)

energy deficit   Condition in which rate of outgoing radiant energy exceeds rate of incoming radiant energy at a given time and place.

energy surplus   Condition in which rate of incoming radiant energy exceeds the rate of outgoing radiant energy at a given time and place.

Entisols   A *soil order* in the *Soil Taxonomy*, consisting of *mineral soils* lacking *soil horizons* that would persist after normal plowing.

entrainment   Drawing in of surrounding air by a rapidly rising, bubblelike air body within a *thunderstorm* cell.

entrenched meanders   Winding, sinuous valley bends produced by *degradation* of a *stream* with trenching of the *bedrock* by downcutting.

environmemtal region   Region of the continental surface that has a unique set of physical environmental factors and supports a distinctive *ecosystem*.

environmental temperature lapse rate   Rate of temperature decrease upward through the *troposphere*; standard value is 6.4 C°/km (3½ F°/1000 ft).

epeirogenic movement   Slow rising or sinking of the *crust* over a large area, without appreciable faulting or folding.

ephemeral annuals   Small desert plants that complete a life cycle very rapidly following a desert downpour.

epipedon   *Soil horizon* that forms at the surface.

epiphytes   Plants that live above ground level out of contact with the *soil*, usually growing on the limbs of *trees* or *shrubs*; also called "air plants."

epoch of geologic time   Geologic time unit, a subdivision of a *period*.

equal-area projections   Class of map projections on which any given area of the earth's surface is shown to correct relative size, regardless of position on the globe.

equatorial air mass   Warm, moist *air mass* with source region over oceans in the *equatorial zone*.

equatorial countercurrent   Narrow *ocean current*, flowing west to east between the two *equatorial currents*.

equatorial current   West-flowing *ocean current* in the belt of *trade winds*.

equatorial easterlies   Upper-level easterly air flow over the *equatorial zone*.

equatorial rainforest   Plant *formation class* within the *forest biome*, consisting of tall, closely set *broadleaf trees* of *evergreen* or *semideciduous* habit.

equatorial trough   Low-pressure trough centered more or less over the equator and situated between the two belts of *trade winds*.

equatorial zone   Latitude zone lying between lat. 10°S and 10°N (more or less) and centered upon the equator.

equinox   Instant in time when the *subsolar point* falls on the earth's equator and the *circle of illumination* passes through both poles. Vernal equinox occurs on March 20 or 21; autumnal equinox on September 22 or 23.

erg   Large expanse of *active dunes* in the Sahara Desert of North Africa.

erosional landforms   Class of the *sequential landforms* shaped by the removal of *regolith* or *bedrock* by agents of erosion. Examples: *canyon*, glacial *cirque*, *marine cliff*.

esker   Narrow, often sinuous embankment of coarse gravel and boulders deposited in the bed of a meltwater *stream* enclosed in a tunnel within stagnant ice of an *ice sheet*.

estuarine delta   *Delta* built into the lower part of an estuary.

Eurasian-Melanesian belt Major *mountain arc* system extending from southern Europe across southern Asia and Indonesia.

eutrophication Excessive growth of algae and other *primary producers* in a *stream* or lake as a result of the input of large amounts of nutrient *ions*, especially phosphate and nitrate.

evaporation Process in which water in *liquid state* or *solid state* passes into the *vapor state*.

evaporites Class of *chemical precipitate sediment* and *sedimentary rock* composed of soluble salts deposited from salt water bodies.

evapotranspiration Combined water loss to the atmosphere by *evaporation* from the soil and *transpiration* from plants.

evergreen leafless plants *Evergreen plants* with fleshy stems but no functional leaves. Example: the cacti.

evergreen plant *Tree* or *shrub* that holds most of its green leaves throughout the year.

evergreen-succulent plants *Evergreen plants* with thick, fleshy leaves or stems.

exfoliation Development of sheets, shells, or scales of *rock* upon exposed *bedrock* or *boulders,* usually in concentric, spheroidal arrangements.

exfoliation dome Smoothly rounded *rock* knob or hilltop bearing rock sheets or shells produced by spontaneous expansion accompanying *unloading.*

exotic river *Stream* that flows across a region of *dry climate* and derives its *discharge* from adjacent uplands where a *water surplus* exists.

exponential decay Program of decrease in a variable quantity at a rate such that the quantity is halved in a constant time period called the *half-life.*

exposed shields Areas of *continental shields* in which the ancient basement *rock,* usually of Precambrian age, is exposed to the surface.

external magnetic field Lines of force of the earth's magnetic field surrounding the earth's solid surface and extending far out into space.

extrusive igneous rock *Rock* produced by the solidification of *lava* or ejected fragmental *igneous rock (tephra).*

Fahrenheit scale Temperature scale in which the freezing point of water is 32°, the boiling point 212°.

fallout *Gravity* fall of atmospheric particles of *particulate matter,* reaching the ground.

fault Sharp break in *rock* with displacement (slippage) of block on one side with respect to adjacent block. (See *normal fault, overthrust fault, transcurrent fault, transform fault.*)

fault block Blocklike *rock* mass lying between two parallel *normal faults.* (See *graben, horst.*)

fault coast *Coast* formed when *shoreline* comes to rest against a *fault scarp.*

fault line Surface trace of a *fault.*

fault-line scarp Erosion scarp developed upon an inactive *fault line.*

fault plane Surface of slippage between two earth blocks moving relative to each other during faulting.

fault scarp Clifflike surface feature produced by faulting and exposing the fault plane; commonly associated with a *normal fault.*

feldspar *Aluminosilicate mineral* group consisting of silicate of aluminum and one or more of the metals potassium, sodium, or calcium. (See *plagioclase feldspar, potash feldspar.*)

fell-field In regions of *tundra climate,* a ground surface littered with rock fragments that may be formed into *stone polygons.*

felsenmeer Expanse of large blocks of *rock* produced by *joint block separation* and *shattering* by *frost action* at high altitudes or in high latitudes.

felsic igneous rock *Igneous rock* dominantly composed of *felsic minerals.*

felsic minerals, felsic mineral group *Quartz* and *feldspars* treated as a *mineral group* of light color and relatively low *density.* (See also *mafic minerals.*)

ferricrete Rocklike surface layer rich in sesquioxide of iron (*hematite*); essentially the same as an exposed layer of *laterite.*

field capacity (See *storage capacity.*)

filmy leaves Leaves that are thin and delicate, as compared with leaves described as *membranous.*

finger lake Long, narrow lake occupying part of the floor of a *glacial trough;* a variety of *trough lake.*

fiord Narrow, deep ocean embayment partially filling a *glacial trough.*

fiord coast Deeply embayed, rugged *coast* formed by partial *submergence* of glacial troughs.

firn Granular old *snow* forming a surface layer in the *zone of accumulation* of a *glacier.*

firn field Field of a *glacier* in which *firn* accumulates.

fixed dunes *Sand dunes* covered by protective vegetation and no longer in the active state.

flocculation The clotting together of colloidal mineral particles to form larger particles.

flood Stream flow at a *stream stage* so high that it cannot be accommodated within the *stream channel* and must spread over the banks to inundate the adjacent *floodplain.*

flood basalts Large-scale outpourings of *basalt lava* to produce thick accumulations of *basalt* over large areas.

flood crest Maximum *stream stage* attained during the passage of a *flood.*

flood current Landward flow of a *tidal current.*

floodplain Belt of low, flat ground, present on one or both sides of a *stream channel,* subject to inundation by a *flood* about once annually and underlain by *alluvium.*

flood stage Designated *stream stage* for a particular point on a *stream,* higher than which overbank flooding may be expected.

flood wave Time sequence of rising and falling *stream stage* during the passage of a *flood.*

fluid Substance that flows readily when subjected to unbalanced stresses; may exist as a *gas* or a *liquid.*

fluid agents *Fluids* that erode, transport, and deposit *mineral* matter and organic matter; they are running water, waves and currents, glacial ice, and wind.

fluvial landforms *Landforms* shaped by running water.

fluvial processes Geomorphic processes in which running water is the dominant *fluid agent,* acting as *overland flow* and *stream flow.*

focus (pl. foci) Point within the earth at which the energy of an *earthquake* is released and from which the *seismic waves* emanate.

foehn (föhn) Warm, dry *wind* produced to the lee of a mountain range as air descends and is adiabatically warmed; essentially the same phenomenon as *chinook winds.*

fog *Cloud* layer in contact with land or sea surface, or very close to that surface. (See *advection fog, radiation fog.*)

folds Wavelike corrugations of *strata* (or other layered *rock* masses) as a result of crustal compression; a form of *tectonic activity.*

food chain Organization of an *ecosystem* into steps or levels through which energy flows as the organisms at each level consume energy stored in the bodies of organisms of the next lower level.

forb Broadleaved *herb,* as distinguished from the grasses.

foredunes Ridge of irregular *sand dunes* typically found adjacent to *beaches* on low-lying *coasts* and bearing a partial cover of plants.

foreshore Sloping face of a *beach* that is within the zone of *swash* and *backwash.*

forest Assemblage of *trees* growing close together, their crowns forming a layer of foliage that largely shades the ground.

forest biome *Biome* that includes all regions of *forest* over the lands of the earth.

formation classes Subdivisions within a *biome* based upon the size, shape, and structure of the plants that dominate the vegetation.

formative elements Syllables used to form the names of *soil orders, suborders,* and *great soil groups* in the *Soil Taxonomy.*

Foucault pendulum Pendulum so designed as to be perfectly free to undergo a steady change in direction of swing as the earth rotates on its axis.

fractional scale Ratio of distance between two points on a map or a globe to the actual distance between the same two points on the earth's surface.

fragipan Dense, moderately brittle layer in the *soil.*

freezing Change from *liquid state* to *solid state* accompanied by release of *latent heat of fusion* becoming *sensible heat.*

freshwater peat Form of *peat* produced by *bog succession;* often identified as the organic *soil horizon* of a *Histosol.*

fringing reef *Coral reef* directly attached to land with no intervening *lagoon* of open water.

front Surface of contact between two unlike *air masses.* (See *cold front, occluded front, polar front, warm front.*)

frost (See *killing frost.*)

frost action *Rock* breakup by forces accompanying the *freezing* of water.

funnel cloud Long, narrow cloud of tubular or funnel shape hanging from the base of a *cumulonimbus cloud;* it represents a *tornado* or a *waterspout.*

gabbro *Intrusive igneous rock* consisting largely of *pyroxene* and *plagioclase feldspar,* with variable amounts of *olivine;* a *mafic igneous rock,* occurs as a *pluton.*

gamma rays High-energy form of radiation at the extreme short *wavelength* (high-frequency) end of the *electromagnetic spectrum.*

gas *Fluid* of very low *density* (as compared with a *liquid* of the same chemical composition) that expands to fill uniformly any small container and is readily compressed.

gaseous cycle Type of *material cycle* in which an element or compound is converted into gaseous form, diffuses through the *atmosphere,* and passes rapidly over land or sea where it is reused in the *biosphere.*

gaseous state *Fluid* state of matter having the properties of a *gas.*

geographic grid Complete network of *parallels* and *meridians* on the surface of the globe, used to fix the locations of surface points.

geographic north (See *true north.*)

geologic norm Stable natural condition in a moist climate in which slow *soil erosion* is paced by maintenance of *soil horizons* bearing a plant community in an equilibrium state.

geomorphology Science of *landforms,* including their history and processes of origin.

geostationary orbit Satellite orbit that holds a fixed position over a selected point on the earth's equator.

geostrophic wind *Wind* at high levels above the earth's surface blowing parallel with a system of straight, parallel *isobars.*

gilgae Small surface relief features of the *soil* that may be knobs and basins or narrow ridges with valleys between, typical of the *Vertisols.*

glacial abrasion *Abrasion* by a moving *glacier* of the *bedrock* floor beneath it.

glacial delta *Delta* built by meltwater streams of a *glacier* into standing water of a *marginal glacial lake.*

glacial drift General term for all varieties and forms of *rock* debris deposited in close association with *Pleistocene ice sheets.*

glacial plucking Removal of masses of *bedrock* from beneath an *alpine glacier* or *ice sheet* as ice moves forward suddenly.

glacial trough Deep, steep-sided rock trench of a U-shaped cross section formed by *alpine glacier* erosion.

glaciation (1) General term for the total process of glacier growth and *landform* modification by *glaciers.* (2) Single episode or time period in which *ice sheets* formed, spread, and disappeared.

glacier Large natural accumulation of land ice affected by present or past flowage. (See *alpine glacier, outlet glacier.*)

glacier terminus Lower end of an *alpine glacier.*

glaciofluvial sediment *Sediment* accumulation formed by meltwater *streams* issuing from a *glacier terminus.*

glaciolacustrine sediment *Sediment* that has accumulated on the floor of a *marginal glacial lake.*

glaze Ice layer accumulated upon solid surfaces by the *freezing* of falling *rain* or *drizzle.*

global water balance Balance among the three hydrologic components — *precipitation, evaporation,* and *runoff* — for the earth as a whole.

gneiss Variety of *metamorphic rock* showing banding and commonly rich in quartz and feldspar.

Gondwana Hypothetical southern hemisphere continent existing through-out Paleozoic and early Mesozoic Eras, consisting of *continental shields* of South America, Africa, India, Madagascar, Australia, and Antarctica; part of *Pangaea.*

gorge (canyon) Steep-sided *bedrock* valley with a narrow floor limited to the width of a *stream channel.*

graben Trenchlike depression representing the surface of a *fault block* dropped down between two opposed, infacing *normal faults.* (See *rift valley.*)

graded profile Smoothly descending *longitudinal profile* displayed by a *graded stream.*

graded stream Stream (or *stream channel*) with *stream gradient* so adjusted as to achieve a balanced state in which average *bed load* transport is matched to average bed load input; an average condition over periods of many years' duration.

graminoid Having long, narrow leaves; for example, the leaves of grasses.

granite *Intrusive igneous rock* consisting largely of *quartz, potash feldspar,* and *plagioclase feldspar,* with minor amounts of biotite and hornblende; a *felsic igneous rock,* occurs as a *pluton.*

granular disintegration Grain-by-grain breakup of the outer surface of coarse-grained *rock,* yielding gravel and leaving behind rounded boulders.

graphic scale Map *scale* shown by a line marked off into length units.

grassland biome *Biome* consisting largely or entirely of herbs, which may include grasses, grasslike plants, and *forbs.*

gravity percolation Downward movement of water under the force of *gravity* through the *soil-water belt* and *intermediate belt,* eventually arriving at the *water table.*

great circle Circle formed by passing a plane through the exact center of a perfect sphere; the largest circle that can be drawn on the surface of a sphere.

great soil groups Third level classification in the *Soil Taxonomy.*

greenhouse effect Accumulation of heat in the lower *atmosphere* through the absorption of *longwave radiation* from the earth's surface.

green revolution Major advance in securing increased agricultural production in the developing nations through improved genetic strains of wheat and rice and the application of *cultural energy.*

Greenwich meridian *Meridian* passing through the Royal Observatory at Greenwich, England, universally accepted as the *prime meridian* of zero *longitude.*

groin Man-made wall or embankment built out into the water at right angles to the *shoreline.*

gross photosynthesis Total amount of carbohydrate produced by *photosynthesis* by a given organism or group of organisms in a given unit of time.

ground moraine *Moraine* distributed beneath a large expanse of land surface covered at one time by an *ice sheet.*

ground radiation *Longwave radiation* emitted by land or water surfaces and passing upward into the overlying *atmosphere.*

ground water *Subsurface water* occupying the *saturated zone* and moving under the force of *gravity.*

ground water recharge Replenishment of *ground water* by downward movement of water through the *unsaturated zone,* or from *stream channels,* or through recharge wells.

grus Loose gravel formed by *granular disintegration* of a coarse-grained *felsic igneous rock,* such as *granite,* in a *dry climate.*

guard cells Cells surrounding the openings of the *stomata* that can open and close those openings to regulate the flow of *water vapor* and other gases.

gullies Deep, V-shaped trenches carved by newly formed *streams* in rapid headward growth during advanced stages of *accelerated soil erosion.*

gypsic horizon *Soil horizon* of accumulation of hydrous calcium sulfate, the *mineral* gypsum.

gypsum *Evaporite mineral,* composition calcium sulfate with water.

gyres Large circular *ocean current* systems centered upon the oceanic *subtropical high-pressure cells.*

habitat Subdivision of the plant environment having a certain combination of *slope,* drainage, *soil* type, and other controlling physical factors.

hachured contour *Contour line* with attached *hachures* (ticks) used to denote a *closed depression* on a topographic map.

hachures Minute, short lines drawn on a topographic map to show direction and steepness of *slope.*

Hadley cell Atmosphere circulation cell in low latitudes involving rising air over an *equatorial trough* and sinking air over *subtropical high-pressure belts.*

hail Form of *precipitation* consisting of pellets or spheres of ice with a concentric layered structure.

hairpin dune Type of *parabolic dune* of highly elongate form.

half-life Time required for an initial quantity at time-zero to be reduced by one-half in an *exponential decay* system or program.

hanging trough Tributary *glacial trough* with floor high above the floor of the main trough that it joins.

hanging valley *Stream* valley that has been truncated by marine erosion so as to appear in cross section in a *marine cliff,* or truncated by glacial erosion so as to appear in cross section in the upper wall of a *glacial trough.*

Hawaiian high Persistent cell of high *atmospheric pressure* located in the *subtropical zone* of the central and eastern North Pacific Ocean.

haze Minor concentration of pollutants or natural forms of *particulate matter* in the *atmosphere* causing a reduction in visibility.

heat (See *sensible heat, latent heat.*)

heat engine Mechanical system in which motion is powered by heat energy.

heat island Persistent region of higher air temperatures centered over a city.

heavy minerals Group of *minerals* having exceptionally high *density,* usually 4 gm/cc and greater, typically occurring in *clastic sediments.*

hematite *Mineral* variety, composition sesquioxide of iron; a product of *chemical weathering* of *mafic minerals;* commonly present in *soil* and *regolith.*

herb Tender plant, lacking woody stems, usually small or low; it may be annual or perennial, *broadleaf (forb),* or *graminoid* (grass).

herbaceous Adjective applied to plants that are *herbs.*

hertz *Wave frequency* of one cycle per second.

hibernation Dormant state of some vertebrate animals during the winter season.

high base status (See *base status of soils.*)

high-latitude climates Group of *climates* in the *subarctic zone, arctic zone,* and *polar zone,* dominated by *arctic air masses* and *polar air masses.*

high water Highest water level reached by the *ocean tide* in a given tidal cycle.

hillslope Sloping land surface between a drainage divide and a *stream channel* on a *landmass* undergoing fluvial denudation.

hillslope profile Profile of a *hillslope* plotted along the path followed by *overland flow* from a drainage divide to the nearest *stream channel.*

histic epipedon Thin surface *soil horizon (epipedon)* consisting of *peat.*

Histosols *Soil order* within the *Soil Taxonomy,* consisting of *soils* with a thick upper layer of organic matter.

hogback Sharp-crested, often sawtooth ridge formed of the upturned edge of a resistant *rock* layer of *sandstone, limestone,* or *lava.*

Holocene Epoch Last epoch of geologic time, commencing about 10,000 years ago; it followed the *Pleistocene Epoch*

and includes the present.

Homolographic projection *Map projection* that is an *equal-area projection* consisting of straight horizontal *parallels* and elliptical *meridians.*

homolosine projection Composite *map projection* consisting of the *sinusoidal projection* between lat. 40°N and 40°S and the *homolographic projection* poleward of those latitudes.

horizonation Degree of development of *soil horizons* as a result of complex combinations of the *pedogenic processes.*

horizon circle Circle representing the positions of all points equidistant from the center point of a *zenithal projection.*

horizontal strata *Strata* lying in approximately horizontal attitude.

horn Sharp, steep-sided mountain peak formed by the intersection of headwalls of glacial *cirques.*

horse latitudes *Subtropical high-pressure belt* of the North Atlantic Ocean, coincident with the central region of the *Azores high;* a belt of weak, variable winds and frequent calms.

horst *Fault block* uplifted between two *normal faults.*

hot spot Center of intrusive igneous and volcanic activity thought to be located over a rising *mantle plume.*

hour circles Imaginary *meridians of longitude,* spaced 15° apart, traveling westward around the globe at 15° per hour, and representing successive hours of the *mean solar day.*

humid climate subtype Subtype of the *moist climate* in which the annual *water surplus* always greater than the annual *soil-water shortage.*

humidity General term for the amount of water vapor present in the air. (See *relative humidity, specific humidity.*)

humification *Pedogenic process* of transformation of plant tissues into *humus.*

humus Dark brown to black organic matter on or in the *soil,* consisting of fragmented plant tissues partly oxidized by *consumer* organisms.

hurricane *Tropical cyclone* of the western North Atlantic and Caribbean Sea.

hydraulic action *Stream erosion* by impact force of the flowing water upon the bed and banks of the *stream channel.*

hydraulic head Difference in level of the *water table* between one point and another, setting up a pressure difference and causing the flow of *ground water* from higher to lower points.

hydrogenic sediments *Sediments* that are *chemical precipitates.*

hydrograph Graphic presentation of the variation in *stream discharge* with elapsed time, based on data of *stream gauging* at a given station on a *stream.*

hydrologic cycle Total plan of movement, exchange, and storage of the earth's free water in *gaseous state, liquid state,* and *solid state.*

hydrology Science of the earth's water and its motions through the *hydrologic cycle.*

hydrolysis Chemical union of water molecules with *minerals* to form different, more stable mineral *compounds.*

hydrosphere Total water realm of the earth's surface zone, including the *oceans,* surface waters of the lands, *ground water,* and water held in the *atmosphere.*

hygrograph Recording *hygrometer;* it produces a continuous record of *relative humidity.*

hygrometer Instrument that measures the *water vapor* content of the *atmosphere;* some types measure *relative humidity* directly.

hygrophytes Plants adapted to a wet environment on the lands.

hygroscopic nuclei *Nuclei* of *condensation* having a high affinity for water; usually salt particles.

iceberg Mass of glacial ice floating in the ocean, derived by *calving* from a *glacier* that extends into tidal water.

icecap Platelike mass of glacial ice limited to the high summit region of a mountain range or plateau; a variety of *glacier.*

ice fall Abrupt steepening of the down-valley gradient of a *valley glacier,* causing the surface ice to break apart in deep *crevasses.*

ice floes Individual patches of *pack ice* detached by the forces of *wind* and *ocean currents.*

ice island Large, thick body of floating freshwater ice, derived from a mass of landfast ice.

Icelandic low Persistant center of low *atmospheric pressure* located over the North Atlantic Ocean and strongly intensified in winter.

ice lobes (glacial lobes) Broad tonguelike extensions of an *ice sheet* resulting from more rapid ice motion where terrain was more favorable.

ice sheet Large thick plate of glacial ice moving outward in all directions from a central region of accumulation.

ice sheet climate Severely cold *climate,* found on the Greenland and Antarctic ice sheets, with *potential evapotranspiration (water need)* effectively zero throughout the year.

ice shelf Thick plate of floating glacial ice attached to an *ice sheet* and fed by the ice sheet and by *snow* accumulation.

ice storm Occurrence of heavy *glaze* of ice on solid surfaces.

ice wedge Vertical, wall-like body of ground ice, often tapering downward, occupying a shrinkage crack in *silt* of *permafrost* areas.

ice-wedge polygons Polygonal networks of *ice wedges.*

igneous rock *Rock* solidified from a high-temperature molten state; rock formed by cooling of *magma.* (See *extrusive igneous rock, felsic igneous rock, intrusive igneous rock, mafic igneous rock, ultramafic igneous rock.*)

illite *Clay mineral* derived by *chemical weathering* from such *silicate minerals* as *feldspar* and muscovite mica.

illuviation Accumulation in a lower *soil horizon* (typically, the *B* horizon) of materials brought down from a higher horizon; a *pedogenic process.*

imagery General term in *remote sensing* for the graphic form of presentation of data obtained by *scanning systems.*

Inceptisols *Soil order* in the *Soil Taxonomy,* consisting of *soils* having weakly developed *soil horizons* and containing weatherable *minerals.*

infiltration Absorption and downward movement of *precipitation* into the *soil* and *regolith.*

infiltration capacity Limiting rate at which falling *rain* or melting *snow* can be absorbed by a *soil* surface in the process of *infiltration.*

influent stream *Stream* that loses *discharge* by seepage through the *channel* floor to recharge the *ground water* body beneath.

infrared emissivity Ratio of *infrared radiation* emission of a gray body to that of a *black body* at the same *absolute temperature.*

infrared radiation *Electromagnetic radiation* in the *wavelength* range of 0.7 to about 200 *microns.*

initial landforms *Landforms* produced directly by internal earth processes of *volcanism* and *tectonic activity.* Examples: volcano, fault scarp.

inner lowland On a *coastal plain,* a shallow valley lying between the first *cuesta* and the area of older *rock* (oldland).

inselberg Small, islandlike hill or mountain rising sharply above a surrounding *pediment* or *alluvial fan.*

insolation Interception of solar energy (*shortwave radiation*) by an exposed surface.

interception Catching and holding of *precipitation* above the ground surface on leaves and stems of plants.

interflow Movement of *soil water* through a permeable *soil horizon* or other shallow permeable layer in a downslope direction parallel with the ground surface; same as *throughflow.*

interlobate moraine *Moraine* formed between two adjacent *ice lobes.*

intermediate belt Zone below the *soil-water belt* too deep to supply capillary water to plants; i.e., too deep to be reached by plant roots.

International Date Line The 180° *meridian of longitude,* together with deviations east and west of that meridian, forming the time boundary between adjacent *standard time* zones that are 12 hours fast and 12 hours slow with respect to Greenwich standard time.

Interrupted sinusoidal projection *Sinusoidal projection* subdivided into a number of sectors (gores), each of which is centered upon a different central *meridian.*

intertropical convergence zone (ITC) Zone of convergence of *air masses* of *tropical easterlies* along the axis of the equatorial trough.

intrusive igneous rock *Igneous rock* body produced by solidification of *magma* beneath the surface, surrounded by preexisting rock.

inversion (See *temperature inversion.*)

inversion lid Top surface of a *low-level temperature inversion,* resisting mixing of colder air below with warmer air above.

ion Atom or group of atoms bearing an electrical charge as the result of a gain or loss of one or more electrons. (See also *cation, anion.*)

ionizing radiation Very short *wavelength electromagnetic radiation,* such as that of *X rays* and *radioactivity,* capable of causing *ionization* of atoms exposed to the radiation.

isobar Line on map passing through all points have the same *atmospheric pressure.*

isobaric surface Surface of equal *atmospheric pressure.*

isogonic line Line drawn on a map to pass through all points having the same *magnetic declination.*

isohyet Line on a map drawn through all points having the same numerical value of *precipitation.*

isostasy Equilibrium state, resembling flotation, in which crustal masses stand at levels determined by their thickness and *density;* equilibrium being achieved by flowage of denser *mantle rock* of the underlying *asthenosphere.*

isostatic compensation Slow rising motion of the *earth's crust* in response to the removal of *rock* during *denudation,* as required by the principle of *isostasy.*

isotherm Line drawn on a map to pass through all points having the same air temperature.

jet stream High-speed air flow in narrow bands within the *upper-air westerlies* and along certain other global latitude zones at high levels.

joint blocks Blocklike masses of *bedrock* bounded by *joints*.

jungle Type of *forest* that is low and dense, consisting partly of *lianas*, bamboo scrub, thorny palms, or thickly branching *shrubs*, usually formed where *tropical rainforest* has been disturbed or destroyed.

joints Fractures within *bedrock*, usually occurring in parallel and intersecting sets of planes.

kame Hill composed of sorted coarse water-laid *glacial drift*, largely *sand* and *gravel*, built into an impounded water body within stagnant ice or against the margin of an *ice sheet*.

kame terrace *Kame* taking the form of a flat-topped *terrace* built between a body of stagnant glacial ice and a rising valley wall.

kaolinite *Clay mineral* typically formed by *hydrolysis* from *potash feldspar* (also from *micas*).

karst Landscape or topography dominated by surface features of *limestone* solution and underlain by a *limestone cavern* system.

katabatic winds (See *drainage winds*.)

Kelvin scale (°K) Temperature scale on which the starting point is absolute zero, equivalent to −273°C.

killing frost Occurrence of below-freezing temperature in the air layer near the ground, capable of damaging frost-sensitive plants.

knob and kettle Terrain of numerous small knobs of *glacial drift* and deep depressions usually situated along the *moraine* belt of a former *ice sheet*.

knot Speed of one *nautical mile* per hour.

lagoon Shallow body of open water lying between a *barrier island* or a *barrier reef* and the mainland.

lag time Interval of time between occurrence of *precipitation* and peak *discharge* of a *stream*.

Lambert conformal conic projection *Map projection* that is a *conformal projection* of the *conic projection* class having two *standard parallels*.

land breeze Local *wind* blowing from land to water during the night.

landform (pedology) General term for the configuration of the ground surface as a factor in *soil* formation; it includes *slope* steepness and *aspect*, as well as *relief*.

landforms Configurations of the land surface taking distinctive forms and produced by natural processes. Examples: hill, valley, plateau. (See *depositional landforms*, *erosional landforms*, *initial landforms*, *sequential landforms*.)

landmass Large area of *continental crust* lying above sea level (base level) and thus available for removal by *denudation*.

landmass rejuvenation Episode of rapid fluvial *denudation* set off by a rapid crustal rise, increasing the *available landmass*. (See also *rejuvenation* of *stream*.)

landslide Rapid sliding of large masses of *bedrock* on steep mountain *slopes* or from high *cliffs*.

langley (ly) Unit of intensity of solar radiation equal to one gram-calorie per square centimeter.

lapse rate (See *environmental temperature lapse rate*, *dry adiabatic lapse rate*, *wet (saturation) adiabatic lapse rate*.)

latent heat Heat absorbed and held in storage in a *liquid* or a *solid* during the processes of *condensation* or *freezing*, respectively; distinguished from *sensible heat*.

latent heat of fusion *Latent heat* released during *melting* or absorbed during *freezing*.

latent heat of vaporization *Latent heat* released during change from *liquid state* to *gaseous state* or absorbed in the reverse change.

lateral cutting Sidewise shifting of a *stream channel* caused by undercutting of steep banks on the outsides of bends.

lateral moraine *Moraine* forming an embankment between the ice of an *alpine glacier* and the adjacent valley wall.

laterite Rocklike layer rich in *sesquioxide of aluminum* and iron, including the *minerals* bauxite and *limonite*, found in low latitudes in association with *Ultisols* and *Oxisols*.

latitude Arc of a *meridian* between the equator and a given point on the globe.

lattice layers A geometric structure of certain *minerals* such as the *clay minerals*, in which extremely thin layers of strongly bonded atoms and *ions* are separated from adjacent layers by weak bonds.

Laurasia Hypothetical northern hemisphere continent existing throughout the Paleozoic and early Mesozoic eras, consisting of what are now *continental shields* of North America and Eurasia; part of *Pangaea*.

laurel forest Variety of *broadleaf evergreen forest* dominated by species of the laurel family (*Lauraceae*).

lava *Magma* emerging on the earth's solid surface, exposed to air or water.

layer silicates *Silicate minerals* having *lattice layer* structure.

leachate Solution of various *ions* and compounds carried down from the waste body of a *sanitary landfill* site to the *ground water* system below.

leaching *Pedogenic process* in which material is lost from the *soil* by downward washing out and removal by percolating surplus *soil water*.

leads Narrow strips of open ocean water between *ice floes*.

liana Woody vine supported on the trunk or branches of a *tree*.

lichens Plant forms in which algae and fungi live together (in a symbiotic relationship) to form a single structure; they typically form tough, leathery coatings or crusts attached to rocks and tree trunks.

life-form Characteristic physical structure, size, and shape of a plant or of an assemblage of plants.

life layer Shallow surface zone containing the *biosphere*; a zone of interaction between *atmosphere* and land surface, and between atmosphere and ocean surface.

lime Soil-conditioning material that may be either calcium oxide or calcium carbonate (*limestone*), applied for the purpose of reducing the *acidity* of the *soil*.

limestone Nonclastic *sedimentary rock* in which *calcite* is the predominant *mineral*, and with varying minor amounts of magnesium carbonate, *silica*, or other minerals, and *clay*.

limestone caverns Interconnected subterranean cavities formed in *limestone* by *carbonation* occurring in slowly moving *ground water*.

limonite *Mineral* or group of minerals consisting largely of iron oxide and water, produced by *chemical weathering* of other iron-bearing minerals.

liquid *Fluid* that maintains a free upper surface and is only very slightly compressible, as compared with a *gas*.

liquid state *Fluid* state of matter having the properties of a *liquid*.

lithification Process of hardening of *sediment* to produce *sedimentary rock*.

lithosphere (1) General term for the entire solid earth realm. (2) In *plate tectonics*, it is the strong, brittle outermost *rock* layer lying above the *asthenosphere*.

lithospheric plate Segment of *lithosphere* moving as a unit, in contact with adjacent lithospheric plates along *subduction zones*, zones of *sea-floor spreading*, or *transform faults*.

Little Ice Age Climatic episode of below-normal temperatures from about −550 to −150 years (A.D. 1450-1850), during which *alpine glaciers* advanced to lower levels.

littoral drift Transport of *sediment* parallel with the *shoreline* by the combined action of *beach drift* and *longshore current* transport.

live dunes *Sand dunes* in an active state of change as *sand* is moved in *saltation*.

loam *Soil-texture class* in which no one

of the three size grades (sand, silt, clay) dominates over the other two.

local time   Mean solar time based on the local *meridian* passing through a given point on the globe.

local winds   General term for *winds* generated as direct or immediate effects of the local terrain.

loess   Accumulation of yellowish to buff-colored, fine-grained *sediment*, largely of *silt* grade, upon upland surfaces after transport in *turbulent suspension* (i.e., carried in a *dust storm*).

longitude   Arc of a *parallel* between the *prime meridian* and a given point on the globe.

longitudinal dunes   Class of *sand dunes* in which the dune ridges are oriented parallel with the prevailing wind.

longitudinal profile   Graphic representation of the descending course of a *stream* or *stream channel* from upper end to lower end; altitude is plotted on the vertical scale, downstream distance on the horizontal scale.

longshore current   Current in the breaker zone and running parallel with the *shoreline*, set up by the oblique approach of waves.

longshore drift   *Littoral drift* caused by action of a *longshore current*.

longwave radiation   *Electromagnetic radiation* emitted by the earth, largely in the range from 3 to 50 *microns*.

low base status   (See *base status of soils*.)

lowland   Broad, open valley between two *cuestas* of a *coastal plain*. (*Lowland* may refer to any low area of land surface.)

low-latitude climates   Group of *climates* of the *equatorial zone* and *tropical zone* dominated by the *subtropical high-pressure belt* and the *equatorial trough*.

low-level temperature inversion Reversal of normal *environmental temperature lapse rate* in an air layer near the ground.

low tide   Lowest water level reached by the *ocean tide* in a given tidal cycle.

loxodrome   Line of constant compass bearing drawn on a map or navigational chart; also known as the *rhumb line*.

macronutrients   Nine elements required in abundance for organic growth, including *primary production* by green plants. (See also *micronutrients*.)

mafic igneous rock   *Igneous rock* dominantly composed of *mafic minerals*.

mafic minerals, mafic mineral group *Minerals*, largely *silicates*, rich in magnesium and iron, dark in color, and of relatively great *density*.

magma   Mobile, high-temperature molten state of *rock*, usually of *silicate*

mineral composition and with dissolved gases (*volatiles*).

magnetic azimuth   *Azimuth* system referred to *magnetic north*.

magnetic declination   Horizontal angle between *geographic north* and *magnetic north*.

magnetic field   (See *external magnetic field*.)

magnetic north   Direction from a given point on the earth's surface following a great *circle* toward the *magnetic north pole*.

magnetic north pole   Surface point toward which the north-seeking point of a magnetic compass is directed.

magnetopause   Outer boundary surface of the *magnetosphere*.

magnetosphere   External portion of the earth's magnetic field, shaped by pressure of the *solar wind* and contained within the *magnetopause*.

mangrove swamp forest   Coastal vegetation of mangrove plants found in shallow muddy-water environments in the *equatorial zone* and *tropical zone*.

mantle   *Rock* layer or shell of the earth beneath the *crust* and surrounding the *core*, composed of *ultramafic rock* of *silicate mineral* composition.

mantle plume   A columnlike rising of heated *mantle* rock, thought to be the cause of a *hot spot* in the overlying *lithospheric plate*.

map projection   Any orderly system of *parallels* and *meridians* drawn on a flat surface to represent the earth's surface.

maquis   Form of dense *sclerophyllous scrub* found throughout the Mediterranean region.

marble   Variety of *metamorphic rock* derived from *limestone* or dolomite by recrystallization under pressure.

marginal glacial lake   Lake impounded between an ice front and rising ground slopes.

marginal ocean basin   Comparatively narrow ocean basin floored by *oceanic lithosphere* and lying between a volcanic island arc and a major body of *continental lithosphere*.

marine cliff   *Rock* cliff shaped and maintained by the undermining action of breaking waves.

marine scarp   Steep seaward *slope* in poorly consolidated *alluvium* or other forms of *regolith*, produced along a coast by the undermining action of waves.

marine terrace   Former *abrasion platform* elevated to become a steplike coastal *landform*.

marine west-coast climate   Cool, *moist climate* of west coasts in the *midlatitude zone*, usually with a sustantial annual *water surplus* and a distinct winter *precipitation* maximum.

maritime air mass   Moist *air mass* developed over an ocean *source region*.

mass wasting   Spontaneous downward movement of *soil, regolith,* and *bedrock* under the influence of *gravity*; does not include the action of *fluid agents*.

material cycle   Total system of pathways by which a particular type of matter (a given element, compound, or *ion*, for example) moves through the earth's *ecosystem* or *biosphere*; also called *biogeochemical cycle* or *nutrient cycle*.

maximum-minimum thermometer   Pair of *thermometers* recording the maximum and minimum air temperatures since last reset.

mean annual temperature   Mean of daily air temperature means for a given year or succession of years.

mean daily temperature   Sum of daily maximum and minimum air temperature readings divided by two.

mean monthly temperature   Mean of daily air temperature means for a given calendar month.

mean solar day   Average time required for the earth to complete one *rotation* with respect to the sun; time elapsed between one *solar noon* and the next, averaged over the period of one *year*.

mean velocity   Mean, or average, speed of flow of water through an entire *stream* cross section.

medial moraine   *Moraine* riding on the surface of an *alpine glacier* and composed of debris carried down-valley from the point of junction of two ice streams.

Mediterranean climate   *Climate* type of the *subtropical zone*, characterized by the alternation of a very dry summer and a mild, rainy winter.

megahertz   *Wave frequency* of one million cycles per second.

melting   Change from *solid state* to *liquid state*, accompanied by absorption of *sensible heat* to become *latent heat*.

membranous leaves   Leaves of the *broadleaf* type that are of normal thickness.

Mercator projection   *Map projection* that is a true *conformal projection* with horizontal *parallels* and vertical *meridians* and with *map scale* rapidly increasing with increase in *latitude*.

mercurial barometer   *Barometer* using the Torricelli principle, in which *atmospheric pressure* counterbalances a column of mercury in a tube.

meridian of longitude   North-south line on the surface of the global *oblate ellipsoid*, connecting the north and south poles.

meridional transport   Flow of energy (heat) or matter (water) across the *parallels of latitude*, either poleward or equatorward.

meridional winds *Winds* moving across the *parallels of latitude,* in a north-south, or south-north direction, along the *meridians of longitude.*

mesa Table-topped *plateau* of comparatively small extent bounded by *cliffs* and occurring in a region of *horizontal strata.*

mesopause Upper limit of the *mesosphere.*

mesophytes Plants adapted to a *habitat* of intermediate degree of wetness and to uniform *soil water* availability.

mesosphere Atmospheric layer of upwardly diminishing temperature, situated above the *stratopause* and below the *mesopause.*

Mesozoic Era Second of three geologic eras following *Precambrian time.*

metamorphic rock *Rock* altered in physical structure and/or chemical (*mineral*) composition in the *solid state* by action of heat, pressure, shearing stress, or infusion of elements, all taking place at substantial depth beneath the surface.

meteorology Science of the *atmosphere;* particularly the physics of the lower or inner atmosphere.

mica group *Aluminosilicate mineral* group of complex chemical formula having perfect cleavage into thin sheets.

microclimate *Climate* of a shallow layer of air near the ground, and including the *soil* surface and plant community within which it is in contact.

micron Length unit; one micron equals 0.0001 cm.

micronutrients Elements essential to organic growth, but only in very small amounts. (See also *macronutrients.*)

microwaves Waves of the *electromagnetic radiation spectrum* in the *wavelength* band from about 0.03 cm to about 1 cm.

midlatitude climates Group of *climates* of the *midlatitude zone* and *subtropical zone,* located in the *polar front zone* and dominated by both *tropical air masses* and *polar air masses.*

midlatitude deciduous forest *Formation class* within the *forest biome* dominated by tall, *broadleaf deciduous* trees, found mostly in the *moist continental climate* and *marine west-coast climate.* (Also called *summergreen deciduous forest.*)

midlatitude zones Latitude zones occupying the latitude range 35° to 55° N and S (more or less) and lying between the *subtropical zones* and the *subarctic (subantarctic) zones.*

midnight meridian Imaginary time meridian opposite to the *noon meridian,* marking the occurrence of midnight and traveling westward around the globe at 15° per hour.

mid-oceanic ridge One of three major divisions of the *ocean basins,* being the central belt of submarine mountain topography with a characteristic *axial rift.*

millibar Unit of *atmospheric pressure;* one-thousandth of a bar. Bar is a force of one million dynes per square centimeter.

mineral Naturally occurring inorganic substance, usually having a definite chemical composition and a characteristic atomic structure. (See *felsic minerals, mafic minerals, silicate minerals.*)

mineral alteration Chemical change of *minerals* to more stable compounds upon exposure to atmospheric conditions; same as *chemical weathering.*

mistral Local drainage *wind* of cold air affecting the Rhone Valley of southern France.

Moho Contact surface between the earth's *crust* and *mantle;* a contraction of Mohorovičić, the seismologist who discovered this feature.

moist climate *Climate* in which the annual *soil-water shortage* is less than 15 cm (5.9 in.).

moist continental climate *Moist climate* of the *midlatitude zone* with strongly defined winter and summer seasons, adequate precipitation throughout the year, and a substantial annual *water surplus.*

moist subtropical climate *Moist climate* of the *subtropical zone,* characterized by a moderate to large annual *water surplus* and a strongly seasonal cycle of *potential evapotranspiration (water need).*

mollic epipedon Relatively thick, dark-colored surface *soil horizon,* or *epipedon,* containing substantial amounts of organic matter (*humus*) and usually rich in *base cations.*

Mollisols *Soil order* within the *Soil Taxonomy,* consisting of *soils* with a *mollic epipedon* and *high base status.*

monsoon forest Plant *formation class* within the *forest biome* consisting in part of *deciduous* trees adapted to a long dry season in the *wet-dry tropical climate.*

monsoon system System of low-level *winds* blowing into a continent in summer and out of it in winter, controlled by *atmospheric pressure* systems developed seasonally over the continent.

monsoon and trade-wind littoral climate *Moist climate* of low latitudes showing a strong rainfall peak in the season of high sun and a short period of reduced rainfall.

montane forest Plant *formation class* of the *forest biome* found in cool upland environments of the *tropical zone* and *equatorial zone.*

montmorillonite *Clay mineral* derived by the chemical alteration of *silicate minerals* in various *igneous rocks.*

moraine Accumulation of *rock* debris carried by an *alpine glacier* or an *ice sheet* and deposited by the ice to become a *depositional landform.* (See *ground moraine, interlobate moraine, lateral moraine, medial moraine, recessional moraine, terminal moraine.*)

morphogenetic regions (See *climogenetic regions.*)

mossy forest Form of upland *forest* (*montane forest*) in highlands of the *tropical zone* and *equatorial zone* characterized by massive accumulations of mosses attached to *tree* branches.

mountain arc Curved (arcuate) segment of an *alpine chain.*

mountain roots Erosional remnants of deep portions of ancient *sutures* that were once *alpine chains.*

mountain winds Daytime movements of air up the gradient of valleys and mountain slopes; alternating with nocturnal *valley winds.*

mud *Sediment* consisting of a mixture of *clay* and *silt* with water, often with minor amounts of *sand* and sometimes with organic matter.

mud flat A nearly flat expanse of fine soft *sediment* (*mud*) deposited in a tidal estuary or bay.

mudflow deposit Deposit left by a *mudflow* and consisting of a mixture of size grades, including *clay, silt, sand,* and boulders.

mudstone *Sedimentary rock* formed by the *lithification* of *mud.*

multiband spectral photography Photography using a combination of several narrow spectral bands within the *visible light* region and the *near-infrared region.*

natric horizon *Soil horizon* having prismatic structure and a high proportion of the sodium *ion.*

natural bridge Natural *rock* arch spanning a *stream channel,* formed by *cutoff* of an *entrenched meander* bend.

natural ecosystem An *ecosystem* that attains its development without appreciable interference by Man and that is subject to natural forces of modification and destruction. (See also *agricultural ecosystem.*)

natural gas Naturally occurring mixture of *hydrocarbon compounds* (principally methane) in the *gaseous state* held within certain porous *rocks.*

natural levee Belt of higher ground paralleling a meandering *alluvial river* on both sides of the *stream channel* and built up by deposition of fine *sediment* during periods of overbank flooding.

nautical mile Unit of distance measurement approximately equal to the average length of one minute of *latitude,* or of one minute of *longitude* measured along the equator; one nautical mile equals 1.85 km (1.15 statute mi), approx.

near-infrared region Narrow *wavelength* band within *infrared region,* immediately adjacent to the red band of the *visible light* region.

needleleaf Leaf form that is very narrow in relation to length, as illustrated by needles of the spruce, pine, or fir.

needleleaf forest Plant *formation class* within the *forest biome,* consisting largely of *needleleaf* evergreen trees. (See also *boreal forest.*)

net photosynthesis *Carbohydrate* remaining in an organism after *respiration* has broken down sufficient carbohydrate to power the metabolism of the organism.

net primary production Rate at which *carbohydrate* is accumulated in the tissues of plants within a given *ecosystem;* units are grams of dry organic matter per year per square meter of surface area.

net radiation Difference in intensity between all incoming energy (positive quantity) and all outgoing energy (negative quantity) carried by both *shortwave radiation* and *longwave radiation.*

névé (See *firn.*)

nimbostratus Low, dense *stratiform cloud* from which *rain* or *snow* is falling.

nitrogen cycle *Material cycle* in which nitrogen moves through the *biosphere* by the process of *nitrogen fixation* and *denitrification.*

nitrogen fixation Chemical process of conversion of gaseous molecular nitrogen of the atmosphere into compounds or *ions* that can be directly utilized by plants; a process carried out within the *nitrogen cycle* by certain microorganisms.

nonclastic sediments Class of *sediments* formed of mineral compounds precipitated from chemical solution or from organic activity.

noon (See *solar noon.*)

noon meridian Imaginary *meridian of longitude* marking the occurrence of *solar noon* and traveling westward around the globe at 15° per hour.

normal fault Variety of *fault* in which the *fault plane* inclines (*dips*) toward the downthrown block and a major component of the motion is vertical.

nuclei (atmospheric) Minute particles of solid matter suspended in the *atmosphere* and serving as cores for *condensation* of water or ice.

nuée ardente Glowing volcanic avalanche; a cloud of incandescent *dust* and *gas;* travels rapidly down the steep side of a *volcano.*

nutrient cycle (See *material cycle.*)

oblate ellipsoid Geometric solid resembling a flattened sphere, with polar axis shorter than the equatorial diameter, elliptical in any polar cross section but circular in any cross section taken at right angles to the polar axis.

oblateness Ratio of difference between length of polar axis and length of equatorial diameter of an oblate ellipsoid to the equatorial diameter, expressed as a simple fraction; also known as the flattening of the poles.

occluded front Weather *front* along which a moving *cold front* has overtaken a *warm front,* forcing the warm *air mass* aloft.

ocean-basin floors One of the major divisions of the *ocean basins,* comprising the deep portions consisting of *abyssal plains* and low hills.

ocean current Persistent, dominantly horizontal flow of ocean water.

oceanic lithosphere *Lithosphere* bearing *oceanic crust.*

oceanic trench Narrow, deep depression in the sea floor representing the *subduction* of an oceanic *lithospheric plate* beneath the margin of a continental lithospheric plate; often associated with an island arc.

oceanography (See *physical oceanography.*)

ocean tide Periodic rise and fall of the ocean level induced by gravitational attraction between the earth and moon in combination with earth *rotation.*

ochric epipedon Surface *soil horizon* (*epipedon*) that is light in color and contains less than 1% organic matter.

offshore That part of a *beach* lying in the zone of shoaling waves and below the level of low *tide.*

offshore bar Low bar of *sand* in the *offshore* region of a *beach.*

old-field succession Form of *secondary succession* typical of an abandoned field, such as might be found in the *moist continental climate* or *moist subtropical climate* of the eastern and central United States; a form of *autogenic succession.*

olivine *Silicate mineral* with magnesium and iron but no aluminum, usually olive-green or grayish-green; a *mafic mineral.*

open folds *Folds* of *strata* in which *dips* are low to moderately steep, without overturning.

orbit Path followed by a planet revolving around the sun, or by a planetary satellite revolving around a planet.

organically-derived sediment *Sediment* consisting of the remains of nonliving plants or animals, or of *mineral* matter produced by the activities of plants or animals.

organic horizon *Soil horizon,* designated as the O horizon, overlying the *mineral* horizons and formed of accumulated organic matter derived from plants and animals.

orogeny Major episode of *tectonic activity* resulting in *strata* being deformed by folding and faulting.

orographic precipitation *Precipitation* induced by the forced rise of moist air over a mountain barrier.

oscillatory waves Water waves in which particles move in vertical orbits, completing one orbital circle with the passage of one wave.

outcrop Surface exposure of *bedrock.*

outgassing Process of exudation of water and other gases from the earth's *crust* through *volcanoes,* to become a part of the earth's *hydrosphere* and *atmosphere.*

outlet glacier Tonguelike ice stream, resembling an *alpine glacier,* fed from an *ice sheet.*

outwash deposit Accumulation of layers of *sand* and *gravel* deposited by meltwater *streams* near the margin of a stagnant *ice sheet* or *alpine glacier.*

outwash plain Flat, gently sloping plain built up of *sand* and *gravel* by the *aggradation* of meltwater *streams* in front of the margin of an *ice sheet.*

overland flow Motion of a surface layer of water over a sloping ground surface at times when the *infiltration* rate is exceeded by the *precipitation* rate; a form of *runoff.*

overthrust fault *Fault* characterized by the overriding of one *fault block* (or *thrust sheet*) over another along a gently inclined *fault plane;* associated with crustal compression.

overwash Movement of storm *swash* entirely across a *barrier island* or barrier *beach* to reach the *lagoon* or *salt marsh* on the inland side.

oxbow lake, oxbow swamp Crescent-shaped lake or swamp representing the abandoned channel left by the *cutoff* of an *alluvial meander.*

oxic horizon Highly weathered *soil horizon* rich in *clay minerals* and *sesquioxides* of low *cation exchange capacity (CEC).*

oxidation Chemical union of free oxygen with metallic elements in *minerals.*

Oxisols *Soil order* in the *Soil Taxonomy* consisting of very old, highly weathered *soils* of low latitudes, with an *oxic horizon* and low *cation exchange capacity (CEC).*

oxygen cycle *Material cycle* in which oxygen moves through the *biosphere* in both gaseous and sedimentary forms.

ozone Gas with a molecule consisting of three atoms of oxygen, O_3.

ozone layer Layer in the *stratosphere,* mostly in the altitude range 20 to 35 km (12 to 31 mi), in which a concentration of *ozone* is produced by the action of solar *ultraviolet rays.*

pack ice   Floating *sea ice* that completely covers the sea surface.

Paleozoic Era   First of three geologic eras comprising all geologic time younger than *Precambrian time*.

Pampa   Geographic region of plains in Uruguay and eastern Argentina, characterized by the dominance of tall-grass *prairie* as the native plant *formation class*.

Pangaea   Hypothetical parent continent, enduring until near the close of the Mesozoic Era, consisting of the *continental shields* of *Laurasia* and *Gondwana* joined into a single unit. (See *continental drift*.)

parabolic blowout dune   Type of *parabolic dune* formed to the lee of a shallow *deflation hollow,* usually found on interior plains in a *dry climate*.

parabolic dunes   Isolated low *sand dunes* of parabolic outline, with points directed into the prevailing wind.

parallel of latitude   East-west circle on the earth's surface, lying in a plane parallel with the equator and at right angles to the axis of rotation.

parallel retreat   Program of *hillslope profile* evolution in which the angle of *slope* remains essentially constant as the slope retreats.

particulate matter   *Solid* and *liquid* particles capable of being suspended for long periods in the *atmosphere*.

passive systems   Electromagnetic *remote sensor* systems that measure radiant energy reflected or emitted by an object or surface.

patterned ground   General term for a ground surface that bears polygonal or ringlike features, including *stone polygons,* and *ice-wedge polygons*.

peat   Partially decomposed, compacted accumulation of plant remains occurring in a bog environment.

ped   Individual natural *soil* aggregate.

pediment   Gently sloping, rock-floored land surface found at the base of a mountain mass or *cliff* in an arid region.

pediplain   Desert land surface of low relief composed in part of *pediment* surfaces and in part of *alluvial fan* surfaces.

pedogenic processes   Group of recognized basic soil-forming processes, mostly involving the gain, loss, translocation, or transformation of materials within the *soil* body.

pedology   Science of the *soil* as a natural surface layer capable of supporting living plants; synonymous with *soil science*.

pedon   *Soil* column extending down from the surface to reach a lower limit in some form of *regolith* or *bedrock*.

peneplain   Land surface of low elevation and slight relief produced in the late stages of *denudation* of a *landmass*.

percentage base saturation (PBS)   Percentage of exchangeable *base cations* with respect to the total *cation exchange capacity (CEC)* of a given *soil*.

perched water table   *Ground water* layer or lens formed over an *aquiclude* and occupying a position above the main *water table*.

perhumid climate subtype   Subtype of the *moist climate* in which the *soil-water surplus* is 60 cm or greater.

periglacial   Pertaining to the physical environment in close proximity to the margin of a continental *ice sheet*.

perihelion   Point on the earth's elliptical *orbit* at which the earth is nearest to the sun.

period of geologic time   Time subdivision of the *era*, each ranging between about 35 and 70 million years in duration.

permafrost   Condition of permanently frozen water in the *soil, regolith,* and *bedrock* in cold climates of subarctic and arctic regions.

permeability   Property of relative ease of movement of *ground water* through *rock* when unequal pressure *(hydraulic head)* exists.

petrocalcic horizon   Hardened *calcic horizon* in the *soil* that does not break apart when soaked in water. (See also *caliche, calcrete*.)

petroleum (crude oil)   Natural liquid mixture of many complex hydrocarbon compounds of organic origin, found in accumulations (oil pools) within certain *sedimentary rocks*.

pH   Measure of the concentration of hydrogen *ions* in a solution; the number represents the logarithm to the base 10 of the reciprocal of the weight in grams of hydrogen ions per liter of water.

photochemical reactions   Chemical reactions occurring in polluted air through the action of sunlight upon pollutant gases to synthesize new toxic compounds or gases.

photoperiod   Duration of daylight on a given day at a given latitude.

photosynthesis   Production of *carbohydrate* by the union of water with carbon dioxide while absorbing light energy. (See *gross photosynthesis, net photosynthesis*.)

phreatophytes   Plants that draw water from the *ground-water table* beneath *alluvium* of dry *stream channels* and valley floors in desert regions.

physical oceanography   Physical science of the *oceans*.

physical weathering   Breakup of massive *rock (bedrock)* into small particles through the action of physical forces acting at or near the earth's surface. (See *weathering*.)

phytogenic dunes   Class of *sand dunes* formed under partial cover of plants.

pingo   Conspicuous conical mound or circular hill, having a core of ice, found on plains of the *arctic tundra* where *permafrost* is present.

pioneer stage   First stage of an *ecological succession*.

plaggen epipedon   Man-made surface *soil horizon (epipedon)* produced by long-continued manuring, incorporating sod or other livestock bedding materials into the *soil*.

plagioclase feldspar   *Aluminosilicate mineral* with sodium or calcium or both.

plane of the ecliptic   Imaginary plane in which the earth's *orbit* lies.

plastic shading   Method of showing *relief* on topographic maps by use of color shades varied according to *aspect* of the *slope*.

plateau   Upland surface, more or less flat and horizontal, upheld by a resistant bed or *formation* of *sedimentary rock* or *lava flows* and bounded by a steep *cliff*.

plate tectonics   Theory of *tectonic activity* dealing with *lithospheric plates* and their activity.

playa   Flat land surface underlain by fine *sediment* or *evaporite minerals* deposited from shallow lake waters in a *dry climate* in the floor of a closed topographic depression.

Pleistocene Epoch   Epoch of the Cenozoic Era, often identified as the Ice Age; it preceded the *Holocene Epoch*.

plinthite   Iron-rich concentrations present in some kinds of *soils* in deeper *soil horizons* and capable of hardening into rocklike material with repeated wetting and drying.

plucking   (See *glacial plucking*.)

plunging fold   *Fold* of *strata* with a descending or rising crest or fold axis.

pluton   Large body of coarse-grained *intrusive igneous rock,* typically of *felsic minerals*. Example: *batholith* of *granite*.

pocket beach   *Beach* of crescentic outline located at a bay head.

point-bar deposits   Deposits of coarse *sediment* accumulating on the insides of growing *alluvial meanders*.

polar air   Air originating in high latitudes and having the characteristics of a *polar air mass*.

polar air mass   A cold *air mass* with *source region* over continents and oceans in lat. 50° to 60° N and S.

polar easterlies   System of easterly surface *winds* at high latitude, best developed in the southern hemisphere, over Antarctica.

polar front   *Front* lying between cold *polar air masses* and warm *tropical air* *masses,* often situated along a *jet stream* within the *upper-air westerlies.*

polar front jet stream   *Jet stream* formed along the *polar front,* where cold *polar air* and warm *tropical air* are in contact.

polar front zone   Broad zone in mid-latitudes and higher latitudes, occupied by the shifting *polar front.*

polar high   Persistent low-level center of high *atmospheric pressure* located over the *polar zone* of Antarctica.

polar low   Persistent center of low *atmospheric pressure* over high latitudes in the upper atmosphere.

polar outbreak   Tongue of cold *polar air,* preceded by a *cold front,* penetrating far into the *tropical zone* and often reaching the *equatorial zone;* it brings rain squalls and unusual cold.

polar zones   Latitude zones lying between 75° and 90° N and S.

pollutants   In air pollution studies, foreign matter injected by Man into the lower atmosphere as *particulate matter* or as *chemical pollutants.*

pollution dome   Broad, low dome-shaped layer of polluted air, formed over an urban area at times when winds are weak or calm prevails.

pollution plume   (1) The trace or path of *leachate* or other pollutant substances, moving along the flow paths of *ground water.* (2) Trail of polluted air carried downwind from a pollution source by strong winds.

polypedon   Smallest distinctive geographic unit of the *soil* of a given area; it consists of *pedons.*

pool of materials   Area or location of concentration of a given material in the *material cycle;* two types are *active pools* and *storage pools.*

potash feldspar   *Aluminosilicate mineral* with potassium the dominant metal.

potential evapotranspiration (water need)   Ideal or hypothetical rate of *evapotranspiration* estimated to occur from a complete canopy of green foliage of growing plants continously supplied with all the *soil water* they can use; a real condition reached in those situations where *precipitation* is sufficiently great or irrigation water is supplied in sufficient amounts.

pothole   Cylindrical cavity in hard *bedrock* of a *stream channel* produced by *abrasion* of a large spherical or discuss-shaped mass of *rock* rotating within the cavity.

prairie   Plant *formation class* of the *grassland biome,* consisting of dominant tall grasses and subdominant *forbs,* widespread in subhumid continental climate regions of the *subtropical zone* and *midlatitude zone.*

Precambrian time   All of geologic time older than the beginning of the Cambrian

Period, i.e., older than 600 million years.

precipitation   Particles of liquid water or ice that fall from the *atmosphere* and may reach the ground. (See *orographic precipitation.*)

pressure cell   Center of high or low *atmospheric pressure,* identified with a *cyclone* or an *anticyclone.*

pressure gradient   Change of *atmospheric pressure* measured along a line at right angles to the *isobars.*

pressure-gradient force   Force acting horizontally, tending to move air in the direction of lower *atmospheric pressure.*

prevailing westerly winds (westerlies)   Surface winds blowing from a generally southwesterly direction in the *midlatitude zone,* but varying greatly in direction and intensity.

primary consumers   Animals that live by feeding on *producers.*

primary minerals   In *soil science,* the original, unaltered silicate minerals of the *igneous rocks* and *metamorphic rocks.*

primary producers   Organisms that use light energy to convert carbon dioxide and water to *carbohydrates* through the process of *photosynthesis.*

primary production   (See *net primary production.*)

primary succession   *Ecological succession* that begins on a newly constructed deposit of *mineral sediment.*

prime meridian   Reference meridian of zero *longitude;* universally accepted as the *Greenwich meridian.*

principal meridian   Reference *meridian* used in the U.S. Land Office Survey.

producers   (See *primary producers.*)

progradation   Shoreward building of a *beach,* bar, or *sandspit* by addition of coarse *sediment* carried by *littoral drift* or brought from deeper water offshore.

psychrometer   (See *sling-psychrometer.*)

Puszta   Lowland region in Hungary, formerly covered by tall-grass *prairie.*

pyranometer   Instrument that measures the intensity of *insolation (shortwave radiation),* including both direct solar beam and indirect sky radiation from *down-scatter.*

pyroclastic sediment   *Sediment* consisting of particles thrown into the air by volcanic explosion in the form of *tephra* or *volcanic ash.*

pyroxene group   Complex *alumino-silicate minerals* rich in calcium, magnesium, and iron, dark in color, high in *density,* classed as *mafic minerals.*

quadrangle   Map area within designated boundary of *meridians* and *parallels.*

quartz   Mineral of silicon dioxide *(silica)* composition.

quartzite   *Metamorphic rock* consisting largely of the *mineral quartz.*

quick clays   *Clay* layers that spontaneously change from a solid condition to a near-liquid condition when disturbed, a process called *spontaneous liquefaction.*

radar   *Wavelength* band within the *microwave* region, beginning at about 0.1 cm and extending to about 100 cm (1 m).

radial pattern   Stream pattern consisting of *streams* radiating outward from a central peak or highland, such as a sedimentary *dome* or a volcano.

radiation balance   Condition of balance between incoming energy of solar *short-wave radiation* and outgoing *longwave radiation* emitted by the earth into space.

radiation fog   *Fog* produced by radiational cooling of the basal air layer.

rain   Form of precipitation consisting of falling water drops, usually 0.5 mm or larger in diameter.

rain gauge   Instrument used to measure the amount of *rain* that has fallen.

raingreen vegetation   Vegetation that puts out green foliage in the wet season, but becomes largely dormant in the dry season; found in the *tropical zone,* it includes the *savanna biome* and *monsoon forest.*

rainshadow desert   Belt of arid climate to lee of a mountain barrier, produced as a result of adiabatic warming of descending air.

raised shoreline   Former *shoreline,* lifted above the limit of wave action; also called an *elevated shoreline.*

ranges   Vertical rows of congressional townships in the U.S. Land Office Survey.

rapids   Steep-gradient reaches of a *stream channel* in which *stream* velocity is high.

recessional moraine   *Moraine* produced at the ice margin during a temporary halt in the recessional phase of *glaciation.*

recharge   (See *ground water recharge.*)

recharge well   A *tube well* used to provide *ground water recharge.*

reclining retreat   Program of *hillslope profile* evolution in which the angle of *slope* decreases with time.

recording thermometer   *Thermometer* equipped with a mechanism for making a continuous record of temperature; same as *thermograph.*

reg   Desert surface armored with a pebble layer, resulting from long-continued *deflation;* found in the Sahara Desert of North Africa.

regolith   Layer of *mineral* particles overlying the *bedrock;* may be derived by *weathering* of underlying bedrock or transported from other locations by *fluid agents.* (See *residual regolith, transported regolith.*)

rejuvenation Onset of an episode of *degradation* (downcutting) by a *stream* in an attempt to reestablish grade at a lower level.

relative humidity Ratio of *water vapor* present in the air to the maximum quantity possible for *saturated air* at the same temperature.

relief Measure of the average elevation difference between adjacent high and low points in a land surface, as for example hilltops and valley bottoms.

remote sensing Measurement of some property of an object or surface by means other than direct contact; usually refers to the gathering of scientific information about the earth's surface from great heights and over broad areas, using instruments mounted on aircraft or orbiting space vehicles.

remote sensors Collective term for detection instruments or devices used in *remote sensing.*

representative fraction (R.F.) *Fractional scale* stated as a simple fraction.

residual regolith *Regolith* formed in place by alteration of the *bedrock* directly beneath it.

residual till Glacial *till* deposited as stagnant ice melts away, leaving the enclosed rock particles behind.

respiration Metabolic process in which organic compounds are oxidized within living cells to yield biochemical energy and waste heat.

retrogradation Cutting back (retreat) of a *shoreline, beach, marine cliff,* or *marine scarp* by wave action.

reverse fault Type of *fault* in which one *fault block* rides up over the other on a steep *fault plane.*

revolution Motion of a planet in its *orbit* around the sun, or of a planetary satellite around a planet.

rhumb line Line of constant compass bearing drawn on a map or navigational chart; also known as a *loxodrome.*

rhyolite *Extrusive igneous rock* of *granite* composition; occurs as *lava.*

ria coast Deeply embayed *coast* formed by partial *submergence* of a *landmass* previously shaped by fluvial *denudation.*

Richter scale Scale of magnitude numbers describing the quantity of energy released by an *earthquake.*

rift valley Trenchlike valley with steep, parallel sides; essentially a *graben* between two *normal faults;* associated with crustal spreading.

rill erosion Form of *accelerated erosion* in which numerous, closely spaced miniature channels (rills) are scored into the surface of exposed *soil* or *regolith.*

roches moutonées Knobs of *bedrock* formed by *glacial abrasion.*

rock Natural aggregate of minerals in the *solid state;* usually hard and consisting of one, two, or more *mineral* varieties.

rock basin Overdeepened section of *bedrock* floor of a *cirque* or *glacial trough* forming a depression, often holding a *tarn.*

rock-defended terrace *Alluvial terrace* protected from further removal by undercutting by the presence of a *bedrock outcrop.*

rockfall Free fall of particles or masses of *bedrock* from a steep *cliff* face, typically accumulating at the cliff base in the form of *talus.*

rockslide Form of *landslide* consisting of the slippage of a *bedrock* mass on a sloping fracture plane.

rock step Abrupt down-step in the rock floor of a *glacial trough.*

rock terrace *Bedrock terrace* remaining after the *degradation* of a *stream channel* induced by a crustal rise or a fall of the sea level.

rock texture (See *texture of rock.*)

rock transformation cycle Total cycle of changes in which *rock* of any one of the three major rock classes — *igneous rock, sedimentary rock, metamorphic rock* — is transformed into rock of one of the other classes.

Rossby waves Horizontal undulations in the flow path of the *upper-air westerlies;* also known as *upper-air waves.*

rotation Spinning of a spherical object around an axis.

runoff Flow of water from continents to oceans by way of *stream flow* and *ground water* flow; a term in the water balance of the *hydrologic cycle.* In a more restricted sense, runoff refers to surface flow by *overland flow* and *channel flow.*

salic horizon *Soil horizon* enriched by soluble salts.

salinization Precipitation of soluble salts within the *soil.*

saltation Leaping, impacting, and rebounding of spherical *sand* grains transported over a sand or *pebble* surface by wind.

salt marsh Peat-covered expanse of *sediment* built up to the level of high tide over a previously formed *tidal flat.*

sand *Sediment* particles between 0.06 mm and 2 mm in diameter.

sand dune Hill or ridge of loose, well-sorted *sand* shaped by wind and usually capable of downwind motion.

sand-ridge desert Area dominated by *longitudinal dunes.*

sand sea Field of *transverse dunes.*

sandspit Narrow, fingerlike embankment of *sand* constructed by *littoral drift* into the open water of a bay.

sandstone Variety of *sedimentary rock* consisting largely of *mineral* particles of *sand* grade size.

sand storm Dense, low layer of *sand* grains traveling in *saltation* over a *sand dune* or *beach* surface.

sanitary landfill Disposal of solid wastes by burial beneath a cover of *soil* or *regolith.*

Santa Ana Easterly wind, often hot and dry, that blows from the interior desert region of southern California and passes over the coastal mountain ranges to reach the Pacific Ocean.

saprolite Layer of deeply decayed *igneous rock* or *metamorphic rock,* rich in *clay minerals,* typically formed in warm, humid climates; a form of *residual regolith.*

saturated air Air holding the maximum possible quantity of *water vapor* at a given temperature and pressure.

saturated zone Zone beneath the land surface in which all pores of the *bedrock* or *regolith* are filled with *ground water.*

savanna biome *Biome* that consists of a combination of trees and grassland in various proportions.

savanna woodland Plant *formation class* of the *savanna biome* consisting of a *woodland* of widely spaced *trees* and a grass layer, found throughout the *wet-dry tropical climate* regions in a belt adjacent to the *monsoon forest* and *equatorial rainforest.*

scale of globe Ratio of size of a globe to size of the earth, where size is expressed by a measure of length or distance.

scale of map Ratio of distance between two points on a map and the same two points on the ground. (See *fractional scale, graphic scale.*)

scanning systems *Remote sensing* systems that make use of a scanning beam to generate images over the frame of surveillance.

scarification General environmental impact term for Man-made excavations and other land disturbances produced for purposes of extracting or processing *mineral* resources.

scattering Turning aside by reflection of solar *shortwave radiation* by *gas* molecules of the *atmosphere.*

schist Foliated *metamorphic rock* in which mica flakes are typically found oriented parallel with foliation surfaces.

sclerophyll forest Plant *formation class* of the *forest biome,* consisting of low *sclerophyllous trees,* and often including *sclerophyllous woodland* or *scrub,* associated with regions of *Mediterranean climate.*

sclerophyllous leaves Leaves of *sclerophylls;* they are hard, thick, and leathery.

sclerophylls Hard-leaved *evergreen*

trees and *shrubs* capable of enduring a long, dry summer.

scree slope (See *talus slope.*)

scrub Plant *formation class* or subclass consisting of *shrubs* and having a canopy coverage of about 50%.

sea breeze Local *wind* blowing from sea to land during the day.

sea cave Cave near the base of a *marine cliff,* eroded by breaking waves.

sea cliff (See *marine cliff.*)

sea-floor spreading Pulling apart of the *crust* along the *axial rift* of the *mid-oceanic ridge,* and representing the separation of two *lithospheric plates* made up of *oceanic lithosphere.*

sea ice Floating ice of the oceans formed by direct freezing of ocean water.

secant cone Cone imagined to pass through two *standard parallels* of *latitude* on the earth's surface.

secant conic projection. Type of *conic projection* using a *secant cone* and having two *standard parallels.*

secondary consumers Animals that feed on *primary consumers.*

secondary minerals In *soil science, minerals* that are stable in the surface environment, derived by *mineral alteration* of the *primary minerals.*

secondary succession *Ecological succession* beginning on a previously vegetated area that has been recently disturbed by such agents as fire, flood, windstorm, or Man.

sections Units of land, one mile square, used in the U.S. Land Office Survey.

sediment Finely divided *mineral* matter and organic matter drived directly or indirectly from preexisting *rock* and from life processes. (See *chemical precipitate, nonclastic sediment, organically-derived sediment.*)

sedimentary cycle Type of *material cycle* in which the compound or element is released from rock by *weathering,* follows the movement of running water either in solution or as sediment to reach the sea, and is eventually converted into rock.

sedimentary dome Up-arched *strata* forming a circular structure with domed summit and flanks with moderate to steep outward *dip.*

sedimentary rock *Rock* formed from accumulations of *sediment.*

sediment yield Quantity of *sediment* removed by *overland flow* from a land surface of given unit area in a given unit of time.

seismic sea wave (tsunami) Train of sea waves set off by an *earthquake* (or other sea floor disturbance) traveling over the ocean surface with a velocity proportional to the square root of the ocean depth.

seismic waves Waves sent out during an *earthquake* by faulting or other crustal disturbance from an *earthquake focus* and propagated through the solid earth.

semiarid (steppe) climate subtype Subtype of the *dry climate* in which *soil-water storage* equals or exceeds 6 cm in at least two months of the year.

semideciduous plants Plants that shed their leaves at intervals not in phase with a season. (See also *deciduous plants.*)

semidesert Plant *formation class* of the *desert biome,* consisting of xerophytic *shrub* vegetation with a poorly developed *herbaceous* lower layer; subtypes are *semidesert scrub and woodland,* and *semidesert scrub.*

semidesert climate subtype Subtype of the dry climate in which soil-water storage exceeds 6 cm in fewer than two months, but is greater than 2 cm in at least one month.

semidesert scrub Subtype of the *semidesert formation class.*

semidesert scrub and woodland Subtype of the *semidesert* plant *formation class.*

sensible heat Heat measurable by a *thermometer;* an indication of the intensity of kinetic energy of molecular motion within a substance.

sequential landforms *Landforms* produced by *external earth processes* in the total activity of *denudation.* Examples: *canyon, alluvial fan, floodplain.*

seral stage Stage in a *sere.*

sere In an *ecological succession,* the series of *biotic communities* that follow one another on the way to the stable stage, or *climax.*

sesquioxide of aluminum Oxide of aluminum with a ratio of two atoms of aluminum to three atoms of oxygen.

shale Fissile, *sedimentary rock* of mud or *clay* composition, showing lamination.

shattering Form of *physical weathering* in which fresh *rock* fractures are produced by powerful mechanical stresses.

shearing Distortion on a *map projection* resulting in the intersection of *parallels* and *meridians* in acute and obtuse angles.

sheet erosion Phase of *accelerated soil erosion* in which thin layers of *soil* are removed without formation of *shoestring rills* or *gullies.*

sheet flow *Overland flow* taking the form of a continuous thin film of water over a smooth surface of *soil, regolith,* or *rock.*

sheeting structure Thick, subparallel layers of massive *bedrock* formed by spontaneous expansion accompanying *unloading.*

shield volcano Domelike accu-

mulation of *basalt lava* flows emerging from long radial fissures on flanks.

shoreline Shifting line of contact between water and land.

short-grass prairie (See *steppe.*)

shortwave radiation *Electromagnetic radiation* in the range from 0.2 to 3 *microns,* including most of the energy spectrum of solar radiation.

shrub Woody perennial plant, usually small or low, with several low-branching stems and a foliage mass close to the ground.

Siberian high Center of high *atmospheric pressure* located over north-central Asia in winter.

side-looking airborne radar (SLAR) *Remote sensing* using *radar* sensor systems that send their impulses toward either side of an aircraft.

silcrete Rocklike surface layer cemented largely with *silica,* widespread in the *tropical zone.*

silica Silicon dioxide in any of several *mineral* forms, including the crystalline form as *quartz* and partly crystalline or amorphous forms such as *allophane* and *chalcedony.*

silicate magma *Magma* from which *silicate minerals* are formed.

silicates, silicate minerals *Minerals* containing silicon and oxygen atoms, linked in the crystal space lattice in units of four oxygen atoms to each silicon atom.

silication *Pedogenic process* in which the proportion of *silica* in a *soil horizon* is increased because the silica remains behind while other materials are removed.

sill *Intrusive igneous rock* in the form of a plate where *magma* was forced into a natural parting in the *bedrock,* such as a bedding surface in a sequence of *sedimentary rocks.*

silt *Sediment* particles between 0.004 mm and 0.06 mm in diameter.

siltstone *Sedimentary rock* consisting of particles largely of *silt* grade.

sine curve Curve on a graph on which the sine of an angle is plotted against the value of that angle from 0° to 360°.

sinkhole Surface depression in *limestone,* leading down into *limestone caverns.*

sinusoidal projection *Map projection* that is a *conformal projection* consisting of straight horizontal *parallels* and of *meridians* that are *sine curves.*

skeletal minerals *Mineral* particles, mostly of *sand* and *silt* grades, that make up the chemically inactive fraction of the *soil,* as distinguished from the clay minerals and related *weathering* products.

slash-and-burn Agricultural system, practiced in the low-latitude rainforest, in which small areas are cleared and the trees burned, forming plots that can be cultivated for brief periods.

slate   Compact, fine-grained variety of *metamorphic rock,* derived from *shale,* showing well-developed cleavage.

sleet   Form of *precipitation* consisting of ice pellets, which may be frozen raindrops.

slickensides   *Soil* surfaces with grooves or striations made by movement of one mass of soil against the other while the soil is in a moist, plastic state.

sling psychrometer   Form of *hygrometer* consisting of a wet-bulb *thermometer* and a dry-bulb *thermometer.*

slip face   Steep face of an active *sand dune,* receiving sand by saltation over the dune crest and repeatedly sliding because of oversteepening.

slope   (1) Degree of inclination from the horizontal of an element of ground surface, analogous to *dip* in the geologic sense. (2) Any portion or element of the earth's solid surface, as in *hillslope.* (3) Verb "to incline."

slump   Form of *landslide* in which a single large block of *bedrock* moves downward with backward rotation upon an upwardly concave fracture surface.

slump block   Block of *bedrock* that moves down to a lower position of rest on a curved slip surface. (See also *slump.*)

small circle   Circle produced when a plane is passed through a sphere, but not passing through the center of the sphere.

small leaf-form   Leaves that are thin, flat, and comparatively wide, but of small overall dimensions as compared with the *broadleaf.*

smog   Mixture of *particulate matter* and *chemical pollutants* in the lower *atmosphere* over urban areas.

snow   Form of *precipitation* consisting of ice particles.

soft layer of mantle   Layer within the *mantle* in which the temperature is close to the melting point, causing the mantle *rock* to be weak; synonym for *asthenosphere.*

soil   Natural terrestrial surface layer containing living matter and supporting or capable of supporting plants.

soil consistence   Quality of stickiness of wet *soil,* the plasticity of moist *soil,* and the degree of coherence or hardness of soil when it holds small amounts of moisture or is in the dry state.

soil creep   Extremely slow downhill movement of *soil* and *regolith* as a result of continued agitation and disturbance of the particles by activities such as *frost action,* temperature changes, or wetting and drying of the soil.

soil enrichment   Additions of materials to the *soil* body; one of the *pedogenic processes.*

soil erosion   Erosional removal of material from the *soil* surface.

soil horizon   Distinctive layer of the *soil,* more or less horizontal, set apart from other soil zones or layers by differences in physical and chemical composition, organic content, structure, or a combination of those properties, produced by soil-forming processes (*pedogenic processes*).

soil moisture   (See *soil water.*)

soil orders   Those 10 *soil* classes forming the highest category in the *Soil Taxonomy* within the *Comprehensive Soil Classification System (CSCS).*

soil profile   Display of *soil horizons* on the face of a *pedon,* or on any freshly cut vertical exposure through the *soil.*

soil science   (See *pedology.*)

soil solum   That part of the *soil* made up of the A and B *soil horizons;* the soil zone in which living plant roots exert a control on the soil horizons.

soil solution   Aqueous solution held in the *soil* as *soil water* and containing dissolved atmospheric *gases* and *ions.*

soil structure   Presence, size, and form of aggregations (lumps or clusters) of *soil* particles.

soil suborders   Second level of classification of *soils* in the *Soil Taxonomy.*

Soil Taxonomy   Taxonomic system of classifying *soils;* part of the *Comprehensive Soil Classification System.*

soil-temperature regime   Characteristic annual cycle of *soil* temperature defined in terms of mean annual soil temperature and the difference between mean temperatures of the warm and cold seasons.

soil-texture classes   Classes of the *mineral* portion of the *soil* based upon varying proportions of *sand, silt,* and *clay,* expressed as percentages.

soil water   Water held in the *soil* and available to plants through their root systems; a form of *subsurface water.* (Synonymous with *soil moisture.*)

soil-water balance   Balance among the component terms of the *soil-water budget;* namely, *precipitation, evapotranspiration,* change in *soil-water storage,* and *water surplus.*

soil-water belt   *Soil* layer from which plants draw *soil water.*

soil-water budget   Accounting system evaluating the daily, monthly, or yearly amounts of *precipitation, evapotranspiration, soil-water storage, water deficit,* and *water surplus.*

soil-water control section   Depth zone in the *soil* recognized for the purpose of establishing the *soil-water regime* and defined in terms of the depth to which a given amount of applied water will moisten a dry soil.

soil-water recharge   Restoring of

depleted *soil water* by *infiltration* of *precipitation.*

soil-water regime   Characteristic type of annual *soil-water budget* adapted to purposes of soil classification and defined in terms of conditions prevailing in the *soil-water control section.*

soil-water shortage   Difference between *potential evapotranspiration (water need)* and *actual evapotranspiration (water use),* representing the quantity of irrigation water that would be required to sustain maximum plant growth.

soil-water storage   Actual quantity of water held in the *soil-water belt* at any given instant; usually applied to a soil layer of given depth, such as 300 cm (12 in.).

solar constant   Intensity of solar radiation falling upon a unit area of surface held at right angles to the sun's rays at a point outside the earth's atmosphere; equal to about 2 gram-calories per square centimeter per minute (2 cal/cm²/min), or 2 *langleys* per minute (2 ly/min).

solar noon   Instant at which the sun crosses the celestial meridian of a given point on the earth; instant at which the sun's shadow points exactly due north or due south.

solar wind   Flow of electrons and protons emanating from the sun and traveling outward in all directions through the solar system.

solids   Substances in the solid state that resist changes in shape and volume, are usually capable of withstanding large unbalanced forces without yielding, but will ultimately yield by sudden breakage.

solid state   State of matter that is dense, has strength to resist flowage, shows elastic properties, and may be a crystalline solid or an amorphous (noncrystalline) substance. Example: *mineral* crystal, *rock.*

solifluction   Tundra (arctic) variety of *earthflow* in which the saturated thawed layer over *permafrost* flows slowly downhill to produce multiple terraces and *solifluction lobes.*

sorting   Separation of one grade size of *sediment* particles from another by the action of currents of air or water.

source region   Extensive land or ocean surface over which an *air mass* derives its temperature and moisture characteristics.

spalling   Form of *weathering* in which concentric, curved *rock* shells are produced; a small-scale form of *exfoliation.*

specific humidity   Mass of *water vapor* contained in a unit mass of air.

spectrum of electromagnetic radiation   Total range of wavelengths and frequencies from *gamma rays* to radio waves.

spheroidal weathering Production of thin, soft concentric shells of decomposed *rocks* as *chemical weathering* penetrates *joints* in *bedrock* under a protective cover of *regolith*.

spine Leaf form that is a hard, sharp-pointed spike.

spit (See *sandspit*.)

splash erosion *Soil* erosion caused by direct impact of falling raindrops on a wet surface of *soil* or *regolith*.

spodic horizon *Soil horizon* containing precipitated amorphous materials composed of organic matter and *sesquioxides* of aluminum, with or without iron.

Spodosols *Soil order* within the *Soil Taxonomy* consisting of *soils* with a *spodic horizon*, an *albic horizon*, with low *cation exchange capacity (CEC)*, and lacking in *carbonate* materials.

spontaneous liquefaction (See *quick clays*.)

spot heights Numerals printed on a map to show the altitude of selected points.

stable air mass *Air mass* in which the *environmental temperature lapse rate* is less than the *dry adiabatic lapse rate* and thus resists being lifted.

stack (marine) Isolated, columnar mass of *bedrock* left standing during retrogradation of a *marine cliff*.

staff gauge Graduated vertical scale used to measure *stream stage*.

standard meridians *Standard time* meridians separated by 15° of *longitude* and having values that are multiples of 15°. (In some cases meridians are used that are multiples of 7½°.)

standard parallels (1) two *parallels of latitude* along which a *secant cone* intersects the surface of the earth or a globe. (2) Parallels of latitude, spaced 24 miles apart, used as secondary *base lines* in the U.S. Land Office Survey.

standard time Time system based upon the *local time* of a *standard meridian* and applied to belts of longitude extending 7½° (more or less) on either side of that meridian.

star dune Large, isolated *sand dune* with radial ridges culminating in a peaked summit; found in the deserts of North Africa and the Arabian Peninsula.

Stefan-Boltzmann law Total energy radiated by a unit area of surface per unit of time varies as the fourth power of the *absolute temperature* (°K).

steppe climate (See *semiarid climate subtype*.)

steppe Plant *formation class* in the *grassland biome* consisting of short grasses sparsely distributed in clumps and bunches and some *shrubs*, widespread in areas of semiarid climate in continental interiors of North America and Eurasia; also called *short-grass prairie*.

stereographic projection *Conformal projection* that is a *zenithal projection* on which any arc of a circle on the globe is shown as an arc of a circle on the map.

stomata Specialized leaf pores that are openings in the outer cell layer through which *transpiration* occurs.

stone polygons Ringlike networks of coarse fragments; same as *stone rings*.

stone rings Linked ringlike ridges of cobbles or boulders lying at the surface of the ground in arctic and alpine tundra regions.

storage capacity Maximum capacity of *soil* to hold water against the pull of gravity; same as *field capacity*.

storage pools Type of *pool* in the *material cycle* in which materials are more or less inaccessible to life. (See also *active pools*.)

storage recharge Restoration of stored *soil water* during periods when *precipitation* exceeds *potential evapotranspiration (water need)*.

storage withdrawal Depletion of stored *soil water* during periods when *evapotranspiration* exceeds *precipitation*, calculated as the difference between *actual evapotranspiration (water use)* and *precipitation*.

storm surge Rapid rise of coastal water level accompanying the onshore arrival of a *tropical cyclone*.

strangler Twining plant or *liana* that surrounds a *tree* trunk, eventually killing and replacing the tree.

strata Layers of *sediment* or *sedimentary rock* in which individual beds are separated from one another along *bedding planes*.

stratified drift *Glacial drift* made up of sorted and layered *clay, silt, sand,* or gravel deposited from meltwater in *stream channels* or *marginal lakes* close to the ice front.

stratiform clouds *Clouds* of layered, blanketlike form.

stratocumulus *Cloud* type of the low-height family consisting of a layer of individual dense cloud masses.

stratopause Upper limit of the *stratosphere*, transitional upward into the *mesosphere*.

stratosphere Layer of *atmosphere* lying directly above the *troposphere*.

stratus *Cloud* type of the low-height family formed into a dense, dark gray layer.

stream Long, narrow body of flowing water occupying a *stream channel* and moving to lower levels under the force of gravity. (See *consequent stream, graded stream, subsequent stream*.)

stream capacity Maximum *stream load* of solid matter that can be carried by a stream for a given *discharge*.

stream channel Long, narrow, troughlike depression occupied and shaped by a *stream* moving to progressively lower levels.

stream deposition Accumulation of transported particles on a *stream* bed, upon the adjacent *floodplain*, or in a body of standing water.

stream erosion Progressive removal of *mineral* particles from the floor or sides of a *stream channel* by drag force of the moving water, or by *abrasion*, or by *corrosion*.

stream flow Water flow in a *stream channel*; same as *channel flow*.

stream gauging Measurement of *stream discharge, mean velocity,* and depth continuously or at intervals over a long period of time at a selected station.

stream gradient Rate of descent to lower elevations along the length of a *stream channel*, stated in m/km, ft/mi, degrees, or percent.

stream load Solid matter carried by a *stream* in dissolved form (as *ions*), in *turbulent suspension,* and as *bed load*.

stream stage Height or elevation of the surface of a *stream* at any given instant.

stream transportation Down-valley movement of eroded particles in a *stream channel* in solution, in *turbulent suspension*, or as *bed load*.

striations Scratches made by *glacial abrasion* on *bedrock outcrops* or the surfaces of rock fragments carried by glacial ice.

strike Compass direction of the line of intersection of an inclined *rock* plane and a horizontal plane of reference. (See *dip*.)

strip mining Mining method in which overburden is first removed from a seam of *coal*, or a sedimentary ore, allowing the coal or ore to be extracted.

structure section Graphic presentation of a vertical cross section from the surface to a given depth, showing the arrangement of *rock* units and other geologic features.

subantarctic low-pressure belt Persistent belt of low *atmospheric pressure* centered about at lat. 65°S over the Southern Ocean.

subantarctic zone Latitude zone lying between lat. 55° and 60°S (more or less) and occupying a region between the *midlatitude zone* and the *antarctic zone*.

subarctic zone Latitude zone between lat. 55° and 60°N (more or less), occupying a region between the *midlatitude zone* and the *arctic zone*.

Subboreal climatic stage Climate stage of the *Holocene Epoch*, with below-average temperatures, spanning the period of about −5000 to −2000 years.

subduction Descent of the downbent edge of a *lithospheric plate* into the

asthenosphere so as to pass beneath the edge of the adjoining plate.

subhumid climate subtype Subtype of the *moist climate* in which annual *soil-water shortage* is greater than zero but less than 15 cm.

sublimation Process of change of *water vapor (gaseous state)* to ice (*solid state*) or vice versa.

submergence Inundation or partial drowning of a former land surface by a rise of sea level or a sinking of the *crust* or both.

subsequent stream *Stream* that develops its course by *stream erosion* along a band or belt of weaker *rock*.

subsidence of air Downsinking of a large mass of air in the central region of an *anticyclone*.

subsidence theory Hypothesis, advanced by Charles Darwin, explaining *atolls* by subsidence of the oceanic crust and continued upbuilding of coral reefs.

subsolar point Point at which solar rays are perpendicular to the earth's surface.

subsurface water Water of the lands held in *soil, regolith,* or *bedrock* below the surface.

subtropical high-pressure belts Belts of persistent high *atmospheric pressure* trending east-west and centered about on lat. 30° N and S.

subtropical jet stream *Jet stream* of westerly winds forming at the *tropopause*, just above the *Hadley cell*.

subtropical zones Latitude zones occupying the region of lat. 25° to 35° N and S (more or less) and lying between the *tropical zones* and the *midlatitude zones*.

succulent Adjective applied to plants with thickened, spongy leaves or stems capable of holding much water.

succulents Plants adapted to resist water loss (*xerophytes*) by means of thickened spongy tissue in which water is stored.

summer berm *Berm* produced by weak waves during the summer season in midlatitudes.

summergreen deciduous forest (See *midlatitude deciduous forest*.)

summer monsoon Inflow of maritime air at low levels from the Indian Ocean toward the Asiatic low pressure center in the season of high sun; associated with the rainy season of the *wet-dry tropical climate* and the Asiatic monsoon climate.

summer solstice *Solstice* occurring on June 21 or 22, when the *sun's declination* is 23½°N.

sun's declination *Latitude* of that *parallel* on which the *subsolar point* is located at any given instant; ranges from lat. 23½°N to 23½°S throughout the *tropical year*.

sun's noon altitude Vertical angle

between the sun and the horizon at the instant of *solar noon*.

sun-synchronous orbit Satellite orbit in which the orbital plane remains fixed in position with respect to the sun.

supercooled water Water existing in the *liquid state* at a temperature lower than the normal freezing point.

surface detention Temporary holding of *precipitation* in minor surface depressions; same as *depression storage*.

surface environment Environment of low pressure and low temperature to which *rock* is exposed near the earth's surface.

surface water Water of the lands flowing exposed (as *streams*) or impounded (as ponds, lakes, or marshes).

surges Episodes of very rapid down-valley movement within an *alpine glacier*.

suspended load That part of the stream *load* carried in *turbulent suspension*.

suture The narrow zone of crustal deformation and alpine structure produced by *continental collision*. Example: Himalayas.

swales Long, narrow depressions lying between *beach ridges*, or between bars of a *point-bar deposit* on a river *floodplain*.

swash Surge of water up the *beach* slope (landward) following collapse of a breaker.

synclinal mountain Ridge or elongate mountain developed by erosion on a *syncline*.

synclinal valley Valley eroded in weak *strata* along the central trough or axis of a *syncline*.

syncline Downfold of *strata* (or other layered *rock*) in a troughlike structure; a class of *folds*. (See also *anticline*.)

taiga (See *cold woodland*.)

talus Accumulation of loose rock fragments derived by *rockfall* from a *cliff*.

talus cone Accumulation of *talus* in the form of a partial cone with apex at top, heading in a ravine or *gully*.

talus slope *Slope* formed of *talus*.

tarn Small lake occupying a *rock basin* in a *cirque* or *glacial trough*.

tectonic activity Crustal processes of bending (folding) and breaking (faulting), concentrated on or near active *lithospheric plate* boundaries.

temperate rainforest (See *broadleaf evergreen forest, laurel forest*.)

temperature inversion Upward reversal of the normal *environmental temperature lapse rate*, so that the air temperature increases upward. (See *low-level temperature inversion, upper-level temperature inversion*.)

tephra Collective term for all size

grades of solid *igneous rock* particles blown out under gas pressure from a volcanic *vent*.

terminal moraine *Moraine* deposited as an embankment at the *glacier terminus* of an *alpine glacier* or at the leading edge of an *ice sheet*.

terrestrial ecosystems *Ecosystems* of land plants and animals found on upland surfaces of the continents.

texture (of rock) Physical property of *rock* pertaining to the sizes, shapes, and arrangements of *mineral* grains it contains, i.e., coarse-grained or fine-grained.

thermal environment Total influence of heat and cold upon living organisms in the *life layer*.

thermal erosion In regions of *permafrost*, the physical disruption of the land surface by melting of ground ice, brought about by removal of a protective organic layer.

thermal infrared *Electromagnetic radiation* in the *infrared radiation wavelength* band approximately from 1 to 20 microns.

thermal pollution Form of water pollution in which heated water is discharged into a *stream* or lake from the cooling system of a power plant or other industrial heat source.

thermal regimes Global set of air-temperature regimes, based upon distinctive annual cycles of *mean monthly air temperature*.

thermocline Water layer in which temperature changes rapidly in the vertical direction.

thermograph (See *recording thermometer*.)

thermometer Instrument measuring temperature. (See *maximum-minimum thermometer, recording thermometer*.)

thermosphere Atmospheric layer of upwardly rising temperature, lying above the *mesopause*.

thorn forest and thorn woodland Subtype of the *thorntree semidesert formation class*.

thorntree-desert grass savanna Subtype of the *thorntree semidesert* plant *formation class*.

thorntree semidesert *Formation class* within the *desert biome*, transitional from *grassland biome* and *savanna biome* and consisting of xerophytic *trees* and *shrubs*. There are two subtypes: *thorn forest and thorn woodland*, and *thorntree-desert grass savanna*.

thorntree-tall grass savanna Plant *formation class*, transitional between the *savanna biome* and the *grassland biome*, consisting of widely scattered *trees* in an open grassland.

throughflow (See *interflow*.)

thrust sheet   Sheetlike mass of *rock* moving forward over a low-angle *overthrust fault.*

thunderstorm   Intense, local convectional storm associated with a *cumulonimbus cloud* and yielding heavy *precipitation,* along with lightning and thunder, and sometime the fall of *hail.*

tidal current   Current set in motion by the *ocean tide.*

tidal inlet   Narrow opening in a *barrier island* or *baymouth bar* through which *tidal currents* flow.

tide   (See *ocean tide.*)

tide curve   Graphical presentation of the rhythmic rise and fall of ocean water because of *ocean tides.*

tide range   Difference in height between one *high water* and the following or preceding *low water* in *ocean tides.*

till   Heterogeneous mixture of *rock* fragments ranging in size from *clay* to *boulders,* deposited beneath moving glacial ice or directly from the melting in place of stagnant glacial ice.

till plain   Undulating, plainlike land surface underlain by glacial *till.*

tombolo   *Sand* bar or narrow *beach* connecting an island with the mainland.

topographic profile   Graph with elevation (altitude) on the vertical scale and distance on the horizontal scale, on which the *relief* of the land surface along a given traverse line is shown by a rising and falling line.

tor   Group of boulders or *joint blocks* forming a small but conspicuous hill.

tornado   Small, very intense wind vortex with extremely low *air pressure* in center, formed beneath a dense *cumulonimbus cloud* in proximity to a *cold front.*

trade winds   Surface *winds* in low latitudes, representing the low-level air flow within the *tropical easterlies.*

transcurrent fault   Variety of *fault* on which the motion is dominantly horizontal along a near-vertical *fault plane.*

transform fault   Special case of a *transcurrent fault* making up the boundary of two moving *lithospheric plates,* usually found along an offset of the *mid-oceanic ridge* where *sea-floor spreading* is in progress.

transpiration   Evaporative loss of water to the atmosphere from leaf pores of plants.

transportation   (See *stream transportation.*)

transported regolith   *Regolith* formed of *mineral* matter carried by fluid agents from a distant source and deposited upon the *bedrock* or upon older regolith. Examples: *floodplain silt,* lake *clay, beach sand.*

transverse dunes   Field of wavelike *sand dunes* with crests running at right angles to the direction of the prevailing wind.

travertine   Carbonate *mineral* matter, usually *calcite,* accumulating upon *limestone cavern* surfaces situated in the *unsaturated zone.*

tree   Large erect woody perennial plant typically having a single main trunk, few branches in the lower part, and a branching crown.

trellis drainage pattern   *Drainage pattern* characterized by a dominant parallel set of major *subsequent streams,* joined at right angles by numerous short tributaries; typical of *coastal plains* and belts of eroded folds.

triangular facets   Steeply inclined *bedrock* surfaces of triangular outline occurring between *canyons* carved into the *fault scarp* near the base of a *fault block* mountain.

tropical air   Air originating in the *subtropical zone* and *tropical zone,* having the characteristics of a *tropical air mass.*

tropical air mass   Warm *air mass* with *source region* over continents and oceans in lat. 20° to 35° N and S.

tropical cyclone   Intense traveling *cyclone* of tropical and subtropical latitudes, accompanied by high winds and heavy rainfall.

tropical easterlies   Low-latitude *wind* system of persistent air flow from east to west between the two *subtropical high-pressure belts.*

tropical easterly jet stream   Upper-air *jet stream* of seasonal occurrence, running east to west at very high altitudes over Southeast Asia.

tropical rainforest   Plant *formation class* similar in structure to the *equatorial rainforest,* but extending into the *tropical zone* along coasts windward to the *trade winds.*

tropical year   *Year* defined as the time elapsing between one *vernal equinox* and the next.

tropical zones   Latitude zones centered on the *tropic of cancer* and the *tropic of capricorn,* within the latitude ranges 10° to 25°N and 10° to 25°S, respectively.

tropic of cancer   *Parallel of latitude* at 23½°N.

tropic of capricorn   *Parallel of latitude* at 23½°S.

tropopause   Boundary between *troposphere* and *stratosphere.*

tropophyte   Plant that sheds its leaves and enters a dormant state during a dry or cold season when little *soil water* is available.

troposphere   Lowermost layer of the *atmosphere* in which air temperature falls steadily with increasing altitude.

trough   (See *glacial trough, hanging trough.*)

trough lake   Lake occupying part of the floor of a *glacial trough.*

true azimuth   *Azimuth* system referred to *true north* (geographic north.)

true north   Direction from a point on the earth's surface to the north geographic pole, following a *meridian* through the point; same as *geographic north.*

truncated spur   Valley spur that has been beveled off by *glacial abrasion* to become part of the sidewall of a *glacial trough.*

tsunami   (See *seismic sea wave.*)

tube well   Drilled water well that makes use of a metal casing.

tundra biome   *Biome* of the cold regions of *arctic tundra* and *alpine tundra,* consisting of grasses, grasslike plants, flowering *herbs,* dwarf *shrubs,* mosses and *lichens.*

tundra climate   Cold *climate* of the *arctic zone* with 8 or more consecutive months of zero *potential evapotranspiration (water need).*

turbidity current   Rapid downslope streamlike flow of turbid (muddy) sea water close to the sea bed, and often confined within a *submarine canyon* on the *continental shelf,* or flowing down the side of an *oceanic trench.*

turbulence   (See *turbulent flow.*)

turbulent flow   *Fluid* flow in which individual water particles move in compex eddies, superimposed on the average downstream flow path.

turbulent suspension   *Stream transportation* in which particles of *sediment* are held in the body of the *stream* by turbulent eddies.

typhoon   *Tropical cyclone* of the western North Pacific and coastal waters of Southeast Asia.

Utisols   *Soil order* in the *Soil Taxonomy,* consisting of *soils* of mesic and warmer *soil-temperature regimes* with an *argillic horizon* and low *base status.*

ultramafic igneous rock   *Igneous rock* composed almost entirely of *mafic minerals,* usually *olivine* or *pyroxene.*

ultraviolet rays   *Electromagnetic radiation* in the *wavelength* range of 0.2 to 0.4 microns.

umbric epipedon   Dark surface *soil horizon (epipedon)* resembling the *mollic epipedon,* but with *percent base saturation (PBS)* less than 50.

unloading   Process of removal of overlying *rock* load from *bedrock* by processes of *denudation,* accompanied by expansion and often leading to the development of *sheeting structure.*

unsaturated zone   Subsurface water zone in which pores are not fully saturated, except at times when *infil-*

tration is very rapid; lies above the *saturated zone.*

unstable air mass *Air mass* with substantial content of *water vapor,* capable of breaking into spontaneous convectional activity leading to the development of heavy showers and *thunderstorms.*

upper-air westerlies System of *westerly winds* in the upper *atmosphere* over middle and high latitudes.

upper-level temperature inversion *Temperature inversion* produced by subsidence of air within an *anticyclone* and occurring at the top of a cool, stable air layer.

uprush (See *swash.*)

valley train Deposit of *alluvium* extending down-valley from a melting *glacier.*

valley winds Air movement at night down the gradient of valleys and the enclosing mountainsides; alternating with daytime *mountain winds.*

Van Allen radiation belt Doughnut-shaped belt of intense ionizing radiation surrounding the earth, within the inner *magnetosphere.*

varves Fine *glaciolacustrine sediment* showing annual banding in the form of alternate light and dark layers.

veldt Region of *steppe* grassland in Orange Free State and the Transvaal of South Africa.

vermiculite Group of *clay minerals* consisting of hydrous *aluminosilicates* rich in iron and manganese; a common product of the weathering of mafic volcanic rocks.

vernal equinox *Equinox* occurring on March 20 or 21, when the *sun's declination* is 0°.

vertical exaggeration Ratio by which the vertical scale on a *topographic profile* is greater than the horizontal scale.

Vertisols *Soil order* of the *Soil Taxonomy,* consisting of *soils* of the *subtropical zone* and the *tropical zone* with high *clay* content, developing deep, wide cracks when dry, and showing evidence of movement between aggregates.

visible light *Electromagnetic radiation* in the *wavelength* range of 0.4 to 0.7 *microns.*

volatiles Elements and compounds normally existing in the gaseous state under atmospheric conditions, dissolved in *magma.*

volcanic ash Finely divided *extrusive igneous rock* blown under gas pressure from a volcanic vent.

volcanic neck Isolated, narrow steep-sided peak formed by erosion of *igneous rock* previously solidified in the feeder pipe of an extinct *volcano.*

volcanism General term for *volcano* building and related forms of extrusive igneous activity.

volcano Conical, circular structure built by accumulation of *lava flows* and *tephra,* including *volcanic ash.* (See *composite volcano, shield volcano.*)

volcano coast *Coast* formed by *volcanoes* and *lava* flows built partly below and partly above sea level.

warm-blooded animal Animal that possesses one or more adaptations to maintain a constant internal temperature despite fluctuations in the environmental temperature.

warm front Moving weather *front* along which a warm *air mass* is sliding up over a cold air mass, leading to production of *stratiform clouds* and *precipitation.*

washout Downsweeping of atmospheric *particulate matter* by *precipitation.*

water deficit Difference between *soil water* present in the *soil* (actual *soil-water storage*) and the *storage capacity* of the soil.

waterfall Abrupt descent of a *stream* over a *bedrock* downstep in the *stream channel.*

watergap Narrow transverse *gorge* cut across a narrow ridge by a *stream,* usually in a region of *folds.*

waterlogging Rise of a *water table* in *alluvium* to bring the *zone of saturation* into the root zone of plants.

water need (See *potential evapotranspiration.*)

waterspout Intense perpendicular vortex of air and water droplets formed beneath a *cumulonimbus cloud* over water; visible as a *funnel cloud.*

water surplus Water disposed of by *runoff, interflow,* or *percolation* to the *ground water zone* after the *storage capacity* of the *soil* is full.

water table Upper boundary surface of the *saturated zone;* the upper limit of the *ground water* body.

water use (See *actual evapotranspiration.*)

wave-cut notch Recess at the base of a *marine cliff* where wave impact is concentrated.

wave cyclone Traveling, vortexlike *cyclone* involving interaction of cold and warm *air masses* along sharply defined *fronts.*

wave frequency Number of waves passing a fixed point in a given unit of time.

wavelength Distance separating one wave crest from the next in any uniform succession of traveling waves.

wave refraction Bending of a wave front as it travels in shallow water, caused by changes in depth of the bottom.

wave theory Ideal model of development of the *wave cyclone* as put forward by J. Bjerknes.

weak equatorial low Weak, slowly moving low-pressure center (*cyclone*) accompanied by numerous convectional showers and *thunderstorms;* it forms close to the *ITC* in the rainy season, or *summer monsoon.*

weather Physical state of the *atmosphere* at a given time and place.

weathering Total of all processes acting at or near the earth's surface to cause physical disruption and chemical decomposition of *rock.* (See *chemical weathering, physical weathering.*)

westerlies (See *prevailing westerly winds, upper-air westerlies.*)

west-wind drift Drift current moving eastward in zone of *prevailing westerlies.*

wet (saturation) adiabatic lapse rate Reduced *adiabatic lapse rate* when *condensation* is taking place in rising air; value ranges from 3 to 6 C°/1000 m (2 to 3 F°/1000 ft).

wet-dry tropical climate *Climate* of the *tropical zone* characterized by a very wet season alternating with a very dry season.

wet equatorial climate Moist *climate* of the *equatorial zone* with a large annual *water surplus,* and with uniformly warm temperatures and high values of *soil-water storage* throughout the year.

wetted perimeter Length of the line of contact between the water of a *stream* and its *channel.*

wilting point Quantity of stored *soil water,* less than which the foilage of plants not adapted to drought will wilt.

wind Air motion, dominantly horizontal relative to the earth's surface.

wind abrasion Mechanical wearing action by wind-driven *mineral* particles striking exposed *rock* surfaces.

windows Certain *wavelength* bands of the *electromagnetic radiation* spectrum within which energy is radiated through the atmosphere to escape into outer space.

wind vane Weather instrument used to indicate *wind* direction.

winter berm *Berm* produced by large, steep storm waves during the winter in midlatitudes.

winter monsoon Outflow of continental air at low levels from the *Siberian high* passing over Southeast Asia as a dry, cool, northerly *wind.*

winter solstice *Solstice* occurring on December 21 or 22, when the *sun's declination* is 23½°S.

Wisconsin Stage Last glacial stage of the *Pleistocene Epoch.*

woodland Plant *formation class,* transitional between *forest biome* and *savanna biome,* consisting of widely spaced trees with canopy coverage between 25% and 60%.

world ocean   Collective term for combined oceans of the globe.

xeric animals   Animals adapted to dry conditions typical of a desert climate.

xerophytes   Plants adapted to a dry environment.

X rays   High-energy form of radiation at the extreme short *wavelength* (high-frequency) end of the *electromagnetic spectrum*.

yazoo stream   *Stream* that enters on the *floodplain* of a larger *alluvial river* and is forced by the presence of the *natural levees* to flow far down-valley before making a junction with the larger stream.

year   Period of time required for one complete *revolution* of a planet in its *orbit* around the sun.

zenithal projection   One of a class of map projections in which the *geographic grid* is centered on a point and has perfect radial symmetry of its geometrical properties.

zone of ablation   Lower portion of *glacier* in which *ablation* exceeds gain of mass by snowfall; zone of wastage of a glacier.

zone of accumulation   Upper portion of a *glacier* in which the *firn* becomes transformed into glacial ice; the zone of nourishment of a glacier.

Index

Number in boldface type gives page on which term is defined or explained.

Cycle of denudation (cont'd)
  in mountainous deserts, 313
Cycle of rock transformation, **181,** 379
  and plate tectonics, 379
Cyclone families, 117
Cyclones, **79,** 109, 112
  traveling, 109
  tropical, **120**
  wave, **112**
  on weather map, 113
Cyclonic storms, **109**
Cylindric projections, **12**

Dams, flood-storage, 303
Dansereau, P., 238
Date Line, International, **34**
Datum plane, **448**
Davis, W. M., 307
Day, mean solar, **20**
Daylight, duration, 186
  and photosynthesis, 186
Daylight saving time, **32**
Debris avalanche, 267
Debris floods, **265**
  man-induced, 265
Decalcification, **209**
Deccan Plateau, 399
Deciduous plants, **228,** 239
Declination, of compass, 451
  magnetic, **451**
  of sun, **28**
Decomposers, **183**
Deep environment, **181**
Deflation, **356,** 363
  man-induced, 363
Deflation hollow, **357**
Deformed strata, 401
Degradation by streams, **299**
Degree, of latitude, 8
  length of, 7, 8
  of longitude, 7
Deicing salt, 173
Delta coasts, **347,** 351
Delta kames, **333**
Deltas, **351**
  arcuate, **351**
  bird-foot, **351**
  cuspate, **351**
  estuarine, **351**

glacial, **331**
Dendritic drainage pattern, **399,** 407
Denitrification, **192**
Density, **169**
Denudation, **256**
  by carbonic acid action, 261
Denudation cycle, **307,** 308-312
  in mountainous desert, 313
  and rejuvenation, 311
  with spasmodic uplift, 310
  stages, 311
Denudation rates, 308
Denudation system, model, 308
Deposition by streams, **292**
Depositional landforms, **289**
Depression contours, **449**
Depression storage, 277
Desalinization, **209**
Desert biome, 232, **241,** 251
Desert climate, 145
Desert formation class, 252
Desert irrigation, 423
Desert landscapes, 312, 313
  mountainous, 313
Desert pavement, **357**
Desert processes, 312
Deserts, 130
  and deflation, 363
  rainshadow, **101**
Desert soils, 224
Desert varnish, **422**
Developable geometric surface, 9
Dew, 92
Dew point, **92**
Diagenesis, 179
Diagnostic horizons, **213**
Diaspore, 174
Diffuse reflection, **51**
Diked salt marshes, 347
Dikes, on floodplains, 303
  igneous, **381,** 385
  on salt marshes, 347
Diorite, **172**
Dip, **397,** 455
Discharge of stream, **279**
Distributaries, **351**

Dokuchaiev, V. V., 211
Doldrums, *80*
Dolomite, 179
Domes, sedimentary, **401,** 402
Dornveldt, 252
Down-scatter, **51**
Drainage basin, **278**
Drainage patterns, annular, **403**
  dendritic, **399,** 407
  trellis, **398**
Drainage systems, **278**
Drainage winds, **82**
Drawdown, **273**
Drift, glacial, **331**
  stratified, **331**
Driftless Area, 328
Drift of sediment, 342
  littoral, **343**
Drifts in mines, **267**
Drizzle, 96
Drumlins, **331**
Dry adiabatic lapse rate, **94**
Dry climates, **133, 144, 157,** 158, 441
  subdivisions, 158, 441
  subtypes, 158, 441
Dry desert, **252,** 421
Dry midlatitude climate, **140, 164,** 432
Dry subtropical climate, **137, 161,** 419, 432
Dry tropical climate, **136, 160,** 416, 419
Dunes, *see* Sand dunes
Dune sand, **177,** 358
Dunite, 172
Duripan, **214**
Dust, atmospheric, 39
  and climate change, 72
Dust Bowl, 434
Dust storms, **357,** 432, 434
  man-induced, 363

Earth, circumference, 4
  curvature, 3
  diameter, 6
  dimensions, 6
  interior, 365
  as magnet, 40

Lumber, 430, 436
Lunar eclipse, 4

Macronutrients, **189**
Mafic minerals, **169**
Mafic rocks, **172**
Magellan, 3, 33
Magma, **169**
  silicate, **169**
Magnesium, in sea water, 42
Magnetic azimuths, **452**
Magnetic declination, **451**
Magnetic field of earth, 40
  external, **40**
Magnetic north, **451**
Magnetic north pole, **451**
Magnetite, 169
Magnetopause, **40**
Magnetosphere, **40**
Malpais, 388
Man as geomorphic agent, 267
Mangrove, 244
Mangrove swamp forest, **244**
Manioc, 413
Mantle of earth, **366**
Mantle plumes, **376,** 385
  and hot spots, 385
Map orientation, 451
Map projections, **9**
  and areas, 10
  classification, 11
  conformal, **10**
  conformal conic, **14**
  conic, **12**
  cylindric, **12**
  equal-area, **10**
  Goode's, 18
  homolographic, **16**
  homolosine, **18**
  interrupted, 18
  Lambert conformal conic, **14**
  Mercator, **15**
  problem, 9
  and scale changes, 10
  secant conic, **14**
  sinusoidal, **17**
  stereographic, **13**
  zenithal, **11**
Map reading, 446, 449

Maps, isobaric, 79
  isogonic, 451
  large-scale, 450
  small-scale, 450
  topographic, 446
Map scale, 9, 449
  fractional, 449
  graphic, **449**
  and map area, 451
Maquis, **249**
Marble, **181**
Marbut, C. F., 211
Marginal glacial lakes, **331**
Marginal ocean basin, 375
Marine cliffs, **339,** 344
  elevated, 353
  raised, 353
Marine ecosystems, 187
  productivity, 187
Marine erosion, 339, 344
Marine scarp, **339**
Marine terraces, **353**
Marine west-coast climate, **139, 163,** 429
Marine west-coast forest environment, **429**
Marshes, 271
  salt, **347**
Mass wasting, **257,** 261-267
  man-induced, 267
  in rainforest, 413
  and weathering, 257
Material cycles, **188**
  gaseous, **188**
  sedimentary, **189**
Maximum-minimum thermometer, **64**
Mean annual temperature, **65**
Mean daily temperature, **65**
Meanders, alluvial, **297,** 301, 303
  entrenched, **304**
Mean monthly temperature, **65**
Mean solar day, **20**
Mean velocity in stream, **279**
Medial moraine, **323**
Mediterranean climate, **139, 162,** 427

Mediterranean environment, **427**
Mediterranean evergreen mixed forest, 249
Megahertz, **46**
Melting, **91**
Mercator, Gerardus, 15
Mercator chart, 15
Mercator projection, **15**
Mercurial barometer, **37**
Meridians, **7**
  principal, **453**
  standard, **31**
  of time, 29
Meridional transport, **57**
  of heat, 57, 76
Meridional winds, **83**
Mesa, **399**
Mesopause, **39**
Mesophytes, **227**
Mesosphere, **39**
Mesothermal climates, 144
Mesozoic Era, **368**
Metamorphic belts, 406
Metamorphic rocks, **169,** 180
Meteorology, **35**
Methane, 180
Mica group, **169**
Microclimate, **234**
Micron, **46**
Micronutrients, **189**
Microthermal climates, 144
Microwaves, **57**
Mid-Atlantic Ridge, 378
Midlatitude deciduous forest, **246,** 431
Midlatitude desert environment, **435**
Midlatitude zones, **50**
Midnight meridian, **29**
Mid-oceanic ridge, **372,** 373
Mile, nautical, **8**
  statute, 9
Military grid, 452
Millibar, **37**
Milliequivalent, 204
Mineral alteration, **174**
Minerals, **168**
  felsic, **169**
  heavy, **176**
  mafic, **169**

SYMBOLIC STATION MODEL

$$\begin{array}{c} \quad \overset{\displaystyle /}{f\!f} \\ T T \, dd \, \overset{C_H}{\underset{C_M}{}} PPP \\ VV \, ww \, \bigcirc{N} \, \pm pp\,a \\ T_d T_d \, C_L \, N_h \, W \, R_t \\ h \quad RR \end{array}$$

SAMPLE PLOTTED REPORT

$$\begin{array}{c} 31 \quad \leftarrow \quad 247 \\ \text{3/4} \; ** \; \bullet \; +28 / \\ 30 \text{---} 6 \; \bullet 4 \\ 2 \quad 45 \end{array}$$

U. S. DEPT. OF COMMERCE
WEATHER BUREAU

SURFACE WEATHER MAP AND STATION WEATHER

EXPLANATION OF SYMBOLS

N — Total amount of cloud 8 = completely covered block **6**

dd — True direction from which wind is blowing 32 = 320° = NW

ff — Wind speed in knots 20 = 20 knots block **9**

VV — Visibility in miles and fractions 12 = 12/16 or 3/4 miles

ww — Present weather 71 = continuous slight snow (block **8**)

W — Past weather 6 = rain (block **11**)

PPP — Barometric Pressure (in millibars) reduced to sea-level 247 = 1024.7 mb.

TT — Current air temperature 31 = 31° F.

N_h — Fraction of sky covered by low or middle cloud 6 = 7 or 8 tenths block **7**

C_L — Cloud type (block **3**) 7 = Fractostratus and/or Fractocumulus of bad weather (scud)

C_M — Cloud type 9 = Altocumulus of chaotic sky

C_H — Cloud type 2 = Dense cirrus in patches

h — Height of base of cloud 2 = 300 to 599 feet

$T_d T_d$ — Temperature of dewpoint 30 = 30° F.

a — Characteristic of barograph trace 2 = rising steadily or unsteadily block **10**

pp — Pressure change in 3 hours preceding observation 28 = 2.8 millibars

RR — Amount of precipitation 45 = 0.45 inches

R_t — Time precipitation began or ended 4 = 3 to 4 hours ago

WW **8** PRESENT WEATHER

↓	Slight or moderate drifting snow, generally low
≡	Fog, sky NOT discernible, no appreciable change during past hour
,	Intermittent drizzle
,,	Continuous drizzle
∞	Haze

h — HEIGHT IN METERS (Approximate) **5**

h	Height
0	0 - 49
1	50 - 99
2	100 - 199
3	200 - 299
4	300 - 599
5	600 - 999
6	1,000 - 1,499
7	1,500 - 1,999
8	2,000 - 2,499
9	At or above 2,500, or no clouds

9 ff KNOTS

Symbol	Knots
◎	Calm
—	1 - 2
⊢	3 - 7
⊢	8 - 12
⊢	13 - 17
⊢	18 - 22
⊢	23 - 27
⊢	28 - 32
⊢	33 - 37
⊢	38 - 42
⊢	43 - 47
⊢	48 - 52
⊢	53 - 57
⊢	58 - 62
⊢	63 - 67
⊢	68 - 72
⊢	73 - 77

N SKY COVERAGE **6** (Total Amount)

○	No clouds
◔	Less than one-tenth or one-tenth
◔	Two-tenths or three-tenths
◑	Four-tenths
◑	Five-tenths
◕	Six-tenths
◕	Seven-tenths or eight-tenths
◕	Nine-tenths or overcast with openings
●	Completely overcast
⊗	Sky obscured

W PAST WEATHER **11**

Symbol	Description
	Clear or few clouds
	Partly cloudy (scattered) or variable sky — Not Plotted
	Cloudy (broken) or overcast
S⃥	Sandstorm or duststorm, or drifting or blowing snow
≡	Fog, or smoke, or thick dust haze
,	Drizzle
•	Rain
*	Snow, or rain and snow mixed, or ice pellets (sleet)
▽	Shower(s)
⍐	Thunderstorm, with or without precipitation

Intermittent rain (NOT freezing), slight at time of observation

Slight rain shower(s)

Moderate or thick freezing drizzle

Intermittent fall of flakes, slight at observation

Continuous rain (NOT freezing), slight at time of observation

Moderate or heavy rain shower(s)

Slight freezing rain

Continuous fall of snowflakes, slight at time of observation

Continuous rain (NOT freezing), moderate at time of observation

Slight snow shower(s)

Ice pellets (sleet, U. S. definition)

Continuous fall of snowflakes, moderate at time of observation

Heavy thunderstorm with hail at time of observation

C_L DESCRIPTION
(Abridged From W.M.O. Code)

Cu of fair weather, little vertical development and seemingly flattened

Cu of considerable development, generally towering, with or without other Cu or Sc bases all at same level

Cb with tops lacking clear-cut outlines, but distinctly not cirriform or anvil-shaped; with or without Cu, Sc, or St

Sc formed by spreading out of Cu; Cu often present also

Sc not formed by spreading out of Cu

St or Fs or both, but no Fs of bad weather

Fs and/or Fc of bad weather (scud)

Cu and Sc (not formed by spreading out of Cu) with bases at different levels

Cb having a clearly fibrous (cirriform) top, often anvil-shaped, with or without Cu, Sc, St, or scud

③ CLOUD ABBREVIATION

St or Fs-Stratus or Fractostratus

Ci-Cirrus

Cs-Cirrostratus

Cc-Cirrocumulus

Ac-Altocumulus

As-Altostratus

Sc-Stratocumulus

Ns-Nimbostratus

Cu or Fc-Cumulus or Fractocumulus

Cb-Cumulonimbus

C_M DESCRIPTION
(Abridged From W.M.O. Code)

Thin As (most of cloud layer semi-transparent)

Thick As, greater part sufficiently dense to hide sun (or moon), or Ns

Thin Ac, mostly semi-transparent; cloud elements not changing much and at a single level

Thin Ac in patches; cloud elements continually changing and/or occurring at more than one level

Thin Ac in bands or in a layer gradually spreading over sky and usually thickening as a whole

Ac formed by the spreading out of Cu

Double-layered Ac, or a thick layer of Ac, not increasing; or Ac with As and/or Ns

Ac in the form of Cu-shaped tufts or Ac with turrets

Ac of a chaotic sky, usually at different levels; patches of dense Ci are usually present also

---

N_h SKY COVERAGE ⑦
(Low And/Or Middle Clouds)

0	No clouds
1	Less than one-tenth or one-tenth
2	Two-tenths or three-tenths
3	Four-tenths
4	Five-tenths
5	Six-tenths
6	Seven-tenths or eight-tenths
7	Nine-tenths or overcast with openings
8	Completely overcast
9	Sky obscured

a BAROMETRIC TENDENCY ⑩

Rising, then falling

Rising, then steady; or rising, then rising more slowly

Rising steadily, or unsteadily

} Barometer now higher than 3 hours ago

Falling or steady, then rising; or rising, then rising more quickly

Steady, same as 3 hours ago

Falling, then rising, same or lower than 3 hours ago

Falling, then steady; or falling, then falling more slowly

} Barometer now lower than 3 hours ago

Falling steadily, or unsteadily

Steady or rising, then falling; or falling, then falling more quickly

C_H DESCRIPTION
(Abridged From W.M.O. Code)

Filaments of Ci, or "mares tails," scattered and not increasing

Dense Ci in patches or twisted sheaves, usually not increasing, sometimes like remains of Cb; or towers or tufts

Dense Ci, often anvil-shaped, derived from or associated with Cb

Ci, often hook-shaped, gradually spreading over the sky and usually thickening as a whole

Ci and Cs, often in converging bands, or Cs alone; generally overspreading and growing denser; the continuous layer not reaching 45° altitude

Ci and Cs, often in converging bands, or Cs alone; generally overspreading and growing denser; the continuous layer exceeding 45° altitude

Veil of Cs covering the entire sky

Cs not increasing and not covering entire sky

Cc alone or Cc with some Ci or Cs, but the Cc being the main cirriform cloud

JUN 6 1970